Algorithms for Intelligent Systems

This book series publishes research on the analysis and development of algorithms for intelligent systems with their applications to various real world problems. It covers research related to autonomous agents, multi-agent systems, behavioral modeling, reinforcement learning, game theory, mechanism design, machine learning, meta-heuristic search, optimization, planning and scheduling, artificial neural networks, evolutionary computation, swarm intelligence and other algorithms for intelligent systems.

The book series includes recent advancements, modification and applications of the artificial neural networks, evolutionary computation, swarm intelligence, artificial immune systems, fuzzy system, autonomous and multi agent systems, machine learning and other intelligent systems related areas. The material will be beneficial for the graduate students, post-graduate students as well as the researchers who want a broader view of advances in algorithms for intelligent systems. The contents will also be useful to the researchers from other fields who have no knowledge of the power of intelligent systems, e.g. the researchers in the field of bioinformatics, biochemists, mechanical and chemical engineers, economists, musicians and medical practitioners.

The series publishes monographs, edited volumes, advanced textbooks and selected proceedings.

All books published in the series are submitted for consideration in Web of Science.

More information about this series at https://link.springer.com/bookseries/16171

Mohit Dua · Ankit Kumar Jain · Anupam Yadav ·
Nitin Kumar · Patrick Siarry
Editors

Proceedings of the International Conference on Paradigms of Communication, Computing and Data Sciences

PCCDS 2021

Springer

Editors
Mohit Dua
National Institute of Technology Kurukshetra
Kurukshetra, India

Ankit Kumar Jain
National Institute of Technology Kurukshetra
Kurukshetra, India

Anupam Yadav
Dr. B. R. Ambedkar National Institute of Technology
Jalandhar, India

Nitin Kumar
National Institute of Technology Uttarakhand
Srinagar, India

Patrick Siarry
Campus Centre de Créteil
Université Paris-Est Créteil
Créteil, France

ISSN 2524-7565 ISSN 2524-7573 (electronic)
Algorithms for Intelligent Systems
ISBN 978-981-16-5746-7 ISBN 978-981-16-5747-4 (eBook)
https://doi.org/10.1007/978-981-16-5747-4

This Springer imprint is published by the registered company Springer Nature Singapore Pte Ltd.
The registered company address is: 152 Beach Road, #21-01/04 Gateway East, Singapore 189721, Singapore

Committees

Patron

Dr. Satish Kumar, Director, NIT Kurukshetra

General Chairs

Mayank Dave, National Institute of Technology Kurukshetra, India
Patrick Siarry, Université Paris-Est Créteil, France
Mohit Dua, National Institute of Technology Kurukshetra, India
Ankit Jain, National Institute of Technology Kurukshetra, India
Anupam Yadav, National Institute of Technology Jalandhar, India
Nitin Kumar, National Institute of Technology Uttarakhand

Organizing Chairs

Harish Sharma, Rajasthan Technical University, Kota
Mukesh Saraswat, Jaypee Institute of Information Technology, India

Finance Chair and Treasurer

Sandeep Kumar, CHRIST (Deemed to be University), Bengaluru

International Advisory Committee

David Asirvatham, Taylor's University
Vicente García Díaz, University of Oviedo, Spain
J. K. Chhabra, NIT Kurukshetra
Mukul V. Shirvaikar, The University of Texas at Tyler, USA
Brahmjit Singh, NIT Kurukshetra
R. K. Aggarwal, NIT Kurukshetra
Michael Sheng, Macquarie University, Sydney, Australia
Ljiljana Trajkovic, Simon Fraser University, Canada
Sheng-Lung Peng, National Dong Hwa University
Jagdish Chand Bansal, SAU, New Delhi
Stefka Fidanova, Bulgarian Academy of Sciences, Bulgaria
Jerry Chun-Wei Lin, Western Norway University of Applied Sciences, Bergen, Norway
Nishchal K. Verma, IIT Kanpur, India
A. K. Singh, NIT Kurukshetra
Mehdi Shadaram, The University of Texas at San Antonio, USA
Marcin Paprzycki, Polish Academy of Sciences, Poland
Xiao-Zhi Gao, University of Eastern Finland, Finland
Sanjiv K. Bhatia, University of Missouri—St. Louis, USA
Rajoo Pandey, NIT Kurukshetra
S. K. Jain, NIT Kurukshetra

Preface

The 2nd International Conference on Paradigms of Communication, Computing and Data Sciences (PCCDS 2021) was jointly organized in virtual format by the Department of Computer Engineering at National Institute of Technology Kurukshetra, India, and Soft Computing Research Society (SCRS), India, from 7 May to 9 May 2021.

The conference aimed to bring together leading academicians, scientists, researcher scholars, UG/PG graduates across the globe to discuss all aspects (current and future) of advanced communications, computing and data science techniques. It enabled the participating researchers to exchange their ideas about applying existing methods in these areas to solve real-world problems. This conference was an attempt towards exploration of amalgamation possibilities of data sciences into computing algorithm with communication proficiency.

This book is an edited volume of the articles presented in the conference. The theme of the book covers the following three major areas such as communication, computing and data sciences. In addition to the contributed papers, four invited keynote speeches were delivered by Dr. Marcin Paprzycki, Systems Research Institute, Polish Academy of Sciences, Poland; Prof. R. K. Agrawal, Jawaharlal Nehru University, New Delhi; Dr. Aruna Tiwari, Indian Institute of Technology Indore; and Dr. Maanak Gupta from Tennessee Technological University, TN, USA. We are grateful to all the keynote speakers for sharing their fruitful and rewarding ideas during the conference.

We express our deep regards to the entire team of PCCDS 2021, all reviewers, authors and participants for their contributions. We are sure that this edited collection will surely fulfil the expectations of the researchers working in different areas of communication, computing and data sciences.

Kurukshetra, India	Mohit Dua
Kurukshetra, India	Ankit Kumar Jain
Jalandhar, India	Anupam Yadav
Srinagar, India	Nitin Kumar
Créteil, France	Patrick Siarry

Contents

About the Editors

Dr. Mohit Dua did his B.Tech. degree in computer science and engineering from Kurukshetra University, Kurukshetra, India, in 2004 and M.Tech. degree in computer engineering from National Institute of Technology, Kurukshetra, India, in 2012. He received Ph.D. in the area of speech recognition from the National Institute of Technology, Kurukshetra, India, in 2018. He is presently working as Assistant Professor in the Department of Computer Engineering at NIT Kurukshetra, India, with more than 16 years of academic experience. He is Life Member of Computer Society of India (CSI) and Indian Society for Technical Education (ISTE). His research interests include speech processing, theory of formal languages, statistical modeling, and natural language processing. He has published more than 60 research papers in various reputed journals and conferences.

Dr. Ankit Kumar Jain is presently working as Assistant Professor in the National Institute of Technology, Kurukshetra, India. He received master of technology from the Indian Institute of Information Technology Allahabad (IIIT), India, and Ph.D. degree from the National Institute of Technology, Kurukshetra. His general research interest is in the area of information and cyber security, phishing Web site detection, Web security, mobile security, IoT security, online social network, and machine learning. He has published more than 35 papers in reputed journals and conferences.

Dr. Anupam Yadav is Assistant Professor, Department of Mathematics, Dr. B. R. Ambedkar National Institute of Technology Jalandhar, India. His research area includes numerical optimization, soft computing, and artificial intelligence, and he has more than ten years of research experience in the areas of soft computing and optimization. He has done a Ph.D. in soft computing from the Indian Institute of Technology Roorkee, and he had worked as a research professor at Korea University. He has published more than twenty-five research articles in journals of international repute, and has published more than fifteen research articles in conference proceedings. He has authored a textbook entitled *An Introduction to Neural Network Methods for Differential Equations*. He has edited three books which are published by AISC, Springer Series. He was General Chair, Convener, and Member of the Steering Committee of several international conferences. He is Member of various research societies.

Dr. Nitin Kumar has been working as Assistant Professor in the Department of Computer Science and Engineering at the National Institute of Technology, Uttarakhand, since 2013. He has obtained his M.Tech. and Ph.D. in computer science and technology from the School of Computer and Systems Sciences, Jawaharlal Nehru University. He has published more than 50 research papers in reputed international journals and international conferences in India and abroad. His current research interests include pattern recognition, biometric recognition and security, image processing, and computer vision.

Prof. Patrick Siarry was born in France in 1952. He received the Ph.D. degree from the University Paris 6, in 1986, and the Doctorate of Sciences (Habilitation) from the University Paris 11, in 1994. He was first involved in the development of analog and digital models of nuclear power plants at Electricité de France (E.D.F.). Since 1995, he is Professor in automatics and informatics. His main research interests are computer-aided design of electronic circuits and the applications of new stochastic global optimization heuristics to various engineering fields. He is also interested in the fitting of process models to experimental data and the learning of fuzzy rule bases and of neural networks.

Communication

Simulation and Implementation of Circular Microstrip Patch Antenna

Hrishikesh Ugale, Shubham Chauhan, and Ashwin Kothari

1 Introduction

In the era of Communication Technology, microstrip patch antennas are very popular since they are used in a various applications in the fields of mobile, military, and satellite wireless communications [1]. These antennas are known for special characteristics such as compactness, light weight low profile and cost effective fabrication on printed circuit boards. A typical Microstrip Patch antenna consists of a metallic patch on top of a dielectric substrate and a ground plane below the dielectric substrate. Maximum radiation in the direction normal to the plane of the patch is the key factor to be considered in the design. Rectangular and circular shaped patch antennas are widely preferred since they are easy of analyze and fabricate, and offer very good radiation characteristics.

5 GHz frequency band is widely used for a large number of commercial communication applications. These applications include rapidly growing IEEE 802.11 ac standard for wireless transmission [2] of data. Many IoT devices also use on 5.1 GHz frequency for communication and use patch antennas [3] for this purpose. Proposed work is based on design of a circular microstrip patch antenna for the applications mentioned above.

Patch antennas come in a wide variety of designs for different frequencies and applications. Design of the antenna aims to maximize the radiated power and minimize the reflected power [4]. Performance of antenna is related to its performance metrics like gain, reflection coefficient, Voltage Standing Wave Ratio (VSWR), etc. Paper presents a brief discussion on each of these performance metrics of the proposed design using CADFEKO simulation tool.

The remainder of the paper is organized as follows. Section 2 presents the literature study of antenna designs proposed by other researchers for different applications.

H. Ugale (✉) · S. Chauhan · A. Kothari
Visvesvaraya National Institute of Technology, Nagpur, India

M. Dua et al. (eds.), *Proceedings of the International Conference on Paradigms of Communication, Computing and Data Sciences*, Algorithms for Intelligent Systems,
https://doi.org/10.1007/978-981-16-5747-4_1

Section 3 briefly describes the design methodology using basic design equations, simple base design and suitable modifications in the base design. Section 4 describes the performance parameters of the antenna as observed in the simulation tool. Section 5 describes the results obtained after implementation and fabrication of the design. Section 6 concludes the work with some remarks about the scope of further improvements and prospects of the proposed work.

2 Related Work

Many designs of patch antennas have been proposed in the literature for different frequencies and applications. The design of the patch antenna has one of the interesting areas of research for many researchers. Rectangular and Circular designs of patch antenna are quite popular in the literature. In [5], a design of a rectangular patch antenna for 4.1 GHz frequency was proposed. Simulations and testing results using HFSS simulation software were presented.

In [6], a Microstrip patch antenna for WLAN applications at 2.4 GHz was designed with a slotted ground plane. The minimum reflection coefficient value was found to be around -19 dB. Simulation of the proposed antenna was carried out using the CADFEKO simulation tool and results were compared with the actual fabricated antenna. A modified design for the same 2.4 GHz frequency was proposed in [7] with slightly improved results.

In [8], authors simulated 4×1 and 8×1 arrays of microstrip patch antennas for resonance at 2.4 GHz used for WLAN applications. Design of rectangular patch antenna for ISM band was proposed in [9]. For the frequencies in the 5 GHz band, a notch based design was proposed in [10]. In [11], the authors proposed the highly compact design of the dual-frequency antenna for both 2.4 and 5 GHz.

Various designs discussed in this subsection were studied. A comparison of the performance of the proposed antenna design with some of these designs is discussed in Sect. 5.

3 Design Methodology

The design of the proposed antenna is briefly discussed in this section. The first subsection describes the primary design equations proposed in the theory for the simplistic design of a circular antenna. The second subsections describe the simple base design and drawbacks observed after the simulation of the base design. The third subsection describes the modifications made in the base design to achieve the desired characteristics.

3.1 Design Equations

This subsection briefly discuss about the use of some basic theoretical equations and design thumb rules for determining the basic parameters of the design like radius of patch, dimensions and material of substrate, and specifications of feedline. Each of them are discussed below.

Radius of Patch (R). The Radius (R) of the patch is the most crucial design parameter to be considered in the designing of the circular patch antenna. Rectangular patch antenna has two design parameters—length (L) and width (W). Both of these parameters need to be carefully adjusted in order to achieve resonance at the desired frequency. In the case of a circular antenna, the only radius needs to be adjusted in order to achieve desired characteristics. This makes the process of designing circular antennas much simpler as compared to rectangular antennas.

According to the theory of Patch Antennas as proposed in [11], the radius of the patch (R) is calculated using the expressions as shown in Eq. 1.

$$R = \frac{F}{\sqrt{1 + \frac{2h}{\pi \epsilon_r F}\left[\ln\left(\frac{\pi F}{2h}\right) + 1.7726\right]}} \tag{1}$$

where,

$$F = \frac{8.791 \times 10^9}{f_r \sqrt{\epsilon_r}}$$

ϵ_r Dielectric constant of the substrate
f_r Resonating frequency in Hertz
h Height of substrate in cm

Desired frequency of operation for proposed antenna is 5.1 GHz. Substituting $f = 5.1$ GHz in Eq. 1, radius of patch (R) comes out to be 8.56 mm.

Specifications of Substrate The dielectric substrate is an important part of the patch antenna. This dielectric substrate materials can be ceramic, semiconductor, ferromagnetic, synthetic, composite, and foams, etc. The important properties of substrate material include its dielectric constant (ϵ_r), permittivity (μ_r), and loss tangent ($\tan \delta$).

In the proposed design, Glass Epoxy substrate (FR4) is selected as the substrate material due to its easy availability. Dielectric constant (ϵ) of this substrate is 4.4. The value of loss tangent($\tan \delta$) for this substrate is 0.02. As a design thumb rule, the size of the substrate is greater than the size of the patch as the patch resides on the top of the substrate. A square substrate of 23 mm $\times$ 23 mm in size is used in the proposed design. The substrate is selected to be 1.6 mm thick with a full layer of the ground plane on the bottom side.

Feedline Specifications Excitation of the antenna is carried out using a microstrip transmission line which is also known as Feedline. The proposed antenna is fed through a feedline with 50 Ω impedance. Width w of feedline for the given impedance value is calculated using the formula stated in Eq. 2 [11, 12].

$$w = \frac{7.48 \times h}{\exp\left(Z_o \frac{\sqrt{\epsilon_r + 1.41}}{87}\right)} - 1.25 \times t \tag{2}$$

where, t - Thickness of copper in the feedline

Z_o Impedance of feedline
h Thickness of substrate
ϵ_r Dielectric constant of the substrate.

Substituting Z_o as 50 Ω in Eq. 2, we get the width of microstrip line equal to 2.95 mm which is approximately 3 mm.

Basic parameters of design like the radius of patch, dimensions and material of substrate, impedance and thickness of feedline are now known to us. Using these parameters, a basic design of the antenna is proposed in the next subsection.

3.2 Basic Design

Initial parameters for circular microstrip patch antenna calculated using thumb rules and formulas discussed in Sect. 2 are illustrated in Fig. 1.

looseness-1Basic circular design of antenna using these initial parameters is proposed in Fig. 1. This design consists of a circular copper patch of radius R on top of a square dielectric substrate. Patch is excited using a feedline having a width W_f. The antenna design shown in Fig. 1 was modelled using CAD tools and simulated using CADFEKO simulation tool. Various performance parameters of the design were determined through simulation and analysis of obtained results was carried out. Some important outcomes that describe the performance and efficiency of the basic design are listed below.

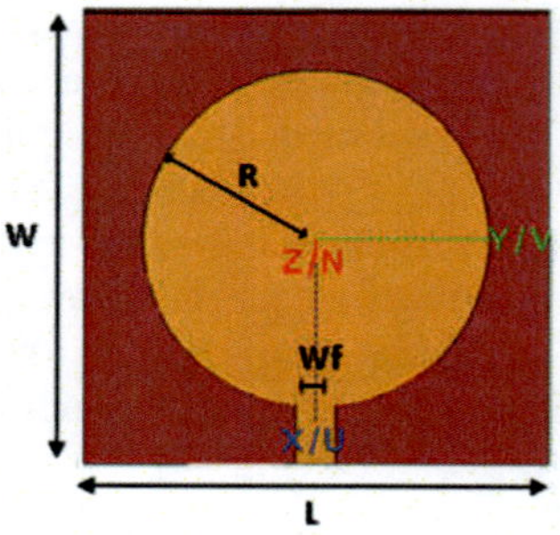

Sr no.	Parameter	Notation	Value
1.	Frequency	f_r	5.1 GHz
2.	Patch Radius	R	8.56 mm
3.	Substrate Thickness	h	1.6 mm
4.	Copper Thickness	M_t	0.035 mm
5.	Feedline Width	W_f	2.95 mm
6.	Dielectric Constant	ϵ_r	4.4
7.	Substrate Length	L	23 mm
8.	Substrate width	W	23 mm

Fig. 1 Basic design in CADFEKO and initial design parameters

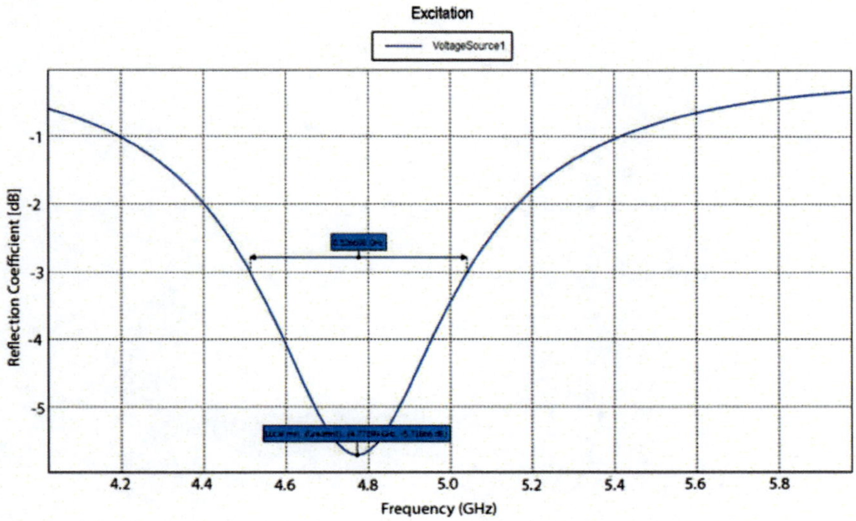

Fig. 2 Variation in reflection coefficient as function of frequency

- Plot of variation in reflection coefficient in dB as a function of frequency is shown in Fig. 2. Since the antenna is intended to be designed for a frequency of 5.1 GHz, it is expected that the minimum value of reflection coefficient should be observed for 5.1 GHz frequency. Simulation results showed that the minimum value of magnitude of reflection coefficient was obtained at the frequency close to 4.7 GHz.
- Corresponding to the frequency of 4.7 GHz, the minimum value of the reflection coefficient was around −5.72 dB. This signifies a large return loss due to improper matching. Desired value for should be below −10 dB for a good performance design.

These observations highlight the lack of efficiency and drawbacks in the basic design which need to be overcome by some modifications. These modifications and their impact have been discussed in the next subsection.

3.3 Modifications in Basic Design

In order to overcome the drawbacks in the basic design certain modifications are made to achieve desired performance metrics. The reflection coefficient of the basic design is found to be −5.72 dB which is quite high. This suggests that the return loss needs to minimized. This high return loss is due to improper impedance matching between the patch and the feedline. Transfer of maximum power from the source to load with minimum or no return loss takes place when the load impedance is made equal to source impedance as per the maximum power transfer theorem.

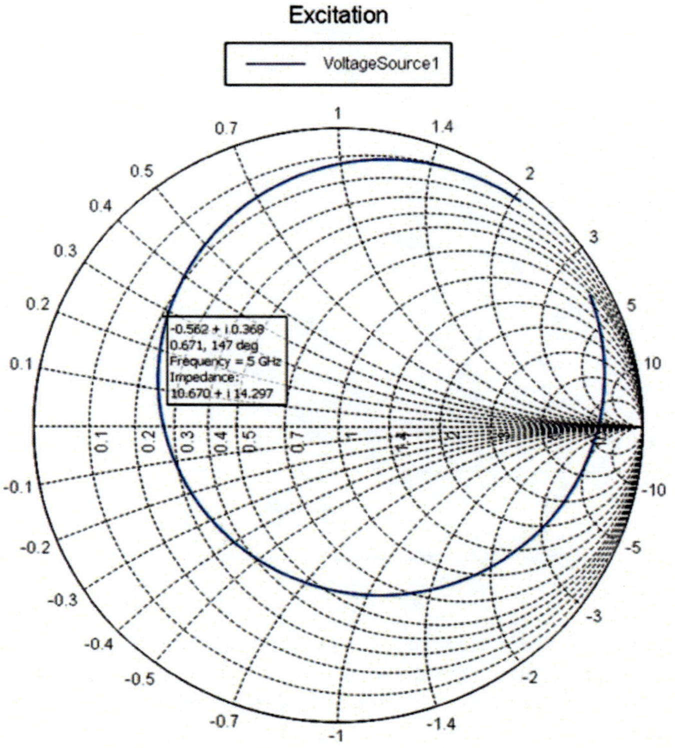

Fig. 3 Smith chart for basic design

The Reactance of the basic antenna design for different frequencies is shown in the Smith chart in Fig. 3. Impedance of 10.67 + j14.297 Ω at the target resonating frequency of 5.1 GHz was observed. This shows the dominance of the inductive component in the impedance of the design and modifications are needed to cancel out this inductive component. This impedance matching can be achieved by adding some slots into the design [13] and other techniques like stub matching. Slots in the patch deviate the direction of the normal flow of current and thus affect the current distribution. In presence of slots, current tends to take a longer path to cross the discontinuity and thus affecting the impedance [13].

Some experimental modifications in the basic design by adding some slots in the patch were made to minimize the return losses. Slots created close to the microstrip provides inset and reduces the distance of excitation from the centre of the patch. Slots are created at another end as well. These slots help to nullify the capacitive and inductive components of impedances and match the impedance of patch with that of the feedline to provide maximum radiation. Final design after a series of experimental modifications is shown in Fig. 4. This design was simulated using CADFEKO and following results were obtained.

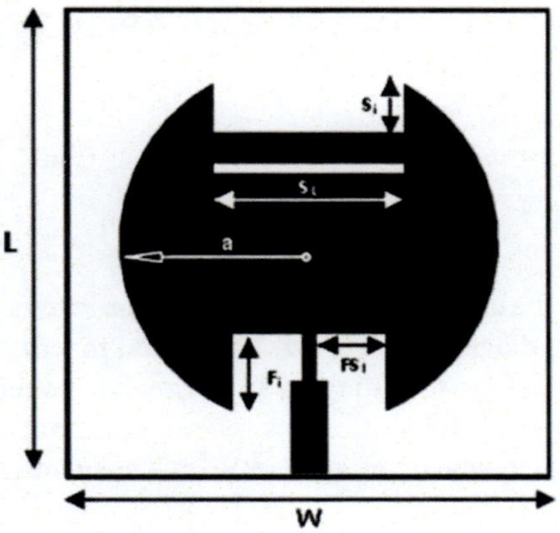

Sr no.	Parameter	Notation	Value
1.	Frequency	f_r	5.1 GHz
2.	Patch Radius	a	9 mm
3.	Substrate Thickness	h	1.6 mm
4.	Copper Thickness	M_t	0.035 mm
5.	Feedline Width	W_f	2.95 mm
6.	Dielectric Constant	ϵ_r	4.4
7.	Substrate Length	L	23 mm
8.	Substrate width	W	23 mm
9.	Slot Length	S_L	10 mm
10.	Slot Thickness	S_w	0.5 mm
11.	Slot Inset	S_i	3 mm
12.	Feedline Inset	F_i	3 mm
13.	Feedline Slot width	FS_i	23 mm

Fig. 4 Final antenna design

- The desired frequency of resonance was obtained i.e. a sharp dip in return loss was observed at 5.1 GHz frequency.
- Desired minimum value of reflection coefficient was achieved. The value was found to be −20.88 dB which is very less than benchmark −10 dB.
- Simulated antenna design showed a maximum antenna gain of 5.9 dBi which is quite good.

The simulation results thus obtained show quite a good performance parameters. The final design and design parameters are shown in Fig. 4.

4 Antenna Performance

This section briefly discusses the performance of the proposed antenna as observed through simulation in terms of various performance metrics like gain, return loss, directivity, etc.

4.1 Gain (G)

The gain of an antenna is defined as the ratio of the power produced by the antenna from a far-field source on the antenna's beam axis to the power produced by a hypothetical lossless isotropic antenna, which is equally sensitive to signals from all directions [11]. FR4 substrate used for the proposed antenna is a lossy material and hence a very large gain cannot be expected. Gain for a good patch antenna with FR4 as the dielectric substrate is typically in the range of 3–6 dBi. When the proposed antenna design was simulated on CADFEKO, the gain was found to be 5.9 dBi. This value is a satisfactorily high value of gain for such antennas.

4.2 Voltage Standing Wave Ratio (VSWR)

The Voltage Standing Wave Ratio signifies the level of mismatch between the antenna and its feed line [11]. Value of VSWR can range from 1 to ∞. For a good practical antenna, VSWR value under 2 is desired. Poor matching of impedance between patch and feedline causes VSWR value to exceed 2. After the simulation, the value of VSWR is found to be 1.2 at the resonant frequency. Variation in VSWR as function of frequency is illustrated in Fig. 5.

4.3 Radiation Pattern

Radiation Pattern of the antenna is a graphical representation of the energy radiated in different directions. Diagrammatic representation of the distribution of radiated energy in space, as a function of direction [11]. This representation can be 2 dimensional (planar) or 3 dimensional. Figure 6 shows the 2D and 3D radiation pattern of the antenna observed through simulation in the YZ plane.

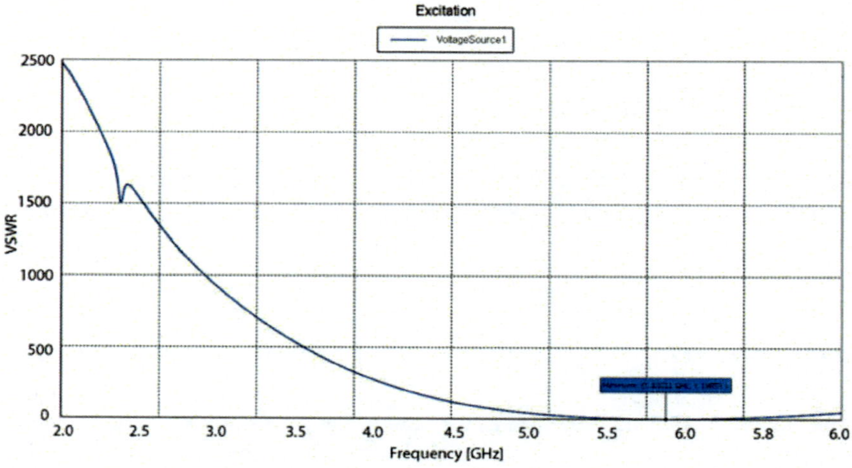

Fig. 5 VSWR plot for different frequencies

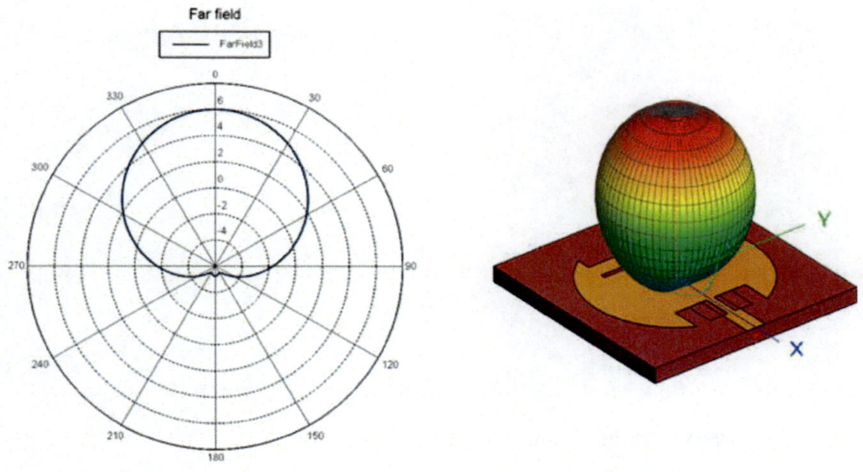

Fig. 6 2D and 3D radiation pattern

4.4 Reflection Coefficient

Reflection Coefficient should be ideally equal to zero as it is desired that the entire electromagnetic wave should get radiated and should not bounce back or get reflected. For a decent performance antenna, the value of the reflection coefficient should be under −10 dB or even under −15 dB for a better performance. Simulation of the proposed antenna design showed that the this value at the desired resonant frequency was −20.88 dB. Figure 7 shows the variation of the reflection coefficient as the function of frequency.

5 Implementation and Results

Fabrication of the proposed design was carried out using manual photolithographic process. In this method, the design is masked on the FR4 substrate having a layer of conducting copper on both top and bottom side. The excess unwanted conducting region is then etched out using ferric chloride (FeCL3) solution. Etching of the patch was done and a SMA connector was soldered at the antenna port. Virtual Network Analyzer (VNA) is a device used for testing characteristics of the antenna. Different characteristics of this fabricated antenna were then measured using a VNA machine by Keysight Technologies also known as VNA Tester. This subsection describes the comparison of results obtained from the VNA testing with the simulation results observed earlier.

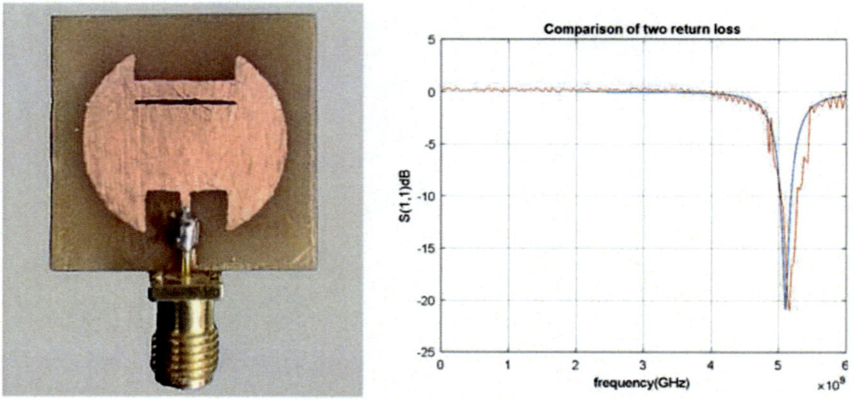

Fig. 7 Fabricated antenna and testing results observed on VNA

5.1 Comparison with Simulated Results

After the fabrication of the proposed design, the performance metrics and parameters like Resonant Frequency (f_r) and Magnitude of Reflection Coefficient were evaluated using a Vector Network Analyzer (VNA). The minimum value of reflection coefficient was found at 5.16 GHz frequency. This minimum value as seen on VNA was −20.48 dB. This clearly indicates that that the resonance was achieved at a frequency close to the target frequency. The magnitude of the reflection coefficient at the resonant frequency was found very close to the simulated value. Thus a good agreement between measured values and simulated values was achieved.

5.2 Comparison with Other Designs

This subsection discuss the comparison of performance of the proposed antenna design with other designs proposed in the literature by other researchers. Table 1 shows the comparison of the proposed design with 3 other designs in the references in terms of shape, frequency of operation, gain and return loss. It can be observed that proposed design has a very good gain and much lesser return loss which is desired for a good antenna.

Table 1 Comparison table

Design	Shape	Frequency (GHz)	Gain (dBi)	Return loss (dB)
Proposed	Circular slotted	5.1	5.9	−21
[6]	Rectangular slotted	2.4	1.75	−14
[14]	Quad	5.1	4.2	−18
[15]	E shaped	5.1	4.1	−21

6 Conclusion

Design of circular-shaped microstrip patch antenna with resonating frequency 5.1 GHz was proposed and simulated using CADFEKO software. Modifications were made in the basic conventional design to obtain the desired performance characteristics. Return loss was reduced by better impedance matching by adding some slots in the design. The maximum gain of 5.9 dBi was achieved which is very good for patch antenna designed with FR4 as substrate. The minimum value of reflection of the coefficient was observed to be −20.88 dB. After obtaining satisfactory simulation results, the design was implemented and fabricated. Results obtained for the simulation are found to be in a very agreement with the results obtained after testing of fabricated design using VNA. As part of further studies, a mathematical model of the proposed design can be proposed and the practical results obtained through simulation and fabrication can be justified and proved theoretically. The gain of the antenna can be further improved by using better substrate and higher quality fabrication techniques.

References

1. NishaBegam, R., Srithulasiraman, R.: The study of microstrip antenna and their applications. In: 2015 Online International Conference on Green Engineering and Technologies (IC-GET), Coimbatore, pp. 1–3 (2015). https://doi.org/10.1109/GET.2015.7453852
2. "5 Ghz IEEE 802.11a for Interference Avoidance" by Motorola Solutions
3. Naik, G., Liu, J., Jerry Park, J.-M.: Coexistence of wireless technologies in the 5 GHz bands: a survey of existing solutions and a roadmap for future research. IEEE Commun. Surv. Tutorials **20** (2018)
4. Liu, Y., Si, L.-M., Wei, M.: Some recent developments of microstrip antenna. Int. J. Antennas Propagation (2012)
5. Werfelli, H., Tayari, K., Chaoui, M., Lahiani, M., Ghariani, H.: Design of rectangular microstrip patch antenna. In: 2016 2nd International Conference on Advanced Technologies for Signal and Image Processing (ATSIP), Monastir, pp. 798–803 (2016). https://doi.org/10.1109/ATSIP.2016.7523197
6. Anusury, K., Dollapalli, S., Survi, H., Kothari, A., Peshwe, P.: Microstrip patch antenna for 2.4 GHz using slotted ground plane. In: 2019 10th International Conference on Computing, Communication and Networking Technologies (ICCCNT), Kanpur, India, pp. 1–6 (2019). https://doi.org/10.1109/ICCCNT45670.2019.8944653

7. Aburgiga, A.M., Shebani, N.M., Zerek, A.R., Kaeib, A.F.: Simulation and analysis of microstrip patch antenna for WLAN applications. In: 2019 19th International Conference on Sciences and Techniques of Automatic Control and Computer Engineering (STA), Sousse, Tunisia, pp. 660–665 (2019). https://doi.org/10.1109/STA.2019.8717278
8. Casu, G., Moraru, C., Kovacs, A.: Design and implementation of microstrip patch antenna array. In: 2014 10th International Conference on Communications (COMM), Bucharest, pp. 1–4 (2014). https://doi.org/10.1109/ICComm.2014.6866738
9. Salai Thillai Thilagam, J., Ganesh Babu, T.R.: Rectangular microstrip patch antenna at ISM band. In: 2018 Second International Conference on Computing Methodologies and Communication (ICCMC), Erode, pp. 91–95 (2018). https://doi.org/10.1109/ICCMC.2018.8487877
10. Khilariwal, S., Verma, R., Upadhayay, M.D.: Design of notch antenna for 5 GHz high speed LAN. In: 2016 International Conference on Wireless Communications, Signal Processing and Networking (WiSPNET), Chennai, pp. 999–1002 (2016). https://doi.org/10.1109/WiSPNET.2016.7566286
11. Balanis, C.A.: Antenna Theory: Analysis and Design, 1st edn (1982)
12. Alisher, B., Fazilbek, Z.: Feed line calculations of microstrip antenna. Int. J. Res. Appl. Sci. Eng. Technol. (IJRASET)
13. Ramamurthy, S., Gopal, D.: Effect of slot size variations on microstrip patch antenna performance for 5G applications. Int. J. Adv. Res. Electron. Commun. Eng. (IJARECE) **7** (2018)
14. Majidi, N., Yaralioglu, G.G., Sobhani, M.R., Imeci, T.: Design of a quad element patch antenna at 5.8 GHz. In: International Applied Computational Electromagnetics Society Symposium (ACES), Denver, CO, pp. 1–2 (2018). https://doi.org/10.23919/ROPACES.2018.8364309
15. Sharma, S., Tripathy, M.R.: Enhanced E- shaped patch antenna for 5.3 GHz ISM band using frequency selective surface. In: 2020 4th International Conference on Trends in Electronics and Informatics (ICOEI) (48184), Tirunelveli, India, pp. 332–336 (2020). https://doi.org/10.1109/ICOEI48184.2020.9142957

Reflection Characteristics Improvement of Wideband Coupled Line Power Divider Using SRR

S. Julius Fusic, T. Sugumari, and S. C. Shivaprakash

1 Introduction

A dual-band Wilkinson power divider is an existing coupled line circuit structure. It is operating at 1.1 and 2.2 GHz frequencies [1]. The design of the dual-band power divider with the variable power dividing ratio, it is simple and compact planar structure with equal power split, and also partial coupled lines are used [2]. Two arbitrary different frequencies models are presented [3]. It is a conventional Wilkinson power divider such as good isolation between the two output ports and all ports should be perfect impedance matching. Compact differential-mode power splitters with dual-band functionality based on composite RLH metamaterial, this splitter is also act as a filter. This is multi-functionality output, where the size is critical [4].

At the midpoints of two coupled line sections with a parallel *LC* circuit [5], so the *LC*-circuit design is should be complicated. Coupled line sections reduce the circuit size, in this paper, there is no need of extra compensation circuits. The Wilkinson original power divider [6]operates at a single band frequency and also it consists of two quarter wavelength lines and then to satisfied the dual-band operations to modified the power divider in [7, 8].

A conventional transmission line is compared with the coupled line in [9] that structures have some advantages, such as compact structures and introducing even- and odd-mode impedances due to flexible design parameters. The electronics engineer Ernest Wilkinson, who initially developed the Wilkinson power divider in the 1960's. In the dual band power divider, a frequency ratio limitation [6] is small. A single or dual-band operations in a coupled line power divider are more research

S. Julius Fusic (✉)
Thiagarajar College of Engineering, Madurai, Tamilnadu, India
e-mail: sjf@tce.edu

T. Sugumari · S. C. Shivaprakash
K.L.N. College of Engineering, Madurai, Tamilnadu, India

M. Dua et al. (eds.), *Proceedings of the International Conference on Paradigms of Communication, Computing and Data Sciences*, Algorithms for Intelligent Systems,
https://doi.org/10.1007/978-981-16-5747-4_2

interests [10]. Recently, metamaterials have been one of the popular research areas in the field of microwaves. The first discovery [11] had been proposed by the Russian physicist Prof. V. Veselago in 1968. The word "Meta-material" is a combination of "meta" and "material" [12], Meta is a Greek word which means altered or changed. It is useful to change the electromagnetic properties. When J. B. Pendry et al. the breakthrough came around the year 2000 [13, 14] that showed an array of circular rings and wire strip could exhibit negative permeability, negative permittivity and hence negative refractive index. From the various structures of split ring resonators are like square, circle, omega shaped, hexagonal-shaped resonators, etc., are used. In terms of a macroscopic permeability function [15, 16] can be expressed by the magnetic response of the artificial material. The circuit parameters may be fully obtained from the geometrical dimensions. *LC*-circuit [17] models can be approached by relatively easy these are the main properties of SRR. In an *LC* resonant circuit, the gap between inner and outer ring acts as a capacitor while the rings themselves act as an inductor [18, 19]. The bandwidth can be improved by using metamaterial in power divider.

2 Split Ring Resonator Design and Equivalent Circuit Analysis

$$f_0 = \frac{1}{2\pi\sqrt{LC}} = 1.65\,\text{GHz} \tag{1}$$

$$\mu_0 = 1.256 \times 10^{-6} \tag{2}$$

$$\varepsilon_0 = 8.854 \times 10^{-12} \tag{3}$$

$$\varepsilon_r = 3.5 \tag{4}$$

To find Inductance:

$$L = \frac{4.86\mu_0}{2}(a - w - d)\left[\ln(0.98/\rho) + 1.84\rho\right] \tag{5}$$

where

$$\rho = \frac{w + d}{a - w - d} \tag{6}$$

To find capacitance:

$$C = (a - 1.5(w + d))C_{\text{pul}} \tag{7}$$

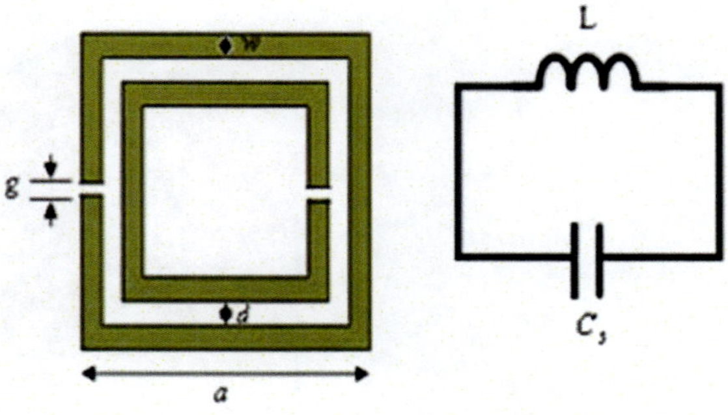

Fig. 1 Structure and equivalent circuit of metamaterial

$$C_{\text{pul}} = \varepsilon_0 * \varepsilon_{\text{eff}} * K \tag{8}$$

$$\text{Where,}\quad \varepsilon_{\text{eff}} = \frac{\varepsilon_r + 1}{2} \tag{9}$$

$$K = \left[\frac{1}{\pi}\ln\left(2\frac{1+\sqrt{k_1}}{1-\sqrt{k_1}}\right)\right]^{-1} \quad \text{for}\quad 0 \le k \le 0.7 \tag{10}$$

$$K = \left[\frac{1}{\pi}\ln\left(2\frac{1+\sqrt{k}}{1-\sqrt{k}}\right)\right] \text{for}\quad 0.7 \le k \le 1 \tag{11}$$

$$\text{where,}\quad k = \frac{d}{d + 2 * w} \tag{12}$$

$$k_1 = \sqrt{1 - k^2} \tag{13}$$

L Inductance = 1.8884e–008 = 18.884 nH
C Capacitance = 2.8050e–014 = 28.05 fF.

The geometry of two split ring resonator is shown in Fig. 2, and then, the simulated scattering parameters are shown in Figs. 1 and 3.

3 Design of Proposed Wideband Power Divider

One of the drawbacks of the existing power divider is only operating at a dual-band frequency. So, to improve the coupled line power divider performance and to broaden the bandwidth, a metamaterial structure has been proposed. Figure 4 shows the proposed to the bandwidth enhancement of coupled line power divider. The circuit layout is a type of metamaterial. The structure is similar to a split ring resonator. The metamaterial structure is located at two output ports. The output port

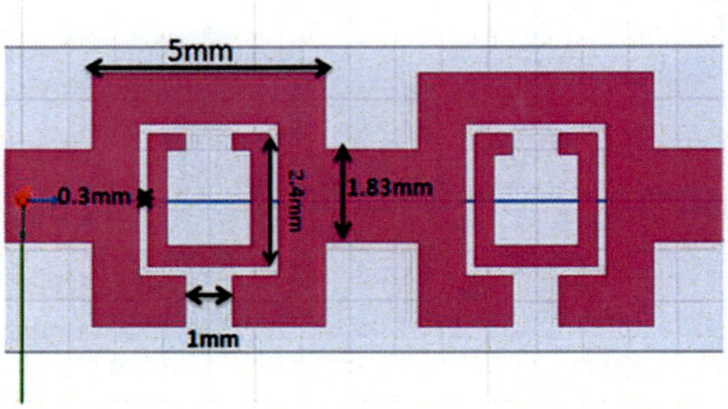

Fig. 2 Geometry of split ring resonator

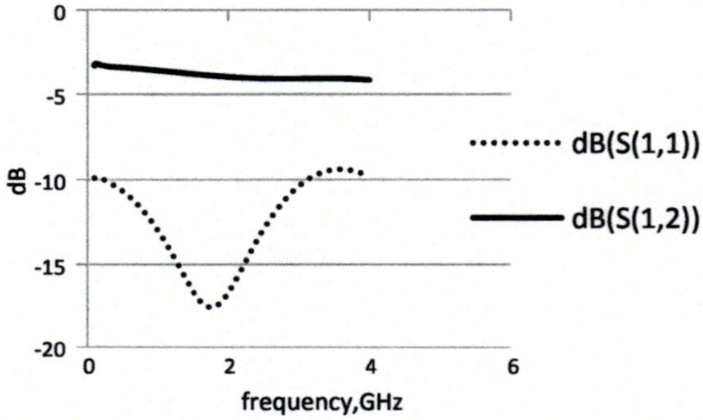

Fig. 3 Simulated scattering parameters

length is 15 mm, to fix the structure of two metamaterial of 5 mm of length. The power divider is fabricated on a 0.8 mm thickness (i.e.,) h = 0.8 mm Taconic RF-35 substrate is used.

Table 1 shows the essential parameters for the design the proposed power divider. $f_0 = 1.65$ GHz $\varepsilon_r = 3.5$ h = 0.8 mm tan $\delta = 0.008$.

4 Comparison of Results and Discussion

The proposed wide band power divider is designed to operate in 0.5–3.0 GHz has been simulated by using the simulator Ansoft 15 High Frequency Structure Simulator (HFSS) and the scattering parameters are analyzed. The power divider is simulated with Taconic-RF35 substrate which has 0.8 mm thick and has a relative permittivity of 3.5. The port impedances are 50, and Fig. 5 shows the different structures of SRR and their detail.

Figures 6, 7, 8, 9, 10 and 11 show the comparison of all the different structures and their results of S_{11}, S_{22}, S_{33}, S_{23}, S_{12}, S_{13}. Compare all these results and select best one is 2srr. Figures 12 and 13 show the best design and result.

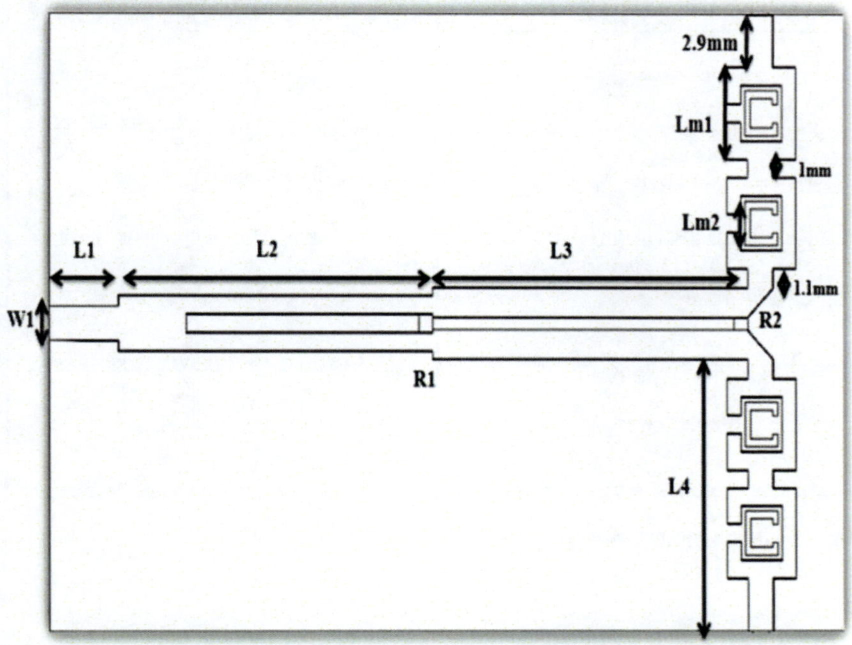

Fig. 4 Layout of proposed power divider

Table 1 Parameters for proposed power divider

Substrate material	Taconic RF-35
Relative permittivity ε_r	3.5
Thickness of the substrate	0.8 mm
$W1$	1.83 mm
$L1$	5 mm
$L2$ and $L3$	23 mm
$L4$	15 mm
Lm1 and Lm2	5 and 2.4 mm
$R1$ and $R2$ resistances value	100 and 300 Ω

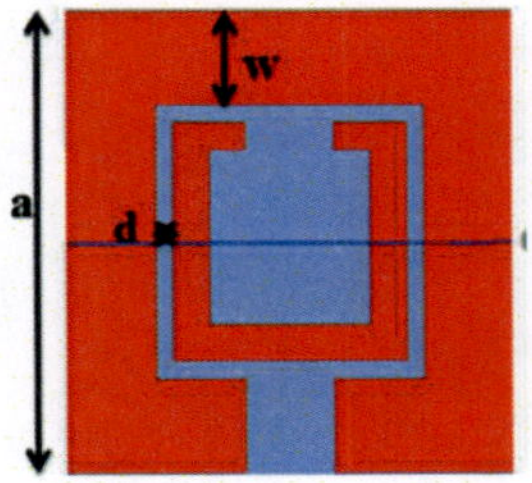

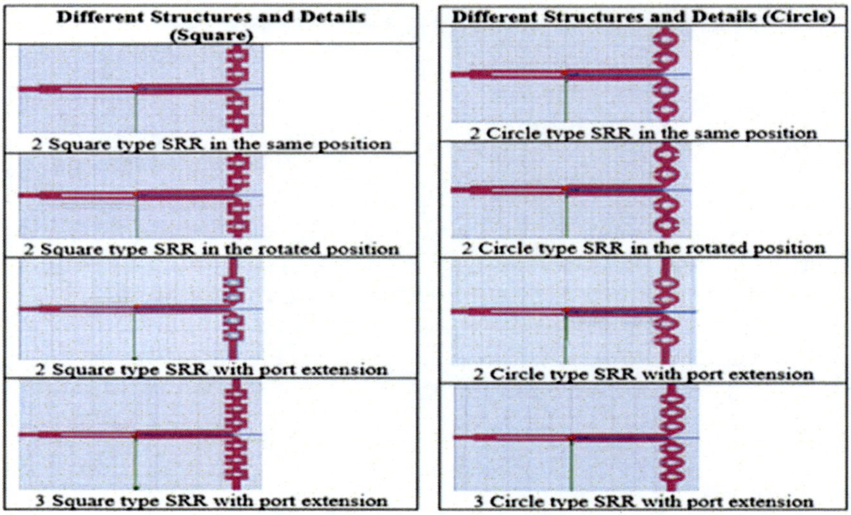

Fig. 5 Different square type and circle type SRR

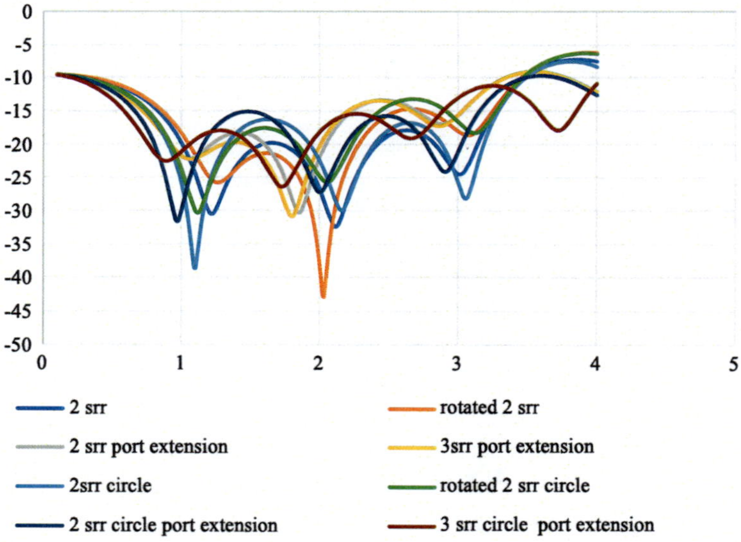

Fig. 6 Comparison of S_{11} results

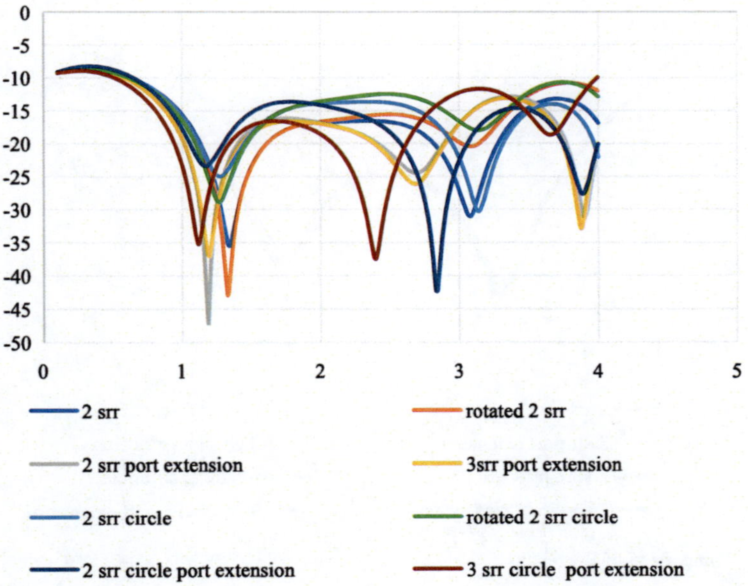

Fig. 7 Comparison of S_{22} results

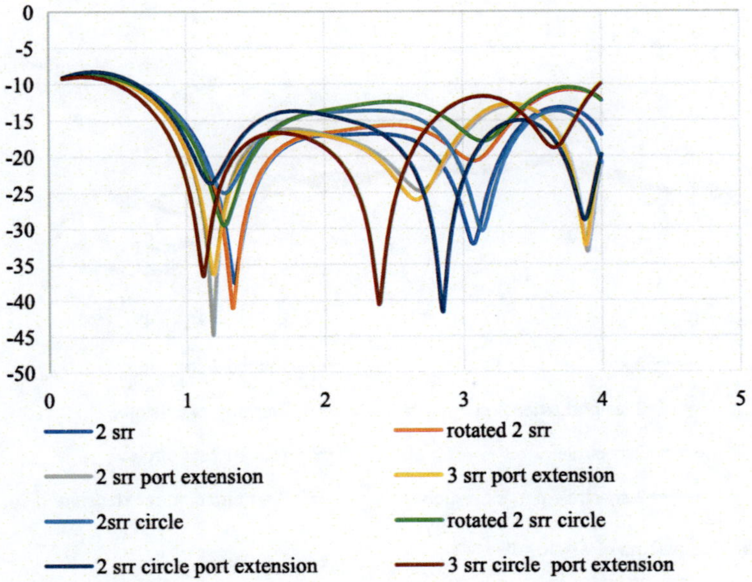

Fig. 8 Comparison of S_{33} results

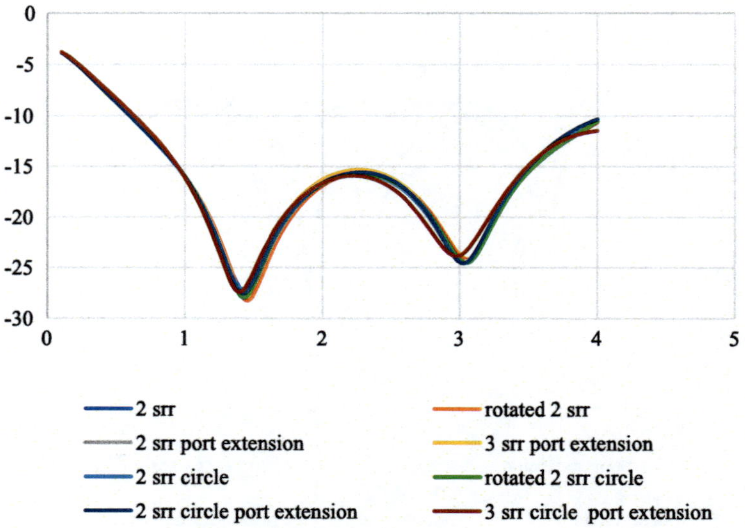

Fig. 9 Comparison of S_{23} results

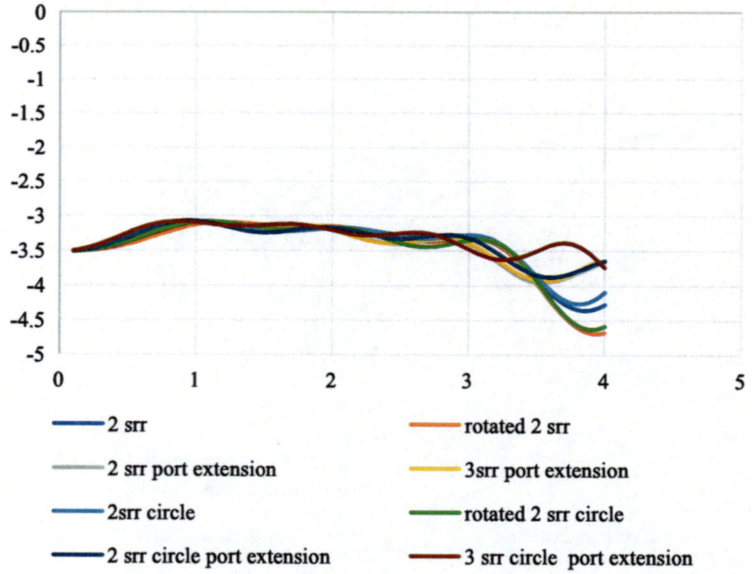

Fig. 10 Comparison of S_{12} results

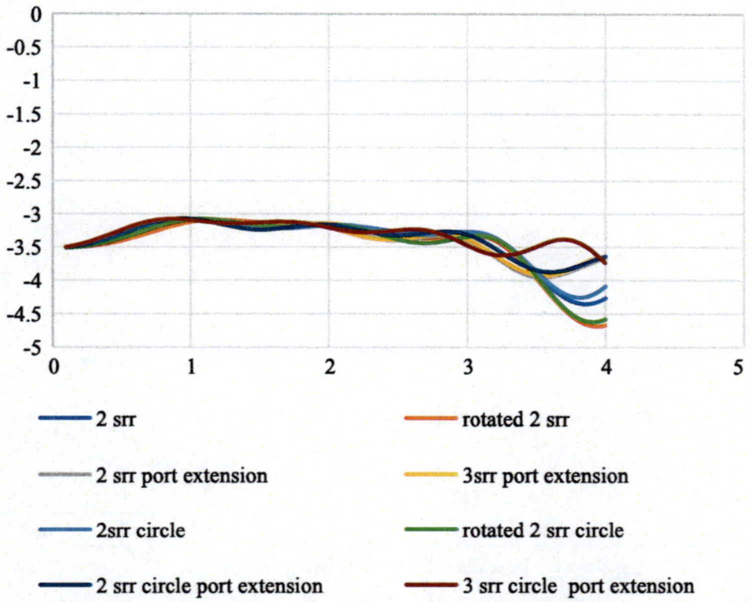

Fig. 11 Comparison of S_{13} results

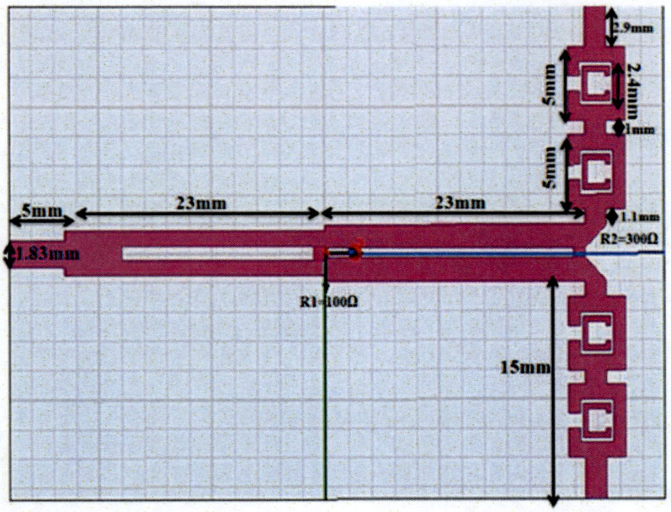

Fig. 12 Final structure of propose power divider

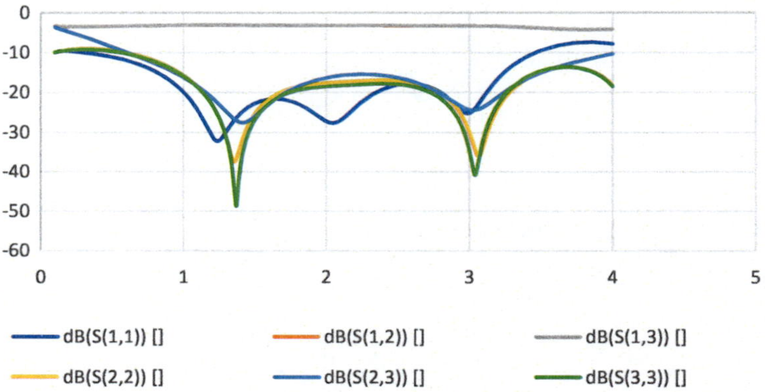

Fig. 13 Final result of proposed power divider

Table 2 Simulated results by using HFSS

S-parameter	Simulated result	dB
S_{11}	Below −10 dB in 0.5–3.5 GHz	−32
S_{22}, S_{33}	Below −10 dB in 0.6–3.7 GHz	−42
S_{23}	Below −10 dB in 0.6–4 GHz	−28
S_{21}, S_{31}	Maintained at −3.1 to −3.45 dB from 0.5 to 3.51 GHz	−3.4

- $a = 5$ mm
- $w = 1$ mm
- $d = 0.3$ mm

where

a the length of the side of the square.
w the width of the conductor.
d the dielectric width between the inner and the outer square.

5 Fabrication and Measured Result

RF 35 of 31mil fabricated by Taconic as per the proposed design was taken for manufacturing the PCB board. Relative permittivity of the fabricated Material is 3.5 and also thickness of the board is 0.8 mm. After the manufacturing of PCB, the connectors and resistors were soldered with their corresponding locations. Here SMA Female Type Edge Connector is used and two resistors are used that is SMD resistors 100 Ω, 200 Ω. Figure 14 shows the prototype of the proposed design. Network analyzer is used to measure the result of fabricated power divider.

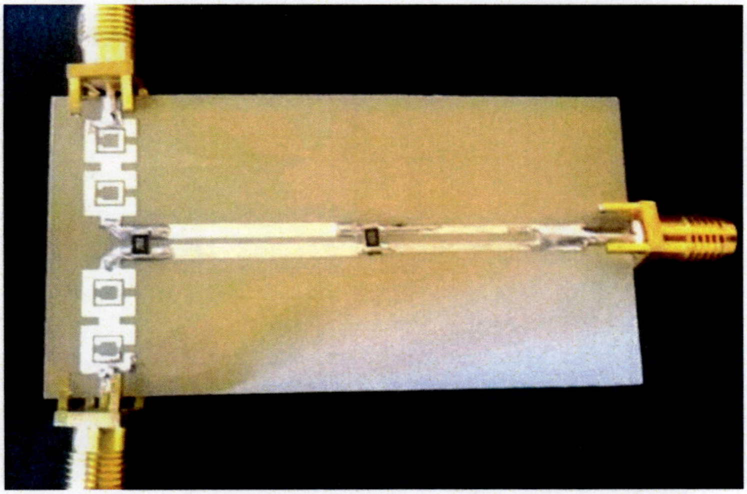

Fig. 14 Photograph of the fabricated power divider

Figures 15, 16 and 17 show the final result of fabricated proposed power divider. Table 3 shows the measured result of fabricated power divider and below Table 4 shows the comparison between the simulate and the fabricated power divider result.

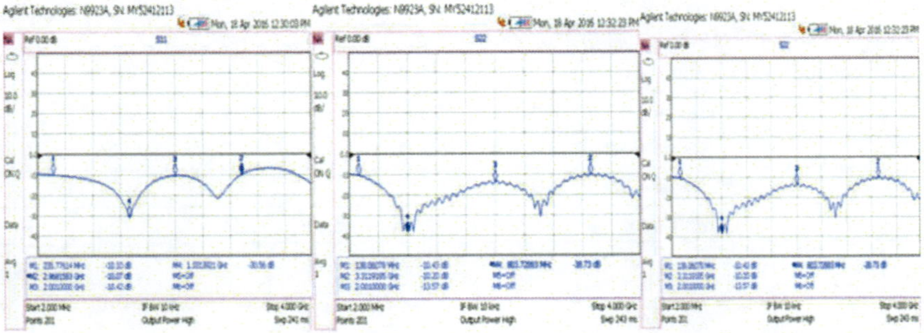

Fig. 15 S_{11}, S_{22} and S_{33} measured result by using vector network analyzer (Return loss)

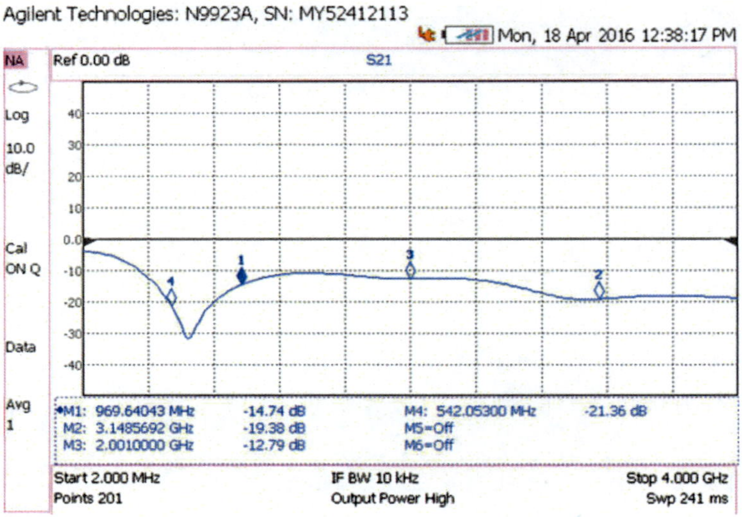

Fig. 16 S_{23} measured result by using vector network analyzer (Isolation)

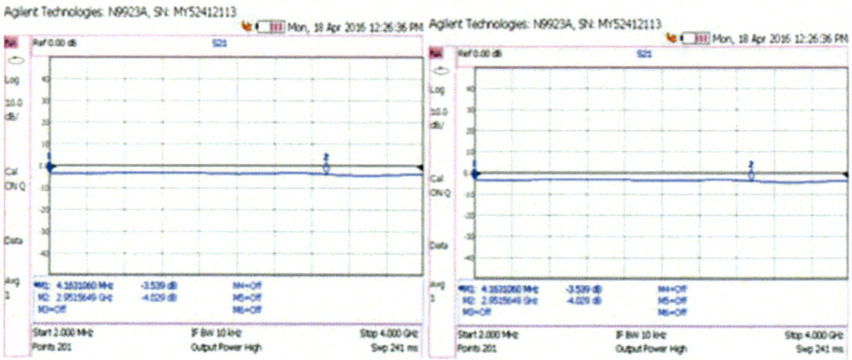

Fig. 17 S_{12} and S_{13} measured result by using vector network analyzer

Table 3 Measured results by using network analyzer

S-parameter	Simulated result	dB
S_{11}	Below −10 dB in 0.5–3.45 GHz	−31
S_{22}, S_{33}	Below −10 dB in 0.5–4.0 GHz	−26
S_{23}	Below −10 dB in 0.5–4 GHz	−22.34
S_{21}, S_{31}	Maintained at −3.8 to −4.5 dB from 0.5 to 3.0 GHz	−3.6

Table 4 Comparison between simulated and fabricated result

S-parameter	Simulated result	Measured result
S_{11}	Below −10 dB from **0.5 to 3.45 GHz**	Below −10 dB from **0.5 to 3.45 GHz**
S_{22}, S_{33}	Below −10 dB from **0.6 to 3.52 GHz**	Below −10 dB from **0.5 to 4.0 GHz**
S_{23}	Below −10 dB from **0.6 to 4 GHz**	Below −10 dB from **0.5 to 4.0 GHz**
S_{21}, S_{31}	Maintained at −3.1 to −3.4 dB from **0.5 to 3.5 GHz**	Maintained at −3.8 to −4.5 dB from**0.5 to 3.0 GHz**

6 Conclusion

The bandwidth of the existing coupled line dual-band power divider operates at 1.1 and 2.2 GHz is improved to 0.5–3.5 GHz is proposed. Hence, the proposed power divider works under the wide range of bandwidth. Split ring resonator structure is used to enhance the bandwidth. The bandwidth of power divider is 3.0 GHz. It has the enhanced the bandwidth compared with the previous dual-band coupled line power divider. It is also compared all the structures and simulated the results chosen the best one. The return loss is also reduced up to −32 dB. Isolation between the ports is obtained −28 dB. Insertion loss is maintained −3.1 dB. So, the proposed power is well suited for *L* Band, *S* band applications, and it is used for mobile phone applications up to 4G. In the future work, the proposed model will implement in MIMO antenna and improve the level bandwidth for 5G mobile antenna applications.

References

1. An analytical approach for a novel coupled-line dual-band Wilkinson power divider. IEEE Trans. Microwave Theor. Tech. **59**(2) (Feb, 2011)
2. Lin, Z., Chu, Q.-X.: A novel approach to the design of dual- band power divider with variable power dividing ratio based on coupled-lines. IEEE Microwave Wirel. Compon. Lett. **22**, 16–18 (2010)
3. Wu, L., Sun, Z., Yilmaz, H., Berroth, M.: A dual-frequency Wilkinson power divider. IEEE Trans. Microw. Theory Tech. **54**(1), 278–284 (2006)
4. Velez, P. (Student Member, IEEE), Duran-Sindreu, M. (Member): Compact dual-band differential power splitter with common-mode suppression and filtering capability based on differential-mode composite right/left-handed transmission-line metamaterials **13** (2014)
5. Wang, X., (Member, IEEE), Sakagami, I., (Life Member, IEEE), Ma, Z., (Member, IEEE), Mase, A., Yoshikawa, M., Ichimura, M.: Miniaturized dual-band Wilkinson power divider with self-compensation structure. IEEE Trans. Compon. Packag. Manuf. Technol. **5**(3) (Mar, 2015)
6. S. B. Cohn.: A class of broadband three-port TEM-Mode hybrids. In: IEEE Transactions on Microwave Theory and Techniques, vol. 16, no. 2, pp. 110–116, (February 1968)
7. Cheng, K.-K.M., Law, C.: A novel approach to the design and implementation of dual-band power divider. IEEE Trans. Microw. Theor. Tech. **56**(2), 487–492 (2008)
8. Yang, T., Chen, J.-X., Zhang, X.Y., Xue, Q.: A dual-band out-of phase power divider. IEEE Microw. Wireless Compon. Lett. **18**(3), 188–190 (2008)

9. Tang, X., Mouthaan, K.: Analysis and design of compact two-way Wilkinson power dividers using coupled lines. Asia–Pacific Microw. Conf. 1319 (7–10 Dec, 2009)
10. Park, M.-J.: Two-section cascaded coupled line Wilkinson power divider for dual-band applications. IEEE Microw. Wireless Compon. Lett. **19**(188–190) (Apr, 2009)
11. Pendry, J.B., Holden, A.J., Robbins, D.J. Stewart, W.J.: Magnetism from conductors and enhanced nonlinear phenomena. IEEE Trans. Microw. Theor. Techn. **74**(11) (Nov, 1999)
12. Castro, P.J., Barroso, J.J., LeiteNeto, J.P.: Experimental study on split-ring resonators with different slit widths. J. Electromagn. Anal. Appl. **5**, 366–370 (2013)
13. Vidyalakshmi, M.R., Dr. Raghavan, S. (National Institute of Technology, Tiruchirappalli): Comparison of optimization techniques for square split ring resonator. Int. J. Microw. Opt. Technol. **5**(5) (Sept, 2010)
14. De ParulDawar, A.: Bandwidth enhancement of RMPA using 2 segment Labyrinth metamaterial at THz. Mater. Sci. Appl. **4**, 579–588 (2013)
15. Smith, D.R. et al.: Composite medium with simultaneously negative permeability and permittivity. Phys. Rev. Lett. **84**, 4184 (Mai, 2000)
16. Ishimaru, A., Seung-Woo, L., Kuga, Y., Jandhyala, V.: Generalized constitutive relations for metamaterials based on the quasi-static Lorentz theory. IEEE Trans. Antennas Propag. (Special Issue) **51**(10), 2550–2557 (2003)
17. Gupta, M., Saxena, J.: Microstrip filter designing by SRR metamaterial. Wirless Pers. Commun. **71**(4), 3011–3022 (Aug, 2013)
18. Cui, T.J., Smith, D., Liu, R.: Metamaterials: Theory, design, and applications (2010)
19. Landau, L., Lifschitz, E.M.: Electrodynamics of Continuous Media. Pergamon, Oxford, U.K. (2000)

All-Optical Frequency Encoded Dibit-Based Half Subtractor Using Reflective Semiconductor Optical Amplifier with Simulative Verification

Surajit Bosu and Baibaswata Bhattacharjee

1 Introduction

In this modern world, photon is a good information transporter. Photon has super fast speed. That is why, it very preferable to the researcher rather than electron [1]. For information transmission, the encoding technique is very needful. The frequency encoding technique [4] is reliable among all other encoding techniques like intensity encoding [2], polarization encoding [18], phase encoding [12], spatial encoding [19], hybrid encoding [9], etc. Here, two different frequencies ν_1 and ν_2 represent the two different digital logic states '0' and '1', respectively. In dibit-based representation technique [4], two consecutive bit positions are chosen to represent a digit. The logic $< 0 >< 1 >$ represents the digital logic state '0' whereas $< 1 >< 0 >$ represents the digital logic state '1'. The two different frequencies ν_1, and ν_2 when placed side by side as $< \nu_1 >< \nu_2 >$ indicate the digital logic state '0' and for $< \nu_2 >< \nu_1 >$ indicate the logic state '1'. Mukhopadhyay [10] first proposed the dibit-based representation technique. MeGeehan et al. [8] proposed a method for simultaneous half subtraction and addition which is based on semiconductor optical amplifiers (SOA) and also uses the periodically poled lithium niobate (PPLN) waveguide for frequency encoding. Rakshit et al. [13] have introduced an all-optical adder and subtractor which is based on the micro-ring-resonator (MRR). Terahertz optical asymmetric demultiplexer-based all-optical half adder and half subtractor has been introduced by Gayen et al. [3]. Rakshit and Roy [14] proposed all-optical adder, subtractor, and comparator circuit using ultrafast switches. These ultrafast switches are based on MRR and circuits are verified by MATLAB software. Singh et al. [16] introduced and simulated half adder, half subtractor, and 4-bit decoder using SOA-MZI con-

S. Bosu (✉)
Bankura Sammilani College, Bankura, West Bengal, India

B. Bhattacharjee
Ramananda College, Bishnupur, Bankura, West Bengal, India

M. Dua et al. (eds.), *Proceedings of the International Conference on Paradigms of Communication, Computing and Data Sciences*, Algorithms for Intelligent Systems,
https://doi.org/10.1007/978-981-16-5747-4_3

figuration. Hu et al. [6] proposed all-optical logic gate and its application as adder, subtractor, etc. There model based on SOA using Quantum-Dot(QD-SOA). Nahata et al. [11] proposed all-optical full adder and full subtractor using SOA and simulated the model in optisystem-16. Mondal et al. [7] introduced designs of optical quaternary adder and subtractor based on high-speed polarization switch and designs are simulated in MATLAB. Swarnakar et al. [17] presented a design of half subtractor circuit using two-dimensional (2D) photonic crystal wave-guides. Optical amplifier and non-linear materials are not used in this model. They simulated their device through finite-difference time-domain and using MATLAB. Logic implementation on the fiber-based is bulky and easily not integrable. TOAD with asymmetrical switching window profile results in an increased crosstalk and reduced switching speed. These limitations can be overcome using frequency encoding in RSOA. PPLN waveguide based frequency encoding and SOA-based frequency encoding are polarization sensitive. In transmission, the state of polarization may change due to reflection and refraction. These problems can be avoided using frequency encoding in RSOA because our proposed designs are based on the cross gain modulation property of RSOA. In this paper, all-optical frequency encoded dibit-based half subtractor using reflective semiconductor optical amplifier (RSOA) is proposed. This devised design performs computation like subtraction at ultra-fast speed. By applying frequency encoding technique and dibit-based logic, this proposed device reduces the bit error problem.

2 Operational Principle

In this communication, a frequency encoded dibit-based half subtractor is devised. The basic components of this design are (a) switching action of RSOA; (b) frequency routing by ADM. The switching action of RSOA and ADM explain below.

2.1 Reflective Semiconductor Optical Amplifier (RSOA)

In RSOA, two facets design in such a way that one facet is highly reflecting coating (HR) and another facet is anti-reflecting (AR) coating [15]. It is low-power efficient, high gain productive with low noise. RSOA has a property that it can transmit a weak probe input signal to the output terminal with the power of the pump signal and it is given in Fig. 1. In these designs, the dotted line acts as the probe signal, and another line indicates the pump signal. The power of RSOA is in between 5 and 20 dBm [5]. The devised designs with RSOA have a great advantage because RSOA consumes very low power and works with a wide power range.

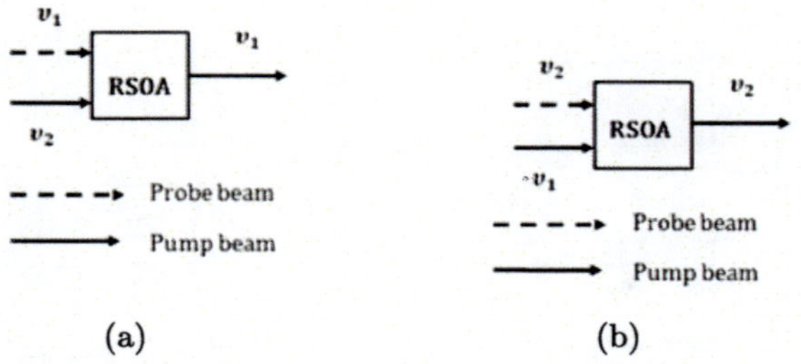

Fig. 1 The working function of RSOA. **a** The probe beam and pump beam are ν_1 and ν_2, respectively, but output gives ν_1. **b** The probe and pump beams are ν_2 and ν_1 respectively but output gives ν_2

2.2 *Add/Drop Multiplexer (ADM)*

Add/drop multiplexer acts as an optical switching device that selects all other frequencies except the biasing frequency at the out port and reflects the particular biasing frequency at the drop port [4]. By changing the biasing frequency, one can get any frequency at the out port and drop port, and it is given in Fig. 2.

3 Scheme of Operation of Frequency Encoded Dibit-Based Half Subtractor

The operation scheme of the proposed frequency encoded dibit-based subtractor is given in the following cases. The schematic block diagram of the proposed half subtractor is given in Fig. 3.

Case 1: Difference (d): In this case, input signal A ($< \nu_1 >< \nu_2 >$) works as dibit input, i.e., $A' = \nu_1$, $A'' = \nu_2$ and also for input signal B ($< \nu_1 >< \nu_2 >$) works as dibit input, i.e., $B' = \nu_1$, $B'' = \nu_2$. All the signals are applied simultaneously. The

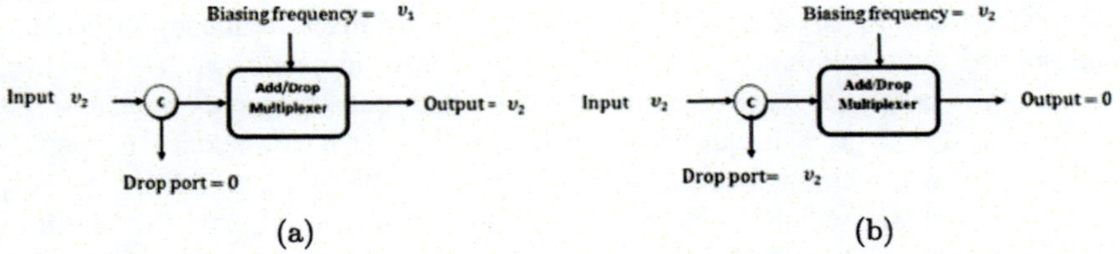

Fig. 2 The working function of the ADM **a** when biasing frequency and input frequency are different then the output port selects the input frequency and reflects nothing at the drop port. **b** When biasing frequency and input frequency are the same then ADM selects input frequency at the drop-port but passes nothing at the output port

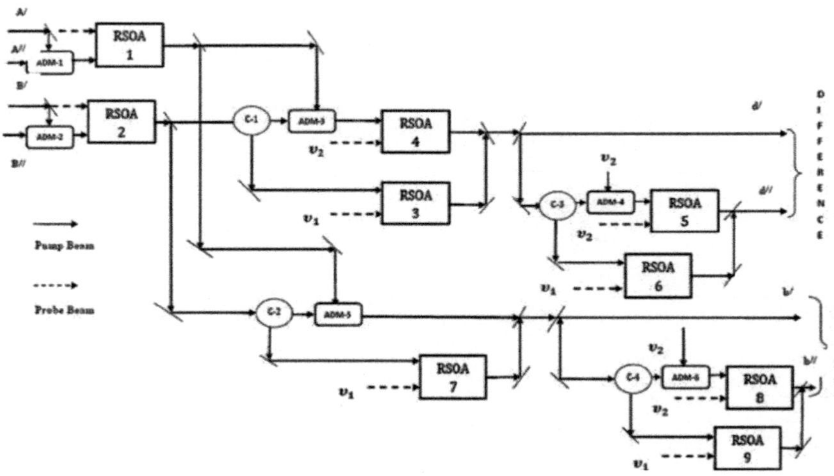

Fig. 3 Schematic diagram of proposed subtractor

real dibit input combination passes through the RSOA-1 and RSOA-2 only because ADM-1 and ADM-2 are used as dibit checking devices. RSOA-1 and RSOA-2 give the output frequency ν_1. The ν_1 frequencies from RSOA-1 and RSOA-2 reach the ADM-3. They act as a biasing frequency and input frequency of ADM-3, respectively. Since biasing and input frequencies are the same, ADM-3 reflects the frequency ν_1. Therefore, the reflected frequency of ADM-3 injects into the RSOA-3 as a pump signal. The RSOA-3 gives the ν_1 frequency at the output d' because ν_1 is the fixed probe signal of RSOA-3, whereas RSOA-4 does not work due to the absence of the pump power. As a result, the output of RSOA-3 reaches the ADM-4 and is selected by ADM-4 due to the biasing frequency ν_2. The selected frequency ν_1 of ADM-4 acts as a pump signal of RSOA-5 which gives ν_2 frequency at the output d'' due to a fixed probe signal ν_2 of RSOA-5. On the other hand, RSOA-6 does not work due to the absence of pump power. Therefore, the output dibit difference (d) is $< \nu_1 >< \nu_2 >$ i.e., $< 0 >< 1 >$ which represents the digital logic state '0'.

For Borrow (b): The output of RSOA-1 and RSOA-2 are ν_1 because probe signal of RSOA-1 and RSOA-2 are frequency ν_1. ADM-5 reflects frequency ν_1 because biasing and input frequency of ADM-5 is the same as frequency ν_1. The reflected frequency ν_1 acts as a pump signal of RSOA-7, and probe frequency is also ν_1. As a result, RSOA-7 gives frequency ν_1 at the final output b'. Frequency ν_1 reaches also the ADM-6 which is biased with ν_2 frequency. Therefore, ADM-6 can easily select ν_1 frequency which injects into the RSOA-8 as a pump signal. The output b'' of RSOA-8 is ν_2 and RSOA-9 does not work due to the absence of pump signal. Therefore, the output dibit borrow (b) is $< \nu_1 >< \nu_2 >$, i.e., $< 0 >< 1 >$ which implies the '0' digital logic state.

Case 2: For Difference (d): In this case, input signal A ($< \nu_1 >< \nu_2 >$) works as dibit input, i.e., $A' = \nu_1$, $A'' = \nu_2$ and also for input signal B ($< \nu_2 >< \nu_1 >$) works as dibit input, i.e., $B' = \nu_2$, $B'' = \nu_1$. All the signals are applied simultaneously. The real dibit input combination passes through the RSOA-1 and RSOA-2 only because ADM-1 and ADM-2 are used as dibit checking devices. Since ν_1 and ν_2 are the probe signal of RSOA-1 and RSOA-2 then ν_1 and ν_2 are the output of RSOA-1 and RSOA-2, respectively. The ν_2 frequency reaches the ADM-3 and ADM-5 but ADM-3 selects the frequency ν_2 because ADM-3 is biased with ν_1 and ADM-5 reflects the frequency ν_2 on the other hand. Therefore, the selected frequency of ADM-3 injects into the RSOA-4 as a pump signal which gives the ν_2 frequency at the output d' because ν_2 is the fixed probe signal of RSOA-4 whereas RSOA-3 does not work due to the absence of pump power. As a result, the output from RSOA-4 reaches to the ADM-4 and which is reflected by ADM-4 due to the same biasing frequency ν_2. The reflected frequency by ADM-4 acts as a pump signal of RSOA-5 which gives ν_1 frequency at the output d''. Therefore, the output dibit difference (d) is $< \nu_2 >< \nu_1 >$, i.e., $< 1 >< 0 >$ which represents the digital logic state '1'.

For Borrow (b): The output of RSOA-1 and RSOA-2 are ν_1, and ν_2, respectively. This frequency ν_2 passes through the ADM-5 because it is biased by the frequency ν_1. Therefore, frequency ν_2 reaches output b'' which reaches also to ADM-6 which is biased with frequency ν_2 then it reflects the frequency ν_2 to the RSOA-9. Frequency ν_1 is obtained at the output b'', because ν_1 is the probe signal of RSOA-9. Here, RSOA-8 does not work due to the absence of the pump signal. Therefore, the output dibit borrow (b) is $< \nu_2 >< \nu_1 >$, i.e., $< 1 >< 0 >$ which implies the '1' digital logic state.

Case 3: For Difference (d): In this case, input signal A ($< \nu_2 >< \nu_1 >$) works as dibit input, i.e., $A' = \nu_2$, $A'' = \nu_1$ and also for input signal B ($< \nu_1 >< \nu_2 >$) works as dibit input, i.e., $B' = \nu_1$, $B'' = \nu_2$. All the signals are applied simultaneously. The real dibit input combination passes through the RSOA-1 and RSOA-2 only because ADM-1 and ADM-2 are used as dibit checking devices. Since ν_2 and ν_1 are the probe signals of RSOA-1 and RSOA-2 then ν_2 and ν_1 are the outputs of RSOA-1 and RSOA-2, respectively. Output frequencies from RSOA-1 to RSOA-2 reach the ADM-3. These frequencies act as biasing frequency and input frequency of ADM-3 respectively. ADM-3 selects the frequency ν_1 because the biasing frequency of ADM-3 is frequency ν_2. Therefore, selected frequency of ADM-3 injects into the RSOA-4 as a pump signal which gives the ν_2 frequency at the output terminal d', because ν_2 is the fixed probe signal of RSOA-4. Here, RSOA-3 does not work due to the absence of pump power. As a result, output from RSOA-4 reaches the ADM-4 and reflected by ADM-4 due to the same biasing frequency ν_2. The reflected frequency by ADM-4 acts as a pump beam of RSOA-6 which gives ν_1 frequency at the d'' terminal. Therefore, the output dibit difference (d) is $< \nu_2 >< \nu_1 >$ i.e., $< 1 >< 0 >$ which represents the digital logic state '1'.

For Borrow (b): Here, the outputs of RSOA-1 and RSOA-2 are ν_2 and ν_1, because probe signals of RSOA-1 and RSOA-2 are ν_2 and ν_1, respectively. This ν_1 frequency passes through the ADM-5 because the biasing frequency of ADM-5 is ν_2. Therefore, ν_1 reaches the output b' and it reaches also to ADM-6 which is biased with frequency ν_2. Here, ADM-6 selects the frequency ν_1 to the RSOA-8. The output of RSOA-8 gives ν_2 at b'' because ν_2 is the fixed probe signal of RSOA-8, whereas RSOA-9 does

Table 1 Truth table of proposed half subtractor

Input						Output					
A'	A''	Digital state	B'	B''	Digital state	d'	d''	Digital state	b'	b''	Digital state
ν_1	ν_2	0	ν_1	ν_2	0	ν_1	ν_2	0	ν_1	ν_2	0
ν_1	ν_2	0	ν_2	ν_1	1	ν_2	ν_1	1	ν_2	ν_1	1
ν_2	ν_1	1	ν_1	ν_2	0	ν_2	ν_1	1	ν_1	ν_2	0
ν_2	ν_1	1	ν_2	ν_1	1	ν_1	ν_2	0	ν_1	ν_2	0

not work due to the absence of pump signal. Therefore, the output dibit borrow (b) is $< \nu_1 >< \nu_2 >$ i.e., $< 0 >< 1 >$ which implies the '0' digital logic state.

Case 4: This case is similar to case-1. Therefore, the output dibit difference (d) is $< \nu_1 >< \nu_2 >$, i.e., $< 0 >< 1 >$ which represents the digital logic state '0' and the output of dibit borrow (b) is $< \nu_1 >< \nu_2 >$ i.e., $< 0 >< 1 >$, digitally it shows '0' logic state. Truth table of half subtractor is given in Table 1.

4 Simulation Results of Proposed Half Subtractor

The proposed half subtractor is verified by simulation in MATLAB Simulink (R2018a) software. As shown in Fig. 4, input dibit signal (A) and signal (B) are applied to the input terminal of the half subtractor. Dibit difference (d) is represented by d', and d''. Similarly, borrow (b) is represented by b' and b''. RSOA and ADM blocks are programmed by MATLAB language to get output. The probe and pump signals are considered as say $\nu_1 = 193.5$ THz (digital logic state '0') and $\nu_2 = 194.1$ THz (digital logic state '1'), respectively. We consider also the probe beam power and pump beam power are -4 dBm, and 8 dBm, respectively, and the injection current of RSOA is 120 mA. We consider the gain saturation of RSOA is 20 dBm. If the probe signal and pump signal are considered as say $\nu_2 = 194.1$ THz and $\nu_1 = 193.5$ THz, respectively, then the output of RSOA is 194.1 THz. ADM block is programmed in such a way that ADM selects all other frequencies at the output except for the biasing frequency which reflects at the drop port. If biasing frequency and input frequency are considered as say $\nu_1 = 193.5$ THz, and $\nu_2 = 194.1$ THz, respectively, then ADM selects $\nu_2 = 194.1$ THz at the output and reflects zero at the drop port. But when the input and biasing frequency are considered the same then ADM reflects the input frequency at the drop port and selects nothing at the output port. If $< 193.5 >< 194.1 >$, $< 194.1 >< 193.5 >$ are applied to the dibit input (A), and input (B) terminal, respectively. As shown in Fig. 4, $< 194.1 >< 193.5 >$, $< 194.1 >< 193.5 >$ obtain at the output terminal of dibit difference (d), and borrow (b). Now all the simulation results of the proposed model are given in Table 2 of the half subtractor.

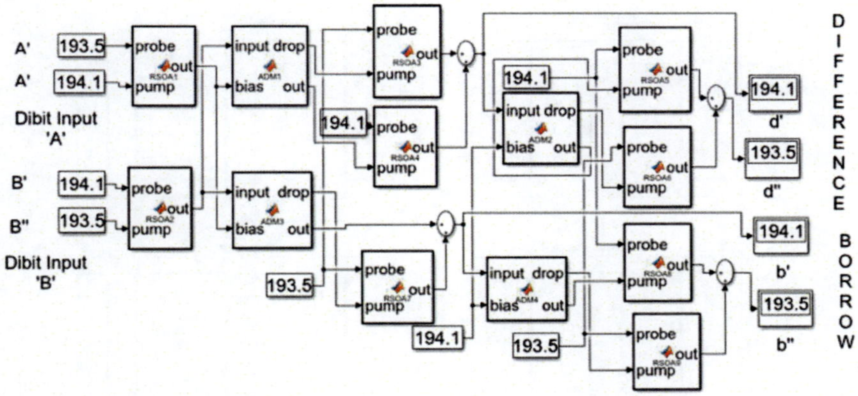

Fig. 4 Simulation model of half subtractor, when $A' = 193.5$ THz, $A'' = 194.1$ THz and $B' = 194.1$ THz, $B'' = 193.5$ THz are applied to the input port

Table 2 Simulation results of frequency encoded half subtractor (all the frequencies are in THz range)

Time (ps)	Input				Output			
	Dibit signal (A)		Dibit signal (B)		Dibit difference (d)		Dibit borrow (b)	
	A'	A''	B'	B''	d'	d''	b'	b''
0–10	193.5	194.1	193.5	194.1	193.5	194.1	193.5	194.1
10–20	193.5	194.1	194.1	193.5	194.1	193.5	194.1	193.5
20–30	194.1	193.5	193.5	194.1	194.1	193.5	193.5	194.1
30–40	194.1	193.5	194.1	193.5	193.5	194.1	193.5	194.1
40–50	193.5	194.1	194.1	193.5	194.1	193.5	194.1	193.5
50–60	194.1	193.5	193.5	194.1	194.1	193.5	193.5	194.1
60–70	194.1	193.5	193.5	194.1	194.1	193.5	193.5	194.1
70–80	193.5	194.1	193.5	194.1	193.5	194.1	193.5	194.1
80–90	193.5	194.1	194.1	193.5	193.5	194.1	193.5	194.1
90–100	194.1	193.5	193.5	194.1	194.1	193.5	193.5	194.1

5 Results and Discussion

In this work, 193.5 and 194.1 THz signals serve as frequency ν_1 and ν_2 which are applied in the time interval of 10 ps. The dibit input signal waveforms are given in Fig. 5a. The dibit output signals for half subtractor are given in Fig. 5b. The input dibit signal waveforms (Fig. 5a) and simulation waveforms of dibit output are verified with the simulation results Table 2 and the truth Table 1 of the proposed design. After

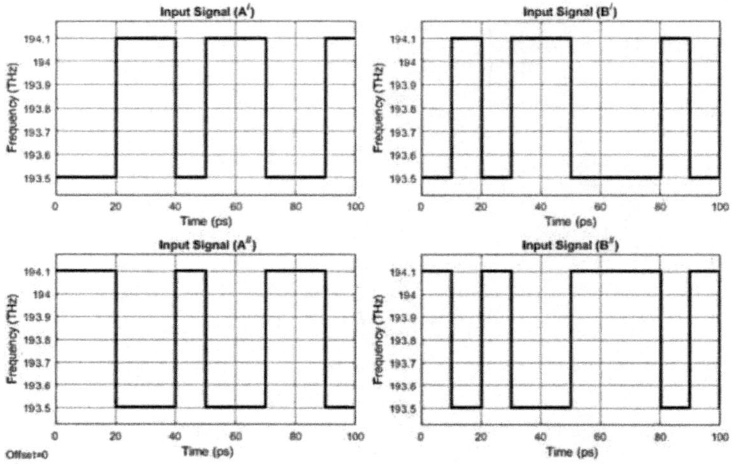

(a)

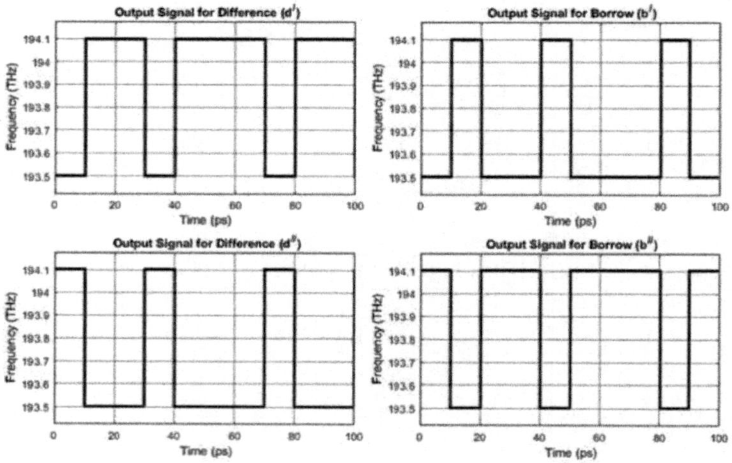

(b)

Fig. 5 **a** Dibit input waveforms of half subtractor. **b** Dibit output waveforms of half subtractor verification of the above data and the simulation results, it may be concluded that the theoretical prediction for output of half subtractor is accurately verified with the simulation output results (Table 2), input and output waveform (Fig. 5a and b), and truth Table 1.

6 Conclusion

In this paper, a different approach for designing a frequency encoded dibit-based half subtractor with RSOA is proposed. The devised design is based on the dibit concept with a dibit checking circuit which can block the errors in the dibit inputs. Bit error problems can be reduced with this devised design which increases the high signal-to-noise ratio. This half subtractor operates at ultra-fast speed and consumes low power. In future, this design can be extended to a full subtractor, and also this design may be used for all-optical computational devices.

References

1. Belhadj, W., Saidani, N., Abdelmalek, F.: All-optical logic gates based on coupled heterostructure waveguides in two dimensional photonic crystals. Optik **168**, 237–243 (2018)
2. Berrettini, G., Simi, A., Malacarne, A., Bogoni, A., Poti, L.: Ultrafast integrable and reconfigurable xnor, and nor, and not photonic logic gate. IEEE Photonics Technol. Lett. **18**(8), 917–919 (2006)
3. Gayen, D.K., Chattopadhyay, T., Bhattacharyya, A., Basak, S., Dey, D.: All-optical half-adder/half-subtractor using terahertz optical asymmetric demultiplexer. Appl. Opt. **53**(36), 8400–8409 (2014)
4. Ghosh, B., Hazra, S., Haldar, N., Roy, D., Patra, S., Swarnakar, J., Sarkar, P., Mukhopadhyay, S.: A novel approach to realize of all optical frequency encoded dibit based xor and xnor logic gates using optical switches with simulated verification. Opt. Spectrosc. **124**(3), 337–342 (2018)
5. Ghosh, B., Hazra, S., Sarkar, P.P.: Simulative study of all-optical frequency encoded dibit-based controlled multiplexer and de-multiplexer using optical switches. J. Opt. **48**(3), 365–374 (2019)
6. Hu, H., Zhang, X., Zhao, S.: High-speed all-optical logic gate using QD-SOA and its application. Cogent Physics **4**(1), 1388156 (2017)
7. Mandal, S., Mandal, D., Mandal, M.K., Garai, S.K.: Design of optical quaternary adder and subtractor using polarization switching. J. Opt. **47**(3), 332–350 (2018)
8. McGeehan, J.E., Kumar, S., Willner, A.E.: Simultaneous optical digital half-subtraction and-addition using SOAs and a PPLN waveguide. Opt. Express **15**(9), 5543–5549 (2007)
9. Mukherjee, K.: Implementation of a novel hybrid encoding technique and realization of all optical logic gates exploiting difference frequency generation alone. Optik **122**(4), 321–323 (2011)
10. Mukhopadhyay, S.: Binary optical data subtraction by using a ternary dibit representation technique in optical arithmetic problems. Appl. Opt. **31**(23), 4622–4623 (1992)
11. Nahata, P.K., Ahmed, A., Yadav, S., Nair, N., Kaur, S.: All optical full-adder and full-subtractor using semiconductor optical amplifiers and all-optical logic gates. In: 2020 7th International Conference on Signal Processing and Integrated Networks (SPIN), pp. 1044–1049. IEEE (2020)

12. Rakheja, S., Kani, N.: Spin pumping driven auto-oscillator for phase-encoded logic–device design and material requirements. AIP Adv. **7**(5), 055905 (2017)
13. Rakshit, J., Chattopadhyay, T., Roy, J.: Design of micro ring resonator based all optical adder/subtractor. Theor. Appl. Phys. **1**, 32–43 (2013)
14. Rakshit, J.K., Roy, J.N.: All-optical ultrafast switching in a silicon microring resonator and its application to design multiplexer/demultiplexer, adder/subtractor and comparator circuit. Opt. Appl. **46**(4), 517–539 (2016)
15. Sarkar, P.P., Satpati, B., Mukhopadhyay, S.: New simulative studies on performance of semiconductor optical amplifier based optical switches like frequency converter and add-drop multiplexer for optical data processors. J. Opt. **42**(4), 360–366 (2013)
16. Singh, P., Singh, A.K., Arun, V., Dixit, H.: Design and analysis of all-optical half-adder, half-subtractor and 4-bit decoder based on SOA-MZI configuration. Opt. Quantum Electron. **48**(2), 159 (2016)
17. Swarnakar, S., Kumar, S., Sharma, S.: Design of all-optical half-subtractor circuit device using 2-d principle of photonic crystal waveguides. J. Opt. Commun. **40**(3), 195–203 (2019)
18. Zaghloul, Y.A., Zaghloul, A., Adibi, A.: Passive all-optical polarization switch, binary logic gates, and digital processor. Opt. Express **19**(21), 20332–20346 (2011)
19. Zhong, L.: Spatial encoded pattern array logic and design of 2d-encoding device for optical digital computing. In: International Conference on Optoelectronic Science and Engineering '90, vol. 1230, p. 12307B. International Society for Optics and Photonics (2017)

An Improved CMOS Ring VCO Design with Resistive-Capacitive Tuning Method

Dileep Dwivedi and Manoj Kumar

1 Introduction

The design challenges in a fully integrated CMOS circuit for a wireless data transmission are to attain low-power and wide frequency range along with low phase noise. In CMOS integrated circuit, voltage-controlled oscillators (VCO) are the main constituent that is used in function generator and high-frequency synthesizer [1–3]. Among different oscillator configurations, ring oscillator and LC oscillator are more widely used configurations to design VCO circuits. Each topology has certain advantages and disadvantages. The ring oscillator is attractive for implementing a compact on-chip VCO and occupies less area. However, its phase noise is larger than the LC oscillator. On the other hand, the LC oscillator has a low-frequency range and requires much larger area due to the use of an inductor [4–9]. Due to the low tuning range, they are not appropriate for wide frequency range applications. Oscillators based on ring topology are usually designed with delay stages such as amplifier and digital logic gate. In delay stages, transition time between input and output controlled by the current that is flowing in the circuit. In the case of amplifier, the gain of an amplifier is modified by controlling the current flowing through it to obtain a wide frequency [10, 11]. In this paper, a VCO based on ring structure is presented to obtain a wide frequency range and improved phase noise performance. In the proposed oscillator, a controllable frequency range is achieved by varying the load capacitance of the delay stage. Varactors are the first option to obtain variable capacitors. Varactors can be designed in the CMOS circuit either by a diode or MOS transistor. Varactors implemented with diode provide a linear frequency range, but

D. Dwivedi (✉) · M. Kumar
University School of Information, Communication and Technology, Guru Gobind Singh Indraprastha University, New Delhi, India

M. Kumar
e-mail: manojtaleja@ipu.ac.in

M. Dua et al. (eds.), *Proceedings of the International Conference on Paradigms of Communication, Computing and Data Sciences*, Algorithms for Intelligent Systems,
https://doi.org/10.1007/978-981-16-5747-4_4

they show adverse effect on the phase noise of VCO due to increase in the internal resistance as compared to MOS varactors. MOS varactors are frequently utilized in CMOS circuits like voltage-controlled capacitors. Different configurations of MOS transistors are used as a variable capacitor [12]. In a conventional type, source, drain, and bulk terminal are tied together ($D = S = B$), and a variable voltage is applied across the source and gate terminal to obtain a variable capacitor. This device can work in the depletion, accumulation, and inversion region. The second structure that is utilized as a MOS varactor in the CMOS circuit is inversion mode MOS (I-MOS). In I-MOS, varactor drain and source ends are shorted ($D = S$) to form one part of the capacitor while poly silicon gate forms the other part. However, the bulk terminal is connected to the highest (lowest) voltage available in the circuit VDD/ground so that it can work in the inversion region only. VCO based on MOS varactor is shown in [13–20]. Further, this paper is structured as follows: In Sect. 2, the structure of the 3-stage ring VCO has been discussed. Results are presented in Sect. 3, and finally, in Sect. 4, conclusions are summarized.

2 VCO Design

The schematic of 3-stage VCO implemented with the proposed delay stage has been presented in Fig. 1. The delay stage used in 3-stage ring VCO consists of the following elements: a CMOS inverter and frequency tuning element. CMOS inverter is designed using transistors $M1$ and $M2$. The output of the CMOS inverter is obtained at the drain terminal of $M1$ and $M2$ transistors and applied to the tuning element. The frequency tuning element is designed with transistors $M3$, $M4$, and $M5$. Transistor $M3$ is utilized as a variable register, controlled by the gate terminal voltage applied to the transistor. Transistor $M4$ and $M5$ are combined in parallel form to make a MOS variable capacitor. MOS varactor capacitance is tuned by the V_{ds}.

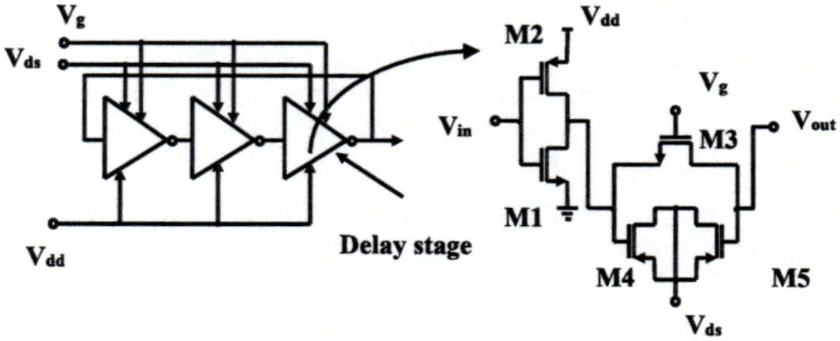

Fig. 1 Schematic diagram of 3-stage VCO

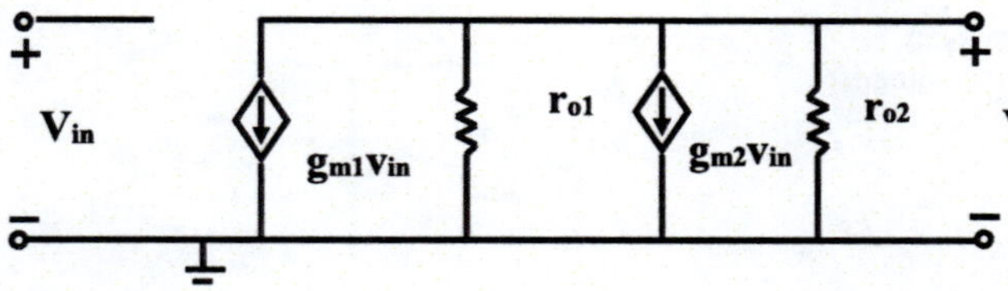

Fig. 2 Small-signal model of CMOS inverter

2.1 Small-Signal Gain and Phase Noise Analysis of VCO Circuit

The small-signal behavior of the CMOS inverter is analyzed in Fig. 2. Consider that both the transistor, $M1$ and $M2$ are in saturation and can achieve the largest voltage gain. Equation (1) is obtained by applying KCL at the output node. Where gm1 and gm2 will represent the trans-conductance of transistors of $M1$ and $M2$, and ro1 and ro2 denote the drain/source resistance of both the transistors. After rearranging Eq. 1, we can get Eqs. (2) and (3). Finally, the small-signal voltage gain (AV) is given Eq. (4).

$$g_{m1}V_{\text{in}} + g_{m2}V_{\text{in}} + \frac{V_{\text{out}}}{r_{o1}} + \frac{V_{\text{out}}}{r_{o2}} = 0 \tag{1}$$

$$V_{\text{in}}(g_{m1} + g_{m2}) = -V_{\text{out}}(r_{o1} \| r_{o2})^2 \tag{2}$$

$$\frac{V_{\text{out}}}{V_{\text{in}}} = -\frac{(g_{m1} + g_{m2})}{(r_{o1} \| r_{o2})^2} \tag{3}$$

$$A_{\text{V}} = -\frac{(g_{m1} + g_{m2})}{(r_{o1} \| r_{o2})^2} \tag{4}$$

The thermal noise performance of the CMOS inverter is analyzed in Fig. 3. The thermal noise of transistors $M1$ and $M2$ is represented by two current sources given by Eqs. (5) and (6). Since both the current sources are uncorrelated, the overall output thermal noise per unit bandwidth of the CMOS inverter is represented by Eqs. (7) and (8). The output noise voltage per unit bandwidth is given by Eqs. (7) and (10). Input-referred noise of CMOS circuit is calculated by Eqs. (11) and (12). Finally, the input-referred noise is represented by Eq. (12). Input-referred noise shows the impact of circuit noise on the input signal.

$$\overline{i_{n1}^2}4KT\left(\frac{2}{3}\right)g_{m2} \tag{5}$$

$$\overline{i_{n1}^2} = 4KT\left(\frac{2}{3}\right)g_{m1} \tag{6}$$

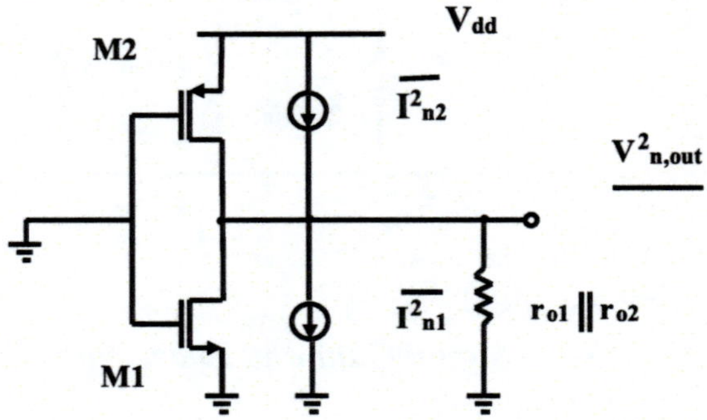

Fig. 3 CMOS inverter circuit diagram including noise sources

$$\overline{i^2_{n,\text{out}}} = \overline{i^2_{n1}} + \overline{i^2_{n2}} \tag{7}$$

$$\overline{i^2_{n,\text{out}}} = 4\,KT\left(\frac{2}{3}\right)(g_{m1} + g_{m2}) \tag{8}$$

$$\overline{V^2_{n,\text{out}}} = \overline{i^2_{\text{out}}} \times (r_{o1} \| r_{o2})^2 \tag{9}$$

$$\overline{V^2_{n,\text{out}}} = 4\,KT\left(\frac{2}{3}\right)(g_{m1} + g_{m2}) \times (r_{o1} \| r_{o2})^2 \tag{10}$$

$$\overline{V^2_{n,\text{in}}} = \frac{\overline{V^2_{n,\text{out}}}}{|A_v|^2} \tag{11}$$

$$V^2_{n,\text{in}} = \frac{4\,\text{KT}\left(\frac{2}{3}\right)(g_{m1} + g_{m2}) \times (r_{o1} \| r_{o2})^2}{(g_{m1} + g_{m2})^2 \times (r_{o1} \| r_{o2})^2} \tag{12}$$

$$V^2_{n,\text{in}} = \frac{4\,KT\left(\frac{2}{3}\right)}{(g_{m1} + g_{m2})} \tag{13}$$

3 Results and Discussion

A three-stage ring VCO having the proposed delay cell is designed in 0.18 μm CMOS technology. The tuning range of oscillation depends on the voltage applied to the load element. Results obtained due to variation in V_{ds} are shown in Table 1 at fixed supply voltage ($V_{\text{dd}} = 1.8$ V) and gate control voltage ($V_{\text{g}} = 1.8$ V) with different values of MOS varactor width (w). As V_{ds} changes from 1 to 1.8 V, the frequency varies from 0.886 to 1.557 GHz. The plot of oscillation frequency versus drain/source voltage is shown in Fig. 4.

Table 1 Generated frequency result with drain/source voltage

Drain/source voltage (V)	Frequency (GHz)		
	$W = 5\ \mu m$	$W = 7\ \mu m$	$W = 10\ \mu m$
1.0	1.557	1.295	1.034
1.1	1.535	1.275	1.014
1.2	1.515	1.251	0.995
1.3	1.485	1.230	0.973
1.4	1.462	1.206	0.954
1.5	1.439	1.186	0.933
1.6	1.418	1.167	0.915
1.7	1.400	1.149	0.900
1.8	1.387	1.133	0.886

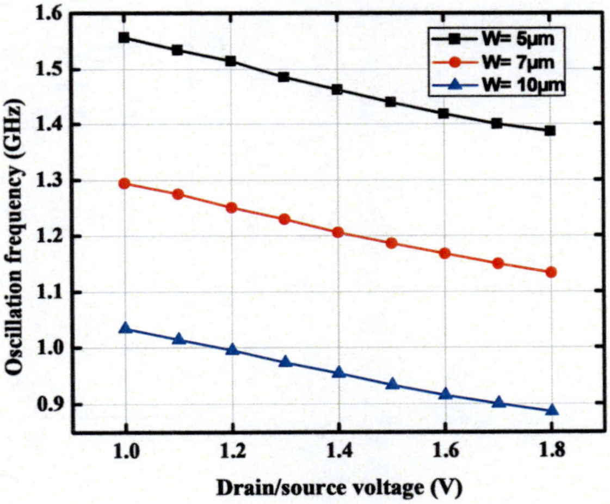

Fig. 4 Frequency sweep with drain/source voltage

Results obtained due to variation in gate control voltage (V_g) are depicted in Table 2 at different values of NMOS transistor width (W). From Table 2, it is evident that with a change in V_g from 1 to 1.8 V, the VCO circuit oscillates from 0.593 to 1.387 GHz. Output oscillation frequency increases with a rise in the gate voltage. Output oscillation curve with variation in gate voltage is presented in Fig. 5.

Variation in supply voltage (V_{dd}) also has been considered, and variation in the oscillation frequency is presented in Table 3. Table 3 provides the information regarding variation in the frequency of VCO design with supply voltage change from 1.5 to 2.5 V. Result of power utilization in the circuit with change in supply voltage is also shown in Table 2. Power dissipation in the VCO circuit varies from 0.164 to 1.822 mW with V_{dd}. Output oscillation frequencies obtained due to change in supply voltage are plotted in Fig. 6.

Table 2 Oscillation frequency variation with change in gate voltage

Gate voltage (V)	Frequency oscillation (GHz)		
	$W = 2\ \mu m$	$W = 4\ \mu m$	$W = 6\ \mu m$
1.0	0.649	0.632	0.593
1.1	0.719	0.689	0.650
1.2	0.792	0.766	0.718
1.3	0.885	0.854	0.804
1.4	1.003	0.974	0.909
1.5	1.125	1.078	1.001
1.6	1.233	1.171	1.076
1.7	1.324	1.232	1.128
1.8	1.387	1.284	1.166

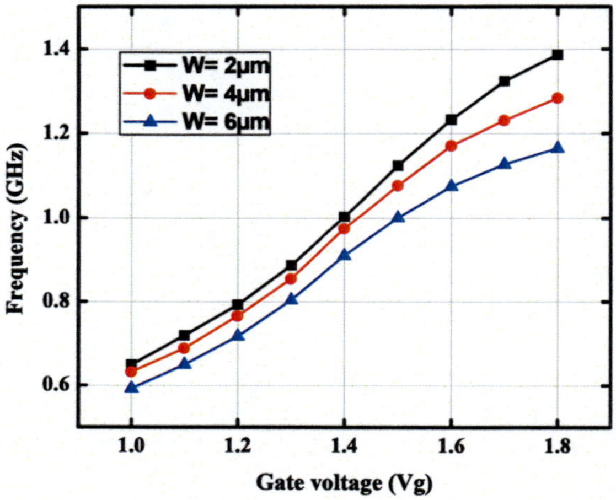

Fig. 5 Frequency variation with gate voltage

Frequencies variation of proposed VCO is shown in Table 4 for different values of temperature range from 10 to 80 °C. From Table 4, it is evident that with a rise in temperature, the frequency decreases due to an increase in temperature carrier mobility change, which results in to decrease in current flowing in the circuit. The dependence between temperature and frequency is presented in Fig. 7.

Phase noise results of the VCO are investigated at different range of gate voltage and drain/source voltage shown in Table 5. From Table 5, it is apparent that with a rise in drain/source voltage and gate voltage, phase noise results have been improved. For the proposed VCO optimized values of performance parameters are reported at fixed 1.8 V supply voltage, drain/source voltage, and the gate voltage.

In Table 6, a comparison of the proposed VCO with other work in the literature has been presented. From Table 6, it is clear that the proposed VCO circuit dissipates less power and shows an enhancement in the figure of merit (FoM) performance

Table 3 Frequency deviation with V_{dd}

Supply voltage (V)	Generated frequency (GHz)			Power dissipation (mW)
	$W = 5\ \mu m$	$W = 7\ \mu m$	$W = 10\ \mu m$	
1.5	1.111	0.711	0.520	0.164
1.6	1.214	0.776	0.568	0.245
1.7	1.309	0.837	0.614	0.344
1.8	1.386	0.887	0.652	0.461
1.9	1.450	0.928	0.682	0.596
2.0	1.486	0.957	0.705	0.751
2.1	1.525	0.978	0.722	0.925
2.2	1.542	1.001	0.737	1.119
2.3	1.577	1.012	0.749	1.133
2.4	1.580	1.022	0.754	1.567
2.5	1.596	1.025	0.759	1.822

Fig. 6 Output frequency variation at with supply voltage

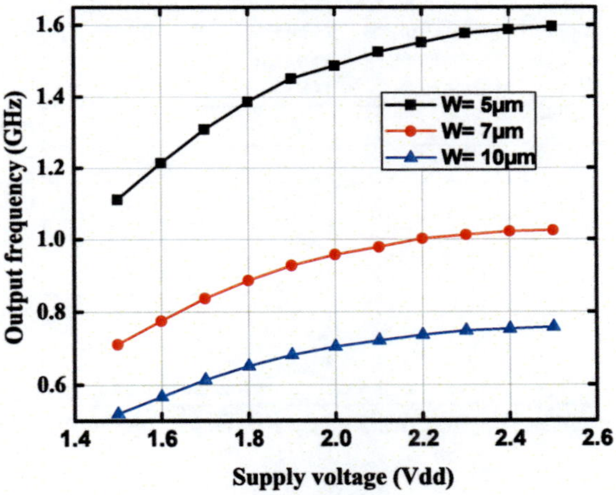

Table 4 Output frequency dependence on temperature

Temperature (°C)	Output frequency (GHz)		
	$W = 5\ \mu m$	$W = 7\ \mu m$	$W = 10\ \mu m$
10	1.477	0.944	0.693
20	1.422	0.910	0.668
30	1.371	0.878	0.642
40	1.322	0.847	0.621
50	1.275	0.817	0.600
60	1.229	0.789	0.580
70	1.188	0.761	0.560
80	1.150	0.734	0.540

Fig. 7 Output frequency variation with temperature

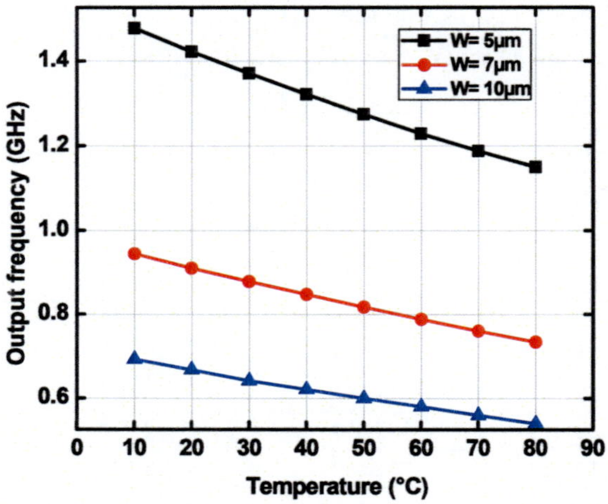

Table 5 Results of performance analysis of proposed VCO

Supply voltage (V)	Drain/source voltage (V)	Gate voltage (V)	PN (dBc/Hz)	PD (mW)	Frequency (GHz)	FoM (dBc/Hz)
1.8	1.2	1.8	−90.399	0.461	1.515	157.36
1.8	1.8	1.8	−96.557	0.461	1.386	162.75
1.8	2.0	1.8	−106.98	0.461	1.307	172.66
1.8	1.8	1.0	−90.934	0.461	0.649	150.54
1.8	1.8	1.2	−100.135	0.461	0.794	161.49
1.8	1.8	1.4	−107.776	0.461	1.002	171.15

Table 6 Comparison of different performance parameter of proposed VCO with other work

References	Tech. (μm)	Frequency range (GHz)	P_{DC} (mW)	PN (dBc/Hz)	FOM (dBc/Hz)
[6]	0.18	0.479–4.09	13	−93.3@1 MHz	154.4
[9]	0.18	0.1–3.5	16	−106 @4 MHz	151.34
[21]	0.18	1.77–1.92	13	−102@1 MHz	156.3
[22]	0.18	5.16–5.93	80	−99.5@1 MHz	155.72
[23]	0.18	0.5–1.2	0.71	−90@1 MHz	151.48
[24]	0.18	4.2–5.9	58	−99.1@1 MHz	156.28
[25]	0.18	8.1–10.5	68.4	−92@1 MHz	153.2
This work	0.18	0.593–1.557	0.461	−96.557@1 MHz	162.75

with the other reported work. Phase noise obtained for the proposed VCO is better than that of references [6, 23, 25] with low-power consumption. The figure of merit (FoM) is calculated by Eq. (14) [25].

$$\mathrm{FoM(dBc/Hz)} = 20\log\frac{f_{\mathrm{osc.}}}{f_{\mathrm{off.}}} - 10\log\frac{P_{\mathrm{diss.}}}{1\,\mathrm{mW}} - \mathrm{PN}(\Delta f) \quad (14)$$

where PN represents the phase noise, $f_{\mathrm{osc.}}$ is center frequency, and $P_{\mathrm{diss.}}$ is power dissipation in the circuit.

4 Conclusions

A low-power VCO with a wide tuning range for radio frequency communication applications has been presented and designed in 0.18 μm standard CMOS technology. Due to reactive tuning load, a wide frequency range for the proposed VCO is obtained, while power consumption in the circuit keeps constant. The proposed VCO generates an oscillating output of 0.593–1.557 GHz corresponding to a control voltage of 1–1.8 V. The proposed VCO observe a phase noise of −96.557 dBc/Hz at 1 MHz offset with 0.461 mW power dissipation at 1.8 V V_{dd}. VCO circuit has a figure of merit of 162.75 dBc/Hz.

References

1. Zhao, B., Lian, Y., Yang, H.: A low-power fast-settling bond-wire frequency synthesizer with a dynamic-bandwidth scheme. IEEE Trans. Circuits Syst. I Regul. Pap. **60**(5), 1188–1199 (2013)
2. Seong, T., Kim, J.J., Choi, J.: Analysis and design of a core-size-scalable low phase noise LC VCO for multi-standard cellular transceivers. IEEE Trans. Circuits Syst. I Regul. Pap. **62**(3), 781–790 (2015)
3. Elabd, S., Khalil, W.: Impact of technology scaling on the tuning range and phase noise of mm-wave CMOS LC-VCOs. Integration **52**, 195–207 (2016)
4. Lee, W.H., Gu, B.J., Nishida, Y., Takao, H., Sawada, K., Ishida, M.: Oscillation-controlled CMOS ring oscillator for wireless sensor systems. Microelectron. J. **41**(12), 815–819 (2010)
5. Nizhnik, O., Pokharel, R.K., Kanaya, H., Yoshida, K.: Low noise wide tuning range quadrature ring oscillator for multi-standard transceiver. IEEE Microwave Wirel. Compon. Lett. **19**(7), 470–472 (2009)
6. Sheu, M.L., Tiao, Y.S., Taso, L.J.: A 1-V 4-GHz wide tuning range voltage-controlled ring oscillator in 0.18 μm CMOS. Microelectron. J. **42**(6), 897–902 (2011)
7. Hafez, A.A., Yang, C.K.K.: Design and optimization of multipath ring oscillators. IEEE Trans. Circuits Syst. I Regul. Pap. **58**(10), 2332–2345 (2011)
8. Kim, J.M., Kim, S., Lee, I.Y., Han, S.K., Lee, S.G.: A low-noise four-stage voltage-controlled ring oscillator in deep-submicrometer CMOS technology. IEEE Trans. Circuits Syst. II Express Briefs **60**(2), 71–75 (2013)
9. Grozing, M., Phillip, B., Berroth, M.: CMOS ring oscillator with quadrature outputs and 100 MHz to 3.5 GHz tuning range. In: ESSCIRC 2004–29th European Solid-State Circuits Conference, pp. 679–682(2003, Sept)

10. Khitouni, N., Boujelben, S., Masmoudi, M.: Sigma delta A/D converter architecture using a current controlled oscillator. In: 2005 12th IEEE International Conference on Electronics, Circuits and Systems, pp. 1–4 (2005, Dec)
11. Hui, Z., Bin, X., Xue-Wen, N., Bang-Xian, M., Zhan-Fei, W.: A low noise current-controlled oscillator with high control linearity. In: Proceedings. 7th International Conference on Solid-State and Integrated Circuits Technology, No. 2, pp. 1551–1554 (2004, Oct)
12. Andreani, P., Mattisson, S.: On the use of MOS varactors in RF VCOs. IEEE J. Solid-State Circuits **35**(6), 905–910 (2000)
13. Jin, J., Yu, X., Liu, X., Lim, W.M., Zhou, J.: A wideband voltage-controlled oscillator with gain linearized varactor bank. IEEE Trans. Compon. Packag. Manuf. Technol. **4**(5), 905–910 (2014)
14. Hegazi, E., Abidi, A.A.: Varactor characteristics, oscillator tuning curves, and AM-FM conversion. IEEE J. Solid-State Circuits **38**(6), 1033–1039 (2003)
15. Danaie, M., Aminzadeh, H., Naseh, S.: On the linearization of MOSFET capacitors. In: 2007 IEEE International Symposium on Circuits and Systems, pp. 1943–1946 (2007, May)
16. Bunch, R.L., Raman, S.: Large-signal analysis of MOS varactors in CMOS-G/sub m/LC VCOs. IEEE J. Solid-State Circuits **38**(8), 1325–1332 (2003)
17. Zhenrong, L., Yiqi, Z., Bing, L., Gang, J., Zhao, J.: A 2.4 GHz high-linearity low-phase-noise CMOS LC-VCO based on capacitance compensation. J. Semicond. **31**(7), 075005 (2010)
18. Sameni, P., Siu, C., Iniewski, K., Mirabbasi, S., Djahanshahi, H., Hamour, M., Chana, J.: Characterization and modeling of accumulation-mode MOS varactors. In: Canadian Conference on Electrical and Computer Engineering 1554–1557 (2005, May)
19. Jangra, V., Kumar, M.: A wide tuning range VCO design using multi-pass loop complementary current control with IMOS varactor for low power applications. Eng. Sci. Technol. Int. J. **22**(4), 1077–1086 (2019)
20. Kumar, M.: Design of linear low-power voltage-controlled oscillator with I-MOS varactor and back-gate tuning. Circuits Syst. Signal Process. **37**(9), 3685–3701 (2018)
21. Chen, Z.Z., Lee, T.C.: The design and analysis of dual-delay-path ring oscillators. IEEE Trans. Circuits Syst. I Regul. Pap. **58**(3), 470–478 (2010)
22. Eken, Y.A., Uyemura, J.P.: A 5.9-GHz voltage-controlled ring oscillator in 0.18-/spl mu/m CMOS. IEEE J. Solid-State Circuits **39**(1), 230–233 (2004)
23. Yoshida, T., Ishida, N., Sasaki, M., Iwata, A.: Low-voltage, low-phase-noise ring voltage-controlled oscillator using 1/f-noise reduction techniques. Jpn. J. Appl. Phys. **46**(4S), 2257 (2007)
24. Danfeng, C., Junyan, R., Jingjing, D., Wei, L., Ning, L.: A multiple-pass ring oscillator based dual-loop phase-locked loop. J. Semicond. **30**(10), 105014 (2009)
25. Liu, H.Q., Goh, W.L., Siek, L.: 1.8-V 10-GHz ring VCO design using 0.18 μm CMOS technology. In: Proceedings 2005 IEEE International SOC Conference, 77–78 (2005, Sept)

Forwarding Strategy in SDN-Based Content Centric Network

Divyanshi Verma, Sharmistha Adhikari, and Sangram Ray

1 Introduction

Internet, as we know it, was intended to be an end-to-end connectivity arrangement centered on packet switching in early 1960s. Initially, it is focused on a point-to-point communication between two hosts. The IP packets that were transferred among hosts contain identified addresses of the nodes. With emergence of World Wide Web, an explosion in number of users occurred along with an even bigger explosion in overall digital content made available through Internet. The frugal and more pervasive hardware motivated by Moore's Law expedites connecting everything to the Internet. There is a significant agreement that the number of addressable devices could cross the likes of billions or even trillions. The IP protocol, initially designed to support the communication between endpoints, is tremendously being used for content propagation. However, the true nature of IP makes the existing Internet architecture a misfit to its prime utilization today. People use Internet to get contents such as web pages, music, or video files, and they only value "what" and are not interested in "where" the contents are actually stored but IP does the opposite and only cares about the "where."

As an attempt to fill the void between "what" and "where," numerous works have come up [1, 2], and it is quite clear from the past literature that naming the data rather than hosts is the way to move forward. In different words, the key behind the propositions is to swap "where" with "what" in the fundamentals of Internet, classically labeled as content centric networking (CCN) or named data networking

D. Verma · S. Adhikari (✉) · S. Ray
National Institute of Technology Sikkim, Sikkim 737139, India

D. Verma
e-mail: b160062@nitsikkim.ac.in

M. Dua et al. (eds.), *Proceedings of the International Conference on Paradigms of Communication, Computing and Data Sciences*, Algorithms for Intelligent Systems,
https://doi.org/10.1007/978-981-16-5747-4_5

(NDN). In 2009, Van Jacobson introduced CCN where host-centric architecture is changed to data-centric architecture [3].

Contrary to IP-based Internet, CCN has only two packet categories [4–10]

i. Interest which is comparable with "get" in HTTP (Hyper Text Transfer Protocol).
ii. Data which is similar to response in HTTP.

Both packets are encoded in an efficient binary XML, and in order to "get" data, a requester or the consumer generates an Interest packet consisting of a name that recognizes the anticipated data. For example, an end user might generate an Interest packet asking for the data named after */documents/papers/divyanshi37.pdf*. A consumer uses the name of a looked-for chunk of data in an Interest packet and transmits it to the network. Routers utilize the same name to propagate the Interest to the producer. After reaching the node having the requested data, the node will reply back with a data packet comprising of both, the name as well as the content, along with a signature of content producer's key that binds the two. The Data/Content packet trails in the reverse path as opted by the Interest in order to reach the requester.

A router manages the data structure called as Pending Interest Table (PIT) in order to record the face that has received the request and also forwards the Interest packet by searching the content name in its Forwarding Information Base (FIB). The FIB maintains a record of the list of faces that could attend to the request. FIB is more or less alike with an IP FIB with one difference that the CCN FIB permits a record of outbound interfaces as well instead of just the incoming interfaces.

The process mentioned above reiterates so that, whenever a cache hit takes place, a content piece is sent back in reverse direction on the path that has been followed by the Interest packet and travels along the PIT information [5–13]. The packets are not just transmitted over hardware connection interfaces but also exchanged between application procedures taking place in a system; thus, the term face instead of interface should be used for specifying the point of incoming and outgoing packets. Once the Interest is received by the node with available data that has wished for data in its CS, a data packet containing the required data's name and content, in addition with a signature generated from key of producer, is reverted back. This data packet retracts the inverse path that was previously followed by the Interest packet to reach the consumer. However, a major drawback in existing CCN is broadcasting of Interest packets all over the network in case there is no matching FIB entry. Broadcasting results in extremely heavy traffic and, thus, low speed transfer especially in large-scale networks, thereby making CCN unsuitable [14, 15].

In this paper, our motivation is to take advantage of the Software Defined Networking (SDN) and design an efficient SDN-based CCN forwarding mechanism that will reduce the unnecessary traffic generated by Interest broadcasting.

The remainder of this paper is organized as follows. Section 2 represents the related work which includes the fundamentals of SDN. Section 3 represents a brief literature survey. The proposed scheme and its performance analysis are given in Sects. 4 and 5, respectively. Finally, Sect. 6 concludes the paper.

2 Related Work

In this section, we briefly discuss the fundamentals of SDN and its existing forwarding mechanisms [14–16]. SDN permits de-unification of the data plane which is responsible for traffic forwarding and the control plane that is responsible for taking decisions pertaining to routing in the network. SDN has an application or programmed unit also known as the SDN controller which is designed in such a way that it is capable of handling the switches and routers and these devices act as just forwarding devices. Thus, the decision maker regarding routing for every switch and router is a SDN controller. The SDN can be modified easily with or without making changes in the physical network because SDN controller is capable of handling forwarding procedures by asserting directions as fed directly into hardware devices.

The SDN controller handles the switches and routers in such a manner that these nodal points act like forwarding expedients only. The routing choices for each and every switch and router are hence maintained and controlled by the controller. SDN processes advance procedures at flow level or packet level by striding rules in physical devices.

Application layer is made up of several modules and managers and among them Content Manager and Table Update Manager take responsibilities for centralized management of the table and contents for forwarding and updating. Once contents are either replaced or cached in the routers, Content List Table, and Content Update Table are accordingly updated by the managers.

Flow Manager, on the other hand, plays a role of actual forwarding management of packets. It manages the flows and routes from source to destination that makes is possible to unicast interest packets rather than broadcasting.

Usually, mesh topology has been considered which is quite often used practically. It has been presumed that the network comprises of groups or clusters and every cluster comprises of many routers. The routers have been clustered using clustering methodology such as K-means clustering which is the clustering technique based on Euclidian distance [17].

Distributed local controllers have been introduced to the network which takes the control over a cluster. In order to forward both Interest packets and data packets, FT (Forwarding Table) has been used instead of PIT and FIB in original CCN. FT is managed by SDN controller application, and the fields and components of the table are as follows:

- **Content Name**: This filed is used to hold the information of the name of contents.
- **RID**: This field is used to hold the information of the router ID where contents are stored.
- **CID**: This filed is used to hold the information of the cluster ID where contents are stored.
- **Requesting Node**: This field is used to hold the information of router ID that requests for content.
- **Requesting Cluster**: This field is used to hold the information of cluster ID that requests for a content

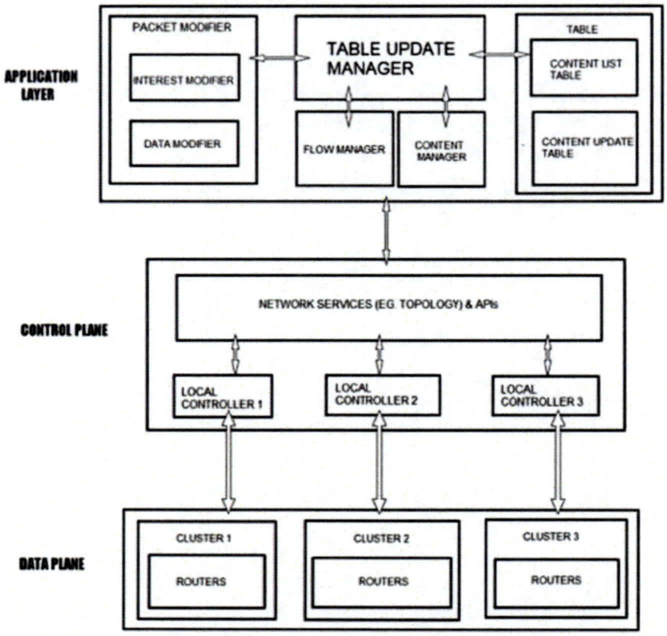

Fig. 1 SDN-based CCN architecture [17]

Based on the above discussion, the basic architecture of SDN-based CCN as proposed by Son et al. [17] is depicted in the following Fig. 1.

The conventional packet structures of CCN are duly modified to support the forwarding strategies of SDN-based CCN [17]. The modified packet structures are given in the following subsections.

Interest Packet: Added three fields to the original CCN Interest packet. When clients request for a certain type of content, the first router in RCID field, the cluster ID that the first router belongs to is recorded in order for data packets to be delivered. Additional header attributes are added to the Interest packet, namely RID, CID, and RCID.

Data Packet: After the requested contents are found, they should be delivered to the clients properly. The structure of data packet has been modified by adding RCID field to the original CCN data packet to find out the origin of the request since local controllers do not know about the content information of others. Additional header attribute RCID is added to the Interest packet also.

Interest packets in this scheme can be unicasted to either servers or intermediate routers which have the requested contents cached in content store (CS). If we consider the existing CCN in a large-scale network, delivery of Interest packets directly to the server may result in a serious problem in terms of network traffic if the size of

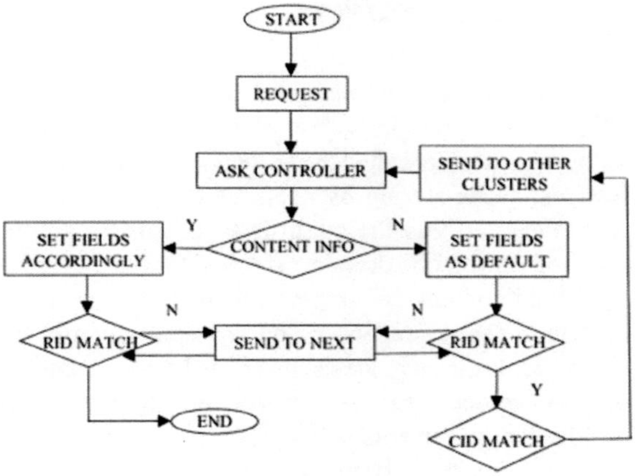

Fig. 2 Flowchart for interest packet forwarding mechanism

the network becomes larger. The Interest packet forwarding algorithm in SDN is described in the following text and depicted with the help of a flow diagram in Fig. 2.

1. Initially, clients request for a certain content by generating Interest packet using the name of the required content.
2. Once the request reaches to a router, the router asks for the location of the content to its local controller. If the controller has the information of the content in the Content List Table, it notifies the router modifying the fields of the Interest packet according to the table. Otherwise, it notifies the router setting up the fields of the Interest packet as default which is predefined in the table. The last row of the tables indicates the default values for unknown contents as shown in Table 1.
3. If local controller knows the location of the content, RID, CID, and RCID field in the Interest packet are modified to the corresponding value.
4. Flow manager defines the flow and router to its destination.
5. If the local controller does not have any information of the content, RID, CID, and RCID field in the Interest packet are modified to the default values of Content List Table. The default router knows that the packet is not for itself but to be delivered to the gateway router which is connected to other clusters of the network.

Table 1 Example of content table in SDN-based CCN

Content name	RID	CID
A	R1	C1
B	R3	C1
C	R5	C1
D	R4	C1
–	R6	0

3 Literature Survey

In this section, a brief background study on the forwarding strategy of SDN-based CCN is conducted. In Kim et al. [18], multi-interface multi-channel (MIMC) technology is used to enhance the performance of a wireless mesh network (WMN). Besides, cache storing by adopting SDN was proposed, i.e., cache decision point was integrated into the controller, and so, the additional overhead in the network required for a cache decision was insignificant. The scheme enhanced the efficiency of cached content distribution where as an alternative of caching everything, and content is cached depending on its popularity and size. The scheme also considers the available space in the cache as well as the resources of the nodes like CPU usage, disk I/O, processing speed, space, etc. In case the candidate node fails to satiate threshold requirements of content being cached, the node is changed to a more suitable one. In case of absence of suitable caching node or when the content is outsized, it will be cached with cache server.

In Son et al. [17], the whole network is divided into clusters by K-means clustering, i.e., clusters are formed on the basis of physical proximity. Each cluster is managed by a local controller which is a software application that stores the table of contents present in the nodes of that cluster. The local controller also manages the flow of Interest packets between neighboring clusters. However, this controller is not aware of the content of any other cluster; thus, in case, the requested content is unavailable in the content table, it forwards the Interest packet to its default routers that are connected to neighboring clusters. Thus, the Interest packets are broadcasted repeatedly, i.e., the Interest packet has to visit each of the adjacent clusters to find the content in a repeated manner. This may lead to infinite loop situation and/or unmanageably heavy traffic in the network.

In Li et al. [16], a group centric greedy *ant colony forwarding* algorithm was used for routing. The scheme divides the topology into several groups/domains where an Internet Service Provider (ISP) manages one or more domains. The scheme has two types of “Ants” with different conducts; Interest Ant and the other is Hello Ant which is created by routers and is used to gather forwarding information. The Hello Ant packet includes information like path overhead, the minimum bandwidth, the round-trip delay, and hops of the whole path. The inter-domain forwarding and intra-domain forwarding are handled differently.

In the inter-domain forwarding progress, after the source node, usually a router, generates the Hello Interest packet, the node randomly chooses a domain name in the inter-domain forwarding Table by the roulette method. Once the domain name is selected, Hello Interest Ant packet is forwarded to all interfaces of the node. It is a major drawback as broadcasting leads to hefty traffic in any network.

According to the scheme, a probability-based forwarding mechanism has been used when internal node of the route receives an Inter-domain Hello Interest Ant. The probability of selection of an interface is dependent on the pheromone value (pheromone is a chemical substance secreted externally by some insects that influences the behavior of other animals of the same species). When the middle node of the path receives an Intra-domain Hello Interest Ant, it forwards the packet to the best of interfaces in the Intra-domain Forwarding Table which has the highest pheromone value. The algorithm focuses on optimization with increasing repeated requests. However, the initial requests could be unsuccessful or time consuming or both.

In Shahid et al. [19], a multicast strategy is introduced which works by forwarding every Interest to all outgoing interfaces by the provided FIB entry as opposed to broadcast strategy which forwards Interests to each interface, in case FIB entry is not present, i.e., executes broadcasting all over the network. Assuming that all routers' FIBs are up to date with all prefaces, the scheme needs regulation from the routing plane. The work also compares implementation of broadcast strategy with multicast strategy in NS-3 centered NDN emulator. However, in place of broadcasting an Interest, the forwarding strategies proposed are: (i) hit and source content from closer nodes and (ii) hit all but source content from closer nodes. Also, the work assumes that content is distributed into same-sized parts known as chunks, and every requesting node generates a different Interest with respect to every chunk it requests. The scheme extends the default Interest packet with addition of sections named P and scope where P is a predefined system parameter with values in the range (0, 100]. Scope holds two objects, "hop count (h) and number of duplicates of an Interest (c)." Default value of P for a chunk as set by each node is in range (0, 100]. The higher the value of P, the deeper the Interest can plunge inside the network, and thus, number of content sources to find reply from with respective data and vice versa for smaller values but not too small as it may lead to DoS (Denial of Service) or delay. Thus, the value of P is altered in dynamic manner as per necessity. Also, $P = 100$ means broadcasting. Additionally, the default data packet is protracted by toting a field titled Copies to ally c's value during the time data packet trips back to the requesting node.

The names of Interest and Data are extended in discovery by adding tags to the default names. These tags identify the category. For instance, a tag "D" (/prefix/D/00) implies that the Interest is a Content Discovery Interest or Content Request Interest, respectively. In search, the bombarding range is adjusted to minimize the overall Interest and Data transmitting.

The Interests that have gone far from the requesting node are rejected. The more the Interest is forwarded outward and replicated to neighbors, the lesser is the value of IFF. Since, a static threshold is utilized, only proximal routers accept and return a reply. The "number of copies" field c helps to halt routers from creating further duplicates of an Interest. Search strategy might present inconsistent Interest and Data, if the value of P and the threshold are not attuned properly. For example, huge difference among P and D might lead to denial of service even if the content is available.

In Lv et al. [20], clusters were based on social relationship of Interest similarity that split community for helping retrieve the content for Interest packets. However, these communities are not temporary, whereas the content keeps on changing which implies the community must also change.

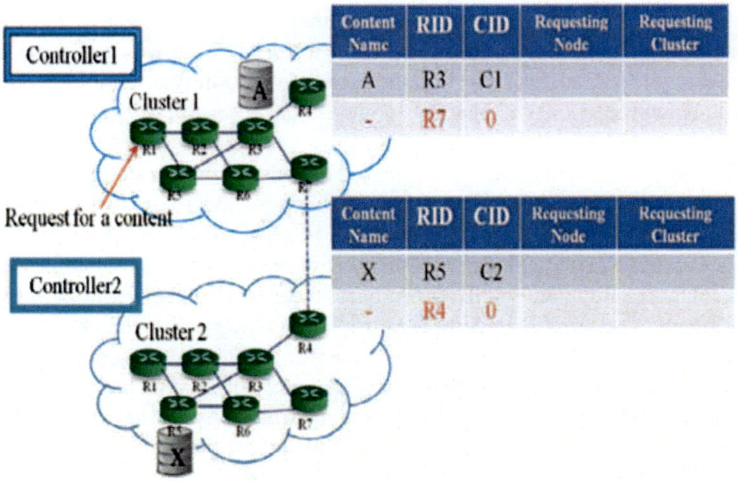

Fig. 3 Example of SDN-based CCN [17]

In Hoque et al. [21], a named data link-state routing protocol was proposed that published the name prefixes by propagating advertisements over the entire network. In [22], a priority-based Interest forwarding strategy was designed, in which Interest packets with higher priority were forwarded prior to those with lower priority. However, for a network to be successful, all requests must be treated equal that would ensure speedy response to the requests.

Our Motivation: The forwarding scheme represented in Son et al. [17] deals with primary stage of how the content is retrieved if it is found in a router's cache. In case the content is not found, the Interest packet is sent to the set default routers (also known as the gateway routers as they are connected to other clusters) of all neighboring clusters (as shown in Fig. 3, router 7 and router 4 are default routers in cluster 1 and 2, respectively). The default router forwards the Interest packet to other default routers (gateway routers), which again is broadcasting. Thus, from this chain of broadcasting, there is always a common set of clusters which are visited repeatedly. Hence, considering the limitations of the above-mentioned schemes in the literature, our motivation is to further improve the Son et al. [17]'s work.

4 The Proposed Scheme

On studying conventional CCN as well as SDN-based CCN, we observed, forwarding can be accurate, efficient and less time-consuming if we work on cache strategies, cache location selection, modification of local controllers and introducing a new concept of "Angel" Clusters.

We propose that local controllers, by the virtue of their high storage capacity store the content that is overwritten on its nodes, i.e., the content deleted from the nodes will be stored by local controller. Moreover, this content will be shared between the clusters that have similar data. Our aim behind sharing the content with other clusters on the basis of content is to assist the Angel Clusters, as will be explained later.

Angel clusters, the problem solvers, are introduced to find the content that the local cluster could not. Angel clusters are clusters formed due to relevance of content of different clusters. It means the clusters that have similar and related content will form a cluster irrespective of their location. When more than 50% of the total content of one cluster matches with that of another cluster, they will act as one Angel cluster.

However, it is necessary to set an upper bound over the number of clusters in an Angel Cluster; else it might happen that majority of the clusters lie in a single Angel Cluster. If that happens, the search time of content will increase manifold. To ensure non-overcrowding of an Angel Cluster, we define an upper bound of 20% of the network's clusters, e.g., if there are 100 clusters altogether, each Angel Cluster can have a maximum of 20 clusters. Also, clusters with low traffic will track if there is any Angel Cluster that has more similar content and will connect with the best content matching Angel Cluster. Thus, Angel Cluster network is not permanent but self-improving. Each Angel Cluster will have a manager controller which is the controller of largest cluster (ties will be broken randomly) within the Angel Cluster and will be responsible for receiving and further forwarding the Interest Packet.

For establishing contact between Angel Cluster and a requesting local controller, we propose to use the mode of advertisements to make local controllers aware of the existence and the content of the Angel Clusters. Thus, when required, the local controller already knows where to find the content. Based on percentage of content similarity of the Angel Cluster with requested content, Angel Clusters is ranked. First, the Interest packet is sent to the best ranked Angel Cluster. Its manager will receive the Interest Packet and match the content with its content store, and if it is found, the fields will be updated accordingly else it will forward the Interest packet to all other clusters of that Angel Cluster. These non-manager clusters revert back to the manager with the updated fields of the Interest packet or a regret message depending on whether the content is found or not. The manager reverts the final response to the requesting cluster which repeats the process with the next best Angel Cluster.

4.1 Proposed SDN-Based CCN Architecture

The proposed SDN-based CCN architecture is depicted in Fig. 4 where clusters B, C, and E form Angel Cluster 1 and clusters A, F and D form Angel Cluster 2. Each cluster has a local controller. Among them local controllers of cluster B and F are the managers of Angel Cluster 1 and 2, respectively, because of highest number of nodes present in them in their respective clusters.

Suppose, the local controller B receives a request for content X. In this case, two following cases are possible:

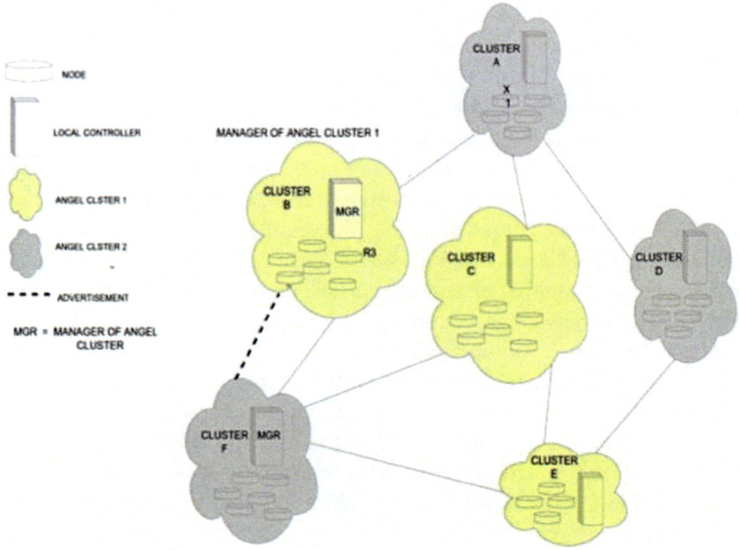

Fig. 4 Example showing proposed modification

Fig. 5 Updated entries of the interest packet

Original CCN packet	RID	CID	RCID
	1	A	B

(i) Local controller has the content with router 3 and that case the entries would be updated.

(ii) Local controller does not have X, and it has received advertisement of Angel Cluster 2. In that case, it sends the Interest packet to the manager of Angel Cluster 2 which checks for the content in cluster F, which does not have X. F sends Interest packet to A and D. D sends back a regret message and A updates the fields RID and CID as 1 and A, respectively, and sends the Interest to F which further sends it to B.

The Interest packet structure is duly modified in our scheme to store the RID, CID, and RCID fields which help in forwarding as shown below in Fig. 5.

4.2 Overall Workflow of the Proposed Scheme

In this section, the overall workflow of the proposed scheme is described step wise as well as depicted in Fig. 6.

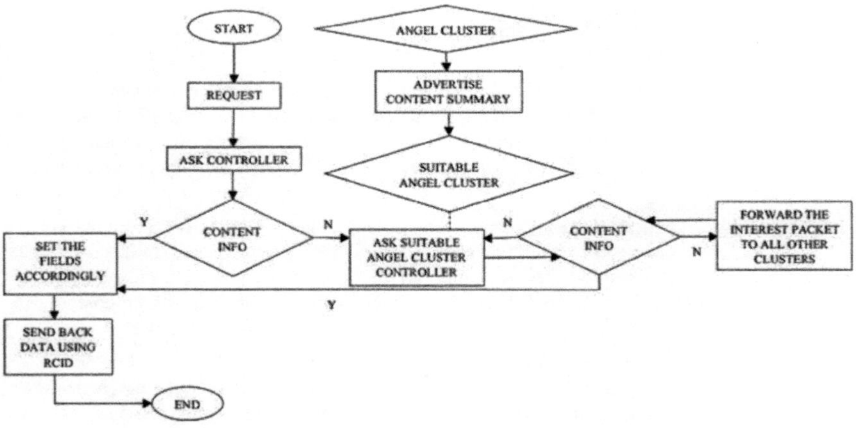

Fig. 6 Flowchart showing proposed scheme

Step 1 Client requests for a certain content by generating Interest packet using the name of the content.

Step 2 Once the request reaches a router, the router asks for the location of the content to its local controller. If the controller has the information of the content in the Content List Table, it notifies the router modifying the fields of the Interest packet according to the table. Otherwise, it notifies the router setting up the fields of the interest packet as default which is predefined in the table.

Step 3 If local controller knows the location of the content, RID, CID, and RCID fields in the Interest packet are modified to the corresponding value.

Step 4 Flow manager defines the flow and router to its destination.

Step5 If the local controller does not have any information of the content, it contacts the Angel Cluster that advertised maximum similar content.

Step 6 If the content is found in that Angel Cluster's Manager, RID, CID, and RCID fields in the Interest packet are modified to the corresponding value.

Step 7 In case the Angel Cluster's Manager does not find the content within its cluster, it sends the Interest packet to all other clusters of that Angel Cluster. The cluster that has the content updates the fields and reverts back to Manager whereas others send a regret message.

Step 8 If entire Angel Cluster did not have the content, the requesting cluster contacts the next best Angel Cluster.

Step 9 The process is repeated till the content is found. If the content is not available with any cluster, the client is sent a regret message.

Table 2 Comparative study of related works

Work	Limitation	Solution in the proposed scheme
Son et al. [17]	There is no record of the clusters that the Interest packet has already been to. It leads to revisiting of clusters leading to heavy traffic	Connected clusters, aware of the content of other clusters through advertisements, in the form of self-improving Angel Cluster leads to non-redundant visiting of clusters
Lv et al. [20]	Clusters based on social relationship of Interest similarity, aka communities, are not temporary, whereas the content is ever changing	Angel Clusters are self-improving. New clusters with similar content can join the existing clusters, whereas clusters with non-similar content can leave the Angel Cluster
Hoque et al. [21]	Interest packets with higher priority are forwarded prior to those with lower priority and that may lead to Interest starvation, i.e., high difference in time taken by different requests	The proposed scheme does not differentiate between the requests; thus, all Interest requests received is treated in the same manner maintaining the monotonicity of the network

5 Performance Analysis

In the proposed scheme, we have successfully overcome the issues discussed in literature review. In our scheme, the clusters are not stand alone but are aware of the content of other clusters through advertisements. We have used content-based clusters which are not permanent but self-improving to keep in tone with the changes in content. However, the concept of clusters on the basis of physical proximity has not been dropped.

Instead of prioritizing the Interest requests that may lead to starvation or longer time to process and respond in case lower priority requests, we have not differentiated between the requests; thus, all requests received will be treated in the same manner maintaining the monotonicity of the network. However, to ensure that the requests are fulfilled speedily, we have prioritized the Angel Clusters that will be approached.

The proposed scheme transforms SDN-based CCN by eliminating broadcasting and using most apt Angel Cluster, thereby, keeps the traffic low in the network.

This section represents a comparative study of the proposed scheme with other relevant schemes discussed in research works and mentions their limitations in terms of Interest packet forwarding. The comparison is shown in the following Table 2.

6 Conclusion

CCN is a networking paradigm which has attracted increasing research interests by various research scholars around the globe. In this paper, we have proposed and analyzed a novel packet forwarding strategy in SDN-based CCN using distributed

SDN controllers and clusters of routers. We have introduced a strategy that enables reduction of the problems that existed in the original CCN and those discussed in literature review. The self-improving Angel Clusters change as per the content and bring about the best of content-based clustering and physical locality-based clustering. We have also overcome the challenge of unnecessary broadcasting and repeated cluster visits of the Interest packets. As a future scope, we are motivated to work on advertisement mechanisms, summarizing content using AI, parameters of matching content, effective management of tables, and compatible protocols between SDN and CCN.

References

1. Ahmed, R., Boutaba, R.: Distributed pattern matching: a key to flexible and efficient P2P search. IEEE J. Sel. Areas Commun. **25**(1), 73–83 (2007)
2. Moskowitz, R., Nikander, P., Jokela, P.: Host identity protocol (HIP) architecture. RFC **4423**, 279–286 (2006)
3. Jacobson, V., Smetters, D.K., Thornton, J.D., Plass, M.F., Briggs, N.H., Braynard, R.L.: Networking named content. In: Proceedings of the 5th International Conference on Emerging Networking Experiments and Technologies, pp. 1–12 (2009)
4. Yaqub, M.A., Ahmed, S.H., Bouk, S.H., Kim, D.: Enabling critical content dissemination in vehicular named data networks. In: Proceedings of the 2018 Conference on Research in Adaptive and Convergent Systems, pp. 94–99 (2018)
5. Adhikari, S., Ray, S., Biswas, G.P., Obaidat, M. S.: Efficient and secure business model for content centric network using elliptic curve cryptography. Int. J. Commun. Syst. 3839 (2018)
6. Adhikari, S., Ray, S., Obaidat, M.S., Biswas, G.P.: Efficient and secure content dissemination architecture for content centric network using ECC-based public key infrastructure. Comput. Commun. Elsevier **157**, 187–203 (2020). https://doi.org/10.1016/j.comcom.2020.04.024
7. Adhikari, S., Ray, S.: A lightweight and secure IoT communication framework in content-centric network using elliptic curve cryptography. In: Recent Trends in Communication, Computing, and Electronics. Springer, Singapore. pp. 207–216 (2019)
8. Jacobson, V., Smetters, D.K., Thornton, J.D., Plass, M.F., Briggs, N.H., Braynard, R.L.: Networking named content. In: Proceedings of the 5th International Conference on Emerging Networking Experiments and Technologies, ACM. pp. 1–12 (2009)
9. Golle, J.P., Smetters, D.: Ccnx access control specifications. Xerox Palo Alto Research Center-PARC, Tech. Rep. (2010)
10. Kuriharay, J., Uzun, E., Wood, C.A.: An encryption-based access control framework for content-centric networking. In: IFIP Networking Conference (IFIP Networking), pp. 1–9 (2015)
11. Udugama, A., Zhang, X., Kuladinithi, K., Goerg, C.: An on-demand multi-path interest forwarding strategy for content retrievals in CCN. In: 2014 IEEE network operations and management symposium (NOMS), pp. 1–6 (2014)
12. Rossini, G., Rossi, D.: Evaluating CCN multi-path interest forwarding strategies. Comput. Commun. **36**(7), 771–778 (2013)
13. Li, C., Okamura, K., Liu, W.: Ant colony based forwarding method for content-centric networking. In: 2013 27th international conference on advanced information networking and applications workshops. IEEE, pp. 306–311
14. Nunes, B.A.A., Mendonca, M., Nguyen, X.N., Obraczka, K., Turletti, T.: A survey of software-defined networking: Past, present, and future of programmable networks. IEEE Commun. Surv. Tutorials **16**(3), 1617–1634 (2014)

15. Son, J., Kim, D., Kang, H. S., Hong, C. S.: Forwarding strategy on SDN-based content centric network for efficient content delivery. In: 2016 International Conference on Information Networking (ICOIN). IEEE, pp. 220–225 (2016)
16. Fundation, O.N.: Software-defined networking: The new norm for networks. ONF White Paper **2**(2–6), 11 (2012)
17. Kim, W.S., Chung, S.H., Moon, J.W.: Improved content management for information-centric networking in SDN-based wireless mesh network. Comput. Netw. **92**, 316–329 (2015)
18. Iqbal, S.M.A.: Adaptive forwarding strategies to reduce redundant interests and data in named data networks. J. Netw. Comput. Appl. **106**, 33–47 (2018)
19. Lv, J., Wang, X., Huang, M., Shi, J., Li, K., Li, J.: RISC: ICN routing mechanism incorporating SDN and community division. Comput. Netw. **123**, 88–103 (2017)
20. Hoque, A.M., Amin, S.O., Alyyan, A., Zhang, B., Zhang, L., Wang, L.: NLSR: Named-data link state routing protocol. In: Proceedings of the 3rd ACM SIGCOMM Workshop on Information-Centric Networking, pp. 15–20 (2013)
21. Aamir, M.: Content-priority based interest forwarding in content centric networks. arXiv preprint arXiv:1410.4987 (2014)
22. Xue, K., He, P., Zhang, X., Xia, Q., Wei, D.S., Yue, H., Wu, F.: A secure, efficient, and accountable edge-based access control framework for information centric networks. IEEE/ACM Trans. Netw. **27**(3), 1220–1233 (2019)

Joint Subcarrier Mapping with Relay Selection-Based Physical Layer Security Scheme for OFDM System

K. Ragini and K. Gunaseelan

1 Introduction

In wireless networks, the communication between legitimate users can simply be overheard by an eavesdropper for intervention because of the broadcast nature of the wireless system. If the intended data transmission channel capacity is greater than the eavesdropping channel capacity, the data can be communicated at a rate close to the intended channel capacity, allowing only the intended receiver (not the eavesdropper) to effectively decode the data. The equivalent of security is defined by the secrecy capacity, which is the capacity difference between the intended data transmission and eavesdropping channels. Conventionally, the security for a communication system is offered at higher layers. Security and efficient transmission with limited resources are great challenges in wireless communication. So that orthogonal frequency division multiplexing (OFDM) is a great promising approach to improve the transmission rate efficiently. If the concept of OFDM is applied to physical layer security, it would be a great notion for reliable communication. OFDM is a multi-carrier system; here, the data bits are encoded to multiple subcarriers at the same time, and it is transmitted simultaneously. The cooperative system (CS) at a wireless node combines to transmit the signal to the destination with the help of relay. This is referred to as relaying. The main advantage of modern relaying system over earlier relaying system is that every node in the system can perform a relay for other nodes. The one-way relaying is a half-duplex system; hence, it needs two times slot to exchange the data between two nods. In every single time slot, the relay changeovers between being a receiver and a transmitter. Duval et al. [1] under power limits for both the base stations and the relay stations, this work investigated

K. Ragini (✉) · K. Gunaseelan
DECE, CEG Campus, Anna University, Tamil Nadu, India

K. Gunaseelan
e-mail: guna_2012@annauniv.edu

M. Dua et al. (eds.), *Proceedings of the International Conference on Paradigms of Communication, Computing and Data Sciences*, Algorithms for Intelligent Systems,
https://doi.org/10.1007/978-981-16-5747-4_6

the cooperative relaying scheme's end-to-end capacity utilizing OFDM modulation. They show that the relay selective scheme outperforms the non-selective relaying by achieving higher capacity. Saman Atapattu et al. [2] proposed the suboptimal relay selection for multiple relay networks and analyzed the performance of the optimal relay selection scheme. In this paper, optimum relay selection for full diversity is achieved with low complexity. The main problem in OFDM-based relay system is power allocation on multiple hops. Guftaar Ahmad Sardar Sidhu et al. [3], the channel gain over different hops may be mutually independent for all subcarriers in a multi-hop network. Unlike the traditional resource allocation problem, the relay node in this case can execute subcarrier permutation over two hops, allowing the signal received on one subcarrier to be sent via a different subcarrier. The goal is to optimize the cognitive radio network's throughput while staying under a limited power budget at the secondary source and relay node. In Shah et al. [4], power allocation is done at the forwarding relay node, together with subcarrier mapping, to optimize the secrecy rate. This ensures that leakage to the eavesdropper is minimized at the forwarding relay node. Zheng et al. [5] investigates at an alternative method that relies on the presence of a full duplex (FD) receiver. The receiver, in instance, transmits jamming noise to degrade the eavesdropper channel while receiving data. The proposed self-protection technique eliminates the requirement for external assistants while still ensuring system stability. The relay selection with subcarrier mapping and optimal power allocation for OFDM-based wireless system with eavesdropper presence is not studied in kinds of literature. Therefore, in this paper, a new joint relay selection with subcarrier mapping, and optimum power allocation is proposed to maximize the secrecy rate of the overall system. The following are the contributions of the proposed work. (1) A new relay selection with subcarrier mapping technique has been proposed to improve secrecy rate performance for the multi-relay assisted OFDM-based wireless networks. (2) Optimum power allocation using closed-form optimization was adapted to maximize the secrecy rate further without nonlinear programming complexity (3) The secrecy rate performances were analyzed with different eavesdropper distances. The remainder of the paper is broken into the following sections. Section 2 discusses our proposed technique and solution, as well as the system model and problem formulation. The results and analysis of our proposed scheme are discussed in Sect. 3. The paper comes to a conclusion with Sect. 4.

2 System Model of OFDM-Based Half-Duplex Relaying

The OFDM-based wireless networks with half-duplex relaying systems have been considered for our study. In this system model, we consider one sender, one receiver, N relay nodes, and one eavesdropper as shown in Fig. 1. The transmission of information has two phases. In the first phase of communication, the source transmits the message to the relay node, and in the second phase, best forwarding relay transmits the received message to the destination. We assume that the distance between

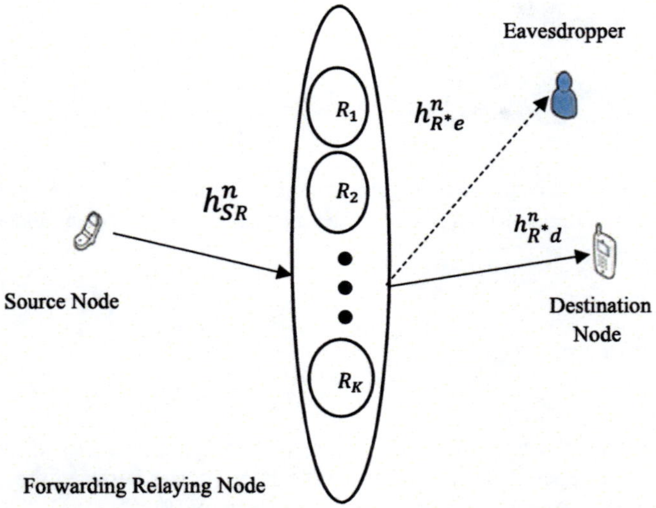

Fig. 1 System model of OFDM-based half-duplex relaying

source and destination is far away, and hence, there is no direct connection between source and destination. Because OFDM is used to access the physical medium, all links have multiple subcarriers. A subcarrier's channel state may change from one hop to the next. Instead of using the same subcarrier to transfer data received at the forwarding relaying node, the channel conditions of the link from the forwarding relaying node to the destination might be taken into account in subcarrier mapping. So, the subcarriers are mapped based on the highest channel gain between the source to relay and also between the relay to destination and transmitted accordingly [6]. Power allocation with maximum transmitter power constraints is done at the source and relay node to utilize efficiently the available power [7].

The main objective of the proposed method is to improve the secrecy rate with optimum power allocation at the source node and relay node in the physical layer security of the wireless network. In the first time slot, the source node transmits its message x_s for N subcarriers to the optimum relay R^*, the received signal at optimum relay is given by

$$y_{SR^*} = \sum_{n=1}^{N} h^n_{SR^*} \sqrt{p^n_S} x^n_s + n_{R^*} \tag{1}$$

where h^n_{s,R^*} denotes the nth sub-channel gain from the source to the optimum relay R^*. p_s represents the transmission power of the source, and n_{R^*} is the additive white Gaussian noise (AWGN). The node relay node applies the amplify-and-forward (AF) protocol. The forwarding relay transmits the received message signal to the destination in the second phase of communication, and the received signal at the destination is given by

$$y_{R^*d} = \left(\sum_{n=1}^{N} h_{R^*d}^{n} h_{SR^*}^{n} \sqrt{p_{R^*}^{n}} \sqrt{p_s^n} x_s^n + \sum_{n=1}^{N} h_{R^*d}^{n} \sqrt{p_{R^*}^{n}} n_{R^*} \right) + n_d \tag{2}$$

where h_{R^*d} represents sub-channel gain from the optimum relay R^* to destination. p_R represents the transmission power of the relay node, and the channel capacity is given by [8]

$$c_d = \frac{1}{2} \log_2 \left(1 + \frac{\left(\sum_{n=1}^{N} |h_{R^*d}^{n} h_{SR^*}^{n}| \sqrt{p_{R^*}^{n}} \sqrt{p_s^n} \right)^2}{N_0 + \left(\sum_{n=1}^{N} h_{R^*d}^{n} \sqrt{p_{R^*}^{n}} \right)^2} \right) \tag{3}$$

In the second phase of communication, eavesdropper capacity is estimated. Here the eavesdropper is considered to be the intermediate between the relay and destination whose channel capacity equation is given by [8]

$$c_e = \frac{1}{2} \log_2 \left(1 + \frac{\left(\sum_{n=1}^{N} |h_{R^*e}^{n} h_{SR^*}^{n}| \sqrt{p_{R^*}^{n}} \sqrt{p_s^n} \right)^2}{N_0 + \left(\sum_{n=1}^{N} h_{R^*e}^{n} \sqrt{p_{R^*}^{n}} \right)^2} \right) \tag{4}$$

where h_{R^*e} is the channel gain between the optimum relay and the eavesdropper. The $\frac{1}{2}$ factor is because of the two-time slots that have been taken for a complete transmission from source to destination. Generally, the parameter metric used for analyze the secrecy performance is the secrecy capacity C_{sec} which is defined as

$$C_{\text{sec}} = [C_d - C_e] \tag{5}$$

where $[.]^+$ identifies that the secrecy capacity cannot be negative, i.e.,

$$C_{\text{sec}} = \max[C_d - C_e, 0]^+ \tag{6}$$

In order to improve the secrecy performance, the optimum relay R^* is obtained by using outage probability analysis for K relays. To select the best relay, the secrecy capacity is calculated for all the K relays based on the outage probability p_{out}, and the equation is given by

$$p_{\text{out}} = p[C_{\text{sec}} < R_{\text{th}}] \tag{7}$$

where p_{out} is the outage probability and R_{th} denotes the threshold rate. The rate of the relays should not exceed the threshold limit.

2.1 Algorithm of Proposed Method for Relay Selection and Subcarrier Mapping

Algorithm for Subcarrier mapping

Input

$$\left|h_{sR_i}^n\right| = \left\{h_{sR_i}^1, h_{sR_i}^2 \ldots, h_{sR_i}^N\right\}$$

$$\left|h_{R_id}^n\right| = \left\{h_{R_i,d}^1, h_{S,R_i}^2 \ldots, h_{R_i,d}^N\right\}.$$

$$\left|h_{R_ie}^n\right| = \left\{h_{R_ie}^1, h_{R_ie}^2 \ldots, h_{R_ie}^N\right\} \text{ where } i \in \left\{r_1, r_2, \ldots r_K\right\}$$

Step 1 Relay Selection using $R^* = arg \min_{r_1, r_2, \ldots r_K} p_{out}$

Step 2 From $S = \left\{h_{SR*}^1, h_{SR*}^2 \ldots, h_{SR*}^N\right\}$ and $C = \left\{h_{R*d}^1, h_{R*d}^2 \ldots\ldots, h_{R*d}^N\right\}$

Step 3 Find $\Lambda_n = \left|\frac{h_{R*d}^n}{h_{R*e}^n}\right|$ where n $\in \{1, 2, \ldots N\}$

$$\left|h_{R*e}^n\right| = \left\{h_{R*e}^1, h_{R*e}^2, \ldots, h_{R*e}^N\right\}$$
$$\Lambda = \{\Lambda_1, \Lambda_2, \ldots \Lambda_N\}$$

Step 4 From $|h_{sR*}|$ find $\left|h_{SR*}^k\right| = \max\{|h_{SR*}|\}$.

Step 5 From Λ find $\left|\Lambda_j\right| = \max\{\Lambda\}$ and select $|h_{R*d}|$ corresponds to $\left|\Lambda_j\right|$

Step 6 $h_{SR*}^k h_{R*d}^j$ i.e. $h_{SR*}^k \leftrightarrow h_{R*d}^j$

Step 7 Remove h_{SR*}^k and h_{R*d}^j from S and C, respectively.

Step 8 If $S \neq \emptyset$ and $C \neq \emptyset$. Go to step 4.

Step 9 Power Allocation for source node and relay node according to (8).

Step 10 End

In step 1, the optimum relay is selected using the R^* equation. Using h_{SR*}^n and h_{R*d}^n, Vector S has been formed, which represents all subcarriers from the source to the forwarding relay node, and vector C has been formed, which represents all subcarriers from the forwarding relay node to the destination. In Step 3, Λ_n is determined by computing the ratio between the gain from the relay node to the destination node and the gain from the relay node to the eavesdropper node for all the subcarriers. In Step 4, kth subcarrier between source and optimum relay is identified by computing $\left|h_{SR*}^k\right| = \max\{|h_{sR*}|\}$ similarly in step 5, jth subcarrier between relay to destination is identified by computing $\left|\Lambda_j\right| = \max\{\Lambda\}$. Then kth subcarrier between source and the optimum relay is mapped with jth subcarrier between optimal relay to destination and removed for the vector S and C. This process is continued until the vector C and S become null. After successful completion of subcarrier mapping, the power allocation is performed in order to optimize the source power and forwarding relay node power. The optimization problem (OP) for power optimization and capacity of the system is given as

$$\text{OP}: \min_{p_s^n, p_{R*}^n, i \in 1,2,\dots,K} C_{\text{sec}} \tag{8}$$

Subject to

i. $\sum_{n=1}^{N} P_s^n = \sum_{n=1}^{N} P_{R^*}^n = P_t$

ii. $P_S^n > 0$ for all $n\ \forall\ n = 1 \ldots$, N

iii. $P_{R^*}^n > 0$ for all $n\ \forall\ n = 1 \ldots$, N

where P_t denotes the total power for is assigned for the source and relay.

3 Results and Discussion

We used MATLAB R2018a to run simulations to test the performance of our proposed technique, assuming Rayleigh fading for the channel coefficients and subcarriers. In the Rayleigh fading model, the path loss exponent is set to 4, illustrating worst-case situations for cellular and long-distance communication. The simulation parameters that we employed in our simulations are listed below in Table 1.

We have compared and analyzed our proposed scheme with three different existing techniques such as (i) baseline scheme [5] where equal power allocation without subcarrier mapping, and without relay selection is done (ii) equal power allocation with relay selection and without subcarrier mapping scheme and (iii) Shah et al. [4] where the sub-carrier mapping with optimal power allocation without relay selection is done.

Table 1 Simulation parameters

Parameters	Value
Wireless channel bandwidth	1 MHz
Noise spectrum density	4.14×10^{21} W/Hz
Number of relays	4
Distance between source and relay	100 m
Distance between relay and destination Distance between relay and eavesdropper for Fig. 5 Distance between relay and eavesdropper for Fig. 4 Distance between relay and eavesdropper for Fig. 3 Number of subcarriers for Figs. 3, 4, and 5	100 m 250 m 200 m 150 m 32

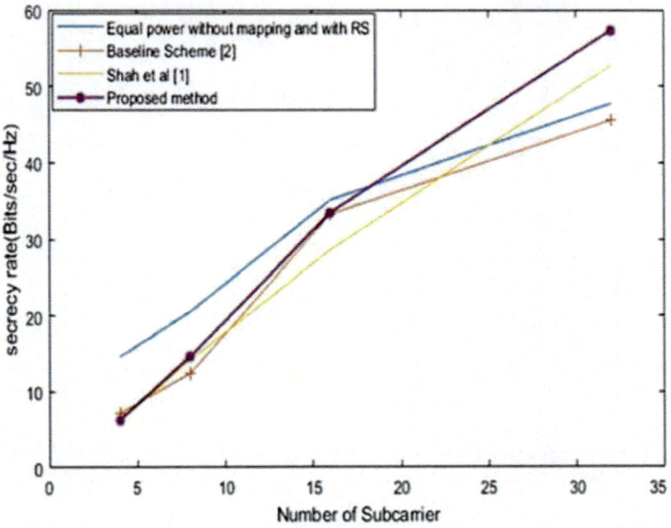

Fig. 2 Effect of number of subcarrier on secrecy rate

Figure 2 shows the impact of the number of subcarriers on the secrecy rate. In this analysis, the maximum transmission power is assumed to be 20 W. Generally, the secrecy rate performance will increase for the more number of subcarriers. However, our proposed algorithm provides a secrecy rate of 58 bps/Hz for the 32 subcarriers which are 12 bps/Hz greater than the baseline method, 10 bps/Hz greater than the equal power without mapping and with relay selection and 2 bps/Hz greater than the [4]. In the baseline scheme, they do not have subcarrier mapping and relay selection. These drawbacks cause poor secrecy rate performance than our proposed method. In the equal power allocation method, equal power is allocated to all the subcarriers which are less complex than the proposed scheme but there is no subcarrier mapping. Despite having a relay selection method, increasing the number of subcarriers does not provide a significant gain due to increased leakage to the eavesdropper. In shah et al. [4], they do not have relay selection which leads to a lower secrecy rate performance than our proposed method.

The effect of relay-eavesdropper distances on secrecy rate is seen in Figs. 3, 4, and 5. The following are the three scenarios:

Case No. 1: The relay network and the eavesdropper are 150 m apart.

Case No. 2: The relay network and the eavesdropper are 200 m apart.

Case No. 3: The relay network and the eavesdropper are 250 m apart.

To analyze the effect of the relay-eavesdropper location on the secrecy rate. In Figs. 3, 4, and 5, the selected relay is located in the same position but the eavesdropper location is varied as mentioned in Table 1. In case No. 1, the secrecy rate of our proposed scheme is approximately 42 bps/Hz for transmission power of 30 W, which is 21 bps/Hz greater than the baseline method, 13 bps/Hz greater than the equal power

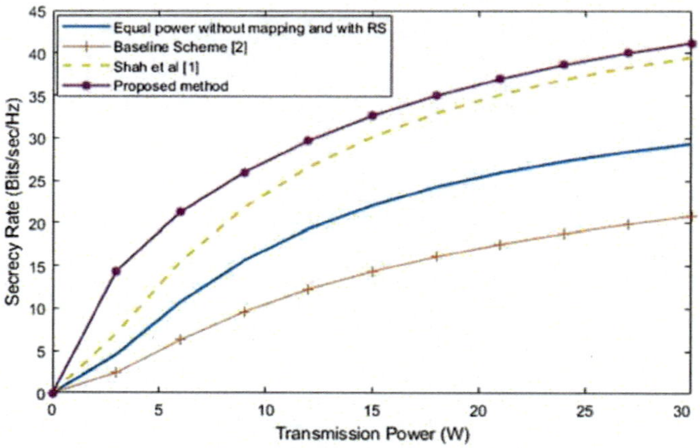

Fig. 3 Effect of transmission power on secrecy rate for case (i)

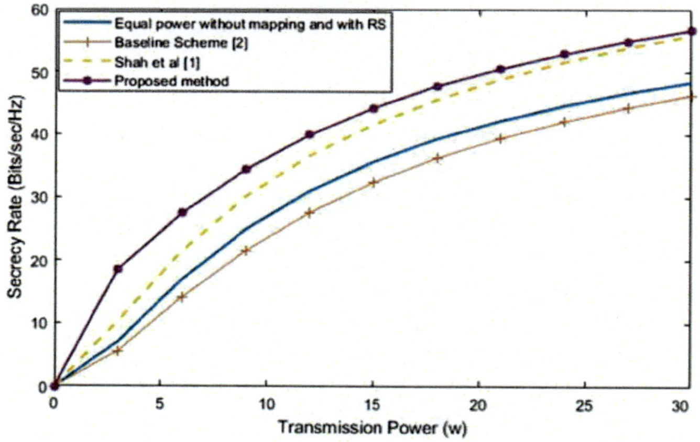

Fig. 4 Effect of transmission power on secrecy rate for case (ii)

without mapping and with relay selection and 2 bps/Hz greater than the Shah et al. [4]. Similarly, in case No. 2 and case No. 3, we inferred that the secrecy rate of our proposed scheme is greater when compared to the other three methods as shown in Figs. 4 and 5. From these analyses, it is inferred that the secrecy rate performance is largely affected when eavesdropper is near to the destination node. Since the proposed algorithm aims at maximizing the secrecy rate and minimizing leakage to the eavesdropper with respect to the distance, it provides better secrecy rate even when the eavesdropper is near to the destination node.

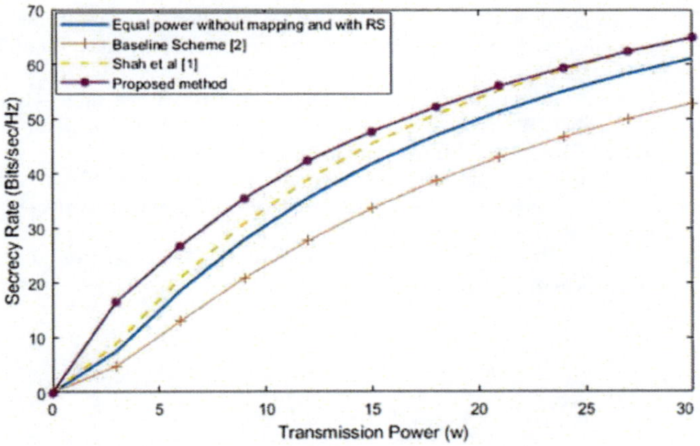

Fig. 5 Effect of transmission power on secrecy rate for case (iii)

4 Conclusion

This paper studied the performance of the physical layer security scheme for the OFDM system assisted with a relay. In order to maximize the secrecy rate, relay selection strategies are incorporated with subcarrier mapping. The optimal power allocation at the source and relay nodes was adjusted to account for channel gain between the source-relay and relay-destination nodes, resulting in minimal leakage to the eavesdropper. Our proposed method eliminates the nonlinear programming complexity using closed-form optimization. The simulation results show that our proposed scheme can significantly enhance the secrecy rate of 42 bps/Hz. The secrecy rate is improved than other existing methods. Also, it maximizes the secrecy rate at low transmitter power and nearby eavesdropper conditions.

References

1. G. Zheng, I. Krikidis et al.: Improving physical layer secrecy using full-duplex jamming receivers. IEEE Trans. Signal Process. (2013)
2. O. Duval, Z. Hasan, E. Hossain, F. Gagnon, V.K. Bhargava: Subcarrier selection and power allocation for amplify-and-forward relaying over OFDM links. IEEE Trans. Wirel. Commun. 1293–1297 (2010)
3. Atapattu, S., Jing, Y., Jing, H. et al.: Relay selection and performance analysis in multiple-user networks, IEE J. Sel. Area Commun. **31**(10), 1517–1529 (2013)
4. Shah, H.A., Koo, I.: A novel physical layer security scheme in OFDM-based cognitive radio networks. IEEE Access Sel. Areas Commun. **6**, 29486–29498 (2018)
5. Choi, Y., Lee, J.H.: A new cooperative jamming technique for a two-hop amplify-and-forward relay network with an eavesdropper. IEEE Trans. Veh. Technol. (2018)

6. Shao, Y. (Graduate Student Member, IEEE), Liew, S.C. (Fellow, IEEE): Flexible subcarrier allocation for interleaved frequency division multiple access. IEEE Trans. Wirel. Commun. **19**(11), 7139–7152 (2020)
7. Goldsmith, A.: Wireless Communications. Cambridge University Press, pp. 1–412 (2004)
8. Zou, Y., Wang, X., Shen, W.: Optimal relay selection for physical-layer security in cooperative wireless networks. IEEE J. Sel. Areas Commun. **31**(10), 2099–2111 (2013)
9. Ca, C., Yang, W., Cai, Y.: Improved relay selection and subcarrier pairing with fairness constraints for OFDM networks. In: 2009 International Conference on Wireless Communications & Signal Processing, pp. 1–4 (2009)
10. Sidhu, G.A.S., Gao, F.: Resource allocation in relay-aided OFDM cognitive radio networks. IEEE Trans. Veh. Technol. **62**(8) (2013)

Prefeasibility Economic Scrutiny of the Off-grid Hybrid Renewable System for Remote Area Electrification

Siddharth Jain, Sanjana Babu, and Yashwant Sawle

Abbreviations

NPC	Net present cost
RF	Renewable fraction (%)
RC	Replacement cost
BT	Battery
WT	Wind turbine
WS	Wind speed (m/s)
CC	Capital cost
COE	Cost of energy (US$/kWh)
EE	Excess energy
HPS	Hybrid power system
GSR	Global solar radiation (kWh/m^2/d)
CT	Converter
BG	Biodiesel generator
DG	Diesel generator

1 Introduction

Electrical energy necessity is increasing rapidly due to rapid industrialization and urban development. India is facing difficulties in supplying electrical energy to sustain its gigantic residents and economic growth. The current requirement for electrical energy surpasses the available resources [1]. In the present day, it is the

S. Jain · S. Babu · Y. Sawle (✉)
Department of Electrical Engineering, Vellore Institute of Technology, Vellore, India

M. Dua et al. (eds.), *Proceedings of the International Conference on Paradigms of Communication, Computing and Data Sciences*, Algorithms for Intelligent Systems,
https://doi.org/10.1007/978-981-16-5747-4_7

phase to bring in unconventional energy sources. Renewable energy sources address an ecological substitute to the electrical energy in a remote location. Unconventional energy sources like PV, hydro-, wind energy, and biomass offer a substantial replacement to the machines that use generators for energy production. Usage of hybrid renewable energy systems varies from the place. There is satisfactory exploration work going on a hybrid renewable energy system. Some of the findings are shown in Table 1. In this current research, the feasibility of PV-wind-biomass-diesel generator-battery-converter for a load of a rural town situated in the western area of Gujarat, India. This study offers a solution for villages and towns with relatable climate and a model that can be used in any nation using unconventional resources.

Throughout this paper, we implement HOMER software to find a way to solve the following issues.

- The aim of this study is to look into the issues and challenges that come with implementing a microgrid system.
- To address scale and cost optimization challenges in remote areas, a novel hybrid renewable energy system (HRES) was created.
- The proposed system emits less pollutants.

This paper is organized with the introduction in Sect. 1, methodology in Sect. 2, resource assessment in Sect. 3, Sect. 4 demonstrates component selection, and Sect. 5 demonstrates HOMER software. The simulation results are presented in Sect. 6, and Sect. 7 highlights the conclusion and discussion.

2 Methodology

The most appropriate site for our study is recognized as a small town in the Tapi district in Gujarat. This study mainly aims to give a minimum total net present cost and cost of energy based on the hybrid system for delivering electrical power to meet the proposed study area's load demands.

2.1 Study Area

The study is conducted in a town, Ukai in the state of Gujarat. It is located at 21 14.0′ N latitude and 73° 35′ E longitude. The town is located in the Taluka of Songadh, in the Tapi district with an overall population of 19,898 [10].

Table 1 Summary of literature review

Authors	Place	Technology	Findings
Shaahid and Elhadidy [2]	Kingdom of Saudi Arabia	PV/DG/BT	Effect of PV/battery penetration on COE, operational hours of diesel gensets
Rajkumar et al. [3]	Barwani district, India	PV/wind/biomass	Result comparison using GA and PSO optimization techniques
Senthil Kumar et al. [4]	Tamil Nadu, India	PV/wind/fuel Cells	Power loss minimization is formulated as a nonlinear problem and optimized by the Nelder Mead-particle swarm optimization algorithm
Baseer et al. [5]	Kingdom of Saudi Arabia	PV/wind/battery/diesel	Save 2800 tons of carbon dioxide emissions
Sharma and Chandel [6]	Khatkar-Kalan	PV	Total yields, comparative yields, as well as production ranges, differed around 1.45–2.84 kWh/kWp-day, 2.29–3.53 kWh/kWp-day, as well as 55–83%
Rehman et al. [7]	Pakisthan	PV/wind/diesel/BSS/converter	Variations are unavoidable due to linear and nonlinear loads, as well as the existence of non-conventional sources of energy
Jahangiri et al. [8]	Iran	PV/diesel	In comparison to a hydrogen-based system, the design for optimizing the micro-power system
Sawle et al. [9]	Ukai	PV/WT/micro- Hydro/DG/BG	The best and most efficient HES, comprising of a 76 kW PV array, a 73 kW 1 WT, a 10 kW DG, a 110 kW BG, 155 BT cells of generic 1kWh lead acid, an 11 kW MH, as well as a 57.2 kW CT

2.2 Load Estimation

The estimation of the study area loads depends on their day-to-day demands and electrical energy requirements for different activities. The total load request was

Fig. 1 Daily profile

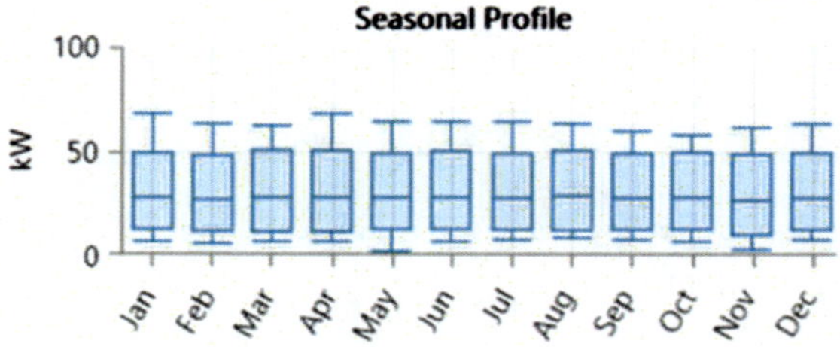

Fig. 2 Seasonal profile

assessed by seeing several demand components, such as residential, industrial, and agricultural load. The community daily load demand is estimated at 686.00 kWh/day with a peak load of 69.49 KW [11]. Figure 1 shows the daily load profile of the location, and Fig. 2 shows the seasonal profile peculiarities.

2.3 System Architecture

The system is created in HOMER software. The proposed system is developed to check the cost effectiveness and optimize hybrid systems. The HOMER software model we developed is shown in Fig. 3.

3 Resource Assessment

We have considered solar, wind, biomass resources in this simulation.

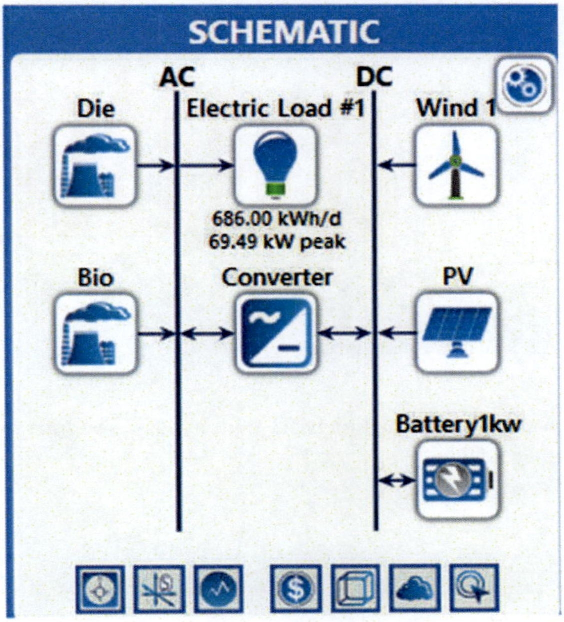

Fig. 3 Proposed hybrid configuration

3.1 Solar

In this research area, there is a surplus amount of solar resource. Solar radiation of Ukai is taken from the NASA Web site [12] as seen in Fig. 4. The monthly

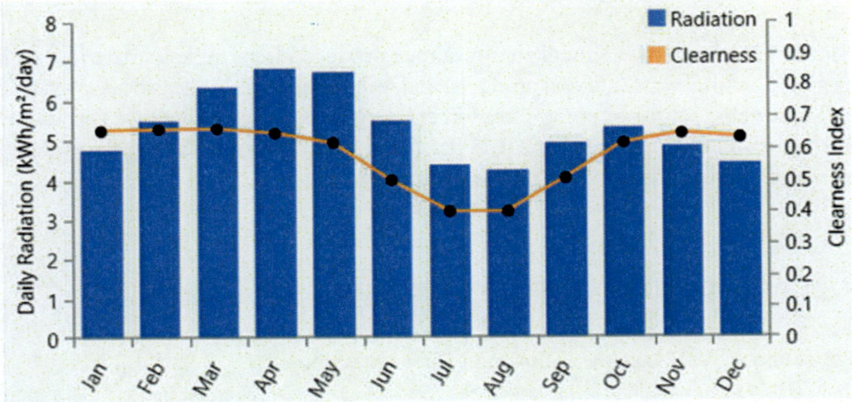

Fig. 4 Solar resource profile

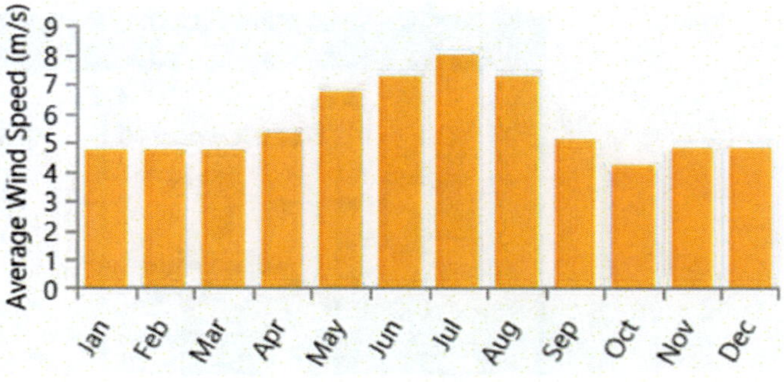

Fig. 5 Wind resource profile

average solar radiation is 5.3 kWh/m^2/day. The solar radiation is the highest in April unavoidable.

3.2 *Wind*

In this study area, wind energy resources are taken from the NASA Web site. The annual average wind speed is 5.63 m/s [12], which can produce a good result throughout the year. July has the highest average monthly wind speed. The detailed data are given in Fig. 5.

3.3 *Biomass*

Biomass is a non-conventional energy source produced from organic materials such as plants, animal wastes, wood products, and dried vegetables. It is also a sustainable and renewable source of energy, used in creating other forms of energy. The annual average availability of biomass is 0.43 tons/day [13–17]. Figure 6 shows the monthly available biomass feedstock. The average price of fuel is 0.67$.

4 Component Selection

For economic analysis, the following values are used.

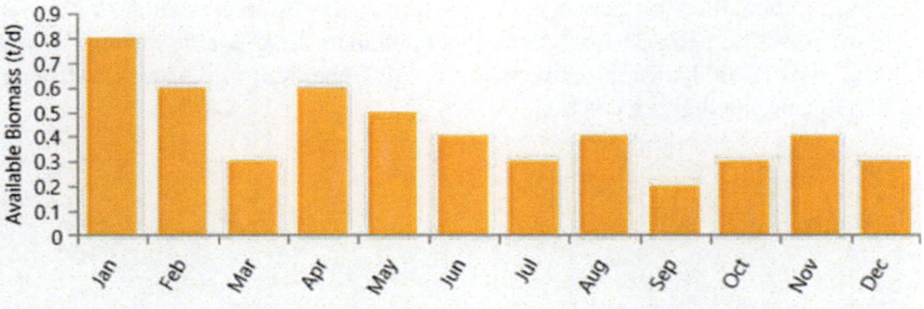

Fig. 6 Biomass resource profile

4.1 PV

Due to technological developments, the cost of solar PV panels has become economically viable and can be used in electricity generation. The capital cost and replacement cost of PV in our study are $1000 and $820, respectively. Operation and management cost is around $ 10 and has a lifetime of 25 years.

4.2 Wind Turbine

Wind energy can be harnessed from the flowing wind through the wind turbine. We have considered the wind turbine's capital cost at $1200, replacement cost at $850, operation and management price at $20. The hub height of the turbine is 17.00 m, and the average lifetime is 20 years.

4.3 Biomass Generator

The pure gaseous form of biomass can be obtained through the biomass gasification process. Initial capital is $500; replacement cost is $250; operation and management cost is $0.030 with the fuel price of $1. The biomass generator has a lifetime of 15,000 h.

4.4 Diesel Generator

The power formation transition increases the flexibility and consistency of the original area to the desired elevations. Distributed creators play an essential role in developing the collection network. Sustainable energy sources similar to solar PV have excessive potential in India as they get abundant solar energy during the year. Capital cost is $1000, the replacement cost is $800, operating and administrative cost is $0.030, and the fuel price is $0.8.

4.5 *Battery*

Batteries store the surplus energy generated and provide backup supply because our renewable energy sources are intermittent. We used a generic 1 kWh lead-acid battery with a capital and replacement cost of $300, operation and management cost $10 per year. The nominal voltage is 12 V and a lifetime of 10 years.

4.6 *Converter*

It is used to perform both as a rectifier and an inverter where AC is converted to DC and vice versa. We used it to keep the power flow between the AC and DC electrical parts. The capital and replacement costs are $300, with an average lifetime of 15 years. The efficiency of both rectifier and inverter is 95% [18].

5 HOMER Software

The United States National Renewable Energy Laboratory (NREL) uses the HOMER software for the technological and economic analyses and optimization of hybrid systems generated and developed. A generation system's physical actions and lifetime cost (including investment and operating costs) are modeled by HOMER software. This software enables the user to analyze several different designs based on technological and economic requirements. This model also helps to recognize the effects of data volatility and user adjustments in inputs to provide the best choice in terms of technological and economic aspects.

In HOMER software, the primary performance of economic calculation is the net present cost (NPC) that can be determined by (1)

$$C_{\text{NPC}} = \frac{C_{\text{Annual,Total}}}{\text{CRF}(i, R_{\text{Project}})} \tag{1}$$

In the above equation, $C_{\text{Annual,total}}$ is a total annual cost, R_{Project} is the project lifetime, and i is the real interest rate.

6 Optimization Results

Table 2 is the complete outcome of cost optimization arranged in the increasing order of cost of energy. As for this study, the result is on line 1 with 139 kW of PV, 112 kW of wind turbine,10 kW of diesel generator, 77 kW of bio generator, 406 kW of lead-acid battery, and 58.2 kW of the converter are selected. The running cost LCOE shows 0.260$/kWh.

Tables 3 and 5 show the complete cash stream for this system. The total net present cost and levelized cost of energy for this model are calculated at $839,675.00 and $0.260 kWh, respectively. The operating cost for this system is $ 29,977/yr.

Table 4 and Fig. 10 show the load consumption of the simulated system. The annual PV module production is at 230,536 kWh/yr.; generic wind turbine is at 212,252 kWh/yr., a diesel generator is at 26,117 kWh/yr. And, biogas generator is at 47,489 kWh/yr. And, the total production is up to 516,394 kWh/yr. Table 4 and Fig. 10 show that the major proportion of electricity is generated from PV and wind as there are a surplus amount of PV and wind sources at the location.

Figure 9 shows the additional electrical power generated. The unmet electric load and capacity shortage are the sum of the energy wasted any of the system's loads

Table 2 Cost optimization of systems

Architecture						Cost			
PV (kW)	WT (kW)	DG (KW)	BG (kW)	BT (kWh)	CT (kW)	NPC ($)	COE ($)	OC ($/yr.)	RF (%)
139	112	10	77	406	58.2	839,675	0.260	29,977	89.6
177	118	10		633	65.9	988,013	0.305	35,461	85.4
205	143	–	77	722	68.9	1.07M	0.332	32,657	88
409	–	10	77	598	82.7	1.19M	0.367	41,330	84.3
240	122	–		1170	84.3	1.27M	0.392	39,061	82
445	–	10		937	66.1	1.43M	0.442	52,732	80

Table 3 System simulation cost

Component	Capital ($)	Replacement ($)	O and M ($)	Fuel ($)	Salvage ($)	Total ($)
DG	1000	1578	10,874.63	91,168.06	62.60	104,558.48
BG	38,500	19,027	52,289.35	1343.17	376.60	110,782.91
WT	134,400	30,350.46	28,957.64	0.00	17,104.4	176,603.67
BT	121,800	107,602.70	52,485.72	0.00	14,589.0	267,299.35
PV	138,980.91	0.00	17,966.78	0.00	0.00	156,947.69
CT	17,466.86	7410.73	0.00	0.00	1394.77	23,482.82
System	452,14.77	165,969.28	162,574.11	92,511.23	33,527.4	839,674.91

Table 4 Proposed system load components

Production	kWh/yr.	%
PV	230,536	44.6
DG	26,117	5.06
BG	47,489	9.20
WT	212,252	41.1
Total	516,394	100

Table 5 Total cost of the system

Total net present cost	$ 839,675
Levelized cost of energy	$ 0.260/kWh
Operating cost	$ 29,977

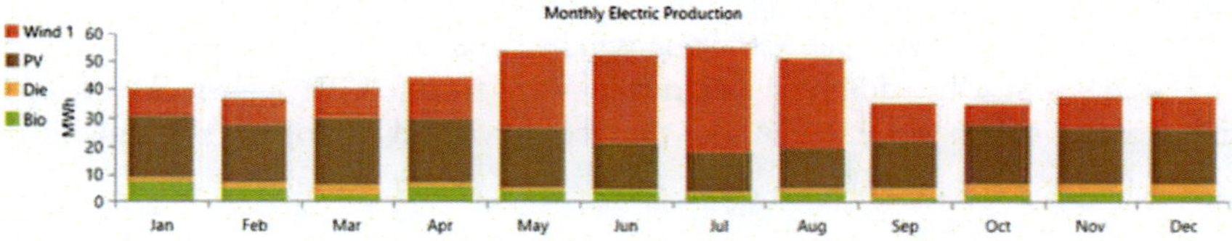

Fig. 7 Per month electrical production

Fig. 8 Overall system load components

Consumption	kWh/yr	%
AC Primary Load	250,246	100
DC Primary Load	0	0
Deferrable Load	0	0
Total	250,246	100

does not utilize that. Figure 10 shows the emissions produced by the system. As we could see that from Fig. 10, carbon dioxide is produced more when compared to other gasses (Figs. 7, 8, 9, and 10; Table 5).

Fig. 9 Excess electricity production

Quantity	kWh/yr	%
Excess Electricity	251,167	48.6
Unmet Electric Load	144	0.0575
Capacity Shortage	239	0.0954

Quantity	Value	Units
Carbon Dioxide	23,054	kg/yr
Carbon Monoxide	177	kg/yr
Unburned Hydrocarbons	6.46	kg/yr
Particulate Matter	10.6	kg/yr
Sulfur Dioxide	56.5	kg/yr
Nitrogen Oxides	200	kg/yr

Fig. 10 Emissions produced by the system

7 Conclusion and Future Scope

Globe is encountering plenty of ecological problems nowadays because of the extensive usage of non-renewable energy resources. Besides, fossil-fueled resources are deteriorating significantly at a shocking level. To conserve non-renewable sources and to protect the atmosphere from its negative impact, it is now a crucial period to rise the usage of non-conventional sources. Solar, wind, and biomass are lavishly in Ukai. As a result of this study, it is seen that NPC, O and M, COEs, RF of proposed HES systems are $839,675.00, $29,977/yr., $0.260 kWh, 89.6%. The costing of the system can be further more decreased based upon the upcoming government policies as well as subsidies provided to villagers. So, the proposed HES can electrify selected locations with basic non-conventional sources of energy.

Acknowledgements I want to thank the Vellore Institute of Technology (VIT) and Prof. Yashwant Sawle for their kind support.

References

1. Markandya, A., Wilkinson, P.: Electricity generation and health. The Lancet **370**(9591), 979–990 (2007)
2. Shaahid, S.M., Elhadidy, M.A.: Technical and economic assessment of grid-independent hybrid photovoltaic–diesel–battery power systems for commercial loads in desert environments. Renew. Sustain. Energy Rev. **11**(8), 1794–1810 (2007)
3. Rajkumar, R.K., et al.: Techno-economical optimization of hybrid PV/wind/battery system using Neuro-Fuzzy. Energy **36**(8), 5148–5153 (2011)
4. Senthil Kumar, J., et al.: Hybrid renewable energy-based distribution system for seasonal load variations. Int. J. Energy Res. **42**(3), 1066–1087 (2018)
5. Baseer, M.A., Alqahtani, A., Rehman. S.: Techno-economic design and evaluation of hybrid energy systems for residential communities: Case study of Jubail industrial city. J. Cleaner Prod. **237**, 117806 (2019)
6. Sharma, V., Chandel, S.S.: Performance analysis of a 190 kWp grid interactive solar photovoltaic power plant in India. Energy **55**, 476–485 (2013)
7. Rehman, S. et al.: Optimal design and model predictive control of standalone HRES: A real case study for residential demand side management. IEEE Access **8**, 29767–29814 (2020)
8. Jahangiri, M., et al.: Feasibility study on the provision of electricity and hydrogen for domestic purposes in the south of Iran using grid-connected renewable energy plants. Energy Strategy Rev. 23, 23–32 (2019)
9. Sawle, Y., Jain, S., Babu, S., Nair, A.R., Khan, B.: Prefeasibility economic and sensitivity assessment of hybrid renewable energy system. IEEE Access **9**, 28260–28271 (2021)
10. https://en.wikipedia.org/wiki/Ukai
11. https://www.gsecl.in/
12. NASA. http://eosweb.larc.nasa.gov
13. Sawle, Y., Gupta, S.C., Bohre, A.K.: Techno-economic scrutiny of HRES through GA and PSO technique. Int. J. Renew. Energy Technol. **9**(1–2), 84–107 (2018)
14. Sawle, Y., Gupta, S.C., Bohre, A.K.: Socio-techno-economic design of hybrid renewable energy system using optimization techniques. Renew. Energy **119**, 459–472 (2018)
15. Sawle, Y., Gupta, S.C., Bohre, A.K.: Review of hybrid renewable energy systems with comparative analysis of off-grid hybrid system. Renew. Sustain. Energy Rev. **81**, 2217–2235 (2017)
16. Sawle, Y., Gupta, S.C., Bohre, A.K.: PV-wind hybrid system: A review with case study. Cogent Eng. J. **3**(1), 1189305 (2016)
17. Sawle, Y., Gupta, S.C., Bohre, A.K.: A novel methodology for scrutiny of off-grid hybrid renewable system (Wiley & Sons). Int. J. Energy Res. **42**, 570–586 (2018)
18. Sawle, Y., Gupta, S.C.: Optimal sizing of photovoltaic/wind hybrid energy system for rural electrification. In: 2014 6th IEEE Power India International Conference (PIICON), pp. 1–4. IEEE (2014)

Wearable Slotted Patch Antenna with the Defected Ground Structure for Biomedical Applications

Regidi Suneetha and P. V. Sridevi

1 Introduction

One of the major diseases causing death and adult disabilities is stroke [1–5], poor blood circulation to the brain causes, death of cells which in turn causes stroke, and it can be of three types ischemic, hemorrhagic, and transient ischemic attack (TIA) or mini-stroke. Medical imaging is the tool used to diagnose this deadly disease. During 1990–2010, the number of strokes decreased by approximately 10% in developed countries and increased by 10% in the developing countries. Around 3.4 million, 6.9 million people got affected by hemorrhagic stroke and ischemic stroke, respectively, in the year 2013. There are around 42.4 million people who are affected by stroke and are alive even today. Around 67% of the strokes occur in adults of 65 and above age. Computed tomography (CT scan) and magnetic resonance imaging (MRI) [6] are currently available techniques for stroke imaging. Unfortunately, the availability of the above machinery is limited due to the cost and hazardous radiation effects due to ionization. Microwave imaging is presenting unconventional and revolutionary results in the medical field [7] at low cost with more ease.

Microwave medical imaging is an advanced rapid inexpensive non-destructive [8] testing tool used to diagnose deadly diseases like cancer [9], hyperthermia, and other treatments also. Microwave imaging [6] uses electromagnetic radiation in the frequency range of 0.3–9.0 GHz. The electromagnetic wave is transmitted [6, 10] into a body under test, and the effects of the microwave signal are measured from the returned signals. A sufficient number of microwave signals are collected at different input frequencies. The inverse scattering problem is solved from the positions of the body obtained, thereby the image of the body under test is reconstructed. Moreover, microwaves have advantages of non-ionizing radiations, non-invasive measurements

R. Suneetha (✉) · P. V. Sridevi
Andhra University College of Engineering, Andhra University, Visakhapatnam, Andhra Pradesh, India

M. Dua et al. (eds.), *Proceedings of the International Conference on Paradigms of Communication, Computing and Data Sciences*, Algorithms for Intelligent Systems,
https://doi.org/10.1007/978-981-16-5747-4_8

[7] which are painless with quite low power requirements. Microwave imaging also finds a wide range of applications in telecommunications, the military, etc.

Recent advances in the study of microwaves and their advantages due to the accessibility and low attenuation of electromagnetic waves and non-ionizing radiation proved that this is the more feasible method of stroke imaging for continuous monitoring and also for tumor detection in various cancers like breast cancer. Low-profile antennas with high-impedance bandwidth in the medium are required as the human body is non-homogeneous, and it is required to use back-scattered signals to differentiate the healthy and affected tissues. The simple design and miniaturized structure of the proposed antenna [10] with omnidirectional [11] radiation pattern performance at the ISM band is well-suited for biomedical applications.

2 Design of the Antenna

The antenna [10, 12–14] is designed and simulated using HFSS software and fabricated on Fr4 substrate with a dielectric constant of 4.4 and loss tangent of 0.02 with 50 Ω microstrip line feeding. The fabricated antenna is tested, and results are obtained from Anritsu vector network analyzer. A rectangular slotted patch [13] antenna with defected ground structure [12, 14–16] with microstrip line feeding is being proposed.

The dimensions of the conventional microstrip patch antenna can be designed using the following equations.

The effective dielectric constant of the substrate is

$$\varepsilon_{\text{eff}} = \frac{\varepsilon_r - 1}{2}\sqrt{\frac{12h + w}{w}} + \frac{\varepsilon_r + 1}{2} \tag{1}$$

$$w = \frac{C}{2\text{fo}\sqrt{\frac{\varepsilon_r + 1}{2}}} \tag{2}$$

where h is the height of the substrate and w is the width of the patch. ε_r is the dielectric constant of the substrate, C is the velocity of light in free space, and fo is the center frequency of the band.

Length of the patch can be obtained from

$$L = L_{\text{eff}} - 2\Delta L \tag{3}$$

where the effective length and ΔL of the patch can be obtained from

$$L_{\text{eff}} = \frac{c}{2fo\sqrt{\varepsilon_{\text{eff}}}} \tag{4}$$

$$\Delta L = \frac{\left(\frac{w}{h} + 0.264\right)(\varepsilon_{\text{eff}} + 0.3)}{\left(\frac{w}{h} + 0.8\right)(\varepsilon_{\text{eff}} - 0.258)} 0.412h \quad (5)$$

2.1 Slotted Rectangular Patch Antenna with Defected Ground Structure

Figure 1 shows an antenna with slots in the patch and defects in the ground plane [14–16] structure along with dimensions. The proposed antenna with dimensions of 24.5 × 15 × 1.6 mm^3 is resonating with triple-band frequency and having impedance bandwidth from 6.3 to 6.9 GHz, from 7.9 to 8.9 GHz, and 9.2 to 9.7 GHz. To maintain the size of the antenna constant without increasing, the addition of slots is done to increase the current path thereby reducing the lowest resonant frequency. The fabricated antenna is as shown in Fig. 2.

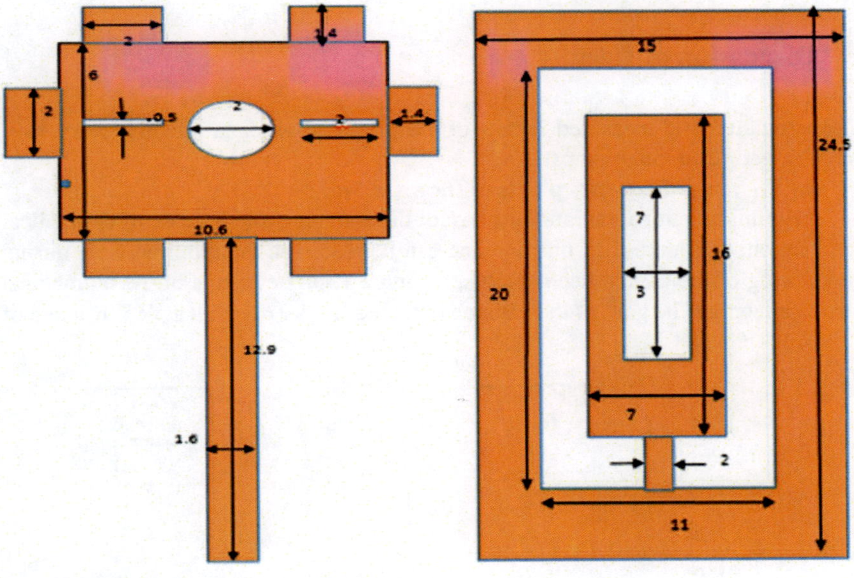

Fig. 1 Proposed antenna design with dimensions **a** Patch **b** Ground

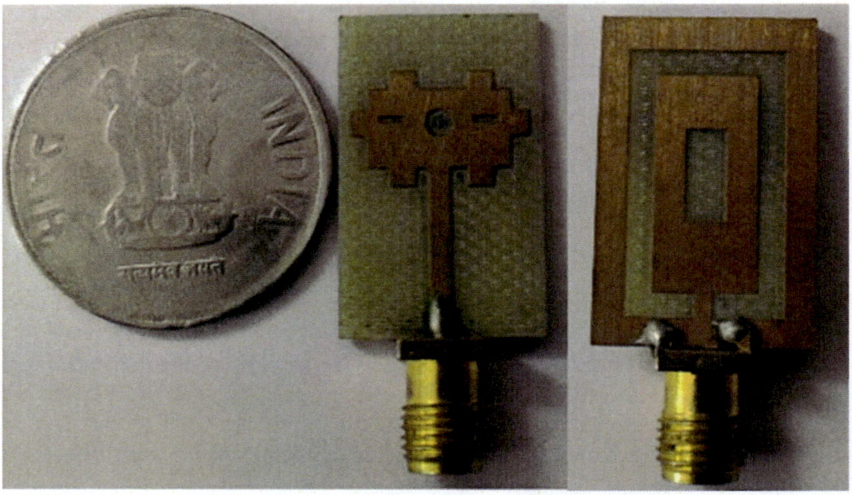

Fig. 2 Fabricated antenna **a** Patch **b** Ground

3 Discussion of Results

The simulated and measured results of the proposed antenna include return loss, VSWR, gain, and radiation pattern.

The S_{11} versus frequency plot is as shown in Fig. 3.

The omnidirectional radiation pattern of the antenna provides ease of biomedical application, and more than one antenna can be placed in the required area, during monitoring of the patient health status, for more accurate results of the biomedical stroke imaging. The gain of the antenna obtained is 1.99 dB with a VSWR value of

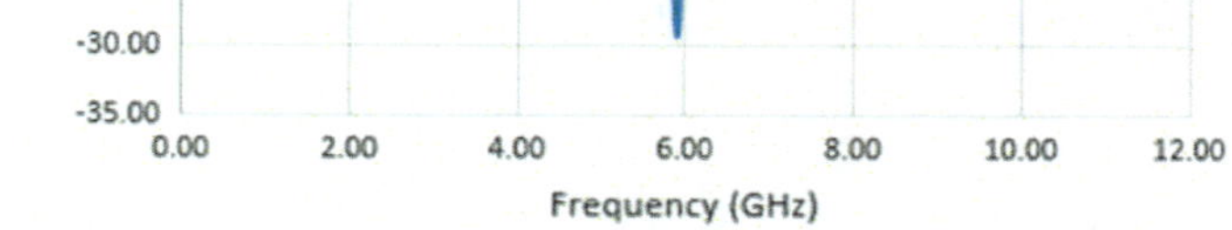

Fig. 3 Simulated and measured S_{11} versus frequency plot of proposed antenna

1.24 at the center frequency as shown in Fig. 4. The measured VSWR vs frequency plot of the proposed antenna using Anritsu network analyzer is as shown in Fig. 5.

To study and maximize the coupling effect between the antenna and human body parts like head, chest, and hand, as the human body is a combination of different layers and materials like blood, bones, muscles, and skin, the antenna is simulated with different values of permittivity, so that the antenna performance can be analyzed for various permittivity values to improve the antenna design for better results. It can be obtained from Fig. 6 that the antenna performs well at the operating frequencies in a different medium. The proposed antenna is suitable for stroke management applications due to the simple structure and size of the antenna as it can be placed comfortably in the helmet as a comparison given in Table 1.

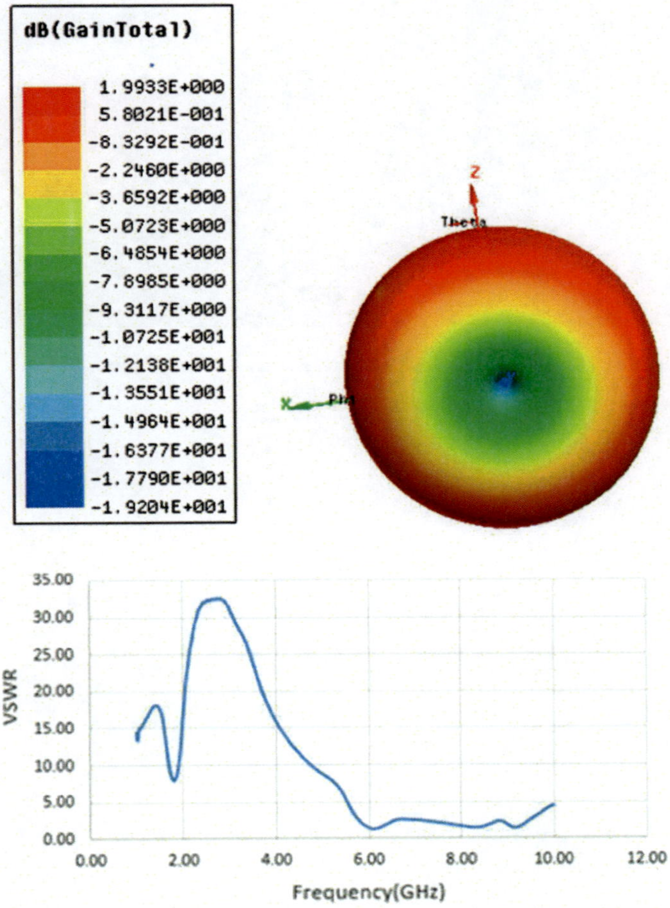

Fig. 4 Simulated gain and VSWR of the proposed antenna

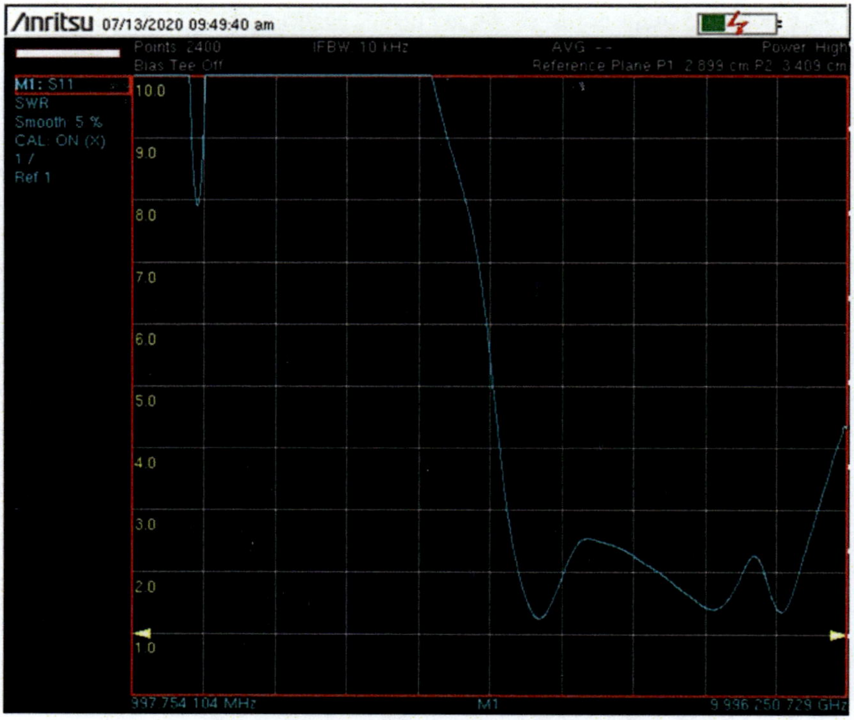

Fig. 5 Measured VSWR versus frequency plot on Anritsu network analyzer

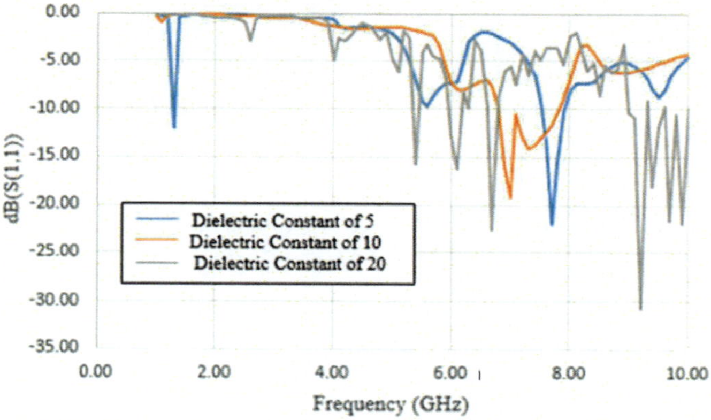

Fig. 6 S_{11} versus frequency plot of simulated results of proposed antenna in a different medium

Table 1 Comparison table

Reference no.	Shape	Substrate	Dimensions (mm^2)	Radiation pattern	Impedance bandwidth	Gain	Structure
Ref. [17]	Vivaldi	Rogers R03010	95 × 120	Directional	(1.1–4)GHz 2900 MHz Single band	4.45–5.25 (dBi)	Unsuitable
Ref. [18]	3D bowtie	Fr4	50 × 50	Directional	(0.5–3)GHz Multi band	NA	Complicated with high profile
Ref. [11]	CPW half circular monopole	Fr4	31 × 61.3	Omnidirectional	(1.5–3) GHz 1800 MHz, 1000 MHz Dual band	5dBi	Suitable
Ref. [11]	CPW half bell-shaped monopole	Fr4	39 × 50	Omnidirectional	(1.4–2.9) GHz 350 MHz, 150 MHz Dual band	6dBi	Suitable
Proposed model	Slotted antenna with defected ground structure	Fr4	24.5 × 15	Omnidirectional	(6.3–9.7)GHz 600 MHz, 1000 MHz, 500 MHz Triple band	1.99 dB	Suitable with simple structure

4 Conclusion and Future Scope

A triple-band antenna for body-centric communication is presented in this paper. The designed antenna characteristics are simulated, and after obtaining the appropriate results, the antenna is fabricated. It can be concluded from the results obtained that the antenna is performing well in the recommended range of frequency. The antenna is resonating at 5.8 GHz ISM band frequencies for biomedical applications like stroke imaging and tumor detection. Particularly, the size of the antenna makes it more feasible to place it in daily wear items like helmets and wallets for continuous monitoring of stroke patients. The presence of ground also reduces the effect of back-scattered signals. Future scope includes that these antennas need to be tested surrounding in different dielectric constants and human phantom models under different conditions for more accurate results.

Acknowledgements The author with unique awardee number MEITY-PHD-2245 would acknowledge Visvesvaraya Ph.D. Scheme, Deity, New Delhi, for providing the fellowship and Andhra University college of Engineering (A), Andhra University, Visakhapatnam, for providing facilities to conduct research activity.

References

1. Liu, Z.G., Guo, Y.X.: Dual band low profile antenna for body centric communications. IEEE Trans. Antenna Propag. **61**(4), 2282–2285 (2013)
2. Hall, P.S., et al.: Antennas and propagation for on-body communication systems. IEEE Antennas Propag. Mag. **49**, 41–58 (07, June)
3. Feigin, V.L., et al.: Global and regional burden of stroke during 1990–2010: Findings from the global burden of disease study 2010. The Lancet **383**, 245–255 (2014)
4. Abtahi, S., Yang, J., Kidborg, S.: New compact multiband antenna for stroke diagnosis system over 0.5–3 GHz. Microw. Optical Technol. Lett. **54**(10), 2342–2346 (2012)
5. Persson, M., et al.: Microwave-based stroke diagnosis making global prehospital thrombolytic treatment possible. IEEE Trans. Biomed. Eng. **62**, 2806–2817 (2014)
6. Bolomey, J.C., Jofre, L.: Three decades of active microwave imaging achievements, difficulties and future challenges. IEEE ICWITS 2010 (1)
7. Stancombe, A.E., Bialkowski, K.S.: Portable biomedical microwave imaging using software-defined radio. In: Asia-Pacific Microwave Conference Proceedings, APMC, 2018 Nov, pp. 572–574 (2019)
8. Hatfield, S.F., Hillstrom, M.A., Schultz, D.N., Werckmann, T.M., Ghasr, M.T., Donnell, K.M.: UWB microwave imaging array for nondestructive testing applications. In: IEEE Instrumentation and Measurement Technology Conference, pp. 1502–1506 (2013)
9. Moloney, B.M., O'Loughlin, D., Elwahab, S.A., Kerin, M.J.: Breast cancer detection—A synopsis of conventional modalities and the potential role of microwave imaging. Diagnostics **10**(2) (2020)
10. Balanis, C.A.: Antenna Theory: Analysis and Design. Wiley India edition, 3rd edn, pp. 811–876
11. Thakkar, Y., Lin, X., Chen, Y., Yang, F., Wu, R., Zhang, X.: Wearable monopole antennas for microwave stroke imaging. In: 2019, 13th European Conference on Antennas and Propagation EuCAP, pp. 1–5 (2019)

12. Shen, W., Yin, S. (Member), Sun, X.: Compact substrate integrated waveguide (SIW) filter with defected ground structure. IEEE Microw. Wirel. Compon. Lett. **21**(2) (2011, Feb)
13. Chen, J., Tsai, J., Row, J.: Wideband circularly polarized slotted-patch antenna with a reflector. In: Proceedings of ISAP 2014, Kaohsiung, Taiwan, 2–5 Dec, 2014, vol. 1, no. c, pp. 615–616 (2014)
14. Fields, C.: Communications asymmetric geometry of defected ground structure for rectangular microstrip: A new approach. IEEE Trans. Antenna Propag. **64**(6) (2016, June)
15. Dwivedi, R.P.: High gain antenna with DGS for wireless applications. In: 2nd International Conference on Signal Processing and Integrated Networks (SPIN), pp. 19–24 (2015)
16. Reddy, M.H., Joany, R.M., Manikandan, G., Nisha, A.S.A.: Design of microstrip patch antenna with multiple slots for satellite communication. In: International Conference on Communication and Signal Processing, pp. 830–834 (2017, April 6–8)
17. Mohammed, B., Abbosh, A., Ireland, D.: Directive wideband antenna for microwave imaging system for brain stroke detection. In: Microwave Conference Proceedings (APMC), 2012 Asia-Pacific, pp. 640–642 (2012)
18. Abtahi, S., Yang, J., Kidborg, S.: New compact multiband antenna for stroke diagnosis system over 0.5–3 GHz. Microw. Opt. Technol. Lett. **54**(10), 2342–2346 (2012)

A Secure and Reliable Architecture for User Authentication Through OTP in Mobile Payment System

Deepika Dhamija and Ankit Dhamija

1 Introduction

A mobile payment system (MPS) can be defined as any payment system that enables financial transactions to be made securely from one organization or individual to another over a mobile network (using a mobile device) [1].

The growth of mobile payment (or m-payment) transactions over the last decade has been largely enabled by increasing speeds of mobile network connections, the rapid proliferation of portable devices (such as smartphones, tablets) and the worldwide penetration of mobile–cellular subscriptions, which reached 96% in 2013 and 40% of the world's population using the Internet. In fact, the global worldwide m-payment market will reach over 450 million users and a transaction value of over USD 721 billion in 2017 [4], USD 897.68 billion in 2018 and is expected to reach a value of USD 3695.46 billion by 2024. [3] Nowadays, there are numerous mobile payment systems available such as Internet banking, credit card, debit card, mobile wallets and unified payments interface (UPI).

When a transaction is under processing through mobile device, the security of the sensitive information entered on the payment page becomes the major concern for the users. This sensitive information can be the user's personal details or the transaction details. Hence, to ensure the security of m-payment transactions, various security techniques have been in use where the majority of those are using cryptography to encrypt the sensitive information before it travels on the network. However, the hackers and intruders continue to innovate new mechanisms of conducting frauds and

D. Dhamija (✉)
Amity College of Commerce, Amity University Gurugram, Gurugram, India
e-mail: ddhamija@ggn.amity.edu

A. Dhamija
Amity Business School, Amity University Gurugram, Gurugram, India
e-mail: adhamija@ggn.amity.edu

M. Dua et al. (eds.), *Proceedings of the International Conference on Paradigms of Communication, Computing and Data Sciences*, Algorithms for Intelligent Systems,
https://doi.org/10.1007/978-981-16-5747-4_9

continue to find ways of compromising the security of m-payment transactions. Over the past few years, there has been a steep rise in the number of failed transactions, security breaches and frauds happening over mobile networks. Hence, there is a pressing need for a secure mechanism which provides secure communication channel for the m-payment transactions.

In this paper, a novel architecture with two-way secure approach is proposed for enhanced security of mobile payment transactions. The approach uses a combination of hashing technique and cryptography to produce a unique one time password (OTP) for every transaction and hence can be a solution to the security issue in m-payments. This approach also protects from interception, reduces the transaction complexity and authenticates the user in a secure manner. The remaining paper is organized as follows: Sect. 2 discusses the related work, literature studied and the shortcomings found. In Sect. 3, the architecture of the proposed technique is presented. Section 4 describes the implementation process. In Sect. 5, results of the technique are explained, in Sect. 6, benefits of the proposed approach are explained, and in section VII, conclusions and remarks are outlined.

2 Literature Review

The literature review was designed and presented in three phases, with each phase presenting existing work on mobile payment architectures, mobile payment models, and mobile payment security techniques. Each of these is presented in Sects. 2.1, 2.2 and 2.3, respectively.

2.1 Mobile Payment Architectures

Zhu et al. [2] proposed a lightweight architecture for secure two party mobile payments (SA2pMP) which also works with the mobile phone with limited resources. The proposed architecture used a cryptographic mechanism, a multi-factor authentication and a distributed transaction log strategy to meet the security requirements of integrity, authentication, confidentiality and non-repudiation. The architecture was implemented in a mobile client that requires JAVA Mobile Edition (ME) and a bank server that requires JAVA Enterprise Edition (EE). B. Shaghayegh [3] proposed a novel service-oriented anonymous architecture for mobile payment where a new protocol called mobicash protocol was proposed which was designed for transactions carried out through smartphones which had features like security, effectiveness and full secrecy to smartphone users. The author tried to establish service-oriented architecture (SOA) in mobicash and removed the problems that occurred in digi-cash. YW Jung et al. [4] proposed an architecture for mobile devices which was designed for virtualization-based trusted execution environment. The architecture provided security in secluded environment through the deployment of smartphone virtualization technology. The key input of this architecture was mobile virtual machine monitor (mVMM) which enabled the user to divide the smartphone platform into different and isolated environment. This application kept sensitive data, its operations and also input and output in secure operating system. The application was also beneficial in mobile banking, streaming and electronic government applications. Guo [5] proposed a four-layered architecture for SMS-based mobile payment systems where the payment was carried out through Bluetooth, cellular networks, wireless LAN and infrared. The payment was executed by sending confirmation through SMS. Security was the limitation in this architecture as SMS messages may be counterfeited by operator or any network insider. The architecture also resulted in loss of data integrity and confidentiality. Zhang et al. [6] proposed an architecture called SIMPA which utilizes the session-based mobile payment system. Peer-to-peer (P2P) payment system is used by the consumer and merchant by using the Session Initiation Protocol (SIP) which increased privacy, integrity and confidentiality during payments. The loophole in this architecture was security as SIP protocol secures from outside attack and not from inside attack. Karnouskos et al. [7] proposed an architecture titled secure mobile payment system (SEMOPS) that aimed to develop a real time and easy to use and operate system for mobile payment point of sale (POS) and peer-to-peer (P2P) transactions.

2.2 *Mobile Payment Models*

Liu et al. [8] proposed a Wireless Access Protocol (WAP)-based model that comprised of six components: purchaser, trading partner, network operator, financial institution like banks, trusted third party (TTP) and data center. The network operator can perform the function of a processor which processes the user payment, besides performing shouldering the responsibility of wireless service provider. The bank is a financial institution where the account of purchaser is operated. Hence, the bank is best fit as the payment processor. TTPs include organizations, such as certifying authority and time-stamping server (TSS), to resolve the disputes if any, by providing notarization service. Data center performs the same function as performed in Karnouskos et al. [7]. It performs the routing function and delivers notifications to addressee payment processor. The authors performed the privacy and non-repudiation analysis, and their results stated that the model is cost efficient.

Song et al. [9] proposed a model for authenticating the third party where both the seller and purchaser are supposed to prove their identities by authenticating themselves with third party. They used private key cryptography between the third party and the seller. Isaac et al. [10] proposed client-centric model with a unique unnamed protocol to be used in a mobile-based payment system which uses a technique focused on self-certified public key-based digital signature together with message recovery. Their model included five major parties: customer, merchant, acquirer, issuer and payment gateway. The model involves no point-to-point communication between the merchant and acquirer, and all the communication is carried out through the customer. The model is more transparent and trustworthy as the customer is kept in loop for all communication and it cannot be bypassed.

Song et al. [11] proposed a commonly agreed authentication model in which the merchant and the customer are supposed to prove their identity to third party before any communication can be initiated. The approach used private key encryption between third party and the merchant and also between the merchant and the customer. The model satisfied the requirement of minimality, manageability and single sign-on.

Meng et al. [12] proposed a wireless access protocol-based mobile payment system where the technique focused on the following aspects: authentication from certifying authority, relationship based on trust, simple architecture and simplicity of transaction steps. The technique is unique as it can be used in both wired and wireless networks. The technique can be useful in providing inter-organizations support to a great extent and enable interoperability to impose dynamic security. The model is independent, interoperable, scalable and robust. The authors used elliptic curve cryptography algorithm to decrease the time taken in key generation in mobile phones. Jianping [13] proposed a SeMoPS model for secure mobile payment service. This model associates the bank with the network operator as the mobile payment handling organization. It uses the Wireless Access Protocol 2.0 which is used for end-to-end security. Roehrs et al. [14] proposed the "4iPAY" model, which supported four ubiquitous commerce assumptions: i–device, carrier, administrator, and physical location independence, and "Pay" to support mobile payment transactions. The model was independent of device, place, provider and cardholder, and it successfully catered to the needs of executing payment transactions in ubiquitous commerce.

2.3 *Mobile Payment Security Techniques*

Ensuring safety of mobile payment transactions is important in the field of mobile commerce. Securing the information whether personal or payment details is necessity in today's world. Yet, it is very difficult to authenticate the users remotely who are using mobile payments and providing a competent level of non-repudiation of payments. Gao et al. [15] proposed a 2D barcode-based mobile payment technique which used two-dimensional barcodes which delivers and supports easy mobile payment transactions and provides a good security. These systems also permitted mobile payments for all type of goods and services that can be recognized by 2D barcodes at any and every location. The system also improved user usability understanding by minimalizing the user inputs in mobile transactions. Javidan et al. [16] proposed a novel algorithm for securing the electronic payments made through mobile phones through the use of one time password with the combination of call back techniques. The proposed method worked by recording the significant information like SIM details, memory chip, mobile device IMEI number and user credentials within the server managing all payments. The program generated a unique encrypted password which was available on both the mobile phone and the payment server. The client or user has one password on execution of program for generating password and inputs the password on the payment page. Along with password, customer SIM card no, mobile phone serial number and user account were checked and verified on the server side.

Everts et al. [17] proposed a payment system named as UbiKiMa: Ubiquitous authentication using a smartphone migrating from passwords to strong cryptography. They leveraged ubiquity of mobile phone and used an application for storing the credentials. The same app also stored the username/password combination which is used as authentication that is further utilized on already prevailing portal where payment takes place that does not offer the stronger, public key-based approach. The approach lacked key management protocols and secure backup strategies which are needed to be developed and integrated in their approach. Rui-Xia [18] proposed an identity--ased cryptography (IBC) mobile payment system where the proposed security technique focused on tampering of payments by adding an identity-based signature together with the communication data. It used the one-time key mechanism for encryption process. The technique ensured the non-repudiation of information during payments. Khan [19] proposed an OTP generation technique using secure hashing algorithm (SHA) by merging more than one unique factor together which in turn makes the OTP different in each time it is generated. The algorithm calculates hash string of 160 bits long, and the algorithm generates a five digit long numerical OTP on the registered mobile number of the user. The approach generated unique OTP and produces unique hash value, and chances of attack are less. The technique can further be improved if used with more cryptographic hash functions for more security. Srinivas et al. [20] proposed an innovative approach for generation of OTP using pictures and images. For executing online monetary transactions, the registered user is required to enter his/her ID and pin number as a first step. Further, as a second step of the authentication, a set of images are made available, and the user has to select any image of his/her choice. Then, the registered user gets one time password generated from the selected image to his registered mobile number. A text box is displayed on the screen where the registered user must enter the received one time password. After entering the OTP into the displayed text box, it is checked on server for successful authentication. Hnaif et al. [21] proposed a new mobile payments system based on RSA public key cryptography. The approach relied upon integer factorization problem. The algorithm was categorized in three subtypes of algorithms: key generation, encryption and decryption algorithms. The proposed method process was divided into authentication, member recognition and payment process. The method was more efficient than other method because it permitted the customers to make payment from their own mobile devices without incurring any additional expenses and that too with more reliability.

The systematic literature review mentioned above revealed the research gap that security is a major concern in mobile payments system and two important requirements must be accomplished when designing a secure mobile payment system:

1. Mobile clients require a payment system in which they feel secure and comfortable.
2. Mobile client requires a payment system in which the transaction gets executed within the stipulated amount of time, low operational cost and user satisfaction. In addition to this, a payment system can be performed on limited functionality mobile phones also.

3 Proposed Architecture

In view of the gap mentioned in previous section, the author proposed novel architecture of mobile payment system integrated with the proposed security technique which increases the security during authentication process.

The proposed architecture also defines the components, participating entities and their relationships. The proposed architecture of mobile payment system is based on the common mobile payment system available in market and is implemented as **N**ovel **A**pproach for **OTP** **G**eneration Using **E**ncryption And **H**ashing **(NAOGEH)** technique (Fig. 1).

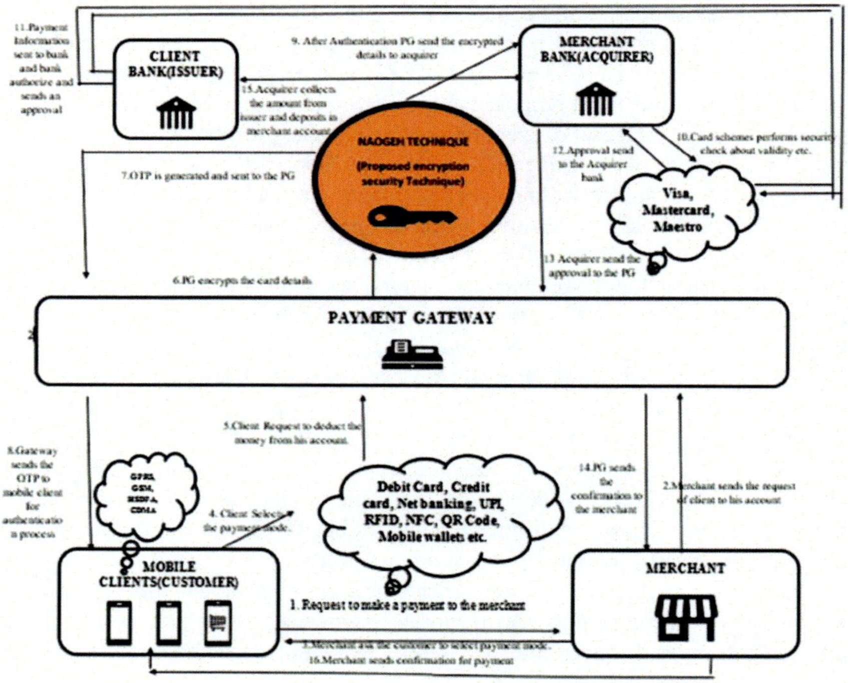

Fig. 1 Proposed architecture with NAOGEH security technique

The steps of the proposed architecture are as follows:

Step 1 The mobile client selects the product or service he wants to purchase from the merchant's site and requests a merchant for making a payment.

Step 2 After that, merchant sends the request to his account.

Step 3 Merchant asks the customer to select the payment mode available on his portal.

Step 4 Mobile client selects the payment modes available like debit card, credit card, net banking, UPI, RFID and mobile wallets. Our payment process works with debit card and credit cards. So, customers enter their credit or debit card details on the payment page. These details include the cardholder's name, 16 digit card number, card expiration date and card verification value (CVV) number.

Step 5 Client makes a request for the money to be deducted from their account, and this information is securely passed to payment gateway.

Step 6 The payment gateway tokenizes or encrypts the card details and performs fraud checks before they send the card data to the acquiring bank. This encryption process is firstly done by card authentication process in which the details entered by the customer are authenticated or not is verified. This checking process is done by sending an OTP to mobile client. Now, to generate a unique OTP, **NAOGEH** technique has been used.

Step 7 OTP is generated with the help of **NAOGEH** technique and sent back to the payment gateway for security check.

Step 8 Payment gateway sends the OTP to mobile client on the registered mobile number of client, which is registered with the client's bank, for authentication process which verifies that the details are entered by an authenticate user.

Step 9 If authentication is successful, then payment gateway sends the encrypted details to the acquirer who is the financial institution of merchant, and if authentication is failed, then payment gateway sends the message to client as authentication failed.

Step10 The acquiring bank sends securely the information to the card schemes (Visa, Mastercard) about validity check, etc.

Step 11 The card schemes perform another layer of fraud check and then send the payment data to the issuer (client's bank) The issuer, after performing fraud screening, authorizes the transaction.

Step 12 The approved or declined payment message is transferred back from the card schemes, then to the acquirer (merchant's bank).

Step 13 The acquirer sends the approval or decline message back to the payment gateway.

Step 14 Payment gateway then transmits the confirmation message to the merchant.

Step 15 If the payment is approved, the Acquirer collects the payment amount from the issuer (client's bank) and deposits the fund into your merchant ac- count deposits the funds into the merchant's account, a process which is known as the settlement; when the actual settlement will occur, depends on the agreement the merchant has with their payment gateway. Based on the message, the merchant may either display a payment confirmation page or ask the customer to provide another payment method.

3.1 Proposed Security Technique

Figure 2 presents a block diagram of the proposed security technique which describes the basic functionality of the approach. Then, a sequence diagram is presented that focuses on the detailed working of the components of block diagram. Finally, the detailed technique has been explained with an example.

Figure 2 shows the functionality of data transfer from mobile client to merchant in the following steps:

Step 1 Mobile client sends his credit card, debit card or other payment mode details to the merchant.

Step 2 The transferred details will be sent in encrypted form with two-way security technique, in which first the hashing technique is applied to reduce the bits and then a unique cryptographic approach for encrypted text is applied.

Step 3 After applying our approach, a unique encrypted code is generated.

Step 4 The unique encrypted code is sent to client in the form of OTP and to merchant site also for verification from both the sides (client and merchant) through payment gateway who is the middleman between client and merchant in this whole payment process.

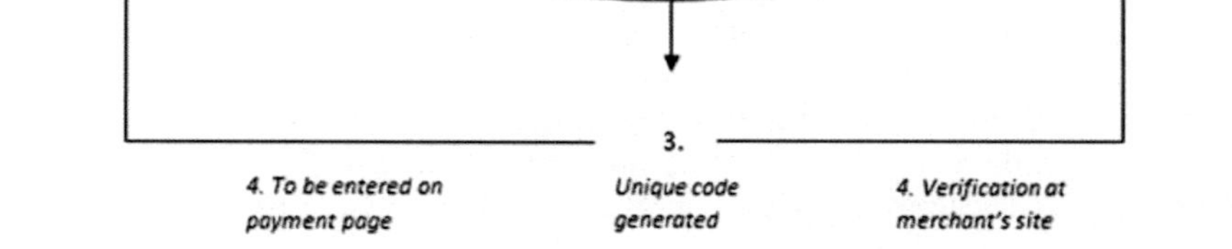

Fig. 2 Basic block diagram

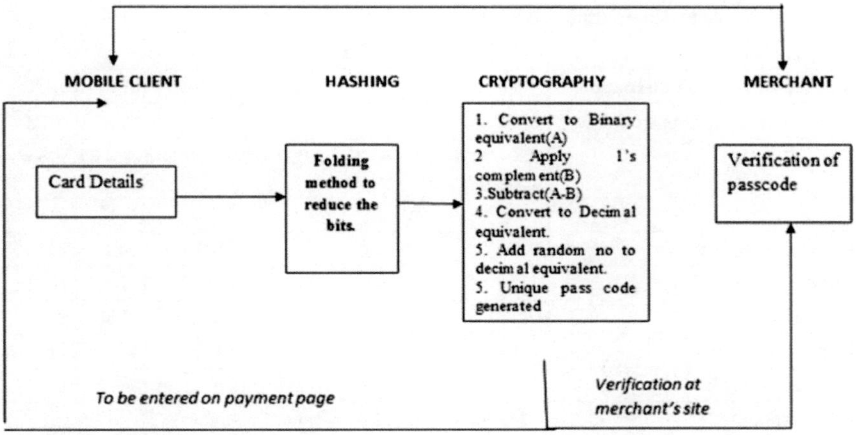

Fig. 3 Sequence diagram of NAOGEH

3.2 Detailed Sequence Diagram

The detailed functionality is explained in Fig. 3. The sequence diagram works as follows:

Step 1 Mobile client sends the 16-digit credit card/debit card number on the payment page.

Step 2 Then, as part of the proposed approach, fold bounded hashing method is applied on the input details to reduce the bits of the data. For reducing the bits, the authors made a group of first four bits/digits and a group of last four bits/digits. The first four bits and last four bits are reversed. Thereafter, all the digits are added. The left most bit is ignored, and last four bits are considered if the total of the bits/digits after addition comes in five bits.

Step 3 Then, the obtained four bit/digit number is converted into its binary equivalent. Thereafter, it is converted into its 1's complement which means that all 0 s bits will become 1 s and all 1 s bits will become 0 s.

Step 4 Subtract the 1's complement number from binary equivalent. If the left most bit is 0, then ignore it, and if it is 1, then consider it.

Step 5 Convert the required result into decimal equivalent and combine this number.

Step 6 Add a random number to each of the digits of decimal number to generate a unique OTP.

Step 7 This unique OTP is sent to client on mobile and email, client is required to enter the same on payment page, authentication process from payment gateway is processed, and payment gateway proceeds further for payment to the merchant which is explained above.

Table 1 Example to demonstrate NAOGEH

Card number	4088-4900-1097-5076			
Breakup of card no into four bits	4088	4900	1097	5076
Fold bounded hashing technique	8804	4900	1097	6705
	8804(folding) + 4900 + 1097 + 6705(folding) = 21,506 = 1506 (discard 2)			
	1	5	0	6
Binary equivalent	0001	0101	0000	0110
1's complement	1110	1010	1111	1001
Subtraction	01101	0101	01111	011
Ignore(0-left most bit)	1101	101	1111	11
Decimal equivalent	13	5	15	3
Add a random no	135,153 + 213,821 (any random number)			
Unique pass code	**348,974**			

Step 8 Merchant verifies the OTP input by client, and finally, the details transferred to merchant's bank and payment will proceed further.

Table 1 presents the detailed example of how the proposed approach is applied to generate the unique passcode which can be entered by the client and which can be used by the merchant for client verification and transaction authentication and approval.

4 Implementation

The technique is implemented on Microsoft Visual Studio 2019 or .Net platform as a Web application which works in mobile browsers and ultimately can be adopted by various mobile applications during payment via mobile. The application works as follows: In this application, the user enters the 16-digit card number, mobile number and email id and clicks on the submit button. After clicking on submit button, a unique OTP is generated which is sent to the user's mobile and email address given in the application.

Figure 4 shows the process that when user enters the card number, their phone number and email address, the application of security technique generates an OTP which is sent on both mobile phone and email address as displayed in Figs. 5 and 6, respectively.

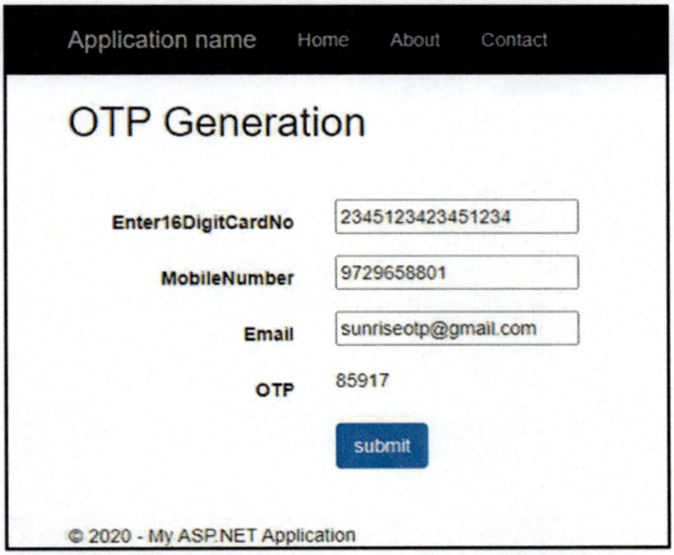

Fig. 4 OTP generation output screen

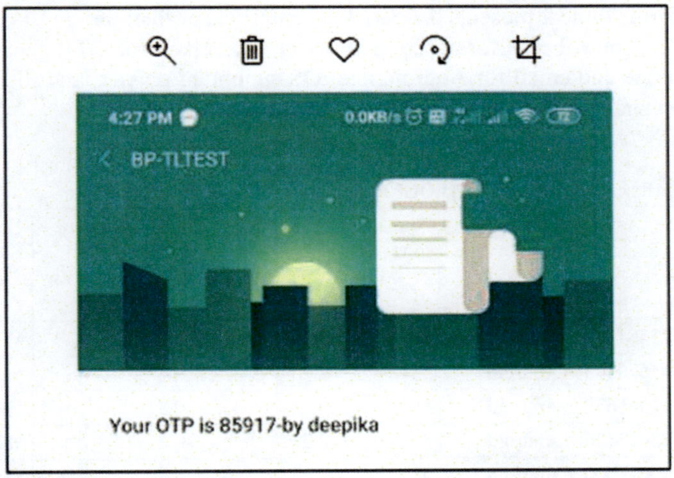

Fig. 5 OTP received on mobile

5 Empirical Evaluation and Results

To evaluate the performance of the proposed NAOGEH technique, an empirical evaluation was carried out where ten different card numbers were taken as input to

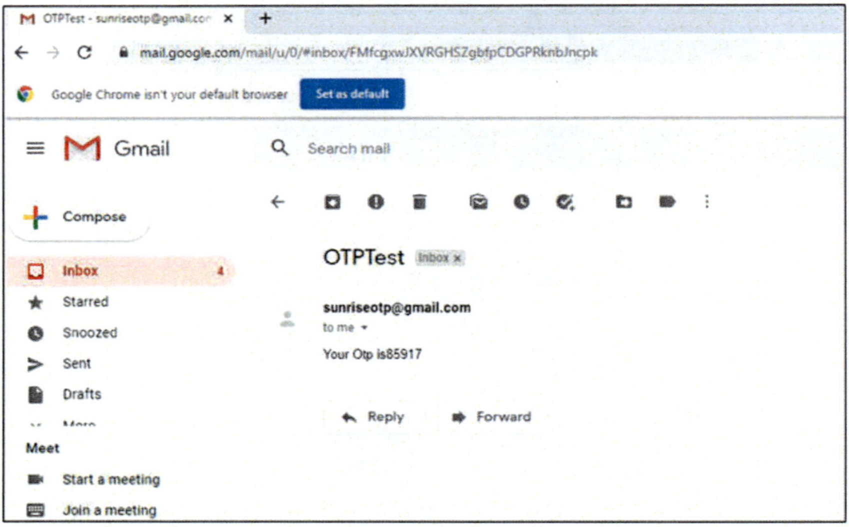

Fig. 6 OTP received on email

the proposed technique. The application was tested multiple times on Android and iOS platform. Table 2 presents the details of empirical evaluation.

Table 2 shows the performance analysis of time taken for OTP generation on mobile phone and email for Android and iOS-supported phones, and Fig. 7 shows the performance analysis graphically.

Table 2 Performance analysis of NAOGEH technique

Card No	Operating system type	Number of time	Time taken for OTP generation on email (s)	Time taken for OTP generation on mobile (s)	Uniqueness
2345123423451234	Android	5	3	2	Yes
2345612345672345	Android	5	2	4	Yes
5467678912345678	Android	5	4	3	Yes
5434900023453456	Android	5	2	2	Yes
6785345610012000	Android	5	4	2	Yes
1434900023452456	Android	5	1	2	Yes
2001870012002001	Android	5	3	2	Yes
2345123423451234	IOS	5	2	3	Yes
2345612345672345	IOS	5	3	2	Yes
5467678912345678	IOS	5	4	3	Yes

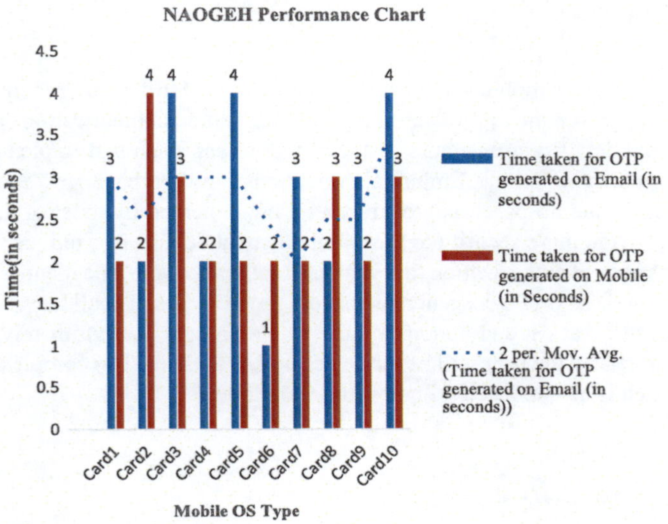

Fig. 7 Performance analysis of NAOGEH technique

From the above empirical evaluation, it is evident that the time taken in OTP generation on email and mobile phones is almost equal for Android and iOS-supported mobile phones. Moreover, the variation is due to network operators only. And also, the OTP generated for the same card number is also unique.

6 Discussion

The proposed technique supports security at high level during authentication process in the mobile payments from the merchant or payment gateway sides. The organizations can also use this technique for authentication of financial transactions.

The technique also has the property of non-repudiation where any user cannot deny of having participated in the transaction. Further, the technique ensures that no tempering in transaction happens. Security and confidentiality of account numbers, passwords and other sensitive information can be ensured by sending one-time OTP technique especially with the NAOGEH technique. The technique also prevents the clients from eavesdropping, cyber-attacks, replay attacks which makes the entire mobile payment process secure.

7 Conclusion

The paper focused on mobile payment security with a goal to design a secure architecture which improves the first requirement of security, i.e., authentication. The paper presented the detailed architecture of the proposed approach and implemented the same as a security technique. Empirical evaluation of the approach gave encouraging results. Hence, the proposed approach can be implemented by most of the mobile clients to provide more security to the client payment details. It is unique in a sense that it takes care of authorization, integrity and authentication. The manner in which unique passcode or OTP gets generated is very much secure, it shall be very difficult for intruders to crack it and understand the whole process, and ultimately, the card number entered by the user will never be accessible to them. The complexity of the new approach is also reduced with hashing technique.

8 Future Direction

The present research work is designed for credit and debit cards only. However, with recent technological advancements, it is essential that this security model is enhanced for other payment modes. So, in this frequently changing scenario, it is also essential to design an integrated secure architecture for mobile payment systems which can be adopted by each payment mode whether its debit card, UPI, mobile wallets, etc. Results in terms of security and time generation in output could be better if such work is used in real-time systems and will be adopted by the payment gateways for transactions.

References

1. Lee, Z.Y., Yu, H.C., Ku, P.J.: An analysis and comparison of different types of electronic payment systems. In: PICMET'01. Portland International Conference on Management of Engineering and Technology. Proceedings Vol.1: Book of Summaries (IEEE Cat. No. 01CH37199), vol. 2, pp. 38–45. IEEE, Portland, OR, USA (2001)
2. Zhu, Y., Rice, J.: A lightweight architecture for secure two-party mobile payment. In: International Conference on Computational Science and Engineering (CSE'09), vol. 2, pp. 326–333. (2009)
3. Shaghayegh, B.: Using service-oriented architecture in a new anonymous mobile payment system. Int. J. Adv. Inf. Sci. Serv. Sci. **3**, 393–396 (2011)
4. Jung, Y., Kim, H., Kim, S.: An architecture for virtualization-based trusted execution environment on mobile devices. In: 2014 IEEE 11th International Conference on Ubiquitous Intelligence and Computing and 2014 IEEE 11th International Conference on Autonomic and Trusted Computing and 2014 IEEE 14th International Conference on Scalable Computing and Communications and Its Associated Workshops, Bali, pp. 540–547 (2014)
5. Guo, W.: Design of architecture for mobile payments system. In: Chinese Control and Decision Conference, CCDC 2008, pp. 1732–1735 (2008)

6. Zhang, G., Cheng, F., Meinel, C.: SIMPA: A SIP-based mobile payment architecture. In: Seventh IEEE/ACIS International Conference on Computer and Information Science (ICIS 2008), Portland, OR, pp. 287–292 (2008)
7. Karnouskos, S., Vilmos, A., Hoepner, P., Ramfos, A., Venetakis, N.: Secure mobile payment–architecture and business model of SEMOPS. Evol. Broadband Serv. (2010)
8. Meng, J., Ye, L.: Secure mobile payment model based on WAP. In: 2008 4th International Conference on Wireless Communications, Networking and Mobile Computing, Dalian, pp. 1–4 (2008)
9. Song, M., Hu, X., Li, J., Deng, L.: An authentication model involving trusted third party for M-commerce. In: International Conference on the Management of Mobile Business (ICMB 2007), Toronto, Ont., pp. 53–53 (2007)
10. Isaac, J.T., Camara, J.S.: Anonymous payment in a client centric model for digital ecosystems. In: 2007 Inaugural IEEE-IES Digital Eco-Systems and Technologies Conference, Cairns, pp. 422–427 (2007)
11. Song, M., Li, J., Wu, X.: A mutual authentication model between merchant and consumer in M-commerce. In: Second International Conference on Innovative Computing, Information and Control (ICICIC 2007), Kumamoto, pp. 489–489 (2007)
12. Meng, J., Ye, L.: Secure Mobile Payment Model Based on WAP, pp. 1–4. https://doi.org/10.1109/WiCom.2008.2121 (2008)
13. Jianping, W.: The analysis and optimization on M-commerce secure payment model. In: 2011 Third International Conference on Communications and Mobile Computing, Qingdao, pp. 41–44 (2011)
14. Roehrs, A., da Costa, C.A., Barbosa, L.: A model for mobile payment in ubiquitous commerce. In: 2011 Simpasio em Sistemas Computacionais, Vitoria, pp. 8–8 (2011)
15. Gao, J., Kulkarni, V., Ranavat, H., Chang, L., Mei, H.: A 2D Barcode-Based Mobile Payment System, pp. 320–329 (2010)
16. Javidan, R., Pirbonyeh, M.A.: A new security algorithm for electronic payment via mobile phones. In: 2010 3rd International Symposium on Applied Sciences in Biomedical and Communication Technologies (ISABEL 2010), Rome, Italy, pp. 1–5 (2010)
17. Everts, M., Hoepman, J.-H., Siljee, J.: UbiKiMa: Ubiquitousauthentication using a smartphone, migrating from passwords to strong cryptography. In: Proceedings of the ACM Conference on Computer and Communications Security, pp. 19–24 (2013)
18. Rui-xia, Y.: Design of secure mobile payment system based on IBC. In: 2015 10th International Conference on Broadband and Wireless Computing, Communication and Applications (BWCCA), Krakow, pp. 422–425 (2015)
19. Hamid, M.: OTP generation using SHA. Int. J. Recent Innov. Trends Comput. Commun. **3**, 2244–2245 (2015)
20. Srinivas, K, Janaki, V.: A Novel Approach for Generation of OTP'S Using Image's. Procedia Comput. Sci. **85**, 511–518 (2016)
21. Hnaif, A., Alia, M.: Mobile Payment method based on public-key cryptography. Int. J. Comput. Netw. Commun. Secur. (2015)

Android Stack Vulnerabilities: Security Analysis of a Decade

Shivi Garg and Niyati Baliyan

1 Introduction

Android mobile operating system (OS) has been continuously attacked by varied malware; however, malware trends over the years show an incomplete picture. According to the AV-test security report [1], there is a slight improvement of malware situation for Android as shown in Fig. 1. Deployment of Android malware surged in 2016 and 2017, while cybersecurity developments and improvements in OS in 2018 and 2019 led to a significant reduction in the simple and easily replicable Android malware.

Despite several security improvements, Android mobile landscape is still far from becoming threat-free. Malware has adopted the new schemes such as code obfuscation, wrappers, variants of old malware, and packers to exploit underlying system vulnerabilities with the same goal of gaining privileges and accessing sensitive information. Android malware has evolved as a long-term intrusive resident in users' devices that can steal sensitive information and gain access. Sophisticated techniques employed in malware make the Android threat landscape difficult to understand [2].

This study is an integral part of research carried out in the field of Android malware. Since Android being the most prevalent mobile OS, it is often attacked by different malware due to underlying vulnerabilities. Many researches in the past have proposed malware detection techniques based on static, dynamic, and hybrid analysis. In one such study, authors detected malicious applications using parallel ensemble classifier *technique using static and dynamic features*. They achieved an accuracy of

S. Garg (✉)
J.C. Bose University of Science and Technology, YMCA, Faridabad, India

S. Garg · N. Baliyan
Indira Gandhi Delhi Technical University for Women, Delhi, India
e-mail: niyatibaliyan@igdtuw.ac.in

M. Dua et al. (eds.), *Proceedings of the International Conference on Paradigms of Communication, Computing and Data Sciences*, Algorithms for Intelligent Systems,
https://doi.org/10.1007/978-981-16-5747-4_10

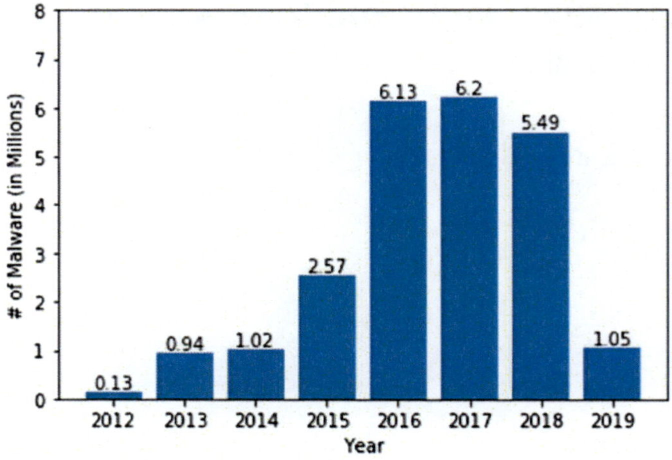

Fig. 1 Number of malware samples across the years

98.27% [3]. Different variants of malware can exhibit similar behavior and characteristics. Therefore, malware can be classified into known malware families. Classifying malware into known families can be fruitful in finding appropriate measure to deal with malware posing similar characteristics. In one such study, researchers classified malware into 71 known families using ensemble classifiers with an accuracy of 94% [4]. Malware can result from the weaknesses or susceptibilities (known or unknown) in an underlying platform. Therefore, it is important to map malware and vulnerabilities. Mapped vulnerabilities can then be assessed on the impact on Android OS layers and different subsystems. Figure 2 shows the pictorial view of the various steps in the Android malware study.

To understand Android threat landscape, it is important to study the underlying vulnerabilities. This paper aims to present an empirical study that analyzes vulnerabilities affecting the Android OS. The main focus is laid on the different subsystems and layers of the Android OS affected by vulnerabilities and impact of these vulnerabilities on confidentiality, integrity, and availability (CIA) triad. Such analysis can

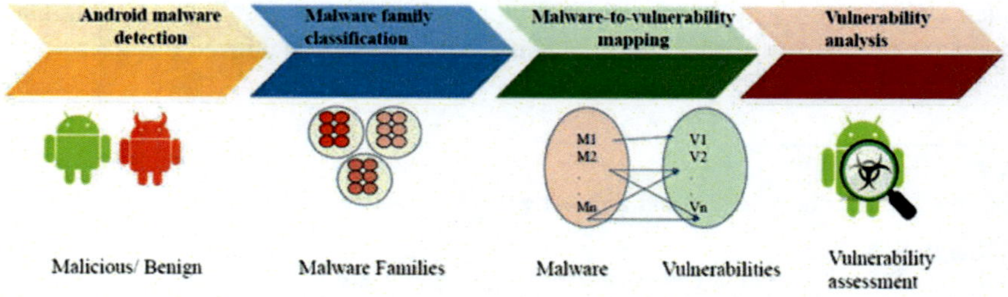

Fig. 2 Different steps in android malware study

help the Android developers to focus on the validation and verification of vulnerabilities and to redesign/improve secure coding procedures of underlying platform and Android apps.

The rest of the paper is organized as follows—Sect. 2 discusses the related work in this area. Section 3 describes the data extraction methodology. Section 4 presents the trends of vulnerabilities in Android between 2016 and 2019. Section 5 discusses general insights and future directions. Finally, Sect. 6 concludes the paper.

2 Related Work

Majority of the works conducted in the past are related to the vulnerabilities in the specific components of Android. Wang et al. [5] discovered six unknown vulnerabilities in three subsystems, i.e., location manager, activity manager, and mount service of Application *Framework* layer. Huang et al. [6] analyzed Android stroke vulnerabilities in Application Framework layer. These vulnerabilities are responsible for DoS attacks and induce soft reboot. In another work by Cao et al. [7], input validation mechanisms used in the Application Framework were analyzed. Ahmed et al. [8] studied privilege escalation vulnerabilities exploiting inter-application communications. Backes et al. [9] studied various security concerns caused by third-party libraries. Ghafari et al. [10] classified a set of 28 code smells related to the security of Android apps in five categories such as insufficient attack protection, security invalidation, broken access control, sensitive data exposure, and lax input validation. Security smells are defined as sets of instructions in the source code and indicate the presence of vulnerabilities. Jimenez et al. [11] identified complex function calls that caused vulnerabilities in Android components by analyzing 42 vulnerabilities from the common vulnerabilities exposure (CVE) database. Linares-Vásquez et al. [12] analyzed 660 Android OS vulnerabilities in terms of their survivability and the impacted subsystems/layers and components of the Android OS.

The aforementioned approaches analyzed vulnerabilities at different layers and components of Android, while this study is all encompassing as it focuses on the entire Android OS stack. This study can be compared to [12] and [13], where there is an increase of +1963 and +1328 respectively, in the number of analyzed vulnerabilities.

3 Design Methodology

The study aims to investigate Android vulnerabilities reported between 2009 and 2019. We have mined 2563 vulnerabilities from CVEdetails.com [14], which is a vulnerability repository that processes Extensible Markup Language (XML) feeds provided by the National Vulnerability Database (NVD) [15]. This study answers the following research questions (RQs):

1. **RQ1**—Which Android layers are most affected by vulnerabilities? This question focuses on the different Android subsystems that are affected by the vulnerabilities so that both Android researchers and app developers can develop better verification and validations tools for secure coding.
2. **RQ2**—How these vulnerabilities impact the CIA of the Android system? This question answers the level of impact (complete, partial, or none) on Android devices.

The study is carried out in two steps: data extraction and analysis.

3.1 Data Extraction Methodology

Android vulnerabilities are extracted from the vulnerability database as illustrated [16] in three main steps.

1. **Identifying the data source**—Vulnerability data is mined from CVEdetails.com. A total of 2563 vulnerabilities are mined from the database between 2009 and 2019.
2. **Deploying web scraping tool**—A web-based scraper tool named Web Scraper 0.4.0 [17] is used for mining vulnerability data from CVEdetails.com. The web scraping starts with the sitemap creation, which is specified with a start Uniform Resource Locator (URL). After this, the data is selected as a tree-type structure, and "Link" selector is specified. Post this, a tabular structure is specified for data extraction, and scraping is started. The scraped data is then exported as a comma-separated value (CSV) file.
3. **Saving the scraped data**—Scraped data is saved as an Excel file and cleaned up for further analysis.

A keyword-based mechanism is used in CVE Details that automatically infers the vulnerability type according to the common weakness enumeration (CWE) and thus prone to errors. Therefore, for verification of each vulnerability type, hierarchical clustering is performed as shown in Fig. 3.

3.2 Analysis

To address RQ1, Android layers and subsystems affected by the different vulnerabilities are identified manually using CVE Details, NVD, and Android issue tracker [18]. The manual analysis yields the distribution of vulnerabilities across Android OS stack.

RQ2 is answered by analyzing CVSS [19] vectors such as severity scores, access levels, attack complexity, and impact on CIA for different vulnerabilities. CVSS 2.0 is considered to analyze these vectors. Confidentiality means the protection of

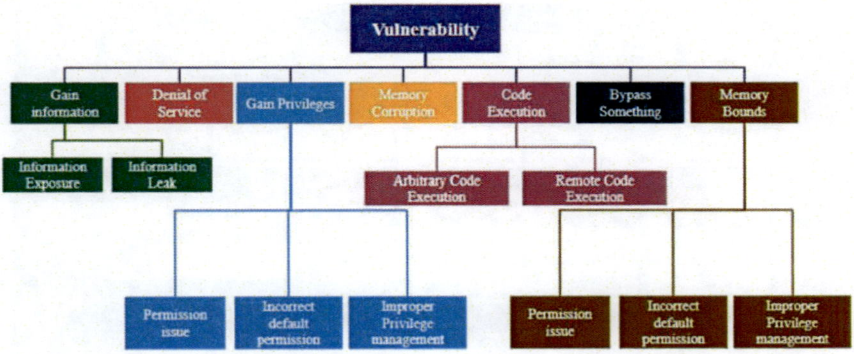

Fig. 3 Hierarchy of vulnerability types

sensitive information from unauthorized access. Integrity ensures that information is modified by any third party, and availability ensures that all the system resources and network services should be readily available to the users.

4 Results

This section presents the quantitative and qualitative results based on the RQs (defined in the Sect. 3). Quantitative data is complemented with the qualitative examples by referring CVE IDs of the vulnerabilities (e.g., CVE-2016-2439).

RQ1 What Android layers are more affected by vulnerabilities?

Heat map shown in Fig. 4 depicts the distribution of vulnerabilities across different Android layers/subsystems. Severity of impact is represented using two color schemes in the heat map—light red (low) to dark red (high) representing the severity impact in different Android layers and light yellow (low) to dark yellow (high) representing the components (internal boxes) in each layer.

It is evident that Linux Kernel is the most impacted layer with 53.3% of vulnerabilities. This is supported by a fact that majority of the modifications are included in the base Kernel by Android Open Source Project (AOSP) to enable mobile features. Seventy-five percent of the vulnerabilities such as memory corruption, buffer overflow, and privilege escalation in this layer affect drivers developed by Original Equipment Manufacturers (OEMs) followed by Kernel subsystem (5%) and boot loader (4%). For instance, CVE-2019-9456 is described as: "*In the Android kernel in Pixel C USB monitor driver there is a possible OOB write due to a missing bounds check. This could lead to local escalation of privilege with System execution privileges needed. User interaction is not needed for exploitation.*"

The second-most impacted layer is Native library with 28.3%. This is due to the *libstagefright* library found in the Media Framework, which constitutes 75% of the

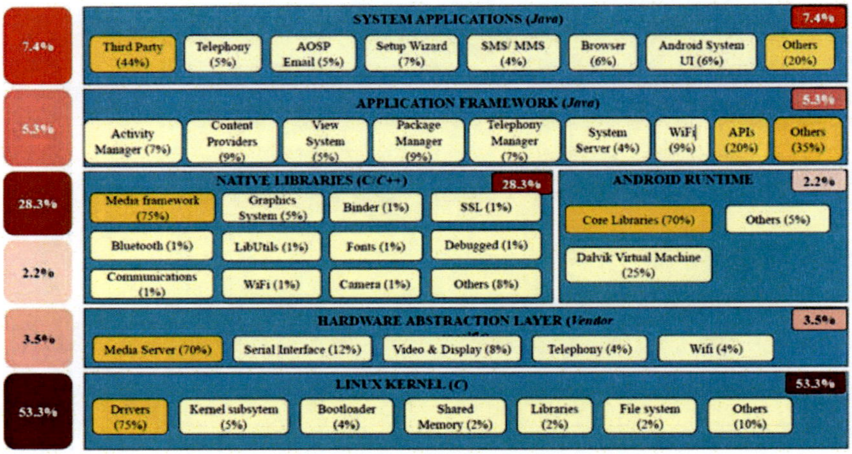

Fig. 4 Distribution of vulnerabilities across different Android layers/subsystems

vulnerabilities. Vulnerabilities found in the Media Framework are buffer overflow, gain information, and code execution. For example, the vulnerability CVE-2019-9416 has the following CVE description [20]: "*In libstagefright there is a possible information disclosure due to uninitialized data. This could lead to remote information disclosure with no additional execution privileges needed. User interaction is needed for exploitation.*"

The System Applications layer is the third-most impacted Android layer with 7.4% of the vulnerabilities with majority of the vulnerabilities (44%) in third-party applications. The applications are written in Java. This layer suffers from vulnerabilities such as DoS, gain information, privilege escalation and bypass something. For instance, CVE-2019-9440 [21] is listed as: "*In AOSP Email, there is a possible information disclosure due to a confused deputy. This could lead to local disclosure of the Email app's protected files with User execution privileges needed. User interaction is needed for exploitation.*"

The next layer impacted by vulnerabilities is the Application Framework with 5.3% vulnerabilities (7.26%). Application Framework mostly suffers from the vulnerabilities such as memory corruption, buffer overflows, privilege escalation, and DoS. For example, CVE-2019-9373 [22] is described as: "*In JobStore, there is a mismatched serialization/deserialization for the 'battery-not-low' job attribute. This could lead to a local denial of service with no additional execution privileges needed. User interaction is not needed for exploitation.*"

Hardware Abstraction Layer (HAL) is the next impacted layer with 3.5% vulnerabilities. Seventy percent of the vulnerabilities belong to media server. HAL suffers from code execution, memory corruption, and DoS vulnerabilities.

The remaining 2.2% of the vulnerabilities are found in Android Runtime (ART). Seventy percent of these vulnerabilities such as gain privileges, gain information, bypass something, and memory corruption are found in core libraries.

Figure 5 shows the different Android layers impacted by the vulnerabilities between 2016 and 2019. It is evident that only the Kernel layer shows a continuous increasing trend of vulnerabilities, whereas other layers show upward and downward trends across the years.

RQ2 How these vulnerabilities impact the CIA of the Android system?

To answer this question, CVSS vectors listed in CVE Details are analyzed for 2563 vulnerabilities. Figure 6 shows the number of vulnerabilities impacting the levels of CIA as complete, partial, or none. It is evident that most of the vulnerabilities have complete impact on CIA.

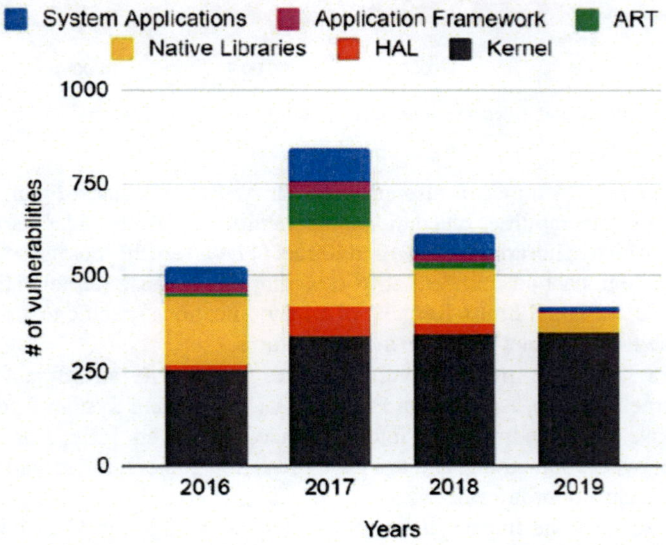

Fig. 5 Different Android layers impacted by the vulnerabilities between 2016 and 2019

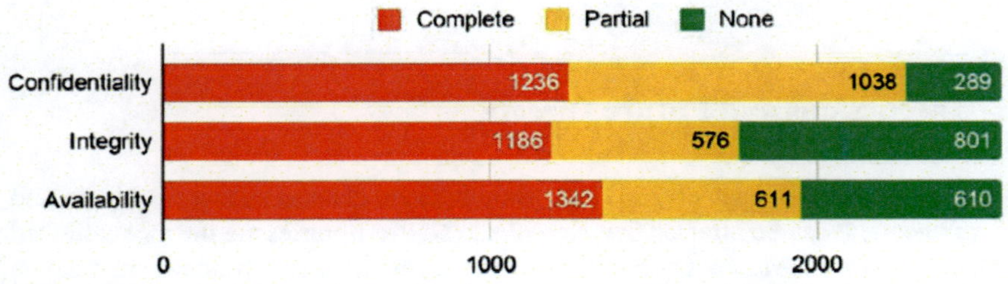

Fig. 6 Number of vulnerabilities impacting the levels of CIA

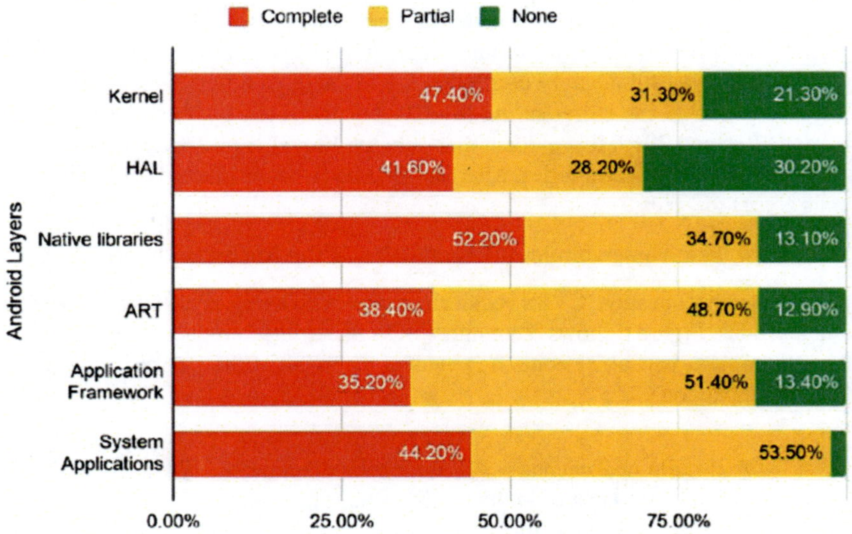

Fig. 7 Impact of vulnerabilities on confidentiality at different Android layers

Furthermore, CIA impact is also analyzed for Android OS stack. Figure 7 shows the impact of vulnerabilities on confidentiality at different Android layers. It should be noted that Native libraries (52.2%) and Kernel (47.4%) are the most affected layers with complete impact on confidentiality. It is primarily because Kernel in the AOSP is a fork of the original Linux Kernel. All the modifications specific to Android are made to the Linux Kernel to enable mobile features.

Figure 8 shows the impact of vulnerabilities on Integrity at different Android layers. Kernel (53.7%), Application Framework (51.2%), and System Applications (50.3%) have complete impact on integrity since these layers suffer from memory corruption, bypass something, and privilege escalation vulnerabilities, and there is a complete disclosure of information.

Figure 9 shows the impact of vulnerabilities on availability. Most layers with complete impact are Kernel (62.3%), Native libraries (57.7%), and Application framework (55.5%), where these layers suffer from DoS vulnerabilities and make the services and resources unavailable to the intended users.

5 Discussions and Future Directions

We analyzed the impact of 2563 vulnerabilities on different Android layers and components. It can be inferred that Kernel and Native libraries are the most affected layers of Android. Moreover, drivers in Kernel layer and media framework of native libraries are most impacted components in these layers (from RQ1).

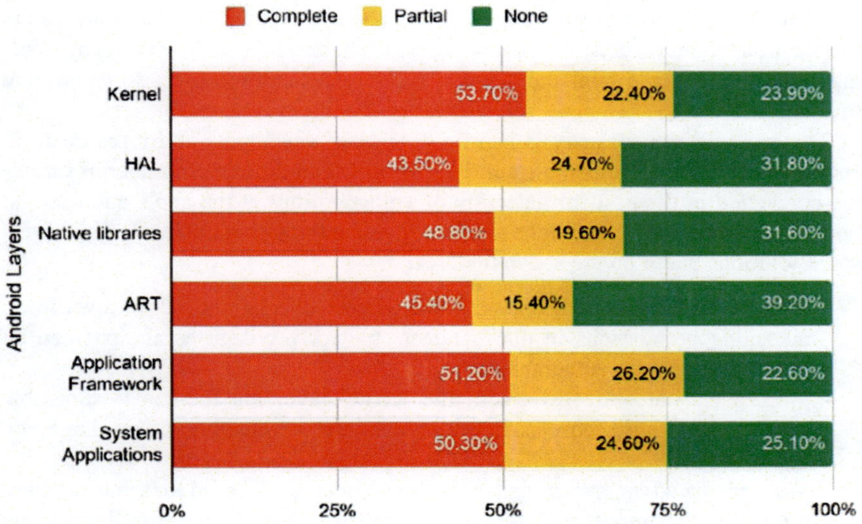

Fig. 8 Impact of vulnerabilities on integrity at different Android layers

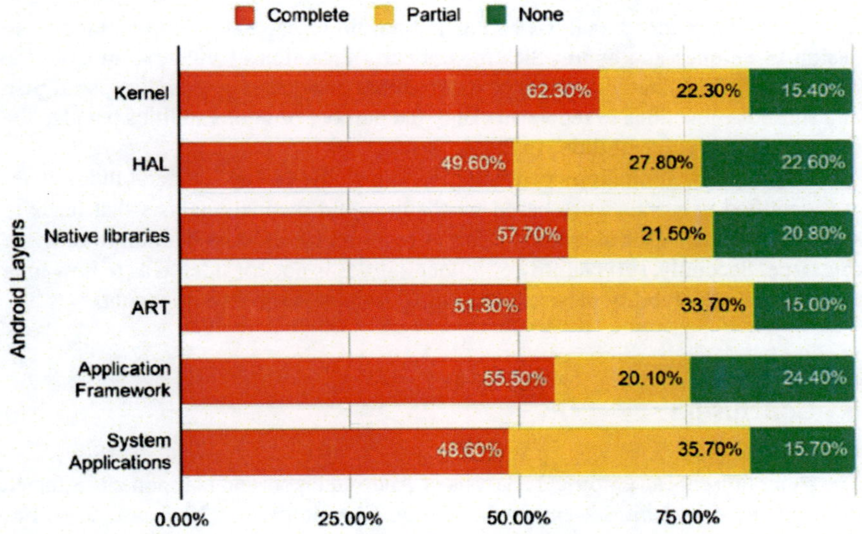

Fig. 9 Impact of vulnerabilities on availability at different android layers

From RQ2, in most of the vulnerabilities, there is a complete disclosure of the information (confidentiality), complete access to attacker to modify/temper any information (integrity), and total shutdown of the services and resources to the system (availability).

The aforementioned analysis can help Android developers and researchers in early identification of vulnerable components of OS stack. Large number of studies are conducted in the past for detection of vulnerabilities at the application levels; however, very few works are done to identify vulnerabilities at OS level. We suggest some solutions based on the above findings:

1. Code transformations such as pre-condition checks for buffer overflows, managing permissions in the manifest file for privilege escalations can be implemented to fix vulnerabilities
2. Third-party hardware drivers in the Kernel layer are the most vulnerable components; thus, more strict validation and verification tasks should be imposed.
3. More advanced programming language practices can be used for secure coding.
4. Assessing the impact of CIA can be helpful in updating Android OS stack architecture. Warning signal to the Android OS designers can be complete impact.

The aforementioned analysis can help Android researchers and practitioners to design techniques and approaches for the early detection of vulnerabilities in the Android OS. This study is a first of its kind that presents a potential research plan and subsequent actions to reduce not only the number of vulnerabilities but also the impact of these vulnerabilities in the Android OS.

There are few limitations pertaining to this study. Firstly, vulnerabilities types are identified using keyword-based mechanism and manual analysis that leads to subjectivity bias. A lot of manual effort is required for cross-validation to mitigate this issue. Secondly, survivability of vulnerabilities is not considered as to how long a particular vulnerability persists in the subsystem or Android component.

6 Conclusion

The paper presents an empirical analysis of Android layers and components affected by different vulnerabilities and its impact on CIA. Impact of 2563 vulnerabilities is analyzed between 2009 and 2019. The insights presented in this paper can help researchers and Android developers to detect most vulnerable components. This can be fruitful in redesigning Android stack with more secure features.

References

1. The AV-Test Security Report 2018/19. https://www.av-test.org/en/news/heightened-threat-scenario-all-the-facts-in-the-av-test-security-report-2018-2019, last accessed 04 Oct 2020
2. Review, Refocus, and Recalibrate The 2019 Mobile Threat Landscape. https://www.trendmicro.com/vinfo/hk-en/security/research-and-analysis/threat-reports/roundup/review-refocus-and-recalibrate-the-2019-mobile-threat-landscape, last accessed 12 Oct 2020
3. Garg, S., Baliyan, N.: A novel parallel classifier scheme for vulnerability detection in android. Comput. Electr. Eng. **77**, 12–26 (2019)
4. Garg, S., Baliyan, N.: Android malware classification using ensemble classifiers. In: Gupta, B.B. (eds 1) MIND 2019. CRC Press, Chap. 10 (2019)
5. Wang, K., Zhang, Y., Liu, P.: Call me back! Attacks on system server and system apps in android through synchronous callback. In: Proceedings of the 2016 ACM SIGSAC Conference on Computer and Communications Security, ser. CCS'16, pp. 92–103. ACM, New York, USA (2016)
6. Huang, H., Zhu, S., Chen, K., Liu, P.: From system services freezing to system server shutdown in android: All you need is a loop in an app. In: 22nd Proceedings of the ACM SIGSAC Conference on Computer and Communications Security, ser. CCS'15, pp. 1236–1247. ACM, New York, USA (2015)
7. Cao, C., Gao, N., Liu, P., Xiang, J.: Towards analyzing the input validation vulnerabilities associated with android system services. In: 31st Proceedings of the Annual Computer Security Applications Conference, ser. ACSAC 2015, pp. 361–370. ACM, New York, USA (2015)
8. Ahmad, W., Kästner, C., Sunshine, J., Aldrich, J.: Inter-app communication in android: Developer challenges. In: 13th Proceedings of the International Conference on Mining Software Repositories, ser. MSR'16, pp. 177–188. ACM, New York, USA (2016)
9. Backes, M., Bugiel, S., Derr, E.: Reliable third-party library detection in android and its security applications. In: Proceedings of the 2016 ACM SIGSAC Conference on Computer and Communications Security, pp. 356–367. ACM (2016)
10. Ghafari, M., Gadient, P., Nierstrasz, O.: Security smells in android. In 2017 IEEE 17th International Working Conference on Source Code Analysis and Manipulation (SCAM), pp. 121–130. IEEE (2017)
11. Jimenez, M., Papadakis, M., Bissyandé, T.F., Klein, J.: Profiling android vulnerabilities. In: 2016 IEEE International Conference on Software Quality, Reliability and Security (QRS), pp. 222–229. IEEE (2016)
12. Linares-Vásquez, M., Bavota, G., Escobar-Velásquez, C.: An empirical study on android-related vulnerabilities. In: 2017 IEEE/ACM 14th International Conference on Mining Software Repositories (MSR), pp. 2–13. IEEE (2017)
13. Mazuera-Rozo, A., Bautista-Mora, J., Linares-Vásquez, M., Rueda, S., Bavota, G.: The Android OS stack and its vulnerabilities: An empirical study. Empir. Softw. Eng. **24**(4), 2056–2101 (2019)
14. CVE details. https://www.cvedetails.com/, last accessed 21 Oct 2020
15. National vulnerability database. https://nvd.nist.gov/, last accessed 22 Oct 2020
16. Garg, S., Baliyan, N.: Machine learning based android vulnerability detection: A roadmap. In: 16th Proceedings of the international conference on information systems security, ICISS 2020, pp. 87–93. India, Springer (2020)
17. Web Scraper. Making web data extraction easy and accessible for everyone. https://webscraper.io/, last accessed 22 Oct 2020
18. Android issue tracker. https://issuetracker.google.com/issues?q=status:open, last accessed 22 Oct 2020
19. A Complete Guide to the Common Vulnerability Scoring System. https://www.first.org/cvss/v2/guide/, last accessed 24 Oct 2020
20. CVE-2019-9416 Detail. https://nvd.nist.gov/vuln/detail/CVE-2019-9416, last accessed 25 Oct 2020

21. CVE-2019-9440 Detail. https://nvd.nist.gov/vuln/detail/CVE-2019-9440/, last accessed 25 Oct 2020
22. CVE-2019-9373 Detail. https://nvd.nist.gov/vuln/detail/CVE-2019-9373/, last accessed 25 Oct 2020

Three-Factor User-Authentication Protocol for Wireless Sensor Networks—A Review

Vaishnavi Mishra and Abhay S. Gandhi

1 Introduction

An assembly for sensing and processing nodes forms a wireless sensor network that is distributed in space to perceive and gather data and handover the observations to centralized repository for statistical analysis and collection of data. The real-time linking of the nodes, constrained with regard to parameters like processing power, memory, energy strength, etc., is required for the exchange of effectual data with external network. They are placed in targeted areas like military war zone, medicinal research, health-condition monitoring, wildfire supervision, smart infrastructure development, etc. to collect environmental parameters. The invasion can manifest as disrupting routing schemes of the nodes, broadcasting fallacious or damaging information, draining the energy of nodes among others. Authentication ensures classified information to be accessed only by the designated entities and is not illegally invaded by trespassers. This review considers the 3-FUA technologies based on knowledge, inherence and possession to mitigate issues related to information and network security.

2 Theory

Wireless sensor networks are resource constrained and lightweight devices. Computing and communication efficiency of sensing nodes are pivotal indexes in designing the validation protocols for WSNs. ECC-based authentication scheme addresses the offline password-guess attack and user-forgery attack, thus enhancing security features as proposed in Sect. 3 (Fig. 1).

Due to limitations like restricted communication, computing and storage capabilities, attackers can tamper the data by launching attacks like MITM, replay, identity-

V. Mishra · A. S. Gandhi (✉)
Department of Electronics and Communication Engineering, Visvesvaraya National Institute of Technology, Nagpur 440010, India
e-mail: asgandhi@ece.vnit.ac.in

M. Dua et al. (eds.), *Proceedings of the International Conference on Paradigms of Communication, Computing and Data Sciences*, Algorithms for Intelligent Systems,
https://doi.org/10.1007/978-981-16-5747-4_11

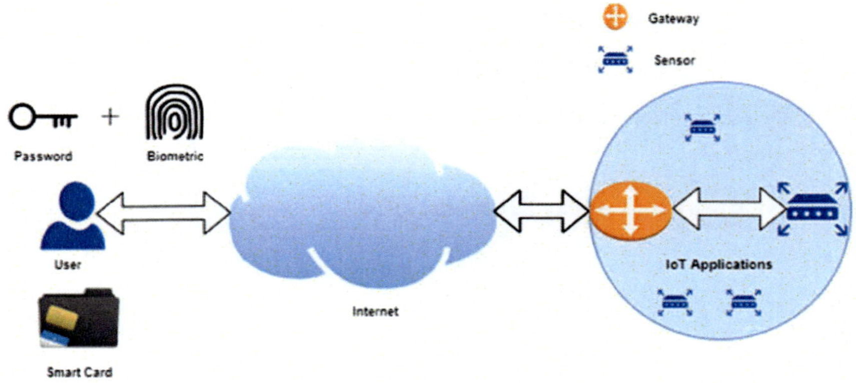

Fig. 1 Generic WSN model

guess and password-guess among others. Three-factor authentication incorporated with honey-list techniques and fuzzy verifiers is a safe way to protect WSNs from such adversaries as proposed in Sect. 4 eliminating elliptic curve cryptography, given the bounded performance capabilities of sensing nodes. The fuzzy extractor is used for biometric recognition.

Security requirements in IoT applications entails considering the CIA (Confidentiality, Integrity and Authenticity) for users and the associated data [7]. Physical unclonable functions are created by instigating variations in the process of IC fabrication for uniqueness. Physical tampering can be detected easily as it changes the internal properties of sensing nodes making them unauthenticated nodes. It is important to create lightweight validation mechanisms because of battery constraints in IoT devices. The protocol proposed in Sect. 5 stores a single challenge–response pair (CRP) for each sensor in place of piling a set. It facilitates lower delay from one end to another along with higher delivery rate for packets making it useful and acceptable for IoT environment.

2.1 Biometric Verification Using Fuzzy Commitment Schemes and Error-Correcting Codes

Error-correcting codes are used to communicate information through an insecure channel. The message is encrypted into a lengthy code which contains redundant information before transmission. It is feasible to reconstruct original message signal from such code even when the noise is present. FCS combines cryptography and error-correcting codes so that the adversary does not get hold of the committed value. It can be accessed only when the indication is close to the original value. Biometrics often contain noise; thus, FCS proves to be of great use.

2.2 Honey-Lists

Honey-list or honey-encryption (HE) algorithm is used for protecting user data by tricking the invalid users when the adversary tries to decrypt the flat text with the aid of incorrect credentials or so-called honey-word. In case of numerous invalid entries or honey-words, the algorithm produces a fake message for validity as per distributed transforming encoding. Wang et al. [3] showed that a fuzzy verifier in addition to honey-list ensures user data security even when 2 out of 3 factors are compromised.

2.3 Fuzzy Extractor

This technology collects user biometric data through data extraction. It is difficult to capture real biometric values due to different noises. The fuzzy extractor helps in the extraction of random strings evenly discarding noise. With the aid of generation, biometric data is used as input, P_i as helper string along a uniformly random string R_i as secret key data. The reproduction algorithm produces the original biometrics R_i when the metric space distance of BIO_i and $\mathrm{BIO}_i^{'}$ lies within specified allowed error-tolerance range.

2.4 Hash Function

Hash function uses a random-length string as input parameter and a value of predetermined length as output [4]. It becomes difficult in a hash function to procure multiple values which furnish the exact result to maintain data integrity, as it is resistant to collusions.

2.5 Physically Unclonable Function

Even when the ICs are manufactured using the same process, they differ slightly from each other because of normal manufacturing variability. A PUF is able to generate the same response for a challenge even when it is utilized multiple times. When it is passed to some other PUF, it produces an entirely varied response R with high probability, where $R = \mathrm{PUF(C)}$.

3 3-FUA on the Basis of Elliptic Curve Cryptography

3.1 Introduction

To realize intelligent ID management practically, IoT allows the objects to access data network by using emerging technologies like RF identification, sensors, etc. WSNs can be used to construct systems to diagnose critical conditions of patients skillfully and in less time. Data transmitted over the wireless channel can be examined by valid entities along with adversaries and can prove fatal if it falls prey to MITM attacks. Validating the user willing to access the sensory data and identity authentication is thus of critical importance when it comes to WSN and WMSN as crucial security mechanism [6].

3.2 Reviewing Protocol Proposed by Li et al.

The authentication schemes in place for WSN and WMSN are susceptible to inability to gain localized password validation alongside shielding theft attacks. In the scheme proposed in [1], the error-correcting codes and fuzzy commitment schemes are utilized for biometric retrieval and EC cryptography to enable forward secrecy. This protocol [2] adopts fuzzy verifiers and honey-list for security even when 2 out of 3 factors are hacked by adversary. Six phases of the protocol are described as below:

System Setup Phase: Gateway node GW initializes the system by selecting an EC $E_p(a, b\ )$ in prime field Z_p with hash function $h\ ()$. Then, it chooses a sub-group G of prime order n of $E_p(a, b)$ and point P. The GW picks a master private key $x \in [1, n-1]$ and determines public key $X = xP$. It also selects (n, k, t) BCH code and corresponding functions ENC, DEC *and* f for imposing security. The gateway GW also declares a moderate integer n_0, such that $2^4 \le n_0 \le 2^8$ for fuzzy verifier. It produces values $\{p, E_p(a, b), P, X, h()\}$, keeping x as secret.

Registering Medical-Professional Phase: U_i needs to register at GW to be considered legitimate for acquiring data of sensor nodes. Figure 2 shows the procedure for the same.

- U_i chooses ID_i, PW_i, a nonce r and extracts the biometric BIO_i from mobile device. Then U_i computes $\mathrm{HPW}_i = h(\mathrm{PW}_i||r)$, and requests the registration $\{\mathrm{ID}_i, \mathrm{HPW}_i, \mathrm{BIO}_i\}$ with private credentials like ID to GW through secure channel.
- Once the request is received, GW first scans for the presence of ID_i in database. When found, U_i prompts for new request. Else, GW chooses one string of k-bits k_i linked to user. Post this, GW calculates $C_i = \mathrm{ENC}\ (k_i)$, $A_i = C_i \oplus B\ \mathrm{IO}_i$, $B_i = h\ ((h\ (\mathrm{ID}_i||k_i) \oplus \mathrm{HPW}_i) \bmod n_0\ )$, $D_i = h\ (\mathrm{ID}_i||k_i||x)$ and $E_i = D_i \oplus \mathrm{HPW}_i$. GW places $\{\mathrm{ID}_i,\ k_i,\ \mathrm{Honey\_List} = \mathrm{Null}\}$ in database, and submits $\{\ A_i,\ B_i,\ E_i,\ X,\ \mathrm{DEC},\ f, n_0\}$ to U_i through safe channel.
- U_i stores r and $\{A_i,\ B_i,\ E_i,\ X,\ \mathrm{DEC},\ f,\ n_0\}$ in mobile device, which includes the parameters $\{A_i, B_i, E_i,\ X,\ \mathrm{DEC},\ f, n_0,\ r\}$.

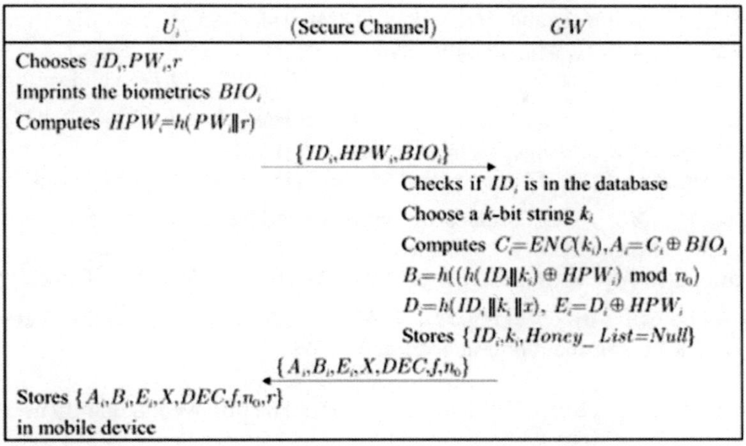

Fig. 2 User registration phase

Registering Patient Phase: With every sensing node S_j having an identity SID_j, the registration center calculates secret key $K_j = h(\text{ID}_{GW}||\text{SID}_j||x)$, and places it in sensing node which is linked to patient.

Login and User-Validation Phase: When U_i wishes to read the sensory data of S_j, he/she must log in to the GWN post which he can acquire the information legally.

- U_i inputs $\text{ID}_i^{'}$, $\text{PW}_i^{'}$ and impresses biometric $\text{BIO}_i^{'}$ to mobile device. It determines $C_i^{'} = f(A_i \oplus \text{BIO}_i^{'}) = f(C_i \oplus (\text{BIO}_i \oplus \text{BIO}_i^{'}))$, $k_i^{'} = \text{DEC}(C_i^{'})$, $cB_i^{'} = h\ \ ((h\ (\text{ID}_i^{'}||k_i^{'} \oplus h\ (\text{PW}_i^{'}||r)) \bmod n_0)$, and validates if $B_i^{'} \overset{?}{=} Bi$. The request is aborted if $B_i^{'} \neq Bi$. It then generates a nonce a and computes $M_1 = aP$, $M2 = aX$, $M3 = \text{ID}_i^{'} \oplus h\ (M1\ ||\ M2)$, $M4 = \text{SID}_j \oplus h\ (M2||M1)$, $D_i^{'} = E_i \oplus h(PW_i{'} \oplus r)$, and $M5 = h(D_i^{'}||\text{SID}_j||M_2)$. At last, U_i forwards the login request $W_1 = \{M_1, M_3, M_4, M_5\}$ to GW.
- While receiving W_1, GW calculates $M_2^{'}{=}xM_1$ and derives $\text{ID}_i^{''}{=}M3 \oplus h(M_1||M_2^{'})$. Then, GW checks for the validity of identity. If found invalid, the session is aborted. Else, GW finds corresponding k_i and calculates $D_i^{''}{=}h\ (\text{ID}_i^{''}||k_i||x)$, $\text{SID}_j = M_4 \oplus h(M_2^{'}||M1)$, $M_5^{'} = h\ (D_i^{''}||\text{SID}_j^{'}||M_2^{'}$, and see if $M_5^{'} \overset{?}{=} M_5$.
- When found unequal, GW places $D_i^{''}$ into Honey_List or discards identity when Honey_List items exceed predefined threshold. Meanwhile, GW discards the session, or request for login stands valid. GW produces random nonce b, and calculates $K_j = h(\text{ID}_{GW}||\text{SID}_j^{'}||x)$, $M_6 = K_j \oplus b$, and $M_7 = h(\text{SID}_j^{'}||K_j||b||M_1)$. At last, GW forwards the message $W_2 = \{M_1, M_6, M_7\}$ to sensing node S_j.
- On reception of W_2, S_j calculates $b^{'} = M6 \oplus K_j$, $M_7^{'} = h(\text{SID}_j\ ||K_j||b^{'}||M_1)$, and checks for the validity of GW by verifying $M_7^{'} \overset{?}{=} M_7$. If valid, S_j generates nonce c and computes $M_8 = c\ P$, $M_9 = cM1.\text{SK}_j = h\ (M1||M_8||M_9)$, $M_{10} =$

$h(\text{SID}_j||K_j||b'||M_8)$, and $M_{11} = h(\text{SID}_j||M_1||M_8\ ||\text{SK}_j)$, then SN_j sends the message $W_3 = \{M_8, M_{10}, M_{11}\}$ to GW.

- On reception of W_3, GW calculates $M_{10}' = h\ (\text{SID}_j||K_j||b||M8)$ and validates S_j by checking $M_{10}' \stackrel{?}{=} M_{10}$. When valid, GW calculates $M_{12}{=}h\ (D_i''||\text{SID}_j'||M_2'||M_8)$ and forwards the message $W_4 = \{M_8, M_{11}, M_{12}\}$ to U_i.
- On reception of W_4, U_i computes $M_{12}' = h(D_i'||SIDj||M_2||M_8)$ and verifies $M_{12}' \stackrel{?}{=} M_{12}.M_{12}' \neq M_{12}$ implies the message is invalid and the session aborts. Else, the validity of GW is confirmed by U_i.
- U_i computes $M_{13}{=}aM_8$, $SK_i{=}h(M_1||M_8||M_{13})$, $M_{11}'{=}h(\text{SID}_j||M_1||M_8||\ \text{SK}_i)$, and validates S_j by checking $M_{11}' \stackrel{?}{=} M_{11}.U_i$. S_j can then use the shared key $\text{SK}_i = \text{SK}_j$ for further communication via GW.

Password Change Phase: Following steps should be followed to modify user password:

- U_i enters ID_i, PW_i and impresses biometric information BIO'_i on his device. The mobile then computes $C_i' = f(A_i \oplus \text{BIO}_i') = f(C_i \oplus (\text{BIO}_i \oplus \text{BIO}_i'))$, $k_i' = \text{DEC}(C_i')$, $B_i' = h((h(\text{ID}_i||k_i') \oplus h(\text{PW}_i||r))\text{mod } n_0)$, and checks $B_i' \stackrel{?}{=} B_i$.
- $B_i' \neq B_i$ leads to the rejection of request. Else, the process continues.
- The device prompts to decide new password, and U_i chooses a new password PW_i^{new}.
- It then calculates $\text{HPW}_i^{\text{new}}{=}h(\text{PW}_i^{\text{new}}||r)$ and $B_i^{\text{new}}{=}h((h(\text{ID}_i||k_i') \oplus h(\text{PW}_i^{\text{new}}||r))$ mod n_0). Lastly, the mobile device replaces B_i with B_i^{new} enabling U_i to modify the password without any help from gateway.

Revocation and Re-register Phase:

When U_i loses his mobile device, he performs these steps:

- U_i submits ID_i along private credentials to GW through safe channel for requesting revocation.
- GW then validates the ID_i and credentials. Post confirmation, ID_i is blocked and the device is revoked by GW making it impossible to login from it.

If U_i wishes to re-register with identity ID_i, he should perform the following:

- U_i decides on a new password PW_i^*, nonce r*, and extracts BIO_i^* from the new device. U_i then figures out $\text{HPW}_i^* = h\ (\text{PW}_i^*\ ||r*)$, and requests for re-registration $\{\text{ID}_i, \text{HPW}_i^*, \text{BIO}_i^*\}$ with personal credentials like ID to GW via secure channel.
- Once $\{\text{ID}_i, \text{HPW}_i^*, \text{BIO}_i^*\}$ is received, GW chooses new string of k-bits k_i^* linked to user. GW goes on to calculate $C_i^* = \text{ENC}\ (k_i^*)$, $A_i^* = C_i^* \oplus \text{BIO}_i^*$, $B_i^* = h\ (\ (h\ (\text{ID}_i||k_i^*) \oplus \text{HPW}_i^*)\ \text{mod } n_0)$, $D_i^* = h(ID_i||k_i^*||x)$, and $E_i^* = D_i^* \oplus \text{HPW}_i^*.GW$ stores $\{ID_i, k_i^*, \text{Honey\_List} = \text{Null}\}$ to database, and submits $\{A_i^*, B_i^*, E_i^*, X, \text{DEC}, f, n_0\}$ to U_i through secure channel.
- U_i stores $\{A_i^*, B_i^*, E_i^*, X, \text{DEC}, f, n_0\}$ to mobile device. U_i can then login to GW with ID_i using new device.

3.3 *Performance and Security Comparisons*

The protocol resists attacks like mobile device loss, gateway node impersonation, user impersonation and sensor node impersonation.

Adversary must resolve ECDLP and ECCDHP to obtain the previous keys, thereby ensuring forward secrecy. The already accepted session keys are safe in the instance of leaking of long-term private keys. Replay attack can be coped up with clock synchronization and random number mechanism. The protocol utilizes the latter and does not face any synchronization problems. It adopts point multiplication on ECC to ensure forward secrecy, so requires more computation cost and time. The communication cost for this is 2,720 bits, and the total cost is '$20T_h + 6T_m$'. Time cost for the user is '$8T_h + 3T_m$', for the GWN is '$8T_h + T_m$' and for the sensor node is '$4T_h + 2T_m$', where T_m, T_h *and* T_s stand for the time period of point multiplication when elliptic curve cryptography, hash operation and symmetric encryption/decryption are used respectively.

4 3-FUA Protocol Using Honey-List

4.1 *Introduction*

This protocol [5] uses hash-functions only prohibiting ECC enabled public key. It uses "honey-lists" and "fuzzy extractors" techniques to preserve security.

4.2 *Reviewing Protocol Proposed by Lee et al.*

Prior to the initiation to registration, gateway forms secret key X_{GWN}.

Registering User and Sensor Phase: To acquire data from WSNs, the user U_i and sensor S_j must register to GW. Figures 3 and 4 show user and sensor registration.

- Registering Users: U_i chooses unique ID_i and PW_i along with biometrics BIO_i post which U_i randomly generates a nonce r_i. U_i calculates $< R_i, P_i >= Ge(\text{BIO}_i)$, $H\ \text{ID}_i = h\ (\text{ID}_i||r_i)$ and $H\quad \text{PW}_i = h\ (r_i||\text{ID}_i\ ||\text{PW}_i)$ and transmits registration request message $\{\ \text{HID}_i,\ \ \text{HPW}_i\}$ to GWN through safe channel which ensures immunity to attacks. Post receiving $\{\text{HID}_i, \text{HPW}_i\}$, GWN ensures whether HID_i is enrolled in database. If not, GWN creates a random string k_i and calculates $a_i = h(\text{HID}_i||X_{\text{GWN}}||k_i)$, $b_i = a_i \oplus \text{HPW}_i$ and $c_i = h(a_i||\text{HPW}_i)$. Post this, GWN feeds HID_i with k_i along HPW_i and stores values $\{b_i, c_i\}$ into a smart card $S\ C$. It then hands over SC to user. Finally, U_i computes $L_i = h(R_i||\text{PW}_i)$ and stores $\{L_i, P_i\}$ into the SC.

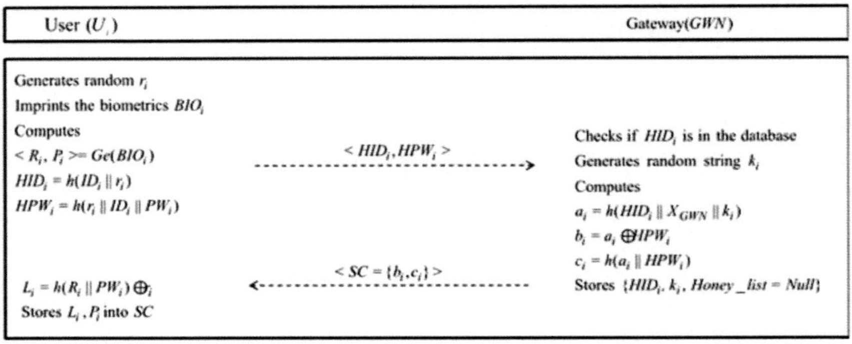

Fig. 3 User registration

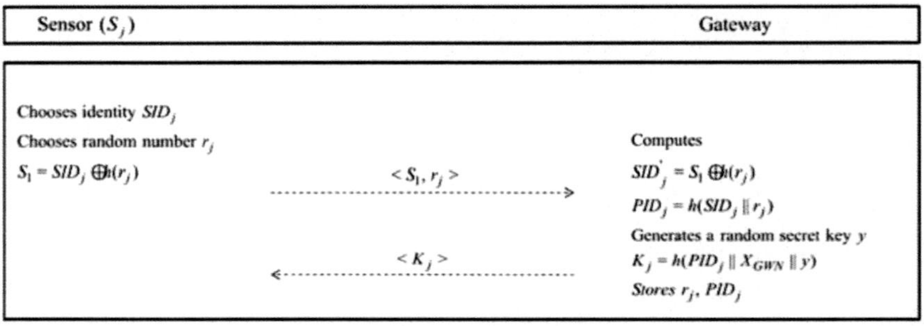

Fig. 4 Sensor registration

- Registering Sensors: Sensor S_j selects identity SID_j and random nonce r_j.S_j computes $S_1 = \text{SID}_j \oplus h(r_j)$ and provides S_1 and r_j to GWN. After receiving request for registration, GWN figures $\text{SID}_j' = S_1 \oplus h(r_j)$ and $\text{PID}_j = h(\text{SID}_j || r_j)$, produces a random secret-key y, calculates $K_j = h\ (\text{PID}_j ||\ X_{\text{GWN}} || y)$ keeping r_j *and* PID_j in memory. It then transmits K_j to sensor.

Login and User Validation Phase: User must login to GW for self-authentication.

- User U_i provides his unique identity ID_i, password PW_i and impresses the biometric BIO_i. U_i then calculates $R_i = Re\,(\text{BIO}_i, P_i)$, $r_i = L_i \oplus h\,(R_i || \text{PW}_i)$, $\text{HID}_i = h\ (\text{ID}_i || r_i)$ and $\text{HPW}_i = h\ (r_i || \text{ID}_i || \text{PW}_i)$. U_i extracts $a_i = b_i \oplus \text{HPW}_i$, computes $c_i' = h(a_i || H\ \text{PW}_i)$ and sees if c_i' and c_i are equal. When they are equal, U_i produces random number N_i and calculates $M_1 = h\ (a_i || \text{SID}_j) \oplus N_i$ and $M_2 = h(a_i || \text{SID}_j || N_i)$. Then, it requests for login by sending <HID_i, M_i, M_2>to GWN.
- On reception of login request, GWN pulls k_i from repository and calculates $a_i' = (\text{HID}_i || X_{\text{GWN}} || k_i)$, $N_i = h(a_i' || \text{SID}_j) \oplus M_1$ and $M_2' = h(a_i || \text{SID}_j || N_i)$ and checks if $M_2 \stackrel{?}{=} M_2'$. If unequal, a_i' adds into $Honey - list$ or suspends iden-

tity when count of items in $Honey - list$ exceeds threshold. Else, GWN calculates $K_j = h\ (h\ (\mathrm{SID}_j\ ||r_j)||\ X_{\mathrm{GWN}}\ ||\ y)$, $M_3 = h(\mathrm{SID}_j||\mathrm{PID}_j||K_j) \oplus N_G$ and $M_4 = h(K_j||\mathrm{PID}_j||N_G)$. Then, GWN sends $<M_3, M_4>$ to a sensor node S_j.

- S_j computes $N_G = h\ (\mathrm{PID}_j||K_j\ ) \oplus M_3$, $M_4^{'} = h(\mathrm{PID}_j||K_j||N_G)$ and compares M_4 with $M_4^{'}$. If same, S_j produces a nonce N_j and computes SK_{ij}=$h\ (\mathrm{PID}_j||K_j||N_G)$ and $M_5 = h\ (\mathrm{SK}_{ij}||K_j||N_G)$. S_j then sends $< M_5 >$ to GWN.
- GWN calculates $\mathrm{SK}_{ij} = h(\mathrm{PID}_j||K_j||N_G)$ and $M_5^{'} = h\ (\mathrm{SK}_{ij}||K_j||N_G)$. GWN checks $M_5 \stackrel{?}{=} M_5^{'}$. When equal, it computes $\mathrm{HID}_{inew} = h\ (N_G||\mathrm{HID}_i)$, $a_{inew} = h\,(\mathrm{HID}_{inew}||X_{\mathrm{GWN}}||N_G)$, $M_6 = (N_i||a_i^{'} \oplus \mathrm{HID}_{inew}$, $M_7 = (N_i||a_i^{'}) \oplus a_{inew}$, $M_8 = (N_i||a_i^{'}) \oplus SKij$ and $M_{\mathrm{gu}} = (SK_{ij}||N_i||a_{inew}||HID_{inew})$. Then, GWN sends $< M_6, M_7, M_8, M_{\mathrm{gu}} > U_i$. On successful session key agreement, GWN modifies HID_i to HID_{inew}. Otherwise, GWN keeps HID_i entact.
- U_i computes $\mathrm{HID}_{inew}^{'} = M6 \oplus (N_i||a_i)$, $a_{inew}^{'} = M_7 \oplus (N_i||a_i)$, $\mathrm{SK}_{ij}^{'} = M_8 \oplus (N_i||a_i)$ and $M_{\mathrm{gu}}^{'} = (\mathrm{SK}_{ij}||N_i||a_{inew}||\mathrm{HID}_{inew})$.$U_i$ verifies whether $M_{\mathrm{gu}}^{'}$ and M_{gu} are same or not. When same, U_i computes $b_{inew} = a_{inew} \oplus \mathrm{HPW}_i$ and $c_{inew} = h(a_{inew}||\mathrm{HPW}_i)$ and updates a_{inew}; b_{inew}; c_{inew} and HID_{inew}. Finally, U_i, GWN and S_j validate each other with same session key.

Change of Password Phase: U_i can change password without any assistance.

- U_i imprints biometrics BIO_i, inputs his/her ID, password and then forwards $\{\mathrm{ID}_i, \mathrm{PW}_i\ and\ \mathrm{BIO}_i\}$ to the smart card.
- The smart card calculates $< R_i,\ \ P_i >= Ge\ (\mathrm{BIO}_i), r_i = L_i \oplus h\ (R_i||\ \mathrm{PW}_i)$, $\mathrm{HPW}_i =\ h\ (r_i||\mathrm{ID}_i||\ \mathrm{PW}_i)$ and $c_i^* = h\ (a_i||\ \mathrm{HPW}_i\ )$. SC compares c_i^* and c_i. When same, U_i is prompted to provide a novel password of his choice.
- After providing new password $\mathrm{PW}_i^{\mathrm{new}}$, he sends it to the SC. Later, it computes $\mathrm{HPW}_i^{\mathrm{new}} = h(r_i||\mathrm{ID}_i||\mathrm{PW}_i^{\mathrm{new}})$, $\mathrm{Lnew}_i^{\mathrm{new}} = h(R_i||_{\mathrm{PW}_i}{}^{\mathrm{new}}) \oplus ri$, $b_i^{\mathrm{new}} = a_i \oplus \mathrm{HPW}_i^{\mathrm{new}}$ and $c_i^{\mathrm{new}} = h(a_i||\mathrm{HPW}_i^{\mathrm{new}})$ and updates $\{L_i^{\mathrm{new}}, b_i^{\mathrm{new}}, c_i^{\mathrm{new}}\}$.

4.3 Performance and Security Comparisons

Formal security analysis built on real-or-random and BAN logic shows that it provides security to various malicious attacks. The adversary cannot speculate the user ID and PW in real-polynomial time. The honey-lists can prevent offline password guess attack. The true ID_i and SID_j are encrypted by a random number. The attacker cannot know the original ID_i and sensor SID_j. Messages are encapsulated with one-way hash function resistant to collision making it secure against forgery attack. While attempting to imitate an entity, the attacker must figure out legitimate messages. But since they are encrypted with random secrets, the protocol is secure to impersonation attacks. The adversary cannot perform desynchronization attack because of mutual authentication. He cannot extract the random nonce making it secure from session key disclosure attack. The adversary cannot guess userś ID and password correctly even when he is a privileged insider without having the biometric secret key preventing it from privileged insider attack because of computational expense. Employing BAN logic, the protocol is proven to provide mutual authentication. The AVISPA simulation tool shows the translation time of the protocol as 0.09 s and search time of 7.89 s for visiting 1040 nodes in On the Fly Mode-Checker (OFMC) analysis making it secure. The computation cost for the protocol is 72.575 ms (T_f+19T_h), and the communication cost is 352 bytes, where T_h, T_f and T_{mul} are the time required to execute hash function, fuzzy extraction and EC point multiplication equal to 0.5, 63.075 and 63.075 ms respectively.

5 3-FUA Using Physically Unclonable Function

5.1 Introduction

The protocol [8] provides security with the aid of physical unclonable function (PUF) and does not require additional phase to modify challenge–response pairs and stores one CRP for every sensor. It utilizes personal biometrics, password and smart card to boost security as compared to 2-FA protocols. It also manipulates basic operations in cryptography, including bit-wise XOR (logical exclusive OR) apart from hash function to attain desired lightweight effect.

5.2 Reviewing Protocol Proposed by Liu et al.

The section below describes the protocol briefly.

Initializing Gateway Node Phase: GWN provides long-term key LTK, utilizing hash function h (.), symm. encryption E_k[.] along decryption D_k[.] algorithms.

Registering Sensor Node: Sensing nodes are enrolled in system via safe channel. Figure 5 shows the same.

- Upon registration with GWN, S_j produces random response C_j, calculates R_j = PUF(C_j) and forwards <SID$_j$, C_j, R_j > to gateway node.
- GWN then computes $(k_j, hd_j) = \text{Gen}(R_j)$, $S_{\text{GWN}-\text{SID}_j} = h(\text{LTK}||\text{SID}_j)$, $V_j = h(C_j||hd_j||k_j)$, stores <SID$_j$, C_j, hd$_j$, Vj > in database and forwards $S_{\text{GWN}-\text{SID}_j}$ to S_j.

User Registration: Users should execute the following steps to acquire information from sensor nodes as shown in Fig. 6.

- U_i sends ID$_i$ in the form of registration request to GWN via secure channel.

Device(SID_j)	GateWay(GWN)
Generate C_j	
Compute $R_j = PUF(C_j)$	
$< SID_j, C_j, R_j >$ $\longrightarrow$	
	$(k_j, hd_j) = Gen(R_j)$
	$V_j = h(C_j \| hd_j \| k_j)$
	Store $< SID_j, C_j, hd_j, V_j >$
	$S_{GWN-SID_j} = h(LTK \| SID_j)$
	$< S_{GWN-SID_j} >$ $\longleftarrow$
Store $S_{GWN-SID_j}$	

Fig. 5 Sensor node registration

User(ID_i)	GateWay(GWN)
Selcet ID_i	
$< ID_i >$ $\longrightarrow$	
	Check ID_i
	$S_{GWN-ID_i} = h(LTK \| ID_i)$
	Generate x, $DID_i = E_{LTK}[ID_i \| x]$
	$< DID_i, S_{GWN-ID_i} >$ $\longleftarrow$
Select PW_i, Imprint Bio_i	
Compute $(\sigma_i, \tau_i) = Gen(Bio_i)$	
$\tau_i^* = \tau_i \oplus h(ID_i \| PW_i)$	
$EID_i^* = S_{GWN-ID_i} \oplus h(ID_i \| PW_i \| \sigma_i)$	
$DID_i^* = DID_i \oplus h(ID_i \| \sigma_i \| PW_i)$	
$C_i = h(S_{GWN-ID_i} \| DID_i \| PW_i \| ID_i)$	
Insert $< \tau_i^*, EID_i^*, DID_i^*, C_i >$ into SC_i	

Fig. 6 User registration

- On reception of registration request, GWN checks whether ID_i is there in repository. When present, U_i qualifies as a valid user. Else, GWN computes $S_{\text{GWN}-\text{ID}_i} = h(\text{LTK}||\text{ID}_i)$, produces x randomly, calculates $\text{DID}_i = E_{\text{LTK}}[\text{ID}_i||x]$ and sends $<\text{DID}_i, S_{\text{GWN}-\text{ID}_i}>$ to U_i.
- U_i then selects password PW_i and impresses the biometric BIO_i. SC_i calculates $(\sigma_i, \tau_i) = \text{Gen}(\text{Bio}_i)$, $\tau_i^* = \tau_i \oplus h\ (\text{ID}_i||\ \text{PW}_i)$, $\text{EID}_i^* = S_{\text{GWN}-\text{ID}_i} \oplus h\ (\text{ID}_i||\ \text{PW}_i||\sigma_i)$, $\text{DID}_i^* = \text{DID}_i^* \oplus h\ (\text{ID}_i||\sigma_i||\ \text{PW}_i)$, $C_i = h\ (S_{\text{GWN}-\text{ID}_i}\ ||\ \text{DID}_i||\ \text{PW}_i||\ \text{ID}_i)$, and saves $< \tau_i^*, \text{DID}_i^*, \text{EID}_i^*, Ci >$ into SC_i.

Login and User Validation Phase: Post mutual authentication, U_i, S_j and GWN compute identical session key.

- User inputs identity ID_i, password PW_i and biometric Bio_i. SC_i of U_i calculates $\tau_i = \tau_i^* \oplus h\ (\text{ID}_i||\text{PW}_i)$, $\sigma_i = \text{Rep}(\text{Bio}_i, \tau_i)$, $S_{\text{GWN}-\text{ID}_i} = \text{EID}_i^* \oplus h\ (\text{ID}_i||\ \text{PW}_i||\sigma_i)$, $D\,\text{ID}_i = \text{DID}_i^* \oplus h(\ \text{ID}_i||\sigma_i||\ \text{PW}_i)$ and examines $C_i' \stackrel{?}{=} C_i$. When C_i' is not equal to C_i present in SC_i, at least 1 out of 3 factors sent by U_i is invalid. SC_i aborts the login phase then; else, continues.

- U_i gets the current timestamp T_1, computes $K_{ug} = h(S_{GWN-ID_i}||DID_i)$ and $M_1 = E_{K_{ug}}[SID_j||T_1]$, and transmits <$DID_i, M_1, T_1$> to GWN.
- On reception of login request, GWN examines the validity of received timestamp T_1 with condition $| T_1^* - T_1| \leq \Delta T$, where T_1^* represents time for receipt of message was and ΔT denotes maximum allowable transmission delay. If the condition holds to be true, GWN pulls ID_i with the help of $ID_i||x = D_{LTK}[DID_1]$ plus the long-term key LKT. Later, GWN computes $S_{GWN-ID_i} = h$ (LTK|| ID_i), being the shared-key of GWN and U_i, $K_{ug} = h(S_{GWN-SID_i}||DID_i)$ and $SID_j||T_1 = D_{K_{ug}}[M_1]$. When the time-stamp in M_1 is invalid, the session discontinues. Else, GWN looks up <SID_j, Cj, hd_j, V_j> in repository, produces nonce number r_g^*, along with current time-stamp T_2. GWN calculates $r_g = h(LTK||r_g^*)$, $S_{GWN-SID_i} = h(LTK||SID_j)$, $M_2 = h(S_{GWN-SID_i}||C_j) \oplus hd_j$, $K_{gs} = h(V_j||S_{GWN-SID_i})$ and $M_3 = E_{K_{gs}}[ID_i||r_g||T_2]$. At last, GWN sends < C_j, M_2, M_3, T_2 > to S_j.
- When getting GWNś message, S_j examines the novelty of T_2. S_j calculates $hd_j = h(S_{GWN-SID_j}||C_j) \oplus M_2$. Subsequently, S_j computes $k_j = Rep(PUF(C_j), hd_j)$ aiding PUF and fuzzy extractor. S_j calculates $V_j = h\ (C_j\ ||hd_j||k_j)$, $K_{gs} = h(V_j||S_{GWN-SID_j})$ and $ID_i||r_g||T_2 = D_{K_{gs}}[M_3]$. Given time-stamp in M_3 as invalid, S_j stops. Else, S_j produces a nonce number r_j, current timestamp T_3, and computes a new challenge $C_j^{new} = h(C_j||r_g)$, a new corresponding response $R_j^{new} = PUF(C_j^{new})$, a session key $SK = h(ID_i||SID_j||r_g||r_j)$, $M_4 = h(SK\ ||r_j||r_g||T_3)$ and $M_5 = E_{K_{gs}}[R_j^{new}||r_j||M_4\ ||T_3]$. Then, S_j submits < M_5, T_3 > to GWN.
- On getting message from S_j, GWN checks for the novelty of T_3. When valid, GWN decrypts $R_j^{new}||r_j||M_4||T_3 = D_{K_{gs}}[M5]$. Post validating the time-stamp of M_5, GWN computes session key $SK = h(ID_i||SID_j||r_g||r_j)$ and checks $M_4 \stackrel{?}{=} h(SK||r_j||r_g||T_3)$. When valid, GWN computes $C_j^{new} = h(C_j||r_g)$, $(k_j^{new}, hd_j^{new}) = Gen(R_j^{new})$, $V_j^{new} = h(C_j^{new}||hd_j^{new}||k_j^{new})$, and modifies the record with <SID_j, C_j^{new}, hd_j^{new}, V_j^{new} >. Then, GWN computes $DID_j^{new} = E_{LTK}[ID_i\ ||r_g']$ as novel identity of ID_i, $M_6 = h(SK||r_j||r_g\ ||T_4)$ and $M_7 = E_{K_{ug}}[DID_j^{new}||r_j||r_g||M_6||T_4]$,and sends <$M_7$, T_4> to U_i.
- On reception of message from GWN, U_i examines novelty of T_4. If it holds, U_i calculates $DID_j^{new}||r_j||r_g||M_6||T_4 = D_{K_{ug}}[M_7]$. After checking the novelty of the time-stamp in M_7, U_i computes a session key $SK = h(ID_i||SID_j||r_g||r_j)$ and checks $M_6 \stackrel{?}{=} h(SK||r_j||r_g||T_4)$. If the equation holds, U_i calculates $DID_i^* = DID_j^{new} \oplus h(ID_i||\sigma_i||PW_i)$ and modifies the old one.

Password and Biometric Change Phase: Assume PW_i^{new} and Bio_j^{new} as the new entities U_i wishes to update. SC_i calculates $(\sigma_i^{new}, \tau_{ji}^{new} = Gen(Bio_j^{new})$, $\tau_i^{**} = \tau_i^{new} \oplus h\ (ID_i\ ||\ PW_i^{new})$, $EID_i^{new} = S_{GWN-ID_j} \oplus h\ (ID_i||PW_i^{new}||ID_i)$.$SC_i$ replaces < τ_i^*, DID_i^*, EID_i^*, C_i > along newly furnished results < τ_i^{new}, DID_i^{**}, EID_i^{new}, C_i^{new} >.

5.3 Performance and Security Comparisons

This protocol has advantages with reference to security, computing costs and functionality. Dynamic identity changes after the user realizes a fruitful communication with GWN and S_j. It can achieve sensing node anonymity and untraceability. Since it guarantees user to be anonymous, attacker cannot determine a valid $M1'$ and secret key in the absence of long-term shared secret making it impossible for the adversary to procure a valid message using the obstructed login message, thus ruling out impersonation attacks. It is secure from GWN impersonation, sensor node impersonation, privileged insider attack, ephemeral-secret-leakage-attack, etc. It is resilient against capture attack of sensing nodes, physical tampering attacks and forward secrecy. In this protocol, CRPs are updated after every successful authentication. It requires lower communication overhead and the cost of implementation increases considering the usage of EC encryption with fuzzy extraction. It achieves superior reliability with enhanced performance features, making it expensive. The computation overhead is of 83.975 ms for user node, 112.075 ms for gateway node, 84.335 ms for sensor node and 290.385 ms is the total overall overhead cost. The communication overhead is of 2,490 bits for 4 messages.

6 Conclusion

In this article, different 3-factor user-authentication techniques are discussed briefly. The first protocol uses elliptic curve cryptography and can accomplish local password change and forward secrecy. It also resists loss of mobile device attacks with the help of error-correcting codes and fuzzy commitment schemes. The forward secrecy is ensured with the help of ECCDHP and ECDLP. With the help of fuzzy verifiers and honey-lists, it can achieve local password change resisting loss of mobile attacks, thus providing a secure, computationally efficient protocol.

To resolve vulnerabilities like simultaneous ID and password guessing attacks, the protocol with honey-list techniques can yield secure mutual authentication with the employment of BAN logic. The widely accepted ROR model shows that it can achieve security of session key. The AVISPA simulation confirms that it can prevent attacks like MITM, replay, etc. It has immunity from imitation, prediction, smart card theft, desynchronization and privileged insider attacks. It is a reliable scheme for in-field WSN environments as it provides mutual validation and user/sensor secrecy.

Compared to protocols in existence for IoT, the protocol with PUF provides most robust security mechanism. It is proven to be resistant to various attacks like physical capture/physical tampering and tracking attacks. As the number of messages sent increases, end-to-end delay rises simultaneously. The throughput increases with the increase in count of sensing devices and user nodes. The quotient between count of packets acquired at the destination and conveyed by the sender gives delivery rate of packets which is in inverse proportion with count of users and sensors. It increases when the count decreases and vice versa.

References

1. Amin, R., Islam, S.H., Biswas, G., Khan, M.K., Kumar, N.: A robust and anonymous patient monitoring system using wireless medical sensor networks. In: Future Generation Computer Systems, vol. 80, pp. 483–495 (2018). https://doi.org/10.1016/j.future.2016.05.032
2. Li, X., Peng, J., Obaidat, M.S., Wu, F., Khan, M.K., Chen, C.: A secure three factor user authentication protocol with forward secrecy for wireless medical sensor network systems. IEEE Syst. J. **14**(1), 39–50 (2020). https://doi.org/10.1109/JSYST.2019.2899580
3. Wang, D., Wang, P.: Two birds with one stone: two-factor authentication with security beyond conventional bound. IEEE Trans. Dependable Secure Comput. **15**(4), 708–722 (2018). https://doi.org/10.1109/TDSC.2016.2605087
4. Paliwal, S.: Hash-based conditional privacy preserving authentication and key exchange protocol suitable for industrial internet of things. IEEE Access **7**, 136073–136093 (2019). https://doi.org/10.1109/ACCESS.2019.2941701
5. Lee, J., Yu, S., Kim, M., Park, Y., Das, A.K.: On the design of secure and efficient three-factor authentication protocol using honey list for wireless sensor networks. IEEE Access **8**, 107046–107062 (2020). https://doi.org/10.1109/ACCESS.2020.3000790
6. Shahi, S., Redestowicz, M., Costadopoulos, N.: Authentication in E-health services. In: 5th International Conference on Innovative Technologies in Intelligent Systems and Industrial Applications (CITISIA), pp. 1–10 (2020). https://doi.org/10.1109/CITISIA50690.2020.9371820
7. Nandy, T., Idris, M.Y.I.B., Noor, R.M., Kiah, L.M., et al.: Review on security of internet of things authentication mechanism. IEEE Access **7**, 151054–151089 (2019). https://doi.org/10.1109/ACCESS.2019.2947723
8. Liu, Z., Guo, C., Wang, B.: A physically secure, lightweight three-factor and anonymous user authentication protocol for IoT. IEEE Access **8**, 195914–195928 (2020). https://doi.org/10.1109/ACCESS.2020.3034219

Trade-Off Between Memory and Model-Based Collaborative Filtering Recommender System

Gopal Behera and Neeta Nain

1 Introduction

The traditional information retrieval process is not an efficient tool or technique for handling the large dataset, so to overcome, a recommendation system is designed in early 1990, and nowadays, recommender system (RecSys) is a more powerful tool in various domains ranging from e-commerce to social media. A RecSys is a filtering technique that recommends the appropriate product or item to a user according to his interest or may recommend a product just reading from the users' profile history. E-commerce company like ebay.in, amazon.com and flip-kart.com [15] uses the recommendation tool for recommending the related items to the user. Based on the filtering approach, a RecSys is broadly categorized into (a) collaborative filtering and (b) content-based filtering (CBF). Generally, a CF technique automatically predicts the user's interest by collecting preferences or taste information from many neighbours; that is, suppose person A and B have the same opinion on an issue. However, they may have a more reasonable opinion on a different issue than a randomly chosen person. For example, a CF predicts TV shows, as per the user's taste of watching.

Many applications use a CF system, which contains vast datasets, such as financial data, user data and environmental data. Further, CF is categorized into memory- and model-based. To generate a recommendation, a memory-based CF uses the entire database's entire data, whereas model-based CF creates a model by using data from the database then that model is used for generating predictions [5]. Further, a memory-based approach of CF consists of item-based and user-based CF. In our work, we have shown the trade-off between memory- and model-based approaches such as

G. Behera (✉) · N. Nain
Department of Computer Science and Engineering, Malaviya National Institute of Technology Jaipur, Jaipur, India
e-mail: 2019rcp9002@mnit.ac.in

N. Nain
e-mail: nnain.cse@mnit.ac.in

M. Dua et al. (eds.), *Proceedings of the International Conference on Paradigms of Communication, Computing and Data Sciences*, Algorithms for Intelligent Systems,
https://doi.org/10.1007/978-981-16-5747-4_12

user-based, item-based and matrix factorization (MF) that includes SVD, SVD++, NMF and PMF with the real datasets.

2 Related Work

Whitnson and Kalakota [9] define a recommender system in the e-commerce point of view as an online system that provides information on a product and enable users to perform a transaction. A recommender engine is helpful to enhance the service and information quality of an online system or in the e-commerce platform, which makes a system a successful model [6]. Also, it improves both service and information quality by providing a personalized product to the user. Model and memory based collaborative recommendation on software library is conducted by McCarey et al. [12], on which they show that memory-based is better than the model-based. Walker and Robillard [13] point out that a recommender system's nature is different in the e-commerce and software engineering domain. Behera and Nain [2] made a comparative study of big mart sales prediction using different machine learning techniques. Breese et al. [5] conducted a comparative study of memory-based technique using vector similarity and Pearson correlation with the model-based techniques using Bayesian network and clustering. Huang et al. [8] compare an item-based, user-based with model-based techniques in e-commerce applications. They focus on recall and precision as evaluation criteria and ignore the standard evaluation metrics like RMSE and MAE. Behera and Nain [3] point out future sales prediction for big mart using an optimization technique. Behera et al. [1] discussed a forecasting model using Holt Winter's technique on univariate data.

To explain the trade-off between memory-based and model-based, this paper is structured as follows. Section 3 describes collaborative filtering and its approaches like various methods of both memory and model-based. Section 4 provides the detailed implementation of both approaches with their evaluation. Section 5 discusses the results and analysis, and finally, conclusion is mentioned in the Sect. 6.

3 Collaborative Filtering Based Recommendation

The idea behind collaborative filtering (CF) is that it makes predictions to similar users on the basis of their likes or dislikes [14]. Let $U = \{u_1, u_2, \ldots, u_m\}$ and $I = \{i_1, i_2, \ldots, i_n\}$ be the set of m users and n items, respectively, a rating [14] is an opinion about items by a user, and then, a CF can be implemented either by using memory-based or model-based.

Memory-Based CF Memory-based CF uses the whole database of user-item for making a prediction. This type of model uses statistical techniques to find a set of related users having similar tastes to an active user; that is, it finds the nearest neighbour of an active user [14] known as user-based CF. Bobadilla et al. [4] elaborate on the processes of the nearest neighbour (NN) method that is (a) selecting similar users to the active user, (b) predicting a rating of item I after aggregating the prediction score of similar users and (c) providing recommendations based on the previous scores. Su and Khoshgoftaar [17] point out the merits of memory-based CF approaches in terms of implementation and accommodation of new data. However, the performance of memory-based CF has decreasing in high sparsity data and has limited scalability for large datasets. Further, memory-based CF is categorized into (a) user-based and (b) item-based CF.

1. User-based CF: User-based recommendation finds similar users based on their preferences and provides the recommendations to the active user who has similar taste, and similarity between users is computed in the form of mean square difference (MSD) and is defined as in Eq. 1.

$$\mathrm{MSD}(u, v) = \frac{1}{|I_{uv}|}\dot{\sum}_{i \in I_{uv}} (r_{ui} - r_{vi})^2 \tag{1}$$

   where u and v are denoted as users, I_{uv} represents item sets that are rated by both user, and r_{ui}, r_{vi} denotes true ratings of user u and v, respectively. The similarity between two users is defined in Eq. 2

$$\mathrm{MSD\_}sim(u, v) = \frac{1}{MSD(u, v) + 1} \tag{2}$$

2. Item-based CF: Item-based CF will find similar patterns of items and provide the recommendation to the active user, and the similarity of items is also determined by using the mean square difference between items defined in Eq. 3.

$$\mathrm{MSD}(i, j) = \frac{1}{|U_{ij}|}\dot{\sum}_{u \in U_{ij}} (r_{ui} - r_{uj})^2 \tag{3}$$

   where i and j is denoted as items, U_{ij} represents a set of users who have rated item i and j, whereas r_{ui}, r_{uj} denotes true ratings of both items by user u. The similarity between two items is defined in Eq. 4

$$MSD\_sim(i, j) = \frac{1}{MSD(i, j) + 1} \tag{4}$$

Model-Based CF Model-based CF generates recommendations by creating a model from ratings [14]. A CF uses both explicit and implicit data, whereas the implicit data are generated by observing users' activity as applications downloaded, music played, Websites visited, etc. [4]. There are two approaches that can be used to develop a model-based CF, which are either a probability or a rating prediction [14]-based approach. Modelling of model-based CF can be done by using machine learning (ML) techniques that includes clustering, classification and rule-based [14]. As model-based CF uses dimensionality reduction strategies, this technique overcomes the memory-based approach's sparsity and scalability problem. However, at the same time, this technique may face loss of information [17].

1. SVD: The singular value decomposition (SVD) [10] is a real or complex matrix factorization. It is the generalization of the eigenvalue-decomposition of a positive semidefinite normal matrix (e.g. symmetric matrix with positive eigenvalues) to any $m \times n$ matrix via an extension of polar decomposition. It has many useful applications in signal processing and statistics. The prediction for SVD is defined in Eq. 5.

$$\hat{R_{ui}} = \mu + b_u + b_i + q_i^T \times p_u \tag{5}$$

where $\hat{R_{ui}}$: is the predicted rating of item i given by user u, q_i and p_u are the feature vectors associated with each item i and user u, respectively. b_u and b_i are the bias term of user u and item i, respectively, and μ is the mean rating of the dataset. So, to estimate all unknown, we need to minimize the regularized squared error using stochastic gradient descent (SGD) as mentioned in Eq. 6

$$\min_{p.q.b} \sum_{R_{ui} \in R_{train}} (R_{ui} - \hat{R_{ui}})^2 + \lambda(\|p_u\|^2 + \|q_i\|^2 + b_u^2 + b_i^2) \tag{6}$$

where λ is the regularized term.

2. SVD++: Sometimes the user provides implicit feedback, like historical rating and browsing data on web pages; no matter what, the user provides a specific rating value. But to a certain extent, this implicit feedback reflects the degree of a user's preference for each latent factor. SVD++ [10] is used to handle the implicit feedback information based on SVD to improve the predictive accuracy of a recommender by adding a factor vector ($y_j \in R^f$) for each item, and these factors describe the characteristics of an item, regardless of whether it has been evaluated or not. Then, a user's factor is modelled so that a better user bias can be obtained. The prediction of SVD++ is defined in Eq. 7.

$$\hat{R_{ui}} = b_u + b_i + \mu + q_i^T \left(p_u + |I_u|^{\frac{1}{2}} \sum_{j \in I_u} y_j \right) \tag{7}$$

where y_j are the set of item factors that capture implicit ratings, and $R(u)$ is the number of items rated by user u. The objective function needs to be minimized to obtain the optimal P and Q and is represented in Eq. 8.

$$\begin{aligned} \min_{p,q} \textstyle\sum_{R_{ui} \in R} [R_{ui} - b_u - b_i - \mu - q_i^T (p_u + |R(u)|^{-\frac{1}{2}} \sum_{j \in R(u)})^2 \\ + \lambda(b_u^2 + b_i)^2 + ||p_u|| + ||q_i||^2] \end{aligned} \tag{8}$$

where λ is the regularization term to prevent overfitting, we use SGD to evaluate the above objective function. Let e_{ui} is the error that is $e_{ui} = R_{ui} - \hat{R_{ui}}$. Let p_u is the element of user matrix P, and q_i is the element of item matrix Q. So, the factors learned iteratively by evaluating error e_{ui} and update the user and the item vector by taking in the opposite direction of the gradient. Then, the user and item vectors are shown in Eqs. 9 and 10, respectively.

$$p_u = p_u + \gamma(e_{ui}q_i - \lambda p_u) \tag{9}$$

$$q_i = q_i + \gamma(e_{ui}p_u - \lambda q_i) \tag{10}$$

3. Probabilistic Matrix Factorization (PMF): PMF is another dimensional reduction model of the CF. Here, the preference matrix of a user is determined with a small number of unobserved factors. Let for M movies and N users, a preference matrix R of $M \times N$ is the product of $M \times D$ and user coefficient matrix U^T of $D \times N$. Training such a model will find the best rank-D approximation to the observed target matrix R of $N \times M$ under the loss function. The PMF [10] models the preference matrix of a user as a product of two lower rank matrix of user and movie and is defined in Eq. 11.

$$p(R|U, V, \sigma^2) = \prod_{i=1}^{N}\prod_{j=1}^{M}[f(R_{ij}|U_i^T V_j, \sigma^2)]^{I_{ij}} \tag{11}$$

where U_i *and* V_j are specific user and item column vectors, respectively. $f(x|\mu, \sigma^2)$ is the probability density function of Gaussian distribution, and I_{ij} is the indicator function equal to 1 if user i rated movie j otherwise 0.
4. Non-negative matrix factorization (NMF): NMF is useful for decomposing multivariate data. NMF [11] is similar to unbiased SVD, but the difference is that in NMF, both user and item features kept as positive. Simultaneously, the multiplicative factor of both item and user features is slightly different in the updation rules of both features. One algorithm can be used to minimize the conventional least-squares error, while the other minimizes the Kullback–Leibler (KL) divergence [19]. The monotonic convergence nature of both algorithms can be proven through auxiliary function analogous. The algorithms interpreted as diagonally rescaled gradient descent (GD), where the rescaling factor is optimally chosen to ensure convergence. The optimization procedure is an SGD [18] with a specific choice of step size that ensures non-negativity of factors. The user and item latent factors are updated using SGD as follows:

$$p_{uf} = p_{uf}\frac{\sum_{i\in I_U} q_{if}\cdot r_{ui}}{\sum_{i\in I_U} q_{if}\cdot \hat{r_{ui}} + \lambda_u|I_U|p_{uf}} \tag{12}$$

$$q_{if} = q_{if}\cdot\frac{\sum_{u\in U_i} p_{uf}\cdot r_{ui}}{\sum_{u\in U_i} p_{uf}\cdot \hat{r_{ui}} + \lambda_i|U_i|q_{if}} \tag{13}$$

Table 1 Sparsity level of dataset

Dataset	Size of dataset	Ratings#	Sparsity(in %)
Movie 100k	1586126	100000	93.69
Movie 1m	23556000	1000209	95.75

4 Implementation

We have use real-world datasets for our experimental setup, and these datasets are movielens 100k[1] and movielens 1m[2] dataset. The datasets with their sparsity level are shown in Table 1. Further, we divided each dataset into training and testing set and evaluate the experiment using threefold cross-validation. In our experiment, we use two accuracy metrics, namely mean absolute error (MAE) and root mean square error (RMSE) for comparing the memory-based with model-based collaborative recommendation system. MAE [7] and RMSE [16] are defined in Eqs. 14 and 15, respectively.

$$\text{MAE} = \frac{1}{N}\sum(r_i - \hat{r}_i) \tag{14}$$

$$\text{RSME} = \sqrt{\frac{1}{N}\sum(p_i - \hat{p}_i)^2} \tag{15}$$

5 Result and Analysis

The evaluation is done for N values ranging from 1 to 20, where N is the number of neighbours. Further, mean square difference measure is used to find the similar users and items in the memory-based approach. We are also using threefold cross-validation to compare the results of both memory-based and model-based approach of CF, which are shown in Tables 2 and 3 for the different dataset of movielens. From Tables 2 and 3 we found that model-based approach is more appropriate to handle the sparse data; here, the movielens 1m dataset is more sparse than the movilens 100k dataset where the sparsity level of these datasets is shown in Table 1.

Similarly, Table 4 shows execution time taken by both model-based and memory-based technique and found that model-based approach takes less time to recommend in both of the datasets as model-based approach reduced the dimensionality of data.

1 https://grouplens.org/datasets/movielens/100k/.

2 https://grouplens.org/datasets/movielens/1m/#_sid=rd0.

Table 2 Comparison of memory-based versus model-based CF for the dataset movielens 1m

Measures	Methods					
	SVD	SVD++	PMF	NMF	item-based	user-based
RMSE	0.8860	**0.8716**	0.8855	0.9200	0.9307	0.9428
MAE	0.6964	**0.6811**	0.6974	0.7267	0.7326	0.7434

Table 3 Comparison of memory-based versus model-based CF for the dataset movielens 100k

Measures	Methods					
	SVD	SVD++	PMF	NMF	Item-based	User-based
RMSE	0.9432	**0.9227**	0.9667	0.9748	0.9944	0.9865
MAE	0.7447	**0.7227**	0.7623	0.7660	0.7868	0.7781

Table 4 Execution time for model and memory-based CF

Methods	Test time (s) on movielens 100k dataset	Test time (s) on movielens 1m dataset
SVD	0.22	3.59
SVD++	2.72	82.52
PMF	0.20	3.28
NMF	0.17	3.24
User-based	3.55	151.82
Item-based	4.32	79.02

Figure 1a shows the threefold comparison of MAE, Fig. 1b shows threefold comparison of RSME of both memory- and model-based CF for dataset movielens 100k, and from the Fig. 1, it is found that SVD++ is better than others as it has the lower MAE and RMSE than other approaches.

Whereas Fig. 2 a and b describes the error rate of the movielens 100k dataset with respect to the value of k and found that the error rate is reduced with the value of K.

Similarly, Fig. 3a and b shows the threefold comparison of MAE and RSME for dataset movielens $1m$, respectively, and found that SVD++ is better than other approaches though the sparsity level is increased. Also, the error rate of movielens dataset$1m$ is shown in Fig. 4 and found that the error rate is reduced with the value of K.

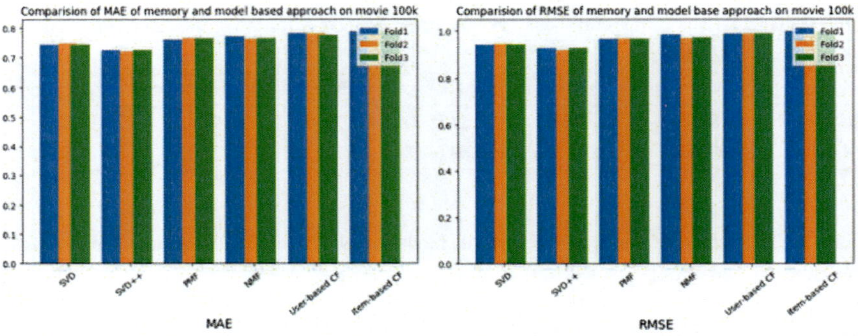

(a) MAE (b) RMSE

Fig. 1 For dataset movielens 100k **a** threefold comparison of mean absolute error; **b** threefold comparison of RMSE of memory-based and model-based CF

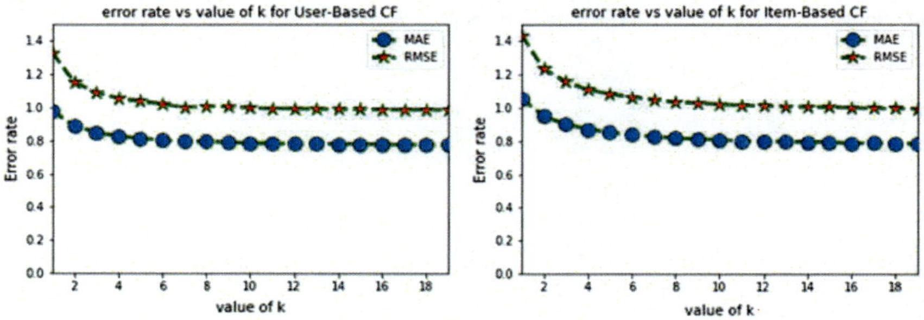

(a) MAE (b) RMSE

Fig. 2 Error rate of memory-based collaborative filtering for movie 100k dataset: **a** error rate of user-based CF with respect to value of K; **b** error rate of item-based CF with respect to value of K

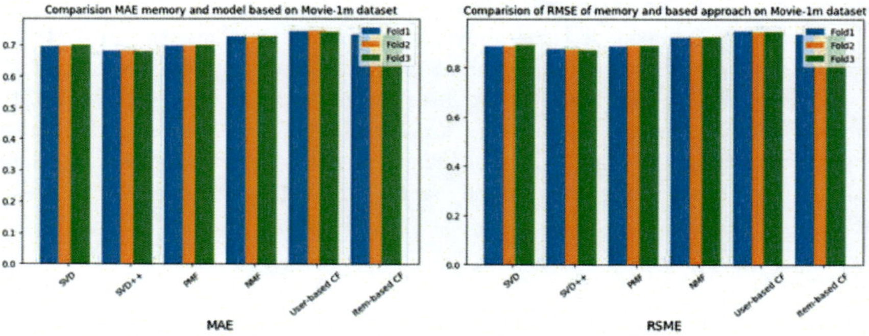

(a) MAE (b) RMSE

Fig. 3 For dataset movielens 1m **a** threefold comparison of mean absolute error; **b** threefold comparison of RMSE; of memory-based and model-based CF

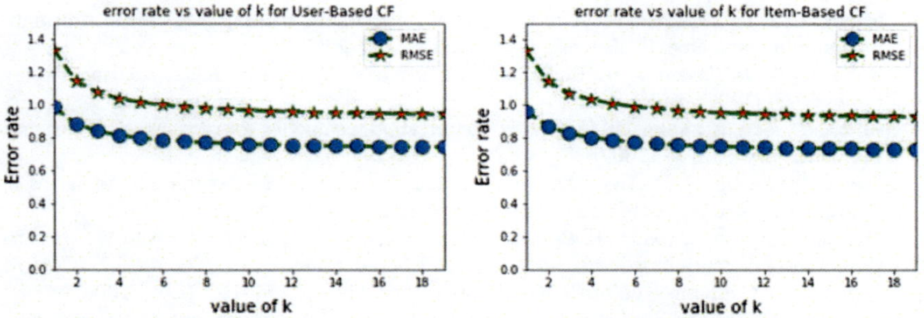

(a) MAE (b) RMSE

Fig. 4 Error rate of memory-based collaborative filtering for movie 1m dataset: **a** error rate of user-based CF; **b** error rate of item-based CF

6 Conclusion

The experimental analysis of both techniques suggests that the model-based approach has produced less error than the memory-based approach of CF. Both prediction errors (RMSE and MAE) are lower than that of the memory-based technique, and hence, it is more accurate. Similarly, the model-based will consume less time than that of the memory-based technique for predicting, and for sparse data, it is found that the model-based is more appropriate to handle than the memory-based. As the model-based techniques reduced the dimensionality, hence there is a chance of information loss. Further, we found that the error rate of the memory-based technique is decreased with respect to the value of neighbours.

References

1. Behera, G., Bhoi, A., Bhoi, A.: UHWSF: univariate holt winters based store sales forecasting. In: International Conference on Machine Learning, Internet of Things and Big Data (ICMIB), pp. 421–432. Springer, Heidelberg (2020)
2. Behera, G., Nain, N.: A comparative study of big mart sales prediction. In: International Conference on Computer Vision and Image Processing, pp. 421–432. Springer, Heidelberg (2019)
3. Behera, G., Nain, N.: Grid search optimization (GSO) based future sales prediction for big mart. In: 2019 15th International Conference on Signal-Image Technology & Internet-Based Systems (SITIS), pp. 172–178. IEEE (2019)
4. Bobadilla, J., Ortega, F., Hernando, A., Gutiérrez, A.: Recommender systems survey. Knowl.-Based Syst. **46**, 109–132 (2013)
5. Breese, J.S., Heckerman, D., Kadie, C.: Empirical analysis of predictive algorithms for collaborative filtering. arXiv preprint arXiv:1301.7363 (2013)
6. Delone, W.H., Mclean, E.R.: Measuring e-commerce success: applying the delone & mclean information systems success model. Int. J. Electron. Commerce **9**(1), 31–47 (2004)
7. Goldberg, K., Roeder, T., Gupta, D., Perkins, C.: Eigentaste: a constant time collaborative filtering algorithm. Inf. Retrieval **4**(2), 133–151 (2001)

8. Huang, Z., Zeng, D., Chen, H.: A comparison of collaborative-filtering recommendation algorithms for e-commerce. IEEE Intell. Syst. **22**(5), 68–78 (2007)
9. Kalakota, R., Whinston, A.B.: Electronic Commerce: A Manager's Guide. Addison-Wesley Professional (1997)
10. Koren, Y., Bell, R., Volinsky, C.: Matrix factorization techniques for recommender systems. Computer 42(8), 30–37 (2009)
11. Lee, D.D., Seung, H.S.: Algorithms for non-negative matrix factorization. In: Advances in Neural Information Processing Systems, pp. 556–562 (2001)
12. McCarey, F., Cinneide, M.O., Kushmerick, N.: A recommender agent for software libraries: an evaluation of memory-based and model-based collaborative filtering. In: 2006 IEEE/WIC/ACM International Conference on Intelligent Agent Technology, pp. 154–162. IEEE (2006)
13. Robillard, M., Walker, R., Zimmermann, T.: Recommendation systems for software engineering. IEEE Softw. **27**(4), 80–86 (2009)
14. Sarwar, B., Karypis, G., Konstan, J., Riedl, J.: Item-based collaborative filtering recommendation algorithms. In: Proceedings of the 10th International Conference on World Wide Web, pp. 285–295 (2001)
15. Schafer, J.B., Konstan, J., Riedl, J.: Recommender systems in e-commerce. In: Proceedings of the 1st ACM Conference on Electronic Commerce, pp. 158–166 (1999)
16. Smyth, B., Cotter, P.: A personalised tv listings service for the digital tv age. Knowl.-Based Syst. **13**(2–3), 53–59 (2000)
17. Su, X., Khoshgoftaar, T.M.: A survey of collaborative filtering techniques. Adv. Artif. Intell. **2009** (2009)
18. Tsuruoka, Y., Tsujii, J., Ananiadou, S.: Stochastic gradient descent training for l1-regularized log-linear models with cumulative penalty. In: Proceedings of the Joint Conference of the 47th Annual Meeting of the ACL and the 4th International Joint Conference on Natural Language Processing of the AFNLP: Volume 1-Volume 1, pp. 477–485. Association for Computational Linguistics (2009)
19. Van Erven, T., Harremos, P.: Rényi divergence and kullback-leibler divergence. IEEE Trans. Inf. Theory **60**(7), 3797–3820 (2014)

RAT Selection Strategies for Next-Generation Wireless Networks: A Taxonomy and Survey

Bhanu Priya and Jyoteesh Malhotra

1 Introduction

In accordance with Cisco visual networking index report [1], 3.6 networked devices per capita and 5.3 billion total Internet users will be reached by 2023. This proliferation of emerging multimedia applications and smart devices has created a conundrum for which it is worthwhile to deliberate upon 5G for high data rate and low latency services. 5G HetNet has emerged as an effective wireless solution that facilitates a coherent communication interface as it consolidates and coordinates numerous radio interfaces, namely LTE, Wi-Fi and 5G-New Radio (NR), Femto cell, LoRa as examples. The convergence of numerous access technologies in 5G HetNet has been one of the most effective means to deliver ever-increasing capacity and better user experience. But the divergence among the RATs in terms of operating frequency, protocols, physical and MAC layer multiple access technologies demands suitable RAT selection for the effective exploitation of 5G HetNets. Therefore, the optimal RAT selection is a vital aspect in the 5G HetNets. The optimal RAT selection mechanism associates a user with a suitable RAT leading to optimal resource utilization in terms of load balancing, spectrum and energy efficiency. However, the emerging 5G applications and scenarios have made the rudimentary network selection solutions ineffective and established diverse challenges and opportunities for the development of advanced user association mechanism.

Numerous sophisticated contributions have surveyed the RAT selection strategies in the 5G HetNets. However, they mainly elaborated the recent advancement in the 5G networks, i.e., key enabling technologies, the architectures, mobility management solutions [2–4]. Therefore, to underline the importance of RAT selection mechanism in the 5G HetNets, this article provides a well-structured taxonomy that classifies the strategies on the basis of different characteristic attributes such as metrics, network,

B. Priya (✉) · J. Malhotra
Guru Nanak Dev University, Regional Campus, Jalandhar, Jalandhar, India

M. Dua et al. (eds.), *Proceedings of the International Conference on Paradigms of Communication, Computing and Data Sciences*, Algorithms for Intelligent Systems,
https://doi.org/10.1007/978-981-16-5747-4_13

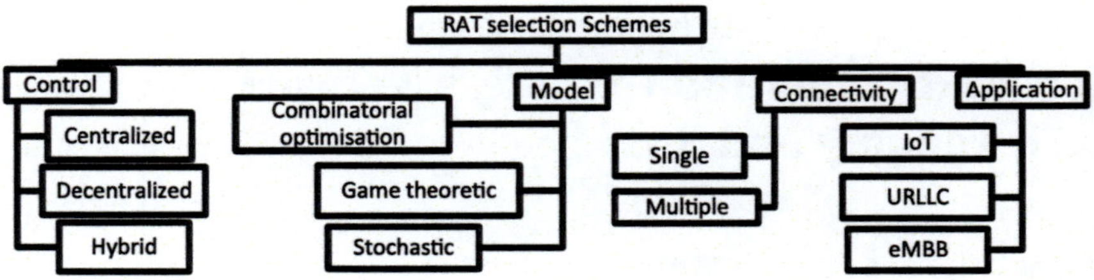

Fig. 1 Proposed taxonomy structure

control, connectivity and methods. Moreover, in comparison with the existing literature, a comprehensive survey including advantages and disadvantages of the recent state-of-the art is presented for the better comprehension to the performance of the RAT selection strategies.

The remainder of this paper is structured as follows. Section 2 provides a hierarchically structured taxonomy followed by a thorough survey of current RAT selection techniques presented in Sect. 3. Section 4 concludes the paper.

2 Taxonomy

A taxonomy is developed for the comprehensive and systematic survey of the current user association algorithms for the 5G HetNets, comprising four characteristic branches: (1) Control (2) Model (3) Connectivity (4) Application as presented in Fig. 1.

On the basis of control, network selection strategies are classified into centralized, decentralized and hybrid approach, whereas combinatorial optimization, game-theoretic and stochastic model-based approaches are classified with reference to the model adopted. Furthermore, network selection mechanism in accordance with single and multi-connectivity is presented. Lastly, in accordance with the application, network selection strategies for Internet of Things (IoT), ultra-reliable low latency (URLLC) and enhanced mobile broadband (eMBB) applications are discussed.

3 Comprehensive Survey of RAT Selection Schemes

3.1 Network Selection on the Basis of Control

The network selection on the basis of control mechanism, namely centralized, decentralized and hybrid, severely affects the computational complexity of the system.

(1) Centralized: All the network selection decisions in the centralized approach are made by the central entity present on the network side that keeps track of the quality of experience (QoE) requirements, resource demand and channel quality from each user and subsequently select a suitable RAT on the basis of collected information. A software-defined network (SDN)-based framework proposed in [5] allows a cross-layer monitoring and centralized management of various RATs for ubiquitous handover in the heterogeneous network. Authors in [6] designed a SDN-based handover management engine in the application layer to determine the best handover decision on the basis of user and network state and quality of service (QoS) requirements of the application. Similarly, Mouwad et al. (2019) presented a SDN-based mobility management solution that exploits the utility theory and game-theoretic approach for handover anticipation to guarantee lower latency and seamless mobility. Authors in [8] presented a novel handover technique that exploits blockchain and SDN techniques to ensure handover with lower delay in the 5G network. Priya et al. (2020) proposed a centralized enabling platform which leveraged a hybrid RAT selection mechanism comprising fuzzy analytical hierarchy process (AHP) and TOPSIS to reduce unnecessary handover and ranking abnormality issue. The centralized approach implemented in [5–9] ensures optimal resource allocation and faster convergence but incurs large amount of signaling.

(2) Decentralized: Decentralized network selection approach allows user equipment (UE) to decide autonomous association to the suitable RAT that maximizes the user's satisfaction and minimizes the signaling overhead. Yan et al. (2018) presented a smart aggregated RAT access strategy that leverages Nash Q-learning for computing set of acceptable RAT selection mechanism while reducing the strategy space to facilitate the long-term throughput with guaranteed QoS. A distributed optimization approach based on multi-agent reinforcement learning is adopted in [11] that leverages dueling deep architecture to achieve the maximum network utility while guaranteeing user QoS in the 5G heterogeneous cellular networks. Authors in [12] proposed a distributed model-driven algorithm that combines machine learning and game theory to learn optimal network selection policy in the 5G HetNets. A smart distributed handover scheme named LESS proposed in [13] utilized the concepts of distributed Q-learning with a minimized action space which facilitates each user to update its own Q-value separately for autonomous handover decisions with guaranteed user QoS. Nevertheless, the work discussed in [10–13] leads to inefficient radio resource utilization as each user acts in their self-interest way and does not consider network preferences and policies.

3.2 Network Selection on the Basis of Models

The choice of model decides the crucial parameters such as convergence rate and stability of the proposed system.

(1) Combinatorial optimization model: A combinatorial optimization problem relates with a general NP-hard in which user associates with a BS to maximize the utility or cost factor. Specifically, utility or cost function quantifies the comprehensive satisfaction offered by each candidate network to conform the service-specific QoS requirements on the basis of specific network metrics. For instance, authors in [5] comprehensively defined fittingness factor that defines the suitability of a particular access point (AP) for a service. The fittingness factor has been formulated by extending the sigmoidal function that accounts the data bit rate demanded by the flow and the bit rate offered by the AP. The utility function combined with the Multiple Attribute Decision Making (MADM), game theory emerged as predominant techniques for the RAT selection. The Jaya algorithm leveraged in the proposed work is modified with the integration of utility function which indicates the efficacy of the network parameters [19]. Authors in [20] presented a traffic differentiated network selection solution that calculates network reputation on the basis of utility function and selects the best network with the help of AHP method. Zhu et al. (2019) proposed an adaptive multi-service network selection model that integrates the utility function with fuzzy logic and MADM technique for the optimal network selection in MEC-enabled 5G HetNet.

(2) Game-theoretic model: The lack of consensus in the attribute evaluation and large number of evaluating criterions restricts the implementation of the methods discussed in [5, 6, 19–21]. Therefore, game-theoretic approach has been adopted and discussed in the literature. For instance, authors in [22] proposed a game theory-based novel network selection mechanism considering APs and vehicles as a player which competes for cellular data offloading and selects the best network on the basis of utility function comprising crucial network parameters. Authors in [23] proposed a game-theoretic auction theory network selection approach for cognitive radio vehicular networks. Moreover, a novel cost function-based MADM scheme is proposed to compute the bidding cost value for various non-safety services. Authors in [24] presented a vertical handover scheme based on non-cooperative game theory for heterogeneous network that maximizes user QoE and network revenues. Similarly, authors in [25] exploited the cooperative and decentralized non-cooperative game theory-based network selection scheme for intra-WBANs and beyond-WBANs, respectively.

3.3 Network Selection on the Basis of Connectivity

To enable reliable transmissions, the network selection on the basis of connectivity plays a major role and hence detailed below:

(1) Single Connectivity: Authors in [30] presented a machine learning-based distributed cognitive RAT association framework that learns terminal experience and user behavior, for optimal single network selection. The network selection problem modeled as Markov decision process (MDP) model is solved with the help of NS-MDP algorithm in [31], which ensures the best network selection for

different type of traffic class with the help of utility function. Authors in [32] proposed a robust handover authentication protocol for all characteristic mobility scenarios generated in heterogeneous network comprising 5G NR and non-3GPP access technologies. The proposed solution utilizes the trapdoor collision property of chameleon hash functions along with the tamper resistance of blockchains to select optimal network in the 5G HetNet.

(2) Multi-connectivity: Multi-connectivity feature allows UE to associate with more than one RATs at a time leading to improved reliability and system throughput as compared to the methods implemented in [30–32]. Wang et al. (2018) leveraged a LSTM-based mechanism that establishes a dual connection on the basis of predictions made from the learned mobility pattern. Authors in [34] presented an SDN-based dual-connectivity mechanism that leverages Lyapunov optimization theory to solve handover overhead problem to ensure ubiquitous communication in 5G-enabled aeronautical network. Poirot et al. (2020) designed a multi-connectivity algorithm to select secondary cell with the help of Max Bitrate-EE and AHP method that aims at improving robustness and performance, while reducing the energy consumption. Ghatak et al. (2020) leveraged Thompson sampling approach with reinforcement learning for elastic multi-connectivity BS-UE association to satisfy dynamic and agile user QoS requirements. Mumtaz et al. (2020) proposed a mobility management framework that leverages the concept of dual-connectivity and strategy-based data split mechanism to execute handover between 4G and 5G radio access technologies.

3.4 Network Selection on the Basis of Applications

The network selection strategies for (i) IoT- (ii) URLLC- and (iii) eMBB-based applications are presented below:

(1) Internet of Things: Mohseni et al. (2019) presented a SDN-based cross-layer handover scheme that efficiently reduces E2E delay and handover latency. The author improvized the classic utility approach to facilitate optimal resource allocation and better QoE with reduced cost. Authors in [39] presented a fully distributed network selection algorithm that utilizes network feedback to guarantee convergence to set of correlated equilibria and low signaling overhead. Authors in [16] designed a stable matching algorithm that facilitates high energy efficiency and reduced data transmission cost, respectively, for the IoT devices and RATs. Sandoval et al. (2019) presented a reinforcement learning-based RAT selection algorithm to attain optimal RAT selection policy that minimizes the transmission delay along with reduced power consumption and operational cost. Goudarzi et al. (2019) presented an intelligent network selection model that hybridizes the biogeography-based optimization and MDP to select the best RAT in the industrial IoT environment.

(2) URLLC: Authors in [41] proposed a conditional make-before-break handover mechanism to achieve zero handover interruption time (HIT) and failure rate

simultaneously in NR for URLLC services. Mahmood et al. (2018) proposed a novel multi-connectivity-based framework for reliability enhancement in 5G-oriented URLLC services and moreover introduced the optimized dual-connectivity parameters to control user operating in this mode. Lee et al. (2019) presented a distributed reinforcement learning-based network selection scheme to learn an optimal network selection policy that configures multi-connectivity mechanism to attain better reliability for emerging URLLC services. Kumar et al. (2019) proposed a D2D-supported handover mechanism to achieve zero millisecond HIT with the help of two modes of operation, i.e., direct and D2D selected by the gNodeB to support seamless mobility.
(3) eMBB: Fan et al. (2020) proposed a deep learning-based handover mechanism in which the online network decouples the throughput maximization of AP and user into two subproblems and leverages the 'Pareto optimal' to compute the best solutions, whereas the deep learning-based offline stage learns from the historical optimization information of the online stage and predicts best solutions. Authors in [46] redesigned the handover mechanism with the help of software-defined ultra-dense network that reduces the E2E delay through revised Markov model considering edge and core networks and removed handover execution state.

The proposed taxonomy and literature survey facilitate the understanding of the diverse aspects of RAT selection strategies in wireless heterogeneous networks, including their respective limitations and performances, and recent state of the art as depicted in Table 1.

Table.1 Categorization of the recent state of the art on the basis of RAT selection criteria

Criteria	Techniques	References
Control	Centralized	[5–9, 20, 34, 38, 40, 45]
	Decentralized	[10–13, 21, 22, 25, 28–30, 36]
	Hybrid	[14–18]
Model	Combinatorial optimization	[5, 6, 19–21, 39]
	Game theoretic	[7, 16, 22–25]
	Stochastic	[9–14, 17, 18, 23–36, 40, 43, 45]
Connectivity	Single	[5–32, 38–42, 44–46]
	Multiple	[33–37, 43]
Application	IoT	[16, 20, 27, 38–40, 46]
	URLLC	[41–44]
	eMBB	[42, 44–46]

4 Conclusion

An optimized RAT selection plays a pivotal role in the 5G HetNet. Therefore, this article presents a well-structured taxonomy that includes the classification of the network selection strategies on the basis of characteristic attributes such as control, model, connectivity and applications from centralized to decentralized, from game-theoretic model to stochastic model, from single to multiple connectivity and from IoT to eMBB applications, respectively. Within each of the classified network selection strategies, their inherent features, limitations and recent state-of-the-art work have been discussed. Indeed RAT selection has to be scrutinized more in-depth in future research to better accommodate the convoluted and random network conditions, in order to realize the full capacity of the next-generation wireless networks.

References

1. Cisco: Cisco Annual Internet Report (2018–2023) White Paper (2018). https://www.cisco.com/c/en/us/solutions/collateral/executive-perspectives/annual-internet-report/white-paper-c11-741490.html. Accessed 22 Dec 2020
2. Rong, B., Zhou, J., Kadoch, M., Sun, G.L.: Emerging technologies for 5G radio access network: architecture, physical layer technologies, and MAC layer protocols. Wirel. Commun. Mob. Comput. **2018**, 1–2 (2018)
3. Pandi, V.S., Priya, J.L.: A survey on 5G mobile technology. In: 2017 IEEE International Conference on Power, Control, Signals and Instrumentation Engineering (ICPCSI), pp. 1656–1659 (2017)
4. Benchaabene, Y., Boujnah, N., Zarai, F.: 5G cellular: survey on some challenging tech- niques. In: 2017 18th International Conference on Parallel and Distributed Computing, Applications and Technologies (PDCAT), pp. 348–353 (2017)
5. Raschella, A., Bouhafs, F., Seyedebrahimi, M., Mackay, M., Shi, Q.: Quality of service oriented access point selection framework for large Wi-Fi networks. IEEE Trans. Netw. Serv. Manag. **14**(2), 441–455 (2017)
6. Gharsallah, A., Zarai, F., Neji, M.: SDN/NFV-based handover management approach for ultradense 5G mobile networks. Int. J. Commun. Syst. **32**(17), e3831–e3831 (2019)
7. Mouawad, N., Naja, R., Tohme, S.: SDN based handover management for a tele-operated driving use case. In: 12th IFIP Wireless and Mobile Networking Conference (WMNC), pp. 47–54 (2019)
8. Yazdinejad, A., Parizi, R.M., Dehghantanha, A., Choo, K.K.R.: Blockchain-enabled au- thentication handover with efficient privacy protection in SDN-based 5G networks. IEEE Trans. Netw. Sci. Eng. 1 (2020)
9. Priya, B., Malhotra, J.: 5GAuNetS: an autonomous 5G network selection framework for Industry 4.0. Soft Comput. **24**(13), 9507–9523 (2020)
10. Yan, M., Feng, G., Zhou, J., Qin, S.: Smart multi-RAT access based on multiagent reinforcement learning. IEEE Trans. Veh. Technol. **67**(5), 4539–4551 (2018)
11. Zhao, N., Liang, Y.C., Niyato, D., Pei, Y., Wu, M., Jiang, Y.: Deep reinforcement learning for user association and resource allocation in heterogeneous cellular networks. IEEE Trans. Wirel. Commun. **18**(11), 5141–5152 (2019)
12. Wang, X., Li, J., Wang, L., Yang, C., Han, Z.: Intelligent user-centric network selection: a model-driven reinforcement learning framework. IEEE Access **7**, 21645–21661 (2019)

13. Sun, Y.: Efficient handover mechanism for radio access network slicing by exploiting distributed learning. IEEE Trans. Netw. Serv. Manage. **17**(4), 2620–2633 (2020)
14. Nguyen, D.D., Nguyen, H.X., White, L.B.: Reinforcement learning with network-assisted feedback for heterogeneous RAT selection. IEEE Trans. Wirel. Commun. **16**(9), 6062–6076 (2017)
15. Alfoudi, A.S.D., Newaz, S.H.S., Ramlie, R., Lee, G.M., Baker, T.: Seamless mobility management in heterogeneous 5G networks: a coordination approach among distributed SDN controllers. In: 2019 IEEE 89th Vehicular Technology Conference (VTC2019-Spring), pp. 1–6 (2019)
16. Arabi, S., Hammouti, H.E., Sabir, E., Elbiaze, H., Sadik, M.: RAT association for autonomic IoT systems. IEEE Netw. **33**(6), 116–123 (2019)
17. Guo, D., Tang, L., Zhang, X., Liang, Y.C.: Joint optimization of handover control and power allocation based on multi-agent deep reinforcement learning. IEEE Trans. Veh. Technol. **69**, 13124–13138 (2020)
18. Wang, D., Sun, Q., Wang, Y., Han, X., Chen, Y.: Network-assisted vertical handover scheme in heterogeneous aeronautical network. In: 2020 Asia-Pacific Conference on Image Processing, Electronics and Computers (IPEC), pp. 148–152 (2020)
19. Munjal, M., Singh, N.P.: Utility aware network selection in small cell. Wirel. Netw. **25**(5), 2459–2472 (2019)
20. Desogus, C., Anedda, M., Murroni, M., Muntean, G.M.: A Traffic type-based differentiated reputation algorithm for radio resource allocation during multi-service content delivery in 5G heterogeneous scenarios. IEEE Access **7**, 27720–27735 (2019)
21. Zhu, A., Guo, S., Liu, B., Ma, M., Yao, J., Su, X.: Adaptive multiservice heterogeneous network selection scheme in mobile edge computing. IEEE Internet Things J. **6**(4), 6862–6875 (2019)
22. Dua, A., Kumar, N., Bawa, S.: Game theoretic approach for real-time data dissemination and offloading in vehicular ad hoc networks. J. Real-Time Image Proc. **13**(3), 627–644 (2017)
23. Kumar, K., Prakash, A., Tripathi, R.: A spectrum handoff scheme for optimal network selection in Cognitive Radio vehicular networks: a game theoretic auction theory approach. Phys. Commun. **24**, 19–33 (2017)
24. Goyal, P., Lobiyal, D.K., Katti, C.P.: Game theory for vertical handoff decisions in het- erogeneous wireless networks: a tutorial. In: Bhattacharyya, S., Gandhi, T., Sharma, K., Dutta, P. (eds.) Advanced Computational and Communication Paradigms, pp. 422–430 (2018)
25. Ning, Z.: Mobile edge computing enabled 5G health monitoring for internet of medical things: a decentralized game theoretic approach. IEEE J. Sel. Areas Commun. **39**(2), 463–478 (2021)
26. Ozturk, M., Gogate, M., Onireti, O., Adeel, A., Hussain, A., Imran, M.A.: A novel deep learning driven, low-cost mobility prediction approach for 5G cellular networks: the case of the Control/Data Separation Architecture (CDSA). Neurocomputing **358**, 479–489 (2019)
27. Sandoval, R.M., Canovas-Carrasco, S., Garcia-Sanchez, A.J., Garcia-Haro, J.: A reinforcement learning-based framework for the exploitation of multiple RATs in the IoT. IEEE Access **7**, 123341–123354 (2019)
28. Ding, H., Zhao, F., Tian, J., Li, D., Zhang, H.: A deep reinforcement learning for user association and power control in heterogeneous networks. Ad Hoc Netw. **102**, 102069–102069 (2020)
29. Mollel, M.S., Abubakar, A.I., Ozturk, M., Kaijage, S., Kisangiri, M., Zoha, A., Imran, M.A., Abbasi, Q.H.: Intelligent handover decision scheme using double deep reinforcement learning. Phys. Commun. **42**, 101133–101133 (2020)
30. Perez, J.S., Jayaweera, S.K., Lane, S.: Machine learning aided cognitive RAT selection for 5G heterogeneous networks. In: 2017 IEEE International Black Sea Conference on Communications and Networking (BlackSeaCom), pp. 1–5 (2017)
31. Tang, C., Chen, X., Chen, Y., Li, Z.: A MDP-based network selection scheme in 5G ultra- dense network. In: 2018 IEEE 24th International Conference on Parallel and Distributed Systems (ICPADS), pp. 823–830 (2018)
32. Zhang, Y., Deng, R., Bertino, E., Zheng, D.: Robust and universal seamless handover authentication in 5G HetNets. IEEE Trans. Dependable Secure Comput. 1 (2019)

33. Wang, C., Zhao, Z., Sun, Q., Zhang, H.: Deep learning-based intelligent dual connectivity for mobility management in dense network. In: 2018 IEEE 88th Vehicular Technology Conference (VTC-Fall), pp. 1–5 (2018)
34. Wang, D., Wang, Y., Dong, S., Huang, G., Liu, J., Gao, W.: Exploiting dual connectivity for handover management in heterogeneous aeronautical network. IEEE Access **7**, 62938–62949 (2019)
35. Poirot, V., Ericson, M., Nordberg, M., Andersson, K.: Energy efficient multi-connectivity algorithms for ultra-dense 5G networks. Wirel. Netw. **26**(3), 2207–2222 (2020)
36. Ghatak, G., Sharma, Y., Zaid, K., Rahman, A.U.: Elastic multi-connectivity in 5G networks. Phys. Commun. **43**, 101176–101176 (2020)
37. Mumtaz, T., Muhammad, S., Aslam, M.I., Mohammad, N.: Dual connectivity-based mobility management and data split mechanism in 4G/5G cellular networks. IEEE Access **8**, 86495–86509 (2020)
38. Mohseni, H., Eslamnour, B.: Handover management for delay-sensitive IoT services on wireless software-defined network platforms. In: 2019 3rd International Conference on Internet of Things and Applications (IoT), pp. 1–6 (2019)
39. Desogus, C., Anedda, M., Murroni, M., Giusto, D.D., Muntean, G.: ReMIoT: reputation- based network selection in multimedia IoT. In: 2019 IEEE Broadcast Symposium (BTS), pp. 1–6 (2019)
40. Goudarzi, S., Anisi, M.H., Abdullah, A.H., Lloret, J., Soleymani, S.A., Hassan, W.H.: A hybrid intelligent model for network selection in the industrial Internet of Things. Appl. Soft Comput. **74**, 529–546 (2019)
41. Park, H., Lee, Y., Kim, T., Kim, B., Lee, J.: Handover mechanism in NR for ultra-reliable low-latency communications. IEEE Netw. **32**(2), 41–47 (2018)
42. Mahmood, N.H., Lopez, M., Laselva, D., Pedersen, K., Berardinelli, G.: Reliability oriented dual connectivity for URLLC services in 5G new radio. In: 15th International Symposium on Wireless Communication Systems (ISWCS), pp. 1–6 (2018)
43. Lee, H., Vahid, S., Moessner, K.: Cognitive Radio-Oriented Wireless Networks. CrownCom. Lecture Notes of the Institute for Computer Sciences. Social Informatics and Telecommunications Engineering 291 (2019)
44. Kumar, N., Kumar, S., Subramaniam, K.: Achieving zero ms handover interruption in new radio with higher throughput using D2D communication. In: 2019 IEEE Wireless Com- munications and Networking Conference (WCNC), pp. 1–8 (2019)
45. Fan, B., He, Z., Wu, Y., He, J., Chen, Y., Jiang, L.: Deep learning empowered traffic offloading in intelligent software defined cellular V2X networks. IEEE Trans. Veh. Technol. **69**(11), 13328–13340 (2020)
46. Erel-Ozcevik, M., Canberk, B.: Road to 5G Reduced-latency: a software defined handover model for eMBB services. IEEE Trans. Veh. Technol. **68**(8), 8133–8144s (2019)

Range-free Localization by Optimization in Anisotropic WSN

Sumit Kumar, Neera Batra, and Shrawan Kumar

1 Introduction

Wireless sensor networks (WSNs) has its wide scope in the various domains like health services, industrial automation, environment monitoring, military services, smart agriculture, and disaster management. In WSNs, the sensor nodes form a network on ad hoc basis for outdoor operations, usually without any fixed infrastructure. The sensor nodes of WSN deployed in the field of interest and draw a random topological arrangement as an anisotropic network in reachably difficult outdoor operations, e.g., deep forest monitoring, stream monitoring, hill rift monitoring, and natural habitat monitoring. The sensor nodes in these outdoor operations are able to communicate their observations to the sink nodes like in IoT [1], however, without knowing the location of the sensors sharing the data carry no significance. Hence, WSN requires to estimate the location of their sensor nodes.

To find out the location of the sensor nodes, global positioning system (GPS) exists as an already well-known system. The GPS necessitates high energy and processing requirements due to its dependency on high-frequency signals [2]. Further, GPS hardware modules become an obligation to carry by the sensor nodes, which affect

S. Kumar (✉) · N. Batra
Department of Computer Science and Engineering, Maharishi Markandeshwar (Deemed To Be University), Mullana, Ambala 133207, India

N. Batra
e-mail: neera.batra@mmumullana.org

S. Kumar
Department of Computer Science and Engineering, Technology Education and Research Institute, Kurukshetra, Haryana 136119, India

S. Kumar
Department of Computer Science, Indira Gandhi National Tribal University, Regional Campus Manipur, Kangpokpi, India

M. Dua et al. (eds.), *Proceedings of the International Conference on Paradigms of Communication, Computing and Data Sciences*, Algorithms for Intelligent Systems,
https://doi.org/10.1007/978-981-16-5747-4_14

the scalability of WSN poorly. It implies that GPS cannot be applied abundantly in outdoor operations of WSN. Therefore, only a few sensor nodes are equipped with GPS modules, and known as anchor nodes. The remaining sensor nodes require to estimate their location from where they are sharing the field observations, and known as unknown nodes.

Therefore, the estimation of the location of the unknown nodes is known as localization. The localization has two broad approaches [3, 4]—range-based and range-free. The range-based localization affects WSN unfavorably, because it requires high energy, more processing power, and sophisticated additional hardware [5]. Therefore, in outdoor operations of WSN, the range-free approach of localization is a more practicable solution.

The range-free localization requires estimating the location of the unknown nodes with the help of anchor nodes. The intended unknown node approximates its distances d from the anchor nodes to estimate its location. The distance d requires obtaining the average hop size of anchor nodes and the hop counts to the anchor nodes as proposed by the benchmark algorithm— DV-hop [6]. Here, hop size and hop count both are erroneously estimated [7–10]. Thus, the distances d $\{d = d_i; \forall i \in n\}$ due to n anchor nodes are estimated imprecisely also [9, 11]. Hence, in this way, the localization of the unknown node of interest leads to a poor estimation.

In this paper, our main contribution includes the proposed algorithm RLO to localize the unknown nodes with high precision without any extra communicational and storage obligations over the sensor nodes. Further, the proposed algorithm maintains high scalability by requiring no extra hardware. The rest of this paper is presented in four sections. Section 2 presents a literature review of some of the existing range-free localization algorithms. Section 3 portrays the proposed work RLO in detail. Then, simulation results and discussion validate the proposed algorithm in Sect. 4. Finally, Sect. 5 concludes the paper with its future scope.

2 Literature Review

Predominantly, DV-hop algorithm is followed in the range-free algorithms in principle for estimative localizations [5, 11, 12]. The algorithms based on DV-hop algorithm improve accuracy of localization by using different approaches. In [13], Wu Wen et al. propose to apply particle swarm optimization (PSO) technique for the localization after improving the hop size. However, PSO is an iterative approach which requires more computations and, thus, more energy also [14]. Kaushik et al. [7] propose to perform localization by using three-hop distant anchor nodes only. However, the limitation of selecting a few anchor nodes out of the existing sensor nodes obstructs its applicability in the sparse network with low sensor node density. Similarly, Kanwar et al. in [8] suggest to apply runner root approach (RRA) for the localization. RRA is a heuristic approach provides an initial viable solution only which must be further improved to attain the final outcome [15, 16]. Therefore, the localization by RRA is an iterative approach which is not suitable for the outdoor battery-ridden disposable networks. Further, Tan et al. put forward genetic algorithm (GA) to apply for distance estimation between the node pairs [17]. The localization results improve by using GA. However, GA attracts high cost of computations and increase time complexity also. Another approach for the precise localization is proposed by Shen et al. [12] by improving the hop size of each anchor node and identifying the most suitable three anchor nodes with respect to the intended unknown node. It implies that the network size is a dominating factor to affect the energy consumption pattern.

In this way, various algorithms are proposed in the literature to estimate the location based on range-free approach. Most of these algorithms consume energy lavishly and, thus, have a resistive acceptance in WSN domain where a network is required to establish by some perishable, tiny size, and feathery weight, battery-ridden sensor nodes. This research gap motivates to propose RLO algorithm, which localizes the unknown nodes precisely by using linear programming without any iterative computations and, thus, keep energy requirements low.

3 Proposed Work: Range-free Localization by Optimization (RLO)

After the network establishment by updating hop tables, RLO requires to estimate distances d $\{d = d_i; \forall i \in n\}$ between n anchor nodes and the intended unknown node based upon hop size and hop counts. The distance equations for d_i suffer by some errors due to imprecise estimation using hop size and hop counts. The error term for every individual distance equation is defined in such a manner so that linear programming can be applied to estimate an optimized location for an unknown node of interest.

In this way, RLO has the following three steps:

(1) Distance estimation
(2) Defining distance error terms
(3) Optimum location estimation.

All the three steps are explained in detail, as follows:

3.1 Distance Estimation

In this step, the intended unknown node estimates its distance d_i from the anchor nodes by obtaining the hop size and hop counts with respect to every anchor node as suggested by DV-hop.

To approximate the hop size, let the n anchor nodes are located at $((x_1, y_1), (x_2, y_2), \ldots, (x_n, y_n))$, then the hop size H is estimated by the following Eq. (1) for each individual anchor node i as follows:

$$H_i = \sum_{j=1, i \neq j}^{n} \sqrt{(x_i - x_j)^2 + (y_i - y_j)^2} / \sum_{j=1, i \neq j}^{n} C_{ij}; \tag{1}$$

where H_i is the hop size of ith anchor node obtained using hop counts C_{ij} between the anchor nodes i and j.

To approximate the distances d_i between an intended unknown node, say b, and the anchor nodes $i\{1 \leq i \leq n\}$, we apply the following Eq. (2).

$$\left.\begin{array}{c} d_1 = H_1 \times C_{1b} \\ d_2 = H_2 \times C_{2b} \\ \vdots \\ d_n = H_n \times C_{nb} \end{array}\right\} \tag{2}$$

3.2 Defining Distance Error Terms

Any distance d_i is an approximation of the exact distance in between the maximum and minimum possible distance ranges due to the anchor node i. Therefore, we establish the $d_{i-\max}$ maximum and $d_{i-\min}$ minimum limits for d_i by Eq. (3) given below.

$$\left.\begin{array}{c} d_{i-\max} = H_{i-\max} \times C_{ib} \\ d_{i-\min} = H_{i-\min} \times C_{ib} \end{array}\right\}; \tag{3}$$

where $H_{i-\max} = H_i^j$ and $H_i{}^j = \sqrt{(x_i - x_j)^2 + (y_i - y_j)^2} / C_{ij}$; $\{\exists j | H_i{}^j \geq H_i{}^k; i \neq j \neq k; 1 \leq k \leq n\}$ is a hop size of the anchor node i with respect to an anchor node j. Similarly, $H_{i-\min} = H_i{}^j$ and $H_i{}^j = \sqrt{(x_i - x_j)^2 + (y_i - y_j)^2} / C_{ij}$; $\{\exists j | H_i{}^j \leq H_i{}^k; i \neq j \neq k; 1 \leq k \leq n\}$ is a hop size of the anchor node i with respect to an anchor node j.

Equation (3) implies that $d_{i-\min} \leq d_i \leq d_{i-\max}$. Therefore, let d_i differ by an error ∂_i from the exact distance with an upper bound of $(d_{i-\max} - d_{i-\min})$. Now, by applying error ∂_i, in Eq. (2), we obtain another Eq. (4), as below:

$$\left.\begin{array}{c} d_1 + \partial_1 = H_1 \times C_{1b} \\ d_2 + \partial_2 = H_2 \times C_{2b} \\ \vdots \\ d_n + \partial_n = H_n \times C_{nb} \end{array}\right\} \tag{4}$$

3.3 Optimum Location Estimation

The distances obtained by Eq. (4) assist to derive the coordinates (x, y) of the unknown node b from the known coordinates (x_i, y_i) of the anchor nodes $i \epsilon n$, by applying following Eq. (5).

$$\left.\begin{array}{c}\sqrt{(x_1-x)^2+(y_1-y)^2}=d_1+\partial_1\\ \sqrt{(x_2-x)^2+(y_2-y)^2}=d_2+\partial_2\\ \vdots\\ \sqrt{(x_n-x)^2+(y_n-y)^2}=d_n+\partial_n\end{array}\right\} \quad (5)$$

In Eq. (5), after squaring and then dividing both sides by $D^2\{D = \sum_{i=1}^{n} d_i\}$ followed by subtracting the last nth equation from the rest of its $(n-1)$ equations, we obtain another Eq. (6) by discarding the trivial terms $\left((\partial_i{}^2 - \partial_n{}^2)/D^2\right)$ with an assumption that $D \gg \partial_n$. The negligence of trivial terms helps to reduce error propagation also.

$$\left.\begin{array}{l}p_1x+q_1y+r_1\partial_1+s\partial_n=D_1\\ p_2x+q_2y+r_2\partial_2+s\partial_n=D_2\\ \vdots\\ p_{n-1}x+q_{n-1}y+r_{n-1}\partial_{n-1}+s\partial_n=D_{n-1}\end{array}\right\}; \quad (6)$$

where the terms p_i, q_i, r_i, s, and D_i are defined for all $i = 1, 2, \ldots, n-1$ as-

$$p_i = 2(x_n - x_i)/D^2$$

$$q_i = 2(y_n - y_i)/D^2$$

$$r_i = -2d_i/D^2$$

$$s = 2d_n/D^2$$

$$D_i = \left(\left(\left(x_n{}^2 + y_n{}^2\right) - \left(x_i{}^2 + y_i{}^2\right) + d_i{}^2\right)/D^2\right) - 1$$

Equation (6) can be written by Eq. (7) in a matrix form as shown below:

$$AB = C; \quad (7)$$

where A, B, and C are as follows:

$$A = \begin{bmatrix} p_1 & q_1 & r_1\, 0 \cdots 0 & s \\ p_2 & q_2 & 0\; r_2 \ldots 0 & s \\ \vdots & \vdots & \vdots\;\vdots \cdots 0 & \vdots \\ p_{n-1} & q_{n-1} & 0\; 0\; \cdots r_{n-1} & s \end{bmatrix}_{(n-1)\times(2+n-1+1)} \qquad B =$$

$\begin{bmatrix} x\; y\; \partial_1\; \partial_2\; \cdots\; \partial_n \end{bmatrix}'_{1\times(2+n)}$, and $C = \begin{bmatrix} D_1\; D_2 \cdots D_{n-1} \end{bmatrix}'_{1\times(n-1)}$.

Equation (7) does not provide a unique solution for the variables x and y due to under-deterministic nature of the set of equations. Therefore, by applying linear programming to solve Eq. (7) with an objective to minimize the magnitude of the error terms ∂_i; $\forall i = 1, 2, \ldots, n$, we get another Eq. (8), as follows:

$$\left.\begin{aligned} &\min \sum_{i=1}^{n} |\partial_i| \\ &such\ that \\ &AB = C \end{aligned}\right\} \tag{8}$$

The objective of Eq. (8) can be achieved by rewriting it in another form by the following Eq. (9):

$$\left.\begin{aligned} &\min\ f'B \\ &such\ that \\ &AB = C \\ &lb \le B \\ &ub \ge B \end{aligned}\right\} \tag{9}$$

where f, lb, and ub are as defined below:

$f = [0\; 0\; 1\; 1 \cdots 1]'_{1\times(2+n)}$, $lb = [-\infty\; -\infty\; 0\; 0 \cdots 0]'_{1\times(2+n)}$,

And, by using Eq. (3), we get ub as

$$ub = \begin{bmatrix} +\infty\; +\infty\; (d_{1-\max} - d_{1-\min})\; (d_{2-\max} - d_{2-\min})\; \ldots\; (d_{n-\max} - d_{n-\min}) \end{bmatrix}'_{1\times(2+n)}$$

Hence, the solution of Eq. (9) yields the values of x and y to localize the unknown node b by optimizing errors.

Equation (9) is significant to analyze the computational cost of the proposed approach. Karmarkar [18] has established that the solution of such equations yields a computational cost of the order of $n^{3.5}$ and established by other localization algorithm ODR [9] also.

4 Simulation Validation

To examine the proposed RLO, simulation is carried by using MATLAB R2014a. The experiments are performed by observing the variations in percentage of anchor nodes, communication range, and density of sensor nodes; on the localization error (L) as suggested [10, 19] by Eq. (10)

$$L = (1/Ru)\left(\sum_{i=1}^{u}\sqrt{(x_{ti} - x_{ei})^2 + (y_{ti} - y_{ei})^2}\right) \times 100\% \tag{10}$$

where (x_{ti}, y_{ti}) and (x_{ei}, y_{ei}) are the true, expected coordinates of the intended unknown node, respectively, R is the communication range, and u is the total number of unknown nodes.

The proposed RLO is analyzed with comparison to DV-hop [6], IDV [12], and ODR [9] algorithms by executing Eq. (10) 500 times for each simulation experimental setup in the anisotropic arrangement of the sensor nodes. The following three experiments are performed:

Experiment 1: localization error vs. percentage of anchor nodes.

Experiment 2: localization error vs. communication range.

Experiment 3: localization error vs. density of sensor nodes.

All the three experiments are performed by considering the following values of their network parameters as shown in Table 1.

4.1 *Experiment 1: Localization Error versus Percentage of Anchor Nodes*

This experiment studies the localization performance with respect to the variation in the percentage of the anchor nodes. The sensor nodes keep the communication

Table 1 Simulation parameters

Parameters	Constant value	Variable value(s)
Area covered (A)(A)(A)(A)(A)	100×100 (m^2)	Not Applicable (NA)
Communication range (R)	NA	15, 20, 25, 30 (m)
Total number of nodes (N)	NA	200, 230, ..., 500
Anchor nodes (m)	NA	5, 10, ..., 25% of$N$
Unknown nodes (u)	NA	$N - m$

(m) stands for meter.

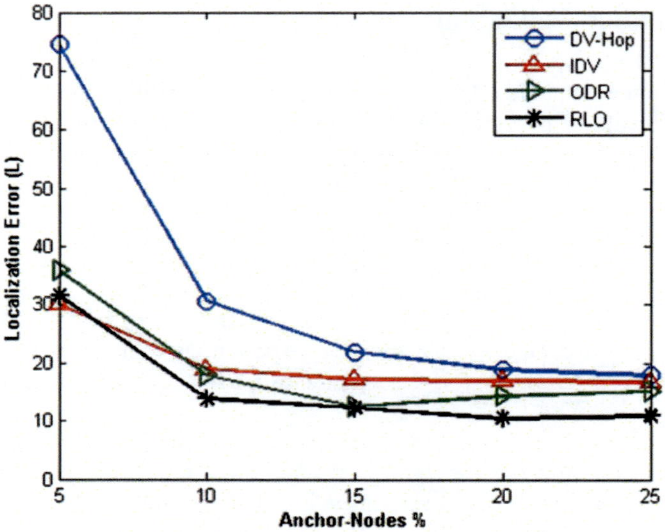

Fig. 1 Localization error versus percentage of anchor nodes

range at 15 units only where the anchor nodes vary as 5, 10, …, 25% of the total 200 sensor nodes. The results of Experiment 1 are shown in Fig. 1.

Figure 1 shows that RLO performs better than DV-hop and IDV algorithms. This improvement is gained due to availability of more reference points in terms of the anchor nodes with rise in the percentage of anchor nodes. Thereby, RLO generates proportionately more distance equations to optimize the location for the unknown nodes, and thus, more precise locations are obtained.

On average, 33, 20, 19, and 16% localization errors are achieved by DV-hop, IDV, ODR, and RLO, respectively, in this experiment.

4.2 Experiment 2: Localization Error versus Communication Range

In Experiment 2, the localization performance is analyzed with respect to the variation in the communication range of the sensor nodes. The sensor nodes vary the communication range with 15, 20, 25, and 30 units where the number of anchor nodes is kept constant at 10%, i.e., 20 of the total 200 sensor nodes. The results of Experiment 2 are shown in Fig. 2.

The results drawn in Fig. 2 show that RLO is able to utilize the communication range in a better way than the remaining two other algorithms. The increase in communication range allows reducing the hop counts for the estimation of the

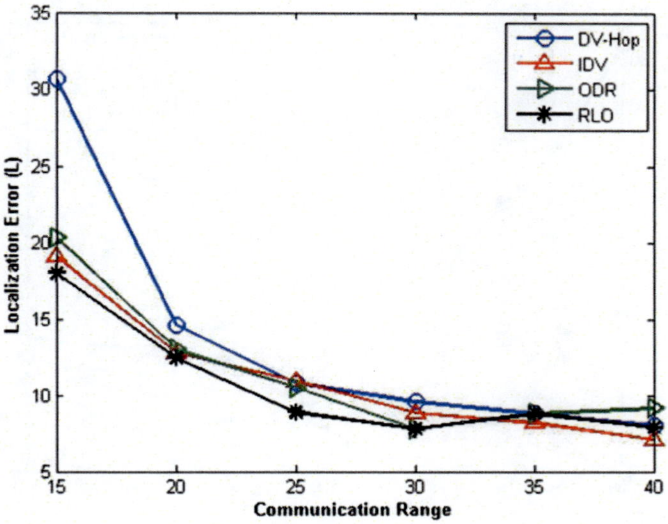

Fig. 2 Localization error versus communication range

distance equations for RLO. The lesser number of hop counts results in more precise distance estimation in between the node pairs, as shown by Table 2 also.

Hence, the localization by RLO is improved with the increase in the communication range.

On average, 16, 13, 12, and 11% localization errors are achieved by DV-hop, IDV, ODR, and RLO, respectively, in this experiment.

Table 2 Relationship among communication range, hop counts, and localization error

Communication range	Hop counts (mean)	Localization error			
		DV-hop	IDV	ODR	RLO
15	4.7	30.68687	19.07862	20.30384	17.99385
20	4.25	14.62948	12.79845	13.02268	12.4634
25	3.15	10.82739	10.92178	10.48949	8.886405
30	3.05	9.564345	8.854545	7.765254	7.765254
35	3.3	8.77704	8.17675	8.815146	8.815146
40	3.2	7.994027	7.061957	9.151466	7.861259

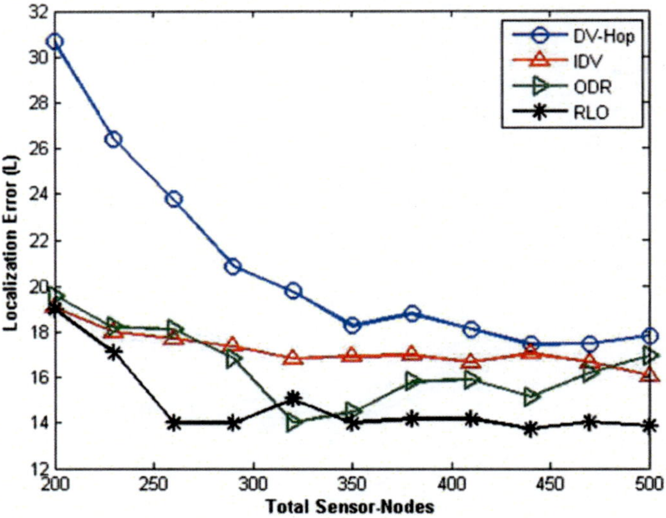

Fig. 3 Localization error versus density of sensor nodes

4.3 Experiment 3: Localization Error versus Density of Sensor Nodes

Here, the localization ability is analyzed with respect to the variation in the density of the sensor nodes. The density varies by raising the number of sensor nodes in the range of 200, 230, …, 500 by keeping the area of observation constant at 100 × 100 sq. units. The sensor nodes keep the communication range at 15 units where the number of anchor nodes is kept at 10% of the total sensor nodes. The results of Experiment 3 are shown in Fig. 3.

This experiment results show that RLO is more consistent than DV-hop and IDV algorithms. As the number of total sensor nodes increases, the availability of the anchor nodes also increases to localize the unknown nodes. Therefore, localization accuracy improves initially and then saturated.

On average, 21, 17, 16, and 15% localization errors are achieved by DV-hop, IDV, ODR, and RLO, respectively, in this experiment.

5 Conclusion

The localization of the unknown nodes in a range-free estimative approach is imprecise due to poor distance estimation between the node pairs. However, the proposed algorithm RLO identified and defined the distance error factors to obtain a precise location of the unknown nodes. The localization problem is transformed in the optimization problem using linear programming in the proposed RLO. This transformation yields the coordinates of the unknown nodes with less localization errors. RLO precisely localize the unknown nodes without any extra hardware, computations, and memory requirements. Further, RLO is an improved algorithm to localize the unknown nodes with its competitors because on average, RLO contributes 14% localization error, whereas DV-hop, IDV, and ODR contribute 20, 17, and 16% localization errors, respectively. Hence, RLO is better to localize the unknown nodes by DV-hop, IDV, and ODR by 6, 3, and 2%, respectively, on average.

In the future, RLO will be extended to the mobile wireless sensor networks (MWSNs) by analyzing a scope to have the mobility parameters.

References

1. Gautam, A.K., Kumar, R.: A comprehensive study on key management, authentication and trust management techniques in wireless sensor networks. SN Appl. Sci. **3**, 50 (2021). https://doi.org/10.1007/s42452-020-04089-9
2. Kudłacik, I., Kapłon, J., Lizurek, G., Crespi, M., Kurpiński, G.: High-rate GPS positioning for tracing anthropogenic seismic activity: the 29, January 2019 mining tremor in Legnica-Głogów Copper District, Poland . Measurement **168** (2021). https://doi.org/10.1016/j.measurement.2020.108396
3. Mohar, S.S., Goyal, S., Kaur, R.: A survey of localization in wireless sensor network using optimization techniques. In: 2018 4th International Conference on Computing Communication and Automation (ICCCA), Greater Noida, India, pp. 1–6 (2018). https://doi.org/10.1109/CCAA.2018.8777624
4. Rai, S., Varma, S.: Localization in wireless sensor networks using rigid graphs: a review. Wirel. Pers. Commun. **96**, 4467–4484 (2017). https://doi.org/10.1007/s11277-017-4397-7
5. Shakshuki, E., Elkhail, A.A., Nemer, I., Adam, M., Sheltami, T.: Comparative study on range free localization algorithms. Procedia Comput. Sci. **151**, 501–510 (2019). ISSN 1877–0509, https://doi.org/10.1016/j.procs.2019.04.068
6. Niculescu, D., Nath, B.: Ad hoc positioning system (APS). In: Proceedings of IEEE Global Telecommunications Conference, vol. 5, pp. 2926–2931 (2001). https://doi.org/10.1109/GLOCOM.2001.965964
7. Kaushik, A., Lobiyal, D.K., Kumar, S.: Improved 3-dimensional DV-hop localization algorithm based on information of nearby nodes. Wirel. Netw. (2021). https://doi.org/10.1007/s11276-020-02533-7
8. Kanwar, V., Kumar, A.: DV-Hop-based range-free localization algorithm for wireless sensor network using runner-root optimization. J. Supercomput. **77**, 3044–3061 (2021). https://doi.org/10.1007/s11227-020-03385-w
9. Optimized distance range free localization algorithm for WSN. Wireless Pers. Commun. **117**, 1879–1907 (2021). https://doi.org/10.1007/s11277-020-07950-7
10. An enhanced DV- Hop localization algorithm for wireless sensor network. Int. J. Wirel. Netw. Broadband Technol. **2**(2), 16–35 (2012)
11. Nemer, I., Sheltami, T., Shakshuki, E., et al.: Performance evaluation of range-free localization algorithms for wireless sensor networks. Pers. Ubiquit. Comput. (2020). https://doi.org/10.1007/s00779-020-01370-x
12. Shen, S., Yang, B., Qian, K., She, Y., Wang, W.: On improved DV-hop localization algorithm for accurate node localization in wireless sensor networks. Chin. J. Electron. **28**(3), 658–666 (2019). https://doi.org/10.1049/cje.2019.03.013

13. Wen, W., Wen, X., Yuan, L., et al.: Range-free localization using expected hop progress in anisotropic wireless sensor networks. J. Wirel. Commun. Netw. **2018**, 299 (2018). https://doi.org/10.1186/s13638-018-1326-8
14. Li, M., Du, W., Nian, F.: An adaptive particle swarm optimization algorithm based on directed weighted complex network. Math. Probl. Eng. **2014**, 7 (2014). Article ID 434972, https://doi.org/10.1155/2014/434972
15. Frankovič, B., Budinská, I.: Advantages and disadvantages of heuristic and multi agents approaches to the solution of scheduling problem. IFAC Proc. Vol. **33**(13), 367–372 (2000). ISSN 1474–6670, https://doi.org/10.1016/S1474-6670(17)37217-8
16. Merrikh-Bayat, F.: The runner-root algorithm. Appl. Soft Comput. **33**, 292–303 (2015). https://doi.org/10.1016/j.asoc.2015.04.048
17. Tan, R., Li, Y., Shao, Y., et al.: Distance mapping algorithm for sensor node localization in WSNs. Int. J. Wirel. Inf. Netw. **27**, 261–270 (2020). https://doi.org/10.1007/s10776-019-00456-5
18. Karmarkar, N.: A new polynomial-time algorithm for linear programming. Combinatorica **4**, 373–395 (1984). https://doi.org/10.1007/BF02579150
19. Kumar, S., Kumar, S., Batra, N.: Optimized distance range free localization algorithm for WSN. Wireless Pers. Commun. **117**, 1879–1907 (2021). https://doi.org/10.1007/s11277-020-07950-7

Polarized Diversity Characteristics Dual-Band MIMO Antenna for 5G/WLAN Applications

Pachiyaannan Muthusamy, Krishna Chennakesava Rao Madaka, P. Krishna Chaitanya, and N. Srikanta

1 Introduction

Nowadays to improve the gain and polarization, the MIMO antenna is utilizing more in modern communication. The dual-band and dual-polarized (DBDP) antenna is essentially used for 5G, WLAN and mobile network application. A diversity characteristic is to reduce the effect of multipath fading because operating bandwidth can cover two different frequency bands and the two orthogonal polarizations. To improve the signal strength and receive the better signal from different direction in ground and space communication, the polarization diversity characteristics antenna is still challenging and focusing toward to improve the gain in DBDP MIMO antenna.

There are many DBDP antenna designed for recent years and reported in two categories. One is based on a single-feed dual-band antenna, and the dual polarization operation is achieved by adding an orthogonal feed to the antenna structure [1–6] which suggests that the resultant DBDP antenna only has two input ports. The other is based on a stacked structure composed of two antenna elements operating at different frequencies, and each element can provide dual orthogonal polarizations through two input ports. In this paper, a dual-band dual-polarized MIMO antenna is proposed by employing single line feed for each element. Author [7] reported 50 × 60 mm antenna 2.45/5.8 GHz with isolation improvement of > −20 dB. By using single-feed and monopole antenna structure, it achieved isolation. In [8], 60 × 95 mm antenna is reported for 2.50–3.25/3.70–4.20 GHz applications. By using C and T shape fed slot >18 dB, isolation is achieved. The coupled feed and two parallel folded monopole antenna with dimension of 20 × 56 mm is reported in [9]. The isolation >-13 dB has been achieved with the help of coupled feed techniques. In another MIMO antenna

P. Muthusamy (✉) · K. C. R. Madaka · P. K. Chaitanya · N. Srikanta
Department of Electronics and Communication Engineering, Advanced RF Microwave and Wireless Communication Laboratory, Vignan's Foundation for Science Technology and Research (Deemed to be University), Vadlamudi, Andhra Pradesh, India

M. Dua et al. (eds.), *Proceedings of the International Conference on Paradigms of Communication, Computing and Data Sciences*, Algorithms for Intelligent Systems,
https://doi.org/10.1007/978-981-16-5747-4_15

20 × 46 mm by using two port and four port, the dual band is achieved [10]. From the above literature, the necessity of dual-band dual polarization is studied. In this work, [11] tried to implement stacked structure consisting of lower part as patch and upper part layer as crossed dipole antenna. The experimental results 10 dB-return bands from 2.4 to 2.48 GHz and 5.15 to 5.85 GHz are used for WLAN applications. Another author [12] discussed MIMO integrated with EBG structure which is useful for WLAN applications. In [13], a compact size with the radiating portion of 0.3 λ_0 × 0.17 λ_0 antenna is proposed for sub-6 GHz wireless applications. The frequency bands 3.29–3.63 GHz and 4.3–5.2 GHz are used for WLAN and 5G applications. A dual-band dual-mode and dual-polarized antenna is reported for WLAN applications [14, 15].

In this way, the proposed MIMO antenna is designed for 5G and WLAN applications. The two elements are kept at orthogonal direction and improve isolation; the decoupling slot is placed at ground structure. The detailed antenna design and simulation result are reported in forthcoming sections.

2 Antenna Geometry

The dimensions of the ground plane are 30 mm × 58 mm, and the substrate dimension is similar to that of ground, but the FR4 substrate height is 1.6 mm with permittivity of 4.4 dielectric constant. The nearly square patch is chosen as the radiator. The dimension of patch is 18.8 mm × 20 mm. The dimensions of the feed line 4.5 mm length and the width of the excitation are also 0.3 mm. There are rectangular diagonally opposite cuts at the boundary of the patch. These are providing the band around 5.6 GHz. The return loss value is improved at WLAN band that is around 5.6 GHz is provided. The two-element patch is orthogonally placed on same substrate and excitation by line fed techniques. The geometrical representation of estimated antenna structure is shown in Fig. 1. To improve isolation between antennas, the decoupling slot is introduced in ground structure with dimension of 27 × 1 mm as shown in Fig. 2. The single-element dimension and design steps are shown in Figs. 3 and 4.

3 Parametric Study

A nearly square patch has been selected at 3.5 GHz. The dimensions of this nearly square patch have been selected for 3.5 GHz. The offset line feeding technique is used for the excitation of the patch. In Fig. 2, Ant-1 and Ant-2 represent evolution of creating two rectangular cuts orthogonal to each other which provides 5.6 GHz resonance, and these are improving the impedance matching further. The non-diagonal cut in the Ant-3 is responsible for circular polarization. The axial ratio of less than 3 dB is obtained. The entire structure has been developed in ANSYS HFSS software. The return loss curve of single-element antenna is shown in Fig. 5.

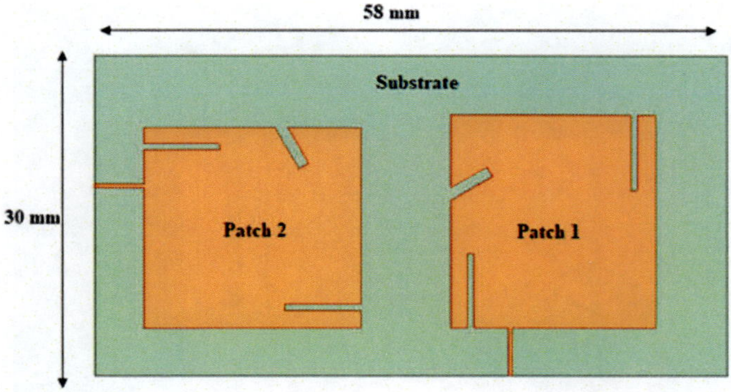

Fig. 1 Geometrical representation of proposed MIMO antenna

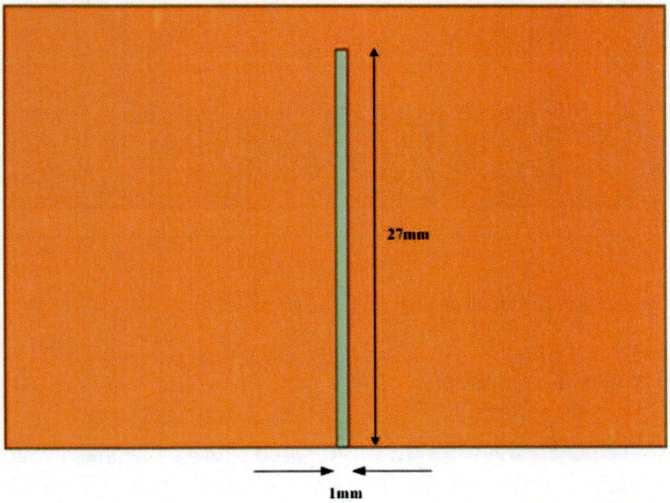

Fig. 2 Ground structure of proposed MIMO antenna

For the antennas Ant-1, 2 and 3, dual-band nature is accomplished around 3.4 GHz and 5.6 GHz. This indicates that the impedance has been matched at two different frequencies 3.4 GHz and 5.6 GHz, so dual-band nature is obtained. For the proposed antenna, the return loss is around -30 dB and -20 dB which are accepted values. The 5G band 3.34–3.45 GHz is covered.

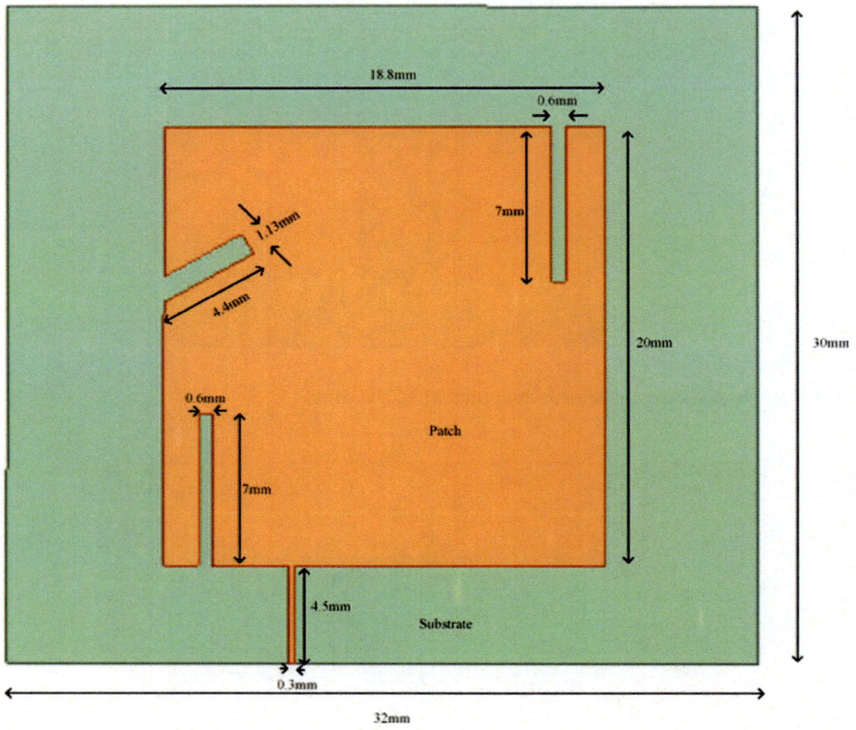

Fig. 3 Geometrical representation of single-element antenna

4 Results and Discussion

This section included various results of proposed MIMO antenna. The return loss curve of S11 and S22 is shown in Fig. 6, which indicates that similar results appeared at both elements, and at -10 dB impedance bandwidth lower band 3.34–3.45 GHz and higher band 5.55–5.7 GHz, the return loss value is showing around –20 dB which is acceptable. In MIMO techniques, the important parameter is isolation improvement. To estimate, the isolation between antenna isolation should be >–15 dB. In this work with the help of decoupling slot in ground structure, the isolation S21 curve showing > –15 dB has been achieved throughout lower band and higher band. The isolation curve is reported in Fig. 6

The VSWR curve is shown in Fig. 7. The VSWR value is less than 2 for two different frequencies which leads to dual-band nature. Figure 8 represents the radiation plot of E field and H field of 3.4 and 5.6 GHz which observe nearly omnidirectional radiation pattern of both bands.

The total gain radiation plot of designed work is shown in Fig. 9 below having moderate gain at 2.6 dB.

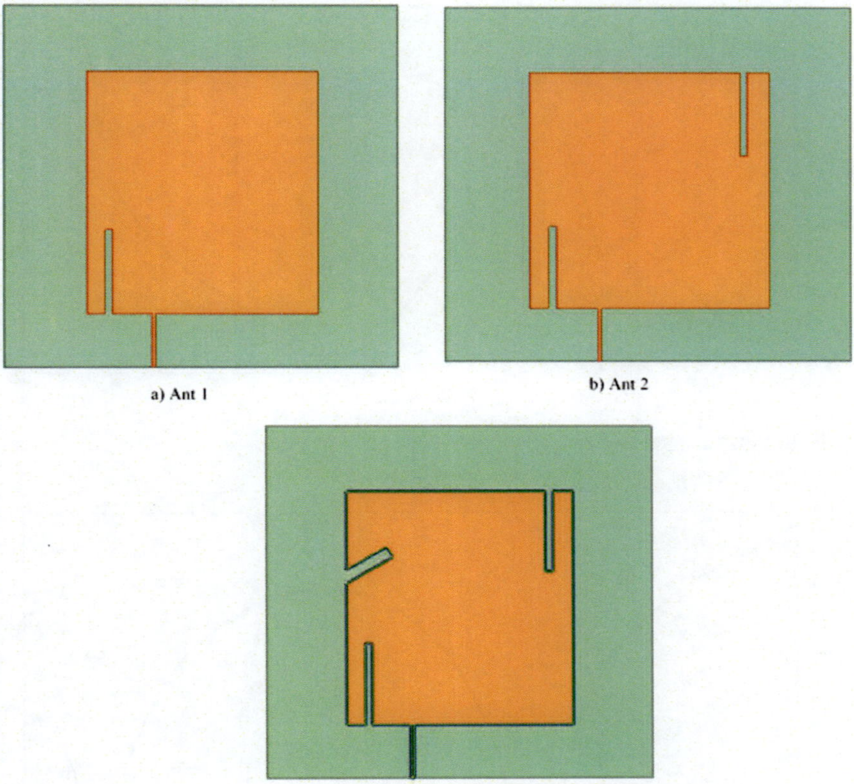

a) Ant 1

b) Ant 2

c) Ant 3 (Proposed Antenna)

Fig. 4 Design steps of single-element antenna

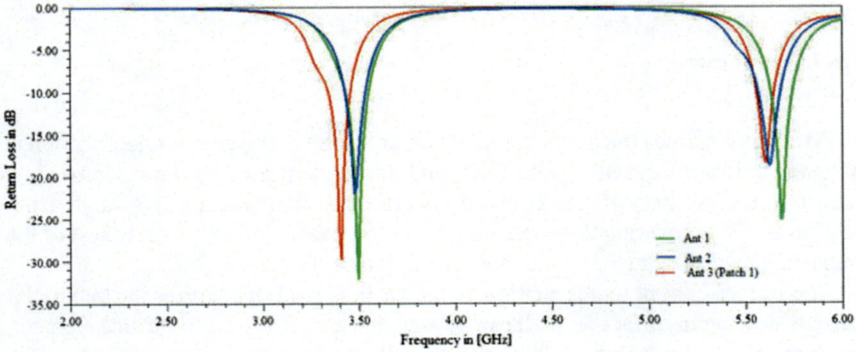

Fig. 5 Return loss curve single-element antenna

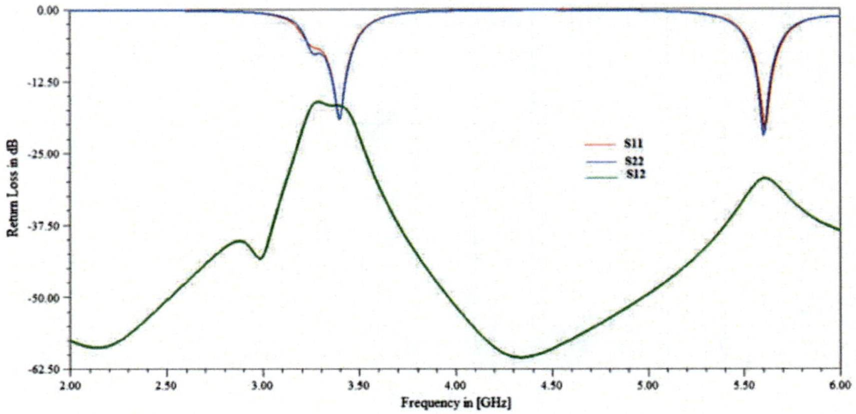

Fig. 6 S11, S22 and S21 curve

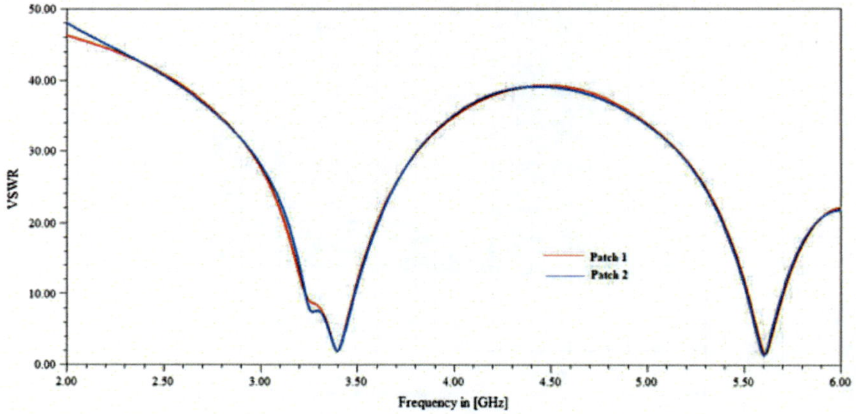

Fig. 7 VSWR curve

Axial ratio defines the polarization nature of antenna. The circular polarization is obtained at lower band 3.34–3.45 GHz, and this is confirmed by the value of axial ratio 1.70 dB. At theta 0°, the axial ratio is estimated in this graph. The angle from –50° to + 50° is covering the circular polarization as shown in Fig. 10. In rest of the frequency band, i.e., higher band, horizontal polarization is available.

The surface current representation is shown in Fig. 11. It analyzes the maximum current flowing in surface level. It can be seen that at patch 1, i.e., right side element, the maximum current flows at both corner side except center patch similarly patch 2 i.e., left side the maximum current flows at entire patch area except that center of patch.

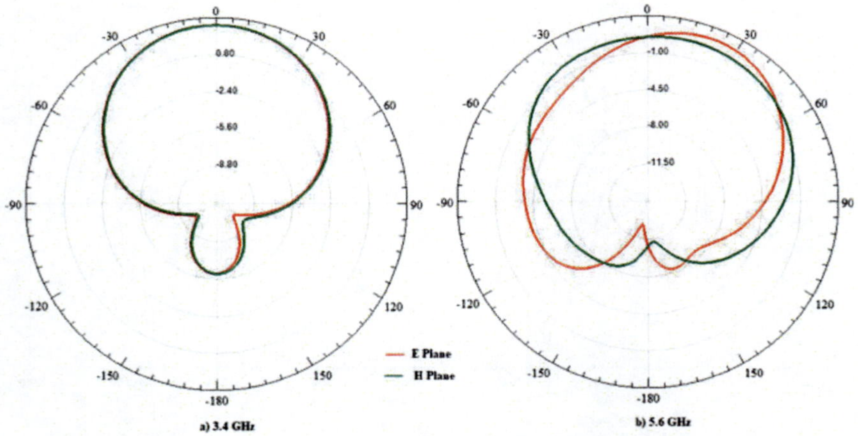

Fig. 8 Radiation plot at **a** 3.4 GHz and **b** 5.6 GHz

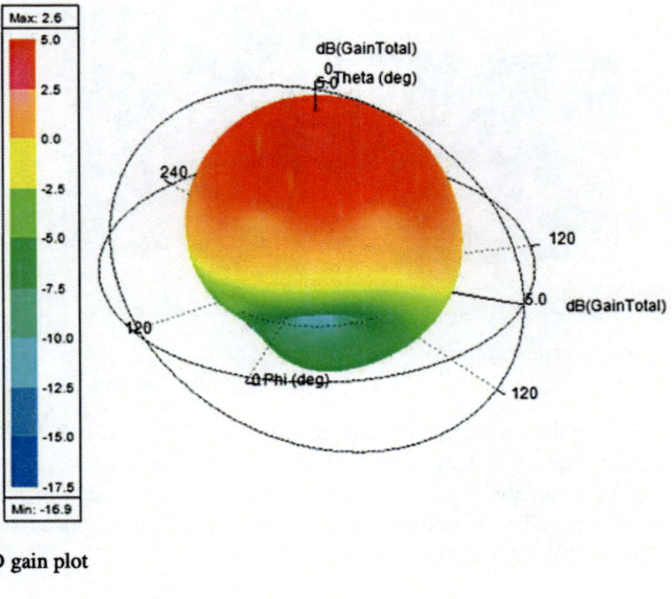

Fig. 9. 3D gain plot

5 Conclusion

The two-element MIMO antenna with dual-band dual polarization diversity characteristics has been designed in this work. This low profile and low-cost antenna having dimension is 30 × 58 mm, and single-element size is 18.8 × 20 mm which is verified and simulated. This is very much essential for modern communication world.

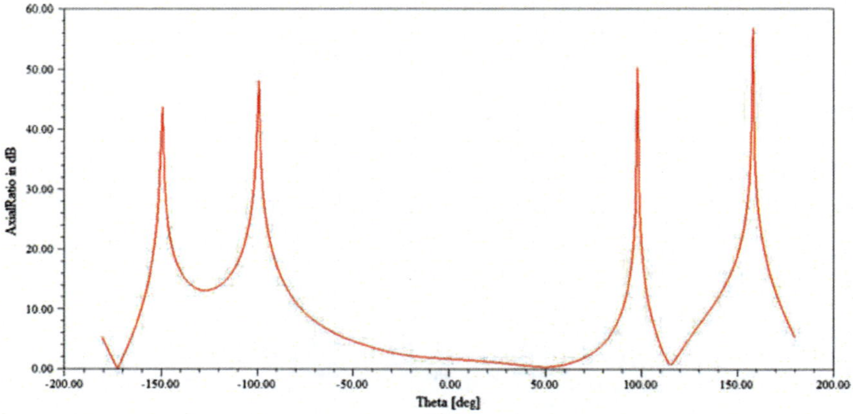

Fig. 10 Theta 0° versus axial ratio in dB

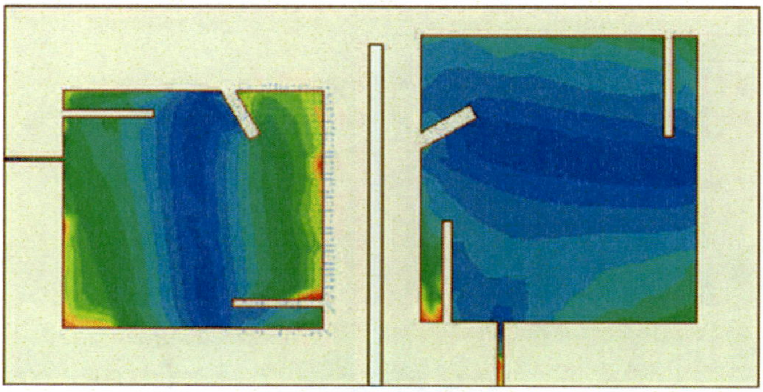

Fig. 11 Surface current representation

At lower band circular polarized −10 dB impedance, bandwidth is 3.34–3.45 GHz which has been achieved, and higher band 5.55–5.7 GHz linear polarization has been achieved. The two-element isolation between antenna is achieved > −15 dB, and it has omnidirectional radiation pattern and moderate gain. This proposed structure is suitable for 5G and WLAN applications.

References

1. He, Y., Pan, Z., Cheng, X., He, Y., Qiao, J., Tentzeris, M.M.: A novel dual-band, dual-polarized, miniaturized and low-profile base station antenna. IEEE Trans. Antennas Propag. **63**(12), 5399–5408 (2015)
2. Sharma, D.K., Kulshrestha, S., Chakrabarty, S.B., Jyoti, R.: Shared aperture dual band dual polarization microstrip patch antenna. Microw. Opt. Technol. Lett. **55**(4), 917–922 (2013)
3. Li, J., Yang, S., Li, B., Nie, Z.: A low profile dual-band dual-polarized patch antenna array with integrated feeding network for pico-base station applications. Microw. Opt. Technol. Lett. **56**(7), 1594–1600 (2014)
4. Kaboli, M., Abrishamian, M.S., Mirtaheri, S.A., Aboutorab, S.M.: High-isolation XX-polar antenna. IEEE Trans. Antennas Propag. **60**(9), 4046–4055 (2012)
5. Sun, Y.X., Leung, K.W.: Dual-band and wideband dual-polarized cylindrical dielectric resonator antennas. IEEE Antennas Wirel. Propag. Lett. **12**, 384–387 (2013)
6. Lai, D.Y., Chen, F.C.: A compact dual-band dual-polarized patch antenna for 1800/5800 MHz cellular/WLAN systems. Microw. Opt. Technol. Lett. **49**(2), 345–349 (2007)
7. Ling, X., Li, R.: A novel dual-band MIMO antenna array with low mutual coupling for portable wireless devices. IEEE Antennas Wirel. Propagation Lett. **10**, 1039–1042 (2011)
8. Cui, S., Liu, Y., Jiang, W., Gong, S.X.: Compact dual-band monopole antennas with high port isolation. Electron. Lett. **47**(10), 579–580 (2011)
9. Wu, Y.T., Chu, Q.X.: Dual-band multiple input multiple output antenna with slitted ground. IET Microwaves Antennas Propag. **8**(13), 1007–1013 (2014)
10. Wang, S.M., Hwang, L.T., Lee, C.J., Hsu, C.Y., Chang, F.S.: MIMO antenna design with built-in decoupling mechanism for WLAN dual-band applications. Electron. Lett. **51**(13), 966–968 (2015)
11. Row, J.-S., You-Jhih, H.: Dual-band dual-polarized antenna for WLAN applications. Microwave Opt. Technol. Lett. **60**(1), 260–265 (2018)
12. Zhang, X.Y., Zhong, X., Li, B., Yu, Y.: A dual-polarized MIMO antenna with EBG for 5.8 GHz WLAN application. Progress Electromagnetics Res. **51**, 15–20 (2015)
13. Ishteyaq, I., Masoodi, I.S., Muzaffar, K.: A compact double-band planar printed slot antenna for sub-6 GHz 5G wireless applications. Int. J. Microwave Wirel. Technol. 1–9 (2020)
14. Lin, J., Qian, Z., Cao, W., Shi, S., Wang, Q., Zhong, W.: A low-profile dual-band dual-mode and dual-polarized antenna based on AMC. IEEE Antennas Wirel. Propag. Lett. **16**, 2473–2476 (2017)
15. Paracha, K.N., Rahim, S.K.A., Soh, P.J., Kamarudin, M.R., Tan, K.G., Lo, Y.C., Islam, M.T.: A low profile, dual-band, dual polarized antenna for indoor/outdoor wearable application. IEEE Access **7**, 33277–33288 (2017)

A Defected Ground Structure Microstrip Antenna for Smart Healthcare Applications

Sujit Tripathy, Pranaba K. Mishro, Bajra Panjar Mishra, and V. Mukherjee

1 Introduction

Detection of tumor cells can be ranked as the most commonly encountered disease, affecting nearly two million individuals per annum. Early diagnosis strategies enable people for better treatment in the early stage and improve the success rate of the treatment. In these days, imaging frameworks are popular in the methods available for the detection of tumor cells. A differentiation analysis of different human organ tumor causing tissue, i.e., benign and malignant cells, is needed. Microstrip patch antenna enables us to fit it into any outer part of the human body which makes it more challenging for the researcher to implement in advanced clinical/medical applications. The properties and advantage of flexible antenna like compact design, low profile, low weight and better gain suggest its implications in smart healthcare system [1].

Different designs for patch antenna are there, and it can be observed that the different types of slots affect the working of the antenna in many ways. Anuja et al. [2] used elliptical patch to design a probe fed antenna and realized U-shaped slot in the patch to increase the bandwidth of the antenna. The evaluated return loss of the antenna shows that the antenna with the single U-slot is yielding a return loss of −11 dB at 6.30 GHz, and antenna with double U-slot is giving the return loss of −11 dB at 4.6–6.25 GHz. Baudha et al. [3] proposed and fabricated an antenna where the bandwidth is increased by realizing a curved slot in the patch. Poovizhi et al. [4] designed a frequency reconfigurable antenna using inset feed technique, and it has been observed that the matching condition is greatly affected by the inset depth and slot width and position to the patch center. The designed antenna is giving a stable

S. Tripathy · B. P. Mishra · V. Mukherjee
SU Institute of Information Technology, Burla, India

P. K. Mishro (✉)
VSS University of Technology, Burla, India

M. Dua et al. (eds.), *Proceedings of the International Conference on Paradigms of Communication, Computing and Data Sciences*, Algorithms for Intelligent Systems,
https://doi.org/10.1007/978-981-16-5747-4_16

radiation pattern. Bakr et al. [5] proposed an antenna with filtering response consisting of four parasitic gap coupled elements, two L-shaped slots and two rectangular slots. The antenna is optimized by manipulating the substrate height, and the filtering response is also achieved. The designed antenna shows great performance for UWB applications.

Çalışkan et al. [6] designed an antenna model using the inset feed technique. Here, FR4 is used as the substrate material. The antenna performance is increased by slotting on the patch and modifying the ground plane, and the quality is improvised. The model is simulated by changing the ground plane and patch structure. The structure is having the same ground plane and microstrip patch with an I-shaped slot. Bohra et al. [7] here considered radar-based microwave imaging for cancer identification method in an early stage. By using fork feed technique and FR4 as the substrate material, a double-sided UWB antenna is designed to work in the frequency range of 5.8–8.8 GHz. Kahwaji et al. [8] proposed an antenna by edge feeding technique and keeping the feed line as 6 mm long and 1.6 mm wide. FR4 epoxy is used as the substrate material with the length of 25 mm and width of 20 mm and thickness of 1.6 mm. FR4 epoxy is used as this is very popular and is easily obtainable. A breast model as a combination of three layers, i.e., skin layer, fat layer and the tumor, has been constructed. The antenna is simulated 6 times over the designed breast model without tumor and with tumor of different sizes.

Ahmed and Imran [9] suggested a cylindrical patch antenna using slotted triangular flared (STF) patch. A curved frequency selective structure (FSS) is installed in the internal surface of the cylindrical substrate material to achieve an in-phase reflection for a wide range of frequencies. Alsharif et al. [10] proposed a compact wearable patch antenna design with improved bandwidth and ultra-wideband property for breast cancer detection. The proposed antenna is showing the operating frequency range from 1.6 to 11.2 GHz. As the antenna is designed as a wearable device, it can be easily integrated into attire. A perceptible methodology of the antenna elements is chosen. The amplitude of the transmitted signal and the received signal of breast model with and without tumor is compared.

According to the theory discussed above, four different microstrip antennas are designed, having the same patch structure and same ground plane but with different substrate material. The permittivity of the material is usually taken in comparison with the free space. This is also called as relative permittivity or dielectric constant ε_r. The substrates with unique dielectric constants offer specific execution of the antenna as shown in Table 1. To support wearability, four different substrate materials those are

Table 1 Substrate materials used for designing wearable microstrip antenna

Material	Thickness (mm)	Permittivity (F/m)
Panama fabric	1.6	2.12
Fleece	1.524	2.22
Dacron	1.624	3
Pure cotton	1.6	1.6

also famous in clothing, i.e., Panama fabric, Pure Cotton, Fleece and Dacron, are used. All the used materials have different properties and offer different specifications. Panama fabric provides good absorbency, good durability and also is widely used substrate material. The material is also a smaller amount of prone to fire. Fleece is pleasing for touching, warming and drying quickly. Further, it easily does not lose its properties when used for a long time and also looks appealing. Dacron material is famous for rigidity; it offers high rigidity. It also has excellent electrical properties. The resulting outcomes reveal the promising characteristic of the defected ground structure with microstrip antenna for smart healthcare implications. This results in a bandwidth of 1.6–11.2 GHz with better return loss and gain. Further, it ensures better performance in comparison with the traditional methods and is a better candidate for real-time applications.

The rest of the paper is organized as follows: Sect. 2 presents the proposed design of the defected ground structure microstrip antenna for smart healthcare applications, and Sect. 3 briefly explains the simulated results and discussion. Finally, Sect. 4 concludes the proposed model.

2 Proposed Model

The suggested model is designed and simulated on a copper ground plane of 70 mm × 60 mm with a workbench setup with HFSS software. Here, the defected ground structure is used in order to get the optimal design [11, 12]. A rectangular dumbbell-shaped slot has been realized in the ground plane below the microstrip line. A microstrip feed line is used to feed the antenna through. Figures below show the top view and side view of the developed antenna.

Figure 1 shows the geometry of the patch of the antenna. Copper in the patch is used with a thickness of 0.1 mm dimensions of the patch as shown in the table. To obtain better, S11 parameter slots in a staircase manner (S1–S13) are introduced in the patch. Dimensions of the patch are shown in Table 2.

In Fig. 2, a breast model is developed with the use of using HFSS simulator. This is a basic virtual human breast. The breast model is designed as with the given specification as skin layer thickness of 3 mm and bulky depth of 22 mm. A fatty tissue region of 22 mm thickness is placed below the skin. A defected cell, i.e., tumor region, is modeled within the fatty tissue layer. The tumor region is sphere shaped having diameter of 20 mm. Table 3 shows the conductivity and dielectric constant of the breast model.

Figure 3 shows the real-time implementation model of the designed antenna. X-rays as plane EM wave need to be provided to impinge into the breast cell as similar to the mammography technique. Now these plane wave X-rays impinge through the breast cell received by the designed MPA placed over the breast. The antenna is being fixed on a protoboard whose connector is attached to a vector network analyzer (VNA). On the VNA screen, we can observe the antenna parameters (S11) variation and can detect the tumorous breast cell.

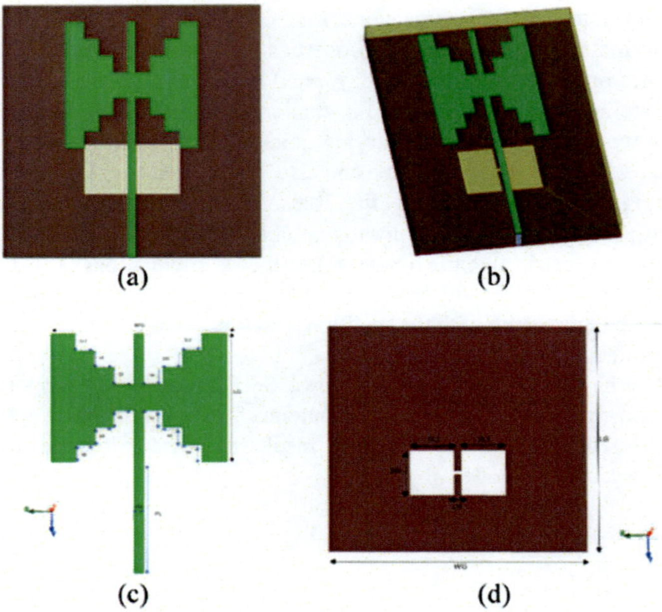

Fig. 1 Proposed antenna model **a** top view **b** side view, **c** geometry of the patch, **d** geometry of the ground plane

Table 2 Patch and ground plane dimensions for different parameters

Parameters	Patch dimension (mm)	Parameters	Ground plane dimension (mm)
LG	30	SL1/SL2/SW	12
WG	36	CN	4
PL	26	LG	60
PW	2	WG	70

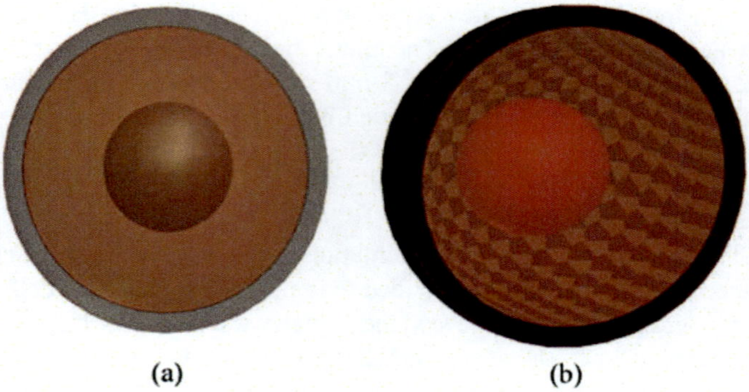

Fig. 2 Breast models **a** top view **b** breast model with tumor

Table 3 Parameter specification of different parameters used in the breast model

Dielectric medium	Conductivity (S/m)	Dielectric constant (F/m)
Normal tissue	0.15	9
Skin	1.1	46.7
Defected tissue	0.7	50

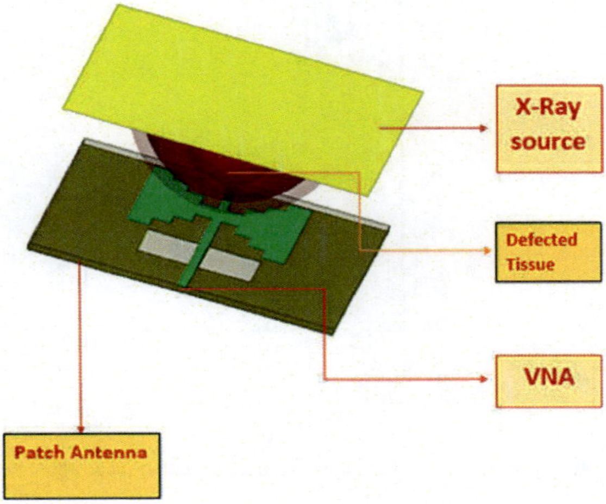

Fig. 3 Implementation of the designed antenna

The measurement on VNA is done for three cases.

- With only microstrip patch antenna
- Microstrip patch antenna over the breast model without defected cell
- Microstrip patch antenna over the breast model with defected cell.

3 Results Discussion

The antenna was simulated four times with four different substrate materials. Figure 4 shows the comparison of the obtained S11 values of the four antennas with different substrate material. From this figure, it can be seen that the design with cotton substrate gives −10 dB return loss with a bandwidth expansion of 3.05–11.2 GHz. A wide band is obtained with center frequency of 8.15 GHz using the proposed antenna model. This gives a maximum value of −32.8 dB at 2.47 GHz for the Dacron substrate. The −10 dB return loss is observed in the entire bandwidth, starting from 1.6–11.2 GHz for Fleece substrate antenna. The −10 dB return loss bandwidth expanding from 2.6 GHz to 11.2 GHz gives the highest value of 28.14 dB at 2.8 GHz with the use of Panama fabric.

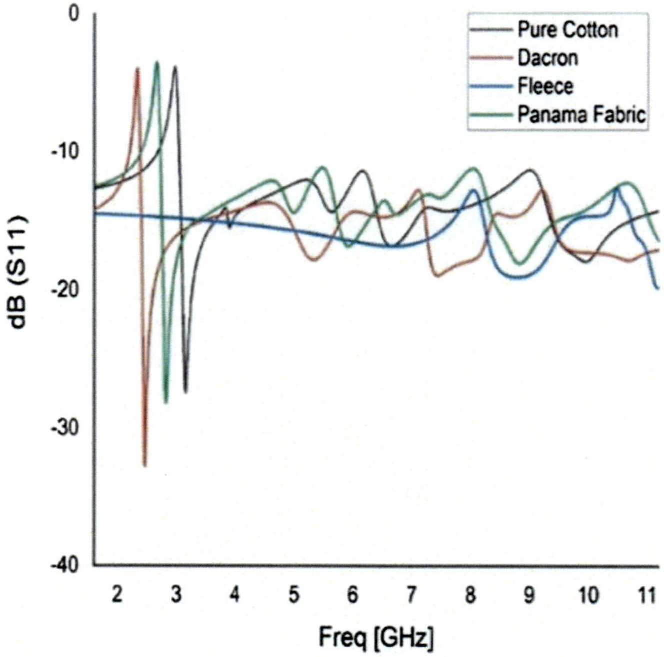

Fig. 4 Comparison of the simulated S11 value of four antennas

From Fig. 4, it can be observed that out of the four designed antennas, the antenna design with Dacron as substrate material is promising better return loss value. So, for the application of the patch antenna over the breast model, the designed antenna with Dacron as substrate is used.

In Figs. 5 and 6, the designed antenna is applied over a breast model. There is a sharp resonance peak of return loss at 6.4 GHz. As shown in Fig. 7, when the designed antenna is applied over breast model with a defected cell in it, there are two sharp resonance peaks of S11 at 6.4 GHz, 8.9 and at 10.7 GHz and a widening of S11 (return loss parameter) for a band of 6.5–11.2 GHz depicting more loss variation due to defected cell. Figure 8 shows the simulated radiation pattern plot of the antenna when applied over the breast model with a defected cell inside. From the above plot, it can be observed that the cross-polarization level overlaps on the co-polarization level signifying the defect in the breast tissues.

Figure 9 presents an assessment of the simulated return loss values of the antenna only, antenna when applied over the breast model with a defected cell in it and antenna when applied over the breast model with a defected cell inside. Antenna response for both the cases is observed and analyzed. From the figure above, it can be seen that when the designed antenna is applied over a breast model, there is a sharp resonance peak of −36.7 dB of return loss at 6.4 GHz. But when the designed antenna is applied over the breast model with a defected cell inside, there are three

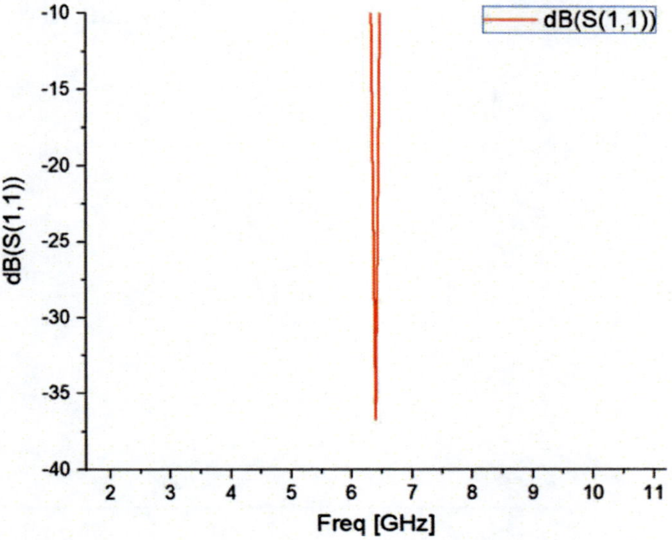

Fig. 5 Return loss of the antenna with the breast model without defected cell

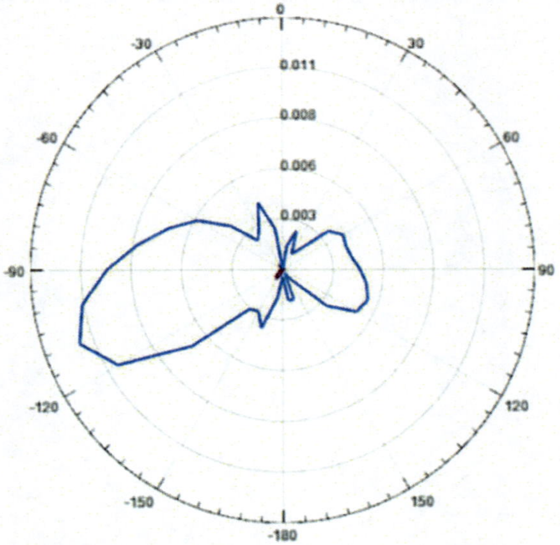

Fig. 6 Radiation pattern of the antenna when applied over a breast model

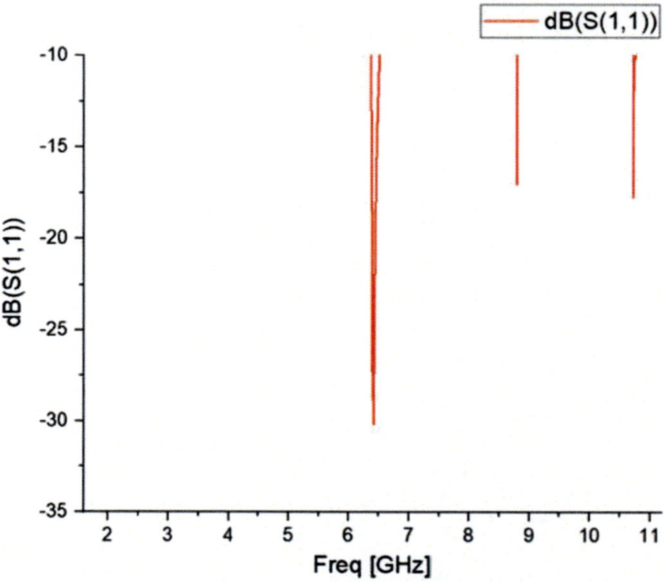

Fig. 7 Return loss of the suggested antenna when applied over a breast model with a defected cells

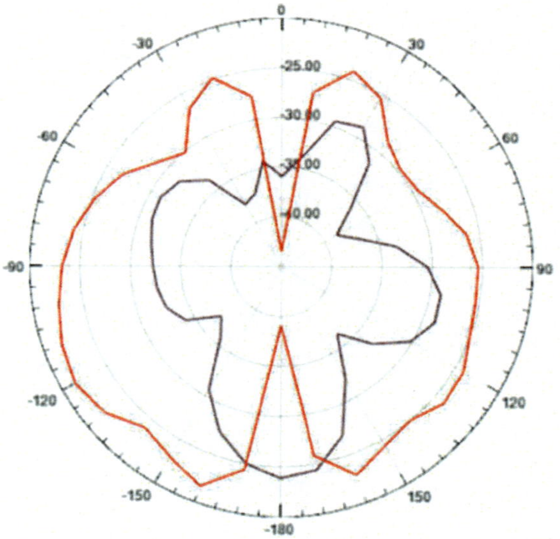

Fig. 8 Radiation pattern of the suggested antenna when applied over a breast model with a defected cells

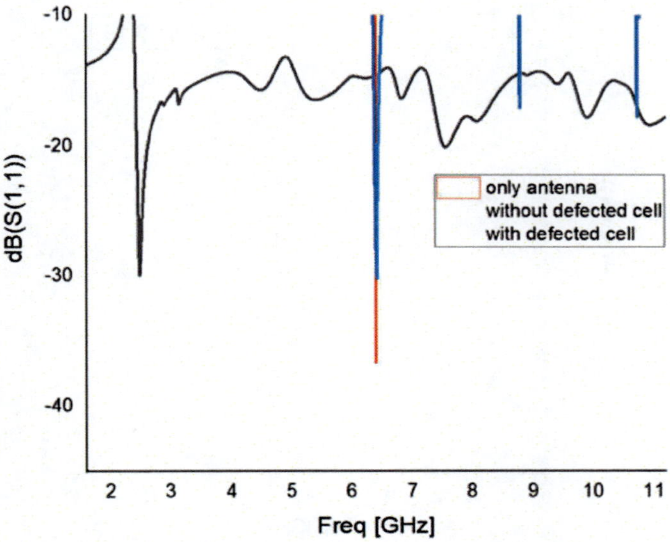

Fig. 9 Assessment of the simulated return loss of antenna with and without defected cell

sharp resonance peaks of return loss at 6.4 GHz, 8.9 and at 10.7 GHz with an observed −30.17 dB return loss at 6.4 GHz and a widening of return loss parameter for a band of 6.3–10.7 GHz depicting more loss variation due to the presence of the defected cell. Also from Fig. 8, it can be seen that the simulated radiation pattern plot of the antenna when applied over the breast model with a defected cell inside the cross-polarization level overlaps on the co-polarization level signifying the defect in the breast tissue. In order to validate the objective, the antenna was simulated several times by changing the size of the defected tissue, i.e., 6 mm radius and 2 mm radius and also by introducing two defected tissue of 2 mm radius and 6 mm radius inside the breast cell at a time. For every simulation, the position of the defected cell was kept different. The simulated results are given below.

Figure 10 shows simulated return loss value of the proposed antenna when applied over a breast model with a defected cell of radius 6 mm inside. It can be seen that there is a sharp resonance peak of −24.5 dB of return loss at 6.4 GHz. Figure 11 shows simulated return loss value of the proposed antenna when applied over a breast model with a defected cell of radius 2 mm inside. It can be seen that there is a sharp resonance peak of −23.2 dB of return loss at 6.4 GHz. Figure 12 shows simulated return loss value of the proposed antenna when applied over a breast model with two defected cells of radius 2 and 6 mm inside. It can be seen that there is also a sharp resonance peak of −27.8 dB of return loss at 6.4 GHz. From all these three figures, it can be seen that whenever a defected cell is present inside the breast volume, a sharp resonance peak of return loss is observed at frequency.

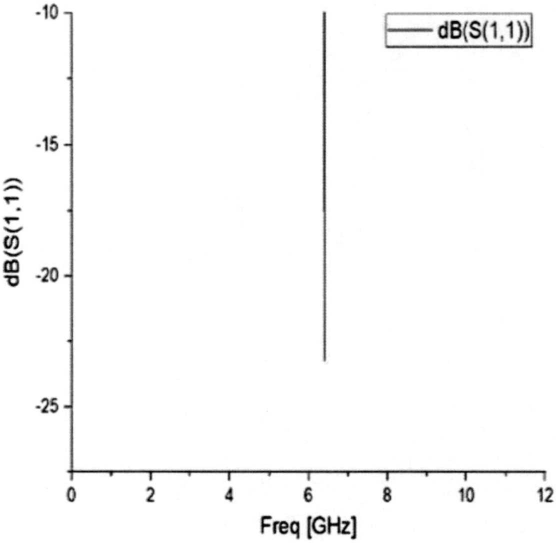

Fig. 10 Simulated return loss of the proposed antenna when applied over a breast model with a defected cell of 6 mm radius

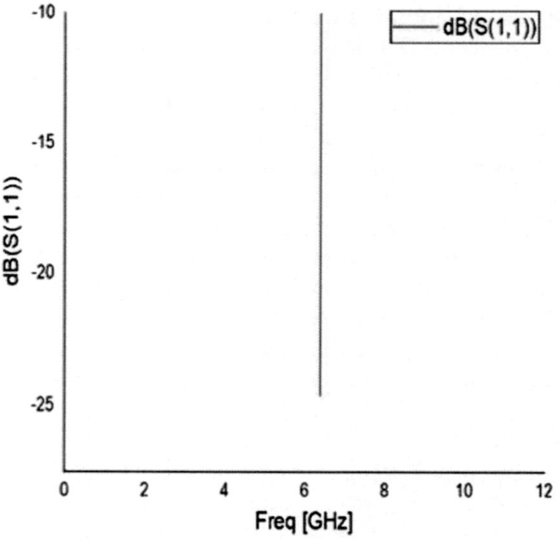

Fig. 11 Simulated return loss of the proposed antenna when applied over a breast model with a defected cell of 2 mm radius

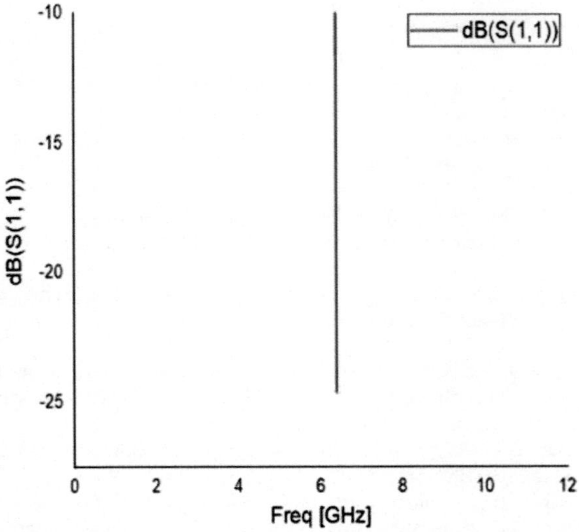

Fig. 12 Simulated return loss of the proposed antenna when applied over a breast model with two defected cells inside

4 Conclusion

The suggested model is used for finding a wearable microstrip UWB antenna to perform disease diagnosis and monitoring in smart healthcare system. Different types of antenna structures are judged to obtain the better return loss parameter. Four different types of substrate materials like Pure Cotton, Dacron, Fleece and Panama fabric were also used and compared to support wearability with the best return loss value. The material used as the substrate for the proceedings of this work is Dacron, as Dacron is promising the best results. The suggested antenna is simulated with a simulated breast model. The performance of the suggested model is tested with three different cases, i.e., only antenna, when applied over a breast model without a defected cell inside, when applied over a breast model having a defected cell in it. The results with different size of the defected cell and in different positions are also taken into consideration. It is observed that the narrow return loss bands get widened when the proposed patch antenna is subjected to the cancerous breast cells. The radiation patterns also reveal that the cross-polarization level increases when antenna is simulated with defected breast cell which is not in the case of the analysis done for antenna without defected cell, which enables us to conclude the presence of tumor. This gives a bandwidth of 1.6–11.2 GHz, minimal cost, better return loss and gain and wearability. This makes it better when compared to traditional methods and is a better candidate for real-time implications. The work can be extended on multiple defected structures and in real-time applications.

References

1. Balanis, C.A.: Antenna Theory: Analysis and Design. Wiley, Hoboken (1997)
2. Anuja, P.S., Kiran, V.U., Kalavathi, C., Murthy, G.N., Kumari, G.S.: Design of elliptical patch antenna with single & double U-slot for wireless applications: a comparative approach. Int. J. Modern Eng. Res. **4**(2), 24–30 (2014)
3. Baudha, S., Vishwakarma, D.K.: Bandwidth enhancement of a planar monopole microstrip patch antenna. Int. J. Microwave Wirel. Technol. **8**(2), 237–242 (2016)
4. Poovizhi, M.: Survey of microstrip patch antenna. Int. J. Sci. Eng. Technol. Res. **6**(2), 223–228 (2017)
5. Bakr, M.S.: Compact broadband frequency selective microstrip antenna and its application to indoor positioning systems for wireless networks. IET Microwaves Antennas Propagation **3**, 1–9 (2019)
6. Çalışkan, R., Gültekin, S.S., Uzer, D., Dündar, Ö.: A microstrip patch antenna design for breast cancer detection. In: Procedia-Soc. Behav. Sci. **195**, 2905–2911 (2015). World Conference on Technology, Innovation and Entrepreneurship
7. Bohra, S., Shaikh, T.: UWB microstrip patch antenna for breast cancer detection. Int. J. Adv. Res. Electron. Commun. Eng. **5**(1), 88–91 (2016)
8. Kahwaji, A., Arshad, H., Sahran, S., Garba, A.G., Hussain, R.I.: Hexagonal microstrip antenna simulation for breast cancer detection. In: Proceedings of International Conference on Industrial Informatics and Computer Systems, pp. 1–4. IEEE (2016)
9. Ahmed, I.I., Elwi, T.A.: A cylindrical wideband slotted patch antenna loaded with frequency selective surface for MRI applications. Eng. Sci. Technol. Int. J. **20**(3), 990–996 (2017)
10. Alsharif, F., Cetin, K.: Wearable microstrip patch ultra wide band antenna for breast cancer detection. In: Proceedings of International Conference on Telecommunications and Signal Processing, pp. 1–5. IEEE (2018)
11. Guha, D., Biswas, S., Kumar, C.: Printed antenna designs using defected ground structures: a review of fundamentals and state-of-the-art developments. Forum Electromagnetic Res. Methods Appl. Technol. **2**, 1–13 (2014)
12. Sh-Ali, J.M.: Performance of defected ground structure for rectangular microstrip patch antennas

A Software-Defined Collaborative Communication Model for UAV-Assisted VANETs

K. S. Arikumar, A. Deepak Kumar, C. Gowtham, and Sahaya Beni Prathiba

1 Introduction

Intelligent transportation system (ITS) is a very important wireless network for VANET system, which offers communication and communication between powerful vehicles. This gives the purpose of usability such as collision avoidance, modification alerts, safety guidance, comfort, and ease of use while driving [1]. There will be no problems such as loss of connectivity over long distances because of its multi-vehicular network, frequent network changes due to traffic congestion. The challenge for VANET is the quality of communication, which can be overcome by using unmanned aerial vehicles (UAVs). UAVs enable fast communication and flexibility between powerful vehicles where infrastructure is not available where it serves as a guide for emergency vehicles [2]. Therefore, UAVs improve VANET connectivity because UAVs are allowed to travel voluntarily due to free environmental issues [3].

Complex communication requirements are satisfied using a swarm of UAVs, i.e., aerial nodes in conjunction with wireless communication. A UAV hybrid network can reduce end-to-end delays between vehicles [4]. So, these wireless connections between cars have access from two bids. Software-defined network (SDN) can also bring benefits to VANETs assisted by multiple UAVs due to the flexibility of network through its intelligent and intermediate control organization [5, 6]. Therefore, the SDN in VANET-enabled model supports automotive services seamlessly [5, 7].

In addition to global networks, other compatible resolutions, like satellite and overhead networks, are also available that can benefit from this model [8, 9]. The space network can have numerous satellites which work as control centers and ground

K. S. Arikumar (✉) · A. D. Kumar · C. Gowtham
St.Joseph's Institute of Technology, Chennai, India

S. B. Prathiba
MIT Campus, Anna University, Chennai, India

M. Dua et al. (eds.), *Proceedings of the International Conference on Paradigms of Communication, Computing and Data Sciences*, Algorithms for Intelligent Systems,
https://doi.org/10.1007/978-981-16-5747-4_17

control stations. In such conditions like disaster management and emergency evacuation, the satellites play a vital role due to the high availability, high reliability, and streaming abilities.

The UAV network in u-VAN acts as a relay network for providing broadband services in between VANETs and core network. The UAV network can be located around 20 km above the earth's surface. Contrasted with base stations in earthly correspondence organizations, UAVs can have huge inclusion to bring to the table administrations on a territorial premise. However, the areas such as rural and highways has less coverage than regular networks. Providing a high network infrastructure for seamless connectivity in sparse environment is not a cost-effective solution. Hence, the proposed u-VAN model, which is based on a SSAGV model, adapts the UAV networking technology as a rapid intermediate for collaborative and automotive communications among vehicles.

In this paper, we proposed a u-VAN network model, which exploits the latest technologies such as SDN and UAVs into VANETs. The SDN in u-VAN supports numerous critical VANET services. Further, SDN enables rapid and flexibility in the network and assists to exploit the available resources effectively. Thus, the u-VAN network model can provide the higher quality of services (QoS) to all the end vehicles with minimal computation and infrastructure cost.

The rest of this paper is structured as follows. The next section provides an overview of the literature survey. Section 3 describes the proposed u-VAN network model and the algorithms associated with it. Section 4 presents the results and discussion. Finally, Sect. 5 concludes the paper with the performance achieved.

2 Literature Survey

The states of wireless solutions are vehicle-to-sensor, vehicle-to-vehicle, vehicle-to-Internet, and vehicle-to-road infrastructure connectivity and focus on the wireless technologies and potential challenges to provide vehicle-to-x connectivity [1]. The control plane in the network is apt for resolving various challenges associated with VANET and moulds the network to meet the necessities to provide high QoS [10]. VANET-based architecture of SDN and it is operating mode to SDN with VANET scenarios [11]. The architecture design uses the concept of SDN by absorbing multi-generation mixing (MGM) and network coding to increase the accuracy and security of data transfer across automotive networks [12].

With vehicle-to-UAV (V2U) communications, complete communication between vehicles is developed, and therefore, the route process is well developed [3]. Connection of vehicles is using by efficiently integrating the communication and networking technologies of UAVs. It has better throughput and delay performance [13]. To improve UAV transmission services, we also suggest a maximum vehicle coverage (MVC) design for two-dimensional UAV movements while data distribution [14]. This type of interaction provides accurate routes, and there are other solutions in the event of a road failure [15].

3 UAV-Assisted VANET Network Model Using SDN

SDN is one of the developing technologies and acts as a higher version of cloud. In addition to cloud-related capabilities, the SDN globally controls the network through programmable and reconfigurable interfaces in between networks. Thus, SDN personifies as a centralized controller with logically adjustable techniques which makes the network more agile, cost-effective, highly dynamic, less management effort, and highly adaptable. When SDN is integrated with VANET networks, the network can be managed in an optimized way, and based on this concept, we propose u-VAN network model to provide high QoS to the vehicles in the network.

3.1 u-VAN Network Architecture

As shown in Fig. 1, the u-VAN network is composed of three layers, namely layer one with space components, layer two with air components, and layer three with ground components. SDN controllers are used to regulate the network behaviors. It also helps to the manage network resources dynamically. Each segment has it is own characteristics to support vehicular services like UAV-assisted VANET network model. To prioritize the decision making of SDN controllers, network slicing is enforced, which avoids the interference of vehicular services with the legacy services. Each layer in the u-VAN network is allocated with each slice for accessing the VANET services with available resources. However, the lower layer units such as infrastructures of VANETs and other local servers assigns the local resources to the vehicles to enable the services. The slices are equipped with timer that the services can be accessed for the requested time period. Thus, the network slicing works dynamically for achieving high QoS with accessing the available resources.

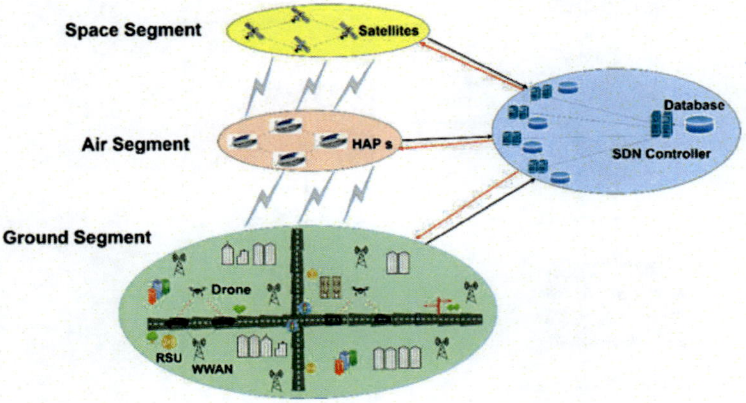

Fig. 1 Software defined space air ground integrated vehicular (u-VAN) networks

The flow of the u-VAN system is discussed in Fig. 2. In u-VAN, the data collection (DC) unit and communication (Comm) unit are designed for each specific spots. DC takes the responsibility of collecting the information regarding the location for each timestamp, and Comm is responsible for transmitting the data to other nodes. Initially, the UAVs are allowed to fly over a particular location to serve as a transmission point.

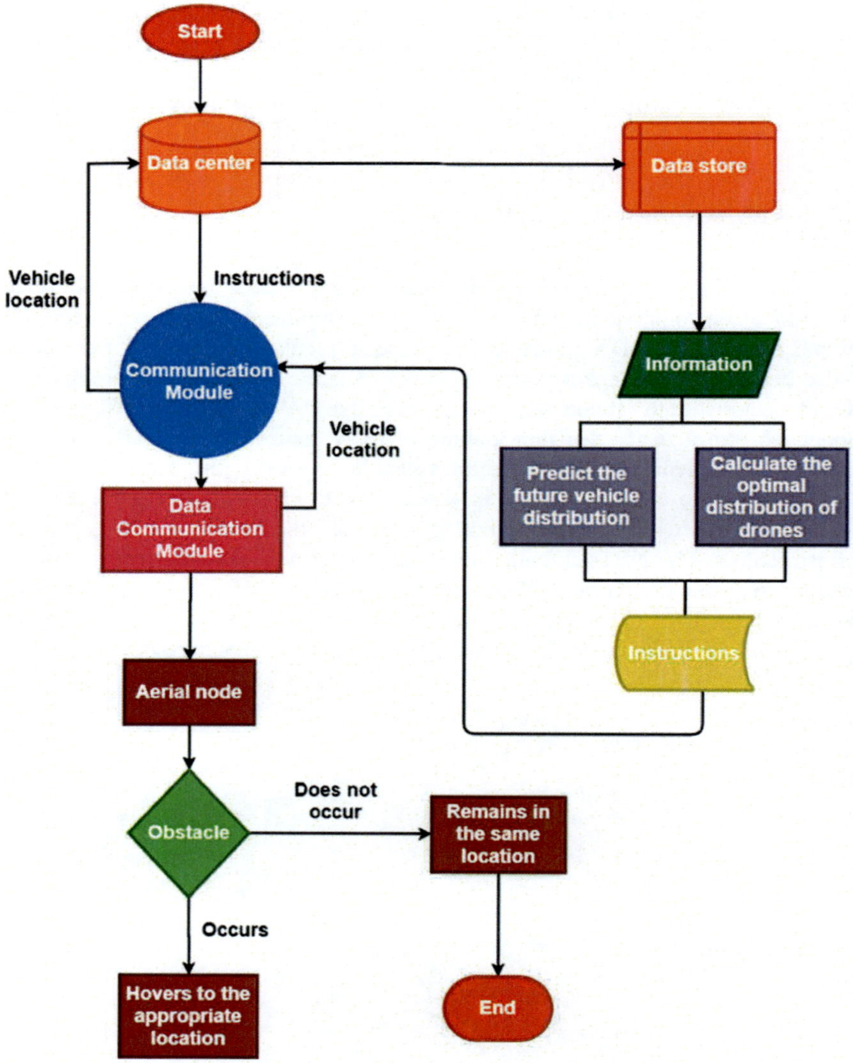

Fig. 2 Flow of the u-VAN system

Then, by utilizing DC and Comm unit, the data are transmitted and processed in u-VAN network. The data center in each base station distributes the UAVs after getting instructions from Comm and DC unit. Further, the data center centrally manages the network. Thus, based on the number of vehicles in a particular location, the UAVs are distributed locally, and the UAVs act as a relay node.

3.2 Layered Structure

The layered structure of u-VAN resembles the SDN architecture. The SDN encompasses three distinct layers, namely application, infrastructure, and control. The infrastructural layer is responsible for processing, storing and communication among resources. The control layer, the pool of SDN-centralized controllers are available and are waiting to target for a particular operation. The controllers are able to communicate and control the available physical resources via the interfaces in between the layers. Thus, the SDN controller is capable of controlling, managing, and organizing the UAVs as the relay nodes in the u-VAN system for offering the VANET services at low cost high QoS.

3.3 Network Settings

UAVs can be used as a relay where the relationship between the satellite and the ground segment is not good due to certain weather conditions. Each UAV is allowed to communicate with the satellite via space layer. It can connect UAVs and vehicle users through optical links and microwave links. LEO satellite network works at height 1,414 km in eight aircrafts, 48 satellites, with different tilt angles. UAVs at a height of 25 km are evenly scattered in the space and are comparatively balanced on Earth. LEO satellites move in a short period of time from 130 min.

3.4 Optimization of Multiple UAVs Assisted VANETs

The u-VAN system deploys multiple number of UAVs on different locations with the help of SDN by considering the emergency demands. Let T be the number of UAVs and the boundary of location be $X\text{max}$ and $X\text{min}$. When the number of UAVs is equal to the boundary of location, the UAVs will not be scheduled, else, T will be less than the boundary for full coverage, then,

$$\text{Tfc} = ((\text{Len}/\text{Rd}) - 1)((\text{Wid}/\text{Rd}) - 1) \tag{1}$$

where Len is the length of the particular location, and Wid is the width of the location. The communication range of UAV is represented as Rd, and the minimum number of UAVs is represented as Tfc.

In u-VAN, the UAV-assisted VANET is considered as a typical multimodal optimization problem, and our proposed u-VAN can be applied for this architecture. When T, the number of UAVs is equal to cover the boundary of location Xmin and Xmax, without any conditions to be applied, the u-VAN system's infrastructure provides the VANET services to the vehicles. If u-VAN finds the excess number of UAVs for full coverage, then the UAVs are mitigated with the consideration of time and energy. For this task, the positioning of UAVs (P-UAV) algorithm finds a sequence to confirm that the UAV moves to the destination provided. This sequence reduces the migrating time of the UAVs in such conditions, as it avoids the UAVs to fly to the original position and allocating to the next destination position. Also, it reduces the energy of the UAVs. The pseudocode of P-UAV algorithm is given in Algorithm 1.

In u-VAN, the UAVs can be in either fly mode or sleep mode. In both the modes, they are capable of communicating with the vehicles, infrastructures, and SDN continuously. Thus, the data traffic of U-VAN is equally distributed, irrespective of the position of the UAVs.

Algorithm 1: P-UAV

Input: The range of simulation area (Xmin; Xmax), the required number of UAVs T;
Output: The positions in time for T UAVs;

1. Randomly initialize T positions as UAVs;
2. **repeat**
3. Predict the distribution of vehicles in next time slice;
4. Built the evaluation function;
5. Perform the function to get T global optima;
6. Assign optimal positions for T UAVs by SA that makes the total path shortest;
7. UAVs hover at the designated position until the current time slice is exhausted;
8. T UAVs respectively fly to their destinations;
9. **until** Terminate condition is satisfied;
10. **end**

4 Results and Discussion

All network simulations are performed by NS-3. To make our simulation closer to the realities of the cities, we embrace the real place. According to 1,000,000 floating car data from the simulation area, we do the math to get moving features, including high-end parking space time, high speed, low speed, and medium speed. These features and map details are submitted to SUMO to make a car traffic model as shown in Table 1.

Table 1 Simulation parameter

Simulation area	2000 m × 2000 m
Mobility generator	SUMO
Number of vehicles	100,200
Number of UAVs	4, 8, 12, 16, 20
Vehicles speed	0–60 km/h
UAV speed	0–90 km/h
Length of time slice	5 s
Vehicles communication range	500 m
UAVs communication range	600 m
UAVs altitude	200 m
Mac protocol	802.11p
Routing protocol	AODV
Radio-propagation model	Two ray ground
Type of traffic	Constant bit rate (CBR)
CBR interval	0.1 s
Packet size	512 bytes

4.1 Packet Delivery Ratio (PDR)

This is the calculation of total number of successfully delivered packets out of total number of packets transmitted. The Eq. (2) calculates the PDR,

$$\mathrm{PR} = P \tag{2}$$

where PR shows the number of packets reaches at the sink vehicle, and P is number of packets transmitted by the source vehicle. The performance of u-VAN is evaluated with the usability of SSAGV method and without it and demonstrated the outcome in Fig. 3. From the figure, it is observed that the proposed u-VAN model has 19% higher packet delivery ratio than the model without SSAGV. The reason behind the higher achievement of PDR is that the u-VAN framework holds the supports of both SDN and UAVs, which assures the delivery of packets with low latency and avoids packet drop. However, the framework without SSAGV has higher probability of packet drop, and thus, the successful delivery ratio may be low.

4.2 End-to-End Delay (EED)

End-to-end delay is the time difference in between the transmission of a data packet. The EED rate is based on the total time taken in number received successfully packets and can be obtained by using the Eq. (3).

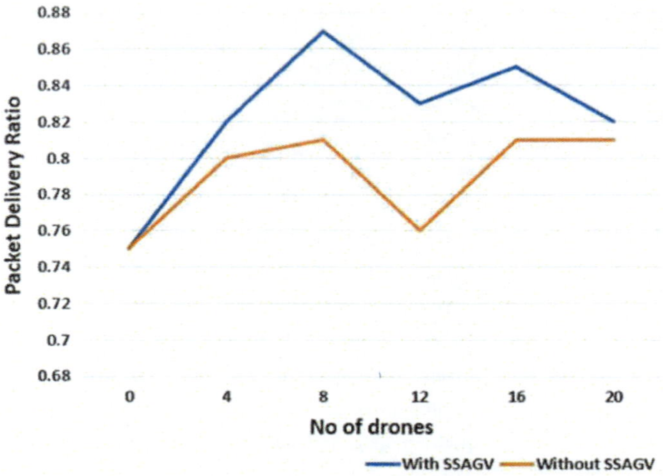

Fig. 3 Analysis of PDR with SSAGV versus without SSAGV

$$\text{EED} = (N * L/R) * P \tag{3}$$

where N be the link, L be the packet length, R be the transmission rate (N links in series, using store and forward between the links), and P packets over N links. Network quality assessed with moderate EED, expected to be lower. According to Fig. 4, UAV-assisted methods do not apply to this metric. The standard EED fluctuates incorrectly in the case of a small car. This situation too occurs in the tightest traffic conditions expected of u-VAN, i.e., reveals a very serious tendency. In complex cases,

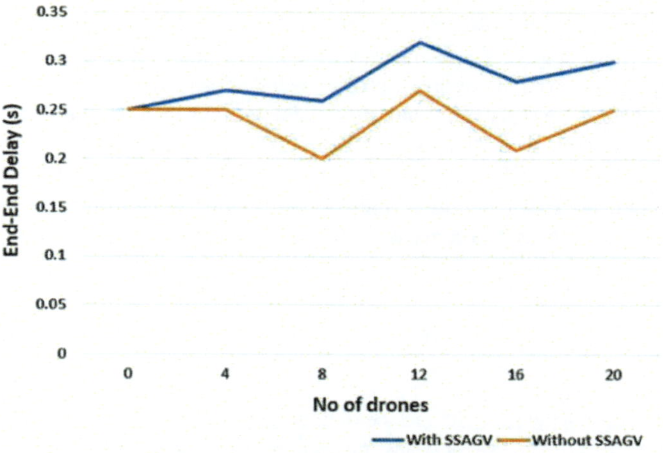

Fig. 4 Analysis of EED with SSAGV versus without SSAGV

u-VAN spends more time on availability but finds an accessible way that enhances the median effects of EED. Due to the nature of u-VAN expending extra time to obtain better accuracy on the route, this is a conflict, and in the middle, the use of time and accuracy causes the u-VAN to fail worse on EEDs. However, this disability can be defeated by protocols based on their geographical route, and the time taken to find a way is often enough to ignore it. End-to-end delay is obtained by using the Eq. (3). Further, the south bound interface of SDN and the faster mobility of UAVs lift up the VANET services with as much lower latency as possible.

4.3 Throughput

It represents the total value of data that can be successfully transferred over time. The throughput efficiency can be calculated by Eq. (4).

$$R = I/T \tag{4}$$

where I be the inventory, R be the rate, i.e., throughput, and T be the time. This test metric shows actual transfer performance of network. Figure 5 gives statistics for each situation. u-VAN achieves the highest exit. With SSAGV method, the pass is well matched via PDR. The higher throughput of u-VAN framework is due to the efficient three-tier architecture like SSAGV and the successful delivery of packets. In traditional VANET, the services are provided with the wireless technology IEEE 802.11p, which has lower bandwidth of 5.9 GHz and cannot support longer range transmissions. In contrast to traditional VANET, the u-VAN is assisted with faster UAVs, which transmits the data in between vehicles with a higher data rate. Thus,

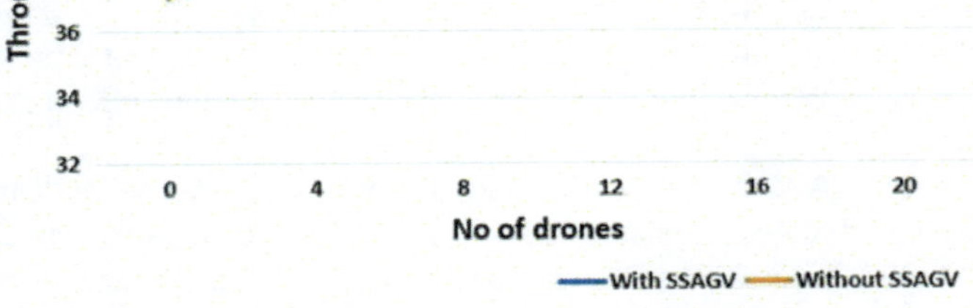

Fig. 5 Analysis of throughput with SSAGV versus without SSAGV

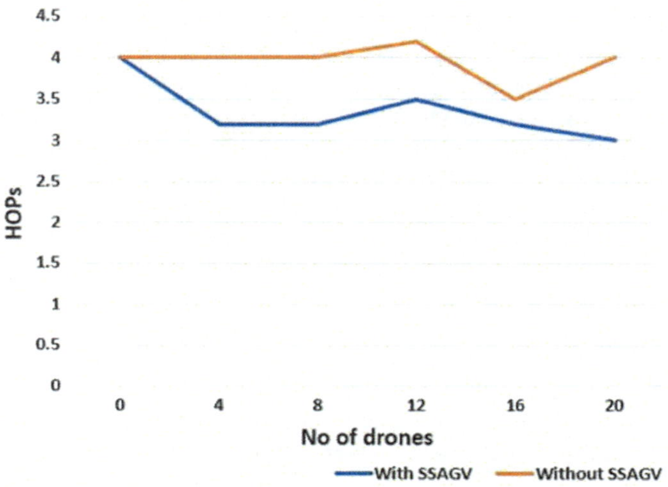

Fig. 6 Analysis of HOPs with SSAGV versus without SSAGV

the u-VAN framework ensures higher throughput than the framework that does not have SSAGV (Fig. 6).

4.4 *The Number of Hops*

The number of HOPs can be defined as the number of MAC layer protocol transmitted from source vehicle to the destination vehicle in the particular timestamp. When the number of transmission is higher, HOPs will be higher and increases the communication cost. Hence, the optimal networks will have reduced number of hops. The u-VAN using model utilizes the existence of UAVs for the specific location to avoid unnecessary hops. When there is more number of vehicles in a particular region, then U-VAN can have two hops, whereas other scenarios can have around six hops.

The outcomes of u-VAN network model are as follows.

- u-VAN simplifies the workload of network by minimizing the management cost, operational cost, initialization cost, and upgradation cost.
- The global view of SDN and continuous network connectivity provision of UAVs in u-VAN provide optimized network functionalities and effective resource utilization.
- On the fly of UAVs, the u-VAN has the ability to agile the network model by adapting the high dynamic characteristics of VANET architectures, which in turn reduces the link failures even in dense network and high dynamic models.

5 Conclusion

In this paper, an SSAGV model-based network model, namely u-VAN, is proposed to support diversity of VANET services in a cost-effective manner. The adaption of SDN and UAVs to support the existing VANET architecture enables the vehicles to experience high QoS with peer-to-peer communication technology. The SDN improves the network with flexible, simplified, and low maintenance. Further, the SDN enhances the interoperability and operation cost. The UAVs act as a relay node to provide continuous communication among vehicles, infrastructures, and core network, which lead to effective utilization of available resources and enable collaborative communication in VANET. The proposed u-VAN architecture is evaluated against the various QoS parameters, and the results prove the superiority of the u-VAN system.

References

1. Lu, N., et al.: Connected vehicles: solutions and challenges. IEEE Internet of Things J. **1**(4), 289–299 (2014)
2. Luo, C., et al.: Unmanned aerial vehicles for disaster management. In: Geological Disaster Monitoring Based on Sensor Networks, pp. 83–107. Springer, Singapore (2019)
3. Oubbati, O.S., et al.: UVAR: an intersection UAV-assisted VANET routing protocol. In: 2016 IEEE Wireless Communications and Networking Conference. IEEE (2016)
4. Wang, X., et al.: VDNet: an infrastructure-less UAV-assisted sparse VANET system with vehicle location prediction. Wirel. Commun. Mob. Comput. **16**(17), 2991–3003 (2016)
5. Lin, N., et al.: A novel multimodal collaborative UAV-assisted VANET networking model. IEEE Trans. Wirel. Commun. (2020)
6. Arikumar, K.S., Natarajan, V., Satapathy, S.C.: EELTM: an energy efficient lifetime maximization approach for WSN by PSO and fuzzy-based unequal clustering. Arab J. Sci. Eng. **45**, 10245–10260 (2020). https://doi.org/10.1007/s13369-020-04616-1
7. Zhou, Y., et al.: Multi-UAV-aided networks: aerial-ground cooperative vehicular networking architecture. IEEE Veh. Technol. Mag. **10**(4), 36–44 (2015)
8. Bertaux, L., et al.: Software defined networking and virtualization for broadband satellite networks. IEEE Commun. Mag. **53**(3), 54–60 (2015)
9. Arikumar, K.S., Natarajan, V.: FIoT: a QoS-aware fog-IoT framework to minimize latency in IoT applications via fog offloading. In: Bhateja, V., Peng, S.L., Satapathy, S.C., Zhang, Y.D. (eds.) Evolution in Computational Intelligence. Advances in Intelligent Systems and Computing, vol. 1176. Springer, Singapore (2021). https://doi.org/10.1007/978-981-15-5788-0_53
10. Smida, K., et al.: Software defined Internet of Vehicles: a survey from QoS and scalability perspectives. In: 2019 15th International Wireless Communications & Mobile Computing Conference (IWCMC). IEEE (2019)
11. Ku, I., et al.: Towards software-defined VANET: architecture and services. In: 2014 13th Annual Mediterranean Ad Hoc Networking Workshop (MED-HOC-NET). IEEE (2014)
12. Bhatia, J., et al.: SDN-enabled network coding based secure data dissemination in VANET environment. IEEE Internet Things J. (2019)
13. Shi, W., et al.: UAV assisted vehicular networks: architecture, challenges and opportunities. IEEE Netw. **32**(3), 130–137 (2018)

14. Zeng, F., et al.: UAV-assisted data dissemination scheduling in VANETs. In: 2018 IEEE International Conference on Communications (ICC). IEEE (2018)
15. Arikumar K. S., Natarajan V.: Fuzzy based dynamic clustering in wireless sensor networks. In: 2016 Eighth International Conference on Advanced Computing (ICoAC), pp. 77–82 (2017). https://doi.org/10.1109/ICoAC.2017.7951749

Comprehensive Survey on Wireless Network on Chips

R. Shruthi, H. R. Shashidhara, and M. S. Deepthi

1 Introduction

Network on Chip (NoC) has emerged in the world of VLSI archetype as a backbone for communication in enabling a higher level of integration in multicore system on chips (SoCs) [6]. Network on chip defines as an architecture that maintains appropriate design solutions for a communication-centric inclination. The major accomplishment of NoC design is the significant utilization of bandwidth between the cores of SoC for a set of seriatim applications. The latest NoC models at first dynamically produce or withdraw the lines present between the SoC components and accomplish the runtime topologies for NoC and in second the congestion management capability at the interconnects to manage reconfigurability with constraint in operations [16]. Nevertheless, the inclusion of dynamically reconfigurable capabilities in an initial network on chips can be used in various models with an area limited overhead concerning the primary NoC pattern [15, 16, 32, 49]. Consequently, the great aids are considerable when compared with custom non-reconfigurable NoC patterns.

SoCs for embedded systems are multifaceted as they are made up of huge numbers of parameters like storage elements, reconfigurable devices, and processing elements, to surge the manageability of SoC available in a diverse atmosphere form. However, the multi-processor system on chip (MPSoC) design is obtained by interconnecting

R. Shruthi (✉) · H. R. Shashidhara · M. S. Deepthi
Department of Electronics and Communication, The National Institute of Engineering, Mysuru, Karnataka 570008, India
e-mail: shruthir@nie.ac.in

H. R. Shashidhara
e-mail: shashidharahr@nie.ac.in

M. S. Deepthi
e-mail: deepthims@nie.ac.in

M. Dua et al. (eds.), *Proceedings of the International Conference on Paradigms of Communication, Computing and Data Sciences*, Algorithms for Intelligent Systems,
https://doi.org/10.1007/978-981-16-5747-4_18

the sub-systems of the SoC components, for the physical scalability and architecture. Processor cores in an embedded system are firstly intended to run the software tasks for various applications, and there will be a trade-off in the optimization of the communication methods. Therefore, the bandwidth capacity for various links should be capable of providing the highest rate of traffic for various tasks in an ultimate SoC design. Secondly, the network, perhaps, undergoes overcrowding of the traffic which intend leads to delay in data streaming, and unacceptable interlink performance, hence, has to be nullified. Consequently, network on chip keenly guarantees bandwidth worst-case conditions for SoC processes with several parallel applications recurrently leading to the outsized topologies and the regular process links of the SoC. In another environment, the growth of innovative methods and frameworks will speed up the time for a differently working initial NoC condition, developed for a variety of applications at each moment, is one among the vital study avenues in a network on chip domain [50].

The runtime reconfigurable framework of NoC can be partially accomplished by the reconfigurable capability of field-programmable gated arrays (FPGA) to familiarize the operation of the network on chip interconnects for exact functional requirement of the SoCs [12]. Then, the dynamically configured SoC is useful in circuit switching communications with enabling other features such as promising reliable high performance, shoot up in run time, reduction in the possible number of hops and reduction in latency. These meet the throughput necessities by the quick dynamic reconfiguration of the routing table with fewer cycles and reduction in the latency for different network topologies [16]. This dynamicity functionality is applicable in wireless network on chips (WiNoCs).

The features and properties that a WiNoC benchmarking environment would possess are: (a) network size (small, medium, large), (b) IP core composition (amount of processing, memory cores, other), (c) topology (regular, irregular), (d) traffic characteristics (spatial and temporal), and (e) QoS requirements (best-effort, guaranteed bandwidth, guaranteed latency). There are other parameters of the WiNoC benchmarks along with the above properties necessary for evaluation of NoC performance. The most vital among them are always being the silicon area, throughput, power/energy dissipation, and latency.

This paper mainly emphasizes on the wide variety of designs and proposals of WiNoC architectures. It is set forth as firstly, the need and progression toward WiNoC architecture in Sect. 1. Next, the performance parameters necessary for the evaluation of the structures in Sect. 2. Followed by this is the detailed information of the tools that are required for the analysis of the designs in Sect. 3. Then, details of the techniques and design proposal along with the results of the evaluation of the designs stated for WiNoC in Sect. 4. Finally, a summary of WiNoC outcomes and the research avenues are indicated in Sect. 5.

2 Performance Analysis Metrics

The most thought-provoking and universally pertinent metrics of network on chip are delay, power consumption, area utilization, bandwidth, and jitter. The performance metrics are classified as Jitter, latency, and bandwidth, whereas the cost factors are area usage and power consumption. These parameters and factors are obtained utilizing media access control (MAC), traffic patterns, topology, routing algorithms, and benchmarks as shown in Fig. 2.

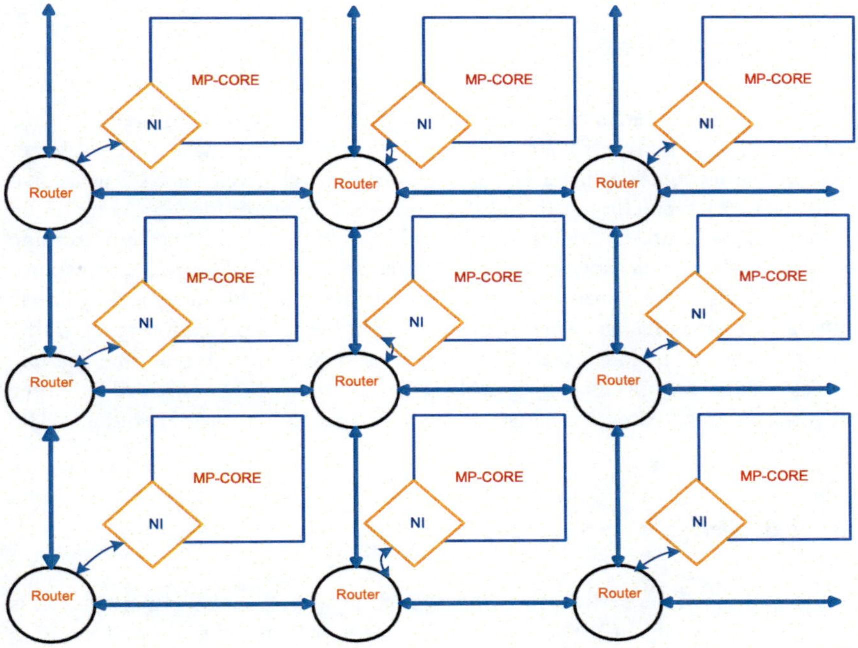

Fig. 1 2D mesh topology of NoC-based multi-processor

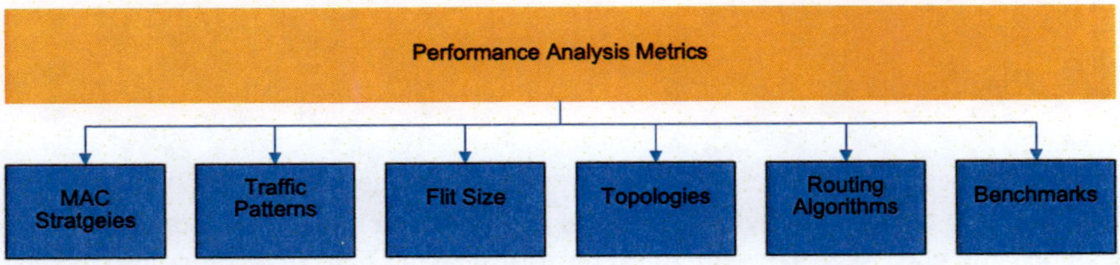

Fig. 2 Metrics used for the analysis of the performance of WiNoC

2.1 MAC Strategies

The challenge lies in the growth of MAC strategies which are necessary in effective sharing of the wireless medium for the increased cores where the uniqueness in the research arena and distinctiveness of the working environment are given. To illuminate the originalities of the new environment, a well-researched analysis of the contexts of MAC protocols that deals with both unique optimization occasions and essential requirements is present in [25].

2.2 Traffic Patterns

For the NoC performance evaluation, the design must be tested for hard real-time scenarios. Traffic workloads are used to provide a well-balanced NoC architecture with the resources available. The traffics can be generated for NoC using two approaches: (a) synthetic traffic and (b) application-based traffic. Synthetic traffic is mainly used in understanding the network's overall performance by informing about the bottlenecks without representing the characteristics of the real applications. Simple to design TGs in hardware or software is the major advantages of synthetic traffic and is also used in many FPGA-based platforms. Application-based traffic is mainly used to obtain the realistic behavior of the network for real-time applications as it gathers results more accurately. The patterns are further divided into three categories (a) trace-based, (b) stochastic, and (c) application cores execution [23].

2.3 Flit Size

For the analysis of the performance, the size of the packets is also important. Flit is the basic unit of transfer between a pair of routers or a unit of flow control within the network. A flit size varying from 64 to 512 bits was used by the investigators of the varied design structures.

2.4 Topologies

In NoC, the design and selection of topologies are considered vital. Based on parameters like hop counts, throughput, latency, and injection rates, many topologies are designed and classified. They are mesh, torus, ring, star, butterfly, spidergon, mixed, hierarchical rings, and trees topologies. Topologies are further categorized as either irregular or regular. The widely used topology is mesh as the path diversity is good; i.e., we get many paths to get from one node to another.

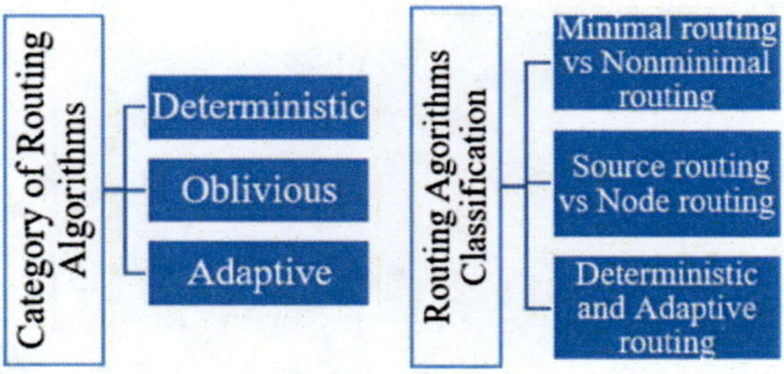

Fig. 3 Routing algorithms characterization

2.5 Routing Algorithms

For the data packets to be routed from one destination to another, a certain path or method has to be followed to deliver the data to the right destination along with the tracking of the shortest path, routing algorithms are necessary. The category of the routing algorithms for NoCs are

(i) Deterministic that chooses the same path for a source–destination pair
(ii) Oblivious that chooses different paths without considering network states
(iii) Adaptive that chooses different paths adapting to the state of the network.

This can be further classified as minimal routing vs. non-minimal routing, source routing vs node routing, and deterministic and adaptive routings [52]. XY routing algorithm [51] is widely used in NoC because of its simplicity. A pictorial representation is given in Fig. 3. There are many types of XY routing algorithms [52], and the selection of the types is dependent on the application or conditions of the environment of NoC architecture. The adaptive XY routing algorithm is opted as the best if the focus is on the utilization of network resources. The wormhole routing segments each data packet into flits. It has the advantage of not storing the complete packet in the switch, and the header flit will wait to route for the next upcoming stages. However, the drawback is the probability of occurrence of livelock and deadlock. But, many researchers have worked in providing the solution to the same [52].

2.6 Benchmarks

The performance parameters of varied architectural designs can be analyzed or compared using certain benchmarks which is a collective set of programs or operations that helps in assessing the relative performance of the design by executing the standard tests and trials of a subsystem. The benchmarks help us in rating the architectures. The benchmarks can be further classified as toy programs, synthetic benchmarks, and real benchmarks as shown in Fig. 4. The toy programs are simple programs with fewer lines of codes. These are not suitable to analyze the performance of the architectures. For example, sorting and matrix multiply. The synthetic

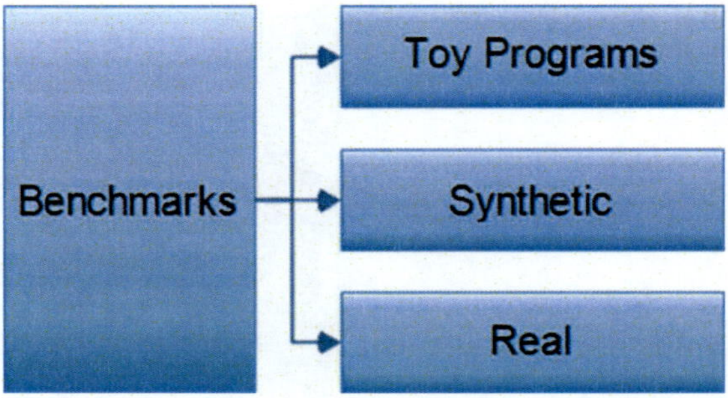

Fig. 4 Types of benchmarks

benchmarks, like dhrystone, are basically artificial programs. They do not represent any real applications. They artificially create fetch, decode, and execution effect. The real benchmark suits like SPEC06 and SPLASH and replicates the standard operations for the analysis of the architecture. The most commonly used benchmarks are PARSEC and SPLASH-2 benchmarks.

3 Tools Used for the Analysis of NoCs

It is significant to evaluate the architectures proposed for NoC among the others. The research community considers area consumption, power utilization, network throughput, and latency as the general criteria for the evaluation of the designs via a simulation process. The other criteria measured for the analysis are packet or wire length, the size of network, the size of buffers, traffic distribution, packet injection ratio, selection strategy, routing algorithm, and finally the packets distribution. Here, a sneak peek at the globally used NoC tools is present with some having extra auxiliary options. The extra options include availability, hardware synthesis, modeling, and simulation [35]. Figure 5 gives us the flow of the steps that takes place in a simulation of WiNoC interconnects [37].

The design and the implementations have to be verified using some kind of tools. There are many tools available for simulation purpose for analyzing the above-mentioned WiNoC parameters. The most popular among the tools is Noxim [39]. It is an open, configurable, extendible, cycle-accurate NoC simulator developed in System C, a system description library written in C++ . This simulator helps in analyzing the power figures and performance of both evolving WiNoC architectures and conventional wired NoC. The XMulator [40] is a cost-effective simulator tool that can be used to evaluate the system design operation. It is an object-oriented listener-based simulation environment. It contains a toolbox of several routing algorithms, network topologies, flexible router models, and switching methods, defined using XML format for the evaluation of multi-computer interconnected networks. Methodology used for the generation of synthetic traffic and also to capture and evaluate the application and cache coherence behavior of NoCs is SynFull [41]. The

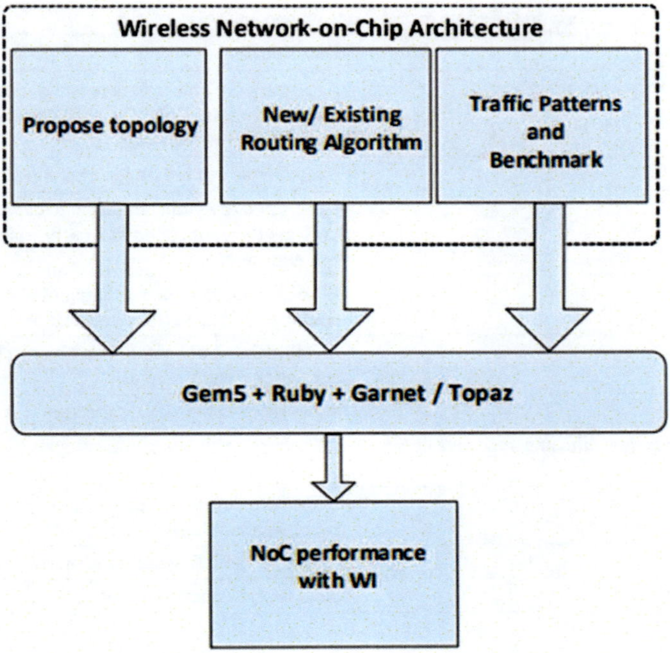

Fig. 5 Flow diagram of simulation setup of WiNoC interconnects

designers can indulge themselves in detailed performance simulations with errors being less than 0.3% and over 50 × speedup on average over the full-system simulation. Another cycle-accurate simulator for NoCs is BookSim [42] which has been designed for flexibility and accurate modeling of network components and metrics such as routing algorithm, topology, router micro-architecture, and flow control, including buffer management and allocation schemes. For vastly parallel and many-core computer systems, an open-source simulation framework based on OMNeT++ is present which is called as MCoreSim [43]. Another simulator highly recommendable for the analysis of computer architecture researchers is the GEM5 [44]. This simulator is designed by merging the best aspects of the M5 [45] that provides a highly configurable simulation framework, multiple ISAs, and diverse CPU models, and GEMS [46], a simulation toolset to evaluate multi-processor architectures. The capabilities of the simulation of GEM5 range from the selection of ISA, CPU model, and coherence protocol to the instantiation of interconnection networks, devices, and multiple systems. The RTL design validation and analysis for any proposed method can be done using the Xilinx ISE design suite and Synopsys design compiler [47]. The problem of congestion that occurs during the detailed routing can be avoided and has turned to be of great help to the RTL designers. The reliability of the proposed systems of wireless communication as well as their network performance can be done using MATLAB modeling. Automation tools for design are provided by NoCs

Table 1 List and description of tools for NoC design and analysis

List of tools	Description
Noxim	Analyzing the power figures and performance of both evolving WiNoC architectures and conventional wired NoC
XMulator	Evaluate the system design operation
SynFull	Generation of synthetic traffic, to capture and evaluate the application and cache coherence behavior of NoCs
BookSim	Flexible and accurate modeling of network components and various network parameters
MCoreSim	Suitable for vastly parallel and many core computer systems analysis
GEM5	Analysis of computer architecture
Xilinx ISE design suite and synopsys design compiler	RTL design validation and analysis
Multi2Sim	Verification of design can be done by running a set of standard benchmarks
HFSS	Antennas modeling and its characterization
CONNECT	NoC model generation tool

Arteris, iNoCs, Silistix [12]. Cadence tools can be used to obtain the analysis area, power, and delays in the designs. To test the design of architecture before the architecture is physically manufactured, the Multi2Sim simulator can be used to test and validate new hardware designs. The verification and proposing a correct design can be done by running a set of standard benchmarks on Multi2Sim. The antennas modeling and its characterization needed for the wireless communication in NoC can be done using high-frequency structural simulator (HFSS). CONNECT framework [54] can be used to acquire the HDL code of several NoC configurations. Apart from the above-mentioned tools, there a few more tools like DARSIM, SunFloor-3D SunFloor, ORION 2.0, NoC Emulation techniques, FlexNoC, and NoCGEN [35] which were proposed by renowned companies and research institutes. A summary of the tools is given in Table 1.

4 Overview of Algorithms and Techniques Used for WiNoC

The design aim of on-chip communication system relies on the data transmission with less latency and higher throughput utilizing minimum resources. An emblematic NoC architecture involves many segments of wires and routers. The architecture provides the necessary communication structure for all the resources of the system. The main objectives of the architectural design are to setup self-sufficient blocks of hardware resources and form the NoC by connecting these blocks to form a network.

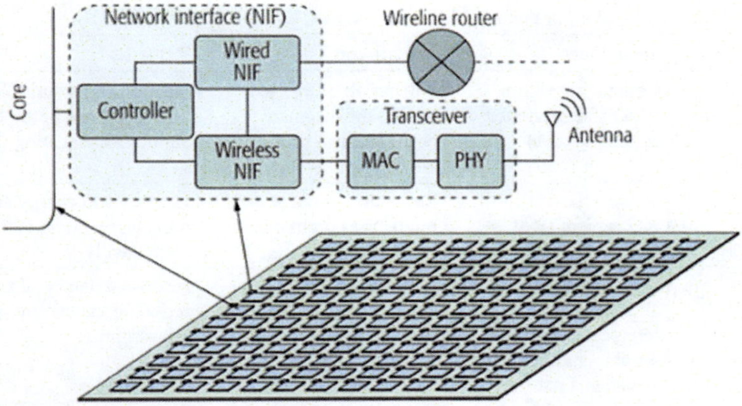

Fig. 6 Schematic diagram of many cores integrated in a hybrid WiNoC

The other objective is mainly on the configurable property and the scalability of the design to make a flexible platform that is adaptable to a variety of workloads for all the scenarios that may occur [24]. The conventional NoCs used multi-hop packet-switched communication. At every hop, the data packet drives through a complexly designed router/switch, which in turn consumed a significant amount of power, throughput, and latency overhead [6]. This drawback was an advantage to WiNoC that has inspired the researchers in proposing, designing, and implementing numerous architectures and routing algorithms for the efficient and the best utilization results in terms of performance, robustness, and energy. Figure 6 gives us the schematic of the hybrid WiNoC. The following section also contains the outcomes and the results obtained for the proposed structures and designs by the researchers as mentioned in the algorithms and techniques section.

Algorithms.
The various types of algorithms adopted by the researchers and analyzed results for different WiNoC architectures are listed in Table 2. As per the literature study, it is found that speed, yield, area, power, efficiency, routing, shortest path, and throughput are the measurable factors for WiNoC.

4.1 Architectures

The various architectures designed and studied by the researchers for WiNoC are summarized in Table 3. The study shows that there is a reduction in the energy, power, area, hop count, routing techniques, and distance, and enhancement is observed in factors like scalability for a large number of cores, speed, throughput, and efficiency/performance.

Table 2 Summary of literature on algorithms for WiNoC

Reference no.	Techniques	Description	Remarks
[2]	Genetic algorithm (GA) and simulated annealing (SA)	Allocation of PEs to reduce the distance between them to the antennas	Genetic algorithm (GA) efficiency was better than simulated annealing (SA)
[31]	Radio access control mechanism (RACM)	Reduction in range and multi-hop communication	Latency and energy are reduced by 30 and 25%, respectively
[7]	Seamless hybrid wired and wireless interconnection for on-chip and off-chip transfer of data	Improving communication efficiency	Increased bandwidth and reducing the energy consumption
[8]	SiESTA	Turning off the radio hubs that are idle	The rise in router buffer size is proportional to the rise in the power consumption of the antenna buffers
[15]	Adaptive WiNoC	Decreasing the traffic at the routers	Along with the reduction in the traffic at the routers, packet latency was also reduced
[17]	Application-specific network on chip (ASNoC)	Reduction in power utilization and area and obtain high-speed communication	Results yield reduction in the area and power with the increase in the speed of operation
[18]	XY, DyAD, odd–even, negative- first, north-last, and west-first	Focuses on the impact of varied schemes of routing	Odd–even routing gave the best results
[30]	Adaptive fault-tolerant wireless routing algorithm	To abide by both permanent and intermittent faults occurring on wireless hubs	Average distance between the nodes was upgraded along with the average delay
[36]	Dynamically configurable routing algorithm	Configuring the router dynamically based on traffic patterns encountered	This dynamicity provided' improved delay and throughput
[22]	HoneyWiN	Challenge the performance of the most suited mesh topology and to prove the efficacy of honeycomb-based WiNoC architectures	HoneyWiN outperforms the mesh-based NoC and is two times faster in operations

(continued)

Table 2 (continued)

Reference no.	Techniques	Description	Remarks
[53]	Secure and dependable routing (SDR)	It is a routing strategy that incorporates security and fault-tolerant applicable for NoC-based MPSoC architecture	Use of SDR shows sound tolerance for faulty links and increase in scalability, security, and dependability

5 Summary

Network on chip covers a broader band of research, extending from issues related to software down to physical level implementation across system topology. WiNoCs are shown to deliver one-hop broadcasts with less delay across the complete chip and can augment point-to-point and multi-hop signaling over outmoded wired network on chip through various research approaches. Thus, in this survey, we have given an overview of different architectures, tools, and technique which are used for state-of-the-art wireless network on chips (WiNoC). The trending research is in exploring the energy benefits and performance of WiNoCs. A modern design practice is foreseen for effective usage of the available resources. In the future, the methodology of layered micro-network design is most preferred field in specializing the complex SoC design paradigm.

Effective solutions to the problems related to design and implementations are still in demand as we go higher from the physical layer. Figure 7 below shows the possible research avenues in the field [20]. The problems related to hardware design get harder as the massive parallelism handling capacity gets increased with runtime techniques. The key challenge also lies in designing the MAC protocol [25] and providing security to the network. These are the areas where the research is being carried out along with the effective WiNoC architecture. The study is required as to be aware of the most likely threats related to eavesdropping, spoofing, and denial-of-service attacks in WiNoC, which can occur due to faulty wireless components and malicious hardware trojans [10, 13].

Table 3 Summary of different architectures for WiNoC

Reference no.	Architecture	Description	Remarks
[14]	Hybrid architecture with the adaptable links	To obtain communication with energy efficiency	WiNoC is compared with electrical NoC designs like mesh, concentrated mesh (CMesh), and flattened butterfly (FB) architectures, and the wireless networks Wcube and iWISE. The area parameter is a trade-off with throughput, speed, and energy and also is scalable to a large number of cores with the best outcomes.
[26]	Hybrid NoC architecture employing both wired and wireless communications and with deadlock-free routing algorithm	To efficiently use wireless links using simulated annealing optimization techniques	Comparable decrease in the hop count for different traffic patterns. Transfer latency has improved along with the reduction in power consumption up to 15%
[1]	Hierarchical wireless-based architecture, HiWA with HoneyWiN routing algorithm	For optimal utilization of wireless resources	Total power consumption is observed to be reduced by 14 and 39, 16, and 19% reduction in latency and 42% and 51% reduction in hop count for conventional NoC and WiNoC, respectively
[3]	Power-efficient fine-grained router architecture (FGRA)	Activation of the receiver antenna only when the signal arrives	In comparison with other approaches like power punch, FlexiBuffer, and NoRD, FGRA power/energy saving of routers static power is 88.76% (base router); 62.5% (wireless interfaces) along with 2.42% area overhead. The network power consumption is decreased to 37.20%
[4]	Routerless NoC	Removal of expensive routers present in traditional NoCs and is grounded on a layered progressive approach	Better performance for all the parameters with a drawback of utilizing more wires in comparison to other designs

(continued)

Table 3 (continued)

Reference no.	Architecture	Description	Remarks
[9]	Four hierarchical WiNoC obtained using obtained using analytic hierarchy process (AHP)	Hybrid hierarchical structures to excel the efficiency	Enhancement in the performance was observed
[19]	ORTHONOC	Pursuing to highlight the ordered broadcast benefit accessible by the wireless technology	Enhancement in the performance was observed for all the parameters
[29]	Hybrid NoC	Hybrid NoC architecture, mesh topology is carried out on different numbers of radio hubs	Sixteen radio hubs give the lowest throughput
[5]	H2WNoC	Reduction of the hardware resources, latency, network cost, and energy utilization	Enhancement in the performance was observed for all the parameters
[28]	Multi-stage interconnection networks (MINs)	Apt for the application of on-chip radio communications technologies	Development in the average delay, saturation, and the energy overhead
[21]	Toroidal folding (TF)-based interconnection architectures	Energy-efficient and multi-modal chip-to-chip communication protocol	Reduction in the network diameters and also the average distance between the nodes
[16]	Argo NI-NoC for GALS architecture	Concentrates on the area efficiency and obtained using combination of TDM scheduling and asynchronous router and NIs micro-architecture in Argo NI-NoC for GALS architecture	Network is found to be 3.5 times smaller than any existing model for the standard functions

References

1. Rezaei, A., Daneshtalab, M., Safaei, F., Zhao, D.: Hierarchical approach for hybrid wireless network-on-chip in many-core era. Comput. Electr. Eng. **51**, 225–234 (2016)
2. Bahrami, B., Jamali, M.A.J., Saeidi, S.: Proposing an optimal structure for the architecture of wireless networks-on-chip. Telecommun. Syst. **62**(1), 199–214 (2016)
3. Mondal, H.K., Gade, S.H., Kishore, R., Kaushik, S., Deb, S.: Power efficient router architecture for wireless Network-on-Chip. In: 2016 17th International Symposium on Quality Electronic Design (ISQED), pp. 227–233. IEEE, March 2016
4. Alazemi, F., Azizimazreah, A., Bose, B., Chen, L.: Routerless network-on-chip. In: 2018 IEEE International Symposium on High Performance Computer Architecture (HPCA), pp. 492–503. IEEE, February 2018
5. Alaei, M., Yazdanpanah, F.: H2WNoC: a honeycomb hardware-efficient wireless network-on-chip architecture. Nano Commun. Netw. **19**, 119–133 (2019)

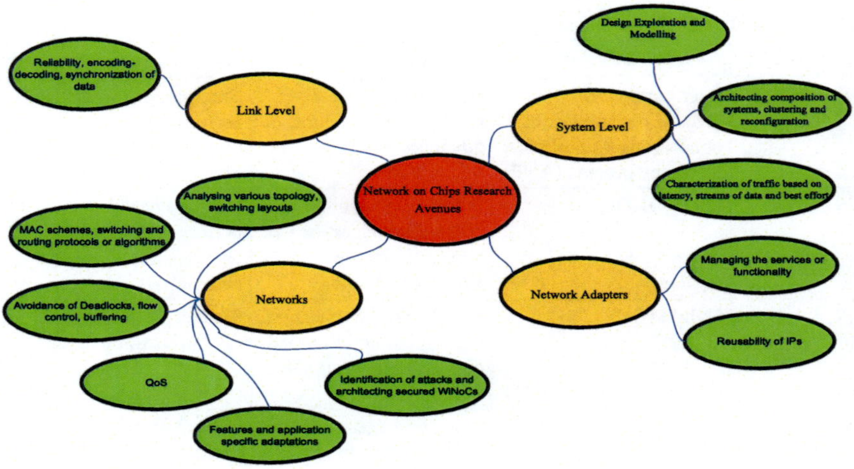

Fig. 7 NoC research avenues classification

6. Ganguly, A., Chang, K., Deb, S., Pande, P.P., Belzer, B., Teuscher, C.: Scalable hybrid wireless network-on-chip architectures for multicore systems. IEEE Trans. Comput. **60**(10), 1485–1502 (2010)
7. Shamim, M.S., Mansoor, N., Narde, R.S., Kothandapani, V., Ganguly, A., Venkataraman, J.: A wireless interconnection framework for seamless inter and intra-chip communication in multichip systems. IEEE Trans. Comput. **66**(3), 389–402 (2016)
8. Catania, V., Mineo, A., Monteleone, S., Palesi, M., Patti, D.: Improving energy efficiency in wireless network-on-chip architectures. ACM J. Emerg. Technol. Comput. Syst. (JETC) **14**(1), 1–24 (2017)
9. Bahrami, B., Jamali, M.A.J., Saeidi, S.: A novel hierarchical architecture for wireless network-on-chip. J. Parallel Distrib. Comput. **120**, 307–321 (2018)
10. Biswas, A.K., Chatterjee, N., Mondal, H., Gogniat, G., Diguet, J.P.: Attacks toward wireless network-on-chip and countermeasures. IEEE Trans. Emerg. Topics Comput. (2020)
11. Deb, S., Ganguly, A., Pande, P.P., Belzer, B., Heo, D.: Wireless NoC as interconnection backbone for multicore chips: promises and challenges. IEEE J. Emerg. Sel. Topics Circ. Syst. **2**(2), 228–239 (2012)
12. De Micheli, G., Seiculescu, C., Murali, S., Benini, L., Angiolini, F., Pullini, A.: Networks on chips: from research to products. In: Proceedings of the 47th Design Automation Conference, pp. 300–305, June 2010
13. Lebiednik, B., Abadal, S., Kwon, H., Krishna, T.: Architecting a secure wireless network-on-chip. In: 2018 Twelfth IEEE/ACM International Symposium on Networks-on-Chip (NOCS), pp. 1–8. IEEE, October 2018
14. DiTomaso, D., Kodi, A., Matolak, D., Kaya, S., Laha, S., Rayess, W.: A-WiNoC: adaptive wireless network-on-chip architecture for chip multiprocessors. IEEE Trans. Parallel Distrib. Syst. **26**(12), 3289–3302 (2014)
15. Shetty, S.S., Moharir, M., Sunil, K., Ahmad, S.F.: Low latency & high throughput wireless-NoC architecture for manycore processors. In: 2018 International Conference on Networking, Embedded and Wireless Systems (ICNEWS), pp. 1–5. IEEE, December 2018
16. Shashidhara, H.R., Prasad, S.N., Prabhudeva, B.L., Kulkarni, S.S.: Design and implementation of Argo NI-NoC micro-architecture for MPSoC using GALS architecture. In: Emerging Trends in Electrical, Communications, and Information Technologies, pp. 451–463. Springer, Singapore (2020)

17. Venkataraman, N.L., Kumar, R.: Design and analysis of application specific network on chip for reliable custom topology. Comput. Netw. **158**, 69–76 (2019)
18. Lit, A., Rusli, M.S., Marsono, M.N.: Comparative performance evaluation of routing algorithm and topology size for wireless network-on-chip. Bull. Electr. Eng. Inform. **8**(4), 1239–1250 (2019)
19. Abadal, S., Torrellas, J., Alarcón, E., Cabellos-Aparicio, A.: OrthoNoC: A broadcast-oriented dual-plane wireless network-on-chip architecture. IEEE Trans. Parallel Distrib. Syst. **29**(3), 628–641 (2017)
20. Bjerregaard, T., Mahadevan, S.: A survey of research and practices of network-on-chip. ACM Comput. Surv. (CSUR) **38**(1), 1-es (2006)
21. Saxena, S., Manur, D.S., Mansoor, N., Ganguly, A.: Scalable and energy efficient wireless inter chip interconnection fabrics using THz-band antennas. J. Parallel Distrib. Comput. **139**, 148–160 (2020)
22. Yazdanpanah, F., AfsharMazayejani, R., Alaei, M., Rezaei, A., Daneshtalab, M.: An energy-efficient partition-based XYZ-planar routing algorithm for a wireless network-on-chip. J. Supercomput. **75**(2), 837–861 (2019)
23. de Lima, O.A., Costa, W.N., Fresse, V., Rousseau, F.: A survey of NoC evaluation platforms on FPGAs. In: 2016 International Conference on Field-Programmable Technology (FPT), pp. 221–224. IEEE, December 2016
24. Kumar, S., et al.: A network on chip architecture and design methodology. In: Proceedings IEEE Computer Society Annual Symposium on VLSI. New Paradigms for VLSI Systems Design. ISVLSI 2002, pp. 117–124. IEEE, April 2002
25. Abadal, S., Mestres, A., Torrellas, J., Alarcón, E., Cabellos-Aparicio, A.: Medium access control in wireless network-on-chip: a context analysis. IEEE Commun. Mag. **56**(6), 172–178 (2018)
26. Hu, W.H., Wang, C., Bagherzadeh, N.: Design and analysis of a mesh-based wireless network-on-chip. In: 2012 20th Euromicro International Conference on Parallel, Distributed, and Network-based Processing, pp. 483–490. IEEE, February 2012
27. Gade, S.H., Rout, S.S., Kashyap, R., Deb, S.: Reliability analysis of on-chip wireless links for many core WNoCs. In: 2018 Conference on Design of Circuits and Integrated Systems (DCIS), pp. 1–6. IEEE, November 2018
28. Mnejja, S., Aydi, Y., Abid, M., Monteleone, S., Palesi, M., Patti, D.: Implementing on-chip wireless communication in multi-stage interconnection NoCs. In: International Conference on Advanced Information Networking and Applications, pp. 533–546. Springer, Cham (2020)
29. Mnejja, S., Aydi, Y., Abid, M.: Exploring hybrid NoC architecture for chip multiprocessor. In: 2018 30th International Conference on Microelectronics (ICM), pp. 307–310). IEEE, December 2018
30. Mortazavi, S.H., Akbar, R., Safaei, F., Rezaei, A.: A fault-tolerant and congestion-aware architecture for wireless networks-on-chip. Wirel. Netw. **25**(6), 3675–3687 (2019)
31. Palesi, M., Collotta, M., Mineo, A., Catania, V.: An efficient radio access control mechanism for wireless network-on-chip architectures. J. Low Power Electron. Appl. **5**(2), 38–56 (2015)
32. Romera, T., Brière, A., Denoulet, J.: Dynamically reconfigurable RF-NoC with distance-aware routing algorithm. In: 2019 14th International Symposium on Reconfigurable Communication-centric Systems-on-Chip (ReCoSoC), pp. 98–104. IEEE, July 2019
33. Salminen, E., Kulmala, A., Hamalainen, T.D.: On network-on-chip comparison. In: 10th Euromicro Conference on Digital System Design Architectures, Methods and Tools (DSD 2007), pp. 503–510. IEEE, August 2007
34. Yuvaraj, M.P.: Design trade-offs for reliable on-chip wireless interconnect in NoC platforms (2014)
35. Achballah, A.B., Saoud, S.B.: A survey of network-on-chip tools. arXiv preprint arXiv:1312.2976 (2013)
36. Xu, S., Meyer, M.C., Jiang, X., Watanabe, T.: A traffic-robust routing algorithm for network-on-chip systems. In: 2019 IEEE 13th International Symposium on Embedded Multicore/Many-core Systems-on-Chip (MCSoC), pp. 209–216. IEEE, October 2019

37. Deb, S., Mondal, H.K.: Wireless network-son-chip: a new era in multi-core chip design. In: 2014 25nd IEEE International Symposium on Rapid System Prototyping, pp. 59–64. IEEE, October 2014
38. Ganguly, A., et al.: Intra-chip wireless interconnect: the road ahead. In: Proceedings of the 10th International Workshop on Network on Chip Architectures, pp. 1–6, October 2017
39. Catania, V., Mineo, A., Monteleone, S., Palesi, M., Patti, D.: Noxim: an open, extensible and cycle-accurate network on chip simulator. In: 2015 IEEE 26th International Conference on Application-Specific Systems, Architectures and Processors (ASAP), pp. 162–163. IEEE, July 2015
40. Nayebi, A., Meraji, S., Shamaei, A., Sarbazi-Azad, H.: Xmulator: a listener-based integrated simulation platform for interconnection networks. In: First Asia International Conference on Modelling & Simulation (AMS 2007), pp. 128–132. IEEE, March 2007
41. Badr, M., Jerger, N.E.: SynFull: synthetic traffic models capturing cache coherent behaviour. ACM SIGARCH Comput. Archit. News **42**(3), 109–120 (2014)
42. Jiang, N., et al.: A detailed and flexible cycle-accurate network-on-chip simulator. In: 2013 IEEE International Symposium on Performance Analysis of Systems and Software (ISPASS), pp. 86–96. IEEE, April 2013
43. Kumar, S., Cucinotta, T., Lipari, G.: A latency simulator for many-core systems. In: SpringSim (ANSS), pp. 151–158, April 2011
44. Binkert, N., et al.: The gem5 simulator. ACM SIGARCH Comput. Archit. News **39**(2), 1–7 (2011)
45. Binkert, N.L., Dreslinski, R.G., Hsu, L.R., Lim, K.T., Saidi, A.G., Reinhardt, S.K.: The M5 simulator: modeling networked systems. IEEE Micro **26**(4), 52–60 (2006)
46. Martin, M.M., et al.: Multifacet's general execution-driven multiprocessor simulator (GEMS) toolset. ACM SIGARCH Comput. Archit. News **33**(4), 92–99 (2005)
47. Pouchak, M.A., Smith, G.A.: U.S. Patent No. 8,418,128. Washington, DC: U.S. Patent and Trademark Office (2013)
48. Tsai, W.C., Lan, Y.C., Hu, Y.H., Chen, S.J.: Networks on chips: structure and design methodologies. J. Electr. Comput. Eng. (2012)
49. Kasapaki, E., Schoeberl, M., Sørensen, R.B., Müller, C., Goossens, K., Jens, S.: Argo: a real-time network-on-chip architecture with an efficient GALS implementation. IEEE Trans. Very Large Scale Integr. (VLSI) Syst. **24**(2), 479–492 (2015)
50. Benini, L., De Micheli, G.: Networks on chips: a new SoC paradigm. Computer **35**(1), 70–78 (2002)
51. Boppana, R.V., Chalasani, S.: A comparison of adaptive wormhole routing algorithms. In: Proceedings of the 20th Annual International Symposium on Computer Architecture, pp. 351–360, May 1993
52. Chawade, S.D., Gaikwad, M.A., Patrikar, R.M.: Review of XY routing algorithm for network-on-chip architecture. Int. J. Comput. Appl. **43**(21), 975–8887 (2012)
53. Fernandes, R., Marcon, C., Cataldo, R., Sepúlveda, J.: Using smart routing for secure and dependable NoC-based MPSoCs. IEEE/ACM Trans. Netw. **28**(3), 1158–1171 (2020)
54. Prasad, B.M.P., Parane, K., Talawar, B.: FPGA friendly NoC simulation acceleration framework employing the hard blocks. Computing 1–23 (2021)

Computing

Giza Pyramids Construction Algorithm with Centroid Opposition-Based Learning

Debolina Bhattacharya and Tapas Si

1 Introduction

The Giza pyramids construction algorithm has come to the literature of artificial intelligence by Harif et al. in the year 2020 [2]. Giza Pyramid is one of the seven wonders of the world. The three pyramids are Khufu Pyramid, Khafre, and Menkaure. Khafre is the largest among all. This construction has taken a lot of effort over time to develop.

Some great ancient principles from the starting of the history of mankind to the beginning of the post-classical era have been inspiring throughout the proceeding of this GPC proposal. The techniques, that were innovated by the ancient builders, have always been cost reducing and optimizing in every way eventually. The ancient management, observation, and antique strategy of well structured agents are the cause of advancement of the civilization and also enhanced the technical strategies of optimization in present era of computing too.

The major problem in the construction of the pyramids is managing the workers. The workers are coolies, slaves, masons, metalworkers, and carpenters. They are directed by a special agent named Pharaoh's special agent. Arranging of the blocks by the workers is done under the management of the Pharaoh's agent. The workers have to maintain their position orderly. The worker who is better in his work will get a sublime rank as and thus there will be a contention to get sublime rank. The best rank is associated with Pharaoh's agent. Generally, when one worker lost energy during the transportation period he may take a rest for some time. When a worker lost too much energy or overtired, he will be replaced by young and energetic workers. To get sub-line rank, there is another competition. They can achieve experience and expertise. The interval between the stone block and the elevation location in the pyramid must be carried to push the stone which is measured by the capability of the workers. Making the block closer to the site in the pyramid, if enough power is available, the workers carry out more shipments. So the ramp gradient, friction force,

D. Bhattacharya (✉) · T. Si
Department of Computer Science and Engineering, Bankura Unnayani Institute of Engineering, Bankura 722146, West Bengal, India

M. Dua et al. (eds.), *Proceedings of the International Conference on Paradigms of Communication, Computing and Data Sciences*, Algorithms for Intelligent Systems,
https://doi.org/10.1007/978-981-16-5747-4_19

and initial velocity encourage the stone block. So this concept has been used by this GPC algorithm.

Though the GPC algorithm showed better performance than its competitive algorithms, it suffers from premature convergence to local optima due to the lack of exploration. The objective of this paper is to improve the GPC algorithm in order to achieve better solutions. As the OBL scheme in metaheuristic algorithms increases the diversification in the search space and enhances the chances to achieve better solutions, we are motivated to incorporate an OBL scheme called COBC in GPC algorithm in this paper. The proposed GPC-COBL is tested on 28 IEEE CEC2013 benchmark function optimization problems. The comparative study is conducted with GPC algorithm. From the experimental results with statistical analysis, it is observed that the proposed algorithm performs better than or equal to the classical GPC algorithm in most of the problems.

The remaining of this paper is organized as follows: Sect. 2 discusses the GPC algorithm. The proposed GPC-COBL is presented in Sect. 3. Section 5 provides the results and discussion. In Sect. 6, the conclusion and potential future works are given.

2 Giza Pyramids Construction Algorithm

For constructing the algorithm, the pyramids were erect by using a straight-on ramp presuming that one ramp is utilized. The angle between the ramp and horizon must be less than 15° and also can be variable. Friction is productive in stone block's displacement but is not contemplate for workers. And in the construction period, some workers are probably to be placed in a new position. The kinetic friction force f_k is defined as:

$$f_k = \mu_k m g \cos\theta \tag{1}$$

Here m is the stone block's mass, g is the earth's gravity, θ is the angle between the ramp and horizon. μ_k is the coefficient of kinetic friction. As we are using Newton's second law, because we are in the X-axis, so we have

$$-mg\sin\theta - f_k = ma \tag{2}$$

where a is acceleration. By using Eqs. 1 and 2, the stone block's acceleration in upward direction has been gained.

$$a = -g(\sin\theta + \mu_k \cos\theta) \tag{3}$$

The stone block's displacement is obtained by:

$$d = \frac{v_0^2}{2g(\sin\theta + \mu_k \cos\theta)} \tag{4}$$

The initial velocity v_0 of the stone block is determined by a uniformly distributed random number in (0, 1) in each iteration. μ_k is obtained as follows:

$$\mu_k = rand[\mu_{k,\min}, \mu_{k,\max}] \tag{5}$$

The worker's new position while the stone block is being pushed:

$$x = \frac{v_0^2}{2g\sin\theta} \tag{6}$$

After calculating the changes of stone block using Eqs. 4 and 6, a new position (i.e., new solution) $\boldsymbol{p}$ is calculated as:

$$\boldsymbol{p} = (p_i + d) * x\epsilon_i \tag{7}$$

Here p_i is the current position and ϵ_i is a vector of the uniform, normal, or Lévy distributed random numbers. When workers got tired or lost energy they are substituted by the energetic workers. It is done to balance the power of ordering. If the primary solution is $X(x_1, x_2, \ldots, x_D)$ with dimension D and generated solution using Eq. 7 is $Y(y_1, y_2, \ldots, y_D)$, then, similar to a uniform crossover operator, the substituting operation is carried out with fifty percent probability as follows:

$$Z = \begin{cases} y_k, if \quad rand[0, 1] \leq 0.5; \\ x_k, otherwise; \end{cases} \tag{8}$$

The algorithm GPC is given in Algorithm 1. In step 1 of Algorithm 1, the initial population of stone blocks or workers is initialized and the evaluation of cost, i.e., objective function values are done in step 2. In step 6, the amount of stone block's displacement is calculated. In step 7, the amount of worker movement is calculated. In step 9, the possibility of substituting workers is checked. The new position and its cost are computed in step 10. If the new cost is better than Pharaoh's agent cost, then Pharaoh's agent cost is updated in steps 11–13. The steps 6–13 are carried out for all N stone blocks or workers. In step 14, the candidate solutions are sorted. The steps 5–14 are repeated until termination criteria are met.

Algorithm 1: GPC Algorithm

```
Generate initial population array of stone blocks or workers (population size)
Generate position and cost of stone block or worker
Determine best worker as pharaoh's agent
for Iteration ← 1 to Maximum iteration do
    for i ← 1 to N do // (all N stones block or workers)
        Calculate amount of stone block displacement
        Calculate amount of worker movement
        Estimate new positions
        Investigate possibility of substituting workers
        Determine new position and new cost
        if new cost < pharaoh's agent cost then
            Set new cost as pharaoh's agent cost
        end
    Sort solution for next iteration
end
```

3 Proposed Method

In this work, we have proposed an improved GPC algorithm by incorporating COBC which is an OBL scheme to achieve better solutions. The concept of OBL and COBC is described in Sects. 3.1 and 3.2, respectively. The proposed algorithm is discussed in Sect. 3.3.

3.1 *Concept of OBL*

Opposition-based learning (OBL), a machine learning scheme [5], has been a successful scheme to developed search algorithms. For performance improvement, the different OBL schemes are successfully hybridized with different metaheuristic algorithms. The *type-I* opposite point is defined below:

Definition 1 *Let* y *be a real number in the interval* $[a, b]$. *The opposite number* $\check{y}$ *is defined as:*

$$\check{y} = a + b - y \tag{9}$$

For example, the opposite $\check{y}$ of $y = 3$ point in the interval $[-10, 10]$ is -3.

Definition 2 *Let* $Y(y_1, y_2, \ldots, y_d)$ *be the D dimensional point. The opposite point* $\check{y}_i$ *of* $y_i \in [a_i, b_i]$ *is defined as:*

$$\check{y}_i = a_i + b_i - y_i \tag{10}$$

For example, opposite $\check{Y}$ of a $2D$ point $Y = (5, 2)$ in the interval $[-10, 10]^2$ is $(-5, -2)$.

3.2 Concept of COBC

Rahnamayan [4] developed the COBC which has been a very useful scheme of OBL. It was first incorporated in DE algorithm resulted in better performance. The entire population of metaheuristic algorithm is considered while computing the centroid opposite points. Let $(X_1, ..., X_N)$ be N points in D-dimensional search space which carries unit mass. Then the centroid of the body is defined as the following:

$$M = \frac{X_1 + X_2 + X_3 + \ldots + X_N}{N} \quad (11)$$

The centroid point in jth dimension is computed as:

$$M_j = \frac{1}{N}\sum_{i=1}^{N} x_{i,j} \quad (12)$$

The centroid opposite point $\check{X}_i$ of X_i is defined as:

$$\check{X}_i = 2 \times M - X_i \quad (13)$$

3.3 Proposed GBC-COBL

The proposed scheme in developing the GPC-COBL algorithm has centroid opposition-based initialization and centroid opposition-based generation jumping. First, the initial population X is generated randomly using uniform distribution in the initialization phase. Then centroid opposite solutions $\check{X}(\check{x}_1, \check{x}_2, \ldots, \check{x}_D)$ get calculated. The dynamic search space range is $[a_j, b_j]$ where a_j and b_j is computed using the following equations:

$$a_j = \min_{\forall i}(x_{ij}) \quad (14)$$

$$b_j = \max_{\forall i}(x_{ij}) \quad (15)$$

where x_{ij} is the jth element of ith solution. If $\check{x}_{ij} > b_j$, then it is re-positioned as follows:

$$\check{x}_{ij} = M_j + (b_j - M_j) \times rand(0, 1) \quad (16)$$

If $\check{x}_{ij} < a_j$, then it is re-positioned as follows:

$$\check{x}_{ij} = a_j + (M_j - a_j) \times rand(0, 1) \quad (17)$$

After calculating the fitness of original and opposite solutions, the best N solutions are selected from $\{X, \check{X}\}$ after sorting based on fitness values. Opposition-based generation jumping using COBC with a generation jumping probability P_{gj} is employed in GPC after opposition-based initialization. The COBC is used to compute the opposite solutions in this phase, and the best N solutions from the original and opposite solutions are selected. The individual positions are updated using the basic position update rule with a probability $(1 - P_{gj})$. The pseudo-code of GPC-COBL is presented in Algorithm 2. In the first step of Algorithm 2, the population X of size N is initialized in the range $[X_{\min}, X_{\max}]$ where $X_{\min}$ and $X_{\max}$ are respectively the lower bound and upper bound. In step 2, opposite solutions are computed using COBC schemes, and best N solutions from original and opposite solutions are selected in step 3. In step 4, the best solutions among all candidate solutions are determined. The steps 5–22 run repeatedly until the termination criteria is met. In steps 7–9, the opposition-based computation is performed, and best N solutions are selected if the condition in step 6 is true. Otherwise, basic operations of GPC are performed in steps 11–19.

Algorithm 2: GPC-COBL Algorithm

```
1  Generate initial population (X) array of stone blocks or workers (population
   size)
2  Calculate the opposite positions of the population array of stone or workers X̌
   using COBC
3  Select best N solutions from {X, X̌}
4  Determine best worker as pharaoh's agent
5  for Iteration ← 1 to Maximum iteration do
6      if rand(0, 1) < P_gj then
7          calculate the opposite positions of the population array of stone or
           workers X̌ using COBC
8          Select best N solutions from {X, X̌}
9          Determine best worker as pharaoh's agent
10     else
11         for i ← 1 to N do // (all N stones block or workers)
12             Calculate amount of stone block displacement
13             Calculate amount of worker movement
14             Estimate new positions
15             Investigate possibility of substituting workers
16             Determine new position and new cost
17             if new cost < pharaoh's agent cost then
18                 Set new cost as pharaoh's agent cost
19             end
20         Sort solution for next iteration
21     end
22 end
```

Table 1 Parameter settings

Parameters	GPC	GPC-COBL
Population size (N)	30	30
Gravity (g)	9.8	9.8
Angle of ramp (θ)	14	14
Minimum friction ($\mu_{\min}$)	1	1
Maximum friction ($\mu_{\max}$)	10	10
Substitution probability	0.5	0.5
Generation jumping probability (P_{gj})	-	0.3

4 Experimental Setup

4.1 Parameter Settings

The parameter settings are given in Table 1.

4.2 Termination Criteria

1. Maximum function evaluations (FEs) = 300,000, or
2. Best-error-runs $E_{best} = |F^* - F| \leq \epsilon$ where F^* and F are respectively the global minima and the best solution achieved by the algorithm, $\epsilon = 1e-8$ is the threshold error.

5 Results and Discussion

The GPC-COBL is tested on 28 IEEE CEC2013 benchmark functions [3]. In this benchmark function suite, f_1–f_5 are unimodal functions, f_6–f_{20} are multimodal functions and f_{21}–f_{28} are the composite functions. The results of GPC-COBL are compared with classical GPC. Each algorithm is executed for 51 independent runs. The same initial population is used for the same run to make a fair comparison. Table 2 presents the mean and standard deviation of the best-error-runs over 51 independent runs, and bold-faced results indicate better. It has been noticed that the mean results of GPC-COBL are better than that of GPC for unimodal functions f_1, f_4, f_5, and again GPC-COBL performs better than GPC for basic multimodal functions for f_{11}, f_{12}, f_{13}, f_{17}, f_{18}, f_{19}, and f_{20}, and also again GPC-COBL performs better than GPC for composite functions f_{22}, f_{23}, and f_{28} whereas GPC performs better than GPC-

Table 2 Mean and standard deviation (in parenthesis) of best-run-errors over 51 independent runs for 30D problems

F	GPC	GPC-COBC	h
1	18571.70987 (2258.8166)	**13544.82044 (2960.4987)**	1
2	**76853821.7 (24282315.99)**	119391103.4 (33974302.87)	−1
3	**39769930305 (10382201979)**	44156896010 (15523368287)	0
4	51874.65978 (4241.8612)	**45913.92632 (4176.7070)**	1
5	1957.609307 (580.9996)	**1173.740523 (283.5266)**	1
6	**740.8203348 (154.7907)**	842.1183672 (361.2197)	0
7	**161.9914925 (46.3346)**	172.8036325 (38.7413)	0
8	20.95163933 (0.0581)	**20.9327948 (0.0552)**	0
9	**33.34487321 (2.8271)**	33.6274813 (2.2654)	0
10	**1577.71504 (338.9570)**	1747.440507 (330.6671)	−1
11	394.9065265 (30.0059)	**333.9010392 (38.0629)**	1
12	476.0318209 (70.7914)	**408.0269684 (46.0983)**	1
13	455.3580995 (61.5310)	**424.8795566 (44.5350)**	1
14	**4875.542293 (277.2015)**	5015.13341 (283.1800)	−1
15	**5361.845452 (582.0225)**	5459.377952 (556.4335)	0
16	**1.969204518 (0.6229)**	2.357596683 (0.4379)	−1
17	466.9667933 (31.9884)	**399.7582165 (54.8985)**	1
18	582.1482068 (85.3223)	**465.4018101 (51.6464)**	1
19	7314.405534 (4076.6506)	**3372.624757 (1754.6402)**	1
20	13.7976766 (0.4644)	**13.06464095 (0.6952)**	1
21	**2013.343231 (34.8244)**	2016.402293 (40.5537)	0
22	5155.60257 (355.5407)	**5251.081981 (294.1639)**	0
23	**6844.966649 (622.9171)**	6986.925265 (718.7758)	0
24	303.086662 (6.3785)	**294.7831289 (9.8940)**	1
25	**313.3670068 (2.9351)**	315.0469941 (3.9387)	−1
26	**208.1592107 (25.1440)**	208.6745065 (4.5634)	−1
27	1203.374402 (69.5318)	**1168.250484 (63.7710)**	1
28	3501.741482 (524.0558)	**3411.976749 (344.7728)**	0

COBL for unimodal function f_2. GPC performs better than GPC-COBL for basic multimodal function f_{10}, f_{14}, f_{16}, and composite function f_{25} and f_{26}. On the other hand, they both got tied for unimodal function f_3, multimodal function f_6, f_7, f_8, f_9, f_{15}, and composite function f_{21}, f_{22}, f_{23}, and f_{28}. We have conducted pairwise Wilcoxon Signed Rank Test [1] with the significance level (α) = 0.05 for analyzing the statistical significance. The test results have been given in the last column of Table 2. $h = 1$ indicates GPC-COBL statistically outperforms GPC, whereas $h = -1$ indicates that GPC statistically outperforms GPC-COBL. GPC-COBL outperforms the GPC for 12 problems, whereas GPC performs better for 6 problems and got tied for 10 problems. As we know from the *No Free Lunch* (NFL) theorem [6] that "if an algorithm performs well on a certain class of problems then it necessarily pays for that with degraded performance on the set of all remaining problems", the proposed GPC-COBL shows poor results on six problems compared to GPC algorithm.

6 Conclusion

We have proposed an improved version of GPC by incorporating the COBC scheme. The GPC-COBL has been tested on IEEE CEC2013 benchmark problems having unimodal, multimodal, and composite functions. The GPC-COBL is compared with classical GPC. The results illustrate that our proposed algorithm GPC-COBL statistically outperforms GPC for most of the problems. In the future, we intend to improve the GPC by incorporating other OBL schemes. We will also further improve the GPC-COBL by using adaptive generation jumping probability.

References

1. Derrac, J., Garcia, S., Molina, D., Herrera, F.: A practical tutorial on the use of nonparametric statistical tests as a methodology for comparing evolutionary and swarm intelligence algorithms. Swarm Evol. Comput. **1**, 3–18 (2011)
2. Harifi, S., Mohammadzadeh, J., Khalilian, M., Ebrahimnejad, S.: Giza pyramids construction: an ancient-inspired metaheuristic algorithm for optimization. Evol. Intell. (2020). https://doi.org/10.1007/s12065-020-00451-3
3. Liang, J., Qu, B., Suganthan, P., Hernández-Díaz, A.G.: Problem definitions and evaluation criteria for the CEC 2013 special session on real-parameter optimization. In: Technical Report 201212, Computational Intelligence Laboratory, Zhengzhou University, Zhengzhou China and Technical Report, Nanyang Technological University, Singapore (2013). https://personal.ntu.edu.sg/EPNSugan/index_files/CEC2013/CEC2013.htm
4. Rahnamayan, S., Jesuthasan, J., Bourennani, F., Salehinejad, H., Naterer, G.: Computing opposition by involving entire population. In: Proceedings of IEEE Congress on Evolutionary Computation (CEC), pp. 1800–1807. IEEE (2014)
5. Tizhoosh, H.: Opposition based learning: a new scheme for machine intelligence. In: International Conference Computational Intelligence for Modelling, Control and Automation and International Conference on Intelligent Agents, Web Technologies and Internet Commerce (CIMCA-IAWTIC'05), pp. 695–701. IEEE (2005)

6. Wolpert, D., Macready, W.: No free lunch theorems for optimization. IEEE Trans. Evol. Comput. **1**(1), 67–82 (1997). https://doi.org/10.1109/4235.585893

Parallelization of Cocktail Sort with MPI and CUDA

C. R. Karthik, Ashwin G. Shanbhag, B. Ashwath Rao, Prakash K. Aithal, and Gopalakrishana N. Kini

1 Introduction

Sorting is a way of arranging the data in a specific format. The sorting algorithm determines how data can be sorted. There are plenty of applications where these sorting algorithms are utilized. Major applications are in searching data in the list and matching entries in the list. Sorting algorithms can be categorized as in-place sorting and not in-place sorting. In-place [1], sorting algorithms are algorithms that do not require any extra space and sorting is intended to happen within the array or list itself. One of the commonly used in-place algorithms is Bubble sort. Not in-place algorithms refer to those algorithms that require equal or extra space based on the elements to be sorted. A general example of a not in-place sorting algorithm is merge sort. Sorting algorithms has a major role in the day-to-day life of programming because of their extensive range of applications.

MPI is an interface between processes. Processes coordinate and communicate via calls to message passing library routines. MPI addresses primarily the message-passing parallel programming model. The message-passing program can exchange local data by using communication operations. MPI consists of a collection of processes. Processes in MPI are heavy-weighted and single-threaded with separate address spaces. On many parallel systems, an MPI program can be started from the command line.

Nvidia Introduced CUDA, a parallel computing platform. CUDA [2] utilizes the power of GPU (graphical processing unit) for computing parallel parts and CPU (central processing unit) for the sequential part of the program. At first, GPU was only used for gaming purposes. Now due to its high computation capability, the use of GPU gradually extended to the field of science, maths, etc., when computation to

C. R. Karthik (✉) · A. G. Shanbhag · B. Ashwath Rao · P. K. Aithal · G. N. Kini
Department of Computer Science and Engineering, Manipal Institute of Technology, Manipal AcademyofHigherEducation, Manipal, Karnataka 576104, India

M. Dua et al. (eds.), *Proceedings of the International Conference on Paradigms of Communication, Computing and Data Sciences*, Algorithms for Intelligent Systems,
https://doi.org/10.1007/978-981-16-5747-4_20

be done is very high and operations are loosely coupled, using GPU increases the performance of the system.

In this paper, we parallelize the cocktail sort [3] algorithm using MPI and CUDA platforms. Before discussing the parallelization of cocktail sort, let's discuss cocktail sort in brief. Cocktail sort, also known as cocktail shaker sort is an in-place sorting algorithm. Like bubble sort [4], it is also an even and comparison-based sorting technique. Cocktail sort sorts the list in two phases per iteration. Firstly, it performs rightward pass and then leftward pass. After the rightward pass, the largest element is moved to the rightmost corner of the list and in the leftward pass, the smallest element is moved to the leftmost corner. The outcome of the first iteration is that the smallest element in the list will be placed on the leftmost corner in the list and the largest element is placed at the rightmost corner. These iterations are continued until the array or list is sorted.

Working of the cocktail sort is illustrated using the below example. Consider an array of integers as shown in Fig. 1. Now in cocktail sort, the initial procedure is to check the first two elements of the array if the first element is greater than the second it swaps the element. At the end of the first forward pass, the largest element will be at the rightmost corner as shown in Fig. 1 (left). Then, it performs a backward pass where at the end the smallest element is settled at the leftmost corner of the array as shown in Fig. 1 (right). Thus, the first iteration completes. If the array is not sorted, then it goes to the second iteration and follows the process same as that of iteration 1 as shown in Fig. 2 (left). But in iteration 2 s forward pass, we can see that at step 3 itself the array is sorted but the algorithm does not know that the array is sorted so it performs the second backward pass as shown in Fig. 2 (right) when no element is swapped in a pass it concludes that the array is sorted.

Fig. 1 First iteration. First rightward pass to take the largest element to the rightmost corner (left). First leftward pass to bring the smallest element to the leftmost corner of the array (right)

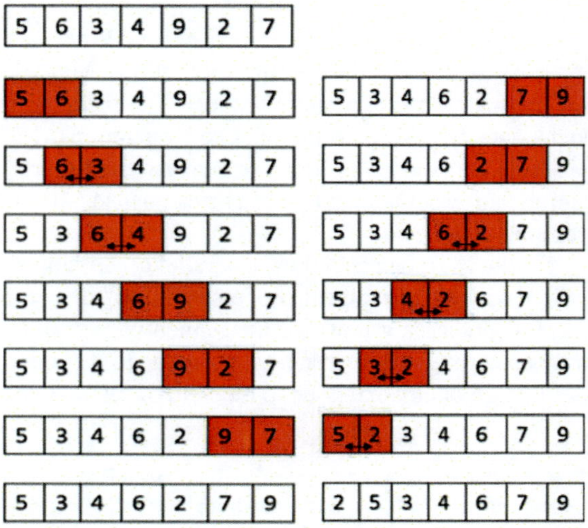

Fig. 2 Second Iteration. Second rightward pass (left). The array is sorted in the second forward pass itself but to confirm second leftward pass (right) is performed when no element is swapped in a pass it concludes that the algorithm is sorted

From the above example, the time complexity for the best case is $O(n)$, when the array is already sorted. But, in the case of average and worst-case when the array is not sorted the time complexity of the algorithm is $O(n^2)$. The rest of the paper is organized as follows: Sect. 2 gives a brief knowledge on previously carried work in the field of sorting as well as parallelizing the algorithm. Section 3 broadly discusses the implementation of the cocktail sort both sequentially as well as parallelly. Section 4 gives a detailed analysis of the results to support the proposed algorithm. Section 5 concludes the paper and Section "References" has the references used in this paper.

2 Literature Survey

In this section, we discuss various previously carried work in the field of sorting as well as parallelizing the algorithms using a parallel computing platform. Beniwal and Grover [5] in their paper compared different sorting algorithms like bubble, quick, heap, merge, insertion sort and derived a conclusion that speed is not the only factor that determines the good sorting algorithm but also it depends on length, the complexity of code, stability, datatype handling capacity, and performance also play a part in determining the good sorting algorithm. Joshi et al. [6] in their work compared the time complexity of different sorting algorithms like bucket, radix, and counting sorting by taken 1, 2, …, 10 as input and showed that non-comparison-based algorithms have $O(n)$ complexity which is better than comparison-based algorithm with $O\ (n \log n)$.

Chhajed et al. [7] in their work compared three sorting algorithms quick, heap and insertion sort, and showed that when compared to the heap and quick sort insertion sort is slower when the input of 10,000 and 30,000 random values were given.

However, the time complexity of all three algorithms is $O(n^2)$. Kumar and Chugh [8] in their paper compared the bubble, quick, merge, heap, and insertion sort based on speed by ranging the number of input values from 100,000 to 1,500,000. They also ranked these algorithms based on speed as follows merge, quick, heap, insertion, and bubble sort. They also showed that when the large data is used merge, quick and heap sort performance is better than the bubble and insertion sort.

Recently, the parallelization of Tim's sort, Super sort, cut sort, and Matrix sort [15–18] have been carried out using MPI and CUDA. Each of these sorting algorithms has been redesigned to run in a parallel environment consisting of CPU and GPU. The results are encouraging and prove that many sorting algorithms can be parallelized with a significant reduction in execution time.

Parallelizing the sorting algorithm increases the performance of the algorithm. But, parallelizing a sorting algorithm is a challenging task. Kumar et al. [9] in their hybrid approach showcased that how a sequential code can be parallelized with function level and block-level parallelization. Tjaden and Flynn [10] in their work proposed approaches for detection and execution of parallel instruction. David [11] in his study showcased the drawbacks of instruction, block, and functional level parallelization. In the case of instruction-level parallelization, there are the chances of high communication overheads. In function level parallelization there are chances that performance is not satisfactory, i.e., if there are two functions that if one function has more loops than the other the speedup of the system may be reduced. Block-level parallelization has better performance when compared with the other two but if there are more than one function calls in a block and they take part in determining the total complexity then running such blocks in parallel may not give good results.

In the case of selection sort [12], there is a need to find out min and max in the unsorted every time due to a decrease in the size of the unsorted list with each iteration. Hence, the recursive sorting algorithms are best suited for parallelization. Tsigas and Zhang [13] proposed an algorithm for parallelizing the quick sort by introducing the multiple threads during the time of partitioning the array into two. Varman et al. [14] introduced the parallel merge sort algorithm where they divided the unsorted array based on the number of threads and each partition was sorted by one thread. After sorting using each thread sorted partitions are merged. Using these concepts, we introduce a novel algorithm for parallelizing the cocktail sort using MPI and CUDA platforms which are discussed in detail in the next section.

3 Methodology

In this section, we discuss how the cocktail sort algorithm can be implemented in sequential as well as parallel using MPI and CUDA platforms. In the sequential implementation of cocktail sort, we just take N number of elements as input and simply perform the operations as discussed in Sect. 1. The general flow diagram for sequential cocktail sort is shown in Fig. 3.

Fig. 3 Flow diagram for sequential cocktail sort

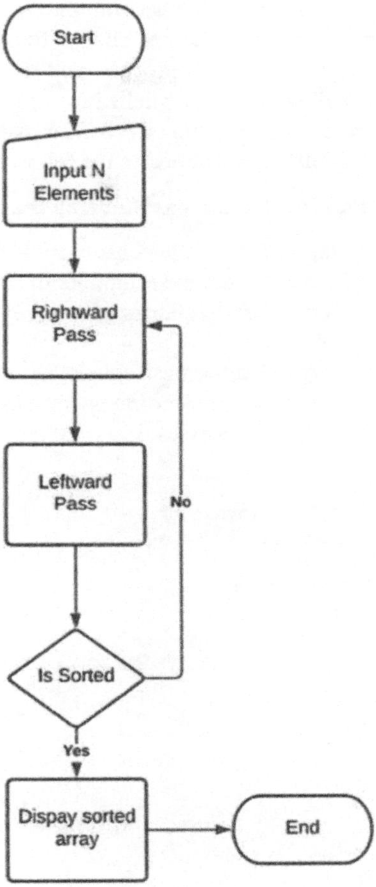

Steps involved in implementing sequential cocktail sort:

Step 1: Take *N* elements as an input that is to be sorted.
Step 2: Check whether the list is sorted or not, if not go to step 3.
Step 3: Perform right forward pass on the list.
Step 4: Perform left forward pass on the list.
Step 5: Perform steps 2 and 3 until the list is not sorted.
Step 6: If no elements in step 4 are swapped then the array is sorted. Hence, exit.

For implementing cocktail sort in parallel the general idea is distributing the number of elements (*N*) to the different processes so that the array is sorted simultaneously in different processes (*P*). Now the major task is to assign the processes with the number of elements (*N*). To do this, we divide the given array based on the number of processes (*N*/*P*) used, and then, we send these chunks to the respective processes. Then, we perform the sorting on these data chunks. After sorting is done, we gather all the sorted chunks for the root process. Later, these sorted chunks are merged in the root process, and the final result is printed. Figure 4 represents the flow diagram for parallelization of cocktail sort using MPI and Fig. 5 represents the implementation using the CUDA platform. Now, the implementation of cocktail sort parallelly is performed in the following steps.

Steps involved in parallelizing the cocktail sort using MPI:

Step 1: Take *N* elements as input that is to be sorted.
Step 2: Define the *P* number of processes.
Step 3: Create chunks of array based on the number of processes *P*. Chunk size is *N*/*P*.
Step 4: Scatter these chunks of array elements to all '*N*' processes.
Step 5: Perform sorting operations on these chunks of data then gather all the data from the processes.

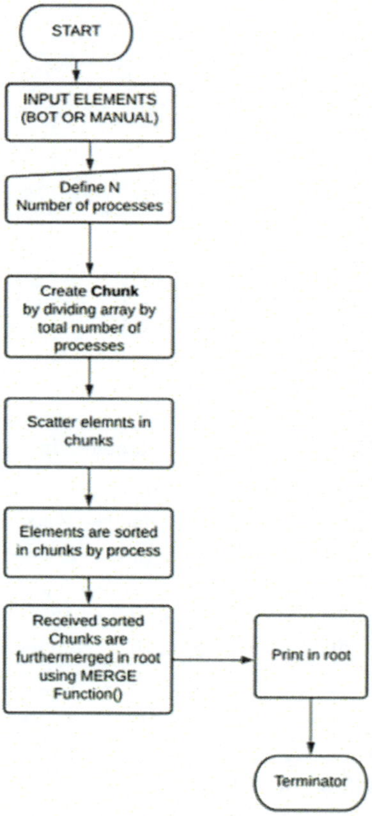

Fig. 4 Flow diagram for parallelizing cocktail sort using MPI

Fig. 5 Flow diagram for parallelizing cocktail sort using CUDA

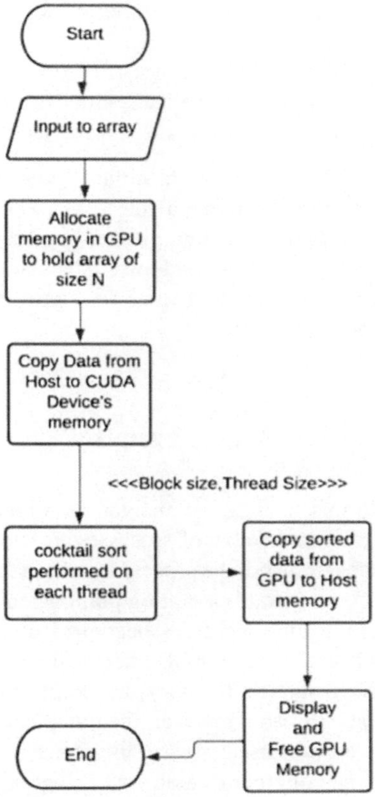

Step 6: Gathered sorted chunks from all the processes are merged in root using the merge function.
Step 7: Print the sorted elements in the root.

From the above steps, we could observe that for the odd number of steps we need $(N/2) - 1$ number of processes and for even-numbered steps require $(N/2)$. Hence, it needs $O(n)$ processes. To perform sequential cocktail sort for N elements time complexity of $O(n^2)$. The time complexity of parallel implementation is $O(N/P)$ where N is the number of elements and P is the number of processes. So, the time complexity of the parallel implementation in MPI can be written as $O(n^2)/O(n)$ resulting in $O(n)$ complexity. Hence, the time complexity of the parallel cocktail sort in MPI is $O(n)$.

Steps involved in implementing cocktail sort in CUDA:

Step 1: Take input N elements into an array to be sorted.
Step 2: Allocate the variable of size array in the CUDA device's memory.
Step 3: Copy all elements from the host's memory to GPU's memory.

Step 4: Calling parallel cocktail sort kernel function with desired blocks and threads.
Step 5: Computation of sorting performed on each thread simultaneously.
Step 6: Copying back sorted array back from the device to the host's memory.
Step 7: Display sorted array with the time taken. Also, free GPU's memory.

For testing the algorithm, using random values of a different range may lead to different results each time tested. This may mislead the performance of the algorithm. To avoid shortcomings and to achieve the best possible results by minimizing the errors, we have made use of a standard dataset [T10I4D100K (.gz)] [19]. Hence, using the standard value for sorting will provide an ideal condition for conducting the experiments.

4 Result and Analysis

In this section, we analyze the time consumed by the parallel algorithm when the different number of processes are used to sort the array of different lengths. In MPI as the number of the process increases the time consumed for computation decreases but after a certain saturation point when the number of the process is increased computation time increases becoming directly proportional to the number of processes. This is because in MPI heavyweight threads are used, due to which each thread will be having its memory; hence, it requires a lot of time for communication among each thread. However, the computation time is purely dependent on the number of processes used. Hence, to get better performance choosing the processes according to the number of elements is a good practice. To test the algorithm, we use a standard dataset [T10I4D100K (.gz)] [19]. Table 1 shows the time taken by the sequential algorithm to sort the N number of elements. Where we can observe that as the number of elements increases the computation time also increases.

Table 2 represents the time taken by the algorithm in MPI to sort the data by ranging the number of processes when the number of elements is 30,000. In MPI, the number of processes has to be specified explicitly while at the time of execution. From Table 2, it can be observed that the number of processes is ranging in between 1 and 64. But after 64, the time gradually starts increasing which means that for 30,000 elements the number of processes required is around 64 to achieve the best possible

Table 1 Computation time taken for different number of elements for sequential algorithm

Number of elements (*N*)	Time (ms)
1000	0.005
5000	0.097
10,000	0.308
30,000	2.069
100,000	22.544

Table 2 Computation time taken by the algorithm when the number of elements is 30,000

Number of processes (*P*)	Time (ms)
1	2.141
2	0.772
4	0.223
8	0.080
16	0.067
32	0.031
64	0.026

Table 3 Computation time taken by the algorithm when the number of elements is 100,000

Number of processes (*P*)	Time (ms)
1	21.547
2	6.168
4	2.533
8	0.967
16	0.452
32	0.238
64	0.154
128	0.111

parallelism. Hence, choosing the proper number of processes based on the given input elements plays a major role in MPI execution. Similarly, Table 3 represents the time consumed by the algorithm in MPI when the number of elements is 100,000, and the number of processes is ranging from between 1 and 128. The graph for processes versus time is plotted when the number of elements is 30,000 and 100,000 which is shown in Figs. 6 and 7, respectively.

$$\text{Speedup} = \frac{\text{Time taken by sequential algorithm}}{\text{Time taken by parallel algorithm}} \tag{1}$$

Considering data from Tables 1 and 3, the speedup can be computed using Eq. (1). The speedup is defined as the ratio of time taken by a sequential algorithm to the time taken by a parallel algorithm. To compute consolidated speedup, time taken by various processes in Tables 2 and 3 has been averaged, respectively. The speedup of MPI against sequential is shown in Table 4.

In CUDA, the user need not specify the number of processes as each thread takes an element and sorting operation is performed. Hence, based on the number of elements, CUDA generates the same number of threads for sorting the data. Table 5 represented the computation time taken by the algorithm when the CUDA platform is used. To test the computation time, the number of elements is ranged between 1000 and 100,000, and the respective time is noted down. The graph for number elements

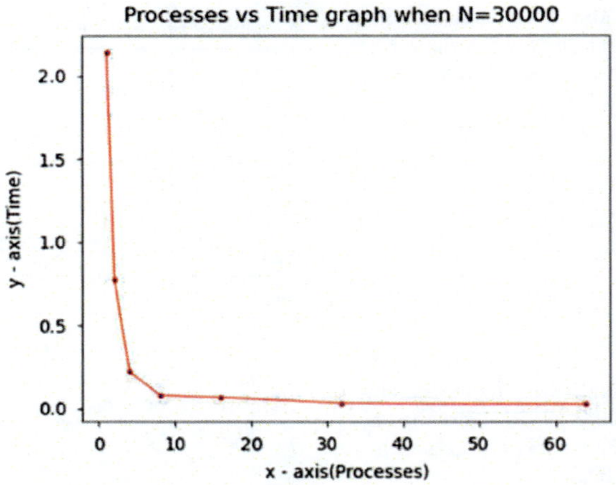

Fig. 6 Graph of processes versus time when the number of elements is 30,000

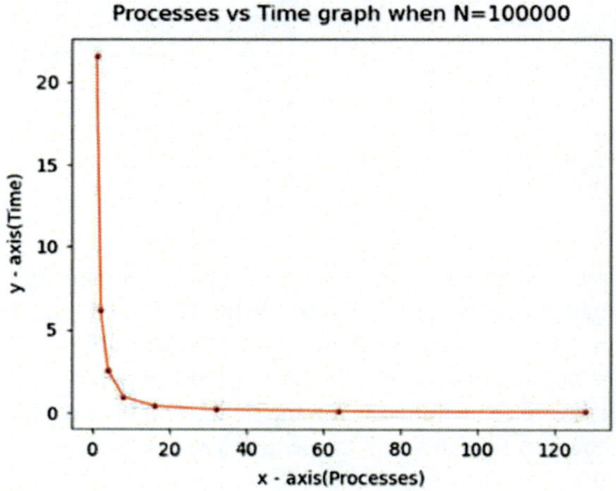

Fig. 7 Graph of processes versus time when the number of elements is 100,000

Table 4 Speedup of MPI against sequential

Number of elements (N)	Speedup
30,000	4.336
100,000	5.606

Table 5 Computation time taken for different number of elements when CUDA platform is used

Number of elements (N)	Time (ms)
1000	0.002765
5000	0.015432
10,000	0.052670
30,000	0.381782
100,000	3.444165

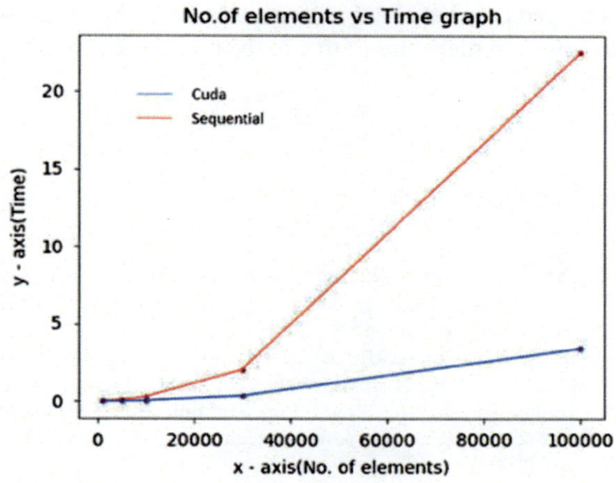

Fig. 8 Graph of the number of elements versus time for both sequential and parallel algorithm using CUDA

Table 6 Speedup of CUDA against sequential

Number of elements (*N*)	Speedup
1000	1.8083
5000	6.2856
10,000	5.8477
30,000	5.4193
100,000	6.5456

versus time is plotted in comparison with the sequential execution time as depicted in Fig. 8. Using Eq. (1), the speedup of the algorithm is calculated which is noted in Table 6.

5 Conclusion and Future Scope

In this paper, we discuss how cocktail shaker sort can be implemented in a parallel manner using computing platforms like MPI and CUDA. We have also analyzed the computation time taken by the sequential as well as a parallel algorithm using a standard dataset so that experimentation could be performed in ideal conditions. Tabulation of empirical results and representation of it in a graphical manner has been accomplished. We also showed that parallelizing a cocktail sort reduces its time complexity from $O(n^2)$ to $O(n)$ where 'n' is the number of elements. Thus, the results of the experiment depict that parallelizing the cocktail sort gives additional strength to the algorithm by increasing the performance of the algorithm. In the future, this parallelized algorithm can be packaged into a library and used in multiple platforms where sorting is essential. This parallel implementation can be containerized and deployed over the cloud as free software so that if anyone intends to perform sorting in bulk, can make use of this in the best possible way as the time taken is decreased exponentially.

References

1. Prajapati, P., Bhatt, N., Bhatt, N.: Performance comparison of different sorting algorithms. Int. J. Latest Technol. Eng. Manag. Appl. Sci. **VI**(Vi), 39–41 (2017)
2. Stojanovic, N., Stojanovic, D.: High-performance processing and analysis of geospatial data using CUDA on GPU. Adv. Electr. Comput. Eng. **14**(4), 109–115 (2014)
3. Elkahlout, A.H., Maghari, A.Y.: A comparative study of sorting algorithms comb, cocktail and counting sorting. Int. Res. J. Eng. Technol. (IRJET) **4**(1) (2017). e-ISSN 2395-0056
4. Astrachan, O.: Bubble sort: an archaeological algorithmic analysis. ACM Sigcse Bull. **35**(1), 1–5 (2003)
5. Beniwal, S., Grover, D.: Comparison of various sorting algorithms: a review. Int. J. Emerg. Res. Manag. Technol. **2** (2013)
6. Joshi, R., Panwar, G., Pathak, P.: Analysis of non-comparison-based sorting algorithms: a review. Int. J. Emerg. Res. Manag. Technol. (2013)
7. Chhajed, N., Uddin, I., Bhatia, S.S.: A comparison-based analysis of four different types of sorting algorithms in data structures with their performances. Int. J. Adv. Res. Comput. Sci. Softw. Eng. **3**(2), 373–381 (2013)
8. Kumar, G., Chugh, H.: Empirical study of complexity graphs for sorting algorithms. Int. J. Comput. Commun. Inf. Technol. (IJCCIT) **1**(1) (2013)
9. Kumar, K.A., Pappu, A.K., Kumar, K.S., Sanyal, S.: Hybrid approach for parallelization of sequential code with function level and block level parallelization. In: International Symposium on Parallel Computing in Electrical Engineering (PARELEC'06), pp. 161–166. IEEE (2006)
10. Tjaden, G.S., Flynn, M.J.: Detection and parallel execution of independent instructions. IEEE Comput. Archit. Lett. **19**(10), 889–895 (1970)
11. Wall, D.W.: Limits of instruction-level parallelism. In: Proceedings of the Fourth International Conference on Architectural Support for Programming Languages and Operating Systems, vol. 19, pp. 176–188 (1991)
12. Jadoon, S., et al.: Design and analysis of optimized selection sort algorithm. Int. J. Electr. Comput. Sci. (IJECS-IJENS) **11**(01), 16–22 (2011)
13. Tsigas, P., Zhang, Y.: A simple, fast parallel implementation of quicksort and its performance evaluation on SUN enterprise 10000. In: Eleventh Euromicro Conference on Parallel, Distributed and Network-Based Processing, 2003. Proceedings, pp. 372–381. IEEE (2003)
14. Varman, P.J., Scheufler, S.D., Iyer, B.R., Ricard, G.R.: Merging multiple lists on hierarchical-memory multiprocessors. J. Parallel Distrib. Comput. **12**(2), 171–177 (1991)
15. Thanagaraja, S., Shanbhag, K., Ashwath Rao, B., Shwetha Rai, N., Kini, G.: Parallelization of Tim sort algorithm using MPI and CUDA. J. Crit. Rev. **7**(9), 2910–2915 (2020)
16. Pereira, S.D., Ashwath, R.B., Rai, S., Gopalakrishna Kini, N.: Super sort algorithm using MPI and CUDA. In: Intelligent Data Engineering and Analytics, pp. 165–170. Springer, Singapore (2020)
17. Yadav, H., Shraddha Naik, B., Rao, A., Rai, S., Kini, G.: Comparison of CutShort: a hybrid sorting technique using MPI and CUDA. In: Evolution in Computational Intelligence, pp. 421–428. Springer, Singapore (2020)
18. Ojha, P., Singh, P., Gopalakrishna Kini, N., Ashwath Rao, B., Rai, S.: Parallel matrix sort using MPI and CUDA. In: RTU TEQIP-III Sponsored 2nd International Conference on Communication and Intelligent Systems (ICCIS 2020), pp. 26–27 (2020)
19. Frequent Itemset Mining Implementation's Repository. http://fimi.cs.helsinki.fi. Accessed 11 Feb 2021

Review of Research Challenges and Future of in DNA Computing Applications

Sapna Jain and M. Afshar Alam

1 Introduction

A DNA PC consists of explicit DNA strands. The unique blend of DNA strands is used in modern applications. Nano computers are another idea to comprehend because these machines use DNA for the capacity of information and to finish muddled estimations. DNA itself works as a structure block in a tremendous scope of zones, for example, nanotechnology and atomic registering. Likewise, it is significant not to be too restricted by the way we imagine a "PC." When we talk about PCs, the more substantial part of us is thinking about our work area PCs or our helpful PC. DNA PCs are an extraordinary idea, although they have applications that could uphold our daily figuring needs. DNA PCs can be minuscule enough to work in the human body, where they may one day perform undertakings, for example, distinguishing unhealthy cells or delivering insulin as needed for a diabetic patient [1, 2]. DNA method is the natural atomic process or sub-atomic figuring. The Dutch Leonard Adelman, United Nations agency, started adding this field [3] at intervals in 1994. His analysis findings focused on using code of deoxyribonucleic acid sub-atomic structure to accomplish estimation. R. Lipton worked [2] on deoxyribonucleic acid and its satisfiability issues. He also and its method to use the required technique to store and utilize it. A deoxyribonucleic acid pc has a high thickness of knowledge warehousing limit, tremendous equal preparing capability, and unprecedented energy productivity.

S. Jain (✉) · M. Afshar Alam
Jamia Hamdard, New Delhi, India
e-mail: drsapnajain@jamiahamdard.ac.in

M. Afshar Alam
e-mail: aalam@jamiahamdard.ac.in

M. Dua et al. (eds.), *Proceedings of the International Conference on Paradigms of Communication, Computing and Data Sciences*, Algorithms for Intelligent Systems,
https://doi.org/10.1007/978-981-16-5747-4_21

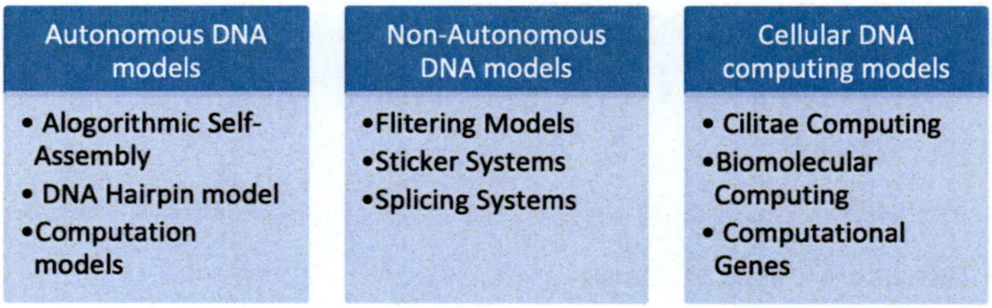

Fig. 1 Commonly used DNA models in research

2 Research Challenges

See Fig. 1.

The research challenges depend on the type of DNA model used in an application as shown in Fig. 1.

2.1 Autonomous DNA Models Research Limitations

The second era of DNA processing focuses on models are atomic scale, independent, and incompletely programmable calculations are determined by the self-gathering of DNA particles and are regulated by DNA-controlling proteins. The possibility of algorithmic self-get together emerged from the combination of DNA figuring, tiling hypothesis, and DNA nanotechnology. Algorithmically self-gathered structures length a decent reach between maximally basic structures (precious stones) and self-assertively intricate informatics tiling's (supramolecular edifices). Algorithmic self-get together is agreeable to test examinations. Permitting the comprehension of included actual wonders. This comprehension may ultimately end in new nanostructured materials and gadgets [4, 5]. DNA self-assembly is another crucial topic that is effectively used through numerous organizations inside the subject, remaining used at least, it is an ambitious undertaking to connect the distance between base up DNA self-assembly with top-down nano-and MICR-creation systems to make high throughput nanodevices. Current advancement can likewise clear the way toward this fundamental reason [6]. Algorithmic self-assembly [7] is agreeable to test examinations allowing the data to included true marvels. This information may also in the end in new nanostructured materials and contraptions. In artificial neural network (ANN), the studying method is fundamentally based totally on the interconnections among the preparing components that speak to the community topology [8].). The Turing contraption idea depends absolutely on taking strolls an appropriately depicted framework that produces changes and head development on a boundless image tape. The image set is limited and characterizes the Turing machine letter set. The top trademark characterizes the new picture to be examined and likely overwritten [9].

2.2 Non-autonomous DNA Models Research Limitations

Early biomolecular registering research was based on lab-scale human-worked DNA models of calculation for tackling complex computational issues. These models produce enormous combinatorial libraries of DNA to give search spaces to resemble separating calculations. A wide range of techniques for library age, arrangement separating, and yield age was there. The presence of shallow atomic weight DNA pieces might be connected to the fracture of nucleosomes from atomic chromatin with an intracellular endonuclease guide during customized cell death toll [10]. Charlton et al. researched the atomic load of DNA from explained, enormous scope, taking care of many mammalian versatile. The way of life is considered in every apoptosis and not resolved in portable tradition conditions, they could not further choose the wellspring of those more modest DNA parts with the fact [11]. Splicing is a fundamental advance of the quality guideline. It permits singular qualities to create different protein stock with explicit frameworks and highlights through the addition or erasure of indispensable useful space names encoded by, on the other hand, joined exons. Varieties in the quality articulation levels of joining administrative segment resolved in masses of malignant growths, and those proteins as often as possible influence the grafting sorts of numerous qualities that trademark ensures most tumors clear natural pathways, including versatile cycle improvement, cell multiplication, and movement, and RNA preparing [12, 13].

2.3 Cellular DNA Computing Models Research Limitations

Ciliates are unicellular eukaryotes chosen in different conditions such as simple comparably to soil for the globe's length that arose more noteworthy than 1 billion years in the past [14]. The computational quality expectation is getting more noteworthy and extra essential for the computerized assessment and explanation of gigantic uncharacterized genomic arrangements. In the past couple of years, numerous quality expectation bundles created. The vast majority of them call to us on the web website. Utilizing went Ian quality revelation in prokaryotic genomes is significantly less intense because of the better-quality thickness of prokaryotes and the nonattendance of introns of their protein-coding territories [15]. The plan of DNA calculation is specifically basically dependent on a "preliminary, what is more, mistakes" technique? The total of the quantitative and prescient data, execution designing format and usage, can consequently scarcely ever be utilized to shape a model-basically based sub-atomic calculation stage indeed. Furthermore, the natural materials, comprising DNA, RNA, or proteins, cannot be reused. At the equivalent time, the DNA arrangements combined for a chose biomolecular figuring might be burned-through or perhaps obliterated at a couple of levels inside the usage of an arrangement of rules. Thirdly, DNA computational analyses are inclined to mistakes. Fourthly, regardless of the truth that biomolecular registering has set up that to find answers for muddled numerical issues, which incorporates np-complete issues, dramatically multiplying the measure of DNA can be required even as the np-whole issues' size develops straight. As a final quit product, it will quick arise as an unimaginable computational endeavor as the measure of hubs for an np-entire issue will increment. Finally, biocomputing sets aside bounty longer effort for every calculation, assessing the silicon-based PC structures. Regularly, executing a calculation to cure a computational issue may require various days or maybe weeks. Simultaneously, as a current introductory circumstance wishes to be analyzed, a similar time-frame is needed for some extraordinary run of the count. Hence, it is miles awkward and costly to impact the biocomputing tests that require rehashed improvement draws near.

3 Future of DNA Applications

3.1 DNA Nanotechnology Application Areas

The streamlining of the DNA meeting approach is moreover of exact awareness to bring DNA nanotechnology into big business cordial use. Furthermore, the fuse of supramolecular associations is symmetrical to DNA base-blending in DNA nanomaterials furnishes new design boundaries with the helpful asset of bringing new get-together modes and highlights.

(a) ***DNA nanostructure-based artificial membrane channel***

DNA nanostructures-based counterfeit film channels highlight opening channels with open closures. It is genuinely preferred to expand engineered layer channels with improvements in responsive opening or last assets. Forty-six to help the circle of biomimetic channels boost responsive DNA nanotechnology was misused to broaden the knowledge of counterfeit film channels. The reproducing film channel with DNA nanotechnology is an engaging liable to recognize questions regarding how channel proteins work. Future examination interest during this region investigates more bendy DNA pore models with different morphology and execution. The gratitude to incorporate DNA-principally based channels with homegrown proteins and different added substances is a couple of other tasks to amass multifunctional film channel that contenders' characteristic ones. With additional DNA nanotechnology advancements, miles expected that over some layer channel occasions would be recreated by utilizing DNA-based nanopores [16].

(b) ***DNA nanostructure-based membrane-floating proteins***

Specifically, mindfulness at the dynamic gets together DNA-copied layer proteins practices and, like this, DNA meeting capacity to substitute the vesicles' morphology. At this stage, the reversible gathering and dismantling ought to be additional bendy and accomplished in a too purchaser control way. Moreover, the elements of these gathering responses should be accelerated to meet practical applications, including drug conveyance and sign transduction [16].

(iii) ***DNA nanostructure-based artificial organelles***

DNA-basically based nanoreactors; researchers additionally fabricated numerous DNA nanocages that work dynamic practices. It is expressed an addressable functionalized DNA box with a curiously sizeable inward pit that transformed into prepared to encasing cargoes. The top becomes altered with a double lock key machine. It opened by adding one of a kind the key oligonucleotides in an incredibly individual oversee manner. More locking devices might counter various cautions, similar to proteins, little atoms, and metal particles, furthermore to temperature or possibly strain. Particular DNA-based manufactured compartmental frameworks could likewise build to emulate the homegrown compartment, including mitochondria, endoplasmic reticulum, lysosomes, and various organelles [16].

(iv) ***DNA nanostructure-based artificial cells***

DNA gel can work the inward cytoskeleton of counterfeit cells, which is fundamentally the same as the cytoskeleton in stay cells, to reinforce layer frameworks. This proposed lipid-DNA cytoskeleton is frequently considered an ideal gadget to make engineered cells and see how cells work. Both the skin and internal cytoskeleton layout strategies give a street to direct the lipid bilayer's development and dynamic changes. Also, deviated and monstrous film vesicles can likewise be worked through a reasonable plan, significant for reproducing cell-acknowledged shape. In a word, the DNA platform might be a successful simple of the cytoskeleton to manage the arrangement of engineered cell shape [16].

3.2 DNA and Cloud Computing

Cloud computing in the organic gadget will exchange the method's situation closer to organic problems with plenty of quicker information acquisition and analysis prices. It could deploy for the development of DNA identifiers based on genome sequences as generation advances. There were many cloud computing applications within genomics and different biological studies and improvement regions in the last few years.

(a) ***Genome Informatics***

These days, the user has the option to establish a record with amazon web contributions is the most punctual transporter guarantor to understand a reasonable distributed computing climate which is one of the diverse business transporters, discharge a computerized gadget model from an enormous kind of conventional and bioinformatics-orientated pics and interface one and all of various huge public genome-arranged datasets. For advanced gadget pictures, we could select previews pre-populated with the system: a unique net-based gadget for acting numerous typical genome examination obligations, Bioconductor: a programming climate joined with the r insights group, browse a genome program, a total arrangement of bioinformatics modules written in the Perl programming language, java cloud bio-Linux: a fixed of bioinformatics gadget which incorporates the Celera constructing agent, and an implication of others. Various photographs that run specific occasions of the US of America. Genome program is underneath advancement.

(b) ***Comparative Genomics***

The torrent of DNA realities presented through the ensuing period, and third-era sequencing procedures is distressing computationally serious application for re-enactment and handling of data which cannot be added by methods for the regular bioinformatic gadget assessment.

(iii) ***Genome Analysis and SNP Detection***

Distributed computing discovered via MapReduce and Hadoop can be utilized to adequately parallelize existing sequential executions of assortment arrangement and genotyping calculations. This mix allows enormous datasets of DNA arrangements to be investigated quickly without forfeiting exactness or requiring sizeable programming program designing endeavors to parallelize the calculation (Langmead et al., 2009). Some other cloud-based pipeline cloud map generously streamlines the examination of freak genome successions. It is miles accessible at the system net stage, and it requires no product program set up when run on the cloud; it could likewise be locally or through amazon's ec2 transporter.

3.3 DNA Fingerprinting

DNA fingerprinting appears as a reliable scientific gadget when pleasantly finished; a couple of researchers have alluded to more extensive testing of human DNA to safeguard that the fragments examined are fabulously crucial for all ethnic and racial associations. It is conceivable to make artificial hereditary examples and use them to mislead measurable specialists, notwithstanding if those examples devised the use of quality enhancement systems, they might be conspicuous from definitive DNA proof.

(a) ***Paleontology***

Know-how molecular diagenesis across geological time scales and spotting preserved biomarkers from the fossil record may additionally aid in our search for evidence of lifestyles on other planets. The search for extra-terrestrial life rests on three possibilities, mainly, existence may additionally never have existed, existence may additionally have existed for a brief time, then gone extinct, or lifestyles can be presently thriving. If the second scenario happens, all that may leave as proof are resistant molecular markers specific to existence. We must be able to apprehend the range of diagenetic alteration of biomolecules across time to come across them on other planets, wherein lifestyles may also become a tenuous start, after which became extinct. Molecular paleontology has an excellent deal to contribute to the look for existence on different planets and add to our know-how of the evolution and extinction of life on this one.

(b) ***Archaeology***

Nuclear DNA has been recovered from skeletal remains hundreds of years vintage and may assist determine a person's sex and, at instances, ancestral historical past. Intriguingly, the artifacts that would have come into touch with DNA-wealthy frame fluids, including saliva, can also be recovered.

3.4 DNA and Medical Diagnostics

DNA-essentially based examination stays in its earliest stages.

(a) ***Heart Disease Prediction***

The occurrence of coronary illness keeps on ascending in the United Kingdom (UK) and numerous world pieces. Coronary disease is a perplexing transaction of the way of life elements and hereditary qualities. As of late, specialists in the UK have found a strategy for recognizing those who have a raised risk of coronary illness. They found that telomeres, which are tiny DNA strands, are represented as closures of chromosomes that help to store essential data that help to keep the coronary illness. They found it limited due to the severe danger of creating coronary disease in men matured 45–64 years of age. The telomeres are estimated in leukocytes, otherwise called white platelets. Scientists accept that as telomere length diminishes, an individual's chromosomes are bound to transform. It identifies with the defensive impact of telomeres, which help to forestall harm to chromosome closes.

(b) ***Brain Disorders Prediction***

In the United States, researchers took a gander at information from individuals with ALS and Parkinson's illness, just as the individuals who did not have the infections. They discovered contrasts in qualities that permitted them to anticipate those people who had an expanded danger for the sicknesses. These distinctions, given to the analysts, examined the axon direction pathway. This pathway includes a convoluted gathering of synthetically interceded messages significant in the cerebrum during fetal development. They work to help and fix the "wiring" of the mind during an individual's whole life [14, 17].

4 Impact of DNA Application

DNA particles can be used as building segments of sub-nuclear robots. Up until this point, regardless, simply clear limits have been cultivated with such DNA robots, for instance, walking around a controlled plan and getting nanoparticles. By and by, Lulu Qian and accomplices from Caltech have developed a DNA robot prepared for performing all the more confounding cargo orchestrating tasks; the heap orchestrating DNA robot is recognized using an essential computation and three sub-nuclear structure blocks [17]. A molecular robot [18] can be best depicted as a coordinated framework shaped through the blend of various atomic parts or gadgets that may fill in as processors or, on the other hand, rationale doors, actuators, and sensors [19]. An atomic robot must be self-sufficient in accepting data from its encompassing and making choices through its capacity of nuclear calculation. As atomic sensors, different DNA-based nanostructures have been created to sense an assortment of signs and convert the data to yield signals for other pieces of the robots [11, 20–28]. Photograph responsive atoms give detecting capacity to the robots [29, 30]. As the data processor DNA-based gadgets, for example, teeter-totter entryways are promising [31, 32]. Control of DNA utilizing photochemical innovation has seemed a promising apparatus to build novel devices for atomic processing [33, 34]. Biomolecular engine proteins, for example, actin-myosin or microtubule-kinesin dynein, have been the most ideal options as actuators for sub-atomic robots [35, 36]. Biomolecular engines can change over compound energy into mechanical work witan astoundingly high productivity and subsequently have been promising as the actuator to drive engineered frameworks [37–40].

5 Conclusion

The field of DNA registering and DNA pc stays alive and promising, at the same time as new difficulties get up. Generally, giant among those is the vulnerability, given the DNA technology, within the computational consequences, and the brilliant enlargement in the number of DNA atoms crucial to address problems of captivating size. As of the overdue announcement of Moore's regulation growing old, the requirement for new registering standards for quicker and elite calculation is in the hunt. DNA writing is the various springing and processing of perfect models expected to replace the traditional silicon desktops. This paper discusses the significance of DNA in engineering and biomolecular contraptions. The territory of DNA registering holds tremendous potential to have been investigated due to its programs in numerous one of a kind fields. However, DNA registering remains at its starting phases, and some constraints have to be defeated earlier than it can indeed supplant silicon computers.

References

1. Sori, A.A.: DNA computer; present and future. J. Eng. Res. Appl. **4**(6), 228–232 (2014) [Online]. Available at: www.ijera.com
2. Lipton, R.J.: DNA solution of hard computational problems. Science (80-) **268**(5210), 542–545 (1995). https://doi.org/10.1126/science.7725098
3. Hameed, K.: DNA computation based approach for enhanced computing power. Int. J. Emerg. Sci. **1**(1), 31–37 (2011)
4. Am. J. Sociol. **53**(9) (2019)
5. Li, D., Huang, H., Li, X., Li, X.: Hairpin formation in DNA computation presents limits for large NP-complete problems. BioSystems **72**(3), 203–207 (2003). https://doi.org/10.1016/S0303-2647(03)00145-X
6. Lin, C., Ke, Y., Liu, Y., Mertig, M., Gu, J., Yan, H.: Functional DNA nanotube arrays: bottom-up meets top-down. Angew. Chem. **119**(32), 6201–6204 (2007). https://doi.org/10.1002/ange.200701767
7. Chhabra, R., Sharma, J., Liu, Y., Rinker, S., Yan, H.: DNA self-assembly for nanomedicine. Adv. Drug Deliv. Rev. **62**(6), 617–625 (2010). https://doi.org/10.1016/j.addr.2010.03.005
8. Lancashire, L.J., Lemetre, C., Ball, G.R.: An introduction to artificial neural networks in bioinformatics—application to complex microarray and mass spectrometry datasets in cancer studies. Brief. Bioinform. **10**(3), 315–329 (2009). https://doi.org/10.1093/bib/bbp012
9. Muraru, M., Popovici, M.-D.: DNA Computing—Modelling and Simulating a Molecular Turing Machine. U.P.B. Sci. Bull. Ser. C **71**(4) (2009)
10. Khanal, O., et al.: DNA retention on depth filters. J. Memb. Sci. **570–571**, 464–471 (2019). https://doi.org/10.1016/j.memsci.2018.10.058
11. Endo, K., Hayashi, K., Inoue, T., et al.: A versatile cis-acting inverter module for synthetic translational switches. Nat. Commun. **4**(1), 2393 (2013). Kabir, A.M.R., et al.: Sci. Technol. Adv. Mater. **21,** 331 (2020)
12. Adleman, L.M.: Adleman1994. Science (80-) **266**, 1021–1024 (1994)
13. Leier, A., Richter, C., Banzhaf, W., Rauhe, H.: Cryptography with DNA binary strands. BioSystems **57**(1), 13–22 (2000). https://doi.org/10.1016/S0303-2647(00)00083-6
14. Namasudra, S., Devi, D., Kadry, S., Sundarasekar, R., Shanthini, A.: Towards DNA based data security in the cloud computing environment. Comput. Commun. **151**, 539–547 (2020). https://doi.org/10.1016/j.comcom.2019.12.041
15. Wang, Z., Chen, Y., Li, Y.: A brief review of computational gene prediction methods. Genomics Proteomics Bioinform. **2**(4), 216–221 (2004). https://doi.org/10.1016/S1672-0229(04)02028-5
16. Shen, H., Wang, Y., Wang, J., Li, Z., Yuan, Q.: Emerging biomimetic applications of DNA nanotechnology. ACS Appl. Mater. Interfaces **11**(15), 13859–13873 (2019). https://doi.org/10.1021/acsami.8b06175
17. https://theconversation.com/organic-computers-made-of-dna-could-process-data-inside-our-bodies-46364
18. Rashedul Kabir, A.M., Inoue, D., Kakugo, A.: Molecular swarm robots: recent progress and future challenges. Sci. Technol. Adv. Mater. **21**(1), 323–332 (2020).https://doi.org/10.1080/14686996.2020.1761761
19. Hagiya, M., Konagaya, A., Kobayashi, S., et al.: Molecular robots with sensors and intelligence. Acc. Chem. Res. **47**(6), 1681–1690 (2014)
20. Tanaka, F., Mochizuki, T., Liang, X., et al.: Robust and photocontrollable DNA capsules using azobenzenes. Nano Lett. **10**(9), 3560–3565 (2010)

21. Yang, Y., Endo, M., Hidaka, K., et al.: Photo-controllable DNA origami nanostructures assembling into predesigned multiorientational patterns. J. Am. Chem. Soc. **134**(51), 20645–20653 (2012)
22. Suzuki, Y., Endo, M., Yang, Y., et al.: Dynamic assembly/disassembly processes of photoresponsive DNA origami nanostructures directly visualized on a lipid membrane surface. J. Am. Chem. Soc. **136**(5), 1714–1717 (2014)
23. Endo, M., Miyazaki, R., Emura, T., et al.: Transcription regulation system mediated by mechanical operation of a DNA nanostructure. J. Am. Chem. Soc. **134**(6), 2852–2855 (2012)
24. Saito, H., Kobayashi, T., Hara, T., et al.: Synthetic translational regulation by an L7Ae-kink-turn RNP switch. Nat. Chem. Biol. **6**(1), 71–78 (2010)
25. Saito, H., Fujita, Y., Kashida, S., et al.: Synthetic human cell fate regulation by protein-driven RNA switches. Nat. Commun. **2**(1), 160, 1–9 (2011)
26. Hara, T., Saito, H., Inoue, T.: Directed evolution of a synthetic RNA–protein module to create a new translational switch. Chem. Commun. **49**(37), 3833–3835 (2013)
27. Ohno, H., Kobayashi, T., Kabata, R., et al.: Synthetic RNA–protein complex shaped like an equilateral triangle. Nat. Nanotechnol. **6**(2), 116–120 (2011)
28. Ohno, H., Osada, E., Inoue, T., et al.: Synthetic RNAprotein nanostructures and their potential applications. In: Guo, P., Haque, F. (eds.) RNA Nanotechnology and Therapeutics, pp. 303–312. CRC Press, Boca Raton, FL (2013)
29. Amrutha, A.S., Sunil Kumar, K.R., Tamaoki, N.: Azobenzene-based photoswitches facilitating reversible regulation of kinesin and myosin motor systems for nanotechnological applications. ChemPhotoChem **3**(6), 337–346 (2019)
30. Qian, L., Winfree, E.: Scaling up digital circuit computation with DNA strand displacement cascades. Science **332**(6034), 1196–1201 (2011)
31. Yoshimura, Y., Fujimoto, K.: Ultrafast reversible photo-crosslinking reaction: toward in situ DNA manipulation. Org. Lett. **10**(15), 3227–3230 (2008)
32. Jacob, G., Murugan, A.: DNA based cryptography: an overview and analysis. Int. J. Emerg. Sci. **3**(1), 36–42 (2013) [Online]. Available at: https://www.researchgate.net/publication/269098843_DNA_based_Cryptography_An_Overview_and_Analysis
33. Howard, J.: Mechanics of Motor Proteins and the Cytoskeleton. Sinauer Associates Inc., Sunderland, MA (2001)
34. Sohal, M., Sharma, S.: BDNA-A DNA inspired symmetric key cryptographic technique to secure cloud computing. J. King Saud Univ.—Comput. Inf. Sci. (2018). https://doi.org/10.1016/j.jksuci.2018.09.024
35. Tanaka, K., Okamoto, A., Saito, I.: Public-key system using DNA as a one-way function for key distribution. BioSystems **81**(1), 25–29 (2005). https://doi.org/10.1016/j.biosystems.2005.01.004
36. Enayatifar, R., Abdullah, A.H., Isnin, I.F.: Chaos-based image encryption using a hybrid genetic algorithm and a DNA sequence. Opt. Lasers Eng. **56**, 83–93 (2014). https://doi.org/10.1016/j.optlaseng.2013.12.003
37. Saper, G., Hess, H.: Synthetic systems powered by biological molecular motors. Chem. Rev. **120**(1), 288–309 (2019)
38. Hess, H., Ross, J.L.: Non-equilibrium assembly of microtubules: from molecules to autonomous chemical robots. Chem. Soc. Rev. **46**(18), 5570–5587 (2017)
39. Liu, H., Schmidt, J.J., Bachand, G.D., et al.: Control of a biomolecular motor-powered nanodevice with an engineered chemical switch. Nat. Mater. **1**(3), 173–177 (2002)
40. Yokokawa, R., Takeuchi, S., Kon, T., et al.: Hybrid nanotransport system by biomolecular linear motors. J. Microelectromech. Syst. **13**(4), 612–619 (2004)

Road Vehicle Tracking Using Moving Horizon Estimation

Gejo Georgesan and K. Surender

1 Introduction

Target Tracking is an important research topic in the field of automated vehicles. Although a number of tracking algorithms have been developed, yet it is still a very challenging task to implement these tracking algorithms in realistic situations. Numerous reasons include high clutter, low visibility of sensors and high target density. Taking such distortions into consideration and using it as constraints can help us achieve relatively better tracking performance. An effective approach for solving such constraints is to incorporate them into any standard filtering algorithm as state constraints, and this process is known as state estimation process.

For most tracking scenarios, Kalman filtering or its derivatives is commonly used to estimate the state of the target (vehicle) based on the state process and corresponding measurement models. Most of the filters in existence focus on either linear equality or linear inequality constraints, but very little research has been conducted focusing on the nonlinear equality/inequality constraints. The basic strategy of MHE in determining the optimal state estimation is to reformulate the estimation problem into an optimization problem using a fixed size estimation window. The MHE is being widely used not only in the field of chemical engineering, but also in hybrid systems, large-scale system distributed network systems and so on. However, the application of moving horizon estimation method in target tracking problems is still an uncharted area. Previous studies suggest that Kalman filter and its derivatives yield extremely good results for linear systems with/without constraints, but in case of non-linear systems the performance of the latter has not been quite satisfactory.

G. Georgesan · K. Surender (✉)
Department of Electronics and Communication Engineering, Visvesvaraya National Institute of Technology, Nagpur 440010, India
e-mail: ksurender@ece.vnit.ac.in

M. Dua et al. (eds.), *Proceedings of the International Conference on Paradigms of Communication, Computing and Data Sciences*, Algorithms for Intelligent Systems,
https://doi.org/10.1007/978-981-16-5747-4_22

2 Moving Horizon Estimation

Moving horizon estimation (MHE) is a state estimation method using a dynamic optimization approach that takes into account a series of measurements observed over time including noise and random variations and thereby generates an estimate of unknown parameters or variables. MHE requires an iterative approach for either linear or nonlinear systems to find a solution. Moving Horizon Estimation is mainly used in the chemical engineering field, and yet, its application in the wireless communication and tracking domain still remains an unexplored field of research. MHE is also known as Receding Horizon Estimation (RHE) and is generally used in the case of a nonlinear system for noisy inputs and outputs. If the system is linear and the inputs and outputs are noisy, then the Kalman filter comes into use. Similarly, if the system is nonlinear and the inputs and outputs are noisy, then the variants of the Kalman filter such as Extended Kalman Filter (EKF) and Unscented Kalman Filter (UKF) come into use. The working principle of MHE can be illustrated using the system mentioned below.

Consider a SISO system with the following input and output equations

$$x(k+1) = f(x(k), u(k)) + v_x \tag{1}$$

$$y(k) = x(k) + v_y \tag{2}$$

where $f: \mathbb{R}^n \rightarrow \mathbb{R}^n$ represents the nonlinear system dynamic function and $h: \mathbb{R}^n \rightarrow \mathbb{R}^n$ represents the nonlinear measurement model function. v_x and v_y are the noise on the system and noise at the sensor respectively. Generally these noises are modeled to represent the uncertainties of the system and are described by independent pdfs. In most of the target tracking systems, these noises are generally modeled as zero mean Gaussian distribution with a constant covariance matrix. $x(k)$ denotes the state vector of the system whereas $u(k)$ and $y(k)$ denote the input and output of the SISO system. The variable k represents the time steps and takes only integer values. The figure below (Fig. 1) shows the input and output measurements over a certain time interval. Now at any time instant k, the current state or the current output of the system can be estimated using MHE.

In Moving Horizon Estimation (MHE), we look backwards for a certain window size of length N_{MHE} and all the past measurements of the input and the output over this window length are observed. Using the principle of MHE, the trajectory which best fits the measurements is generated and is basically called the state trajectory. Based on the state trajectory the output at any time instant k can be determined (Fig. 2).

The basic idea behind MHE depends on three basic steps: Predict, Measure and Correct. Now this can be achieved by minimizing the cost function, $J_{N_{MHE}}$ represented as

$$J_{N_{\text{MHE}}}(x, u) = J_{\text{output}} + J_{\text{input}} \tag{3}$$

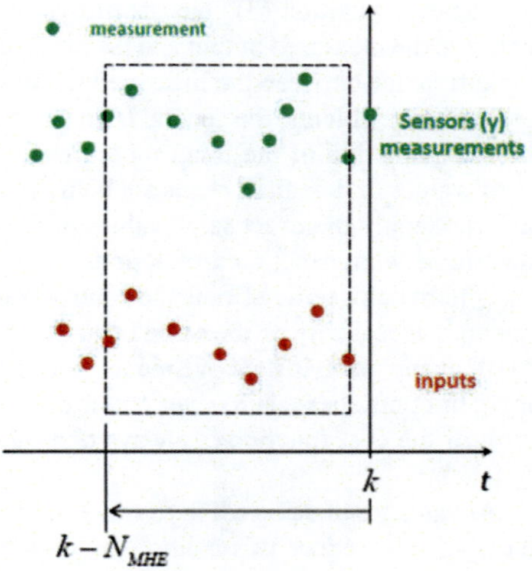

Fig. 1 Input and output measurement for a SISO system

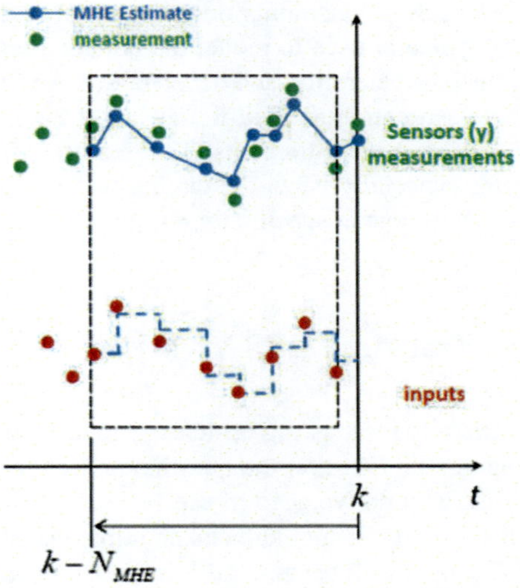

Fig. 2 Estimating state trajectory in a SISO system

where,

$$J_{\text{output}} = \sum_{i=k-N_{\text{MHE}}}^{k} \left\| \hat{y}(i) - y(i) \right\|^2 \tag{4}$$

$$J_{\text{input}} = \sum_{i=k-N_{\text{MHE}}}^{k-1} \left\| \hat{u}(i) - u(i) \right\|^2 \tag{5}$$

The cost function (3) consists of two terms, The first term (4) is the difference between the measured output and the predicted output, whereas the second term (5) is the difference between the measured input and the predicted input. For various state estimation problems, the second term can be excluded as it takes into consideration the cost function of the input measurements. But for certain systems wherein the input values or the initial states are being fed to the system, the lower level controller may not apply the exact same values to the system. Thus, the second term helps us determine what exact control actions went through the system. In other words, the J_{input} term helps us determine the control actions that go in as input to the system that may not necessarily be the same control actions sent by the user. By the principle of MHE at any given instant of time the output state would depend only on the previous input or control state and never on the current input or current state. Thus the output term of the cost function is always one step longer than the input term of the cost function.

Advantages of using MHE to solve target tracking models (state estimation) could be significant. Since this method is optimization based, constraints in target tracking problems can be naturally handled by MHE as additional (non)linear and/or (in)equality constraints under consideration. In vehicle tracking, state constraints are typically used to model the bounded disturbance on vehicle movement such as vehicle acceleration and deceleration. Another major advantage of using MHE as a state estimation method in any target tracking model is that it considers a window of N latest measurements. This feature is very significant in target tracking problems especially when the targets are occluded by each other leading to no reliable measurement at specific time steps.

3 Tracking: GPS VS MHE

Generally for tracking vehicles on roads, the most common used technology is tracking via the GPS. So the question arises, given the widespread and dominant use of GPS, why do we need to use MHE for vehicle (target) tracking? While answering this question it should be taken into consideration that one cannot directly compare GPS and MHE because GPS is a sensor used for target tracking, whereas MHE is a filter used to reduce the noise from the sensory data and measure the hidden states that sensors don't directly measure. Therefore, we can't have a direct comparison between MHE and GPS because the former is a filtration method whereas the latter is a sensor. The GPS being a sensor does not usually measure the intermediate states between two output states, this is where filters like EKF and MHE gain importance. Moreover, the GPS range and measurements are also affected by several types of random errors and biases. Errors arise from a variety of sources and cause fluctuations in the position estimation while tracking an object using the GPS (Global Positioning System) sensor. Listed below are a few major causes of error caused during the process of tracking using a GPS.

- Satellite clock errors (1.5–3.5 m)
- Satellite orbital errors (<1 m)
- Atmospheric Propagation delay (1.5–1.7 m)
- Multipath errors (0.5–1.3 m)
- Receiver Noise (0.7–1.5 m).

4 Constrained Target Tracking Using MHE

4.1 System Specification

The general dynamic model for target tracking problem

$$x_{k+1} = f(x_k) + w_k \tag{6}$$

The observation equation

$$y_k = h(x_k) + v_k \tag{7}$$

where k takes integer values, $h: \mathbb{R}^n \rightarrow \mathbb{R}^n$ is a nonlinear measurement model and $f: \mathbb{R}^n \rightarrow \mathbb{R}^n$ is a nonlinear system function. $x_k \in \mathbb{R}^n$ is the state vector and $y_k \in \mathbb{R}^n$ is the vector of available measurements. Also the vectors $w_k \in \mathbb{R}^n$ and $v_k \in \mathbb{R}^n$ are Gaussian noise of the process and the measurement respectively, described by its independent pdfs represented as $p(w_k) = \mathcal{N}(0, Q)$ and $p(v_k) = \mathcal{N}(0, R)$ respectively, where both Q and R are covariance matrices.

4.2 Road Width Constraints

As discussed before, when moving along a road network the ground targets are constrained. Therefore, the knowledge of road maps and terrain database can be used as constraints and thereby incorporated into the tracking algorithm. In most of the existing techniques, the road map constraint target motion is in a one-dimensional physical space (road width is ignored) and are then incorporated as equality constraints. This is a fairly good approach only when an observer is far away from moving objects (unmanned aircraft tracking any ground vehicle). However in the case of automated vehicles, only objects in proximity are of interest. Therefore here we consider the road network information as road width inequality constraints and the motion of the vehicle is restricted by these physical constraints in both straight and curved road segments.

Nonlinear State Inequality Constraints: Consider a curved or a circular road segment as shown in Fig. 3, represented by two arcs with the upper and the lower

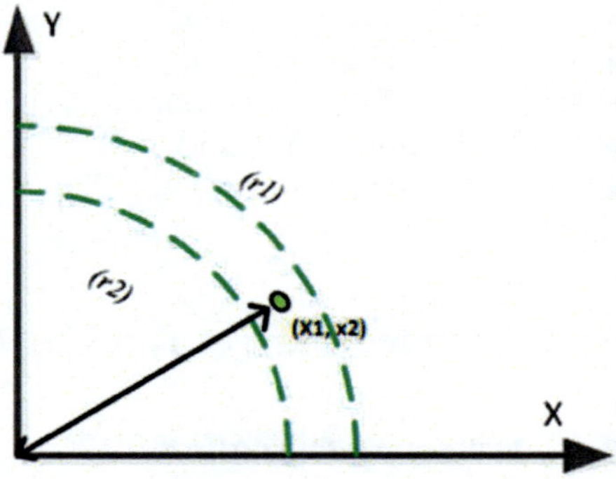

Fig. 3 Non linear inequality constraint on a curved road

boundaries of the road having radius equal to r_1 and r_2 respectively when measured with respect to the center of origin of the cartesian coordinate system.

The nonlinear inequality constraint is defined by

$$r_1 \leq \sqrt{x_{1,k}^2 - x_{2,k}^2} \leq r_2 \tag{8}$$

where for each time step k, the vehicle position states $x_{1,k}$ and $x_{2,k}$ are subject to nonlinear constraint as defined in Eq. (8).

Linear State Inequality Constraints Consider a and b to be the two side boundaries of a straight road and at every time step k, the state vector x_k is subject to the following linear inequality constraint:

$$a_k \leq C_k(x_k) \leq b_k \tag{9}$$

where $a_k, b_k \in \mathbb{R}^c$, $C_k : \mathbb{R}^n \rightarrow \mathbb{R}^c$ and the inequality ($\leq$) holds for all the elements of the vector and $a_k < b_k \forall\ k$.

5 Simulation and Results

In this section, the study and simulation of the Moving Horizon Estimation Algorithm is majorly segregated into two categories, viz.

- Vehicle tracking along a straight road
- Vehicle tracking along a curved/circular road.

5.1 Vehicle Tracking Along a Straight Road

Consider a moving vehicle on a straight road section defined by two outer boundaries with measurements along the x-axis to be $a = 30$ m and $b = 40$ m respectively, centered at the origin of the Cartesian coordinate system.

The vehicle dynamics is defined by a white noise motion model.

$$x_{k+1} = \begin{bmatrix} 1 & T & 0 & 0 \\ 0 & 1 & 0 & 0 \\ 0 & 0 & 1 & T \\ 0 & 0 & 0 & 1 \end{bmatrix} x_k + \begin{bmatrix} T^2/2 & 0 \\ 0 & T^2/2 \\ T & 0 \\ 0 & T \end{bmatrix} w_k \tag{10}$$

where the state vector is defined as

$$x_k = [x_{1,k}, x_{2,k}, x'_{1,k}, x'_{2,k}]^T \tag{11}$$

The road width is considered to be equal to 10 m and the speed limit for this road segment is considered to be equal to 5 m/s (18 km/hr) with a sampling time of $T = 1$ s (Fig. 4).

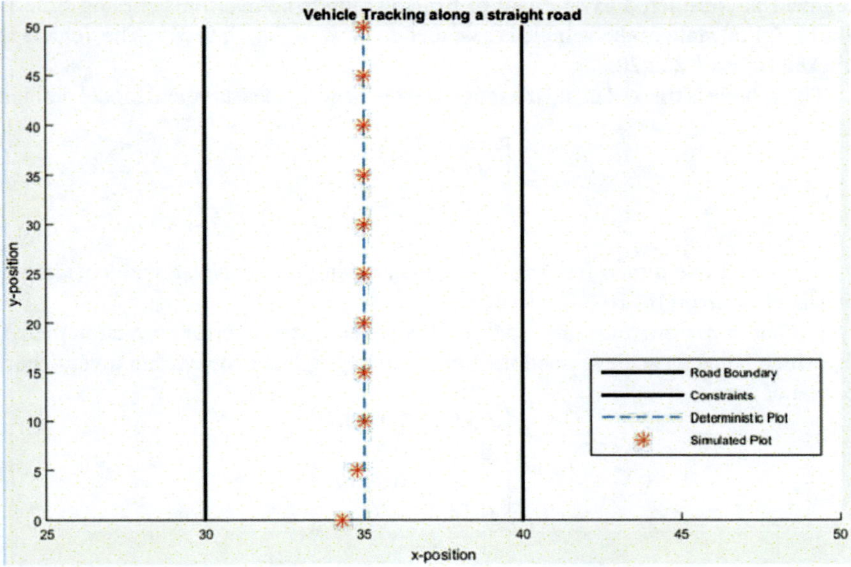

Fig. 4 Tracking scenario on a straight road

The Mean Square Error (MSE) defined as the average squared difference between the actual and the predicted values in the case of vehicle tracking along a straight road using MHE is measured and noted.

5.2 Vehicle Tracking Along a Curved Road

Consider a moving vehicle on a curved/circular road section defined by two arcs with radius $r_1 = 90\,\text{m}$ and $r_2 = 100\,\text{m}$ centered at the origin of the Cartesian coordinate system. The vehicle dynamics is defined by a white noise motion model.

$$x_{k+1} = \begin{bmatrix} 1 & T & 0 & 0 \\ 0 & 1 & 0 & 0 \\ 0 & 0 & 1 & T \\ 0 & 0 & 0 & 1 \end{bmatrix} x_k + \begin{bmatrix} T^2/2 & 0 \\ 0 & T^2/2 \\ T & 0 \\ 0 & T \end{bmatrix} w_k \tag{12}$$

where the state vector x_k is defined as in Eq. (11).

The state vector x_k consists of the vehicle position and velocity along x and y directions respectively. w_k is a two dimensional zero mean Gaussian Process Noise with covariance matrix $Q = diag\{1, 1\}$ where $diag\{\}$ means a diagonal matrix. The road width is again considered to be equal to 10 m and the speed limit for this road segment is considered to be equal to 5 m/s (18 km/hr). Also the sampling time is $T = 1$ s. Initial state of the vehicle is assumed to be $x_0 = [95, 0, 0, 5]^T$. The vehicle is tracked for $k = 1,2....,20$.

The vehicle is tracked using the range and the bearing measurement model defined as:

$$z_k = \begin{bmatrix} \sqrt{x_{1,k}^2 - x_{2,k}^2} \\ \arctan\left(\frac{x_{2,k}}{x_{1,k}}\right) \end{bmatrix} + v_k \tag{13}$$

where v_k is a two dimensional zero mean Gaussian Process Noise with covariance matrix $R = diag\{16, 10^{-2}\}$.

For better performance of the MHE filter we consider an error correction matrix E_r which helps us achieve good tracking with negligible errors within a very short period of time.

$$E_r = \begin{bmatrix} 0.4001 & 0.02276 \\ 0.1209 & 0.3055 \\ 0 & 0.2159 \\ 0.1438 & 0 \end{bmatrix} v_n \tag{14}$$

where v_n=1 is the number of vehicles (target) being tracked. Figure 5 shows the tracking of a vehicle along a curved/semi-circular road. It can be observed that despite a vague starting point being provided as the initial condition by the observer, MHE does achieve a perfect tracking within a short span of time.

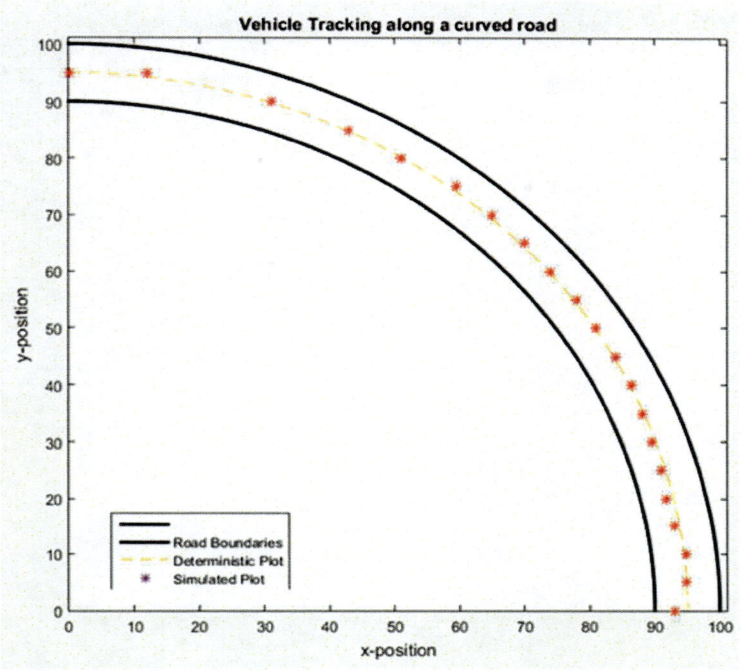

Fig. 5 Tracking scenario on a curved/circular road

The error in tracking can be determined by measuring the deviation between the measured output and the actual output. To determine the effectiveness of the system designed, the initial position of the vehicle (target) is considered far from the actual position and the performance of the MHE filter is monitored over the MHE window length. The quicker the target (vehicle) traces itself to the ideal tracking condition, the better the system design. Consider the initial point by the observed to be 6 m away from the actual point along the y-axis and 2 m away from the actual point along the x-axis, meaning the state vector is assigned the initial value of $[89, 2, 3, 7]^T$.

Figures 6 and 7 show the error in the measurement along the MHE horizon length. The error in the y position is represented by the plots along the North and South directions. Similarly the error in x-position is represented by the plots along the East and West directions. The basic idea behind MHE working principle is the state estimation. State Estimation is defined as a process of estimating the state of a system by eliminating the errors and inaccuracies from measurement data. State Estimation is commonly referred as the heart of moving horizon estimation, since the principle of predicting and correcting the future outputs or the state of the system entirely depends on the property of state estimation. Figures 8 and 9 graphically represents the comparison of the estimated and the actual value of the state vector. It can be observed that the vague values provided as the initial state by the user quickly converges to

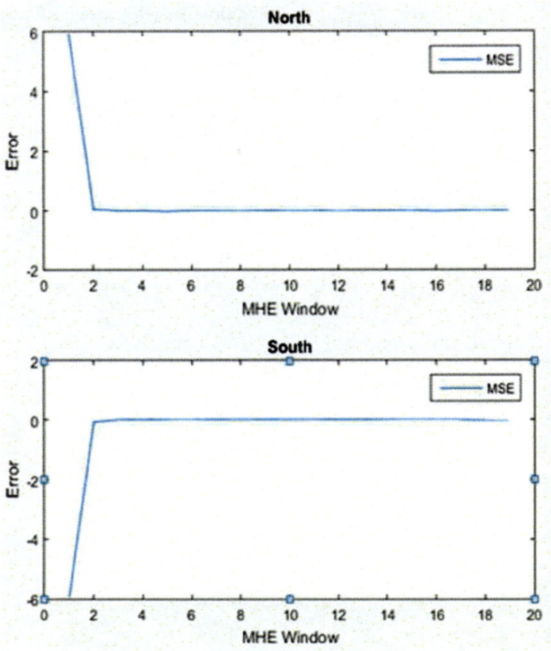

Fig. 6 Error in tracking along y-position

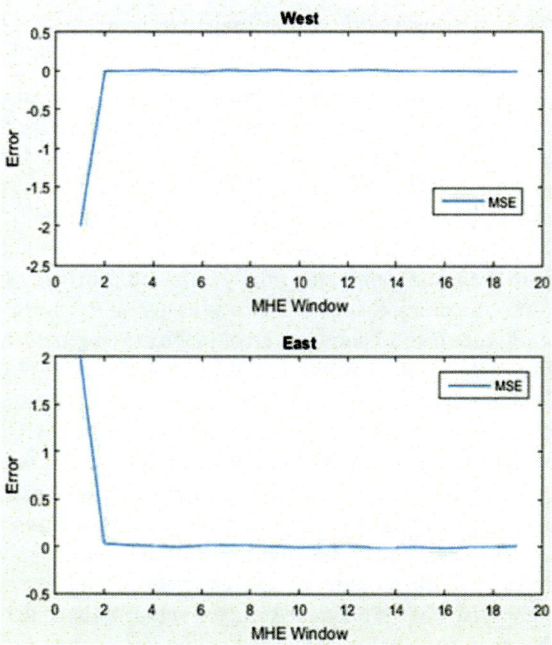

Fig. 7 Error in tracking along x-position

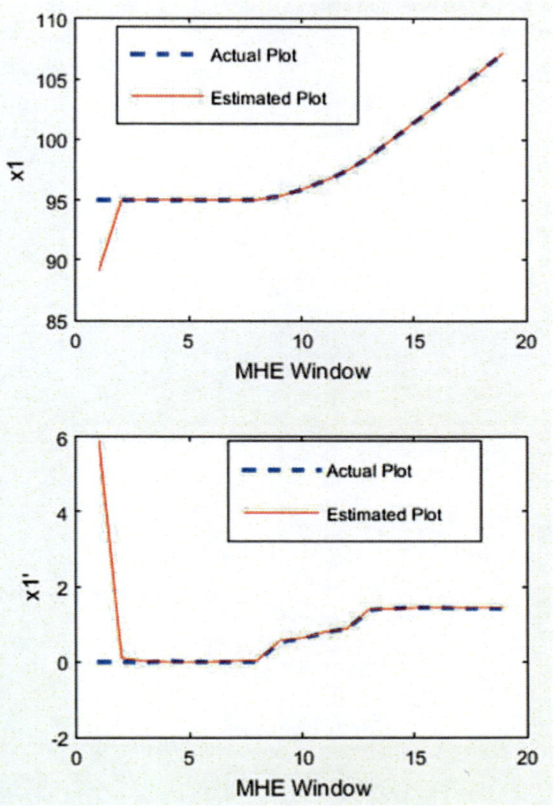

Fig. 8 Comparison of state vector coefficients

the actual value using MHE and thus minimizing the error in the state vector thereby provides accurate tracking results.

For the state vector as represented in Eq. (11) the state estimate comparison for the first two variables is represented in Fig. 8 and the same for the latter two is represented in Fig. 9. The over-lapping of the two graphs indicates the efficiency of MHE filter used in this system.

The performance of MHE filter can be determined by measuring the Mean square error (MSE)

$$\text{MSE} = \frac{1}{(2(K+1))^{-1}} \sum_{k=0}^{K} \sum_{i=1}^{2} \left\| \hat{x}(i,k) - x(i,k) \right\|^2 \qquad (15)$$

where K is the horizon length or window size. Various research and study conducted till date have proved the efficiency of MHE over other conventional filters. Here we compare the performance of MHE with different horizon length (*i.e.* 4, 8, 15 and 20) for the given system (Table 1).

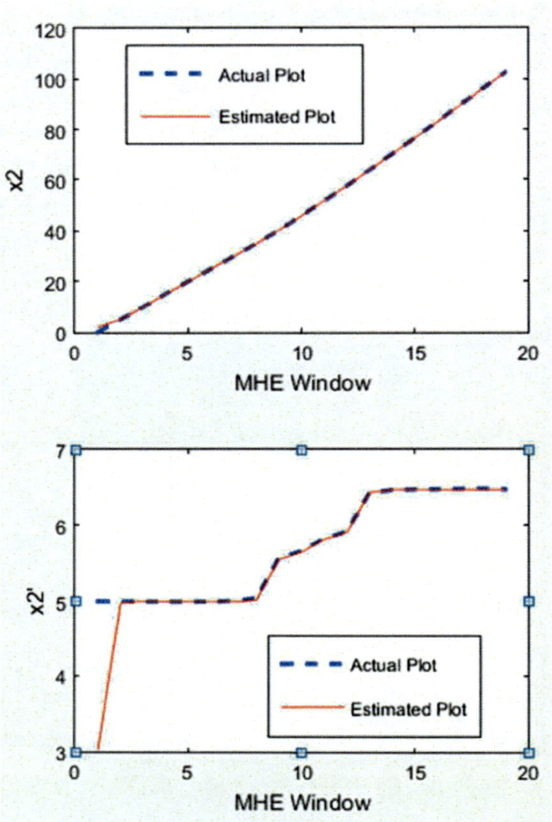

Fig. 9 Comparison of state vector coefficients

Table 1 Estimation performance of filters across different window lengths

N_{MHE}	MSE	
4	1.014	
8	0.426	
15	0.329	
20	0.297	
EKF	2.62	

It can be observed that the proposed MHE algorithm provides reasonably good performance with increased window size. But it should be taken into consideration that increased window size also increases the simulation time and thereby would increase the computational cost of the system.

Among various target tracking algorithms [1, 2], the moving horizon estimation (MHE) has a good efficiency in being able to accommodate different types of constraints [3, 4]. In an experiment carried out, [5] several conventional filters namely the divided difference filter (DDF), Gaussian mixture filter (GMF), unscented kalman filter (UKF) and the constrained particle filter (CPF) are compared for the same target tracking model. After analyzing the simulation results it is evident that the MHE filter out performs the unconstrained conventional filters and yields results at a comparatively lower computational cost. The proposed MHE [5] also verifies better performance of MHE for increased window size. Similarly in another experiment [6], a performance comparison is carried out between EKF, MHE and MHT for the same tracking model using the same initial condition. The numerical simulation for these algorithms is performed using the RMSEs through a hundred Monte Carlo simulations for the same scenario. MHE based tracking provides better tracking results for reduced computational cost.

6 Conclusion and Future Scope

In this paper, a new ground model based vehicle tracking method has been proposed considering both linear and nonlinear constraints in a comprehensive way. Initial simulation studies indicate the efficiency of the proposed algorithm when compared to various conventional algorithms mainly the Extended Kalman Filter. The mean square error achieved for the proposed algorithm is quite appreciable compared to various other algorithms across different horizon lengths. To further study the benefits and sustainability of the proposed algorithm, various real world experiments with actual sensor measurements would be considered as future work.

References

1. Simon, D.: Kalman filtering with state constraints: a survey of linear and nonlinear algorithms. In: IET Control Theory Application, vol. 4, no. 8, pp. 1303–1318 (2010). https://doi.org/10.1049/iet-cta.2009.0032. ISSN 1751-8644
2. Straka, O., Dunik, J., Šimandl, M.: Truncation nonlinear filters for state estimation with nonlinear inequality constraints. Automatica (J. IFAC) **48**(2), 273–286 (2012). https://doi.org/10.1016/j.automatica.2011.11.002
3. Rao, C.V., Rawlings, J.B., Mayne, D.Q.: Constrained state estimation for nonlinear discrete-time systems: stability and moving horizon approximations. IEEE Trans. Automatic Control **48**(2), 246–258 (2003). https://doi.org/10.1109/TAC.2002.808470
4. Simon, D., Chia, T.L.: Kalman filtering with state equality constraints. IEEE Trans. Aerospace Electron. Syst. **38**(1), 128–136 (2002). https://doi.org/10.1109/7.993234

5. Ding, R., Yu, M., Chen, W.-H.: A multiple target tracking strategy using moving horizon estimation approach. In: 24th International Technical Conference on the Enhanced Safety of Vehicles (ESV) (2005)
6. Ding, R., Yu, M., Oh, H., Chen, W.: New multiple-target tracking strategy using domain knowledge and optimization. IEEE Trans. Syst. Man Cybern.: Syst. **47**(4), 605–616 (2017). https://doi.org/10.1109/TSMC.2016.2615188

Hybrid End-to-End Architecture for Hindi Speech Recognition System

A. Kumar, T. Choudhary, M. Dua, and M. Sabharwal

1 Introduction

Automatic speech recognition (ASR) task is to convert speech utterances into corresponding text sequences. The conventional ASR system calculates the posterior distribution, $p(W/X)$, of most probable word sequence W by using the speech feature X. By applying Baye's rule, this term is further decomposed into $p(W)$ and $p(X/W)$, which denotes the language model and acoustic model respectively. This conventional procedure requires many step-wise refinements [1] and a complex decoding process [2]. Statistical HMM-based acoustic modeling has been widely used to perform this task. DNN-based acoustic modeling also relies on state alignments obtained after GMM-HMM acoustic modeling. End-to-end ASR objective is to make more straightforward training and decoding processes by introducing a single-network architecture of deep learning framework. This single-network architecture is trained without any previous refinement steps and calculates the sequential letter probabilities given a speech signal [1, 3].

End-to-end ASR can be done using two major approaches; (1) connectionist temporal classification (CTC) [4]), and (2) attention-based approach [3]. CTC performs the frame-based classification by utilizing the Markov assumption to tackle the sequential problem using dynamic programming [5, 6]. Attention-based models do the sequence modeling using the encoder–decoder network, in which the attention layer acts as the interface. The encoder includes the different recurrent neural network (RNN) layers to model acoustics, and the decoder contains at least one

A. Kumar (✉) · M. Dua
National Institute of Technology, Kurukshetra, Haryana, India

Present Address:
A. Kumar
KIET Group of Institutions, Ghaziabad, India

T. Choudhary
GLA University, Mathura, India

M. Sabharwal
Galgotias University, Greater Noida, India

M. Dua et al. (eds.), *Proceedings of the International Conference on Paradigms of Communication, Computing and Data Sciences*, Algorithms for Intelligent Systems,
https://doi.org/10.1007/978-981-16-5747-4_23

RNN layer to calculate the output word sequence. These approaches treat the language model, acoustic model, and pronunciation model as a whole instead of treating them separately. Attention-based models are more flexible and do not require conditional independence assumption, but they need more information about the sequential arrangement of input features [1]. It suffers from the alignment issue, which can be handled by different methods [7, 8].

To take advantage of both approaches, hybrid CTC/attention architecture was introduced [1]. The combination of two network architectures in one increases the complexity and improves the ASR performance [8, 9]. In the hybrid CTC/attention model, a multiobjective learning method is used in which the CTC objective function is attached with an attention-based encoder network as regularization [8]. In decoding, joint-decoding was performed by combining both model scores in rescoring algorithm [9].

This work implements the hybrid CTC/attention architecture using the ESPnet speech recognition toolkit [10] and Hindi speech database.[1] This paper provides the detailed formulation of an end-to-end ASR system (Sect. 2). Section 3 describes the corpus details, and Sect. 4 presents the experimental setup and results. Section 5 concludes the paper with the final findings.

2 Related Work and Contribution

ASR field has witnessed rapid demand in the last few years. However, only a fraction of languages having speech recognition capabilities among the 7000 dialects spoken around the globe [11]. A reliable ASR system generally requires an enormous amount of speech and text data, which could undoubtedly require a year for one language. Hindi is a widely spoken language with millions of speakers, but due to the unavailability of open-source speech corpus, very little research has been done for Hindi ASR. In 2020, IIT Madras conducted a Hindi ASR challenge and released a dataset of 50 h. In the past, few works have been reported for the Hindi language but on different datasets [12–17]. The majority of works are based on statistical modeling, and a few of them are based on deep learning techniques. End-to-end models are not properly investigated for the Hindi language, and this work attempts to investigate the end-to-end models for Hindi ASR on limited resource conditions. A detailed comparison with state-of-the techniques on the same dataset can be found on the ASR challenge site.[2]

[1] IITM Hindi Speech Corpus: a corpus of native Hindi Speech Corpus—*Speech signal processing lab, IIT Madras.*

[2] https://sites.google.com/view/asr-challenge/leaderboard?authuser=0.

3 End-to-End ASR

3.1 Connectionist Temporal Classification (CTC)

CTC is based on Baye's theory. The CTC uses K-length letter sequence, $L = (l_k \in \upsilon | k = 1, 2, \ldots, K)$ and "blank symbol" in addition to handle repetation of symbols. A modified letter sequence along with blank symbol can be denoted as:

$$\begin{aligned} L' &= \{<b>, l_1, <b>, l_2, <b>, \ldots, l_k, <b>\} \\ &= \{l'_k \in \upsilon \cup \{<b>\} | k = 1, 2, \ldots, 2k+1\} \end{aligned} \tag{1}$$

The posterior distribution $p(L/X)$ can be defined as:

$$p(L/X) = \sum_{z} p(L/Z, X) p(Z/X) \tag{2}$$

By assuming the conditional independence assumption, we can write Eq. 2 as:

$$\approx \sum_{z} p(L/Z) p(Z/X) \tag{3}$$

Here, $p(Z/X)$ denotes the CTC acoustic model, and $p(L/Z)$ denotes the letter model. The term $p(Z/X)$ can be further decomposed with the help of conditional independence and chain rule as:

$$p(Z/X) = \prod_{t=1}^{T} p(z_t/z_1, \ldots, z_{t-1}, X) \tag{4}$$

$$\approx \prod_{t=1}^{T} p(z_t/X) \tag{5}$$

$p(z_t/X)$ is the frame-wise posterior distribution, which can be modeled by different RNN like bidirectional long short-term memory (BLSTM) [18, 19].

$$p(z_t/X) = \text{softmax}(\text{Lin.}(h_t)) \tag{6}$$

$$h_t = \text{BLSTM}_t(X) \tag{7}$$

In Eq. 6, Lin.(.) denotes the linear layer and softmax(.) denotes the softmax activation function. The CTC letter model $p(L/Z)$ can be defined with the help of Baye's role as:

$$p(L/Z) = \frac{p(Z/L) p(L)}{p(Z)} \tag{8}$$

$$\approx \prod_{t=1}^{T} p(z_t/z_{t-1}, L)\frac{p(L)}{p(Z)} \tag{9}$$

In Eq. 9, $p(z_t/z_{t-1}, L)$ is the state-transition probability, $p(L)$ is language model, and $p(Z)$ is the state-prior probability. By combining Eqs. 5 and 9, the posterior probability $p(L/X)$ can be represented by the objective function as:

$$p(L/X) \approx \underbrace{\sum_{z}\prod_{t=1}^{T} p(z_t/z_{t-1}, L)p(z_t/X)\frac{p(L)}{p(Z)}}_{\triangleq p_{\text{ctc}}(L/X)} \tag{10}$$

Equation 10 is computed with the help of dynamic programming. The CTC closely matches HMM/DNN, except that it applies Baye's rule to $p(L/Z)$ instead of $p(W/X)$.

3.2 Attention-Based Encoder Decoder

Attention-based model directly calculates the posterior probability $p(L/X)$ with the help of chain rule and does not relies on conditional independence assumption.

$$p(L/X) = \prod_{t=1}^{T} p(l_k/l_1, \ldots, l_{k-1}, X) \tag{11}$$

$$\triangleq p_{\text{att}}(L/X) \tag{12}$$

The posterior probability is calculated with the help of following:

$$h_t = \text{Encoder}(X) \tag{13}$$

$$a_{kt} = \begin{cases} \text{Content-Attention}(q_{k-1}, h_t) \\ \text{Location-Attention}(\{a_{k-1}\}_{t=1}^{T}, q_{k-1}, h_t) \end{cases} \tag{14}$$

$$r_k = \prod_{t=1}^{T} a_{kt}.h_t \tag{15}$$

$$p(l_k/l_1, \ldots, l_{k-1}, X) = \text{Decoder}(r_k, q_{k-1}, l_{k-1}) \tag{16}$$

Equations 13 and 16 denote the encoder and decoder networks. Equation 14 denotes the attention mechanism, and r_k denotes the hidden vector. Encoder network converts the input feature vector X into frame-wise hidden vector h_t.

$$\text{Encoder}(X) \triangleq \text{BLSTM}_t(X) \tag{17}$$

There are different attention mechanisms available like location-aware attention [20], dot-product attention [21], additive attention [22], multihead attention [23], and coverage attention mechanism [24]. Decoder is the another recurrent network based on past output l_{k-1}, hidden vector q_{k-1}, and letter-wise hidden vector r_k.

$$\text{Decoder}(.) \triangleq \text{softmax}(\text{Lin.}(\text{LSTM}_k(.))) \tag{18}$$

In Eq. 18. $\text{LSTM}_k(.)$ is the unidirectional LSTM model, which calculates the hidden vector q_k as:

$$q_k = \text{LSTM}_k(r_k, q_{k-1}, l_{k-1}) \tag{19}$$

3.3 *Hybrid CTC/Attention*

The hybrid CTC/attention [1] architecture takes the advantages of both architecture in training and decoding.

3.3.1 Multiobjective Training

The multiobjective training is based on the multiobjective learning framework [8], in which we train the attention-based encoder with the help of CTC objective function as an auxiliary task. The logarithmic linear combination of CTC and attention objective function increases the robustness and convergence speed [10] as:

$$\pounds_{\text{MOL}} = \lambda \log p_{\text{ctc}}(L/X) + (1-\lambda) \log p_{\text{att}}(L/X) \tag{20}$$

Here, λ is the tuning parameter, which satisfy $0 \le \lambda \le 1$ condition.

3.3.2 Joint-Decoding

The hybrid CTC/attention architecture linearly combines the CTC and attention-based sequence probabilities to perform joint-decoding. The joint-decoding objective function can be represented as:

$$\widehat{\pounds} = \text{argmax}_{\ell \in \upsilon}\{\lambda . p_{\text{ctc}}(L/X) + (1-\lambda) \log . p_{\text{att}}(L/X)\} \tag{21}$$

Table 1 Details of the Hindi database

Hindi	# utterances	Duration (in hours)
Train-set	27,131	40
Dev-set	3330	5
Eval-set	3349	5

4 Database

The database was released by IIT, Madras,[3] under the Hindi ASR challenge. It consists a total of 50 h of speech utterances. Speech utterances were recorded for the Hindi language with male and female speakers of different age groups. 40 h of speech utterances were used as a training set, and 5 h of speech utterances were used as a development set. The remaining utterances were used for testing purposes. Table 1 shows the statistic about the database.

5 Experimental Setup and Results

The baseline system was released by IIT, Madras, using the Kaldi toolkit [25]. GMM-HMM-based acoustic modeling was used to prepare the baseline model with n-gram language modeling. The baseline model was built using LDA + MLLT + SAT training over 39-dimensional MFCC features. In this article, we developed the end-to-end ASR model using ESPnet [10] toolkit. For the end-to-end ASR model, we used the hybrid CTC/attention mechanism. The data preparation was done using the Kaldi data directory. For feature extraction, 80-dimensional log Mel Filterbank (FBANK) features along with three-dimensional pitch features were used in training. ESPnet compressed all the information into JSON file as a data preparation step for end-to-end modeling. We used a projected BLSTM encoder network in this work. We used four layers for the encoder and one layer for the decoder network. The hybrid CTC/attention model with $\lambda = 0.5$ (equal contribution) was used to train and decode the network. Other hyperparameters details are shown in Fig. 1.

Table 2 presents the comparison of different end-to-end ASR models. The four-layer BLSTM was used as an encoder network with 320 cells in each layer. The linear projection layer with 320 cells is also used. The decoder was a one-layer model with 300 cells. The conventional beam search algorithm with beam size 20 was used in decoding. We observed that hybrid CTC/attention model $m = 0.5$ performs better with the lowest CER 30.61%. Except for this improvement, the learning alignment

[3] IITM Hindi Speech Corpus: a corpus of native Hindi Speech Corpus - *Speech signal processing lab, IIT Madras.*

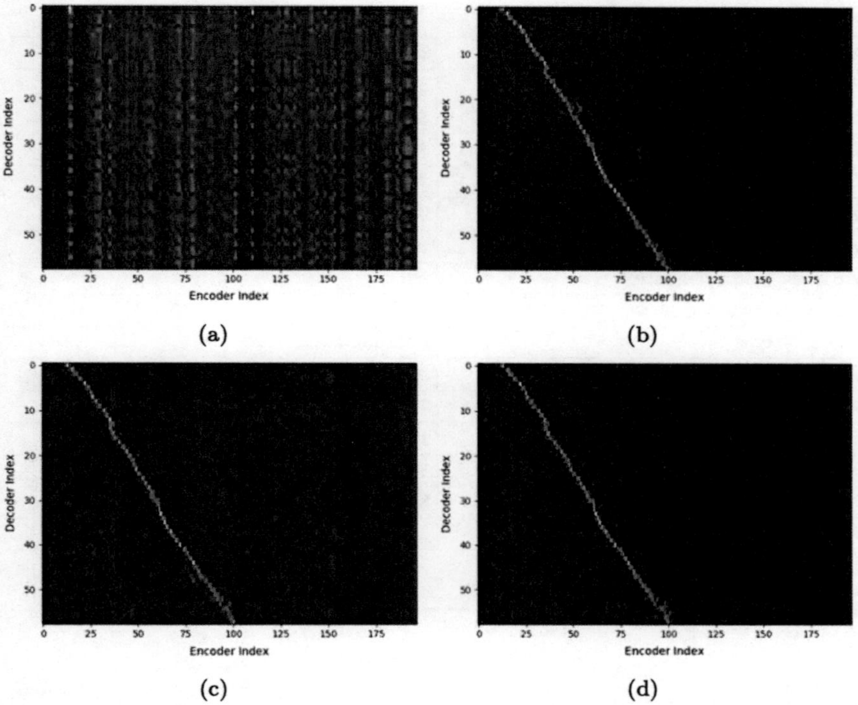

Fig. 1 Comparison of learning alignment speed between character (y-axis) and acoustic frames (x-axis) of Hybrid CTC/Attention ($\lambda = 0.5$) model over the training epochs (1, 5, 10, 20). Alignments are shown for the speech utterance (ahd_28_long_265_hin-005000-008808-1-1). **a** Hybrid CTC/Attention 1 epoch; **b** hybrid CTC/Attention 5 epoch; **c** hybrid CTC/Attention 10 epoch; **d** hybrid CTC/Attention 20 epoch

is also very fast in the hybrid CTC/attention model in comparison to other models. Figure 1 shows the learning alignment speed in different epochs of the hybrid CTC/attention model. It shows the learning rate reduction in different epochs of the hybrid CTC/attention model (Fig. 2; Table 3).

All the experiments have been done using the Ubuntu 18.04 OS, 8 GB RAM with 4 GB NVIDIA graphic support. The training time for one model is approx 48 h, and decoding takes additional 4 h. The evaluation set includes 3349 sentences, and the development set includes 3330 speech sentence utterances. Both are having over 0.2 million characters set.

Table 2 Common hyperparameters details

# of encoder BLSTM cell	320
# of encoder projection units	320
Encoder subsampling	Skip every *n* frame from input to *n*th layer
# of decoder LSTM cells	300
Optimization	Adadelta
Adadelta ρ	0.95
Adadelta $\in$	10^{-8}
Adadelta $\in$ decaying factor	10^{-2}
Maximum epoch	20
Attention type	Location-aware
Attention dimension	320
# convolutional filters	10
Convolutional filter widths	100
Batch-size	30
Max. input length	800
Max. output length	150

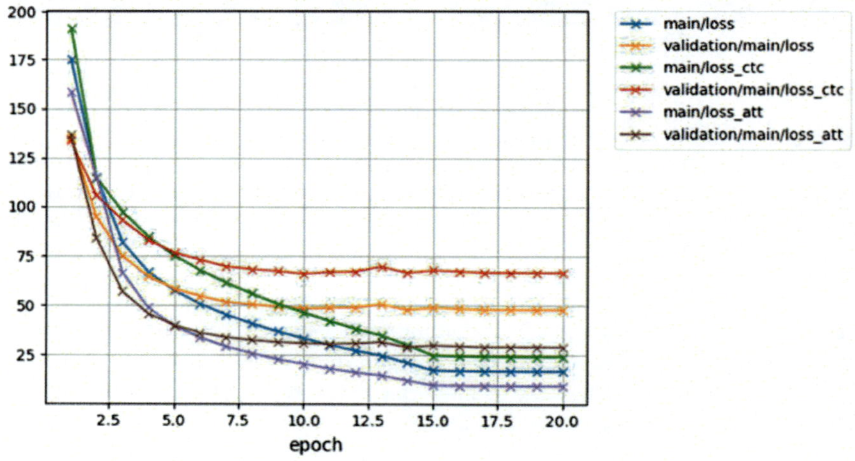

Fig. 2 Loss rate

Table 3 Common hyperparameters details

Model	Matric	Dev	Eval
CTC	CER(%)	39.21	34.61
Attention (content-based)	CER(%)	39.61	35.41
Attention (+ location-based)	CER(%)	38.41	33.61
Hybrid CTC/Attention ($\lambda = 0.2$)	CER(%)	36.61	31.10
Hybrid CTC/Attention ($\lambda = 0.5$)	CER(%)	35.21	30.61
Hybrid CTC/Attention ($\lambda = 0.8$)	CER(%)	37.41	33.21

6 Conclusion

This work implements the end-to-end Hindi ASR using hybrid CTC/attention architecture without an additional language model. End-to-end models do not require any alignments generated by HMM-based acoustic modeling, DNN pre-training, and complex search during decoding. This mechanism helps to reduce the complex ASR pipeline structure but does not perform well in comparison to statistical models without trained language models. This work can be extended to use the language model and other attention mechanisms to improve the performance of the existing ASR model.

References

1. Watanabe, S., Hori, T., Kim, S., Hershey, J.R., Hayashi, T.: IEEE J. Sel. Topics Signal Process. **11**(8), 1240 (2017)
2. Kürzinger, L., Watzel, T., Li, L., Baumgartner, R., Rigoll, G.: In: International Conference on Speech and Computer, pp. 258–269. Springer (2019)
3. Chan, W., Jaitly, N., Le, Q., Vinyals, O.: In: 2016 IEEE International Conference on Acoustics, Speech and Signal Processing (ICASSP), pp. 4960–4964. IEEE (2016)
4. Graves, A., Fernández, S., Gomez, F., Schmidhuber, J.: In: Proceedings of the 23rd international conference on Machine learning, pp. 369–376 (2006)
5. Chorowski, J., Bahdanau, D., Cho, K., Bengio, Y.: arXiv preprint arXiv:1412.1602 (2014)
6. Graves, A., Jaitly, N.: In: International Conference on Machine Learning, pp. 1764–1772. PMLR (2014)
7. Chorowski, J., Jaitly, N.: arXiv preprint arXiv:1612.02695 (2016)
8. Kim, S., Hori, T., Watanabe, S.: In: 2017 IEEE international conference on acoustics, speech and signal processing (ICASSP), pp. 4835–4839. IEEE (2017)
9. Hori, T., Watanabe, S., Hershey, J.R.: In: Proceedings of the 55th Annual Meeting of the Association for Computational Linguistics, vol. 1: Long Papers, pp. 518–529 (2017)
10. Watanabe, S., Hori, T., Karita, S., Hayashi, T., Nishitoba, J., Unno, Y., Soplin, N.E.Y., Heymann, J., Wiesner, M., Chen, N., et al.: arXiv preprint arXiv:1804.00015 (2018)
11. Grézl, F., Karafiat, M., Janda, M.: In: 2011 IEEE Workshop on Automatic Speech Recognition & Understanding, pp. 359–364. IEEE (2011)
12. Kumar, A., Aggarwal, R.: J. Intell. Syst. **30**(1), 165 (2021)
13. Kumar, A., Aggarwal, R.K.: Comput. Sci. **21**(4) (2020)

14. Kumar, A., Aggarwal, R.K.: Int. J. Speech Technol. 1–12 (2020)
15. Kuamr, A., Dua, M., Choudhary, A.: In: 2014 International Conference on Electronics and Communication Systems (ICECS), pp. 1–5. IEEE (2014)
16. Aggarwal, R.K., Dave, M.: Telecommun. Syst. **52**(3), 1457 (2013)
17. Passricha, V., Aggarwal, R.K.: In: Intelligent Speech Signal Processing, pp. 5–37. Elsevier (2019)
18. Hochreiter, S., Schmidhuber, J.: Neural Comput. **9**(8), 1735 (1997)
19. Graves, A., Jaitly, N., Mohamed, A.R.: In: 2013 IEEE Workshop on Automatic Speech Recognition and Understanding, pp. 273–278. IEEE (2013)
20. Chorowski, J., Bahdanau, D., Serdyuk, D., Cho, K., Bengio, Y.: arXiv preprint arXiv:1506.07503 (2015)
21. Luong, M.T., Pham, H., Manning, C.D.: arXiv preprint arXiv:1508.04025 (2015)
22. Bahdanau, D., Cho, K., Bengio, Y.: arXiv preprint arXiv:1409.0473 (2014)
23. Vaswani, A., Shazeer, N., Parmar, N., Uszkoreit, J., Jones, L., Gomez, A.N., Kaiser, L., Polosukhin, I.: arXiv preprint arXiv:1706.03762 (2017)
24. See, A., Liu, P.J., Manning, C.D.: arXiv preprint arXiv:1704.04368 (2017)
25. Povey, D., Ghoshal, A.,Boulianne, G.,Burget, L.,Glembek, O.,Goel, N.,Hannemann, M.,Motlicek, P., Qian, Y., Schwarz, P., et al.: In: IEEE 2011 Workshop on Automatic Speech Recognition and Understanding. IEEE Signal Processing Society (2011)

Smart Home Infrastructure with Blockchain-Based Cloud IoT for Secure and Scalable User Access

Sangeeta Gupta, Kavita Agarwal, and M. Venu Gopalachari

1 Introduction

Applications ranging from industrial to banking transaction-based domains incorporate blockchain internally, with a goal to reduce the commission against the interest paid towards the intermediaries. It is also used in different scenarios for these applications varying confined to the requirements. For example, in financial sector, it is also useful in tracking the records in a transparent way across all the users in the network. In retail applications, it is used to track the procurement and placement of goods and services by automation of supplier and buyer details instead of dealing with all the users in a manual way as depicted in Fig. 1. This helps in reduction of time and money. However, when using smart devices, it is efficient to integrate blockchain with IoT to yield productive outcomes to the people involved at different levels of interaction. It also enables to provide security for data via smart contract code that records the transactions inside the blocks through the sensors via permissioned and permission less consensus mechanisms [1].

Utilization of multiple technologies to ensure secure smart home application development and sustainability is the growing trend in modern world. A wide set of techniques ranging from numerical correlation-based analysis for traffic pattern inference to a range of evaluation metrics are explored and implemented efficiently to minimize various data affordability and privacy concerns. However, as most of the houses are provided with smart devices to monitor the inflow and outflow movement around, there is a need to devise mechanisms that ensure safety against unintended

S. Gupta (✉) · K. Agarwal
Department of Computer Science and Engineering, Chaitanya Bharathi Institute of Technology, Gandipet, Hyderabad 500075, India

M. Venu Gopalachari
Department of Information Technology, Chaitanya Bharathi Institute of Technology, Gandipet, Hyderabad 500075, India

M. Dua et al. (eds.), *Proceedings of the International Conference on Paradigms of Communication, Computing and Data Sciences*, Algorithms for Intelligent Systems,
https://doi.org/10.1007/978-981-16-5747-4_24

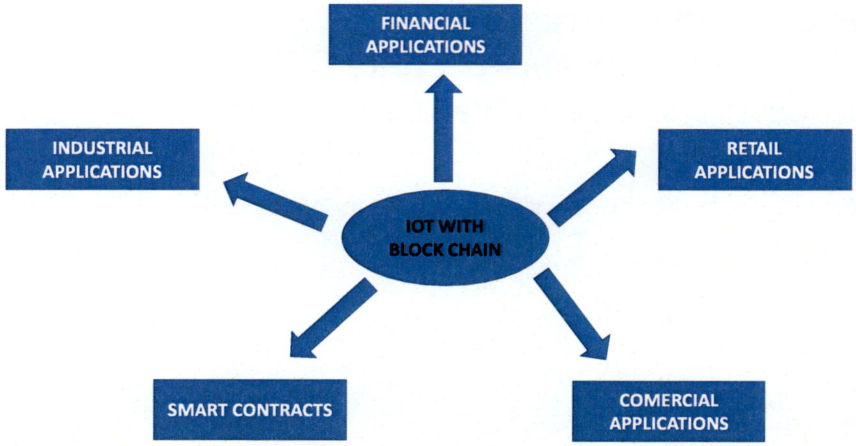

Fig. 1 Blockchain application areas

access by the intruders. Also, though cloud signifies availability and restricted access to signify non-malicious users, it is centralized in nature. This incurs an ease of access to the intruders due to shared key access provision [2, 3].

In secure smart home, it is necessary to consider about numerous security attacks that should be zeroed out, rather than working in a one-way-based approach. Moreover, the number of resources consumed in span of resource utilization also needs to be addressed to maintain a balanced cost-based outcomes. It is also essential to process incoming data in any format such as audio, video or text to be recorded in the transaction logs of the blockchain.

The use of an appropriate blockchain type among public, private and consortium relies on the user requirement-based application specification to be deployed. For example, if a similar level of security is the target functionality to be implemented among sister organizations, then consortium is a preferable choice. At the other end, if security is to be incorporated to provide in-house firewall access, then private blockchain is suitable, and if availability is the dominant characteristic to be attained, then public blockchain is a suitable environment. Hence, before moving to the development phase, it is essential to identify and articulate the requirements with a thorough analysis to utilize resources effectively [4]. In addition, based on the appropriate blockchain category selected to proceed with experimentation, the nature of encryption algorithm also plays a prominent role to strengthen security across the nodes.

2 Literature Survey

The utilization of IoT technology is surrounded by a wide set of challenges related to the cost incurred in the placement of multiple devices within a stipulated range and also privacy-based mechanisms due to storage of all records in a master-based storage model and non-standard models to build the real-time architecture set-up. Also, there is a great dependency of the users on the manufacturers of the devices deployed at their smart homes to identify log of actions in the event of theft. However, these difficulties can be zeroed out by integration of blockchain with IoT that results in cost reduction, distributed data access and measured security. Works in literature describe the public access feature of the blockchain offered to infinite users that is quite unsafe.

In existing works discussed in literature, there is no light throw on the evaluation parameters used to assess aspects related to security and scalability. Also, public blockchain is used to depict a distributed scenario between the set of nodes, where anonymous user's data cannot be captured in an efficient manner. It is emphasized in existing works that the data captured via sensor device is saved to local storage, which is unsafe; for example, whenever robbery takes place, the thief always focuses on the visible cupboards than invisible ones as they target many items in limited time. Also, cyber hackers are also focusing on local storage retrieval initially to gain access in a quick time. In addition, there is no representation for increased number of user access facility preserving the security aspects [5].

With the advent of Internet of Things, it became effortless to manage the devices in smart homes through gateways. Blockchain-based smart home gateway is proposed to overcome privacy issues faced by centralized gateways in smart homes. Blockchain is used to a maximum extent to ensure data integrity, authentication and efficient communication. It provides distributed and trust-free solutions, where online-distributed ledgers are being used. In [6], IoT devices and gateways have been assigned IDS for communication, and they have enough computing power to execute encryption. First the ID verification of devices and gateway is done using pre-shared keys. Then, SHA encryption algorithm key is applied to communicate via message between gateway and devices. The information gathered from the IoT devices by the gateway is stored in blocks of blockchain along with their hash using SHA-3 hash algorithm. A wide number of environments such ad ethereum, dapps and truffle were used to generate results. Performance evaluation based on data traffic was carried out to obtain optimized response time recordings via sensors. The architecture developed has some limitation in terms of additional computational complexity to perform blockchain operations.

A secure architecture for secure transmission and verification of devices using master-less approach based on Merkley tree was used for data verification [7]. Raspberry Pi model 3b and ARM-based devices were used in experiments to derive the advantages such as low cost, limited storage facility and processing capability. Blockchain with Base64 encoding was utilized in implementation. However, the methodology does not confirm robustness against a wide set of attacks that take place at once without any indication of performance when multiple devices are added to the network.

Mutual authentication-based secure system can be used in smart homes that include an integration of blockchain, signature and message authentication code to efficiently authenticate home gateway, authenticate group members anonymously and also helps in connection-oriented auditing of user's access history. The communication architecture in a sensor-based smart environment can also include residential users who uses Internet to connect wireless devices to remotely communicate with the home gateway. Upon receiving access order request from the users, the home gateway captures the actions with the relevant home devices. Here, the home gateway is triggered via network connectivity across the nodes at home being interconnected through the wireless sensor networks. There is a challenge of securing the devices and the communications in resource constrained devices. This results in increase of the computation expense-based delivery due to the recorded results into smart contract leading to additional smart contract invocations [8].

With the increased demand of smart technology, issues related to privacy have also increased. Also, if the security is based on the miner computations, then there is a tremendous increase in security and storage issues. Hence, there is a necessity to integrate multiple technologies in such a way that leads to reduction in cost, power consumption and storage space. An essential component to achieve the required features is to reduce the number of nodes across the blockchain network. To provide solutions to the aforementioned problems, graph theory plays a prominent role in the placement of the nodes in the ledger and world state components [9]. However, the arrangement of nodes should be in such a way that there is no dense or sparse placement of data across, but a proper balance is to be ensured. Various applications have been designed using IoT like smart homes and cities, smart energy, security and surveillance. IoT is also a target for cyber criminals, and they are using lots of tools and technologies to carry out cyber-attacks. Cyber forensics is required for the investigation of attack in order to get the evidence. Heterogeneous nature of IoT environment makes it difficult to conduct forensic investigation in IoT. It relies on collecting digital evidence from service providers which leads to evidence contamination. Blockchain-based IoT forensic model was proposed that prevents the admissibility of tampered logs into evidence [10].

3 Proposed Work

In applications like smart home, sensors are deployed to monitor all electronics installed in a house and also to observe the inflow and outflow of people into the house. The central coordinator is preferred to be cloud environment that records all the storage necessities associated with the transactions performed in the blocks. It also stores the valid and invalid attempts made to access the ledger data. However, the entire monitoring system, if in the hands of a central controller, will lead to security issues. To overcome this, an integration of IoT with blockchain is essential

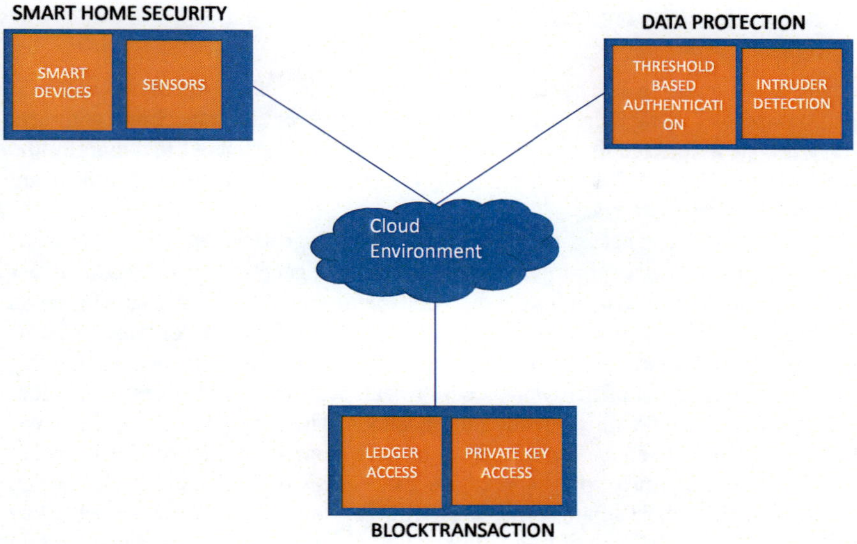

Fig. 2 Multi-user cloud access

that distributes the monitoring of access to items in a house consistently across all the members in the network. Smart contracts with various access right possibilities based on the user roles can be developed.

In the proposed work, a threshold value is integrated with the secret key such that when the upper limit is reached, warning is raised to identify thefts or invalid access made to unlock any secure device, whose data is in turn recorded in a particular block, assigned with an Id. The access is now distributed by diverting the transactions captured via sensors to multiple cloud locations, and retrieval is permitted based on the privileges assigned to different user groups. The block diagram is as shown in Fig. 2. In general, the data captured via sensors for a smart home provides a single-key based access to multiple users [11].

In the proposed framework, numerous user accesses to multiple keys stored in secure private cloud is developed. Also, the data captured is stored in the blocks on a permanent basis, and the editable data is kept in the cloud environment. To prevent unexpected access, a threshold range is set to verify the authenticity of the users as shown in Fig. 2. Smart contracts are represented to execute the transactions and also to capture the information logs records access.

4 Experimental Evaluation

If there exists multiple user access to multiple device access in a smart home, then it becomes tedious to track the actual origination in the event of theft. Smart contracts with a private access facility are deployed along with IoT gateways that monitor the local devices. This reduces the dependency of devices on gateways, and deploys are being done using contracts.

Moreover, user can access the code based on privilege assigned to them. In addition, the use of consensus algorithm based on mining or generic access functionalities enables to make decisions with respect to mutable/immutable block access based on user assigned roles. It is also necessary to identify the level of software or hardware significance at which security is to be refined or fine-tuned to enhance security aspects. Physical level of hardware security deals with biometric access provision, and logical one deals with soft code modifications. The access rights can be provisioned to different users based on sharing of keys among themselves. For instance, members of same family will prefer to use a single key to unlock the devices when returning back to their smart home. This will help the intruder to retrieve the key easily. Addition or deletion of user is linked with the associated ledgers that contain most recent and historic transaction information records stored in them as showcased in Fig. 3.

Integration of IoT with blockchain enables multiple security issues to be resolved. However, it is necessary to specify the classes of security based on multiple access levels by multiple user groups to strengthen security. The primary objectives achieved through the work are to

1. Provide read-only access permission to the data saved in blocks of a blockchain ledger.
2. Release editable data in cloud via private key to record log of all actions.
3. Threshold range is set to verify the authenticity of user to track intruders.
4. Overcome space utilization for multi-key access storage in cloud network.

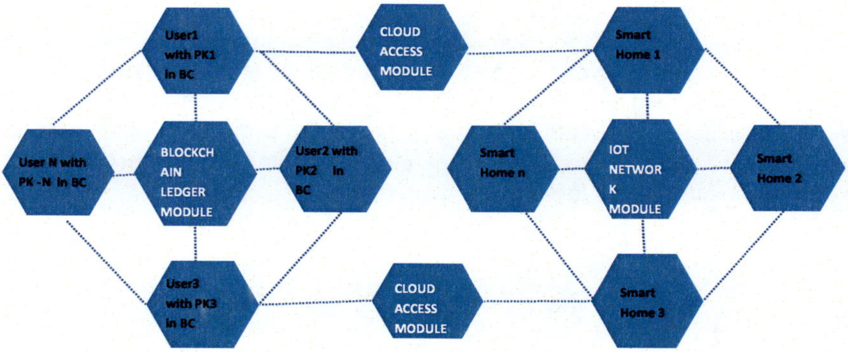

Fig. 3 Blockchain IoT ledger architecture

Table 1 User permissions and access time for smart home-based evaluation

Type of user	Access permission	Single key-based time to access (s)	Multi-key-based time to access (s)
Owner	Read/write/update/delete	0.05	0.06
Guest	Read/write	0.08	0.09
Stranger	Malicious attempt, hence access denied	0.10	0.11
Intruder	Malicious attempt, hence access denied	0.11	0.12

5. Assist scalable nodes for secure objectives established for steps 1 through 4.

Users at different levels will be given access to the smart devices at home after verification of the access time with the threshold value. The access time is measured in milliseconds, captured via sensors based on the number of people and number of keys used to access the entire system. As sensors record the time intervals, the log is collected and stored in a database where the analysis is carried out based on the number of accesses. The data is initially stored in a local database and then moved to a read-only and immutable ledger. If a single address is generated for the record stored in blockchain ledger, then it is categorized as a single-key access, and if there are more than 1 addresses generated, then it is categorized as a multi-key access. If the threshold value is set at 0.09 s that incorporates the average time to grant access, then any attempt that fails to meet the required threshold will be reported as a malicious one, and hence, the user is denied to access the smart device. As shown in Table 1, the users at owner level possess permissions to view, write, delete and update the ledger contents, while the guest level users can view and write their data into the ledger, but they cannot modify or delete the existing contents. At the other end, strangers and intruders who try to gain entry into the house will be aborted from doing so due to the single-key access time and multi-key access time exceeding the threshold value. Also, it is observed from Table 1 that though the time taken to access via multiple keys is more than the single key, it ensures security to keep malicious users away from unintended consequences.

5 Conclusion

Integration of IoT with blockchain enables multiple security issues to be resolved. However, it is important to specify the classes of security depending on different access levels by multiple user groups to build up security. In the proposed framework, multiple keys being accessed by multiple users stored in secure private cloud are presented. Also, the data captured is stored in the blocks on a permanent basis, and the editable data gets stored in the cloud environment. The access rights can be provisioned to different users based on sharing of keys among themselves. Also, it is observed from the proposed work evaluation that though the time taken to access via multiple keys is more than the single-key-based access, it ensures security based on threshold level parameter to keep the malicious users away from unintended consequences.

References

1. Dang, T.L.N., Nguyen, M.S.: An approach to data privacy in smart home using blockchain technology. In: 2018 International Conference on Advanced Computing and Applications (ACOMP), Ho Chi Minh City, 2018, pp. 58–64. https://doi.org/10.1109/ACOMP.2018.00017
2. Singh, S., Ra, I.-H., Meng, W., Kaur, M., Cho, G.H.: SH-BlockCC: a secure and efficient Internet of things smart home architecture based on cloud computing and blockchain technology. Int. J. Distrib. Sens. Netw. **15**(4) (2019). https://doi.org/10.1177/1550147719844159
3. Gupta, S., Aluvalu, R.: Twitter based capital market analysis using cloud statistics. Int. J. Socioecol. Knowl. Dev. (IJSKD) **11**(2), 54–60 (2019). https://doi.org/10.4018/IJSKD.2019040104
4. She, W., Gu, Z., Lyu, X., Liu, Q., Tian, Z., Liu, W.: Homomorphic consortium blockchain for smart home system sensitive data privacy preserving. IEEE Access **7**, 62058–62070 (2019). https://doi.org/10.1109/ACCESS.2019.2916345
5. Rahim, K., Tahir, H., Ikram, N.: Sensor based PUF IoT authentication model for a smart home with private blockchain. In: 2018 International Conference on Applied and Engineering Mathematics (ICAEM), Taxila, 2018, pp. 102–108. https://doi.org/10.1109/ICAEM.2018.8536295
6. Lee, Y., Rathore, S., Park, J.H., Park, J.H.: A blockchain-based smart home gateway architecture for preventing data forgery. HCIS **10**(1), 9 (2020). https://doi.org/10.1186/s13673-020-0214-5
7. Saxena, U., Sodhi, J.S., Anwar, R.: Augmenting smart home network security using blockchain technology. Int. J. Electron. Secur. Digit. Forens. **12**(1), 99 (2020). https://doi.org/10.1504/IJESDF.2020.10025326
8. Lin, C., He, D., Kumar, N., Huang, X., Vijayakumar, P., Choo, K.R.: Home chain: a blockchain-based secure mutual authentication system for smart homes. IEEE Internet Things J. **7**(2), 818–829 (2020). https://doi.org/10.1109/JIOT.2019.2944400
9. Qu, C., Tao, M., Yuan, R.: A hypergraph-based blockchain model and application in internet of things-enabled smart home. Sensors **18**, 2784 (2018). https://doi.org/10.3390/s18092784
10. Agbedanu, P., Jurcut, A.D.: BLOFF: a blockchain-based forensic model in IoT. In: Singh, S., Jurcut, A.D. (eds.) Revolutionary Applications of Blockchain-Enabled Privacy and Access Control, pp. 59–73. IGI Global (2021). https://doi.org/10.4018/978-1-7998-7589-5.ch003
11. Gupta, S., Godavarti, R.: IoT data management using cloud computing and big data technologies. Int. J. Softw. Innov. (IJSI) **8**(4), 50–58 (2020). https://doi.org/10.4018/IJSI.2020100104

Hybrid Computing Scheme for Quasi-Based Deployment in the Internet of Things

Ansh Mehta, Shubham Pabuwal, and Saurabh Kumar

1 Introduction

The prominent use of the Internet of Things (IoT) has been observed in different fields in the last few years. This is mainly due to the reason that the IoT environment tries to realize the ubiquitous communication, wherein the computation and control are viewed as two of the most crucial tasks of the devices deployed in the region [1, 2]. There is a need to perform the real-time computation and controlling by the devices as the events may occur at any time, at any place, and these events may be heterogeneous in their characteristics. An event may be the data and information which are sensed and actuated upon [3]. In the IoT environment, the devices gather these data and information from different sources and communicate them to a device which may be located at remote or inaccessible places [4]. Thus, there is a need to address the issues of connectivity, reachability, and coverage in the IoT environment which consists of a significant amount of heterogeneous networks.

To address the issues discussed above, there is a need to implement a deployment strategy which supports all the phases of deployment, i.e., pre-deployment, post-deployment, and re-deployment [5]. This is due to the reason that in practical scenarios, there is a possibility of non-functioning of a few of the devices. In such cases, either additional devices must be deployed or the existing devices must be reconfigured to act on the behalf of non-functioning devices. This improves the connectivity and reachability of the network. In the literature, authors have discussed the advantages and limitations of the different techniques of deterministic and random deployments [6–8]. One of the limitations is the ineffectiveness of these techniques

A. Mehta · S. Pabuwal · S. Kumar (✉)
Department of Computer Science & Engineering, The LNM Institute of Information Technology, Jaipur 302031, Rajasthan, India
e-mail: saurabh.kumar@lnmiit.ac.in

A. Mehta
e-mail: 17ucs027@lnmiit.ac.in

S. Pabuwal
e-mail: 17ucs157@lnmiit.ac.in

M. Dua et al. (eds.), *Proceedings of the International Conference on Paradigms of Communication, Computing and Data Sciences*, Algorithms for Intelligent Systems,
https://doi.org/10.1007/978-981-16-5747-4_25

to implement the re-deployment phase of deployment in irregular terrains. In this context, the quasi-based deployment strategy [5] serves the purpose by implementing a technique which is a mix of both the deterministic and random strategies. The authors in [9, 10] have proven its better efficiency in comparison with random and deterministic strategies in terms of connectivity, coverage, and response time.

Furthermore, in the IoT environment, there is a need to process the different applications with their distinct communication and computational requirements. Although the heterogeneous devices in the IoT environment are assumed to be distributed in a highly complex scenario, the information from these devices are required to process the different tasks for the existing applications in the region of interest. The events, as and when occur, are needed to be reported by one or more of the deployed devices [4, 11]. There is a possibility that a certain information is required from a specific set of deployed devices. This emphasizes the importance of clustering in the IoT environment. The clustering helps in the organization of the deployed devices in terms of information-driven, context-driven, or computation-driven scenarios [12, 13].

The devices in a cluster communicate the sensed event, data, and information to the cluster head which has more communication and computation potentials. The communication between the devices and cluster head may follow either the centralized or distribution computing models based on the parameters such as the time spent in transmission, overhead time, and processing time [14, 15]. The processing center in a centralized model has the set of services, resources, and know-how parameters. The clients report the occurrence of events to the processing centers for further processing. While the processing center in a distributed model has the know-how services only and the clients have the resources to process the events. In a clustered IoT network, the clients are considered as the devices sensing the events and the processing center is the cluster head, which is responsible to receive and aggregate the event information from the deployed devices in the network [12].

The centralized computing model has limitations in the case of IoT environment. The network lifetime gets severely affected by more of the computation and communication performed by the IoT devices resulting in consumption of more energy in terms of storage and computing capabilities [1]. Due to the large amount of deployed devices, the centralized model creates a significantly higher network traffic and overhead. Due to the transmission of large amount of data to the processing center and more use of local file accesses, the total bandwidth requirement may exceed the available bandwidth, which may result in the degraded performance of the whole system. On the other hand, the distributed computing model suffers from longer network delays with significantly smaller networks [13]. However, there are certain advantages of both the models for different densities of devices in the network. Since the IoT networks may have varying densities, there is a need to combine the characteristics of both the models to focuses on dynamically choosing the type of computation scheme in real-time IoT environment.

In this context, the paper proposes a hybrid computing scheme for the efficient quasi-based deployed devices in the IoT environment. The proposed scheme utilizes the advantages of both the centralized and distributed models of computation and helps in dynamically choosing the type of computation required for a cluster in the network. The novelty of the proposed work may be outlined by two facts. First, the proposed hybrid model addresses both the inter-cluster and intra-cluster communication hierarchy. Second, the proposed work is implemented for quasi-based deployment technique which is a first attempt by authors in this paper. The proposal is implemented and evaluated on the IoT-based Cooja platform [16] to evaluate its efficacy.

The rest of the paper is organized as follows. Section 2 presents the system model for proposal of the algorithm discussed in Sect. 3. The results are discussed in Sect. 4. Finally, Sect. 5 concludes the work.

2 System Model

In this section, the basic terminologies are presented along with the model of the system used for the proposal of the algorithm. The system assumes a layered IoT framework [17] to be used for the organization of the devices in the network. The layered framework uses two types of devices, namely, sensor devices and IoT devices. The sensor device is responsible to sense the event and report the sensed event by forwarding its information to the IoT device. The IoT device is responsible to receive and aggregate the information received from the sensor devices, either to perform data analysis at its end or forward it further to other IoT devices using the IP-based communication.

Further, it is assumed that the IoT environment consists of a huge amount of devices and are clustered. The clustering of devices is performed in such a way that the IoT device serves as the cluster head (CH) of each cluster. There is a possibility of the presence of one or more clusters in the network. Moreover, it is also possible that the devices in one cluster need the information from other clusters. For instance, the devices of a cluster may require the location information of other devices in the network. In such a case, there are two types of communication which may be performed, namely, inter-cluster and intra-cluster. The intra-cluster communication is performed among the devices in the same cluster. On the other hand, the inter-cluster communication is performed among the CHs of each cluster.

The proposed work utilizes our previously proposed dynamically distributed collaborative computing scheme (DDCCS) for cluster-based communication [18] among the devices having the benefits of both the centralized and distributed models. The scheme 1 of DDCCS assumes that the number of nodes within a cluster is more and number of CHs is less. The intra-cluster communication in this scheme uses distributed model, and the inter-cluster communication uses centralized model. Similarly, the scheme 2 of DDCCS assumes that the number of nodes within a cluster is less and number of CHs is more. The intra-cluster communication in this scheme uses centralized model, and inter-cluster communication uses distributed model. With the use of above-mentioned system assumptions, this paper addresses to dynamically choose the type of computation model in the real-time IoT environment. The proposed algorithm is discussed in the following section.

3 Proposed Algorithm

In this section, the proposed algorithm of hybrid computing scheme is presented. The proposed algorithm works in two steps. In the first step, the proposed hybrid computing scheme is employed for dynamically choosing the type of computation in a cluster. Then, the centralized and distributed algorithms used for computation in the proposed approach is discussed for an event-driven scenario.

Initially, the devices are deployed on the terrain using the quasi-based deployment [5] technique. The deployed devices form a layered framework, wherein the clustering of devices is performed using the pragmatic clustering approach for IoT as proposed in [12]. Each cluster is assigned with a unique identification represented by $cluster\_id$. In each of the clusters, the number of devices is counted. Further, a *threshold* is defined for each of the clusters. The threshold signifies the total number of devices to exist in a particular cluster. The value of threshold is decided based on the coverage requirement of the terrain. A significant advantage of quasi-based deployment is that it allows to control the extent of deployment. Thus, based on the terrain coverage requirement, the deployment can be performed. Also, based on the extent of processing needs in different applications, the number of clusters are formed. It is not necessary to have equal number of devices in a cluster. Therefore, using the concept of *threshold*, the type of computation is decided for a particular cluster. If the number of devices is more than the threshold in a cluster, then the distributed computing is performed. Similarly, if the number of devices in a cluster is less than the threshold, then the centralized computing is performed. The same scheme applies for the communication among the CHs of different clusters in the network, as explained in the DDCCS scheme in Fig. 1. The proposed hybrid computing scheme is summarized in Algorithm 1. The algorithm takes the number of clusters and list of deployed devices ($deviceList$) as inputs and provides output as the type of computation, i.e., centralized or distributed, to be performed in a cluster.

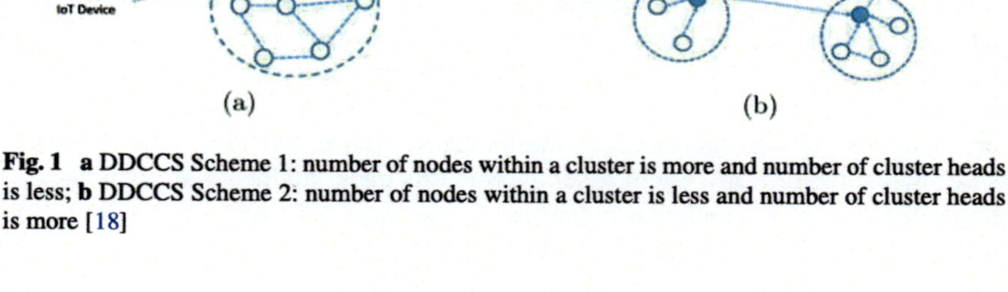

Fig. 1 **a** DDCCS Scheme 1: number of nodes within a cluster is more and number of cluster heads is less; **b** DDCCS Scheme 2: number of nodes within a cluster is less and number of cluster heads is more [18]

Algorithm 1 Hybrid Computing Scheme Algorithm

Require: Number of clusters, List of deployed devices (*deviceList*).
Ensure: Type of Computation (*typeOfComp*).

```
set_comp_params(deviceList); /*Set the paramaters of devices*/
Map clusterDeviceCount - KEY = Cluster_Id, VALUE = Count;
Map clusters - KEY = Cluster_Id , VALUE = List of devices in a cluster; /*Generate the cluster IDs*/
for device in deviceList do
  if device.cluster_id not in clusters then
    clusters[device.cluster_id] = []/*not defined*/
  end if
  clusters[device.cluster_id].append(device);
  clusterDeviceCount[device.cluster_id] = clusterDeviceCount(device.cluster_id,0) + 1; /*Count the devices in a cluster*/
end for
Assign a threshold for number of devices in a cluster.
for cluster_Id,count in clusterDeviceCount.items() do
  for device in clusters[cluster_Id] do
    if count > threshold then
      Perform distributed computing in the cluster.
    else
      Perform centralized computing in the cluster.
    end if
  end for
end for
```

Further, for an event-driven scenario, it is assumed that n number of events are occurring in the terrain and is represented as $E = \{e_1, e_2, \ldots, e_n\}$. These events are sensed by the sensor devices or IoT devices, in whose communication range the events have occurred. If the events are sensed in the communication range of a sensor device, then the event information is forwarded to the CH of its own cluster. However, if the event is sensed by an IoT device, it can either use the information for processing it or forward it to the other IoT devices in other clusters. The data forwarding and computation, if needed either at the sensor or IoT devices end, use the concept of hybrid computation scheme, i.e., either the centralized or distributed computing is used.

The proposed algorithm uses a simple client-server paradigm for performing the computation using the centralized model. The IoT device acting as the CH of a cluster acts as the server and the sensor devices play the role of a client. Similarly, if the communication is between IoT devices (CHs) of all the clusters, then the CHs act as the client and the IoT device, which processes the data received from each cluster, act as the server. However, the distributed communication is performed by using the concept of a *token*. As and when there is a need to process the information by a device within a cluster, the CH generates a Mobile Agent (MA) and circulates it in the network. The MA is a piece of code which tempts the devices to compute and find the solution to its problems at its own site. The processing by a device in the cluster can only be performed if it possess the MA. This ensures mutual exclusion and avoids

Algorithm 2 Distributed Computing Algorithm

Require: Clustered IoT environment, Sensor devices, IoT devices (CH).
Ensure: Distribution of Mobile Agent (MA) in the network.
1: *EventGenerator* generates an event at location (x, y).
2: Events sensed by the devices in communication range.
3: Devices receive the event information on *recieve_event* function by listening on $port = id + constant$.
4: Stores the event, creates a mobile agent (MA) with the event information.
5: Add current device to MA's history.
6: Finds next device not in MA history and in the communication range to current device. Sends the token to the next device.
7: Next device receives the token, perform step 5 & 6.
8: If no other device to send the MA, then it is returned to the CH.
9: CH receives the MA and checks its history if someone is left or not.
10: If a device is left, then CH sends the MA. Client repeats from Step 5.
11: If processing of all devices is finished, CH computes total propagation cost of circulation, reachability, and connectivity.

unnecessary wastage of resources for processing in the network. Periodically, the server circulates the MA among all its devices, on the reception of which, the devices execute its requirements and then passes the MA to the other devices in the network. The last device returns back the code to the CH. The proposed approach uses the simple token bus protocol for the communication of MA in the network. This mechanism is performed for both the inter-cluster and intra-cluster communication in the network. The proposed distributed computing algorithm is described in Algorithm 2.

The proposed algorithm focuses on two important aspects. First, it discusses the dynamic choice between centralized and distributed computing scheme for the clustered IoT environment using quasi-based deployment of devices. The proposed hybrid computing scheme assumes the use of layered IoT framework, which distinguishes the communication and computing capability of sensor and IoT devices in the network. Second, it discusses the implementation of centralized and distributed scheme of computing for both the intra-cluster and inter-cluster communication environment based on the density of the cluster. This helps in the analysis of performance of the distributed scheme in an event-driven scenario. The results obtained on implementation of the proposed work is discussed in the following section.

4 Results and Discussion

In this section, the results obtained on implementing the proposed algorithms is discussed. The proposed work is implemented in two phases. In the first phase, the proposed algorithms are implemented using python platform. The results obtained from the implementation in the first phase is fed as input in the IoT-based Cooja platform. The implementation on Cooja is evaluated using a database application.

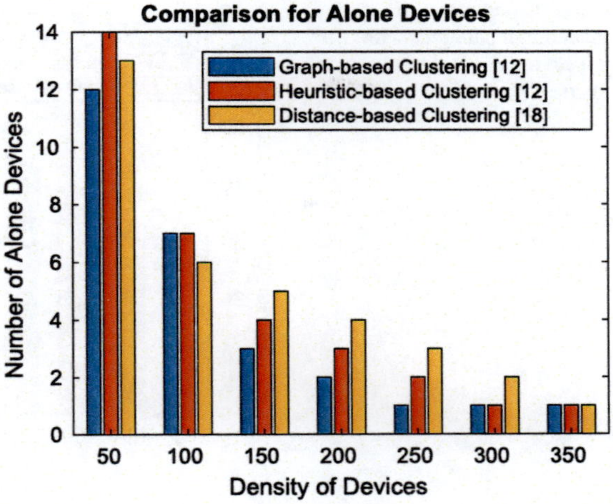

Fig. 2 Comparison of clustering algorithms for alone devices in the network

Initially, a set of D={50, 100, 150, 200, 250, 300, 350, 400, 450, 500} devices are deployed on a terrain of size $100 \times 100\,\text{m}^2$ using quasi-based deployment technique. The Halton discrepancy sequence [9] is used for the quasi-based deployment. The devices are clustered using the pragmatic clustering algorithm [12] for the layered IoT framework. In this algorithm, there are two approaches used for clustering, namely, heuristic and graph-based. Both the approaches use the concept of graphs and location of devices for clustering of sensor and IoT devices in the network. These approaches are compared with the distance-based clustering algorithm. The distance-based approach clusters on the basis of communication range of these devices.

A comparison of these clustering algorithm for quasi-based deployment of devices is performed. The comparison is done on the basis of the number of alone devices left, upon clustering, in the terrain. The alone devices are those which are left being not clustered. There is a need to reduce the number of alone devices in the network to increase the connectivity and reachability. Figure 2 shows the comparison of clustering algorithm for a set of the deployment of D={50, 100, 150, 200, 250, 300, 350} number of devices. It is observed from figure that as the number of devices increases, the number of alone devices decreases in the network. The number of alone devices are more for lesser device density in the region. It is also observed that with an increase in the density of devices, the heuristic and graph-based algorithms perform better than the distance-based approach.

It is known that there is a trade-off between the number of clusters and computational efficiency. It is more suitable to have optimized number of clusters depending on the type of application being processed in the region. This helps in optimizing the computational efficiency of devices in the terrain. A comparison is performed for the clustering algorithms with respect to the number of clusters formed in the region, as depicted in Fig. 3. It can be observed from figure that as the density of devices in the region increases, the number of clusters also increases for all the three algorithms.

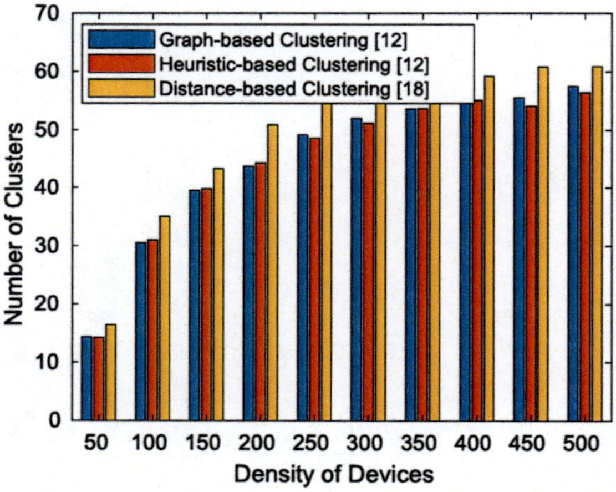

Fig. 3 Comparison of clustering techniques with respect to the number of clusters

However, the distance-based approach produces more number of clusters in comparison with the heuristic and graph-based approaches. Also, the heuristic approach performs better than the graph-based approach.

Further, the proposed hybrid computing scheme is implemented for deployment using random, Halton, and a combination of random and Halton deployment techniques. For the combined random and Halton deployment, 75% of the devices are deployed using random technique and 25% of the devices are placed in the region using Halton technique. A plot of the percentage of coverage with respect to the different densities of deployed devices is shown in Fig. 4. It is observed that the quasi-based deployment using Halton strategy provides more coverage in the network in comparison with the random technique of deployment. It is also observed that the combined deployment using random and Halton techniques provides more coverage than the random deployment technique.

The proposed hybrid computing scheme is then implemented for event-driven scenario on the IoT-based Cooja platform using same set of devices on the $100 \times 100\,\text{m}^2$ terrain. The in-built database application named Antelope is used for the implementation. The events are assumed to attain mobility in the terrain. Both the single instance and multiple instances of events are considered for the implementation. In the case of multiple instances of events, it is possible that the events are generated at the same spatial location. For the clustered environment, the non-detection of events is evaluated for different densities of devices in the network. As shown in Fig. 5, the quasi-based deployment using Halton strategy drops lesser number of events as compared to the random deployment. Also, the combined Halton and random technique performs better than the random technique in terms of detection of events in the region. However, the pure quasi-based strategy detects more number of events as the coverage of the region is maximum using this strategy, as discussed and shown in Fig. 4.

Furthermore, the reachability of devices in the region is evaluated on implementing the hybrid computing scheme in the terrain. The evaluation is performed for different

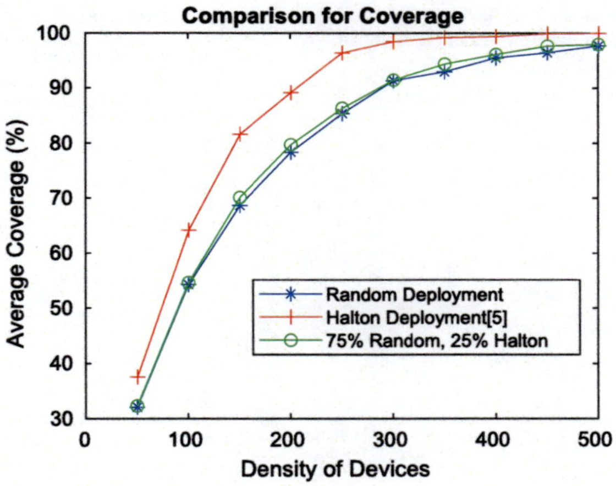

Fig. 4 Comparison of coverage for different deployment techniques using the proposed hybrid computing scheme

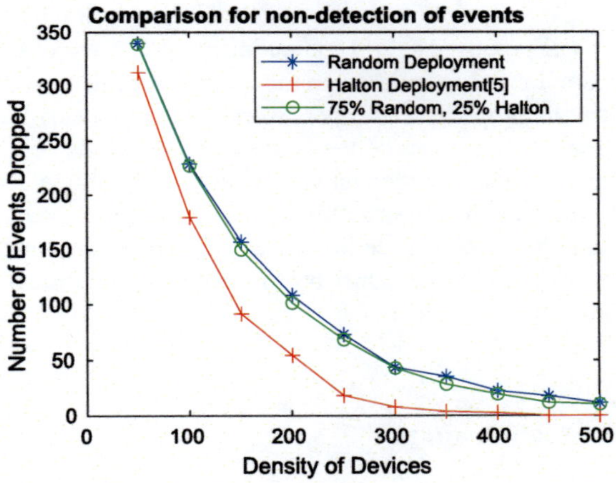

Fig. 5 Comparison of non-detection of events for different deployment techniques

degrees of devices in the region. The degree of a devices specifies the number of edges with which the device is connected to other devices in the network. It is observed that the fully reachable network easily approaches to a fully connected network. There is an influence of node degree on the reachability and connectivity of devices in the network. The average reachability of the devices is compared for random, Halton, and combined random and Halton deployment strategies, as depicted in Fig. 6. It is observed that the devices deployed using random technique has lesser reachability as compared to the devices deployed using quasi-based technique using Halton strategy. It is evident from the figure that the devices deployed using the combined approach of random and Halton tries to achieve the performance of quasi-based technique using Halton strategy.

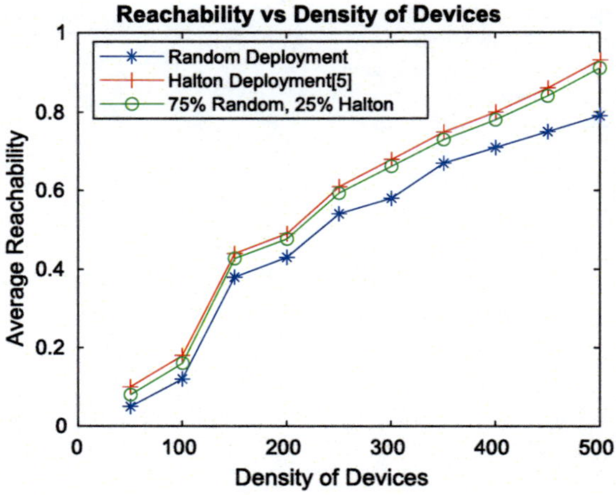

Fig. 6 Plot of average reachability versus density of deployment for different deployment techniques

The proposed hybrid computing scheme dynamically chooses the type of computation for a cluster based on the density of devices in the cluster. The results suggest that the proposed algorithms works best with quasi-based Halton deployment technique. It is because of the better coverage of the deployment technique. At certain times, certain applications demand that the random and quasi-based deployment techniques must be combined to provide better coverage in the terrain. To understand the performance in such a scenario, the proposed algorithm is evaluated and it is found that the values try to achieve the performance as observed in the case of quasi-based approach.

5 Conclusions

The IoT environment assumes a huge amount of heterogeneous devices deployed in the region of interest. In such a scenario, there is a need of clustering of these devices to provide context-oriented services to different applications in the network. There is a need to dynamically choose the type of computing to be used for processing the sensed event in the region of interest. In this context, this paper addresses the use of a hybrid scheme of computing, which dynamically choose the type of computing between centralized and distributed based on the density of deployment in the terrain. The proposed computing model is evaluated for different densities and techniques of deployment. The results suggest that the proposed algorithm works better with quasi-based deployment and provides better coverage and reachability in the terrain. In future, the proposed algorithm can be tested for its efficacy on a real testbed for different communication parameters.

References

1. Koshizuka, N., Sakamura, K.: Ubiquitous id: standards for ubiquitous computing and the internet of things. IEEE Pervas. Comput. **9**(4), 98–101 (2010)
2. Salim, F., Haque, U.: Urban computing in the wild: a survey on large scale participation and citizen engagement with ubiquitous computing, cyber physical systems, and internet of things. Int. J. Hum.-Comput. Stud. **81**, 31–48 (2015)
3. Kumar, S., Zaveri, M.: Event localization based on direction of arrival using quasi random deployment in internet of things. In: Proceedings of SAI Intelligent Systems Conference, pp. 170–188. Springer (2018)
4. Pandey, S.K., Zaveri, M.A.: DoA-based event localization using uniform concentric circular array in the IoT environment. Comput. J. **62**(10), 1403–1425 (2019)
5. Pandey, S.K., Zaveri, M.A.: Quasi random deployment and localization in layered framework for the internet of things. Comput. J. **61**(2), 159–179 (2018)
6. Deif, D.S., Gadallah, Y.: Classification of wireless sensor networks deployment techniques. IEEE Commun. Surv. Tutor. **16**(2), 834–855 (2013)
7. Farsi, M., Elhosseini, M.A., Badawy, M., Ali, H.A., Eldin, H.Z.: Deployment techniques in wireless sensor networks, coverage and connectivity: a survey. IEEE Access **7**, 28940–28954 (2019)
8. Priyadarshi, R., Gupta, B., Anurag, A.: Deployment techniques in wireless sensor networks: a survey, classification, challenges, and future research issues. J. Supercomput. 1–41 (2020)
9. Niederreiter, H.: Quasi-Monte Carlo methods and pseudo-random numbers. Bull. Am. Math. Soc. **84**(6), 957–1041 (1978)
10. Pandey, S.K., Zaveri, M.A.: Optimized deployment strategy for efficient utilization of the internet of things. In: Proceedings of the International Conference on Advances in Electronics, Communication and Computer Technology (ICAECCT), pp. 192–197. IEEE (2016)
11. Pandey, S.K., Zaveri, M.A.: Event localization in the internet of things environment. Procedia Comput. Sci. **115**, 691–698 (2017)
12. Kumar, J.S., Zaveri, M.A.: Clustering approaches for pragmatic two-layer IoT architecture. Wirel. Commun. Mob. Comput. **2018** (2018)
13. Jabeur, N., Yasar, A.U.H., Shakshuki, E., Haddad, H.: Toward a bio-inspired adaptive spatial clustering approach for IoT applications. Future Gener. Comput. Syst. **107**, 736–744 (2020)
14. Corman, F., D'Ariano, A., Pacciarelli, D., Pranzo, M.: Centralized versus distributed systems to reschedule trains in two dispatching areas. Public Transp. **2**(3), 219–247 (2010)
15. Cao, X., Chen, J., Xiao, Y., Sun, Y.: Building-environment control with wireless sensor and actuator networks: centralized versus distributed. IEEE Trans. Ind. Electron. **57**(11), 3596–3605 (2009)
16. Eriksson, J., Österlind, F., Finne, N., Tsiftes, N., Dunkels, A., Voigt, T., Sauter, R., Marrón, P.J.: Cooja/mspsim: interoperability testing for wireless sensor networks. In: Proceedings of the 2nd International Conference on Simulation Tools and Techniques, pp. 1–7. ACM (2009)
17. Pandey, S.K., Zaveri, M.A.: Localization for collaborative processing in the internet of things framework. In: Proceedings of the Second International Conference on IoT in Urban Space, pp. 108–110. ACM (2016)
18. Pandey, S.K., Zaveri, M.A.: A graph-based communication algorithm using collaborative computing scheme in the internet of things. In: Proceedings of the Second International Conference on Advanced Computational and Communication Paradigms (ICACCP), pp. 1–6. IEEE (2019)

IoT Based Health Alert System Using 8051 Microcontroller Architecture and IFTTT Implementation

Mohit Taneja, Nikita Mohanty, Shrey Bhiwapurkar, and Sumit Kumar Jindal

1 Introduction

With the constant advancement in technology and increasing mobility specifications in systems, health care has been made commercially available and a patient-centric method is approached over the conventional orientations [1]. Its demand has surged recently due to the novel Coronavirus which has pushed for driven isolation and an independent lifestyle [2].

Remote patient monitoring (RPM) works on the principles of having access to the patient's dynamic information with regards to their health without being around traditional tools and facilities to do so. This provision enhances health services and makes it affordable [3]. It is made possible with the application of sensors and wireless networks and has led to the expansion of healthcare in various manners. Heartbeat and body temperature have always been vital for diagnosis. Without a medical background, it is often difficult for patients or people related to them to keep track of abnormal conditions which is now made possible in the comfort of one's home because of RPM [4]. Hospital availability also increases with an increase in patients shifting toward RPM. However, many doctors have raised concerns regarding this method due to inaccuracy which may lead to misdiagnosis. Hence, the system is recommended for those who do not require heavy medical care till further advancements [5].

IoT is built on sensors, gateway, and wireless network which allows users to communicate and access data. In the health industry, this development lets professionals connect with the patients and identify the best treatment process for patients and reach the expected outcome. It has significantly led to reduced travel, cost, and time in long-term monitoring [6]. Insights and predictions based on data collected can also be provided using the model. In general, the patients and their families are

M. Taneja · N. Mohanty · S. Bhiwapurkar · S. K. Jindal (✉)
School of Electronics Engineering, Vellore Institute of Technology, Vellore, Tamil Nadu, India

M. Dua et al. (eds.), *Proceedings of the International Conference on Paradigms of Communication, Computing and Data Sciences*, Algorithms for Intelligent Systems,
https://doi.org/10.1007/978-981-16-5747-4_26

kept involved and in control of the situation but a drawback of the system is that it is very specific and needs to be programmed accordingly since different age groups have different safe ranges [7].

This work presents a health monitoring system to keep periodical checks as well as an alert system to execute an impulsive response to critical conditions using an 8051 microcontroller, NodeMCU, and IoT implementation. Serial communication plays a key role for the transfer of data [8]. The entire work is structured concerning the proposed system's architecture which is defined in Sect. 2 inclusive of its working and hardware setup. The outcome and its reasoning are illustrated in Sect. 3, while Sect. 4 compares and contrasts existing literature. It also emphasizes the diversity in systems. Lastly, the inference with the future scope is elaborated in Sect. 5.

2 Proposed System

2.1 System Architecture

Figure 1 depicts the design of the system proposed. The Arduino Uno is converted into an in-system programming (ISP) programmer to burn the code onto the microcontrollers instead of using a programmer board, and this is a practical and economical method to load the bootloader. It is also used to supply power to the system.

The sensors are interfaced with the microcontroller to process the data and display on the LCD for data availability on the user end which is refreshed at regular intervals as programmed with the help of timers and the oscillator. This implies that there is real-time regulated monitoring.

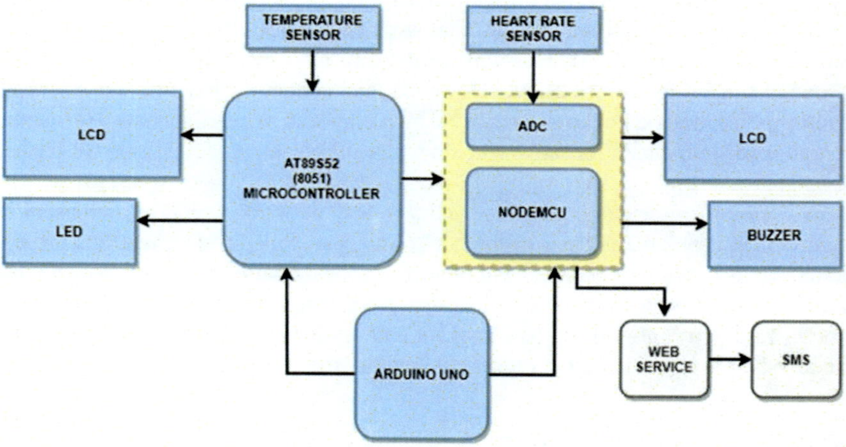

Fig. 1 Block diagram of the proposed system

On meeting the conditions set to indicate a critical state, the LED and buzzer attached are activated. A connection is also built between the 8051 microcontrollers and NodeMCU for serial communication of patient's health data. Thus, an alert system is successfully set and functioning.

The ESP8266 (NodeMCU) is a wireless fidelity (Wi-Fi) module that is configured as a station to connect to the Wi-Fi router and deliver a message by connecting to a host, IF this then that (IFTTT) web service aids in doing so.

2.2 Flowchart

Figure 2 elaborates on the workflow of the proposed model. The temperature sensor is primarily made of semiconductor ceramics and polymers. It varies the negative temperature coefficient transistor to change resistance to detect temperature. On the other hand, the heart rate sensor is constructed with a light-emitting diode and light detector and uses the optical power variation of the light emitted and absorbed on contact with the surface to indicate a change in blood flow.

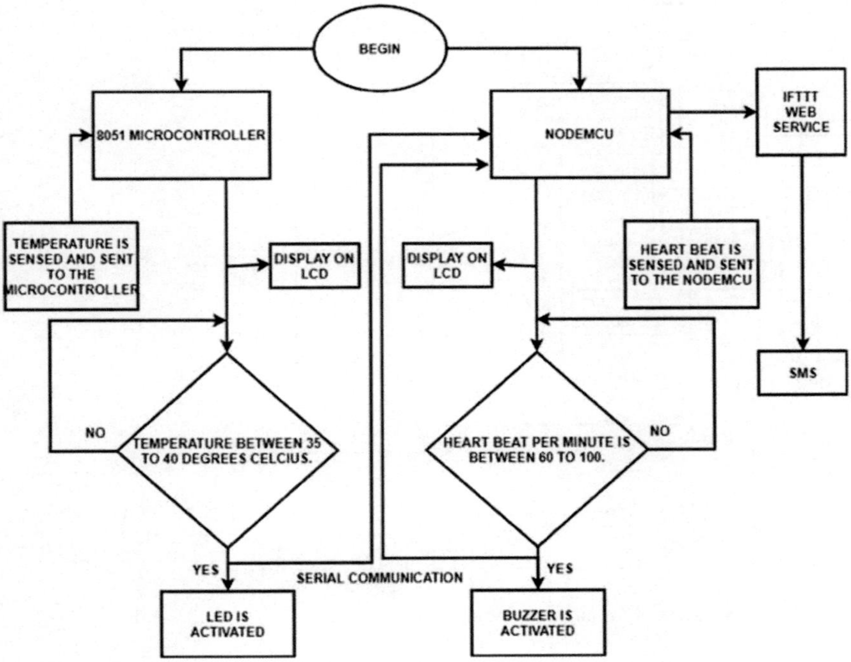

Fig. 2 Flow chart of the proposed system

The data collected by the sensors is processed and checked for abnormal levels in the parameters within the microcontroller. Human body temperature should lie in the range of 35–40 °C and the resting heart rate between 60 and 100 beats per minute.

If the parameters are not within the range, the buzzer and LED are turned on to alert the people nearby. Alongside this, the NodeMCU connects to the IFTTT web service using the unique resource locator (URL) generated and pushes for a predefined SMS to be sent to the registered mobile number.

2.3 *Experimental Setup*

Figure 3 shows the configuration of elements in the model. The system is cost-effective and built using readily available, inexpensive components. The implementation is not very complex either. In the modern-day, maximum utilization of resources in an efficacious manner is sought after and simulation helps us achieve that by checking the feasibility of hardware and testing the model.

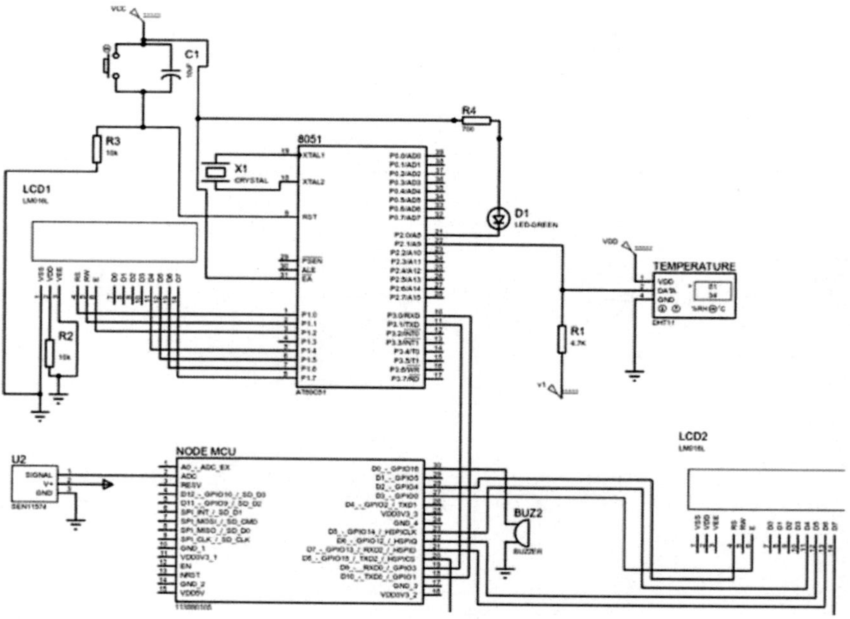

Fig. 3 Circuit diagram of the proposed system

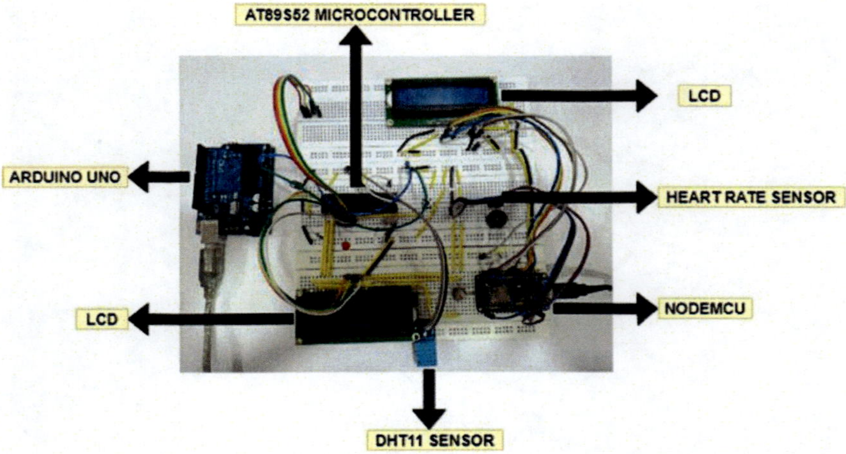

Fig. 4 The hardware setup of the proposed system

2.4 Hardware Used

Figure 4 exhibits the hardware and connections. The Arduino Uno is employed to program the microcontrollers by converting it into an ISP programmer. An AT89S52 of the 8051 microcontroller family and a NodeMCU is used to establish communication between the said system and the users. A heart rate or pulse sensor and DHT11 sensor are utilized to detect heartbeat patterns and temperature. The temperature sensor is a digital sensor hence its values are read by the 8051 instantly. On the other hand, the heart rate sensor is an analog sensor which means it can't be processed directly and needs an analog to digital converter (ADC), the NodeMCU has an inbuilt ADC, thus we attach the above sensor to it instead of the 8051. Conditions are assimilated with the help of LED and buzzer. While the 16×2 LCD visualizes and provides a simplified comprehension of the working of the system.

3 Result and Discussion

Once power has been supplied to the system, data is immediately sensed and processed. Figure 5 displays the case where the body temperature falls within the range defined so the temperature value is displayed on the LCD but the LED doesn't glow as an indication that there is no abnormality and the vitals are kept in check.

Figure 6 indicates when the body temperature falls or rises in comparison to the safe levels, and this suggests that the patient might be in a critical condition. In the given case, we pushed the temperature down to mimic hypothermia. The temperature value is displayed on the LCD, the LED glows to signify the same. This alerts the

Fig. 5 Body temperature lies in the safe range

Fig. 6 Body temperature is out of the safe range

Fig. 7 Heart rate and IFTTT request

people around the patient of their condition. Using serial communication, the 8051 microcontroller and NodeMCU interact with one another and exchange data, and this is implied by another glowing LED. The same is followed when the temperature of the patient goes up the range defined thus suggesting the health condition known as hyperpyrexia also commonly called fever.

In Fig. 7, the LCD portrays the heart rate sensor's output as it did for the temperature sensor. When the heart beats per minute crosses the extremes, a buzzer sets off to make anyone near the patient aware of the situation because they need immediate attention. The second line depicts the NodeMCU requesting to connect to the web service using a distinctive URL to announce the same with the help of a messaging service if there's no one around the patient to take measures to help them out.

While building the system, a mobile number is registered on the IFTTT site to be linked to the server. The working principle of IFTTT is to define a condition, and an action to be taken if the condition is true. In this work, once the internet connection is established which is the initial condition set, the service triggers a predefined message from the server through the linked mobile number to the registered mobile number just like how it is shown in Figs. 8 and 9.

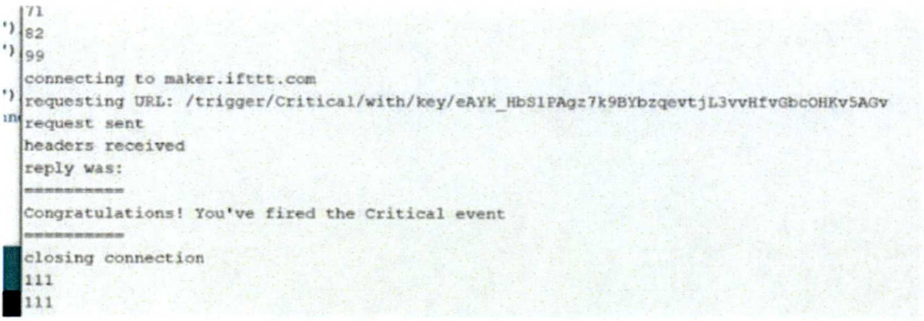

Fig. 8 The linkage between NodeMCU and IFTTT web service

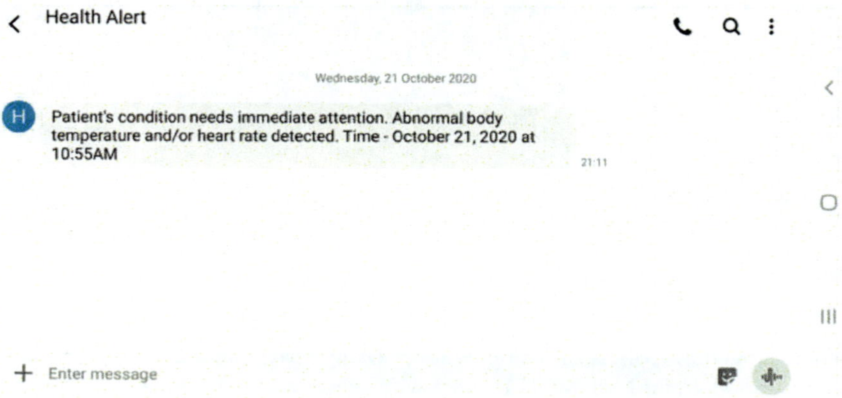

Fig. 9 Message received in case of an emergency

4 Study of Work with the Existing Literature

Stokke R. studied numerous use test cases to delve into the positives and negatives of personal emergency response system (PERS) and deduced that the systems encourage independent living with a sense of security, the systems need to be compact in order to be user-friendly [9]. De San Miguel et al. worked on analysing if PERS is effective enough and inferred that those who use these systems feel more safe and secure than those who don't [10]. Priyanka Das et al. discussed the developments in such systems and its impact on the healthcare industry, keeping both the professionals and the patients at an advantage [11]. Harshitha Bhat et al. elaborated on the expansion of wireless technology in healthcare with usage not limited to a class of users making the model flexible [12]. These works provided motivation to work on a system which focuses on keeping the user as the top priority.

Table 1 Comparative study with existing literature

No.	Title	Result	Comparison
1	iCare: A mobile health monitoring system for the elderly [13]	Wireless patient health monitoring and assistance system for the elderly hence targeting selected users	Consists of features similar to the proposed work but additionally store the previous data for record purposes
2	Health monitoring system based on GSM [14] and design and simulation of microcontroller based wireless patient intelligent health monitoring system with GSM alert technology [15]	Implementation of a health monitoring system using GSM module, microcontroller and pulse sensor	Both of them use GSM for location services, while the proposed work is IoT based and utilizes the IFTTT server to send messages instead
3	Zigbee and GSM based Patient health monitoring system [16]	Implementation of a patient monitoring system using Zigbee, SMS and GSM	Utilizes Zigbee for wireless transmission for data, reducing latency, this is also IoT based but the proposed system works closely with NodeMCU and IFTTT
4	Healthcare blockchain system using smart contracts for secure automated remote patient monitoring [17]	It implements patient monitoring by using sensors calling contracts and writing event records on Ethereum based blockchain protocol	Utilizing blockchain contracts to secure patient data. The proposed system has no such record and secure system
5	Smart health monitoring system based on IoT and cloud computing [18]	It implements AES encryption on sensor data and uses cloud for data storage and NodeMCU for processing	The patient data including location is encrypted using AES for privacy. The work proposed has no secure system because data is not being stored but patient information is safely transmitted with a unique API key used to connect to the server

Table1 discusses various existing literatures to compare and contrast models proposed from which it can be inferred that the proposed system targets an instantaneous response system rather than a record system.

5 Conclusion

Health monitoring systems are extremely beneficial to the medical field. With growing technology, innovation isn't limited to professionals and access to better facilities is made possible. We analyze and implement a health monitoring and alert system with the help of 8051 microcontrollers and NodeMCU. The main goal was to make an efficacious system with minimal expenses which was achieved by converting the Arduino into an ISP programmer instead of using a programmer board as well as using NodeMCU and IFTTT instead of buying additional modules to attach to the 8051 microcontroller. IFTTT is easy to work with and understandable hence it doesn't restrict its usage to a certain community but is open to everybody thus making this system very user-friendly. The system is flexible, cost-effective, and reliable, while being elementary to operate. The software, simulation, and hardware were implemented, debugged, and scrutinized constantly for the best given results.

This system can further be improved and polished by adding multiple other sensors to check other vitals. A global positioning system (GPS) could be implemented to attain the location of the patient as well. We can integrate an alarm system as a regular reminder for the patient to take their medicines. The development of an application or website for the same will enhance the novelty of the proposed system. Thus, expanding on a large scale can be in the works with the continuous addition of modules and advancing technology.

References

1. Saranya, M., Preethi, R., Rupasri, M., Veena, S.: A survey on health monitoring system by using IoT. Int. J. Res. Appl. Sci. Eng. Technol. **6**(3) (2018)
2. Shin, W., Rho, J.: A study on the design of real-time monitoring system using IoT sensor in respirator. Int. J. Adv. Smart Converg. **9** (2020)
3. Parihar, V.R., Tonge, A.Y., Ganorkar, P.D.: Heartbeat and temperature monitoring system using Arduino. Int. J. Adv. Eng. Res. Sci. **4**(5) (2017)
4. Mansor, H., Shukor, M.H.A., Meskam, S.S., Rusli, N.Q.A.M., Zamery, N.S.: Body temperature measurement for remote health monitoring system. In: 2013 IEEE International Conference on Smart Instrumentation, Measurement and Applications (ICSIMA), Kuala Lumpur (2013)
5. Arora, S.: IoMT (Internet of Medical Things): Reducing Cost While Improving Patient Care. IEEE pulse, Oct 2016
6. Patil, P., Waichal, R., Kumbhar, U., Gadkari, V.: Patient health monitoring system using IOT. Int. Res. J. Eng. Technol. **4**(1) (2017)
7. Shelar, M., Singh, J., Tiwari, M.: Wireless patient health monitoring system. Int. J. Comput. Appl. **62**(6) (2013)
8. Ali Mazidi, M., Mazidi, J.G., McKinlay, R.D.: The 8051 Microcontroller and Embedded Systems using Assembly and C, 2nd edn.
9. Stokke, R.: The personal emergency response system as a technology innovation in primary health care services: an integrative review. J. Med. Internet Res. **18**(7), e187 (2016)
10. De San Miguel, K., Lewin, G., Burton, E.L., Howat, P., Boldy, D., Toye, C.: Personal emergency alarms: do health outcomes differ for purchasers and nonpurchasers? Home Health Care Serv. Q. **36**(3–4), 164–177 (2017)
11. Das, P., Deka, R., Sengyung, S., Nath, B.K., Bordoloi, H.: A review paper on patient monitoring system. J. Appl. Fundam. Sci. (JAFS) **1**(2) (2015). ISSN 2395-5554 (Print), ISSN 2395-5562 (Online)
12. Bhat, H., Shetty, N., Shetty, A.: A review on health monitoring system using IoT. Int. J. Eng. Res. Technol. (IJERT), ICRTT—2018 **06**(15)

13. Lv, Z., Xia, F., Wu, G., Yao, L., Chen, Z.: iCare: a mobile health monitoring system for the elderly. In: 2010 IEEE/ACM International Conference on Green Computing and Communications & International Conference on Cyber, Physical and Social Computing, Hangzhou, 2010
14. Jaiswal, H., Safwan, M., Rajput, H.: Health monitoring system based on GSM. Int. Res. J. Eng. Technol. **5**(3) (2018)
15. Ezeofor, C.J., Okeke, R.O., Ogbuokebe, S.K.: Design and simulation of microcontroller based wireless patient intelligent health monitoring system with GSM alert technology. Int. J. Eng. Trends Technol. (IJETT) **24**(2) (2015)
16. Purnima, P.S.: Zigbee and GSM based patient health monitoring system. In: 2014 International Conference on Electronics and Communication System (ICECS-2014)
17. Griggs, K.N., Ossipoval, O., Kohlios, C.P., Baccarini, A.N., Howson, E.A., Hayajneh, T.: Healthcare blockchain system using smart contracts for secure automated remote patient monitoring. J. Med. Syst. (2018)
18. Siam, A.I., Elazm, A.A., El-Bahnasawy, N.A., El Banby, G., Abd El-Samie, F.E.: Smart health monitoring system based on IoT and cloud computing. In: International Conference on Electronic Engineering, Menoufia University, Egypt, ICEEM2019, 7–8 Dec 2019

Role of Speech Separation in Verifying the Speaker Under Degraded Conditions Using EMD and Hilbert Transform

M. K. Prasanna Kumar and R. Kumaraswamy

1 Introduction

Speaker verification efficiency drops significantly under conditions like background speakers and noise. This can be improved by adding a preprocessing stage like speech separation. Speech separation task can be classified into determined, under-determined and over determined cases depending on the count of microphones and speakers as explained in [1]. There is a challenging case of underdetermined source separation where we have only one mixed signal from two or more than two source signals as described in Eq. 1, with $z(n)$ being the mixed signal, $y_1(n)$ and $y_2(n)$ being the first speech signal and second speech signal, respectively.

$$z(n) = y_1(n) + y_2(n) \tag{1}$$

1.1 Single Channel Speech Separation (SCSS)

SCSS algorithms are most suited at the front end of speaker verification as it deals with one microphone. Therefore, in this paper, we concentrate on SCSS-based algorithms

M. K. P. Kumar (✉)
Department of Electronics & Telecommunication, B.M.S. College of Engineering, Bangalore, Karnataka, India
e-mail: prasannamk.tce@bmsce.ac.in

R. Kumaraswamy
Department of Electronics & Communication, Siddaganga Institute of Technology, Tumkur, Karnataka, India

M. Dua et al. (eds.), *Proceedings of the International Conference on Paradigms of Communication, Computing and Data Sciences*, Algorithms for Intelligent Systems,
https://doi.org/10.1007/978-981-16-5747-4_27

instead of multi-channel speech separation algorithms [2]. A well-known classification of SCSS falls under supervised and unsupervised methods. Supervised and unsupervised methods are also known as guided and unguided methods, respectively.

1.2 Supervised Speech Separation

Supervised methods of separating the speech signals from the mixture include hidden Markov model (HMM) and Gaussian models [3]. Efficiency of the algorithms depends on training model and availability of original source signals. An algorithm based on speech production mechanism and vocal tract model is given in [4]. For independent component analysis (ICA)-based method, readers can refer to [1]. ICA-based models work well on determined and over-determined cases and synthetic mixtures of audio and speech signals. Their performance decreases for real recorded mixtures and underdetermined case.

1.3 Unsupervised Speech Separation

Computational auditory scene analysis (CASA) dominates in unsupervised speech separation [5] and uses time frequency analysis to model harmonicity and modulation. The drawback of CASA-based system is in its inability to model the entire auditory system as it is. Non-negative matrix factorization (NMF) also plays a major role in this category of source separation up to some extent. It divides the entire mixing matrix into two submatrices with respect to temporal and spectral properties. It gives maximum result if at least one of the sources is known in advance and can be classified under semi-supervised speech separation [6]. Speech separation by estimating the mixing parameters and using speech specific information is described in [2]. Supervised and unsupervised speech separation based on fundamental frequency and formant frequencies is discussed in [7]. Separation of speech mixtures using clustering techniques is mentioned in [8]. Estimating number of sources in single channel speech mixture and speech separation is described in [9]. Combined empirical mode decomposition (EMD) and speech-specific information are discussed in [10]. Single channel speech separation based on EMD and Hilbert–Huang transform was given in [11].

1.4 Speaker Verification

Speaker verification deals with identity claim of a person [12]. Efficiency drops significantly under unwanted background conditions [13]. Most of the algorithms work based on voice activity detection, and it suffers under low SNR levels [14].

RASTA filtering methods have been discussed in [15]. There is still scope in developing algorithms which can significantly improve the efficiency of verification system under noisy and multi-speaker environment.

1.5 Combined Speech Separation and Speaker Recognition

Harmonicity-based algorithms are discussed in [16] which makes use of the difference in pitch values. Sequential grouping-based methods [17] use corrupted speech frames by the interference signal and passing it through a model for further processing. Usable speech detection-based methods are described in [18] which work based on target to interference ratio being above 20 dB. Bayesian approach is also used in joint speech enhancement and speaker verification [19].

1.6 Recent Developments in Speech Separation

Recent developments are toward supervised and semi-supervised algorithms under multi-channel condition. Most of the methods assume that few labeled data sets are available for training the supervised model. A semi-supervised method based on frame-wise phonetic labels and sample level speaker labels is discussed in [20]. A hybrid method based on facial expressions and lip movements combined with acoustic speech is discussed in [21]. In this method, it is assumed that the facial expressions of a speaker will not change as compared to changes in acoustic features of speech signal. Recent research has also shown the possibility of combining trained speech separation, speaker diarization and speaker recognition in order as discussed in [22]. A toolkit is developed on combined speech enhancement and source separation with respect to dereverberation, denoising and source separation at the front end as discussed in [23].

1.7 Organization of the Paper

Rest of the paper is divided as follows. Section 2 gives the background of EMD. Section 3 describes the database used and proposed speaker verification system. It also explains the denoising procedure using EMD and its effect on speaker verification with experimental results. Section 4 explains the speech separation algorithm for single channel condition and its effect on speaker verification system with experimental results. Finally, Sect. 5 concludes the paper.

2 Empirical Mode Decomposition (EMD)

The EMD technique decomposes the original signal into different frequency components known as intrinsic mode functions (IMFs) starting from high frequency to low frequency as shown in Fig. 1. The EMD decomposition is modeled using Eq. 2.

$$z(n) = \sum_{k=1}^{M} c_k(n) + r_M(n) \tag{2}$$

In Eq. 2, $c(n)$ is the IMF function, and $r(n)$ is known as the residue signal which has no significance left with respect to modulation in the signal. The details of EMD algorithm can be obtained in [24].

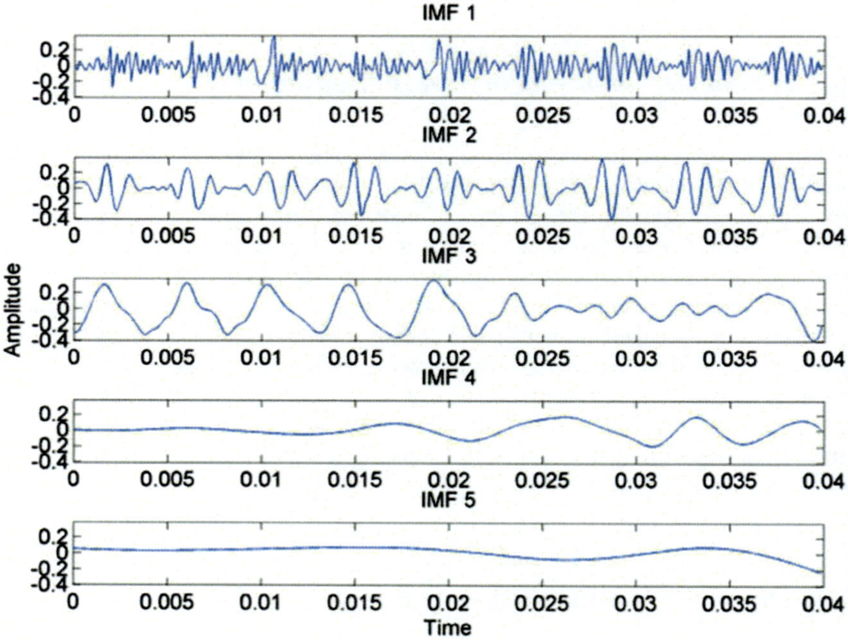

Fig. 1 Illustration of IMFs

3 Proposed Approach

3.1 Database

Speaker verification system was evaluated on a NIST database consisting of 80 speakers with 59 male and 21 female speakers. Each speaker has ten utterances, where eight utterances were used for training and two for testing. Noisy signal is simulated by adding white noise from − 10 to 30 dB. Experiments were also conducted on a database consisting of 50 speakers with 30 male and 20 female speakers under real noisy background with each speaker having ten utterances. Multi-speaker data is collected in real noisy background using one microphone and two speakers talking simultaneously. The initial study was on reducing the effect of noise using EMD to improve the speaker verification system. Finally an unsupervised source separation algorithm based on EMD and Hilbert transform is applied at the front end to deal with multi-talker speaker verification.

3.2 Block Diagram of the Proposed Approach

Proposed approach is as shown in Fig. 2. EMD is used for denoising and speech separation at the front end for effective speaker verification under degraded conditions. The denoising procedure and speech separation for single channel are described in later sections. Mel frequency cepstral coefficients (MFCC) used for feature extraction as shown in Eq. 3.

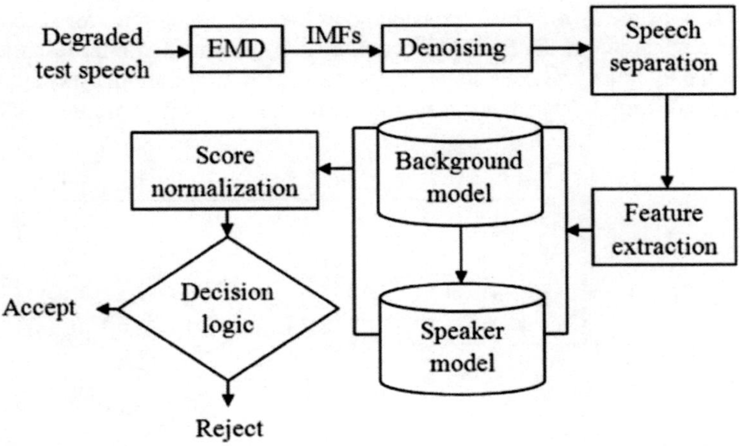

Fig. 2 Combined speech separation and speaker verification

$$d_n = \sqrt{\frac{2}{L}} \sum_{k=1}^{L} (\log V_k) \cos[n(k - \tfrac{1}{2})\tfrac{\pi}{L}], \quad n = 0, 1, ..., L - 1 \tag{3}$$

where d is the cepstral coefficient, V represents output of the filter, and L is the count of filters. GMM model is used for modeling of the feature vectors as shown in Eq. 4. Here, M represents the count of GMM densities.

$$p(x_i|\lambda_j) = \sum_{i=1}^{M} w_j b(x_i|\lambda_j) \tag{4}$$

In Eq. 4, b is the likelihood function. The details of GMM modeling are given in [25].

3.3 Improved Speaker Verification Under Noisy Conditions

We propose a successive EMD decomposition method to eliminate the noise components and retain IMFs of higher order as it contains excitation source information. This procedure is explained in Fig. 3.

Noisy speech is decomposed into two IMFs, namely $c_1(n)$ and $c_2(n)$, respectively, rather than decomposing into many IMFs and discarding initial few IMFs as done in most of the existing methods. Where $c_1(n)$ is of high frequency, and $c_2(n)$ is of low frequency. Generally, first IMF component is of high energy, and second IMF is of low energy. Also, the first IMF contains most of the noise present in it. Therefore, $c_1(n)$ is decomposed further into two new IMFs. Repeat this procedure until the energy of $c_1(n)$ falls below $c_2(n)$. This is considered as the stopping criteria for successive EMD decomposition of the first IMF component. $c_2(n)$ obtained after last stage is less affected by noise and consists of excitation source information. $c_2(n)$ after last stage is less distorted. Objective evaluation is done using equal error rate (EER).

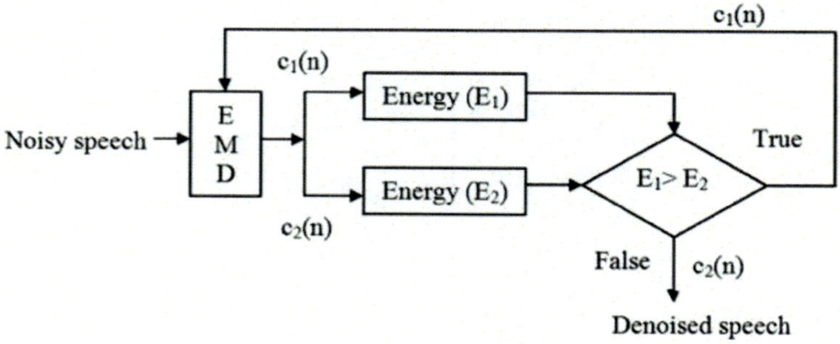

Fig. 3 Noise suppression using EMD

Figures 4, 5, 6 and 7 show the EER in percentage for speech degraded with − 5 dB, 0 dB, 5 dB and 10 dB white noise, respectively. The performance of speaker verification improves after separating speech from noise with EMD-based approach at the front end as shown in Fig. 8 and Table 1.

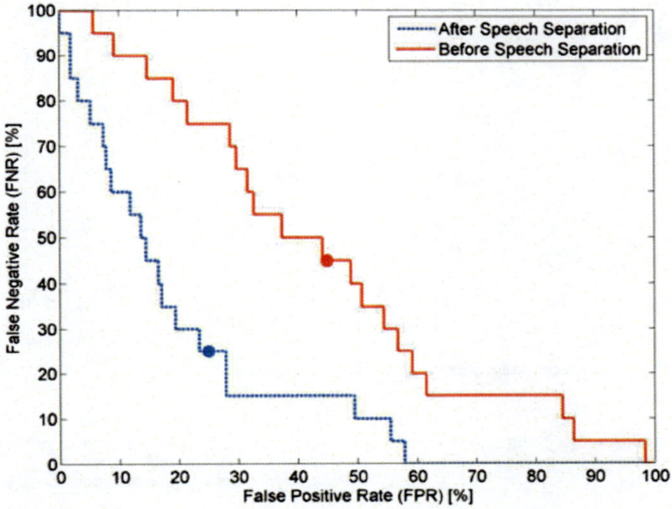

Fig. 4 EER (%) for speech degraded with − 5 dB white noise

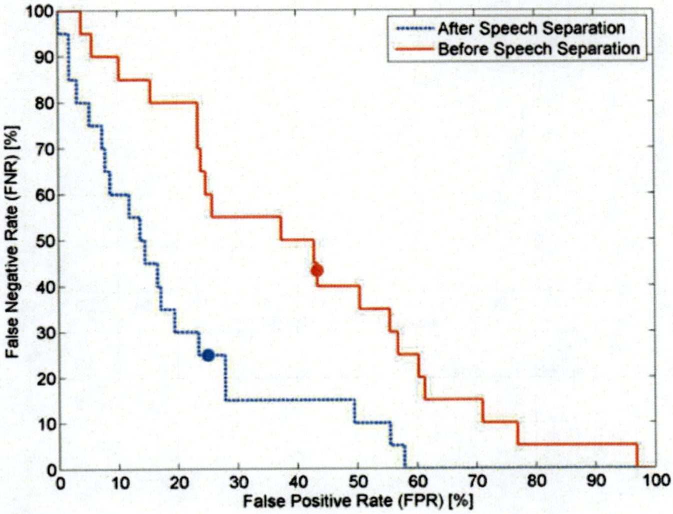

Fig. 5 EER (%) for speech degraded with 0 dB white noise

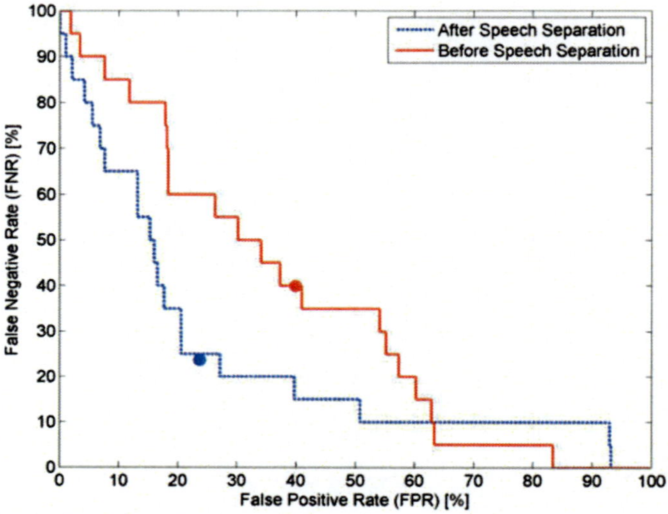

Fig. 6 EER (%) for speech degraded with 5 dB white noise

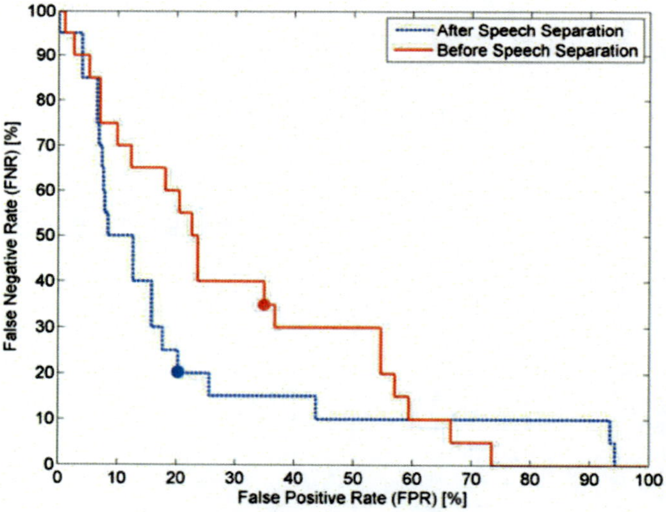

Fig. 7 EER (%) for speech degraded with 10 dB white noise

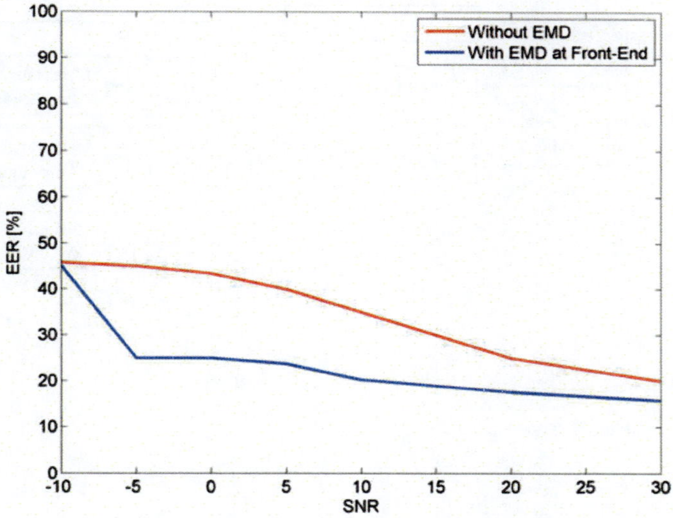

Fig. 8 Performance of speaker verification in EER (%) for speech degraded with white noise at various levels

Table 1 Performance of speaker verification in EER (%) with 256 GMMs (under noisy environment)

SNR (dB)	EER (%) without EMD	EER (%) with EMD
− 10	45.78	45
− 5	45	25
0	43.42	25
5	40	23.68
10	35	20.26

4 Improved Speaker Verification Under Multi-Speaker Environment

In this paper, we consider single channel separation of two speakers as front end for effective performance of speaker verification system. Figure 9 shows the proposed approach.

Mixed speech signal is decomposed into two IMFs. We further frame them into matrices with duration of each frame as small as 5 ms to minimize the interference of one speaker with another speaker. In Fig. 9, Yf, C1 and C2 are matrix representation of mixed speech signal, IMF1 and IMF2, respectively. Then, we estimate Hilbert transform of each frame as shown in Eq. 5.

$$H[s(t)] = \frac{1}{\pi} \int_{-\infty}^{\infty} \frac{s(\tau)}{(t-\tau)} d\tau \quad (5)$$

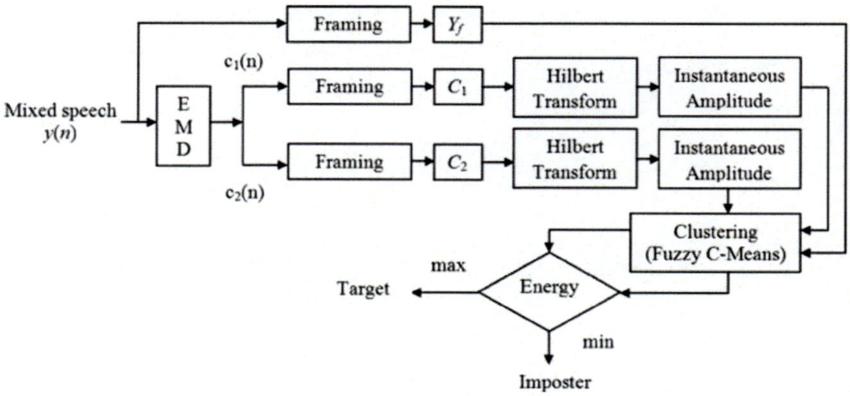

Fig. 9 Single channel speech separation

Analytic signal is given by

$$\begin{aligned} z(t) &= s(t) + jp(t) \\ &= a(t)e^{j\theta(t)} \end{aligned} \quad (6)$$

where a is the instantaneous amplitude. The Hilbert transform is applied to C1 and C2 to estimate instantaneous amplitudes. Next, it is necessary to cluster them into two groups to separate the speakers. The grouping can be done using either K-means clustering or fuzzy C-means clustering. Once the groups ids are created by clustering, the corresponding speakers are assigned with speech frames from Yf as shown in Fig. 9. That is corresponding speakers are assigned with speech frames from the mixed signal to retain high level of speech intelligibility. The target and imposter speakers are identified from the energy content of speech signals obtained after clustering. Later, it is given to the speaker verification module. The detailed algorithm and fuzzy C-means clustering can be found in our previous research given in [8]. To further illustrate the contribution of single channel speech separation method discussed here, we compare the separation results with other existing algorithms. This comparison is as shown in Table 2.

In Table 2, M + M represents male and male speech mixture, M + F represents male and female speech mixture, and F + F represents female and female speech mixture. The objective measures used for comparison are signal to artifact ratio (SAR), signal to interference ratio (SIR) and improvement in signal-to-noise ratio (ISNR). These objective measures are commonly used in evaluating single channel source separation and are discussed in detail in [8]. The proposed approach is compared with non-negative matrix factorization-based algorithm and Itakura–Saito NMF (ISNMF) and pseudo-stereo mixture-based algorithms which are discussed in the literature and in [8]. It is observed that proposed approach produces promising

Table 2 Comparison of single channel speech separation algorithms (all measures are in dB)

Speakers	Proposed approach			Non-negative matrix factorization			Itakura–Saito NMF			Pseudo-stereo mixture and 2D histogram		
	SAR	SIR	ISNR	SAR	SIR	ISNR	SAR	SIR	ISNR	SAR	SIR	ISNR
M + M	4.8	9.1	3.5	1.7	0.4	0.6	2.9	2.5	0.6	2.6	1.2	2.7
M + F	4.2	8.5	1.8	1.5	0.3	0.7	3.8	5.9	0.7	2.1	3.9	0.6
F + F	7.1	7.2	3.3	1.1	0.1	0.3	3.5	2.4	0.3	1.1	1.3	0.3

results when compared to other algorithms and is hence used in front end of speaker verification system.

Table 3 shows EER (%) under multi-speaker environment with 256 GMMs. In Fig. 10, EER (%) under multi-speaker environment with 80 speakers from NIST database is shown. Similar experiment was also conducted on database of 50 speakers collected in real noisy background, and the result is as shown in Fig. 11. From both figures, we can note that the EER reduces to a greater extent after speech

Table 3 Performance of speaker verification in EER (%) under multi-speaker environment with 256 GMMs

	NIST database (80 speakers)		Database collected in real noisy environment (50 speakers)	
Number of sentences used for training	Before speech separation	After speech separation	Before speech separation	After speech separation
1	53.13	32.05	53.1	28.57
2	53.13	22.27	52.32	25.72
3	52.32	19.05	52.38	21.43
4	52	18.75	52.15	21.43
5	48.38	18.75	52.13	21.43
6	48	17.34	52.12	21.43
7	46.34	16.23	50.81	19.22
8	46.3	15.92	50	17.83

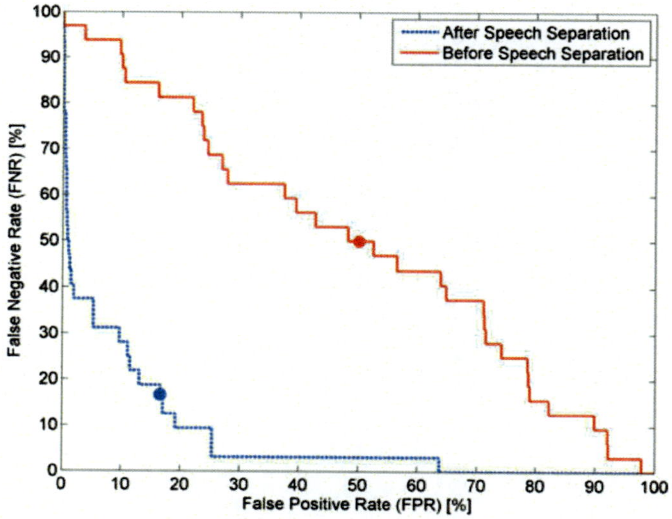

Fig. 10 EER (%) under multi-speaker environment (with 80 speakers from NIST database)

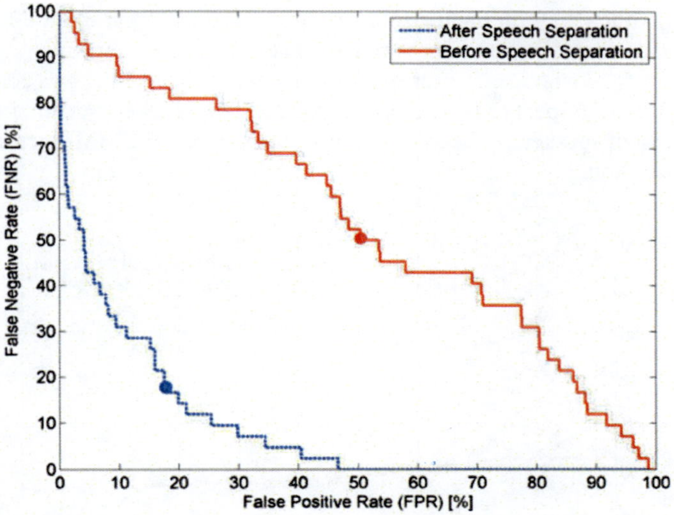

Fig. 11 EER (%) under multi-speaker environment (with 50 speakers from database collected in a real noisy environment)

separation at the front end using proposed approach. In Fig. 12, we compare the EER before and after speech separation with varying number of sentences used for training. For all cases, EER reduces to a greater extent after speech separation.

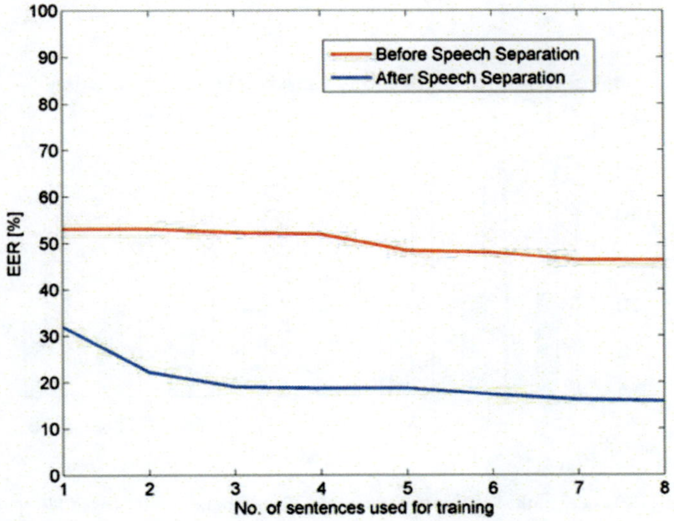

Fig. 12 EER (%) with respect to sentence count used for training

In Fig. 13, EER (%) with number of speakers is evaluated for about 80 speakers from NIST database. The speaker verification system performance improves as the number of speakers increases after speech separation. Figure 14 shows comparison of EER (%) with respect to number of GMMs used for speaker modeling. It shows improvement of speaker verification system as the number of GMMs increases after speech separation.

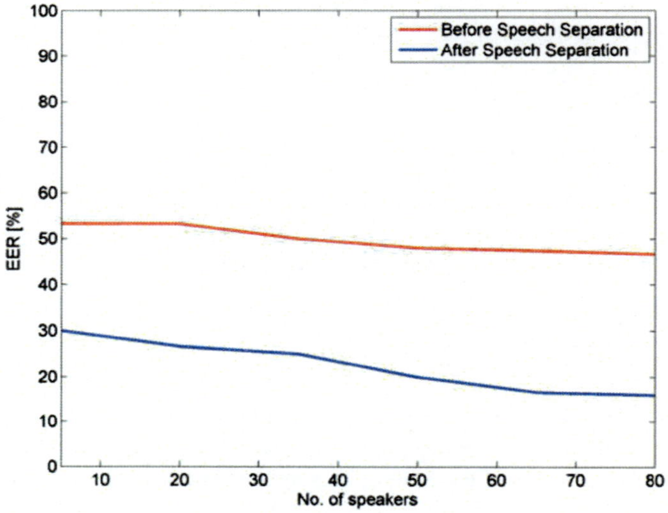

Fig. 13 EER (%) with respect to speaker count

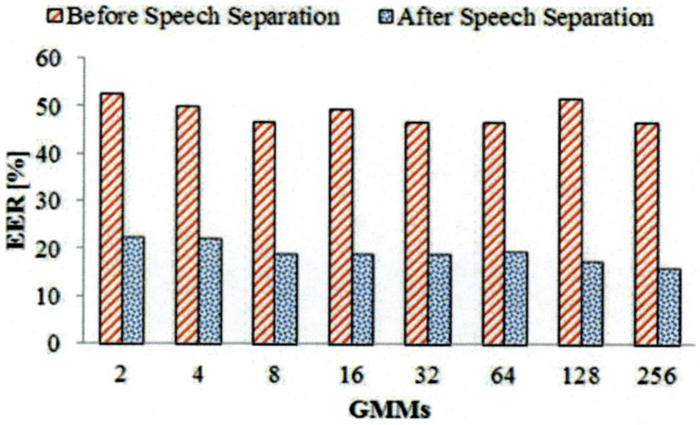

Fig. 14 EER (%) under multi-speaker environment with number of GMMs used (80 speakers from NIST database)

5 Conclusion

In this paper, we presented a method to improve the efficiency of speaker verification system under noisy and multi-speaker environment by adding a speech separation system. The results obtained show significant improvement of speaker verification system when EMD is used at the front end for noise suppression and speech separation. The improved performance is due to the filtering of noise performed by EMD and its data-driven philosophy. The strategy of using higher order intrinsic mode functions which mostly contain excitation source information eliminates most of the noise and improves performance under noisy conditions, and the strategy of combining EMD and Hilbert transform in separating single channel speech mixture provides encouraging results under multi-speaker environment.

References

1. Karhunen, J., Oja, E.: Independent Component Analysis, 1st edn. Wiley (2001)
2. Kumaraswamy, R., Yegnanarayana, S.: Determining mixing parameters from multi speaker data using speech specific information. IEEE Trans. Audio Speech Lang. Process. **17**, 1196–1207 (2009)
3. Ellis, D.: Model based scene analysis. Computational Auditory Scene Analysis, 1st edn. Wiley, Pittsburgh, PA (2006)
4. Michael, S., Michael, W.: Source filter based single channel speech separation using pitch information. IEEE Trans. Audio Speech Lang. Process. **19**, 242–254 (2011)
5. Li, P., Guan, Y.: Monaural speech separation based on computational auditory scene analysis and objective quality assessment of speech. IEEE Trans. Audio Speech Lang. Process. **14**, 2014–2023 (2006)
6. Schmidt, M., Olsson, R.:. Single channel speech separation using sparse non negative matrix factorization. In: 9th International Proceedings on Spoken Language Processing, Pittsburgh, pp. 2614–2617 (2010)
7. Prasanna Kumar, M.K., Kumaraswamy, R.: Supervised and unsupervised separation of convolutive speech mixtures using f0 and formant frequencies. Int. J. Speech Technol. **18**(4), 649–662 (2015)
8. Prasanna Kumar, M.K., Kumaraswamy, R.: An unsupervised approach for co channel speech separation using Hilbert Huang transform and fuzzy c means clustering. Int. J. Speech Technol. **20**(1), 1–13 (2017)
9. Prasanna Kumar, M.K., Kumaraswamy, R.: Single channel speech separation using empirical mode decomposition and multi pitch information with estimation of number of speakers. Int. J. Speech Technol. **20**(1), 109–125 (2017)
10. Prasanna Kumar, M.K., Kumaraswamy, R.: Single channel speech separation using combined EMD and speech specific information. Int. J. Speech Technol. **20**(4), 1037–1047 (2017)
11. Prasanna Kumar, M.K., Kumaraswamy, R.: Single channel speech separation based on empirical mode decomposition and Hilbert transform. IET Signal Process. **11**(5), 579–586 (2017)
12. Rudramurthy, M., Kumaraswamy, R.: Speaker verification under degraded conditions using empirical mode decomposition based voice activity detection algorithm. J. Intell. Syst. **23**, 359–378 (2014)
13. Rosenberg, A., Snoog, F.: Recent research in automatic speaker recognition. Adv. Speech Signal Process. 701–738 (1991)

14. Prasanna, S., Pradhan, G.: Speaker verification by vowel and non-vowel like segmentation. IEEE Trans. Audio Speech Lang. Process. **21**, 854–867 (2013)
15. Shao, Y., Wang, D.: Model based sequential organization in co channel speech. IEEE Trans. Audio Speech Lang. Process. **14**, 289–298 (2006)
16. Decheveigne, A., Kawahara, H.: Speech separation for speech recognition. J. Phys. **24** (1994)
17. Wajdi, G., Amel, B.: Improved EMD usable speech detection for co channel speaker identification. In: NOLISP, pp. 184–191. LNAI (2013)
18. Krishnamachari, K., Yantorno: Spectral autocorrelation ratio as a usability measure of speech segments under cochannel conditions. IEEE Intell. Signal Process. Commun. Syst. (2000)
19. Maina, C., Walsh, J.:. Joint speech enhancement and speaker identification using Monte Carlo methods. In: International Proceedings of Inter Speech (2009)
20. Du, Y.: Semi supervised multichannel speech separation based on a phone and speaker aware deep generative model of speech spectrograms. In: 28th European Signal Processing Conference, Amsterdam, pp. 870–874 (2021)
21. Michaelsanti, D.: An overview of deep learning based audio visual speech enhancement and separation. IEEE Trans. Audio Speech Lang. Process. 1–9 (2021)
22. Raj, D.: Integration of speech separation, diarization and recognition for multi speaker meetings In: IEEE Spoken Language Technology Workshop, Shenzhen, pp. 897–904 (2021)
23. Li, C.: End to end speech enhancement and separation toolkit designed for ASR integration. In: IEEE Spoken Language Technology Workshop, Shenzhen, pp. 785–792 (2021)
24. Huang, N., Shen, Z.: The empirical mode decomposition and Hilbert spectrum for nonlinear and non-stationary time series analysis. Int. Proc. R. Soc. Lond. 903–995 (1998)
25. Reynolds, D., Quateri, T.: Speaker verification using adapted Gaussian mixture models. Digital Signal Process. **10**, 19–41 (2000)

Comparative Analysis of Two Hardware-Based Square Root Computational Algorithms

Prince Choudhary, Atishay Jain, Alankrit Agrawal, and Poornima Mittal

1 Introduction

Square root computation is an important nonlinear operation which is often required in various applications. In digital signal processing, these computations form an integral part of root mean square calculations, instantaneous envelope estimation in AM demodulations and 3-D graphics algorithm implementations. Therefore, the idea of hardware implementation, of square root calculations so as to produce fast computations, seems very lucrative. However, the algorithms for square root computations are rather complex and difficult to implement in the hardware unlike other basic arithmetic operations.

There are two approaches to finding the square root of a number: one is using a digit-by-digit calculation method, and the other is to use piecewise-polynomial approximations. The digit-by-digit methods include the Restoring and Non Restoring [6] algorithms which produce one correct bit in each iteration. In each iteration, simple operations of addition and subtraction are performed based on the previous iteration results. The other class of algorithms includes the algorithms like Newton Raphson method [3, 7] and Taylor Series Expansion method [4]. These methods prove to be more efficient for processors. The Newton Raphson approximation is widely used in square root computations. An initial seed approximation of the root is iterated in this method to get a value closer to the actual root.

In this paper, the attention is given to two algorithms for 32-bit square root calculations, one is based on the Newton Raphson Inverse method and the other based on the Non Restoring method. Each system is considered starting with its mathematical niceties and to their hardware implementations.

P. Choudhary · A. Jain · A. Agrawal · P. Mittal (✉)
Delhi Technological University, Rohini, Delhi, India
e-mail: poornimamittal@dtu.ac.in

M. Dua et al. (eds.), *Proceedings of the International Conference on Paradigms of Communication, Computing and Data Sciences*, Algorithms for Intelligent Systems,
https://doi.org/10.1007/978-981-16-5747-4_28

The remaining part of this paper is organized as follows: Sect. 2 describes the Newton Raphson inverse algorithm and its implementation; Sect. 3 presents the Non Restoring algorithm and its implementation; Sect. 4 presents the results of the implementations in terms of physical characteristics, and Sect. 5 concludes the paper.

2 Newton Raphson Inverse Algorithm

Newton–Raphson method is a numerical solution of mathematical problem which involves making an estimate of the solution, and then using this estimated solution to refine a subsequent estimate of the solution. As per this method, the following equation estimates an approximate zero of a nonlinear function $f(y) = 0$:

$$y_{n+1} = y_n - \left[f(y_n)/f'(y_n)\right] \tag{1}$$

For the nonlinear function $y^2 - A = 0$, (1) leads to:

$$y_{n+1} = 1/2(y_n + A/y_n) \tag{2}$$

This approach proves to be an efficient technique for square root computations. It is based on the process of successive approximation. However, since (2) has the division by y_n operation, it is a computationally expensive equation. Hence, to avoid the division operation, the algorithms are modified to calculate the inverse of square root of A. The iteration (1) is performed on an inverse square root function:

$$f(z) = 1/z^2 - A \tag{3}$$

The iteration yields an inverse square root estimate:

$$y_{n+1} = z_n/2\left(3 - Az_n^2\right) \tag{4}$$

Thus, Newton Raphson Inverse method takes in an initial approximation of the inverse of square root performs its iteration of (4) on it and returns a new approximation closer to the actual inverse. With each iteration, the accuracy of the result is doubled as (4) has a near to quadratic convergence. After two iterations, an enough accurate value of the inverse of square root of A is obtained, $z_2 = 1/\sqrt{A}$. It is then multiplied by A to compute the final square root value of A.

The accuracy of the initial seed approximation affects the accuracy of the final result. Now with only simpler operations of multiplication, subtraction and shifting involved in the method, strategy to determine an accurate initial approximation call for attention. One way of making an initial approximate has been suggested in [1] which gives the equation:

$$z_0 = 1/[(2A/3) + 0.354167] \tag{5}$$

Table 1 Normalization of input to the system requirements

Base 10	4000000
Initial Input (32-bits)	00000000001111010000100100000000
	↓↓↓↓↓↓↓↓↓↓↓↓↓↓↓↓↓↓↓↓↓↓↓↓↓↓↓↓↓↓↓↓
Index	31 30 29 28 27 26 25 24 23 22 21 20 19 18 17 16 15 14 13 12 11 10 9 8 7 6 5 4 3 2 1 0
Normalization (32-bits)	$1.1110100001001000000000000000000 \cdot 2^{21}$
Final input (24-bits)	1.11101000010010000000000

Using this strategy, A is first normalized by multiplying 4^n until it is in the range of $[0.25, 1)$ and then subsequently used in Eqs. (5) and (4). However, again to avoid the expensive division operation, the strategy proposed in [2] is used to make an accurate initial approximation. A 32-bit number is taken as input. This binary number ($A = a_0\ a_1\ a_2\ \ldots\ a_k\ a_{k+1}\ a_{k+2}\ \ldots\ a_{32}$) is normalized and truncated to 24 bits, as shown in Table 1, to give input to the system.

The left most bit that holds 1 is identified and then the radix point is shifted by changing the exponent of 2 to the index value of the left most 1. The final normalized input is always in the range of $[1, 2)$. If k is the index with left most 1, then we intend to compute:

$$y = \surd\left(a_k \cdot a_{k+1}a_{k+2}a_{k+3} \ldots a_{k+23}2^k\right) \tag{6}$$

$$y = \surd\left(M * 2^k\right), \quad \text{where } M = a_k.a_{k+1}a_{k+2} \ldots a_{k+23} \tag{7}$$

Therefore,

$$\surd A = \surd M * 2^{k/2} \;\; \text{when } k \text{ is even} \tag{8}$$

and

$$\surd A = \surd M * 2^{k-1/2} * \surd 2 \;\; \text{when } k \text{ is odd} \tag{9}$$

Here, we work toward obtaining the value of $\surd M$ using the Newton Raphson Inverse method. The seed estimate (z_0) is given by (10) with the help of Table 2 as suggested in [2]:

$$z_0 = \text{Magic constant} - 0.5 * M \tag{10}$$

This value is fed into (2) to get a new, more accurate value of the inverse of square root. The final value obtained after two iterations is multiplied by A to obtain the square root.

Table 2 Magic constants for different ranges

From (base 10)	To (base 10)	Magic constant (base 2)
1.00	1.25 (open)	1.10000001100001001111010
1.25	1.375 (open)	1.10000100110111010000111
1.375	1.5 (open)	1.10001010111111001100111
1.5	1.625 (open)	1.10010001011001000011000
1.625	1.75 (open)	1.10011001000100100100010
1.75	1.875 (open)	1.10100010100010100001111
1.875	2.0 (open)	1.10110101011111011100101

The hardware implementation of the algorithm consists of multipliers, adder/subtracters and shifter units. The shifter unit takes in the normalized and truncated M as input and shifts it to right by one bit, thus giving $0.5 * M$. The subtracter unit executes (10) using the magic constants already stored in the lookup table.

Subsequently, result from the subtracter is squared by a multiplier unit. The squared value is then taken in by another multiplier and multiplied with $0.5 * M$. A second subtracter unit takes the result from second multiplier and subtracts it from binary constant 1.5 producing a value which is multiplied with the result of first subtracter in a third multiplier. The result from the third multiplier is then looped back in the circuit pertaining to (4) in case more iterations are needed. After 2 iterations, an accurate value, z_2, for the inverse of square root is obtained. Finally, another multiplier unit multiplies z_2 and M and returns the square root $\sqrt{M}$.

3 Non Restoring Algorithm

Non Restoring algorithm is an iterative digit-by-digit method which generates a correct quotient bit in each iteration. Partial remainder generated each time is the primary focus of the algorithm. The number taken as input is divided in pair of bits in both directions starting from the radix point. Beginning with the most significant pair as the partial dividend, 01 is subtracted from it. If the result of subtraction turns out to be non-negative, the most significant bit of the quotient is set as 1 else it is set to 0. For each subsequent iteration, the partial remainder from the previous iteration and the next most significant pair of bits clubbed together form the partial dividend. In the case when partial remainder from last iteration turned out to be negative, bits 01 are appended to the current quotient bits and subtracted from the current partial dividend and the same criterion discussed above is used to determine the next quotient bit. Whereas in the case when partial remainder turned out negative, bits 11 are appended to the current quotient bits and added to the current partial dividend. If the result obtained has an overflow, implying the partial remainder to be positive, the next quotient bit is set to 1 else the quotient bit is set to 0 and again algorithm proceeds with the case of negative partial remainder. The remainder from the last iteration is taken as the final precise value in case it is non negative, else addition operation is performed on it to obtain the final value. An example of Non Restoring algorithms for an 8-bit number is shown in Fig. 1.

The hardware circuitry of the discussed algorithm for a 32-bit operand is shown in Fig. 2 [2]. The implementation requires 3 registers, one each for the input radicand,

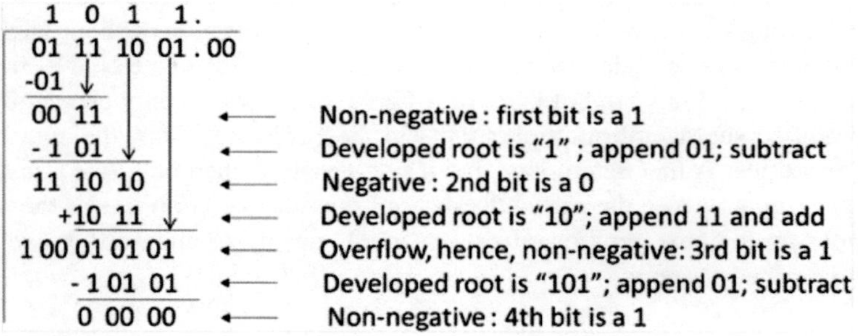

Fig. 1 Example of square root using Non Restoring algorithm

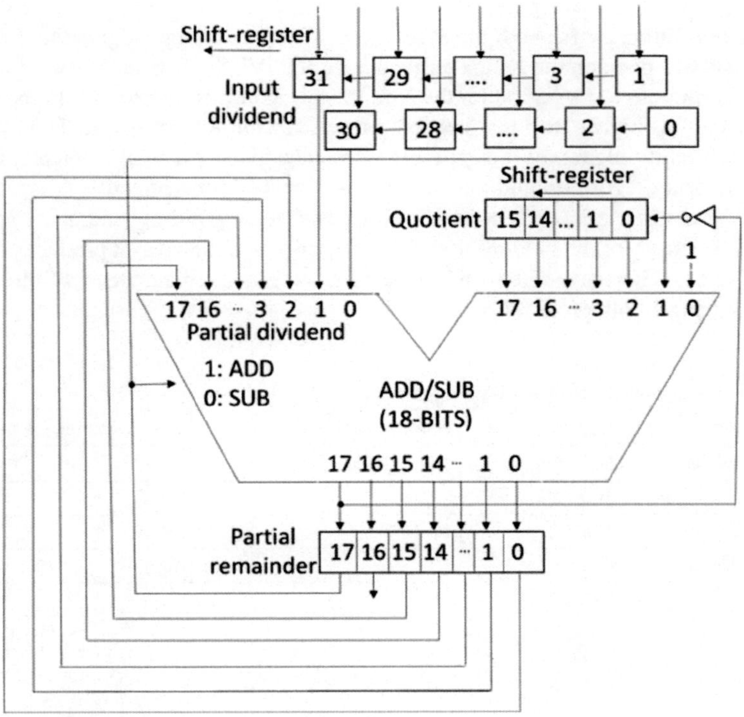

Fig. 2 Iteration of the Non Restoring algorithm

partial remainder and the quotient, and an adder/subtracter unit which adds if control input is 1, else it subtracts.

For an input 32-bit radicand, the algorithm goes through 16 iterations to obtain the final square root. In hardware, each iteration translates to a clock cycle. In each cycle, the operand register is shifted by 2 bits to the left and the quotient register initialized to 0 is shifted by 1 bit. The register for partial remainder is also initialized to 0.

Another alternative to appending 11 and addition operation without having to restore the partial remainder even if it is negative (hence, the name Non Restoring) has been suggested in [10]. Sutikno [10] offers the solution to only append 01 and use controlled subtraction multiplex units in the hardware. Thus, the sign of the partial remainder is first determined and if non-negative, then only it is transferred to the next iteration else the partial dividend of previous iteration acts as the partial remainder for the next iteration. However, in this paper, we choose to go with the earlier described strategy.

4 Results

The device summary for both the systems in percentage usage is given in Table 3. Synthesis was done on the Xilinx target device XC5VLX50T from Virtex5 family.

The minimum clock period for the Non Restoring unit operation is 5.19 ns which is only 0.04 ns higher than that needed for the Newton Raphson unit. The register usage is slightly higher, by a negligible difference given the total available, in the Non Restoring algorithm and also this method makes more effective usage of the LUT-FF pairs in use. In each cycle, the Non Restoring algorithm generates a correct bit. The accuracy of the final result from it, however, is reduced as it produces only a 16-bit value. Whereas the Newton Raphson Inverse method generates a 24-bit result, hence having a high precision.

Table 3 Utilization summary for both algorithms

Parameters	Number available	Newton Raphson (% used)	Non Restoring (% used)
Slice registers	28,800	0.08	0.39
Slice LUTs	28,800	0.69	0.63
LUT-FF pairs	196	12.24	49.49
Bonded IOBs	480	10	10.6

5 Conclusion

This paper has detailed two popular square root computation strategies. Though the results show that both the designs have almost similar performances in terms of the clock period, the Newton Raphson Inverse method has the obvious advantage of providing both the square root and inverse of square root values simultaneously. However, in terms of the complexity of the circuit, Non Restoring algorithm is relatively simpler as it only employs basic hardware components like adder/subtracters and shift registers. Also, Newton Raphson Inverse method require some pre-processing of normalization of the input number and a lookup table access to make a close initial estimate before it is given as input to the square root computation circuit. This does add some delay in the whole process. Therefore, both algorithms being well suitable for VLSI implementation can be chosen to perform the square root computations depending on the requirements.

References

1. Allie, M., Lyons, R.: A root of less evil [digital signal processing]. IEEE Signal Process. Mag. **22**(2), 93–96 (2005)
2. Hasnat, A., Bhattacharyya, T., Dey, A., Halder, S., Bhattacharjee, D.: A fast FPGA based architecture for computation of square root and Inverse Square Root. In: Devices for Integrated Circuit (DevIC), pp. 383–387. Kalyani (2017)
3. Kabuo, H., Taniguchi, T., Miyoshi, A., et al.: Accurate rounding scheme for the Newton-Raphson method using redundant binary representation. IEEE Trans. Comput. **43**(1), 43–51 (1994)
4. Kwon, T., Draper, J.: Floating-point division and square root implementation using a Taylor-series expansion algorithm with reduced look-up tables. In: 51st Midwest Symposium on Circuits and Systems, Knoxville, TN, pp. 954–957 (2008)
5. Kumar, B., Raj, K., Mittal, P.: FPGA implementation and mask level CMOS layout design of redundant binary signed digit comparator. Int. J. Comput. Sci. Netw. Secur. **9**(9), 107–115 (2009)
6. Li, Y., Chu, W.: A new non-restoring square root algorithm and its VLSI implementations. In: Proceedings International Conference on Computer Design. VLSI in Computers and Processors, Austin, TX, USA, pp. 538–544 (1996)
7. Ramamoorthy, C.V., Goodman, J.R., Ki, K.H.: Some properties of iterative square-rooting methods using high-speed multiplication. IEEE Trans. Comput. **C-21**(8), 837–847 (1972)
8. Rathod, A.P.S., Lakhera, P., Baliga, A.K., Mittal, P., Kumar, B.: Performance comparison of pass transistor and CMOS logic configuration based de-multiplexers. In: International Conference on Computing Communication and Automation (ICCCA), pp. 1433–1437. Galgotias University, Noida, India (2015)
9. Rawat, G., Rathod, K., Goyal, S., Kala, S., Mittal, P.: Design and analysis of ALU: Vedic mathematics approach. In: International Conference on Computing Communication and Automation (ICCCA), pp. 1372–1376. Galgotias University, Noida, India (2015)
10. Sutikno, T.: An efficient implementation of the non restoring square root algorithm in gate level. Int. J. Comput. Theory Eng. 46–51 (2011)

Confluence of Cryptography and Differential Privacy: A Hybrid Approach for Privacy Preserving Collaborative Filtering

S. Sangeetha, G. Sudha Sadasivam, V. Nithesh, and K. Mounish

1 Introduction

Recommendation systems are used to suggest relevant products to users to choose from millions of items in e-commerce websites. collaborative filtering (CF) is the most popular recommendation mechanism that uses machine learning techniques to accurately predict the users interest. To provide such accurate recommendation, massive amount of user data is used which creates a threat to individual privacy. In particular, neighborhood based collaborative filtering is prone to privacy attacks. Literature survey suggests that continual monitoring of covariance matrix or recommendation results helps to identify the user ratings [1]. In this paper, we aim to tackle the privacy preserving issue by blending cryptography and differential privacy based mechanism.

Privacy in recommendation system can be achieved using cryptography, obfuscation and perturbation. Obfuscation based technique is easy to understand and implement but completely degrades the recommendation results. Cryptography based mechanism requires high-computational cost. Perturbation is implemented by adding differentially private noise that masks the accurate ratings to protect users.

McSherry [2] did a seminal work on introducing differential privacy into the covariance matrix for recommendation. Further Friedman [3] demonstrated differentially private framework for matrix factorization. Zhu et al. [4] created a recommendation aware sensitivity and private neighbor selection for efficient recommendation results. Badsha et al. [5] proposed a cryptography based recommendation system that computes suggestion on the encrypted user data.

Even though differential privacy provides a rigorous mathematical guarantee, it still has some drawbacks and research barriers. In particular, the existing works have the following limitations:

S. Sangeetha (✉) · G. Sudha Sadasivam · V. Nithesh · K. Mounish
PSG College of Technology, Peelamedu, Coimbatore, Tamil Nadu, India

M. Dua et al. (eds.), *Proceedings of the International Conference on Paradigms of Communication, Computing and Data Sciences*, Algorithms for Intelligent Systems,
https://doi.org/10.1007/978-981-16-5747-4_29

- Differential Privacy induces large noise to mask the individuals which in turn affects the recommendation results. Such noise addition is made to mask the presence or absence of user. Such noise is calibrated based on sensitivity. In recommendation, sensitivity is based on the presence or absence of user in rating matrix which leads to higher sensitivity.
- Cryptography based methods used in privacy preservation makes true information inaccessible to any third party and the recommendation engine. Such techniques encrypt data and perform all the computations on the encrypted form of data. Finally, the results are send to the user who can only decrypt the data to get the results. Such end-to-end computation on encrypted data incurs larger computational cost.

To overcome these drawbacks, we propose a hybrid private recommendation system. With respect to neighborhood based recommender system both the rating and neighbors have to be masked. Existing techniques add excess noise since the noise addition happen in two stages, namely masking the rating and masking the neighbor. Secondly, the differentially private Laplace mechanism adds huge amount of noise in existing algorithms. Recent advancement of differential privacy is considered in our work to avoid the addition of huge noise which in turn improves the accuracy of the recommendation algorithm.

1.1 Differential Privacy

Differential privacy introduced by DWork [6] was initially used for private statistical data release. Its promising mathematical guarantee made Differential Privacy popular in industries and academia. The intuition behind Differential Privacy is that the aggregated query release should mask the presence or absence of individual user in the dataset.

Definition 1 (ϵ, δ) Differential Privacy [7, 8]. A randomized algorithm A provides ϵ, δ-differential privacy if for any pair of neighboring datasets D and D', and for every set of outcomes $S \subseteq \mathbb{R}$, A satisfies:

$$\Pr[A(D) \in S] \leq \exp(\epsilon) \cdot \Pr[A(D') \in S] + \delta, \quad (1)$$

where ϵ is the privacy budget. When $\delta > 0$ provides approximate differential privacy. When $\delta = 0$ it is termed as real differential privacy.

To ensure ϵ, δ-differential privacy we used analytical differential privacy proposed by Balle and Wang [9]. The analytical differential privacy uses ℓ_2 sensitivity. It is a state-of-the-art improvement over standard gaussian mechanism. In the standard gaussian mechanism, the ϵ value must be in the range (0,1], whereas in analytical DP it ranges from $(0,\infty]$.

Definition 2 (*Analytical Gaussian Mechanism*) Let $f : \mathbb{X} \rightarrow \mathbb{R}^d$ be a function with global ℓ_2 sensitivity Δ. For any ϵ 0 *and* $\delta \in [0, 1]$, the gaussian output perturbation mechanism $M(x) = f(x) + Z$ with $Z \sim \mathcal{N}(0, \sigma^2 I)$ (ϵ, δ)- DP if and only if

$$\Phi\left(\frac{\Delta}{2\sigma} - \frac{\varepsilon\sigma}{\Delta}\right) - e^{\varepsilon}\Phi\left(-\frac{\Delta}{2\sigma} - \frac{\varepsilon\sigma}{\Delta}\right) \leq \delta \tag{2}$$

Differential privacy provides a rigourous mathematical guarantee to protect individual in the dataset. However, the performance of recommendation is affected with the excess noise addition. Hence, private algorithms designed with minimal noise addition is necessary to provide accurate recommendation results. The major issues addressed in this paper include:

- high computational time of cryptography based privacy preservation techniques.
- high noise addition in existing DP recommendation system.
- handling KNN attack on recommendation system.
- handling privacy issues due to observation of similarity matrix.

A brief introduction on sensitivity and the local sensitivity used in the proposed algorithm (Algorithm 1) is explained in the following section.

1.2 Sensitivity

In general sensitivity measures, the change in dataset when we remove one user's record from the input dataset [8]. Gaussian noise scales the noise based on ℓ_2 sensitivity instead of ℓ_1 in laplace noise.

Definition 3 (*ℓ_2 Sensitivity*) The ℓ_2 sensitivity of a function $f: \mathbb{N}|\mathcal{X}| \rightarrow \mathbb{R}^k$ is:

$$\Delta_2(f) = \max_{\substack{x,y \in \mathbb{N}|\mathcal{X}| \\ \|x-y\|_1 = 1}} \|f(x) - f(y)\|_2$$

In differential privacy, there are two types of sensitivities that can be used one is local sensitivity and other is global sensitivity. Global sensitivity considers the maximal changes in all the neighboring datasets.

Definition 4 (*Global Sensitivity*). For $f : R \rightarrow \mathbb{R}$, the global sensitivity of f is defined as

$$\mathrm{GS}_f = \max_{D,D'} \|f(D) - f(D')\|_2,$$

where D and D' are the neighboring datasets.

Since global sensitivity considers maximal changes, it yields high noise which affects the performance of recommendation results. Local sensitivity is introduced by Nissim et al. [10] which calibrates record based sensitivity instead of global sensitivity and was able to satisfy differential privacy.

Definition 5 (*Local Sensitivity*). For $f : R \to \mathbb{R}$, the local sensitivity of f is defined as

$$\mathrm{LS}_f = \max_{D'} \|f(D) - f(D')\|_2,$$

where D and D' are the neighboring datasets.

Compared to global sensitivity, it generates less noise which help to improve the performance of the algorithm. The local sensitivity for Algorithm 1 is calculated as the maximum change in the similarity when we remove a user's rating record.

To address the issues, major objectives of the paper are:

- Cryptography based techniques yields high accuracy since it does not add noise like differential privacy approaches. But these techniques are subject to huge computation time. Larger time is consumed when the model training is performed on the encrypted data. By decrypting the similarity matrix, the proposed approach reduces the computational time.
- To design a robust recommender that does not ask the user to submit the plaintext at data collection stage. Our algorithm submits the ciphertext to the server and eliminates the dependency on a trusted third party server.
- Existing recommender preserve privacy by adding noise to rating matrix and neighbors. Cryptography helps to mask the rating and eliminates noise addition to rating matrix.
- Even though cryptography help in masking original rating it cannot tackle KNN attacks [1]. KNN attack will identify the individual user with the help of final recommendation results. A private algorithm with differentially private gaussian noise is designed to tackle KNN attack.
- To tackle privacy attack that is launched by continuous observation of the changes in similarity matrix [1]. In our private crypto DP algorithm, the differentially private noise is added to the encrypted similarity matrix, and then it is decrypted for further computations. Hence, privacy attack on similarity matrix can be overcome.

The rest of the paper is organized as follows. Proposed privacy algorithms are explained in Sect. 3. Experimental results are presented in Sect. 4 followed by conclusion in Sect. 5.

2 Related Work

A seminal work on applying differential privacy for private Recommendation is proposed by McSherry [2]. A differentially private laplace noise is added to the covariance matrix and the noise added matrix is used for further recommendation.

Further Zhu et al. [4] used Laplace noise with recommendation aware sensitivity. The primary focus of the private collaborative filtering algorithm is to handle KNN attack proposed by Calandrino et al. [1]. Existing algorithms can be used for either trusted or untrusted recommender setting. Also Laplace mechanism adds large noise that degrades the performance of recommender. In this paper, we combine cryptography and differential privacy so that it can be used in trusted as well as untrusted recommendation environment. Recent advancement in differential privacy namely analytical Gaussian mechanism is used to avoid excess noise addition. Li et al. [17] propose a private algorithm for shared collaborative filtering using differential privacy.

Several works are carried out using homomorphic encryption for privacy preservation [16, 18]. Basu et al. [12] propose a homomorphic encryption based recommendation system for cloud environment. The work eliminates the need for a trusted third party server to collect the user ratings. Erkin et al. [11] developed an end to end content based recommendation system which performs all the operations on encrypted data. Homomorphic encryption is used to encrypt the user rating and all the computations are performed on the encrypted data. Kikuchi et al. [14] proposed a cryptographic protocol to protect individual rating matrix. Further Kikuchi et al. [13] implemented collaborative filtering using partial homomorphic encryption. The private data is shared between two honest but curious parties and learnt privately with homomorphic encryption.

Elnabarawy et al. [15] present a comprehensive survey on various privacy preserving collaborative filtering algorithms. The survey elaborates various vulnerabilities in collaborative filtering and the techniques to solve it. Friedman et al. [19] brief on privacy risks and mechanisms to handle the privacy risks.

2.1 Summary

The major issues, while preserving the privacy in recommender system, include privacy preservation of user rating and resistance against KNN attacks. Conventional privacy preserving mechanisms do not address both these issues. Cryptography based techniques have better performance with increased computation cost. Also cryptography based methods cannot provide prevention from serious privacy violations like KNN. On the other end, differential privacy provides a rigid privacy guarantee against KNN attack but noise addition at multiple stages leads to unsatisfactory performance.

Hence, in this paper, we propose hybrid approach that integrates the cryptography and differential privacy based mechanisms to strike a balance between privacy and attack resistance. Such integration provides a comprehensive privacy guarantee as well as increased recommendation accuracy.

3 Hybrid Approach

In this section, the overview of the methodology is discussed. Further the proposed private CryptoDP collaborative filtering algorithm and the privacy parameters used on the algorithm are elaborated. The major phases of recommendation system include: Data collection, data publication, and data prediction. The overview of the hybrid approach is depicted in Fig. 1.

3.1 Data Collection

Each user encrypts their rating and sends it to the recommendation server. Cryptensor [21] is an encrypted torch tensor that supports secure computation. The computations are performed on the encrypted data to ensure privacy preserved learning. The cryptensor implements a cryptographic mechanism called secure multiparty computation (MPC). The computation result thus obtained on encrypted data is identical to the output from plain (unencrypted) data. The proposed system uses cryptensor to protect the rating provided by user. In the data collection phase, each user encrypts the ratings and sends it to the server.

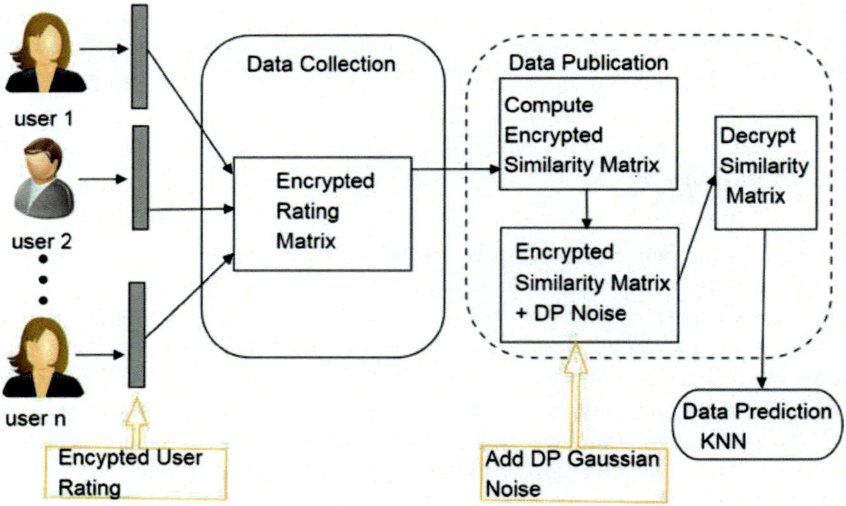

Fig. 1 Overview of the hybrid approach for privacy preserving recommender system

3.2 Data Publication

Encrypted ratings from all users collected by the server is obtained from data collection phase. The main objective is to design a robust system that prevents the KNN attack. Hence, KNN recommendation algorithm is used for prediction. After obtaining the encrypted ratings, the next step in KNN is to compute the similarity matrix. Such matrix can be computed with two well known similarity metrics cosine (COS) or Pearson (PCC). In this paper, both the metrics are used for similarity matrix computation. The major challenge is to compute the similarity matrix on encrypted rating matrix. Hence, cryptensor is used to compute similarity matrix from encrypted rating. The KNN attack is made by observing the changes in similarity matrix and the prediction results obtained from the recommender. Cryptography helps to prevent the attacks on similarity matrix but it cannot handle the attack on prediction result. Moreover, end-to-end recommendation performed using encrypted data incurs high-computational cost. Therefore, the DP gaussian noise is added to the similarity score to prevent the attack from prediction results. Hence, the proposed hybrid approach is resistant to attacks on similarity matrix as well as the prediction results.

To prevent inference from prediction results, the proposed CryptoDP algorithm (Algorithm 1) is used. The algorithm outputs a perturbed encrypted similarity matrix. Since the recommendation performed on encrypted similarity matrix incurs high-computational cost, the similarity matrix is decrypted. Further k nearest neighbor algorithm is applied on the decrypted similarity matrix obtained from Algorithm 1.

Primarily neighborhood based methods are classified into user based and item based. In user based method, the item rated by the active user have to be protected. In item based method, the similarity matrix is observed continuously; hence, the objective is to protect user's identity. In the proposed algorithm, the similarity matrix is decrypted only after applying the differential privacy noise. Hence, the proposed CryptoDP can deal with both cases.

Algorithm 1: CryptoDP Collaborative Filtering

Input: S = $E\{s(i,j)\}$ -Encrypted similarity matrix
ϵ - Privacy Parameter
δ - Privacy Parameter for Gaussian
Output: S' = $E\{s(i,j)\}$ -Encrypted similarity matrix with perturbation

1. The encrypted Similarity Matrix S is obtained
2. Compute the gaussian noise $g = \mathcal{N}(0, \delta\epsilon\Delta(LS))$
3. Encrypt the gaussian noise and obtain $E(g)$
4. Perturbation: Perturb the simalrity matrix by adding gaussian noise S' = $E\{s(i,j)\} + E(g)$

In Algorithm1 $\Delta(LS)$ is the local sensitivity, which was explained in Sect. 1.2. Compared to global sensitivity, local sensitivity adds less noise which helps improving the performance of recommendation. The changes in removing a users record is used to determine the sensitivity in Algorithm 1. Such sensitivity is calculated on the encrypted similarity matrix as follows:

$$\Delta(LS) = \max_{i,j} E(\|s(i,j) - s'(i,j)\|)$$

where $i, j \in I$ for item to item similarity. It is $i, j \in U$ for user to user similarity. $s(i, j)$ and $s'(i, j)$ measures the similarity score with and without user, respectively. Using Algorithm 1 encrypted similarity matrix with DP noise perturbation is obtained. The final step in data publication is the decryption of similarity matrix.

3.3 Data Prediction

The decrypted similarity score from data publication is used for data prediction. The KNN user to user and item to item recommendation algorithms are trained and evaluated using the similarity score. The performance and privacy of the proposed algorithm are evaluated in the following results section.

4 Results

In this section, the proposed algorithm performance is evaluated. The analysis is performed based on the identified research issues.

1. How does the hybrid approach of cryptography and Differential Privacy beneficial? A comprehensive comparison of the baseline with the proposed CryptoDP is performed. The performance is evaluated using various metrics like MAE, RMSE, Precision@10 and Recall@10 is evaluated.
2. How does CryptoDP perform related to the existing privacy preserving algorithms? We compare the performance of the proposed algorithm with existing privacy preserving algorithms.

Also the neighbor selection and execution vary based on the user to user or item to item approach, and the similarity metric used. Hence, a comprehensive analysis using both Pearson (PCC) and cosine (COS) is evaluated.

4.1 Dataset

The MovieLens dataset is collected by GroupLens Research which is the standard benchmark dataset to test the Recommendation algorithm [20]. Movielens 100K and Movielens 1M datasets are used from GroupLens. The Movielens consists of the ratings provided by different user's for various movies. The Movielens 100K dataset contains 100,000 ratings by 943 users on 1682 items. Each user has rated at least 20 movies. Users and items are numbered consecutively from 1. Movielens 1M file contain 1,000,209 anonymous ratings of approximately 3900 movies made by 6040 MovieLens users.

4.2 Metrics

To measure the performance of recommender Precision, Recall, Root Mean Square Error (RMSE) and Mean Absolute Error (MAE) are used.

$$\text{MAE} = \frac{1}{|D|} \sum_{a,i \in D} \left| r_{ai} - \hat{r}_{ai} \right|$$

$$\text{RMSE} = \sqrt{\frac{1}{|D|} \sum_{a,i \in D} \left| r_{ai} - \hat{r}_{ai} \right|}$$

where D is the test dataset. r_{ai} is the actual rating and $\hat{r}_{ai}$ is the predicted rating. For a user u, Precision P_u@N and Recall R_u@N are computed as $\frac{|\text{Rel}_u \bigcap \text{Rec}_u|}{|\text{Rec}_u|}$ and $\frac{|\text{Rel}_u \bigcap \text{Rec}_u|}{|\text{Rel}_u|}$ respectively. Let Rec_u denote the set of N items recommended to user u and Rel_u denote the set of N items relevant to user u.

4.3 Analysis

In this section, the performance of CryptoDP and intuitive comparison to the existing private algorithm is performed.

Table 1 highlights the experimental results using user and item based KNN. The similarity matrix is computed using cosine similarity. It can be observed that the proposed CryptoDP results are closer to the baseline results for precision and recall. The error measures MAE and RMSE of the proposed algorithm are comparable to the baseline results.

Table 2 compares the results of the proposed algorithm using Pearson calculation.

In Fig. 2 the proposed algorithm is compared to the existing private algorithm [4]. The MAE value for baseline, Zhu et al. and CryptoDP are plotted. Specifically, when $k = 30$ CryptoDP achieves an MAE of 0.726 which is much better than 0.73 of Zhu et al. It can be observed from the figure that the proposed CryptoDP algorithm outperforms the existing private algorithm for all the values of k. This is achieved with the proposed privacy preserving hybrid algorithm.

Table 1 Comparison on COS ML 100K

	k	MAE		RMSE		Precision@10		Recall@10	
		Non private	Crypto DP	Non private	Crypto DP	Non private	Crypto DP	Non private	Crypto DP
Item-based	5	1.0010	1.1093	1.5086	1.5172	0.9922	0.9925	0.9906	0.9879
	10	0.7939	0.8682	1.0045	1.0569	0.9876	0.9826	0.9846	0.9813
	15	0.7631	0.8603	0.9306	1.0038	0.9843	0.9862	0.9797	0.9782
	20	0.7434	0.8477	0.9064	0.9728	0.9809	0.9871	0.9756	0.9769
	25	0.7410	0.8238	0.9095	0.9701	0.9777	0.9761	0.9722	0.9719
	30	0.7396	0.8139	0.8897	0.9673	0.9747	0.9738	0.9685	0.9673
	35	0.7381	0.8102	0.8825	0.9609	0.9718	0.9708	0.9649	0.9631
	40	0.7373	0.8118	0.8811	0.9579	0.9683	0.9687	0.9620	0.9603
	45	0.7343	0.8013	0.8736	0.9481	0.9652	0.9685	0.9589	0.9561
	50	0.7313	0.7983	0.8691	0.9287	0.9615	0.9593	0.9559	0.9569
User-based	5	0.3643	0.6298	1.0514	1.1318	0.2494	0.3023	0.2531	0.3238
	10	0.4959	0.6038	0.3662	0.4719	0.2436	0.3198	0.2466	0.3182
	15	0.4195	0.5138	0.2672	0.4231	0.2402	0.3103	0.2434	0.3163
	20	0.3913	0.4823	0.2352	0.3328	0.2378	0.2992	0.2407	0.2983
	25	0.3750	0.4719	0.2218	0.3281	0.2359	0.2951	0.2384	0.2913
	30	0.3643	0.4719	0.2131	0.3123	0.2343	0.2908	0.2370	0.2899
	35	0.3593	0.4517	0.2097	0.3029	0.2327	0.2893	0.2353	0.2871
	40	0.3547	0.4509	0.2056	0.2999	0.2313	0.2891	0.2339	0.2861
	45	0.3520	0.4418	0.2028	0.2983	0.2301	0.2873	0.2328	0.2851
	50	0.3496	0.4407	0.2002	0.2918	0.2290	0.2851	0.2317	0.2842

5 Conclusion

Collaborative filtering algorithms have been widely used by multiple e-commerce platforms to provide accurate recommendation to its users'. The privacy of users' involved in such systems is of utmost importance, hence, a hybrid approach with cryptography and Differential Privacy is proposed in this paper. Such hybrid approach provides an efficient solution to strike a balance between performance and privacy of the user. Important contributions of this paper is to protect the user rating and similarity matrix with cryptography mechanisms. Also, to protect the user from KNN attack using Differential Privacy mechanism. Extensive experimental results on two benchmark datasets confirm that the proposed solution show outstanding results on MAE, RMSE, precision, and recall. It should be noted that the proposed solution exceeds the performance of the existing private algorithm.

Table 2 Comparison on PCC ML 100K

	k	MAE		RMSE		Precision@10		Recall@10	
		Non private	Crypto DP	Non private	Crypto DP	Non private	Crypto DP	Non private	Crypto DP
Item-based	5	1.3061	1.3801	2.5002	2.9392	0.4513	0.4728	0.3690	0.3719
	10	0.8402	0.9238	1.1409	1.2392	0.4535	0.4623	0.3903	0.3939
	15	0.7691	0.8739	0.9361	1.1893	0.4542	0.4603	0.3975	0.4029
	20	0.7542	0.8538	0.8984	0.9830	0.4546	0.4599	0.4017	0.4013
	25	0.7469	0.8431	0.8724	0.9732	0.4550	0.4583	0.4049	0.4193
	30	0.7430	0.8401	0.87	0.9703	0.4552	0.4573	0.4080	0.4182
	35	0.746	0.8237	0.8748	0.9639	0.4555	0.4593	0.4105	0.4281
	40	0.7424	0.8203	0.8596	0.9603	0.4558	0.4583	0.4134	0.4281
	45	0.7424	0.8192	0.8616	0.9539	0.4560	0.4573	0.4157	0.4283
	50	0.7398	0.8163	0.8533	0.9503	0.4562	0.4563	0.4178	0.4317
User-based	5	0.9347	0.9939	1.1681	1.2312	0.1646	0.1723	0.1648	0.1673
	10	0.4959	0.6039	0.4229	0.5338	0.1650	0.1673	0.1652	0.1683
	15	0.4195	0.5282	0.3159	0.4237	0.1658	0.1703	0.1659	0.1683
	20	0.3913	0.4921	0.2797	0.3729	0.1665	0.1689	0.1666	0.1692
	25	0.3750	0.4819	0.2638	0.3823	0.1672	0.1683	0.1673	0.169
	30	0.3643	0.4702	0.2507	0.3631	0.1677	0.1693	0.1679	0.1698
	35	0.3593	0.4621	0.2454	0.3412	0.1682	0.1690	0.1684	0.1703
	40	0.3547	0.4589	0.2380	0.3418	0.1687	0.1691	0.1688	0.1702
	45	0.3520	0.4519	0.2358	0.3348	0.1691	0.1683	0.1692	0.1703
	50	0.3775	0.4639	0.2332	0.3363	0.1695	0.1702	0.1696	0.1719

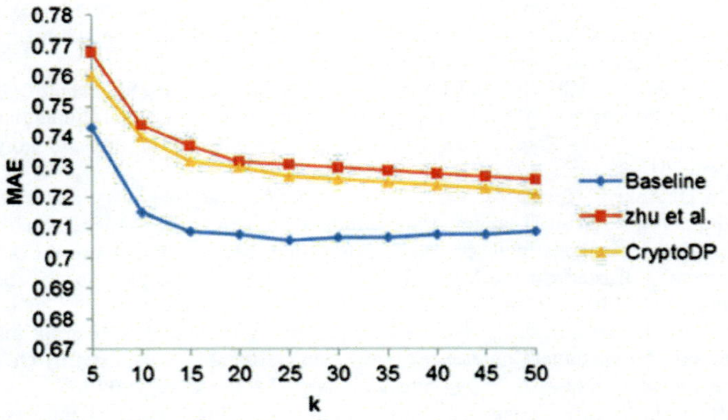

Fig. 2 PCC item on ML1M comparison to existing private algorithm

References

1. Calandrino, J.A., Kilzer, A., Narayanan, A., Felten, E.W., Shmatikov, V.: 'You might also like:' privacy risks of collaborative filtering. In: 2011 IEEE Symposium on Security and Privacy, Oakland, CA, USA, pp. 231–246 (2011). https://doi.org/10.1109/SP.2011.40
2. McSherry, F., Mironov, I.: Differentially private recommender systems: building privacy into the netflix prize contenders. In: Proceedings of the 15th ACM SIGKDD International Conference on Knowledge Discovery and Data Mining (KDD), pp. 627-635. Association for Computing Machinery, Inc., Paris, France (2009). https://doi.org/10.1145/1557019.1557090
3. Friedman, A., Berkovsky, S., Kaafar, M.A.: A differential privacy framework for matrix factorization recommender systems. User Model. User-adapt. Interact. **26**, 425–458 (2016)
4. Zhu, X., Sun, Y.: Differential privacy for collaborative filtering recommender algorithm. In: IWSPA 2016—Proceedings of the 2016 ACM on International Workshop on Security and Privacy Analytics, New Orleans, Louisiana, USA, pp. 9–16 (2016). https://doi.org/10.1145/2875475.2875483
5. Badsha, S., Yi, X., Khalil, I.: A practical privacy-preserving recommender system. Data Sci. Eng. **1**, 161–177 (2016)
6. Dwork, C.: Differential privacy. In: Bugliesi, M., Preneel, B., Sassone, V., Wegener, I. (eds.) Automata, Languages and Programming, ICALP 2006. LNCS, vol. 4052, pp. 1–12. Springer, Berlin, Heidelberg (2006)
7. Zhu, T., Li, G., Zhou, W., Yu, P.S.: Differential Privacy and Applications, 1st edn. Springer, Heidelberg (2017)
8. Dwork, C., Roth, A.: The Algorithmic Foundations of Differential Privacy. Now Publishers Inc., Hanover, MA, USA (2013)
9. Balle, B., Wang, Y.X.: Improving the Gaussian mechanism for differential privacy: analytical calibration and optimal denoising. In: Dy, J., Krause, A. (eds.) 35th International Conference on Machine Learning (ICML 2018), pp. 678–692. PMLR (2018)
10. Nissim, K., Raskhodnikova, S., Smith, A.: Smooth sensitivity and sampling in private data analysis. In: Proceedings of the Annual ACM Symposium on Theory of Computing, pp. 75–84. ACM Press, San Diego, CA, USA (2007). https://doi.org/10.1145/1250790.1250803
11. Erkin, Z., Beye, M., Veugen, T., Lagendijk, R.L.: Privacy-preserving content-based recommendations through homomorphic encryption. In: 33rd WIC Symposium on Information Theory, Boekelo, The Netherlands, pp. 71–77 (2012)
12. Basu, A., Vaidya, J., Kikuchi, H., Dimitrakos, T.: Privacy-preserving collaborative filtering for the cloud. In: Proceedings—2011 3rd International Conference on Cloud Computing Technology and Science (CloudCom), Athens, Greece, pp. 223–230 (2011). https://doi.org/10.1109/CloudCom.2011.38
13. Kikuchi, H., Aoki, Y., Terada, M., Ishii, K., Sekino, K.: Accuracy of privacy-preserving collaborative filtering based on quasi-homomorphic similarity. In: Proceedings—IEEE 9th International Conference on Ubiquitous Intelligence and Computing IEEE 9th International Conference on Autonomic and Trusted Computing (UIC-ATC 2012), Fukuoka, Japan, pp. 555–562 (2012). https://doi.org/10.1109/UIC-ATC.2012.131
14. Kikuchi, H., Kizawa, H., Tada, M.: Privacy-preserving collaborative filtering schemes. In: Proceedings—International Conference on Availability, Reliability and Security (ARES 2009), Fukuoka, Japan, pp. 911–916 (2009). https://doi.org/10.1109/ARES.2009.148
15. Elnabarawy, I., Jiang, W., Wunsch, D.C.: Survey of Privacy-Preserving Collaborative Filtering. arXiv 1–8 (2020)
16. Canny, J.: Collaborative filtering with privacy. In: Proceedings—IEEE Symposium on Security and Privacy 2002-January, Berkeley, CA, USA, pp. 45–57 (2002)
17. Li, J., et al.: Enforcing differential privacy for shared collaborative filtering. IEEE Access **5**, 35–49 (2017)
18. Erkin, Z., Beye, M., Veugen, T., Lagendijk, R.L.: Privacy enhanced recommender system. In: Thirty-first Symposium on Information Theory in the Benelux, Rotterdam, Netherlands, pp. 35–42 (2010)

19. Friedman, A., Knijnenburg, B.P., Vanhecke, K., Martens, L., Berkovsky, S.: Privacy aspects of recommender systems. In: Recommender Systems Handbook, 2nd edn., pp. 649–688. Springer, USA (2015)
20. Dataset. https://grouplens.org/datasets/movielens/. Accessed 29 Mar 2021
21. CrypTensor—CrypTen 0.1 documentation n.d. https://crypten.readthedocs.io/en/latest/cryptensor.html. Accessed 29 Mar 2021

Review on Recent Developments in the Mayfly Algorithm

Akash Jain and Anjana Gupta

1 Introduction

Conventional methods used for optimization purposes are based on the analytical properties of the function (like continuity, differentiability, convexity, Lipschitz, etc.) [1]. Even the numerical methods give the approximate sense of derivative by specifying boundary conditions. But in both analytical and numerical methods, we are computing derivative and the problem of optimization based on derivative search technique is that it often gets trapped in the local optima. Nonlinearity, non-convex problems are unsolvable by the use of these methods as their time complexity goes beyond the bounds of a polynomial (or in other words exponential time complexity). Unfortunately in the real world, we have to deal with such complicated functions so there is a need to go for derivative-free optimization techniques or bio-inspired techniques. As we all are living in the age of artificial intelligence and machine intelligence so bio-inspired optimization became popular and significant. As a product of efforts in this area, MA was developed in 2020 by Zervoudakis and Tsafarakis [2]. MA is a unique algorithm as it is the only algorithm that combines the goodness of three major algorithms viz. Genetic Algorithm (GA), Particle Swarm Optimization (PSO), and Firefly Algorithm (FA).

The primary objective of this review paper is first to give a brief to the readers about the MA. Then, after discussing its pros and cons along with its applications this review paper further extends to all the latest developments undergone in the area of this algorithm. We will demonstrate that how researchers modified this MA to introduce

A. Jain (✉) · A. Gupta
Department of Applied Mathematics, Delhi Technological University, Bawana Road, New Delhi 110042, India

A. Gupta
e-mail: anjanagupta@dce.ac.in

M. Dua et al. (eds.), *Proceedings of the International Conference on Paradigms of Communication, Computing and Data Sciences*, Algorithms for Intelligent Systems,
https://doi.org/10.1007/978-981-16-5747-4_30

a better version of MA with more robustness, compatibility, and applications. Then finally, its future aspects along with concluding remarks.

1.1 Motivation

A hybrid algorithm that combines the goodness of evolutionary adaptation and swarms behavior intelligence was proposed by combining genetic algorithm and particle swarm optimization (PSO) by Garg [3]. MA combines the GA, PSO, and FA. FA can be treated as a special case of PSO [2]. There is a very well-known theorem in machine learning, i.e., No FREE Lunch Theorem [2, 4] which says that not a single algorithm is perfect in itself. It can be just relatively better than others. That means it can be good on certain parameters but it can't satisfy all at same time. In light of that every algorithm requires some timely modifications and improvements. So this gives us an inspiration to do a critical review on the recent developments that happened in MA.

Contribution
The main contributions of this review paper are:

- Review paper on MA and its related follow-up algorithms for the first time to the best of our knowledge.
- Specifically deals with applications and future aspects of this algorithm.
- In this paper, we are dealing with MA [2], hybrid MA-HS (mayfly and harmony Search) [4], Improved MA with opposition based learning rules [5], Improved MA [6], Negative MA [7], Improved MA with Chebyshev maps [8], Regrouping MA [9], multi-start MA [10], and heterogeneous MA [11].

2 Mayfly Algorithm (MA)

MA was recently proposed by Konstantinos Zervoudakis and Tsafarakis in 2020. It was inspired by the mating and dancing pattern of mayflies (male and female in the swarm) [2]. Before moving to the theory of MA, let's first see this flowchart given in Fig. 1 based on the classification of evolutionary algorithms [12].

2.1 Methodology of MA

The location of every mayfly in range space signifies a possible result to the problem.

Primarily one pair of mayflies is arbitrarily created indicating the male and female one separately. Every mayfly regulates its route toward its best personal position

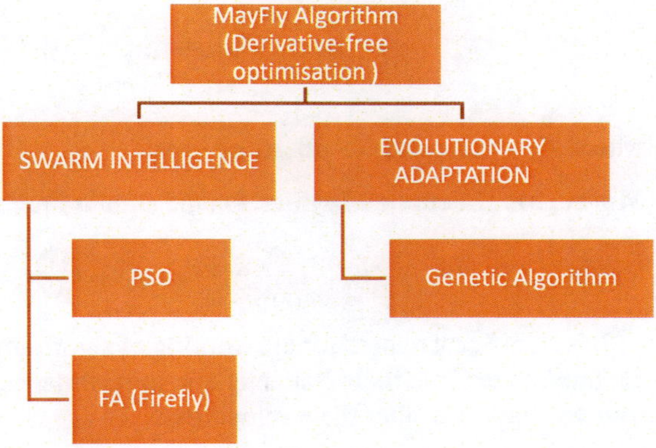

Fig. 1 Flowchart representing components of Mayfly Algorithm (MA)

(pbest) to that point, as well as the global best position to that point (gbest). Now, we will see about movement and mating of mayflies [2].

Movement of Male Mayflies:

$$x_h^{t+1} = x_h^t + v_h^{t+1} \tag{1}$$

Assuming x_h^t is the present position of hth mayfly in the range space at time count t; the position is altered by the addition of a velocity v_h^{t+1} to the present position.

Velocity of Male Mayflies:

Here, we consider modified velocities which are obtained by imposing gravity coefficient g which works similar to PSO inertia weight [13].

$$v_{hj}^{t+1} = g * v_{hj}^t + a_1 e^{-br_p^2}(\text{pbest}_{hj} - x_{hj}^t) + a_2 e^{-br_g^2}(\text{gbest}_j - x_{hj}^t) \tag{2}$$

Here, v_{hj}^t is the velocity of hth mayfly in dimension $j = 1, 2, \ldots n$ at time step t. x_{hj}^t is the position of hth mayfly in dimension j at time step t. a_1 and a_2 are attraction constants for scaling the influence of cognitive and social component, respectively.$pbest_h$ is the finest position of mayfly h^{th} had ever visited. Consider the problem of minimization of the fitness function

$$\text{pbest}_i = \begin{cases} x_h^{t+1}, & \text{if } f(x_h^{t+1}) < f(\text{pbest}_h) \\ \text{Same}, & \text{otherwise} \end{cases} \tag{3}$$

To restrict a mayfly's visibility to others, we used b which is the visibility coefficient in (2).r_p is the Cartesian distance between x_h and $p\text{best}_h$. While r_g is the distance between x_h and $g\text{best}$. Then, the calculation of distances is done by (4)

$$||x_h - X_h|| = \sqrt{\sum_{h=1}^{n} (x_{hj} - X_{hj})^2} \quad (4)$$

where X_h corresponds to $p\text{best}_h$ or $g\text{best}$

Movement of Female Mayflies. On the similar lines of males, we take the analog as

$$y_h^{t+1} = y_h^t + v_h^{t+1} \quad (5)$$

It is significant to note for the working of algorithm superior mayflies in the group continue to execute their characteristic nuptial dance. Hence, the superior mayflies have to constantly alter their velocities as

$$v_{hj}^{t+1} = g * v_{hj}^t + d * m \quad (6)$$

where d is the nuptial dance coefficient and m is the random value picked from the interval $[-1, 1]$.

Velocity of Female Mayfly:

$$v_{hj}^{t+1} = \begin{cases} g * v_{hj}^t + a_2 e^{-br_{mf}^2}(x_{hj}^t - y_{hj}^t) & \text{if } f(y_h) > f(x_h) \\ g * v_{hj}^t + fl * m & \text{if } f(y_h) \le f(x_h) \end{cases} \quad (7)$$

where fl is the random walk coefficient, m is random value in the range $[-1, 1]$, and r_{mf} is the Cartesian distance between male and female.

MATING: With the help of crossover operation, we can generate offsprings as

$$\begin{aligned} \text{offspring}_A &= \varphi * \text{male} + (1 - \varphi) * \text{female} \\ \text{offspring}_B &= \varphi * \text{female} + (1 - \varphi) * \text{male} \end{aligned} \quad (8)$$

where φ is the random value within a specific range.

Improvements In Terms Of Convergence:. can be done by setting velocity limits ($V_{\max}$), imposing gravity coefficient (g), reduction of nuptial dance, and random walk [by scaling down with some parameter say δ lies (0, 1)] and by Mutation of genes of offsprings [2].

Mutation: To deal with the problem of premature convergence, there is addition of random mutation to a proportion of the population (generally, we add γ which is a standard normal distribution variate or Gaussian distribution) in Eq. (8).

2.2 *Mayfly in Multiobjective Optimization*

MA is supplied with a depository of solutions that upholds the finest non dominated solutions. By adding crowding distance and do the necessary adjustment in the male and female equations MA may be well-suited for multiobjective as well.

Male case in multiobjective optimization: Remains same as in the case of a single objective.

Female case in multiobjective optimization:

$$v_{hj}^{t+1} = \begin{cases} g * v_{hj}^{t} + a_2 e^{-br_{mf}^2}(x_{hj}^{t} - y_{hj}^{t}) & \text{if male leads female} \\ g * v_{hj}^{t} + fl * m, & \text{otherwise} \end{cases} \tag{9}$$

3 Flow Chart of Mayfly Algorithm

See Fig. 2.

4 Comparison of MA with Previous State-Of-The-Art Algorithms

BenchMark Functions

See Table 1.

Comparison tables of MA with Firefly (FA), PSO, and GA

See Tables 2 and 3.

Observations. From above tables, the following observations can be made:

- MA is superior than Firefly as it has better time complexity in all cases (average, best, and worst) when compared on Rastringin function which is very well-known multimodal function.
- MA has better average run compared to PSO and GA on sphere and Rastringin function. But PSO has slightly better avg run on Rosenbrock function.

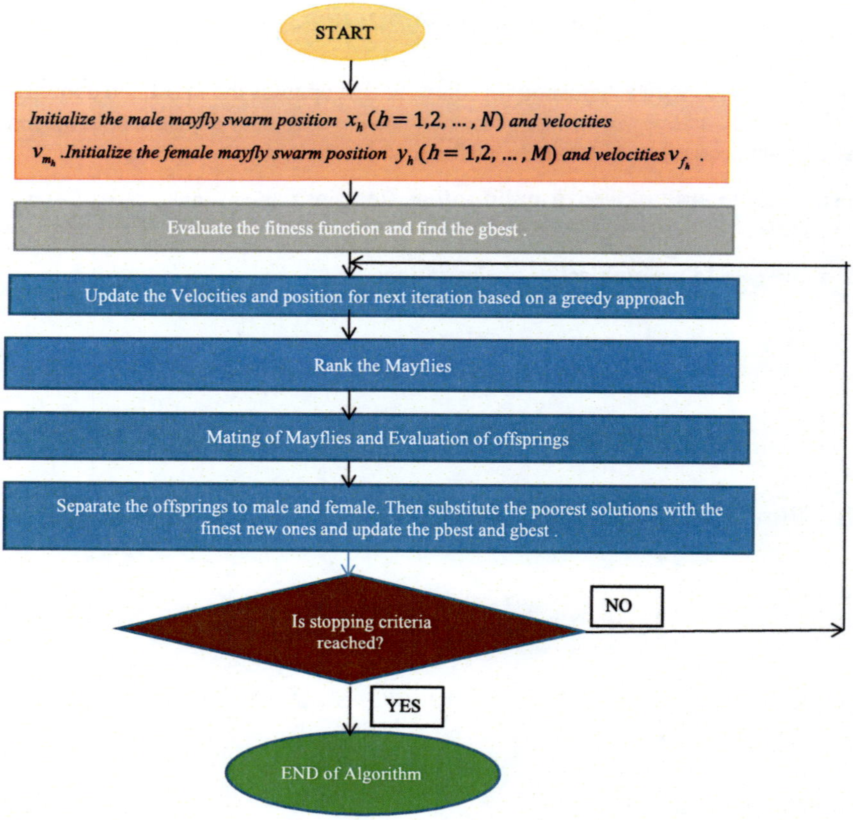

Fig. 2 Flowchart depicting working of MA

Table 1 Benchmark functions (both unimodal and multimodal). Here, global minimum or $f_{\min}$ is 0 for all test functions

Test function Id	Function name	Mathematical exp	Range	
T_1	Sphere (unimodal)	$T_1(x) = \sum_{h=1}^{\gamma} x_h^2$	$-10 \leq x_h \leq 10$	0
T_2	Rosenbrock (unimodal)	$T_2(x) = \sum_{h=1}^{\gamma-1} [100(x_{h+1} - x_h^2)^2 + (1 - x_h)^2]$	$-5 \leq x_h \leq 10$	0
T_3	Rastringin (multimodal)	$T_3(x) = A_n + \sum_{h=1}^{\gamma} [x_h^2 - A * \cos(2\pi x_h)]$ where $A = 10$	$-5.12 \leq x_h \leq 5.12$	0

Table 2 Comparison of MA with PSO and GA

Function ID	GA	PSO	MA
T_1	1.73e−02	1.63e−07	1.17e−07
T_2	1.82e+02	6.33e+01	6.77e+01
T_3	2.83e+01	8.18e+01	1.19e+01

Here, only Avg run is compared (for more details about this refer [2])

Table 3 Comparison of MA with Firefly

Statistics	Firefly (FA)	Mayfly (MA)
Best run	1.34e+02	5.96e+00
Avg run	1.78e+02	1.19e+01
Worst run	2.48e+02	2.18e+01

Here, only Rastringin function is considered (for more details refer [2])

5 Advantages and Challenges of MA

5.1 Advantages of MA

- Bypass the problem of getting caught in local optima due to exploration and exploitation of complete search space [2].
- Suitable for both single objective and multiobjective optimization problems [2]. In multiobjective problems finding, Pareto front is feasible with the help of a bio-inspired algorithm [14].
- Multiobjective MA can better handle the Pareto front as compare to NSGA-II. So it can be widely deployed in those problems having two or more objective functions.
- Even though it seems difficult for other landmark algorithms like GA, PSO, etc., to locate global optima. MA located superior values on various state-of-the-art test functions including both unimodal as well as multimodal benchmark functions.
- With the same resources, it has better efficiency, consistency, and convergence rate as compared to previous algorithms (GA, FA, PSO, etc.).
- It can tackle both continuous and discrete optimization problems. E.g.: flow shop scheduling one.

5.2 Challenges of MA

- Premature convergence.
- Problem of feature selection.
- Problem of stucking to local optima still prevails.

- Velocity updation may cause stability issues due to change in existing solutions.

6 Recent Developments in MA

Hybrid MA-Harmony Search (HS) [4]

This is fundamentally a hybrid algorithm developed by combining the goodness of MA and HS. This work [4] deals with the feature selection problem. By compressing the dimension of the data set by deleting unused features is called feature selection. This helps to shrink the time required to train an algorithm as well as reduces the space requirement by excluding the unused features. In this, authors have taken the help of S–shaped transfer functions to convert continuous search space to a binary one. Shortcomings can be observed in terms of earlier convergence and weaker mutations which can be overcome with the help of using different shapes of transfer functions. Later, this work can be extended to other numerical optimization problems of soft computing and artificial intelligence. For pseudocode and experiments refer to [4].

MA with Opposition Based Rules (OBL) [5]

It combines the goodness of MA and opposition based rules based on Ying-Yang philosophy [5]. It gives significant results as residual errors dropped drastically over certain benchmark functions, while performing Monte Carlo simulations.

Negative MA [6]

In this, male mayflies will improve their velocities pertaining to the poorest mayflies together with their poorest trajectories. Not like as in the original algorithm, the male mayflies will run away from their poorest trajectories and the global poorest candidates. Simulation tests proved that both the Negative MA and MA algorithms would work well in optimizing both types of benchmark functions (multi as well as unimodal). Also, it holds good for the non-symmetric one. But the shortcoming of Negative MA is that it fails to work better in unimodal benchmark functions since MA provides much simplicity over there.

Improved MA with velocity updation [7]

Significant updation in the velocity was carried out in this thus helping in better movement and locating a better optimal position in the search space. When the mayflies are distant from each other, they have to improve their velocities with larger rates and when they are close to each other, velocities will be improved with smaller rates.

Therefore, Eq. (2) must be revised as follows:

$$v_{hj}^{t+1} = a_h e^{-\frac{b}{r_p}} (p_{hj} - x_{hj}^t) \tag{10}$$

Simulation results show that it has a better efficiency rate both in unimodal (sphere function) and multimodal (S Tang function) state-of-the-art benchmark functions.

MA with Chebyshev maps [8]

It is very interesting to note that chaotic maps play an important role in metaheuristics algorithms coming from the theory of nonlinear dynamics modeling and simulations. As with the help of these chaotic maps, we can replace random coefficients involved in bio-inspired algorithms. Also, chaotic maps gave good simulation results when applied to the Whale optimization algorithm [15] and Firefly algorithm. In this, researchers take the help of Chebyshev maps one of the kinds of chaotic maps to perform simulation of modified MA. But results are not satisfactory but that doesn't mean chaotic maps fail their relevance. It may be due to the choice of Chebyshev map but changing chaotic maps might help in improved simulated results.

7 Applications

See Table 4.

8 Concluding Remark and Future Directions

After doing a review of the MA and the recent developments that happened in MA we are in a position to say that MA is a better algorithm as compared to the previous landmark algorithms like PSO, GA, and FA. By the virtue of the No Free Lunch theorem, it becomes essential to dig further so to improve the loopholes of the present algorithm and make it more robust to wider applications. So we reviewed recent hybrid algorithms of MA like MA-HS, MA-OBL, improved MA with velocity adaptation, Negative MA, and finally MA with Chebyshev maps. But the scope of further improvements and research hasn't been finished yet.

FUTURE PROSPECTS: We can also go for a hybrid of MA with other landmark group behavior algorithms like ABC (Artificial Bee Colony) [12], social group optimization (SGO) [16], etc. By replacing the demerits of MA on certain parameters with merits of the latter one, for e.g., even one can also go into quantum computing. For that, they can replace the PSO-based position and velocity equations of MA with Quantum-PSO [17] equations to bring more robustness, stability, and better convergence. Again for future perspectives multiobjective MA can be used to solve various types of multiobjective optimization-based engineering and real-life problems as this method is more robust than NSGA-II [2]. Also, the dynamic alteration of parameters involved in the velocity and position equations with the help of fuzzy reasoning can increase the efficiency of the original algorithm. With this, we would like to conclude by saying that this paper presents a brief review of the original MA and its follow-up hybrid algorithms. As a result, it will help the research community to investigate further on MA and related algorithms.

Table 4 Summary of algorithms discussed with applications

Algorithm	Properties	Application
Mayfly Algorithm [2]	Combines GA, PSO, and FA. Applicable for both continuous and discrete problems	Continuous optimization problems and in combinatorial application is flow shop scheduling problem
MA-HS Algorithm [4]	Combines MA and HS uses S shape transfer function to change the continuous objective function into a binary one	Feature selection in AI and Ml
MA-OBL [5]	Combines the MA with opposition-based rules	Better convergence rate with both multimodal and unimodal benchmark function as compared to MA
Negative MA [6]	Considers the worst position of a swarm and tries to implement the Mayfly in a negative approach	Better simulation result only in multimodal and non-symmetric benchmark functions
Improved MA [7]	Velocity updation of original MA	useful for both uni and multimodal objective functions
MA Chebyshev map [8]	Based on the idea to replace random coefficients with chaotic maps. Here, Chebyshev maps are used	Useful to avoid stagnation during the iterations of MA
Regrouping MA [9]	Based on the regrouping of a swarm of mayflies	useful to avoid stagnation during the iterations of MA
Multi-start MA [10]	Based on idea to incorporate the Multi-start initialization of mayflies	solves the problem of stucking to local optima to some extent
Heterogeneous MA [11]	Multiple ways to update their position in this heterogeneous type of MA	Increased the efficiency of the original algorithm

References

1. Deep, K., Bansal, J.C.: Mean particle swarm optimisation for function optimisation. Int. J. Comput. Intell. Stud. **1**(1), 72–92 (2009)
2. Zervoudakis, K., Tsafarakis, S.: A mayfly optimization algorithm. Computers Ind. Eng. **145** (2020). https://doi.org/10.1016/j.cie.2020.106559
3. Garg, H.: A hybrid PSO-GA algorithm for constrained optimization problems. Appl. Math. Comput. **274**, 292–305 (2016)
4. Bhattacharyya, T., Chatterjee, B., Singh, P.K., Yoon, J.H., Geem, Z.W., Sarkar, R.: Mayfly in harmony: A new hybrid meta-heuristic feature selection algorithm. IEEE Access **8**, 195929–195945 (2020)
5. Gao, Z.M., Zhao, J., Li, S.R., Hu, Y.R.: The improved mayfly optimization algorithm with opposition based learning rules. J. Phys. Conf. Ser. **1693**(1), 012117 (2020). https://doi.org/10.1088/1742-6596/1693/1/012117
6. Zhao, J., Gao, Z.M.: The negative mayfly optimization algorithm. J. Phys. Conf. Ser. **1693**(1), 012098 (2020)

7. Gao, Z.M., Zhao, J., Li, S.R., Hu, Y.R.: The improved mayfly optimization algorithm. J. Phys. Conf. Ser. **1684**(1), 012077 (2020)
8. Zhao, J., Gao, Z.M.: The improved mayfly optimization algorithm with Chebyshev map. J. Phys. Conf. Ser. **1684**(1), 012075 (2020)
9. Zhao, J., Gao, Z.M.: The regrouping mayfly optimization algorithm. In: 7th International Forum on Electrical Engineering and Automation (IFEEA), Hefei, China, pp. 1026–1029. IEEE (2020)
10. Zhao, J., Gao, Z.M.: The multi-start mayfly optimization algorithm. In: 7th International Forum on Electrical Engineering and Automation (IFEEA), Hefei, China, pp. 879–882. IEEE (2020)
11. Zheng-Ming, G.A.O., Su-Ruo, L.I., Juan, Z.H.A.O., Yu-Rong, H.U.: Heterogeneous mayfly optimization algorithm. In: 2nd International Conference on Machine Learning, Big Data and Business Intelligence (MLBDBI), pp. 227–230. IEEE (2020)
12. Fan, X., Sayers, W., Zhang, S., Han, Z., Ren, L., Chizari, H.: Review and classification of bio-inspired algorithms and their applications. J. Bionic Eng. **17**, 611–631 (2020)
13. Bansal, J.C., Singh, P.K., Saraswat, M., Verma, A., Jadon, S.S., Abraham, A.: Inertia weight strategies in particle swarm optimization. In:Third World Congress on Nature and Biologically Inspired Computing, Spain, pp. 633–640. IEEE (2011)
14. Deb, K.: Multi-Objective Optimization using Evolutionary Algorithms, Ist edn. Wiley, England (2001)
15. Mirjalili, S., Lewis, A.: The whale optimization algorithm. Adv. Eng. Softw. **95**, 51–67 (2016)
16. Satapathy, S., Naik, A.: Social group optimization (SGO): a new population evolutionary optimization technique. Complex Intell. Syst. **2**, 173–203 (2016). https://doi.org/10.1007/s40747-016-0022-8
17. Yang, S., Wang, M.: A quantum particle swarm optimization. In: Proceedings of the Congress on Evolutionary Computation, Portland, USA, vol.1, pp. 320–324. IEEE (2004). https://doi.org/10.1109/CEC.2004.1330874

Comparative Analysis of Dynamic Malware Analysis Tools

Mohamed Lebbie, S. Raja Prabhu, and Animesh Kumar Agrawal

1 Introduction

Nowadays, the main source of interaction and communication is through the internet. The present Covid pandemic saw a steep rise in the usage of internet and more people had joined the internet bandwagon. And expectedly, malicious actors are using this upward trend effectively to target users with malware regardless of which operating system and hardware platform one is using. With the rapid rise in the adoption of technology and the digitization of major business process like payment processes, banking transaction, information sharing, procurement, etc., there has reportedly been increase in the attacks on businesses using malware of various types ranging from backdoors, rootkits, Trojans, ransomware, etc.

This had increased the need for malware analysis and to reduce the time line to carry out the same. The malware analysis is broadly classified into two categories viz., static and dynamic analysis. The techniques that are used to detect a malware by identifying the malware signatures is known as static analysis, and analysing their behavior is known as dynamic analysis. These techniques can, with some certainty, determine if the samples/files are malicious or not. There are certain limitations in the static malware analysis as it would not be able to discern packed or encrypted malware, self-modifying codes, dynamically loaded modules, etc. This encouraged the researchers to move to dynamically analysing a malware. In the authors' opinion, there is no dearth of malware analysis tools; in fact, the real problem is quite the opposite—the problem of many. There are so many tools available that intimidate the malware analyst as to which one to select for which task. With the wide range of tools and sandboxes available, clearly determining which tools or sandboxes to use determines success.

M. Lebbie (✉) · S. R. Prabhu · A. K. Agrawal
National Forensic Sciences University, Gandhinagar, Gujarat, India

M. Dua et al. (eds.), *Proceedings of the International Conference on Paradigms of Communication, Computing and Data Sciences*, Algorithms for Intelligent Systems,
https://doi.org/10.1007/978-981-16-5747-4_31

2 Related Work

Malware is a general name for malicious software or codes. Malware does various nefarious activities ranging from demanding a ransom to stealing sensitive data [1]. Most malware is written in a middle level language and once the code is completed, and it is compiled in such a way that the hardware and/or software can be able to execute it [2]. Each type of malware gathers information about the infected device without the knowledge or authorization of the user.

Malware analysis is the process of learning how a malware functions and any potential damages that can be caused by a given malware [3]. Malware analysis is the process used to determine and understand the malware type, nature, attacking methodologies and more. There are two types of malware analysis viz., static (code) analysis and dynamic (behavioral) analysis. Static analysis is the process of debugging a software or code without actually executing the malware [4, 5]. This analysis is further divided into basic static analysis and advanced static analysis. Basic static analysis involves submitting the code to tools like virus total [6] where it will be analysed by various anti-virus solutions, checking the string of the code and using tools like PEiD [7]. And for advanced static analysis, a reverse engineering of the code is done. Dynamic analysis is a quick way of analysing a malware. This type of analysis observes the behavior of the code when it is executed. The two basic ways used are comparative approach and runtime behavior analysis. Comparative approach involves comparing the state or changes of a system before and after running the malicious code on the system, while runtime behavior analysis involves using specialized tools to constantly monitor the activities of the malware when it is been run in the system. Sandbox analysis is another technique generally used by malware analysts. Sandbox is an isolated environment used for testing. It enables researchers to execute programs or files without harming or destroying the system, software or application that they are running on. Security professionals use sandboxes to test for malicious software or codes. They are used to protect organization's critical infrastructure from the suspicious code [8].

It was definition have discussed in detail about the various types of malwares and different viruses which are a threat to the computer. Along with that they have reviewed obfuscation techniques, limitations and the techniques of memory-based analysis. The conclusion was drawn based on the fact that to conduct any malware research, and it is important to collect the malware samples and acquire them [9].

A definition and explanation about the malware analysis and the benefits were stated. It is explained in their article how it can be implemented regardless of how complex it is. The results obtained are effective during the identification phase and its analysis. Their methodology includes iterative mechanism which increases the identification of malwares. The conclusions were elaborate and has been described in depth about the whole process [10].

A zero-day attack and detection by using cuckoo sandbox tool was developed, and it was examined under the UNIX system. In the experimental study, they incorporated their client's PC with their personalized python code, and then UNIX- based cuckoo sandbox was used to prevent attacks. The study outcome revealed that it had enhanced results and was very effective in identifying malwares and also showed that the time for the analysis increased when the malware samples were more [11].

3 Proposed Work

There are literally dozens of malware analysis tools available in the market (Refer Appendix for some of the tools), which adds to the woos of the malware analyst in selecting the appropriate tools for the detection of required information. The aim of this research paper is to help the malware analyst in selecting the appropriate tools for the task at hand. This is achieved by comparing the detection capabilities of the various dynamic malware analysis tools with respect to the various features of the malware. However, to be able to complete the research in the limited available time, the following two criteria are established:

- A malware sample set with only 10 malware samples were selected which represents various malware families viz., rootkits, remote administration trojans, worms, etc.
- Only seven malware features are analysed which, in the authors' opinion, represents some of the most important qualities of the malware for basic static and dynamic analysis.

3.1 Malware Features

The followings are the malware features that are considered for analysis in this research. This list is not exhaustive but it captures the most important features of the malware belonging to various families.

File System Changes Detection. Any changes in the file system give an insight into what the malware is intended to do in the infected device. This process lets us know which files were either accessed or altered or exfiltrated by the malware.

Registry Changes Detection. Registry holds the key to most of the activities that happen in the windows operating system. Malware uses registry keys especially to enable persistence so that they can survive reboots. Registry keys also help in analysing the URLs accessed, files/folders opened, most recently used applications and many more.

Imports Detection. Imports are functions that a software or malicious code calls from other location especially a dynamically linked file or DLL. The import address table or the IAT gives hint about the functionality of the malware. Some malwares could dynamically load the DLL which would necessitate dynamic analysis.

Hooks. Hooks are techniques used by cyber criminals to change the behavior of application, operating system or malicious code. The process of hooking is used to monitor function calls, analyse the function parameters and monitor the information flow. Cybersecurity teams can detect the hidden traces of API hooking techniques through memory analysis frameworks like volatility [12].

Network Analysis. This process identifies and capture any attempt of TCP or UDP connection that is established by the malware which helps us to understand what the malware is doing and where it is connecting to.

Trace. Trace is the resulting product of a hooking function. It contains information such as the parameters accessed and processed and the extended functions called by the monitored function.

Packer. Packers are used by malware to hide the malicious code and to evade the anti-virus detection without compromising the functionality of the malware. When packed, the malware code is either compressed or encrypted and a decompression or decoding stub is appended to the malware to retrieve the original code on the fly.

3.2 Dynamic Malware Analysis Tools

There are many malware analysis tools available, both free ware and commercial versions. For this research, the following free ware is used based on their popularity.

Regshot. This is a tool used to compare the amount of registry changes made during the malware execution [13]. It is an open-source program, it compares the registry at two given points by creating a snapshot of the registry before any program is added, removed or modified and then take another snapshot after the modification and compare them.

Process Explorer. Process explorer is an open-source task manager and system monitor tool that shows information about which handles and DLLs have been opened or loaded by an executable or program [14]. It shows the path from which a given process is running. This helps to identify a malicious code even if it has a generic name like any legitimate process running on the system.

Process Monitor (Procmon). This is an advanced monitoring tool which combines two sysinternals utilities (Filemon and Regmon) to perform and display real-time file system, registry and process/threads activities [15].

Cuckoo Sandbox. Cuckoo Sandbox is an open-source software for automating the analysis of suspicious files. To do so it makes use of custom components that monitors how the malicious processes behave, while running in an isolated environment. Cuckoo Sandbox is compatible with Windows OS, macOS, and Linux [16]. Cuckoo sandbox retrieves win32 API calls, details about created, deleted and downloaded files by the malware and also the network traffic trace in PCAP format. It takes input

Table 1 Analysis of samples using Regshot

Samples	System file changes	Registry changes	Import	API hooking	Network analysis	Trace	Packer
Vcffipzmnipbxzd.exe	–	Yes	–	–	–	–	–
rootkit.exe	–	Yes	–	–	–	–	–
131.exe	–	Yes	–	–	–	–	–
njRAT.exe	–	Yes	–	–	–	–	–
Build.exe	–	Yes	–	–	–	–	–
Cerber.exe	–	Yes	–	–	–	–	–
dumped.exe	–	Yes	–	–	–	–	–
fake intel (1).exe	–	Yes	–	–	–	–	–
yfoye–dump.exe	–	Yes	–	–	–	–	–
BOTBINARY.EXE	–	Yes	–	–	–	–	–

like binary and URL and gives the result of file system changes, registry changes, mutexes created, memory dumps and PEID details [16]. Cuckoo is also highly extensible and offers a lot of additional content made by contributors in the community. It includes an utility to download this content using 'community.py.' Malware today are designed to check if it's executed inside a virtual environment or sandbox, and if it realizes that it was executed in such an environment; then it may hide its malicious behavior or stop execution to prevent analysis. The analysis of such malware may fail, while trying to analyse them with cuckoo [17].

4 Result Analysis

4.1 Analysis of Regshot Results

Regshot is a very powerful open-source tool used in most dynamic analysis, and the result of analysis of our malware sample set is given in Table 1. It can be seen that Regshot was able to detect the registry changes successfully in all the malware samples.

4.2 Analysis of Process Explorer Results.

Process explorer used by malware analyst to have a deeper analysis on malware. We analysed samples of malware and got the following results (Refer Table 2). The process explorer was able to detect imports in all the samples successfully; however,

Table 2 Analysis of samples using process explorer

Samples	System file changes	Registry changes	Import	API hooking	Network analysis	Trace	Packer
Vcffipzmnipbxzd.exe	Yes	Yes	Yes	–	–	–	–
rootkit.exe	–	–	Yes	–	–	–	–
131.exe	Yes	–	Yes	–	Yes	–	–
njRAT.exe	–	Yes	Yes	–	–	–	–
Build.exe	Yes	Yes	Yes	–	Yes	–	–
Cerber.exe	–	Yes	Yes	–	–	–	–
dumped.exe	Yes	–	Yes	–	Yes	–	–
fake intel (1).exe	Yes	Yes	Yes	–	Yes	–	–
yfoye–dump.exe	Yes	–	Yes	–	Yes	–	–
BOTBINARY.EXE	Yes	Yes	Yes	–	Yes	–	–

the file system changes, registry changes and network analysis were successful only in certain samples. It could be because the samples did not have this property or the process explorer was not able to detect them successfully. But, it can be noted that Regshot detected registry changes in all our samples. This indicates certain shortcomings in the process explorer.

4.3 Analysis of Procmon Results

Procmon also known as process monitor is one of the widely used open-source tools for dynamic malware analysis. Using this tool, we analysed the malware samples, and the results showed that Procmon was able to detect registry changes successfully in all the samples but failed to detect file system changes in one of the samples leaving a question mark on the integrity of this tool (refer Table 3).

4.4 Analysis of Cuckoo Sandbox Results

Using cuckoo sandbox [18], we analysed malware samples to identify the different artifacts that can be detected by the sandbox and the results are shown in Table 4. It can be noted here that cuckoo sandbox was able to detect imports successfully in all the samples. However, the various other artifacts viz., file system changes, registry changes, API hooking, network analysis, trace analysis and packers are only partly successful suggesting that cuckoo sandbox may not be complete in all respects when it comes to malware analysis.

Table 3 Analysis of samples using process monitor

Samples	System file changes	Registry changes	Import	API hooking	Network analysis	Trace	Packer
Vcffipzmnipbxzd.exe	Yes	Yes	–	–	–	–	–
rootkit.exe	Yes	Yes	–	–	–	–	–
131.exe	Yes	Yes	–	–	–	–	–
njRAT.exe	Yes	Yes	–	–	–	–	–
Build.exe	Yes	Yes	–	–	–	–	–
Cerber.exe	Yes	Yes	–	–	–	–	–
dumped.exe	Yes	Yes	–	–	–	–	–
fake intel (1).exe	–	Yes	–	–	–	–	–
yfoye–dump.exe	Yes	Yes	–	–	–	–	–
BOTBINARY.EXE	Yes	Yes	–	–	–	–	–

Table 4 Analysis of samples using Cuckoo Sandbox

Samples	System file changes	Registry changes	Import	API hooking	Network analysis	Trace	Packer
Vcffipzmnipbxzd.exe	Yes	Yes	Yes	Yes	–	–	Yes
rootkit.exe	–	–	Yes	–	–	–	–
131.exe	Yes	–	Yes	–	Yes	–	Yes
njRAT.exe	–	Yes	Yes	–	–	–	–
Build.exe	Yes	Yes	Yes	Yes	Yes	–	–
Cerber.exe	–	Yes	Yes	Yes	–	Yes	–
dumped.exe	Yes	–	Yes	Yes	Yes	–	–
fake intel (1).exe	Yes	Yes	Yes	–	Yes	–	–
yfoye_dump.exe	Yes	–	Yes	Yes	Yes	Yes	–
BOTBINARY.EXE	Yes	Yes	Yes	Yes	Yes	–	–

4.5 Comparative Analysis

After analysing the results of all the selected tools, the authors did a comparative analysis of these tools to arrive at their capabilities in detecting the malware features that were discussed. The result of the comparative analysis is depicted in Table 5. The analysis shows the following:

- Regshot is reliable in detecting registry changes.
- Process explorer is not completely reliable for system file changes and registry changes. More investigation is required for network analysis using process explorer, while import detection is found reliable.

Table 5 Comparative analysis of dynamic malware analysis tools

Tools	System file changes	Registry changes	Import	API hooking	Network analysis	Trace	Packer
Regshot	–	Yes	–	–	–	–	–
Process Explorer	Partly	Partly	Yes	–	Yes*	–	–
Procmon	Partly	Yes	–	–	–	–	–
Cuckoo Sandbox	Partly	Partly	Yes	Yes*	Yes	Yes*	Yes*

Yes Results are reliable, *Yes* Results are reliable but may need more investigation, *Partly* Result are only partly reliable. Need to use multiple tools to do complete and reliable investigation

- Procmon is not completely reliable when it comes to system file changes but it is successful in detecting registry changes.
- Cuckoo sandbox is only partly successful in detecting file system changes and registry changes, while it is successful in detecting imports and network analysis. More investigation is needed to study the reliability of trace analysis and packer detecting using cuckoo sandbox.

5 Conclusion and Future Work

The dilemma faced by the malware researcher or analyst is not the dearth of analysis tools but the availability of varied tools with differing capabilities. This has made the selection of tools a personal choice. This will reduce the analyst to use only the tool that he/she is familiar with rather than exploring other tools which has got similar or more capabilities. In this research, malware samples belonging to different malware families are tested for seven most important malware features using five regularly used malware analysis tools. The results are compared that shows the capabilities of each tool. This would help the analyst in selecting the appropriate tool for the task at hand which in turn may result in efficient analysis and reduced time frame. The future research may include more samples from different malware families, more malware features and the complete list of tools mentioned in the Appendix.

Appendix

1	Md5deep	14	IDA Pro
2	Strings	15	OllyDbg
3	PEiD	16	WinDbg

(continued)

(continued)

4	Dependency Walker	17	x64Dbg
5	PEView	18	Capture BAT
6	Resource Hacker	19	CFF Explorer
7	PE Explorer	20	Hexeditor
8	Procmon	21	ImportREC
9	Process Explorer	22	Memoryze
10	Regshot	23	OfficeMalScanner
11	Apate DNS	24	GMER
12	Inetsim	25	Fiddler
13	Wireshark	26	Detect It Easy (DIE)

References

1. What is Malware? Forcepoint (2020) Retrieved from https://www.forcepoint.com/cyberedu/malware
2. RIPS: RIPS—automate security testing and manage your risks (2020). Retrieved from https://www.ripstech.com/product/tour/
3. Comodo News and Internet Security Information: What Is Malware Analysis? Malware Analysis Techniques (2020). https://blog.comodo.com/malware/different-techniques-for-malware-analysis/
4. Aslan, Ö., Samet, R.: Investigation of possibilities to detect malware using existing tools. In: 2017 IEEE/ACS 14th International Conference on Computer Systems and Applications (AICCSA), pp. 1277–1284. IEEE (2017)
5. Jamalpur, S., Navya, Y.S., Raja, P., Tagore, G., Rao, G.R.K.: Dynamic malware analysis using cuckoo sandbox. In: 2018 Second international conference on inventive communication and computational technologies (ICICCT), pp. 1056–1060. IEEE (2018)
6. Ijaz, M., Durad, M.H., Ismail, M.: Static and dynamic malware analysis using machine learning. In: 2019 16th International Bhurban Conference on Applied Sciences and Technology (IBCAST), pp. 687–691. IEEE (2019)
7. Infosec Resources: Static malware analysis—infosec resources (2020). Retrieved from https://resources.infosecinstitute.com/topic/malware-analysis-basics-static-analysis/
8. SearchSecurity: What Is Sandbox (Software Testing and Security)? (2020). Definition From Whatis.Com. Retrieved from https://searchsecurity.techtarget.com/definition/sandbox
9. Sihwail, R., Omar, K., Ariffin, K.A.Z.: A survey on malware analysis techniques: static, dynamic, hybrid and memory analysis. Int. J. Advan. Sci. Eng. Inform. Technol. **8**(4–2), 1662 (2018)
10. Bermejo Higuera, J., Abad Aramburu, C., Bermejo Higuera, J.R., Sicilia Urban, M.A., Sicilia Montalvo, J.A.: Systematic approach to malware analysis (SAMA). Appl. Sci. **10**(4), 1360 (2020)
11. Al-Rushdan, H., Shurman, M., Alnabelsi, S.: On Detection and prevention of zero-day attack using cuckoo sandbox in software-defined networks (2020)
12. Yin, H., Liang, Z., Song, D.: HookFinder: identifying and understanding malware hooking behaviors (2008)
13. Code.google.com.: Google Code Archive—long-term storage for google code project hosting (2021). Retrieved from https://code.google.com/archive/p/regshot

14. Docs.microsoft.com.: Process Explorer—Windows Sysinternals (2021). Retrieved from https://docs.microsoft.com/en-us/sysinternals/downloads/process-explorer
15. Docs.microsoft.com.: Process monitor—windows sysinternals (2021). Retrieved from https://docs.microsoft.com/en-us/sysinternals/downloads/procmon
16. "Cuckoo Sandbox—Automated Malware Analysis". Cuckoosandbox.org. https://cuckoosandbox.org/
17. Malwarebytes Lebs: Automating malware analysis with Cuckoo Sandbox (2016). Available at: https://blog.malwarebytes.com/threat-analysis/2014/04/automating-malware-analysis-with-cuckoo-sandbox/ (Accessed: 2021)
18. Bai, J., Shi, Q., Mu, S.: A Malware and variant detection method using function call graph isomorphism. Security and Communication Networks (2019)

Bebop Drone GCS Forensics Using Open-Source Tools

Rishi Dhamija, Pavni Parghi, and Animesh Kumar Agrawal

1 Introduction

A drone is a remotely piloted aerial vehicle which is also known as unmanned aerial vehicle (UAV) and is a subpart of a broader system called unmanned aerial system (UAS) which consists of drones and any other component required to control drone operations. Being relatively new drone technology is still able to get all the attention a technology may need to propagate. This electronic device was initially used for hobby flying, entertainment, and drone photography in the civil world. However, because of its low costs and ease of availability, non-state actors have also started showing interest in drones to be able to use them for malicious purposes like cyber espionage, cybercrimes, etc.

To control any drone operation, GCS plays a vital role as it is the only source of control and communication with the drone. A GCS consists of mission planning facility, displays for sensor data and drone's location, and is capable of monitoring the drone in real time as per [1]. In commercially available recreational drones, GCS also serves as data storage where all the data related to the drone in flight such as telemetry, media captured by drone, way-points, flight log, and GPS coordinates is transferred to and stored on the GCS linked with the drone making it a rich source of information for a forensic examiner. A GCS can be a hardware device, software, or a combination of both. Generally, for commercially available recreational drones, a mobile app serves the functionality of the GCS. Considering the kind of data GCS stores, it is a potential goldmine of data required to conduct a sound drone forensic examination. However, to gain access to that data, a depth analysis of GCS is a must. Drone forensics also brings a lot of challenges because of the non-availability of forensically sound and validated tools, no published guidelines for carrying out forensics, no standard OS, drone peripherals, and storage format [2].

R. Dhamija (✉) · P. Parghi · A. K. Agrawal
National Forensic Sciences University, Gandhinagar, Gujarat, India

M. Dua et al. (eds.), *Proceedings of the International Conference on Paradigms of Communication, Computing and Data Sciences*, Algorithms for Intelligent Systems,
https://doi.org/10.1007/978-981-16-5747-4_32

Taking the extensive capabilities of a GCS into account along with its importance in aerial operations of a drone and high forensic value data stored in it makes GCS forensics a must. A forensic image of the controlling device of the Parrot Bebop 2 drone was obtained from VTO laboratories [3]. In this case, it was a Samsung mobile device that had an app to serve the functionalities of a GCS. FreeFlight 3 app installed on a mobile phone was the GCS of the drone [4].

2 Literature Review

Research in [5] presents a generalized UAV forensic framework by classifying them into different categories based on their class and weight. The framework proposed is based upon existing frameworks for network forensics, objective-based digital forensics, and event-based digital forensics. The proposed framework takes customization, Wi-Fi, Bluetooth, geo-location, storage media, and other sensor-related data into account. Work in [6] presents categories in which the drones can be broadly classified and their mode of flying. Along with classification, the paper also presents the most common vulnerabilities that can be easily exploited by a hacker in commercial drones by taking 3DR's smart drone "SOLO" as a case study. The main emphasis is given on the GPS spoofing vulnerability in with the hacker spoofs the civil GPS signals so that now the drone will fly according to the coordinated supplied by a hacker. The paper [7] presents a modular approach to develop a virtual universal GCS for UAVs by mentioning certain key functionalities of UAVs and GCS itself to make it feasible to add new and remove old modules to gain desired functionalities freely. It also points out the basic functionalities that a UAV's GCS must have to be efficient enough to work with any kind of UAV. Research in [8] presents a tool "GRYPHON" to carry out forensics of data flash and telemetry log files stored on GCS created automatically after the drone has been armed. ArduPilot was used as firmware for GCS from which the presented tool was able to extract data to find anomalies, GPS coordinates mapping, unexpected altitude variations, determine whether a CMD command was executed, data-specific extraction errors, AC/DC measurements analysis, and check CRC from repositories in addition to timeline analysis. Researchers in [9] carried out work using open-source tools, and some basic scripts have been developed to aid forensic analysis of two drones for identification of operators and extraction of data from drones. In [10], the issue of drone forensics being an uncovered area has been brought out. So, the authors have presented an introductory discussion of unmanned aerial vehicle analysis and provided a digital forensic investigation report of a test Parrot Bebop unmanned aerial vehicle. Researchers in [10] have presented the extraction and identification of artifacts from the open-source tools and some basic scripts developed for the analysis of two popular drone systems the DJI Phantom 3 Professional and Parrot AR. The challenge of extracting data from the mobile app from an unrooted phone was brought out in [11]. Authors in [12] carried out the extraction and interpretation of important artifacts found in recorded flight logs on both the internal memory of the UAV and the controlling application for effective analysis of drones. The work depicts in [13] is to explore drone forensics in terms of challenges, forensic investigation procedures, and experimental results through a forensic investigation study performed on a Parrot AR drone 2.0. Researchers [14] proposed a robust digital drone forensic application with log parameters of drones through a graphical user interface (GUI) developed using JavaFX 8.0 which allows users to extract and examine onboard flight information. The paper [15] consists of a report of potential attacks against the Parrot Bebop 2 drone. In [16], authors focus on drone forensics and present a methodology of investigating physical and digital evidence to acquire all the possible data from drones. The research in [17] takes the capabilities of GCS to another level by proposing a single surveillance system for multiple UAVs and displaying their real-time operation information in 3D with a decentralized architecture. Work in [18] also aims to propose a mobile ground control station to aid in local surveillance by enhancing the capabilities of GCS by integrating various sensors and receiving data from multiple sources to aid in the identification of paths, people, vehicles, etc.

3 Methodology

The study in this paper focuses on the extraction of data from Parrot Bebop drone, like telemetry, waypoint, multimedia files, GPS coordinates, user details, firmware details, etc. This methodology enumerated in Fig. 1 contains the following phases.

Phase 1 (Information Gathering): The process shall begin with gathering information about the incident, potential actors involved, type of drone which caused the incident, type of GCS used, and any other information leading toward deciding a sound approach to carry on with the rest of the investigation process.

Phase 2 (Preparation): Followed by information gathering, an investigator needs to prepare the resources that will be required for the seizure of the device (GCS) forensically to prevent the intervention of any kind and to keep the data integrity intact. Taking pictures of the drone and its visible components along with any other detail visible may also take place in this phase.

Phase 3 (Identification): Next thing will be the identification of the components of the seized GCS. Those components can be flash storage devices, mobile devices, remote controllers, electric parts, etc. This step also includes the identification of the physical and digital evidence.

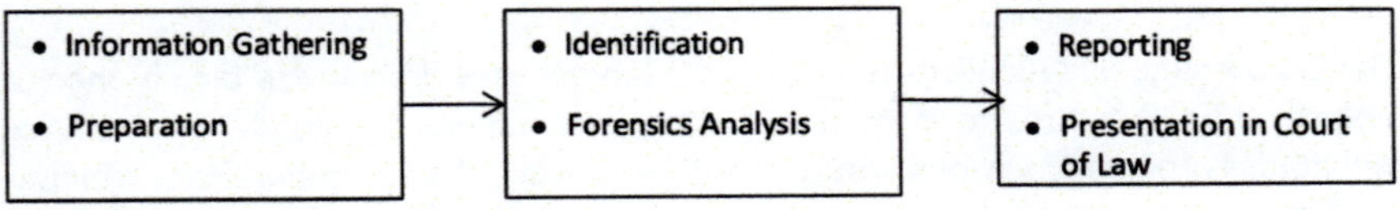

Fig. 1 Methodology

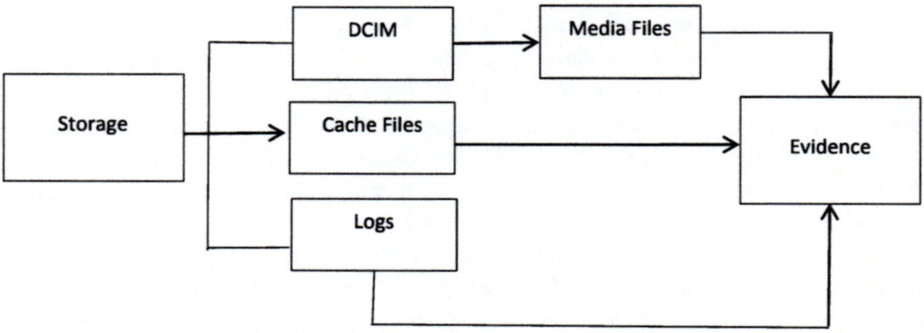

Fig. 2 Samsung phone data storage location

Phase 4 (Forensic Analysis): Next step will be to make a forensic image of the device by copying the data bit-by-bit and then creating an exact copy of that image so that examination of the evidence can be done as part of forensic analysis. Data like serial number, MAC address, photos/videos, telemetry data, etc., can be obtained in this step. Open-source tools would preferably be used for the analysis, maintaining data integrity at all times. Data like serial number, MAC address, photos/videos, telemetry data can be obtained in this step.

Phase 5 (Reporting): After the thorough analysis extracted artifacts to be reviewed and, on that basis, a report is to be generated stating all the events that took place from receiving the request for examination till the end of the analysis. The report should mention all the artifacts found and other correlating facts and figures that need to be presented in the court of law.

Phase 6 (Presentation): Final step in GCS forensics will be the presentation of the forensic examination report in the court of law to validate the findings and identify and prosecute the perpetrator of the crime.

This paper focuses on the forensics of GCS, which is an android application installed on a Samsung android phone for the case under study. Data storage location for GCS on Samsung phone illustrated in Fig. 2 was studied at length. The image was analyzed using the following open-source forensic tools.

- Autopsy 4.13.0 [19]
- HxD editor 2.4.0.0 [20].

4 Results and Discussion

The GCS image downloaded from the VTO laboratories Web site in the absence of a physical GCS controller of the Bebop drone was analyzed using the open-source tool autopsy to gather various artifacts of the drone and the controller. HxD which is another open-source hex editor was used to open the GCS image in hex format and

_id	name	type
1	vto.drone@gmail.com	com.google
2	dummyAccount	com.parrot.freeflight3.release

Fig. 3 Account associated with the FreeFlight3 app

gather the artifacts by studying the hex values. After analyzing the image file, the following artifacts could be obtained.

A. *User credentials*
A Gmail account that was used to log in to the GCS (FreeFlight app in case Parrot Bebop drone) for controlling and giving commands to the drone was found in autopsy while analyzing the image file. Details for the same are shown in Fig. 3.

B. *Database*
Figure 4 shows the structure of the database being used by the FreeFlight app to store the data. To extract this information, both autopsy and HxD editor were used.

C. *Details of components*
Figures 5 and 6 depict the details of the components associated with the UAV itself like its OS, GPS version, Android version of the device comprising the application-based GCS, etc.

D. *Wi-Fi info*
Figure 7 shows information about the Wi-Fi being used by the GCS app for communication and details like its driver and firmware information. This data was extracted using HxD editor.

E. *Host Name and Language*

```
00000160  6E 63 65 0C 43 52 45 41 54 45 20 54 41 42 4C 45  nce.CREATE TABLE
00000170  20 70 72 65 73 65 6E 63 65 20 28 5F 69 64 20 49   presence (_id I
00000180  4E 54 45 47 45 52 20 50 52 49 4D 41 52 59 20 4B  NTEGER PRIMARY K
00000190  45 59 2C 20 67 61 69 61 5F 69 64 20 54 45 58 54  EY, gaia_id TEXT
000001A0  20 4E 4F 54 20 4E 55 4C 4C 2C 20 72 65 61 63 68   NOT NULL, reach
000001B0  61 62 6C 65 20 49 4E 54 20 44 45 46 41 55 4C 54  able INT DEFAULT
000001C0  28 30 29 2C 20 72 65 61 63 68 61 62 6C 65 5F 74  (0), reachable_t
000001D0  69 6D 65 20 49 4E 54 20 44 45 46 41 55 4C 54 28  ime INT DEFAULT(
000001E0  30 29 2C 20 61 76 61 69 6C 61 62 6C 65 20 49 4E  0), available IN
000001F0  54 20 44 45 46 41 55 4C 54 28 30 29 2C 20 61 76  T DEFAULT(0), av
00000200  61 69 6C 61 62 6C 65 5F 74 69 6D 65 20 49 4E 54  ailable_time INT
00000210  20 44 45 46 41 55 4C 54 28 30 29 2C 20 73 74 61   DEFAULT(0), sta
```

Fig. 4 Structure of database

```
Offset(h) 00 01 02 03 04 05 06 07 08 09 0A 0B 0C 0D 0E 0F  Decoded text

00000000  7B 20 22 76 65 72 73 69 6F 6E 22 3A 20 22 31 2E  { "version": "1.
00000010  32 22 2C 20 22 73 6F 66 74 77 61 72 65 5F 76 65  2", "software_ve
00000020  72 73 69 6F 6E 22 3A 20 22 34 2E 32 2E 31 22 2C  rsion": "4.2.1",
00000030  20 22 68 61 72 64 77 61 72 65 5F 76 65 72 73 69   "hardware_versi
00000040  6F 6E 22 3A 20 22 48 57 5F 30 32 22 2C 20 22 64  on": "HW_02", "d
00000050  61 74 65 22 3A 20 22 32 30 31 37 2D 30 38 2D 32  ate": "2017-08-2
00000060  39 54 31 31 33 38 30 30 2D 30 36 30 30 22 2C 20  9T113800-0600",
00000070  22 70 72 6F 64 75 63 74 5F 69 64 22 3A 20 32 33  "product_id": 23
00000080  31 36 2C 20 22 73 65 72 69 61 6C 5F 6E 75 6D 62  16, "serial_numb
00000090  65 72 22 3A 20 22 50 49 30 34 30 33 37 36 41 42  er": "PI040376AB
000000A0  36 43 34 32 31 33 35 33 22 2C 20 22 70 72 6F 64  6C421353", "prod
000000B0  75 63 74 5F 6E 61 6D 65 22 3A 20 22 42 65 62 6F  uct_name": "Bebo
000000C0  70 20 32 22 2C 20 22 75 75 69 64 22 3A 20 22 42  p 2", "uuid": "B
```

Fig. 5 Internal details of drone with serial no. and product name

```
00000170  22 3A 20 2D 32 2C 20 22 67 70 73 5F 61 76 61 69  ": -2, "gps_avai
00000180  6C 61 62 6C 65 22 3A 20 74 72 75 65 2C 20 22 67  lable": true, "g
00000190  70 73 5F 6C 61 74 69 74 75 64 65 22 3A 20 33 39  ps_latitude": 39
000001A0  2E 39 36 31 32 31 34 2C 20 22 67 70 73 5F 6C 6F  .961214, "gps_lo
000001B0  6E 67 69 74 75 64 65 22 3A 20 2D 31 30 36 2E 32  ngitude": -106.2
000001C0  31 36 33 36 32 2C 20 22 63 72 61 73 68 22 3A 20  16362, "crash":
000001D0  30 2C 20 22 6A 75 6D 70 22 3A 20 6E 75 6C 6C 2C  0, "jump": null,
000001E0  20 22 72 75 6E 5F 74 69 6D 65 22 3A 20 33 31 35   "run_time": 315
000001F0  38 30 30 2C 20 22 74 6F 74 61 6C 5F 72 75 6E 5F  800, "total_run_
00000200  74 69 6D 65 22 3A 20 33 31 35 39 37 30 2C 20 22  time": 315970, "
```

Fig. 6 GPS details with late long and total run time

```
5B 77 69 66 69 20 64 72 69 76 65 72 20 3A 20 32  [wifi driver : 2
33 31 34 39 39 5D 20 5B 66 69 72 6D 77 61 72 65  31499] [firmware
20 3A 20 4D 41 52 4C 49 4E 5F 35 2E 31 52 5F 57   : MARLIN_5.1R_W
31 36 2E 35 31 2E 35 5F 44 20 64 61 74 65 20 3A  16.51.5_D date :
20 30 78 32 30 31 36 31 32 31 36 5D               0x20161216]
```

Fig. 7 Wi-Fi details

While investigating the image file of the drone, the hostname of the Parrot Bebop drone was found along with the language used (English) for sending and receiving commands as shown in Figs. 8 and 9, respectively.

F. *Flight Log*

Figure 10 shows the flight log of the drone which gives the product ID, serial number of the drone, controller model, etc.

Fig. 8 Language

```
android_metadata

locale
en_US
```

Fig. 9 Hostname used


```
"id": 0,
"date": "20170829113800",
"product_id": 2316,
"serial_number": "PI040376AB6C421353",
"uuid": "BA3C3DBB35927B2B08228E755FA0F759",
"controller_model": "SkyController",
"controller_application": "nap",
"product_style": -2,
"gps_available": true,
"gps_latitude": 39.961214,
"gps_longitude": -106.216362,
"run_time": 315800,
"total_run_time": 315970,
"run_origin": 0,
"crash": 0,
"jump": 0,
```


Fig. 10 Flight log

Table 1 gives the comparison between the tools used and the pieces of evidence that were found using that tool. It is noticed that autopsy along with the HxD editor was able to find out most of the data.

Table 1 Comparative analysis of extracted artifacts

Artifact Name	Autopsy	HxD
User details	YES	–
Database	–	YES
Drone components	–	YES
Flight log	YES	–
Update logs	YES	–
Wi-Fi details	–	YES
Language	YES	–

5 Conclusion and Future Work

A lot of research is already going on the forensics of drones and other unmanned vehicles; however, there is not much focus on the controller being used to operate them. Hence, this research presents a methodology to carry out forensic analysis of the ground control station of a drone. The steps that are part of the proposed methodology can be used to investigate controllers of other drone families. The methodology highlights that data can be obtained from the ground controller of the drone with the use of openly available tools. The research brings out an important fact that how a drone can be linked to a specific controller based on serial no. found during the investigation, thereby ensuring non-repudiation. While a few expensive commercial tools are available in the field of GCS forensics, this work focused on using open-source tools. The work relies more on manual analysis to get the forensic artifacts from the GCS which is time consuming and needs patience. As part of future work, a combination of manual analysis and automated tools needs to be used to extract already known file formats and to manually analyze new file formats. Efforts can also be made to decrypt the encrypted proprietary file types to extract more detailed information about the drone's flight as part of future work.

References

1. Hong, Y., Fang, J., Tao, Y.: Ground control station development for autonomous UAV. In: Xiong, C., Liu, H., Huang, Y., Xiong, Y. (eds.) Intelligent Robotics and Applications (ICIRA 2008). Lecture Notes in Computer Science, vol. 5315 (2008)
2. Roder, A., Raymond Choo, K.-K., Le-Khac, N.-A.: Unmanned aerial vehicle forensic investigation process: Dji phantom 3 drone as a case study (2018)
3. https://www.vtolabs.com/drone-forensics. Last accessed 31 Nov 2020
4. Kamoun, F., Bouafif, H., Iqbal, F.: Towards a better understanding of drone forensics: a case study of Parrot AR drone 2.0. Int. J. Digital Crime Forens. 1–23 (2019)
5. Jain, U., Rogers, M., Matson, E.T.: Drone forensic framework: Sensor and data identification and verifica-tion. In: 2017 IEEE Sensors Applications Symposium (SAS), Glassboro, NJ, USA, pp. 1–6 (2017)
6. Arteaga, S.P., Hernández, L.A.M., Pérez, G.S., Orozco, A.L.S., Villalba, L.J.G.: Analysis of the GPS spoofing vulnerability in the Drone 3DR Solo. IEEE Access **7**, 51782–51789 (2019)
7. Walter, B., Knutzon, J., Sannier, A., Oliver, J.: Virtual UAV ground control station. In: AIAA 3rd "Unmanned Unlimited" Technical Conference, Workshop, and Exhibit (2004)
8. GRYPHON: Drone Forensics in Dataflash and Telemetry Logs: Evangelos Mantas and Constantinos Patsakis (2019)
9. Barton, T.E.A., Hannan Bin Azhar, M.A.: Forensic analysis of popular UAV systems. In: 2017 Seventh International Conference on Emerging Security Technologies (EST), Canterbury (2017), pp. 91–96
10. Horsman, G.: Unmanned aerial vehicles: a preliminary analysis of forensic challenges. Digit. Investig. **16**, 1–11 (2016)
11. Azhar, M.A., Bin, H., Barton, T., Islam, T.: Drone forensic analysis using open source tools. J. Digital Forens. Secur. Law (2018)
12. Agrawal, A.K., Sharma, A., Khatri, P., Sinha, S.R.: Forensic of unrooted mobile device. Int. J. Electron. Secur. Digital Forens. **12**(1), 118–137 (2019)

13. Barton, T., Azhar, H.: Open source forensics for a multi-platform drone system. In: Matousek, P., Schmiedecker, M. (eds.) 9th EAI International Conference on Digital Forensics & Cyber Crime, pp. 83–96. Springer (2018)
14. Renduchintala, A., Jahan, F., Khanna, R., Javaid, A.: A comprehensive micro unmanned aerial vehicle (UAV/drone) forensic framework. Digital Invest. (2019)
15. Iqbal, F., et al.: drone forensics: examination and analysis. Int. J. Electron. Secur. Digital Forens. 245–264 (2019)
16. Bouafif, H., Kamoun, F., Iqbal, F., Marrington, A.: Drone forensics: challenges and new insights. In: 2018 9th IFIP International Conference on New Technologies, Mobility and Security (NTMS), Paris, pp. 1–6 (2018)
17. Segor, F., Bürkle, A., Partmann, T., Schönbein, R.: Mobile Ground Control Station for Local Surveillance. In: 2010 Fifth International Conference on Systems, Menuires, France, pp. 152–157 (2010)
18. Perez, D., Maza, I., Caballero, F., et al.: A ground control station for a multi-UAV surveillance system. J. Intell. Robot. Syst. **69**, 119–130 (2013)
19. Autopsy home page. https://www.autopsy.com/. Last accessed 15 Nov 2020
20. HxD editor home page, https://mh-nexus.de/en/hxd/. Last accessed 10 Nov 2020

Statistical Analysis on the Topological Indices of Clustered Graphs

Sambanthan Gurunathan and Thangaraj Yogalakshmi

1 Introduction

Clustering is used in genetic technology to group the genes with similar patterns [8]. Ustumbas [10] used tripartite graph clustering in social network as a cluster of three types of data objects simultaneously and produce useful information for a recommended system.

A computer network is known as digital telecommunication network. In this network, graphs are structured into clusters. Electronic links are employed within the same cluster where as optical links are employed between the cluster communications. These data links may be wire or optic cables or wireless media. Interconnection networks are strenuous to work with abstract terms. This motivated many researchers to propose new improved network graphs arguing the benefits and performance evaluation in different contexts. Interconnection network is modeled mathematically as a graph having vertices as cluster nodes and edges as links between them. The topology of such graphs determined certain properties of it.

In recent days, optical transpose interconnection system networks provide efficient connectivity for new optoelectronic computer architectures with technologies of optics and electronics. It has many parallel paths between nodes of network which accelerate large amount of data transfers. Thus, the network, having maximum parallel paths, is sturdy [2].

Topological indices find applications in networking, multiprocessor interconnection networks connect thousands of processor, memory pairs which paid attention

S. Gurunathan · T. Yogalakshmi (✉)
Department of Mathematics, SAS, VIT, Vellore, India
e-mail: yogalakshmithangaraj@vit.ac.in

S. Gurunathan
e-mail: gurunathan.s@vit.ac.in

M. Dua et al. (eds.), *Proceedings of the International Conference on Paradigms of Communication, Computing and Data Sciences*, Algorithms for Intelligent Systems,
https://doi.org/10.1007/978-981-16-5747-4_33

due to the accessibility of inexpensive powerful microprocessor, and microchips. These network families have topologies which indicate the communication pattern of various problems will be resolved by the topological indices. Randic [6] introduced first degree-based index in 1975, Bollabas and Erdos [1] extended it in 1998 to general Randic index. First and second Zagreb indices were introduced in the year 1972 by Gutman[5]. Vukicevic [11][12] originated geometric arithmetic and inverse sum indeg indices. In 2011, Ranjini et al.[7] calculated the explicit expressions for the Shultz indices.

In this paper, clustered graphs have been originated from an algorithm. Degree-based indices such as first and second Zagreb, first and second multiplicative Zagreb, Hyper Zagreb, atom-bond connectivity (ABC), and geometric indices of the clustered graphs have been computed. Moreover, the regression analysis on Randic, SCI, ABC, ISI, and GA indices of the clustered graphs has been established which interprets that those indices are highly correlated with $R = 0.99$.

2 Basic Terminology

A graph G consists of a set V of vertices and a collection E of unordered pair of vertices called edges. The degree of a vertex u, d_u is the number of edges incident with it. If uv is an edge of G, then u and v are adjacent vertices. A complete tripartite graph, G is a graph if the vertex set $V(G)$ can be partitioned into three subsets V_1, V_2, V_3 such that every vertex of $V_i (i = 1, 2, 3)$ is adjacent with all other vertices of $V_j (j \neq i)$.
The first degree-based topological index is Randic index [6]

$$R_{1/2}(G) = \sum_{uv \in E(G)} \frac{1}{\sqrt{d_u \cdot d_v}} \tag{1}$$

Bolloas and Erdos [1] defined general Randic index

$$R_\alpha = \sum_{uv \in E(G)} (d_u \cdot d_v)^\alpha \tag{2}$$

The sum-connectivity index is invention by Zhou and Nenad Trinajstic [13] and defined as

$$\mathrm{SCI}(G) = \sum_{uv \in E(G)} \frac{1}{\sqrt{d_u + d_v}} \tag{3}$$

And the general sum-connectivity index $\chi_a(G)$ was introduced in 2010 [14],

$$\chi_a(G) = \sum_{uv \in E(G)} (d_u + d_v)^a \tag{4}$$

In 1972, Gutman [5] introduced first and second Zagreb index and defined as

$$M_1(G) = \sum_{v\in V(G)} d_v^2 = \sum_{uv\in E(G)} [d_u + d_v]; \quad M_2(G) = \sum_{uv\in E(G)} [d_u \cdot d_v] \tag{5}$$

In 2012, Ghorbani et.al [4] defined first multiple Zagreb index $PM_1(G)$, second Zagreb index $PM_2(G)$, and these indices are defined as

$$PM_1(G) = \prod_{uv\in E(G)} [d_u + d_v]; \quad PM_2(G) = \prod_{uv\in E(G)} [d_u \cdot d_v] \tag{6}$$

Recently, Shirdel et al. [9] proposed the hyper Zagreb index as

$$\text{HM}(G) = \sum_{uv\in E(G)} [d_u + d_v]^2 \tag{7}$$

The most useful connectivity topological index was atom-bond connectivity (ABC) index introduced by Estrada et al. [3] and defined as

$$\text{ABC}(G) = \sum_{uv\in E(G)} \sqrt{\frac{d_u + d_v - 2}{d_u \cdot d_v}} \tag{8}$$

Vukicevic and Furtula [11] introduced the geometric arithmetic (GA) index and defined as

$$\text{GA}(G) = \sum_{uv\in E(G)} \frac{2\sqrt{d_u \cdot d_v}}{d_u + d_v} \tag{9}$$

D.Vukicevic and M.Gaperov [12] introduced inverse sum indeg index (ISI) as

$$\text{ISI}(G) = \sum_{uv\in E(G)} \frac{d_u \cdot d_v}{d_u + d_v} \tag{10}$$

3 Algorithm of Clustered Graphs

The clustered graph is constructed by the following steps:

step-1: Take any complete tripartite graph $K_{r,s,t}$ with $r \leq s \leq t$.
step-2: Introduce a vertex of degree 2 in each edge of a complete tripartite graph.
step-3: Treat the edges of the obtained graph in step-2 as the vertices of the new graph.
step-4: Two vertices of the new graph in step-3 are adjacent if and only if they have common vertex in the graph of step-2.

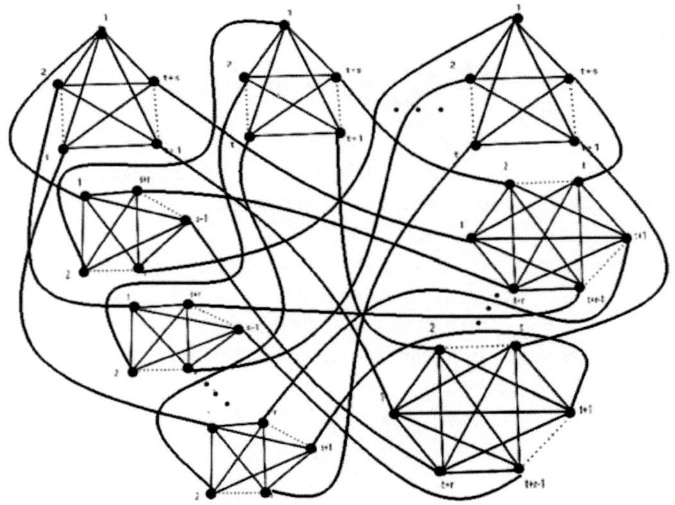

Fig. 1 Clustered graph $CL(K_{r,s,t})$

step-5: Finally, the obtained new graph in step-4 is the clustered graph which is shown in Fig. 1, denoted by $CL(K_{r,s,t})$

Possible degree of the vertices of the clustered graph are $r+s, s+t, r+t$. This clustered graph has $2rst(\frac{1}{r}+\frac{1}{s}+\frac{1}{t})$ vertices and $\frac{1}{2}[(r+s+t)(rs+st+tr)+3rst]$ edges.

4 Topological Indices of the Clustered Graphs

Throughout this section, the clustered graph $CL(K_{r,s,t})$ is represented as H

Theorem 1 *The general sum-connectivity index χ_α and the general Randic index R_α of the graph H are*

$$R_\alpha(H) = \frac{1}{2}[r(s+t)^{2\alpha+1}(s+t-1)+s(r+t)^{2\alpha+1}(r+t-1)+t(r+s)^{2\alpha+1}(r+s-1)+2sr(s+t)^\alpha(r+t)^\alpha+2rt(s+t)^\alpha(r+s)^\alpha+2ts(r+t)^\alpha(r+s)^\alpha]$$

$$\chi_\alpha(H) = 2^{\alpha-1}(s+t)^{\alpha+1}r(s+t-1)+2^{\alpha-1}(r+t)^{\alpha+1}s(r+t-1)+2^{\alpha-1}(r+s)^{\alpha+1}t(r+s-1)+(r+s+2t)^\alpha sr+(r+2s+t)^\alpha tr+(2r+s+t)^\alpha ts$$

Proof Proof is clear from the definition.

Corollary 1 *The Randic and sum-connectivity indices of clustered graph H are*
$R_{1/2}(H)=\frac{1}{2}[r(s+t-1)+s(r+t-1)+t(r+s-1)+\frac{2sr}{\sqrt{(s+t)(r+t)}}+\frac{2rt}{\sqrt{(s+t)(r+s)}}+\frac{2ts}{\sqrt{(r+s)(r+t)}}]$

$SCI(H) = 2^{-3/2}(s+t)^{1/2}r(s+t-1) + 2^{-3/2}(r+t)^{1/2}s(r+t-1) + 2^{-3/2}(r+s)^{1/2}t(r+s-1) + sr(r+s+2t)^{-1/2} + rt(r+2s+t)^{-1/2} + st(2r+s+t)^{-1/2}$

Proof It is proved by putting $\alpha = \frac{-1}{2}$ in Theorem 1.

Theorem 2 *Let H be the clustered graph. Then the family of Zagreb indices are*

$$M_1(H) = r(s^3+t^3) + s(r^3+t^3) + t(r^3+s^3) + 6rst(r+s+t)$$
$$M_2(H) = (s^4-s^3)(r+t) + (r^4-r^3)(s+t) + (t^4-t^3)(r+s) + 2rst(4r^2 + 4s^2 + 4t^2 + 3st + 3rt + 3rs) + 2(r^2s^2 + t^2r^2 + t^2s^2)$$
$$PM_1(H) = r^3s^3t^3(r+s)^2(s+t)^2(t+r)^2(r+s+2t)(r+2s+t)(2r+s+t)(r+s-1)(s+t-1)(r+t-1)$$
$$PM_2(H) = \frac{1}{8}r^3s^3t^3(r+s)^5(s+t)^5(t+r)^5(s+t-1)(r+t-1)(r+s-1)$$
$$HM(H) = r^2(s+t)^4(s+t-1) + s^2(r+t)^4(r+t-1) + t^2(r+s)^4(r+s-1) + s^2r^2(s+r+2t)^2 + t^2r^2(r+2s+t)^2 + t^2s^2(2r+s+t)^2$$

Proof Proof is obvious from formulae (5), (6) and (7).

Theorem 3 *The atom-bond connectivity and geometry-arithmetic indices of H are*

$$ABC(H) = \frac{1}{\sqrt{2}}[r(s+t-1)^{3/2} + s(r+t-1)^{3/2} + t(r+s-1)^{3/2}] + rs\sqrt{\frac{r+s+2t-2}{(r+t)(s+t)}} + rt\sqrt{\frac{r+t+2s-2}{(r+s)(t+s)}} + st\sqrt{\frac{s+t+2r-2}{(s+r)(t+r)}}$$

$$GA(H) = \frac{1}{2}[r^2(s+t) + s^2(t+r) + t^2(r+s) - 2rs - 2st - 2rt + 6rst]$$

Proof Proof is obvious.

Theorem 4 *The inverse sum indeg (ISI) index of H is*

$$ISI(H) = \frac{1}{4}[6rst(r+s+t-1) + (r+t)(s^3-s^2) + (r+s)(t^3-t^2) + (s+t)(r^3-r^2)] + \frac{rs(r+t)(s+t)}{r+s+2t} + \frac{rt(r+s)(t+s)}{r+2s+t} + \frac{st(r+s)(r+t)}{2r+s+t}$$

Proof The result is easily obtained.

5 Interrelation on Indices

Correlation coefficients are highly correlated among all such indices with $R = 0.9$. The scatter plot of Randic, SCI, ABC, GA, and ISI indices are exhibited in Figs. 2 and 3. It is clear that the correlation between those indices are perfectly positive by computing the Pearson's correlation coefficient. To exhibit a linear relationship between indices, the general linear regression model is of the form

$$y = k_0 + k_1x \tag{11}$$

where y is the unknown value(responder), x is the known value (predictor), and k_0, k_1 are linear regression coefficients. R-square, residual standard error and F-value for

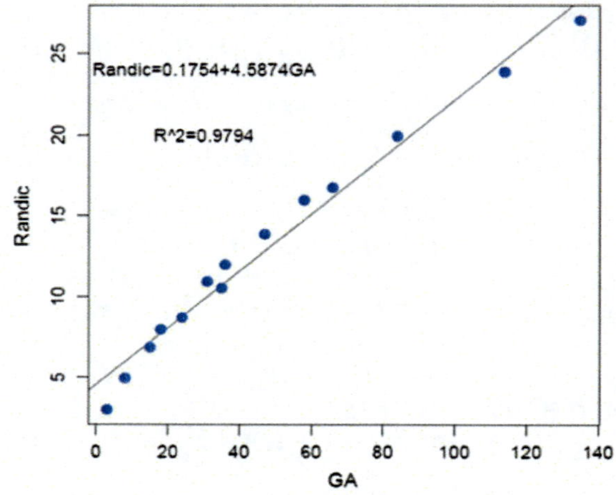

Fig. 2 Correlation between Randic and GA indices

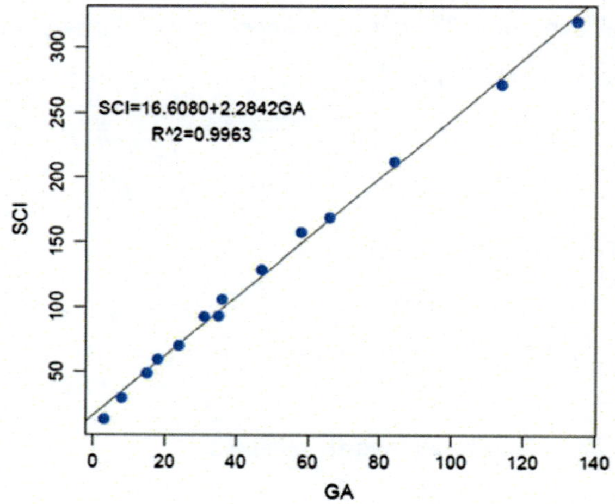

Fig. 3 Correlation between SCI and GA indices

the pair of indices have been computed. The best fit will occur when the R-square value is near 1 and error value near to 0 with larger F-value.

From the linear regression equation (11) and Table 1, the relationship between Randic index and GA index is obtained as

$$R(H) = 4.5874(\pm 0.4517) + 0.175425(\pm 0.007349)GA(H) \tag{12}$$

The regression coefficients k_0 and k_1 in linear regression equation (12) are significant. The coefficient of determination between Randic index and GA index is $R^2 = 0.9794$. This indicates that there is a good relation between Randic and GA indices. The scatter plot between Randic and GA is shown in Fig. 2. Using data given

Table 1 Numerical values of indices of clustered graphs

Name of complete tripartite graph	r	s	t	Randic	SCI	M_1	M_2	ISI	ABC	GA
Triangle graph	1	1	1	3	13.5	24	48	6	3.621	3
Diamond graph	2	1	1	4.966	29.79	70	190	17.3	7.886	8
5-Wheel graph	2	2	1	7.988	59.31	172	576	42.86	15.39	18
Octahedral graph	2	2	2	12	106.1	384	1536	96	27.24	36
(3,2)-Fan graph	3	1	1	6.871	48.74	152	536	37	12.9	15
(3,3)-Fan graph	3	2	1	10.95	92.63	334	1360	82.86	23.92	31
(4,3)-Cone graph	3	2	2	15.98	157.4	692	3256	172.7	40.27	58
9-Circulant graph	3	3	3	27	319.6	1944	11664	486	81.21	135
(4,2)-Fan graph	4	1	1	8.73	69.94	282	1242	67.93	18.54	14
(4,3)-Fan graph	4	2	1	13.87	129.2	574	2800	141.5	33.34	47
(4,4)-Cone graph	4	2	2	19.93	212.7	1120	6144	278.4	54.34	84
(5,2)-Fan graph	5	1	1	10.55	93.11	472	2512	113	24.72	35
(5,3)-Fan graph	5	2	1	16.76	168.6	910	5226	223	43.52	66
(4,5)-Cone graph	5	2	2	23.85	271.6	1692	10704	418.9	69.36	114

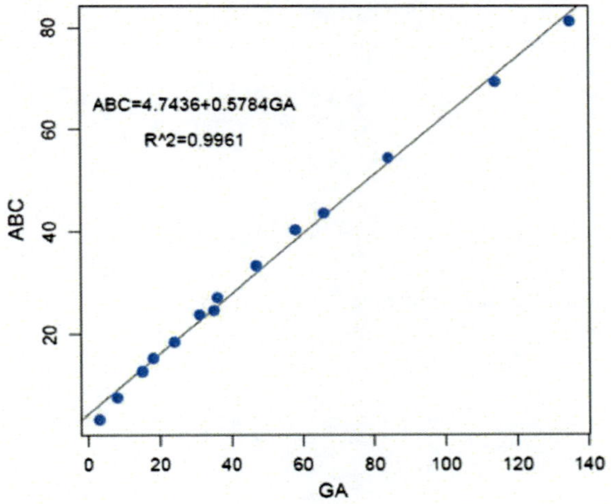

Fig. 4 Correlation between ABC and GA indices

in Table 1, the linear regression equations of SCI and ABC indices on GA index are obtained as (Fig. 4)

$$\text{SCI}(H) = 16.6080(\pm 2.4810) + 2.2842(\pm 0.0404)\text{GA}(H) \tag{13}$$

$$\text{ABC}(H) = 4.7437(\pm 0.6422) + 0.5784(\pm 0.0104)\text{GA}(H) \tag{14}$$

From Eqs. (13) and (14), the coefficient of determination for SCI and ABC are $R^2 = 0.9963$ and $R^2 = 0.9961$, respectively. This shows that the established correlation is best. Similarly, the regression equations of ISI, M1, and M2 are obtained.

$$\mathrm{ISI}(H) = -23.3327(\pm 5.2579) + 3.7240(\pm 0.0855)GA(H) \quad (15)$$

$$M1(H) = -89.112(\pm 22.000) + 14.940(\pm 0.358)GA(H) \quad (16)$$

$$M2(H) = -1086.902(\pm 279.145) + 93.488(\pm 4.542)GA(H) \quad (17)$$

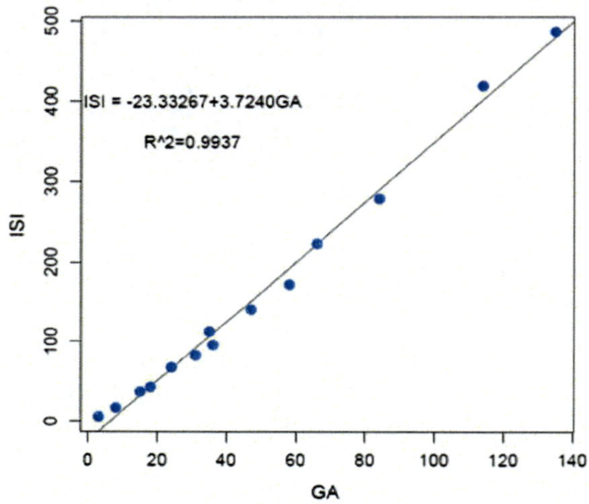

Fig. 5 Correlation between ISI and GA indices

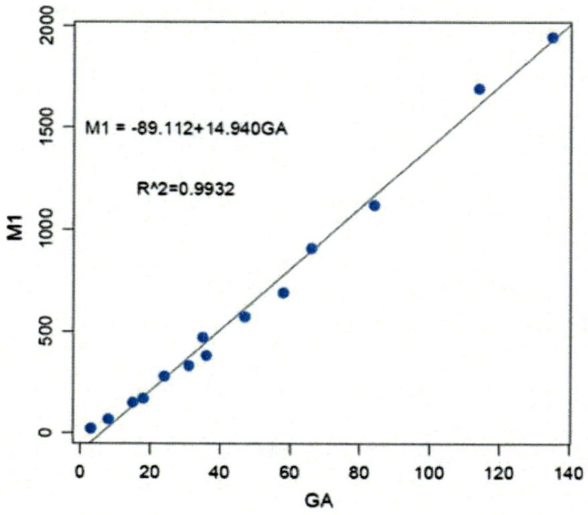

Fig. 6 Correlation between M1 and GA indices

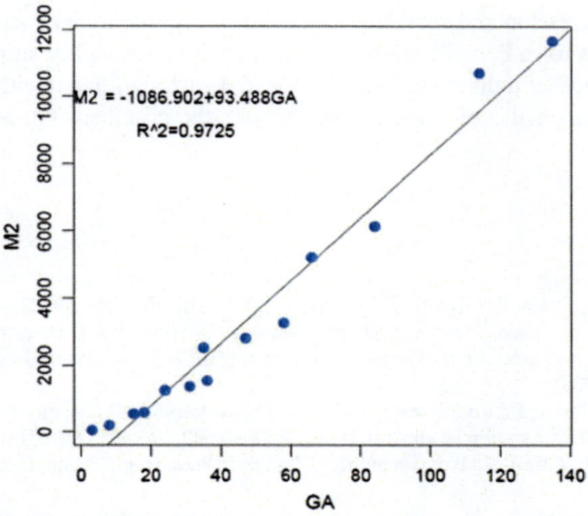

Fig. 7 Correlation between M2 and GA indices

Equations (15)–(17) give R^2 values are 0.9937, 0.9932, and 0.9725, respectively. These values indicate ISI, first and second Zagreb indices having positive correlation with residual error, shown in Figs. 5, 6 and 7.

Algorithm to obtain correlation and regression coefficients

Input: m values of Predictor(x) and Response(y)
Output: Correlation coefficient between x and y, Regression equation
Step 1: Start
Step 2: Declare variables $j, m, x(m), y(m), S_x, S_y, S_{x^2}, S_{y^2}, S_{xy}, V_r, COV(x, y), R, k_0, k_1$
step 3: Enter m
Step 4: Repeat until $j \leftarrow m$ Read $x(j)$
Step 5: Repeat until $j \leftarrow m$ Read $y(j)$
Step 6: Put $S_x, S_y, S_{x^2}, S_{y^2}, S_{xy} == 0$
Step 7: For($j = 1, j = m, j = j + 1$)
$S_x \leftarrow$ `Mean` $(x(j))$
$S_y \leftarrow$ `Mean` $(y(j))$
$S_{x^2} \leftarrow$ `Mean` $((x(j))^2)$
$S_{y^2} \leftarrow$ `Mean` $((y(j))^2)$
$S_{xy} \leftarrow$ `Mean` $(x(j) * y(j))$
Step 8: Calculate R. $R \leftarrow \frac{S_{xy} - (S_x * S_y)}{\sqrt{(S_{x^2} - (S_x)^2) * (S_{y^2} - (S_y)^2)}}$
Step 9: Compute V_r. $V_r \leftarrow S_{x^2} - (S_x)^2$
Step 10: Calculate k_1. $k_1 \leftarrow \frac{S_{xy} - (S_x * S_y)}{V_r}$
Step 11: Calculate k_0. $k_0 \leftarrow S_y - k_1 * S_x$
Step 12: Print R
Step 13: Print $y \leftarrow k_0 + k_1 * x$

6 Conclusion

Clustered graph $CL(K_{r,s,t})$ is constructed in Sect. 3. Various degree-based indices such as Randic, Zagreb, hyper Zagreb sum-connectivity, ABC, ISI, and GA have been found in Sect. 4. In Sect. 5, the inter-relation between the indices are discussed through the regression and correlation analysis. It is evident that Randic, SCI, ABC, and ISI indices form best fit with GA index that they are highly significant to each other, whereas other indices such as M_1, M_2, HM are significant with residual error. In future, the properties of clustered graphs with these indices will be established.

References

1. Bollobos, B., Erdos, P.: Graphs of extremal weights. ARS Combin. **50**, 225–233 (1998)
2. Chen, Weidong, Xiao, Wenjun, Parhami, Behrooz: Swapped (OTIS) networks built of connected basis networks are maximally fault tolerant. IEEE Trans. Parallel Distrib. Sys. **20**(3), 361–366 (2008)
3. Estrada, E., Torres, L., Rodriguez, L., Gutman, I.: An atom-bond connectivity index: modelling the enthalpy of formation of alkanes. Indian J. Chem. **37**, 849–855 (1998)
4. Ghorbani, M., Azimi, N.: Note on multiple Zagreb indices. Iran. J. Math. Chem. **3**(2), 137–143 (2012)
5. Gutman, I., Trinajstic, N.: Graph theory and molecular orbitals total p-electron energy of alternant hydrocarbons. Chem. Phys. Lett. **17**, 535–538 (1972)
6. Randic, M.: On characterization of molecular branching. J. Am. Chem. Soc. **97**, 6609 (1975)
7. Ranjini, P.S., Lokesha, V., Rajan, M.A.: On the Shultz index of the subdivision graphs. Adv. Stud. Contemp. Math. **21**(3), 279–290 (2011)
8. Shen, C., Liu, Y.: A tripartite clustering analysis on microRNA, gene and disease model. J. Bioinform. Comput. Biol. **10**(01), 1240007 (2012)
9. Shirdel. G.H, Pour. H.R, Sayadi. A.M.: The HyperZagreb index of graph operations. Iran. J. Math. Chem. **4**(2), 213–220 (2013)
10. Ustunbas, Y., Guducu, S.G.: A recommendation model for social resource sharing systems based on tripartite graph clustering. In: 2011 European Intelligence and Security Informatics Conference, pp. 378–381(2011)
11. Vukicevic, D., Furtula, B.: Topological index based on the ratios of geometrical and arithmetical means of end vertex degrees of edges. J. Math. Chem. **46**, 1369–1376 (2009)
12. Vukicevic, D., Marija, G.: Bond additive modeling 1. Adriatic indices. Croatica Chemica Acta **83**(3), 243–260 (2010)
13. Zhou, B., Trinajstic, N.: On a novel connectivity index. J. Math. Chem. **46**, 1252–1270 (2009)
14. Zhou, B., Trinajstic, N.: On general sum-connectivity index. J. Math. Chem. **47**, 210–218 (2010)

A Reliable and Tamper-Free Double-Layered Vaccine Production and Distribution: Blockchain Approach

R. Mythili, Revathi Venkataraman, Neha Madhavan, H. Gayathree, and R. Balasubramaniam

1 Introduction

Twentieth century has vaccine as a great achievement. Serious harms can be the cause to the vaccinator's body, if the injection has a problem. Hence, it may cause the vaccinator unable to resist many infectious diseases, if the problematic vaccine such as expired/corrupted is injected. The health of the patients can be harmed due the diseases some may even cause to death. The young children and infants have less immunity; they need vaccination for safety. Meantime, vaccine industry, in recent years, experiences many unexpected vaccination failures as expiration, medical frauds, and other mischievous activities, which should be strictly supervised and banned. Nearly, 100 children were died due to Shanxi' invalid vaccine. Enormous harm caused due to these incidents have occurred at many places. Correct supervision is required for the vaccine production and further actions. Food and Drug Administration (FDA) is responsible primarily for determining vaccine releases list to medical agencies. Further, the local FDA is answerable for assisting the localization of the vaccine. Inside blockchain, vaccine data are combined with the timestamp for better traceability grit.

R. Mythili (✉) · N. Madhavan · H. Gayathree · R. Balasubramaniam
SRM Institute of Science and Technology, Ramapuram, India
e-mail: mythilir2@srmist.edu.in

R. Venkataraman
SRM Institute of Science and Technology, Kattankulathur, India

M. Dua et al. (eds.), *Proceedings of the International Conference on Paradigms of Communication, Computing and Data Sciences*, Algorithms for Intelligent Systems,
https://doi.org/10.1007/978-981-16-5747-4_34

1.1 Significance of Blockchain Technology

Bitcoin [1, 2] is a dispersed digital mode of money, without a bank or single controller, which can have transactions from user-to-user without an intermediate person. Bitcoins involve mining. It is also been predicted as the next generation of currencies and has been gaining popularity among people. They do not involve normal cash and are a system which can be viewed by everyone. Bitcoin involves a blockchain system which has a decentralized system with the help of blockchains.

The most basic unit of the blockchain is transactions. Input and output are the two parts of bitcoin system. The input depends on the prior block. The output of the chain is addressed and currency value. Transfers are represented by transactions. Transactions are to be initiated by someone at every point. After that, it will go to the blockchain network [3, 4]. The nodes inside the blockchain contain the transaction pools. A list of transaction information received by other nodes is known as transaction pool. The transaction is combined in these pools at a certain time. At the last, one block will be added in the blockchain. Chain consisting of many blocks is called a blockchain. The data of the after block are to be changed due to the later behavior of the block. The changes are reflected in the latest block. Once the destroyer modifies the data from one block to the latest block, then the modification will be discovered.

1.2 Double-Layered Blockchain Structure

Double-layered blockchain is designed for the main objective of safety against privacy leakages and better traceability. First layer includes the private data of vaccine enterprises with determined access control. Production records are main in vaccine production process which consist the private data. The corresponding hash values are recorded by each production and the hash of previous process record. Next layer of public data leads to any node participating in the blockchain can see it due to public blockchain storage visibility. Corresponding electronic signatures, vaccine information (e.g., expiration date and manufacture date), and primarily of production records of hash are public blockchain data. Vaccine enterprises can submit production records in a timely manner while does not leak the private data. Tampering with production records are avoided through the non-tampering attribute of the blockchain. It also prevents the enterprises from falsely producing production data by limiting production capacity of each record and the existence of the timestamps on the blockchain.

1.3 New Multi-node Cooperate Consensus

New multi-node cooperate consensus mechanism involves the major role of supervisory nodes. Sort services for other nodes and correctness verification of the blockchain replica are provided by the primary supervisory node. The packaged block is sent to the primary supervisory node, and the primary supervisory node ordering broadcasts the block to the blockchain network. The other nodes only receive the blocks sent by the primary supervisory node and are in the correct order. Other nodes can also connect to the primary supervisor node and verify that the blockchain replica they saved is correct based on the blockchain data it holds. When necessary, an ordinary supervisory node is replaced as the primary supervisory node.

1.4 Timestamp-Based Data Cutting

The proposed vaccine blockchain system combines the following new vaccine data cutting mechanism. The supervisor node sets a cut flag for each block. It determines whether the vaccine can be timestamp-based cut and the vaccine validity stored in the block. It periodically checks if the validity period of the vaccine represented by a block expires. If it expires, the cut flag will be set true. When the vaccination agency sends a signal to the supervisory node, it indicates that a certain batch of vaccine has been used. The cut marks for the blocks representing this batch of vaccines will also be set true. Eventually, these blocks will be discarded.

The overall structure of this paper is as follows: Sect. 2 of existing systems and Sect. 3 with proposed architecture details and design. And, Sect. 4 describes the system implementation and evaluation. Finally, the paper is concluded with the conclusion.

2 Existing Systems

Blockchain system contributes data security revolutions recently in various daily life applications like supply chain management [5, 6], social [7], and health care [8–11] applications. This section explains some of the major blockchain systems that attain data access control in the aforementioned applications.

Toyoda et al. [12] designed a novel product ownership management based on blockchain for the purpose of anti-counterfeits in the supply chain management process. The system has achieved supply accuracy and business economic cost with good responsiveness. Even then, the system lacks in complexity and cooperation integrity.

Kaur et al. [13] have identified the latest silico-peptide-based vaccine identification system for fighting with swine influenza virus. The system has achieved relatively improved data storage efficiency and privacy protection over the earlier systems with variation mercifulness. Meanwhile, the system is not suitable for high-level degree of risk control, complete information acquisition, and successful product quality assurance.

Tang et al. [14] have developed a micro-needle array for vaccine patch applications in tumor diagnosis and treatments. The system has performed better in cost performance, threshold flexibilities, and robustness but leads to failure in case of big payloads.

Fu et al. [15] have introduced a new blockchain-based endogenous risk management supply management system that is capable of achieving supply chain operation efficiency. But, it downs the performance of complex and resource/energy dilapidation.

Conditions. Food [16] and vaccination [17, 18] systems are inculcated nowadays with the modern blockchain technology for data identifiable, thereby human safety.

Hjálmarsson et al. [19] have developed a blockchain system for e-voting system that satisfies the requirements of the legislature which helps in nationwide elections. Sakoshi Nakamoto [20] developed an electronic cash system with peer-to-peer facility makes the transactions safe by hashing them into ongoing chain of work. Casino et al. [21] have discussed about the current status, classification, and the issues of the blockchain system. Buterin [22] has developed a decentralized smart contract system that shows how a smart contract coding works. Rouhani and Deters [23] explain about the application of smart contract and their use.

3 Proposed System

The proposed vaccine blockchain system enhances the existing system by taking care of the distribution and the endpoint of the vaccine distribution. It provides a system that keeps records of the vaccine and which patient has bought that particular vaccine. This keeps track of the vaccine from production until distribution. This helps us understand its distribution status. A double-layered blockchain structure is used for data storage and its access. First layer consists of private data about the production of vaccine and circulation details with hash. Next higher layer is public data, including production records hash, circulation records hash, and vaccine information. Hashing method is used to improve the reliability and data retrieval. Each block is used to contain the entire vaccine record. In the proposed system, each block mainly consists of three transaction, which are spread by companies, the lot release agency, and centers for disease control and prevention (CDC). These transactions are made up of five data fields, a timestamp, the sender, the recipient, the amount, and data records. The timestamp contains the transaction time. For the information, the records of recipient and amount are set as null, and the records are stored in the "data" field. The records are arranged as a hash table structure, which is an organization of unique keys and their values. With the hash table format, the records could be formatted when kept in the blockchain. Once a vaccine fault occurs, the vaccine blockchain is used

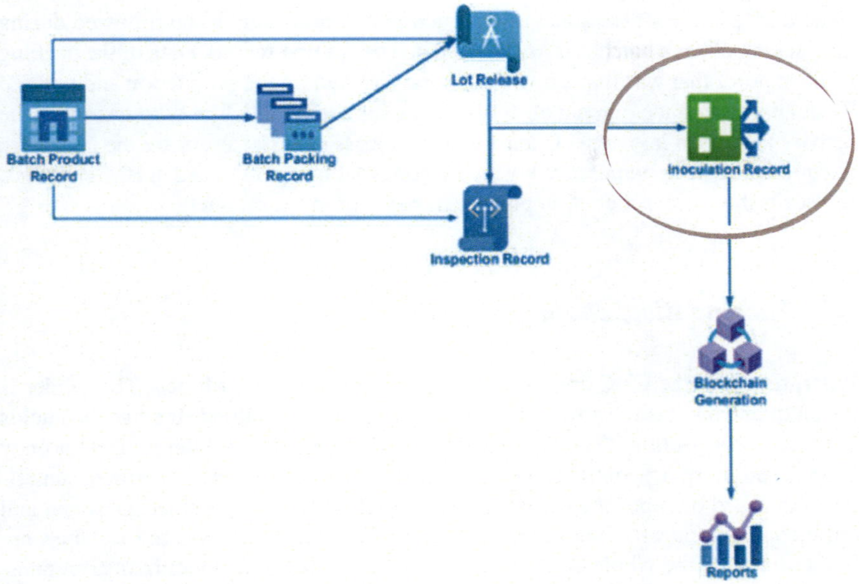

Fig. 1 Tamper-free vaccine blockchain system with inoculation

to track the process of vaccine distribution to find responsibility for the accident. The specification of the query is stored in the data field. Then, the system judges the identity via the specifications. The inoculation record can be quired which brings out vaccine receptions another type that is the vaccine circulation record. It also provides stage of users for this and what information can be viewed by them. All the information in the database can be viewed by the admin only, and only certain records can be viewed by the users. These records can be viewed by the doctors also to get information about the remedy for a patient. This system addresses the problems of vaccine expiration and vaccine record fraud, supports vaccine traceability, addresses trust and security issues by providing access control for various roles, maximizes efficiency, and improves the operational efficiency.

The proposed vaccine blockchain system is executed by the following modules of BPR/BMR records creation, vaccine blockchain generation, and vaccine accident investigation as depicted in Fig. 1.

3.1 BPR/BMR Records Creation

Batch packaging record (BPR) contains all information of the batch and requirements of GMP documentations of the packaging process. Storing process contains everything about BPR. Written document of the batch from dispensing to dispatch stage that talks about the stepwise instruction and procedure to be followed during the packing of each batches is known as batch packaging record. Data of the packing is like a proof that was made by batches and checked by the production and verified by quality assurance personnel. It also contains the details like who has done the activity and when it is done. Chemical and process manufacturers are the important documents of batch manufacturing record (BMR). From start to end of BMR process, it records the entire production process of batch of given product.

3.2 *Vaccine Blockchain Generation*

Enterprises and the CDC are the three types of transaction of block. The sender is the recipient, and data records content, including a timestamp, are the five data fields of types of transactions. The vaccine blockchain based on the Ethereum technology is designed by smart contracts. Open source tools software platform and decentralized are based on blockchain technology that allows users to use decentralized and build applications with blockchains are known as Ethereum. Expired vaccines are broadcasted by the whole vaccine supply chain to the corresponding institution as remainder; for forcibly injecting expired vaccines, the inoculation institution is not rewarded. So, the enterprises do not profit from providing expired vaccines.

3.3 *Vaccine Accident Investigation*

The process of entire vaccine circulation of vaccine blockchain is used to affix responsibility and investigate for the accident. Vaccine query functions are provided by vaccine blockchain system which are also implemented. Ethereum platform provides that the smart contracts are launched by message calls. The data field of the messages is used to store the parameters of the query and identifies and system judges via the parameters. The inoculation record, which is invoked by vaccine recipients, is one type, and the vaccine circulation record, which can be made by all three vaccine, is another type of vaccine record that can be queried.

4 Implementation and Evaluation

The proposed vaccine blockchain system is implemented with the real-time data feed of vaccine and other personal details. The system has the potential capability of making the records.

4.1 Double-Level Blockchain Structure

A double-layered vaccine blockchain structure is designed as follows. The first level with private data about the production and circulation details and corresponding hash. The next level is public data, including production records hash, circulation records hash, and vaccine information.

4.2 Nodes

One of the most important concepts in blockchain technology is decentralization. None of the computers or persons can own the chain. Instead, it is a distributed system via the nodes connected to the main chain. Nodes maintain temporary copies of the blockchain and keep the network functioning.

4.3 Blocks

Every chain consists of multiple blocks, and each block has basic elements and functions:

- The data in the block.
- A 32-bit whole number
- The hash which is a 256-bit number.

The system consists of blocks of information the person is feeding about the vaccine.

4.4 Frontend Software

Web Technologies
Web technologies are used to display and perform operations of what is needed and how they will be shown to the user.

CSS
To enhance the overall look of the web, cascading style sheet is used.

Bootstrap
Bootstrap is used to create buttons and form where some records can be created and is subsequently added to the blockchain.

4.5 Backend Software

Python

Pycharm Python language is used to write the code which consists of various modules like production, packing, inspection, inoculation, blockchain system, and logout. The Pycharm-integrated development environment is used to develop and write the code.

The proposed system is implemented under the aforementioned computer environment and exhibits the manufacturing/packing/inoculation data, storage, and vaccine blockchain generation as in Fig. 2.

The proposed double-layered blockchain-based vaccine system is implemented by an Ethereum smart contract code written in solidity language. Some of the sample smart contract codes are illustrated below:

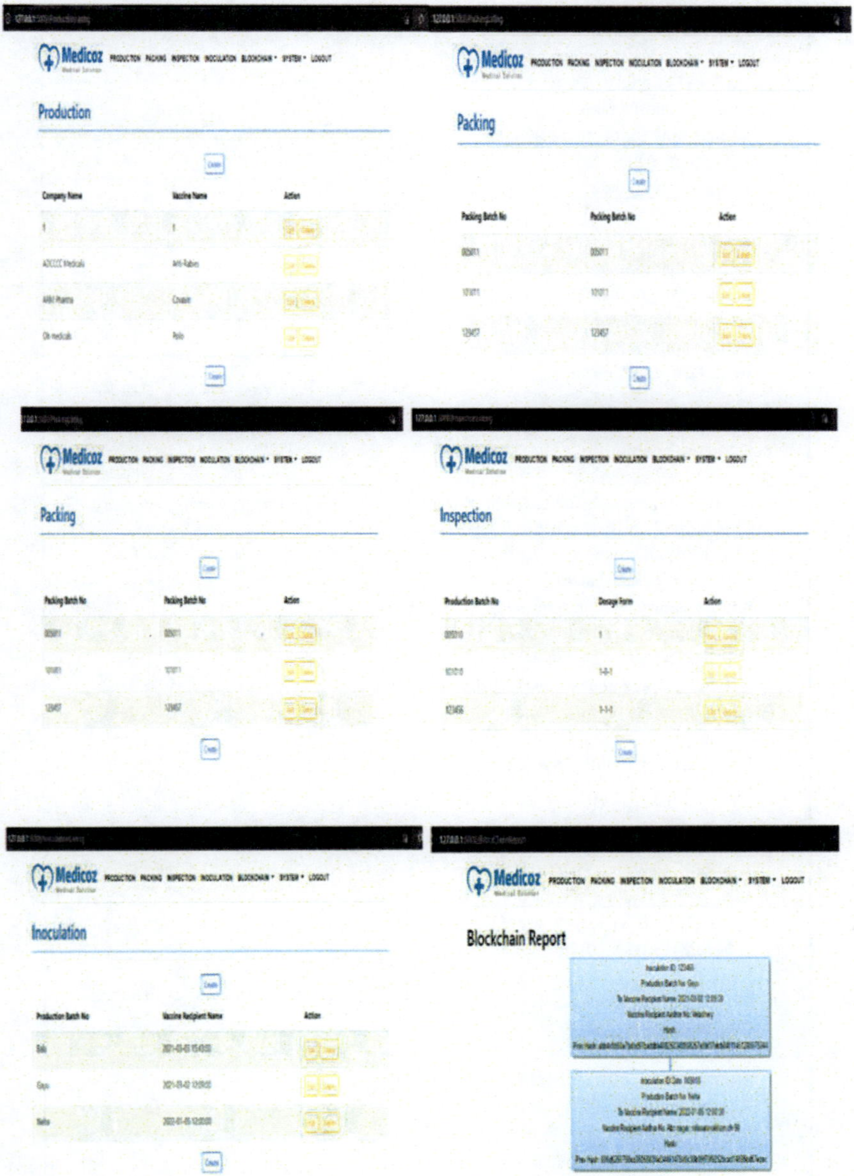

Fig. 2 Vaccine blockchain system—data storage and blockchain generation

```
pragma solidity >=0.4.21 <0.6.0;
contract ProductManagement {
  struct RawMaterial{
    address manufacturer;
    string serial_number;
    string part_type;
    string creation_date;
  }
  struct Vaccine{
    address manufacturer;
    string serial_number;
    string vaccine_type;
    string expiry_date;
    bytes32[6] rawMaterials;
  }
  mapping(bytes32 => Vaccine) public vaccines;
  constructor() public {
  }
  function concatenateInfoAndHash(address a1, string memory s1, string memory s2,
string memory s3) private returns (bytes32){
    //First, get all values as bytes
    bytes20 b_a1 = bytes20(a1);
    bytes memory b_s1 = bytes(s1);
    bytes memory b_s2 = bytes(s2);
    bytes memory b_s3 = bytes(s3);
    //Then calculate and reserve a space for the full string
    string memory s_full = new string(b_a1.length + b_s1.length + b_s2.length +
b_s3.length);
    bytes memory b_full = bytes(s_full);
    uint j = 0;
    uint i;
    for(i = 0; i < b_a1.length; i++){
      b_full[j++] = b_a1[i];
    }
    for(i = 0; i < b_s1.length; i++){
      b_full[j++] = b_s1[i];
    }
    for(i = 0; i < b_s2.length; i++){
      b_full[j++] = b_s2[i];
    }
    for(i = 0; i < b_s3.length; i++){
      b_full[j++] = b_s3[i];
    }
    //Hash the result and return
    return keccak256(b_full);
  }
```

```
    function buildRawMaterial(string memory serial_number, string memory part_type, string
memory expiry_date) public returns (bytes32){
        //Create hash for data and check if it exists. If it doesn't, create the part and return the
ID to the user
        bytes32 part_hash = concatenateInfoAndHash(msg.sender, serial_number, part_type,
expiry_date);
        require(rawMaterials[part_hash].manufacturer == address(0), "RawMaterial ID al-
ready used");
        RawMaterial memory new_part = RawMaterial(msg.sender, serial_number,
part_type, expiry_date);
        rawMaterials[part_hash] = new_part;
        return part_hash;
    }
    function buildVaccine(string memory serial_number, string memory vaccine_type, string
memory expiry_date, bytes32[6] memory part_array) public returns (bytes32){
      //Check if all the rawMaterials exist, hash values and add to vaccine mapping.
      uint i;
      for(i = 0;i < part_array.length; i++){
        require(rawMaterials[part_array[i]].manufacturer != address(0), "Inexistent part used
on vaccine");
      }
      //Create hash for data and check if exists. If it doesn't, create the part and return the ID
to the user
      bytes32 vaccine_hash = concatenateInfoAndHash(msg.sender, serial_number, vac-
cine_type, expiry_date);
        require(vaccines[vaccine_hash].manufacturer == address(0), "Vaccine ID already
used");
        Vaccine memory new_vaccine = Vaccine(msg.sender, serial_number, vaccine_type,
expiry_date, part_array);
        vaccines[vaccine_hash] = new_vaccine;
        return vaccine_hash;
      }
      function getRawMaterials(bytes32 vaccine_hash) public returns (bytes32[6] memory){
        //The automatic getter does not return arrays, so lets create a function for that
        require(vaccines[vaccine_hash].manufacturer != address(0), "Vaccine inexistent");
        return vaccines[vaccine_hash].rawMaterials;
      }
    }
```

After implementing the proposed system, the inferences derived through various Figures 3, 4, and 5 are as follows: Fig. 3 depicts the time taken for each transaction per second, as linear. Figure 4 shows the plotting of the blockchain created (in process) vs. the blockchain updated in the vaccine record, as approximately 42% of blocks updated. And, Fig. 5 shows the number of transactions per block size, as almost of linear relationship.

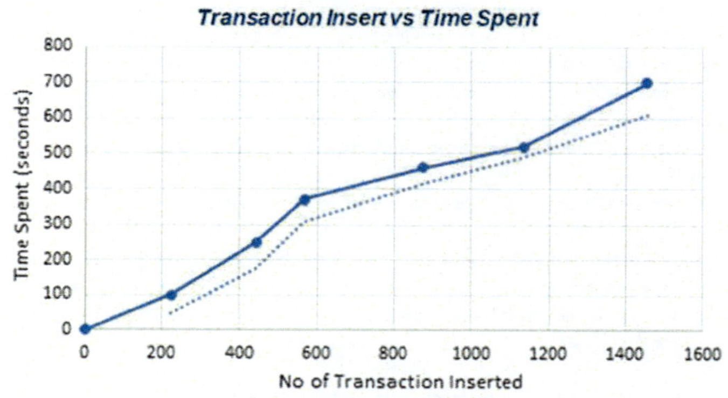

Fig. 3 Transaction insert versus time spent

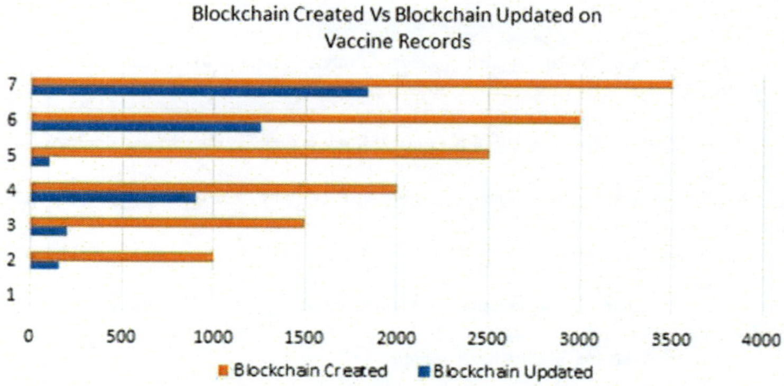

Fig. 4 Blockchain created versus blockchain updated on vaccine records

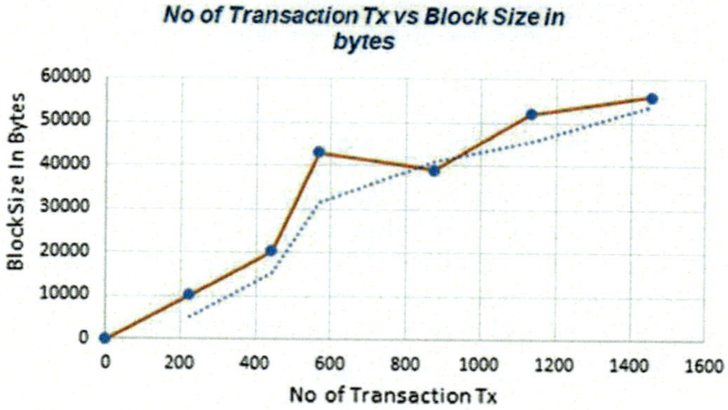

Fig. 5 No. of transaction (Tx) versus block size in bytes

5 Conclusion

The proposed system exhibits the problem of applying blockchain technology to safe vaccine production and distribution. An intelligent system for vaccine supervision is developed as vaccine blockchain system. Meanwhile, smart contracts are used for querying personal inoculation records, and vaccine circulation was designed for consumers, vaccine institutions, and the government, to trace vaccine operation records. The system has been demonstrated, and the issue of vaccine accountability thereby DoS attacks is resolved using the blockchain system. Additionally, Ethereum smart contracts are used to detect expired vaccines, while remainders of expired vaccines can be automatically sent to regulators and institutions of vaccine supply chain. This takes care of the problem of vaccine expiration, high operation costs, internal and external audit, and possibility of fraud and third-party interference. Overall, the proposed novel vaccine blockchain systems provide much better environment for safe vaccine production and distribution with improved traceability.

References

1. Jang, H., Lee, J.: An empirical study on modeling and prediction of bitcoin prices with Bayesian neural networks based on blockchain information. IEEE Access Practical Innov. Open Solutions **6**(99), 5427–5437 (2018)
2. Garay, J.A., Kiayias, A., Leonardos, N.: The bitcoin backbone protocol: analysis and applications. In: Springer Annual International Conference on the Theory and Applications of Cryptographic Techniques, pp. 281–310 (2015)
3. Hughes, L., Dwivedi, Y.K., Misra, S.K., Rana, N.P., Raghavan, V., Akella, V.: Blockchain research, practice and policy: applications, benefits, limitations, emerging research themes and research agenda. Int. J. Inf. Manage. **49**, 114–129 (2019)
4. Beck, R., Avital, M., Rossi, M., Thatcher, J.B.: Blockchain technology in business and information systems research. Bus. Inf. Syst. Eng. **59**(6), 381–384 (2017)
5. Dolgui, A., Ivanov, D., Potryasaev, S., Sokolov, B., Ivanova, M., Werner, F.: Blockchain-oriented dynamic modelling of smart contract design and execution in the supply chain. Int. J. Prod. Res. 1–16 (2019)
6. Kentaroh Toyoda, P. Mathiopoulos, T., Sasase, W., Ohtsuki, T.: A Novel Blockchain-Based Product Ownership Management System (POMS) for Anti-Counterfeits in the Post Supply Chain (2017)
7. Akter, S., Wamba, S.F.: Big data and disaster management: a systematic review and agenda for future research. Annals Oper. Res. 1–21. Andoni, M., Robu, V., Flynn, D.: Blockchains: crypto-control your own energy supply. Nature **548**(7666), 158 (2017)
8. Mettler, M.: Blockchain technology in healthcare: the revolution starts here. In: IEEE 18th International Conference on Health Networking, Applications and Services (Healthcom), pp. 1–3 (2016)
9. Ahmed, S., Broek, N.T.: Food supply: Blockchain could boost food security. Nature **550**(7674), 43 (2017)
10. Kim, M.G., Lee, A.R., Kwon, H.J., et al.: Sharing medical questionnaries based on blockchain. In: 2018 IEEE International Conference on Bioinformatics and Biomedicine (BIBM). IEEE (2018)
11. Kurki, J.: Benefits and guidelines for utilizing blockchain technology in pharmaceutical supply chains: case Bayer Pharmaceuticals. Bachelor thesis, Aalto University (2016)

12. Archa, B.A., Achuthan, K.: Trace and track: enhanced pharma supply chain infrastructure to prevent fraud. In: Springer International Conference on Ubiquitous Communications and Network Computing, pp. 189–190 (2017)
13. Kaur, R., Kaushik, V.: Silico Peptide based Vaccine Identification against Swine Influenza Virus (2018)
14. Tang, Y.-H., Lin, Y.-H., Huang, T.-T., Wang, J.-S., Hu, Y.-C., Shiao, M.-H.: Development of micro-needle array for Tumor vaccine patch applications (2019)
15. Fu, Y., Zhu, J.: Big Production Enterprise Supply Chain Endogenous Risk Management Based on Blockchain (2019)
16. Behnke, K., Janssen, M.F.W.H.A.: Boundary conditions for traceability in food supply chains using blockchain technology. Int. J. Inform. Manage. (2019). https://doi.org/10.1016/j.ijinfomgt.2019.05.025. Christi-dis, K., Devetsikiotis, M.: Blockchains and smart contracts for the internet of things. IEEE Access **4**, 2292–2303 (2016)
17. Dhandapani, J., Uthayakumar, R.J.A.: An EOQ model for a high cost and most wanted vaccine considering the expiration period. J. Anal. **27**(1), 55–73 (2019)
18. Pass, R., Seeman, L., Shelat, A.: Analysis of the blockchain protocol in asynchronous networks. In: Springer Annual International Conference on the Theory and Applications of Cryptographic Techniques, pp. 643–673 (2017)
19. Hjálmarsson, F.Þ., Hreiðarsson, G.K., Hamdaqa, M., Hjálmtýsson, G.: Blockchain-based E-voting system. In: 2018 IEEE 11th International Conference on Cloud Computing (CLOUD), San Francisco, CA, USA, pp. 983–986 (2018). https://doi.org/10.1109/CLOUD.2018.00151
20. Satoshi Nakamoto satoshin@gmx.com www.bitcoin.org "Bitcoin—Peer to peer electronic cash system"
21. Casino, F., Dasaklis, T.K., Patsakis, C.: A systematic literature review of blockchain-based applications: current status, classification and open issues. Telematics Inform. **36**, 55–81 (2019)
22. Buterin, V.: A Next-Generation Smart Contract and Decentralized Application Platform (2015)
23. Rouhani, S., Deters, R.: Security, Performance, and applications of smart contracts: a systematic survey. IEEE Access **7**, 50759–50779 (2019)

Utilizing Stage Change of Subjects for Event Discovery in Online Social Networks

Sanjeev Dhawan, Kulvinder Singh, and Amit Batra

1 Introduction

There are several categories of social networks fulfilling several objectives, for example, friendship systems (Facebook, Instagram), career development systems (LinkedIn) between others. Social media nowadays is indispensable place which helps in sharing of knowledge, social communication, and other tasks. Twitter is the usually employed social networking site. Twitter is used as a microblogged social networking site that permits individuals to transmit and obtain messages in texts with number of characters limited to 280 termed as "tweets." Whereas few common subjects are updates of an individual state, a noteworthy part of these messages is a reaction caused by events like political events (e.g., elections [1], protests [2]), natural disasters [3] (e.g., tsunami and earthquakes [4]), and so forth. These social networks particularly generate vast amounts of individual-produced data pertinent to scenario reports and can be enormously employed for several category of applications, for example, social network analysis [5], event stream distribution [6], information gathering [7], recommendation systems [8], networks modeling [9], and origin recognition of rumors [10]. The discovery of an event can be illustrated as observing a particular trend in information aggregation (particularly a stream of information). The objective will be to recognize the main narrative to talk about a specific instance that occurred at a particular period of clock and location. Trend or a deformity in the information can be considered as event. To consider the case of social networking

S. Dhawan · K. Singh · A. Batra (✉)
Department of Computer Science and Engineering, University Institute of Engineering and Technology, Kurukshetra University, Haryana, India

S. Dhawan
e-mail: sdhawan2015@kuk.ac.in

K. Singh
e-mail: ksingh2015@kuk.ac.in

M. Dua et al. (eds.), *Proceedings of the International Conference on Paradigms of Communication, Computing and Data Sciences*, Algorithms for Intelligent Systems,
https://doi.org/10.1007/978-981-16-5747-4_35

sites, it is pertaining to the recognition of the main narrative on a subject of concern by continuously keeping an eye on news streams. Dou et al. [11] explain an activity as "a happening resulting in a modification in the volume of text information that talk about the concerned subject at a particular period of time." As a matter of fact, Twitter can be observed as an indispensable origin of information beneficial to know about a variety of events occurring across the world, where each individual is considered as a possible informer. In spite of the fact that Twitter is an indispensable origin of precious information in contrast to conventional media news, the examination of Twitter data brings few problems, as mentioned below: Since there is a limit on the number of characters a tweet can contain, a tweet particularly represents a content that is short comprising of amorphous sentences and abbreviations. Tweets produce a great volume of information that need extensible techniques to interpret their content. Time-sensitivity of tweets, i.e., these are posted in synchronous manner and have a major association with the clock these are shared. Sometimes, tweets do not represent beneficial data. The processing of these tweets not only incurs a cost in terms of time needed moreover also deteriorate the standard of the outcomes. In particular, as per "Parikh and Karlapalem [12], few of the tweets are fruitless and do not represent any useful information." Sometimes the same are pertinent to private content change of individuals, however, even are spam. We noticed that currently employed techniques are susceptible to break down whenever the basic model modifies, and the earlier estimated model is not sound anymore. Discovering an event is a cumbersome job because of the inherent features of tweets. Most techniques take into consideration the task to observe outliers (anomalies) and furthermore examine if these outliers are associated with events [13–15]. If an assumption is made that a prototype modifies, the job of discovering an activity is depicted as a stage change discovery. Few subtle problems for the aforesaid category of technique will be to make an attempt to distinguish a modification in framework active elements from an abnormality or malformation. If dynamic change is observed, the model cannot be considered as stationary and, thus, should be handled employing adaptive approaches [16]. Therefore, happening of an event can be associated to a modification of the active element of the basic framework. This signifies tweets pertinent to a happening have features of a novel framework and, therefore, cannot be taken as an anomaly. If the random anomalies are detected (outlier discovery) [17], the discovery approaches take the system as stationary. The modification of active elements in non-static surroundings comprise of a consequence termed as covariate shift. The modification in the dynamics of information dissemination through the course of time can be noticed in several course of actions. Adaptable study pertains to modifying extrapolative techniques online in the course of their functioning to respond to covariate shifts [16]. This can happen quickly (for instance, as shifting takes place from one detector to the other with a distinct accuracy) or gradual; a detector is obviously having calibration steadily reduced for a period of clock. Our technique is centered on the determination of the information about the significant words withdrawn from the text of the tweets to categorize the significant subject as a happening. This work suggests an innovative technique, which takes the advantage from the phase change discovery approaches to discover events on Twitter. This suggestion is an addition in the earlier work done in [18]. This inclusion enhanced the subject of rendering outcomes. Furthermore, six datasets are added by us, out of which one was earlier tagged and the remaining untagged, resulting in additional in-depth capability analysis. In contrast to the earlier work done, an attempt is made to improve and enhance the topic cum subject withdrawal steps by suggesting a summary graph partitioning approach illustrated in Part 3.2. Ultimately, an attempt is made to suggest a systematic technique of the active elements of rumor proliferation in social media to prove our technique. Toward the aforesaid approach, the major works are depicted underneath: An attempt is made to provide a proof that the phase changes are performed continually; i.e., these are categorized as second-order changes. As far as our knowledge is concerned, no work has been proposed in the existing literature which gives formal proof pertaining to this theory. An attempt is made to represent the event discovery task as a stage change detection task, and experimental proof of this stage change is provided and revealed a standard representation of the aforesaid reaction. By employing such representation, we were in a situation to fine adjust the boundary variables of our event discovery technique, contracting the exploration room of aforesaid variables. With the help of this, we can recognize the modification in the system dynamics. An attempt is made to suggest a graph to outline major significant words pertinent to a happening and employ a network subdivision approach to take out keywords of a particular happening. The aforesaid task is structured as given below: Part 2 depicts the pertinent study for discovering event employing "Twitter." In order to carry out this task, an attempt is made to form groups of significant words that are more associated to each other, and we also eliminate unimportant words that in the end can be allocated to a group. Section III depicts the method and approach to analyze the data; Section IV depicts the major outcomes, besides few discussions on phase changes; and Section V depicts the conclusions and future research directions. An attempt is made to employ a technique of the dynamic of rumor proliferation in social networks. The set of equations illustrated in this paper was solved through the Monte Carlo technique. This technique was featured by three differential equations defined in the SIR model. By employing the Monte Carlos outcomes, we were in a position to represent the phase change available in the SIR technique, and we observed that the changes in actual datasets have identical features as the changes observed in the artificial technique, therefore, supporting the proof for theory presented in our work.

2 Related Work

Suggested in [19], TEDAS is an approach that discovers and examines activities in "Twitter." This approach serves the following responsibilities: discovery of recent activities, categorization of activities as per their significance, and spatial (or temporal) production of a trend for the event. In [20], the authors examined content posted on Twitter by withdrawing characteristics that depict the surrounding factors. The authors discovered an earthquake with a high probability. They formulated a probabilistic spatiotemporal technique, assuming each individual as an equipment and utilized a bit reordering to obtain the center and path of place of happening. In a similar fashion, in [21], the authors suggested an irregularity classification utility that describe detector irregularities employing social media content, for example, automotive congestion collisions. In [22], the authors suggested an approach centered on extraction of the text with machine learning techniques to discover activities synchronously pertinent to trafficking on Twitter. The aforementioned investigations [19–22] pertain to a category of approaches that takes into consideration a particular context of the information. EDCoW [23] (Event Discovery using "Clustering of Wavelet-based Signals") is a suggestion that depicts keywords in the form of signals, centered on [24], by employing ripple examination of the keywords. Afterward, it excludes insignificant keywords by observing their respective wave associations. The left-over keywords are afterward grouped to constitute activities using a modularity-centered network subdivision approach. If any category of event is considered, because of the necessity to employ a transformation centered on wavelets along with modularization of the graph, EDCoW needs a huge volume of calculation, thereby, making it an unscalable choice. Like in several applications pertinent to NLP, the techniques which are particular to a specific area usually do the best than the techniques that are general or open-domain; however, their utilization is restricted only to the suggested subject like conditions of the weather, traffic collisions, and others. Dissimilar to this category of technique, our proposal suggests a novel approach for discovering an event on Twitter that takes into consideration any event irrespective of its scenario. Our suggestion takes into consideration that concurrent events can happen simultaneously; therefore, an attempt is made to assume that the summary graph subdivision approach employed can make a distinction between events. Additionally, because of the cross-correspondence matrix, it takes into consideration sets of activities, and the suggestion is not able to make a distinction among several events that occur at the same time, for example, 2 cricket matches that happen at the same time. Mathioudakis and Koudas [25] suggested the Twitter Monitor. Apart from discovering events on Twitter in real-time, they have given significant examination that synthesizes an explanation of each topic. Authors in [27] depicted Tweetvent, a technique that discovers occurrence of tweets at intervals in short within a predetermined time window as events. They recognized trending keywords and clustered them as per their concurrence, utilizing a context withdrawal technique centered on [26]. In [29], the authors suggested an approach centered on effective Bayesian Systems [30]. The author's approach utilizes the information pertaining to tweets and examine the subject spreading job and observes two major features of an apparent subject, namely key-node and attractiveness. They clustered them centered on their concurrence employing DBN-centered approach. Additionally, they group event slices into events by taking into consideration both their similarity in content and frequency distribution employing a variation of the Jarvis-Patrick approach [28]. Every group or cluster is matched with Wikipedia topics to recognize actual-life events. In [31], researchers forecasted riots utilizing Twitter employing few characteristics obtained from posts (spatial, temporal, and text) in the Twitter. They made an assumption that happenings could occur in the same place for a particular period of clock or several happenings in differing places. In [33], the authors suggested a method that employs a measure centered on term "frequency-inverse document frequency of the bigram" in a particular period of clock and builds a categorization of the topmost likely activities. The author's technique presumes. As preliminary data that a predetermined specified number of activities are occurring in a span of clock and gives a categorization of the topmost likely happenings with corresponding significant words. In [32], the authors suggested a novel approach for discovery of an event, by taking into consideration every individual of a social networking site as a detector. As a matter of fact, since they take into consideration the independence of detectors, the researchers established an evaluation score that relates the uncommon happening in a social information flow. In [34], the authors suggested a data entropy-centered incident discovery approach to recognize incidents and the place of their occurrence by grouping the large density of tweets employing Twitter information. The Shannon entropy of selected individuals, place, clock "intervals, and hashtags" are predicted to evaluate the distribution of happenings employing the "entropy" optimization speculation technique. The geo-labeled tweets are drawn out over a particular clock interval to recognize the place of a happening, and the events are visualized in maps of geography. In [33], the authors suggested a prototype of probable incidents, i.e., the count of subjects is specified, and the technique gives the topmost probable *k* subjects over a provided time interval. This category of technique needs non-automatic involvement to tag whether the obtained subjects are incidents or not. In contrast, our suggested method suggests a classifier that instinctively determines whether an event happened. To analyze the suggestion, they employed entropy, Event Discovery Hit, Cluster Score, and False Panic Rate in the course of four significant disaster events, which are recognized to depict the efficacy of the suggestion. The result of the experiment decides the scope and remarkable spreading path of recognizing incidents from a novel viewpoint that illustrates 96% of enhanced incident discovery correctness. We observe that entropy-centered techniques for discovering incidents on twitter as illustrated in [34] are considerably employed in NLP employed to incident discovery in twitter [35–38], slthough no rationalization has been observed pertaining to this certainty. In this proposal, a guess is made that twitter possess a stage change of second order. By employing this portrayal, we are in a situation to fine-tune the variables of our technique with more efficacy. Investigations as aforementioned depict that discovering Twitter events employing emerging topics is achievable. It is significant to observe that the phase change supposition is powerful than recognizing outliers. The present proposal is motivated by the utilization of different techniques as aforementioned, for example, bigrams and their likelihood; however, in contrast to the work already done, a benefit of the truth that entropy is a metric of amount of data to depict the happening of an incident is taken, and utilizing this, a new technique is employed that discovers the stage change of the entropy of subjects. As a matter of fact, an attempt is made to discover the modification in model active elements and appropriately categorize a subject pattern as a happening.

3 Methodology

3.1 Datasets

An attempt is made to prove our proposal by employing two datasets gathered by the authors in [33], and they collected tweets pertaining to incidents happening worldwide in 2012:

The FA Cup Final dataset: FA Cup is the significant contest of football and pertains to the older football league. This dataset comprises of tweets pertaining to the FA Cup. In 2012, "Chelsea and Liverpool" participated in a final match, where Ramirez and Drogba scored one goal each for "Chelsea and Carrol" scored for Liverpool. Therefore, Chelsea defeated Liverpool. The match was played for 90 min along with half time break of 15 min. Figure 1 depicts the curve of time versus gathered tweets pertaining to the FA Cup. Every specimen comprises of tweets in sixty seconds. General elections of Mexican: Elections in Mexico conducted in 2012. For this election, the voters made their use of votes. SXSW-2012: It is a group of film, melody, and engineering carnival conducted each year in USA—Cyprus kidnapping (2016): A man of Egypt kidnapped an airplane in Cyprus. The Super Tuesday Primaries dataset: It pertains to Tuesdays earlier in United States presidential election when the primary elections hold in the highest number of states. The president in USA is chosen in an indirect election and decided by the "electors of the Electoral College." In 2012, Super Tuesday occurred on March 6, 2012, with 419 delegates. An attempt is made to depict the curve of time versus gathered tweets pertaining to this occasion, where every specimen comprises of the count of tweets. In [33], the researchers built the above datasets employing the formal incident hash labels. The authors built the empirical evidence by examining the popular press news to recognize major subjects for every information. The description of the dataset can be obtained from their manuscript. They recognized twenty-two and thirteen significant subjects for Super

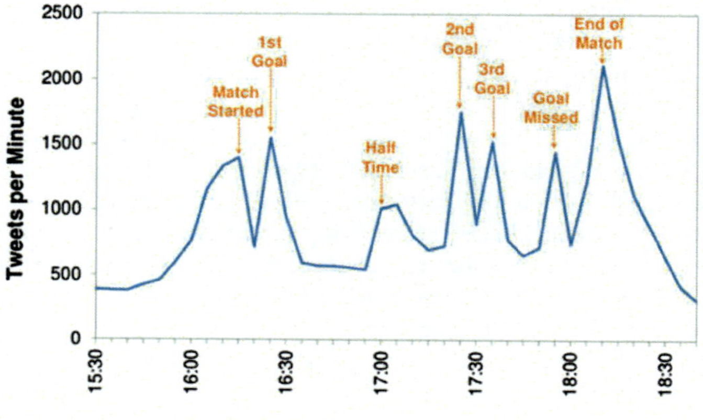

Fig. 1 Tweets pertaining to FA Cup

Tuesday and FA Cup datasets, individually. Apart from the aforementioned datasets, few non-tagged datasets gathered have been employed. It comprises of tweets for five actual incidents. Ebola spread (2014): One disastrous event in the history was spread of Ebola virus, resulting in major casualty of lives and socioeconomic disturbance in Africa. The focal points were majorly in the countries of Liberia, Guinea, and Sierra Leone—Sismo Ecuador: A disastrous earthquake occurred in Ecuador in 2016. The magnitude of earthquake was observed as 7.8 Mw. The regions that were nearer to the epicenter were destroyed. These five datasets were selected to prove our assessment, like general elections and melody carnival, irrespective of the dialect employed (English or Spanish). Going along with Aiello et al., an attempt is made to carry out data cleansing in the dataset, eliminating stop words (e.g., pronouns and prepositions), mentions, punctuation marks, and URLs. Furthermore, all the letters are lowered, so as to normalize the writing.

3.2 *Stage Change Centered Entropy (SCc Entropy)*

The concept of tweet slice has been suggested for labeled element identification, not pertinent to incident discovery. An attempt is made to suggest a technique to recognize actual incidents on Twitter by discovering the subjects of significance introduced in a particular clock period, where our approach is centered on the determination of information about the time series that sum up the rate of occurrence of the significant words withdrawn from content of tweets by using a concurrence algorithm. In a same fashion, we take into consideration a "tweet" to be a group of "bigrams." A bigram is a series of two adjoining components from a sequence of symbols, which are particularly syllables, words, or letters. This approach can be employed to discover any event since no prior data pertaining to the event is taken into consideration. As a matter of fact, every tweet is divided into words (tokenize). Thereafter, we construct the bigrams for every tweet. Our approach comprises of six phases or steps illustrated in sections outlined below:

Determining Bigrams. Assume f: G - > Y is an operation which transforms the bigram $g \in G$ to the count set Y for all time periods in the dataset. The function f is provided by $f(g) = \{y_t \in Y | y_t = \#g_t\}, \forall\, t \in \{s_0, s_1, \ldots, s_M\}$, t is group of all time periods in a observations set of dimension M, $\#g_t$, is the frequency of the bigram g at every time t in the dataset. All bigrams are contained in the set G.

Building the Time Series. Since the proposal has been designed for real-time processing, an attempt is made to gather firstly a group of observations within a slot, and then the slot is advanced ahead to analyze additional information. The window is defined by $X^{j:m} = \{Y_q\}$, where $q \in \{s_j, s_{j+1}, \ldots, s_{j+m-1}\}$ is a time slot within X, s_j is the beginning time, and m denotes the count of components of X. Y_q defines the group of number of bigrams at interval q. Therefore, G^X, Y^X, and $f^X(g)$ are firstly determined for the 1st window, and these sets are determined continuously for the succeeding windows. The window of size m is slide by single time slice, in a manner that the gap between the successive slots is precisely one perception. Moreover, for every acquired bigram, we determine the likelihood of its happening in a particular time slot. It is noteworthy that everywhere in our approach, the value of m is fixed. An attempt is made to construct the time sequence of frequencies $F^X(g)$ = $\{f_q(g)\}$ constrained to a window X as

$$f_q\ (g) = \frac{y_q(g)}{\sum_{k \in q} y_k(g)} \tag{1}$$

where $f_q(g)$ depicts a perception of the time sequence of rate of occurrences related to every bigram in the respective time slot at interval q. Equation (1) transforms $Y^X \subset Y$, the group of numbers constrained to X, into a "vector of frequencies" F^X. The likelihood cum probability of occurrence of the bigram g over a particular time slot q is depicted by $\hat{f}_q = f_q$.

Determining the Entropy. In order to determine the information about a bigram time sequence, we in the first place observe $\hat{f}_q$ as the guess of the likelihood about frequency of the bigram g at every time period q in X. Hence, assume $\hat{F}^X$ as the trajectory of the guess of likelihoods of frequency of every bigram in X. Therefore, the Shannon entropy for every X is illustrated as

$$E_g^X = \sum_{k \in q} -\hat{f}_k(g)\ \ \log\left(\hat{f}_k(g)\right) \tag{2}$$

Analysis of the Entropy. The entropy E_g^X gives an estimate of the amount of data at every window X. Therefore, "windows" that represent large "entropies" are probably to depict the happening of an incident. Centered on the worth of the entropy, it can be inferred that if a time slot represents an irregular behavior, i.e., an incident, for a particular bigram g. In this scenario, it is concerned with discovering a stage change within a time slot that did not represent an incident and a time slot that represents an incident. An attempt is made to talk about how we discover the phase change in Section IV. If an incident is discovered in the time slot X, then the incident is allocated to the time s_{j+m-1}, the rearmost perception in X.

Building the Summary Graph. For this, an attempt is made to suggest an alternative form of the frequent pattern-growth approach initially incorporated to conform the approach to employ bigrams. By taking into consideration that if a stage change is discovered, i.e., an incident, about a particular bigram g_j at clock period s, it extends the presentation of the bigram employing an approach usually utilized in frequent itemset mining. The FP-growth algorithm is well organized and extensible approach for extraction that employs an expansion of the prefix-tree arrangement for compact information repository termed as frequent-trend tree. A concurrence data of the elements is returned by the algorithm; however, in our suggestion, we choose the happenings on the second height of the tree, i.e., bigram. Using the approach, sets of bigrams are found which are pertinent to g_j, i.e., these probably occur in the "tweet" as g_j. "Levenshtein distance" can be seen as a computation measure to calculate the dissimilarity between two series. Furthermore, to mitigate the chosen bigrams to a small-scale group with "bigrams" which are nearer to every other, i.e., are largely pertinent, an attempt is made to reject the bigrams that are more faraway to g_j. More particularly, for word-based components, we can specify the distance as the minimal count of edits which are single-letter edits (deletion, insertion, or exchange) needed to convert one word-based component to other. In order to carry out this, assume leven as the Levenshtein distance, an attempt is made to specify the distance between the bigrams $\beta = (\beta 1, \beta 2)$ and $\delta = (\delta 1, \delta 2)$ as

$$\text{Distance}(\beta, \delta) = \text{minimum}\ \{\text{leven}(\beta 1, \delta 1) + \text{leven}(\beta 2, \delta 2), \text{leven}(\beta 1, \delta 2) + \text{leven}(\beta 2, \delta 1)\} \quad (3)$$

where $\beta 1$, $\beta 2$, $\delta 1$, and $\delta 2$ are words of the bigrams β and δ. But an attempt is made to utilize the aforementioned distance specified in Eq. (3) in the bigrams acquired by using the alternate form of frequent pattern-growth in the time slot X, and we acquire all groups of bigrams with distance to bigram g_j having value lesser than μ, a random value of threshold. By utilizing this group of bigrams, a graph is built where words are represented by the vertices, and an edge is drawn between two words when they pertain to the alike bigram. In our proposal, we employed $\mu = 8$, as it will be illustrated in Sect. 4. In the end, a graph is returned by the algorithm pertaining to the aggregation of the incidents linked with the period s.

Summarized Network Partitioning. We divide the summarized network to eliminate unnecessary keywords that in due course occurs in the earlier stage. This task assists to recognize the incidents accurately. Next to the division of the network, an attempt is made to take into consideration that every resulting associated component having more than two elements is related to an event. To carry out the division, we employed a task known as "MCL." The approach illustrates a series of "stochastic matrix" tasks (expansion, "inflation") termed as operant. Hence, we can recognize more than one event within the same window using our suggested approach. The major plan of the MCL approach is to imitate the due course in a normalized network. In order to perform this, it employs an arbitrary walker and determine the count of number of times the walker moves by every link (current). The approach enhances the flow wherever strong current is there (the node obtained several traverse) and reduces the flow wherever current is weak (the node obtained less traverse). The approach employs the operant to enhance/reduce the flow. The growth operant prefers the routes which are shorter, i.e., arbitrary traversals with smaller number of stages, stimulating the traversal to new sets. Hence, the growth operant is accountable for permitting the flow to link several areas of the network. The aforementioned operant links new likelihoods to all sets of vertices, reducing the likelihood for lengthy traversals and enhancing the same for shorter traversals. The inflation operant is accountable for both making stronger and weaker the flow of current. The consequence of inflation will be of increasing the odds of rides with in the groupings and reducing the rides among groupings. An attempt is made to utilize this algorithm since it permits two vertices within two groups individually. This is attained without any early information about the structure of the groups. Put it differently, although if two different incidents have same words, the approach has the capability to divide them.

3.3 The SCc Entropy Algorithm

The algorithm determines the entropy of a time slot and determines if this time slot pertains to a stage change amid the presence and absence of an incident. The pseudocode of our suggested approach is illustrated in Procedure 1. In order to carry out this, it obtains the index of the beginning time (j_0) of X as input and can use the tweets that relate to every window, and a sequence of discovered significant words is returned for every incident discovered in X. As a first step, the set G^X is initialized mining the bigrams in X. For every window, it determines number of bigrams in X and ranks in descending sequence. The procedure visits G to determine the entropy and likelihood vector of X.

Procedure 1 The suggested approach for Event Discovery in Online Social Networks Centred on Stage Changes

```
Data: j0, the index designating the beginning period in X
Outcome: A sequence of discovered keywords for every event discovered
for j = j0 to j0+m-1   do
    initialize G^X
    for g ∈ G^X  do
        for q ∈ {s_j, s_j+1,…., s_j+m-1}   do
        /*building Y^X                              */
            y_q = f^X(g_q)
        end
    end
    rank (G^X, category= 'reduce', by= 'y_q')
    list_bigrams = ∅
    for g ∈ G^X_0:199   do  /* 200 bigrams in G^X    */
        for q ∈ {s_j, s_j+1,…., s_j+m-1}   do
        /* Determining the likelihood vector F^X(g)*/
            f̂_q = y_q / Σ_{k∈q} y_k
        end
        E^X_g = 0
        for q ∈ {s_j, s_j+1,…., s_j+m-1}   do
        /*determining the entropy of X*/
            E^X_g = E^X_g + (−f̂_q log(f̂_q))
        end
        if 0.1 < E^X_g < 0.7 and y_{j+m−1} > 10 and
        max(F̂^X) = = f̂_j+m-1 then
            list_bigrams = add (list_bigrams, g)
            list_return = FP-growth (g, G^X, X, μ = 8)
            list_bigrams = add (list_bigrams, list_return)
        end
    end
    network = buildGraph (list_bigrams)
    network = MCL (network)
    keepIncident (network)
end
```

We restrict to 200 more representative bigrams to eliminate litter. For every bigram in $G^X_{0:199}$, it builds the group of likelihood of the bigram at every window in X and determines the successive time slot entropy. It discovers if a time slot represents a stage change to an incident. In the matter of discovery, we take into consideration for the network building, the analyzed and the associated bigram, employing the approach illustrated in Part 3.2.5. We select the entropy discovery time period of $0.1 < E^X_g < 0.7$ to discover the incident by examining the Receiver Operating Characteristic curve of our discovery approach, as illustrated in Part 4. At the end, the procedure comprises of the building of the summarized network, which is built centered on the sets of bigrams provided by the frequent pattern-growth approach in X. The guideline $\max\left(\hat{F}^X\right) = \hat{f}_{j+m-1}$ make sure that the information about the rearmost window has the greatest value, giving the indication of happening of a stage change. This phrase circumvents false positive discovery for incidents that were recognized earlier. The MCL is employed in the summarized network. The complexity in terms of time of our suggested approach is $O(|G^X|^3\ |X| + |G^X|\ |X|\ |g|^2 + S(2\ |G^X|^{2.807}))$, where $|G^X|$ is the quantity of bigram for a corresponding time slot X, $|X|$ is the count of windows, S is the count of stages needed for convergence for the network "clustering" approach (MCL), and $|g|$ is the count of "letters" of the greatest "bigram" g in time slot X. But, practically since S is particularly small (less than 30), and $|g| < |G^X|$, hence, it can be deduced that our approach has "time complexity" of $O\ (|G^X|^3\ |X|)$. In spite of the fact that the time complexity is cubic on $|G^X|$, that is, the count of bigrams in X, as depicted previously, we take into consideration at most 200 bigrams, as we experimentally examined that the rate of occurrences of bigrams is little (particularly lesser than 5) whenever count of bigrams taken into consideration is larger than 150, considering all sets of information examined in this proposal. These bigrams are not probably to be pertinent to incidents.

3.4 Analysis

The ground truth comprises of a set of significant words that specify a subject. The suggested approach returns a group of keywords with which the comparison of ground truth is made. To examine our outcomes, we employ the same group of measurement metrics depicted in [33]. We select the aforementioned measures since we employed the alike dataset, so the outcomes are straight away equivalent to their outcomes. The measures are as follows: Subject recall (S-Rec): % of ground truth incidents efficiently discovered, that is, the true positive rate for incident discovery. S-Rec = (ground truth Subject ∩ Discovered Subject incidents)/(ground truth Subject incidents)—Significant word Accuracy (SW-Accuracy): % of accurately discovered significant words over the sum of significant words for a provided ground truth incident, that is, the true negative rate for significant word discovery. SW-Accuracy = (ground truth Significant words ∩ discovered significant words)/ (discovered significant words) – Significant word Recall (SW-Rec): % of accurately discovered significant words over the sum of significant words for a particular ground truth incident, that is, the true positive rate for significant word discovery. SW-Rec = (ground truth significant words ∩ discovered significant words)/(ground truth significant words) – F1-Score (K-Score): For a good contrast among the approach, we employed the F1-score for significant words measures. K-Score = 2. (SW-Rec.SW-Accuracy)/(SW-Rec + SW-Accuracy) It is noteworthy that an attempt is made to determine the values of the above measures for every window.

4 Outcomes and Discussion

4.1 Stage Change

In this approach, an analogy of an epidemic contamination with resistance is employed that illustrates whenever a person is infectious with a disease, afterward a particular period of time, the person is healed, and immunity is restored. In the beginning, to examine the phase change, an attempt is made to represent the data distribution behavior in a social network. This behavior is caught by SIR technique. Therefore, it is taken into consideration that different persons in a social network can be in three different situations pertinent to a slice of data, that is, a presentation of the incident: Condition 0 (V): It is not known about the occurrence of an event by an individual (Vulnerable).

Condition 1 (I): The person has information pertaining to occurrence of an incident, and the individual was infectious (infected); therefore, the individual begins to make posts pertaining to it, resulting in propagation of the information pertaining to the incident in the graph (infectious). Condition 2 (R): The data is known to the individual, but due course of time, the information has no impact on the behavior of an individual (obsolete), and the individual does not disclose it in the graph (recuperated). The approach comprises of group of actions outlined below:

$$V \xrightarrow{c} I \xrightarrow{\gamma} R \tag{4}$$

where c represents the rate of contamination, it is taken into consideration the likelihood of suffering from illness when a vulnerable person comes in touch with an infectious individual, and γ is the rate of recovering from the disease, that is, if the tenure of the disease is T, then $\gamma = 1/T$, because a person gets recovered in T slices of time. The VIR approach is depicted by three differential equations:

$$\frac{\mathrm{d}V}{\mathrm{d}t} = \frac{-bIV}{C} \tag{5}$$

$$\frac{\mathrm{d}I}{\mathrm{d}t} = \frac{bIV}{C} - \gamma I \tag{6}$$

$$\frac{dR}{dt} = \gamma I \tag{7}$$

where C is the count of individuals in the graph. This approach for examining rumor has already been employed by Zanette, but the work did not examine correspondences with actual information. Because our objective is to examine changes in social networks, an attempt is made to employ the small-world graph approach that comprises of an intricated abstraction approach for social interactivities. Hence, we examined the stage changes present in the VIR approach illustrated aforementioned, put in an application to several occurrences of small-world graphs that depict the social interactivities. The group of Eqs. (5), (6), and (7) depicts the framework, however, cannot have straight away solution in case of particular aspect. Therefore, we can have good understanding of how these phase changes conduct in actual datasets. For this, we represent a pertinent crucial stationary exponent. Therefore, we employed "Monte Carlo" abstraction to predict the state R, which is responsible for the count of recuperated topics whenever there exist no more subjects that may be infectious (static rule). The system is modeled in the Monte Carlo simulation as a small-world graph, such that every node depicts an entity, and the links between entities depict contacts. We executed the program number of times so that numeric convergence is reached, and the results obtained are statistically relevant. In this manner, every subject can be represented by one of three conditions depicted by the VIR approach, and the procedure execution is done till the static condition is achieved, that is, till there remains no infectious persons. With help of outcomes of the Monte Carlo procedure execution, we predict the order variable, the static density of recovered persons, as $\sigma\ (c, S) = \langle C_R\ (c, S)\rangle/S$, where $\langle C_R\ (c, S)\rangle$ is defined as the average of the count of persons in condition 2 (R) in a small-world graph of size S in the static condition, that is, whenever the count of persons in condition 1 (I) becomes zero. In this proposal, it is assumed $\gamma = 0.25/25$ as in earlier research. Moreover, we observe in existing research that the selection of several values for γ brings a modification at the crucial point; however, the behavior of the framework remains to stand unchanged, that is, pertain to the alike category of existence everywhere. We observe that this approach comprises of a continual stage change because the density develops continually from the absorptive condition (small density) to the functional condition (large density). We observe the development of the framework via Monte Carlo execution, where an attempt is made to employ different dimensions of network $S = \{103, 203, 503, 1003, 2003\}$. The trial is repeated 15,000, 10,000, 7000, 5000, and 2000 times, respectively. We select aforementioned values of S to depict three infectious persons arbitrarily chosen between 103 persons. We give an account of our dataset for the order variable variations, $\Delta\ \sigma\ (c, S) = [\langle C_R\ (c, S)^2\rangle - \langle C_R(c, S)\rangle^2] \cdot S$, predicted with Monte Carlo executions, and it depicts the differing order variable variation at a significant point. This characteristic is usually illustrated in the existing research, and it is feasible to create depiction pertaining to phase change, like differing order variable variation, a non-finite association length, and a power-law deterioration nearby crucial state. For obtaining the accurate place of the significant point b_C, we determine the proportion of the "second moment" and the "square of the first moment" of the count of persons in condition 2 (R), depicted as $m_S = \langle C_R(c, S)^2\rangle/\langle C_R\ (c, S)\ \rangle^2$. m_S versus b can be observed. The singular behavior nearby the significant point can be illustrated as a power sequence. The significant exponent β is described as $\sigma\ (c, S)\ _{c\text{->}cd}\ \alpha\ (c - c_d)^{\beta}$. The log–log graph of the σ (c, S) versus $(c - c_d)$ is depicted, and it is observed that $\beta = 0.471$, $S = 5003$, and 1000 reiterations. It has been examined now in what manner information about the bigram with time slot acts with information of the tweets. With this work, we represent the stage change depicted by the VIR employing the Monte Carlo approach obtaining the numerical solution. As illustrated in Part-3.2, aim is to discover the stage changes about the information among successive time slots. The active elements of the entropy of the typical bigram (the bigram having many happenings) of the FA cup dataset are considered. As we determined the "entropy" of a time slot, we take into consideration that "the entropy of all time slots" within a window remains unchanged. Note that afterward time 5000, the model begins to represent other active element at which point the entropy is larger than earlier. We observe that the resemblance is of curved form (S-form) and represents the features of second-order stage changes. It is assumed as order variable σ equals E_g^X. We saw few disconnections because the suggestion requires the time to be discrete such that posts are tweeted to build the time sequence of rate of occurrence of bigram. The exponent β in the neighborhood of the significant point $s_d = 5000$ as the slant of the shaped line $E_g^X(s)\ \alpha\ (s - s_d)^{\beta}$ in a log–log graph of $E_g^X\ (s)$ versus $(s - s_d)$ is predicted. The significant exponent $\beta = 0.4788$ tells about the continual stage change, as depicted by the researchers. The parameter β is near to the outcome acquired in "Monte Carlo" execution. This significant exponent is represented by a steady transient that could be an outcome of the inherent active element of the social network commenting for events alike live games. Hence, the stage changes represent alike behavior. In such scenarios, few individuals post instantly after the happening of an incident, and few individuals need small period of time prior to commenting; therefore, the model modifies continually and at a slow pace. To discover steady transients, we adopt that when the entropy goes from small to large within the min and max (second-order characteristic), a change can be discovered. Therefore, the entropy discovery period should not include the maximum value. Note that we are not interested in discovering the transient only whenever the entropy achieves its largest value. Hence, because the approach is dependent on some parameters (entropy span, sliding window size m, summary network threshold μ), an examination is carried out to calculate what value of this parameter will optimize precision. A work suggested an approach termed as "Bayesian Optimization," which comprises of optimizing tasks like a "black box." Commonly, this estimation is done through a Gaussian process because of some features (expanding to some points and not parametric). The approach comprises of, with few familiar points, deciding the shape of the function utilizing regression. Therefore, centered on the regression of the Gaussian task, a utility function can be specified which comprises of determining the next candidate for the variables focusing at the improvement of particular measure. In this proposal, a division is carried out for the variables m and μ. An attempt is made to employ an arbitrary beginning point, and the algorithm is executed five times, observing the parameters $m = 15$, entropy period [0.1, 0.7], and $\mu = 8$. The aforementioned parameters are utilized in the research. Additionally, the examination of the Receiver Operating Characteristic curve for "FA Cup" example confirms with the aforesaid period of the entropy determined through the aforementioned Bayesian approach. A good outcome, with AUC = 0.807, was achieved with $m = 15$ and entropy discovering period [0.1, 0.7]. The Receiver Operating Characteristic curve was built as the "Sensitivity" (true positive rate) versus "1-Specificity" (false positive rate) for differing values of $m \in \{7, 10, 15\}$. Every place of the Receiver Operating Characteristic curve was determined altering the entropy discovering period from [0.1, 0.4] to [0.1, 1.7] in stages of 0.1 slices. With our representation of event discovery as a stage change task, we ensure that the incident lies within the time slot, as we adjust the parameter like size of the window with the recognition of the stage change features. Furthermore, an attempt is made to concurrently fine-tune the entropy period as we are familiar a priori, because of our representation that the event will probably occur within the window.

4.2 Evaluation of Event Discovery

Our approach comprises of catching the stage change of the entropy whenever it alters from nearby 0 to a middle value, that is, the minute a bigram appears to a minute we take into consideration that it is an incident. On the grounds of demonstration, it depicts few discovered bigrams employing the suggested approach in the examined dataset. The "bigrams" ({Chelsea, goal}, {yellow, card}, {Liverpool, goal}) illustrate the 3 goals and 2 yellow cards, depicting the estimated period and significance of every incident also. Observe that the recognized period is not the accurate period when the incidents happened or continued because the individuals require a span of time to react to the incident. These outcomes also give proof that the proposed research has better sensitivity to time: the yellow cards happened in a small interval, however, suggested work organized to make a difference between them. In contrast to ground truth, it can be observed that the discovered incidents are near to the actual incident, having approximate four seconds of interval among them. On the grounds of demonstration, few discovered subjects with the respective ground truth are observed. To assess our outcomes, we employed the measures illustrated in Part 3.4. We made a contrast between the achieved outcomes and few approaches from the existing work. The only outcomes achieved by the researchers in [32] have been gathered from the actual manuscripts and pasted into the chart of outcomes. It can be seen that the proposed technique is the second finest in respect of the SW-Accuracy and SW-Rec metrics and achieves the best outcome in the S-Rec metric. It can be seen that the proposed technique is the second finest when in respect of the S-Rec metric, SW-Accuracy, and SW-Rec for the "FA Cup" dataset, however, it achieves the finest outcome in respect of "K-score." Again, our suggested approach achieves the best outcomes in terms of K-score. An attempt is made to scale the assessment of the proposed technique by employing the above five untagged datasets. In this assessment, ourselves mostly focus on recognition of recall metric of pertinent events. For these instances, the proposed procedure has been instructed (observe the variables set by employing the Receiver Operating Characteristic curve), and ourselves employed the untagged datasets as an examination. In order to perform this, we determine the actual-life incidents and incidents which are discovered by the proposed technique. For every untagged dataset, we take into consideration that every specimen comprises of tweets for a duration of 1 h. To determine the recall, we gathered all incidents discovered by the proposed approach, afterward, created a manual connection of incidents. To achieve the above, it is manually explored to observe that if the significant words obtained from the approach had any relationship to the actual-life incidents. In order to perform this, we examined in particular news blogs the significant words recognized by the approach so as to create a representation of the recognized event. On the grounds of demonstration, lines tagged as 1, 2, and 3 depict three specimens of incidents discovered by the proposed approach, two of them pertain to incidents such that we were in a position to observe a correlation in actual-life incidents, and one of them depicts an unrecognized incident. This creation, along with significant words and the time that proposed approach recognized the incident. Apart from this, we determined the proportion of incidents recognized for every dataset, as we can observe at the conclusion of every chart.

5 Conclusion and Future Research Directions

In the proposed research, we employed entropy to depict the happening of an incident in social networks. It is detected that while the happening of an incident, the entropy of the bigrams drawn from the social networks modifies its active elements, and we found a continual stage change of the entropy active elements. To achieve this, we characterize the active elements of rumor dissemination in social media, and we observe and examine the characteristic of the stage change observed in the framework. Henceforth, we suggested a new approach to discover incidents in Twitter centered on time sequence constituted by the likelihoods of the significant words drawn from the tweets post. We observed proof that the stage change (using Monte Carlo) has the similar behavior of the change found in the Twitter dataset. Our approach, in spite of the fact that not taking into consideration any earlier information pertaining to the tweet content, can recognize any incident (also in differing dialects). By taking into consideration, the phase changes, we represent robust practical and theoretical proof of the existent of tweet changes, as well as behavior pertaining to them. By employing such representation, we were in a position to adjust the variables of our incident discovery approach, mitigating the area of search of all the aforementioned variables. Moreover, we give some proof that the proposed approach is responsive to discover incidents that occur in short-time intervals. The suggested approach represents adequate results in contrast to standards and represented best outcomes in contrast to few approaches in existence for tagged and untagged datasets. As future research directions, we plan to carry out a representation of more significant exponents, apart from carrying out an examination on the active characteristics of the VIR technique. We expect to be feasible a good comprehension of the stage changes as well as characteristics of the active elements of the framework. Furthermore, we intend to make a difference between first-order and second-order stage changes to utilize more appropriate approaches to discover every category of incident.

References

1. Tumasjan, A., Sprenger, T.O., Sandner, P.G., Welpe, I. M. (2010). Predicting elections with twitter: What 140 characters reveal about political sentiment. In: Proceedings of 4th International AAAI Conference Weblogs Social Media, vol. 10, no. 1, pp. 178–185
2. Tufekci, Z., Wilson, C.: Social media and the decision to participate in political protest: observations from tahrir square. J. Commun. **62**(2), 363–379 (2012)
3. Zahra, K., Imran, M., Ostermann, F.O.: Automatic identification of eyewitness messages on Twitter during disasters. Inf. Process. Manage. **57**(1), Art. no. 102107 (2020)
4. Peary, B.D.M., Shaw, R., Takeuchi, Y.: Utilization of social media in the east japan earthquake and tsunami and its effectiveness. J. Natural Disaster Sci. **34**(1), 3–18 (2012)
5. Kim, J., Hastak, M.: Social network analysis: Characteristics of online social networks after a disaster. Int. J. Inf. Manage. **38**(1), 86–96 (2018)
6. Jin, H., Lin, C., Chen, H., Liu, J.: QuickPoint: Effciently identifying densest sub-graphs in online social networks for event stream dissemination. IEEE Trans. Knowl. Data Eng. **32**(2), 332–346 (2020)
7. Wu, X., Zhu, X., Wu, G.-Q., Ding, W.: Data mining with big data. IEEE Trans. Knowl. Data Eng. **26**(1), 97–107 (2014)
8. Bothorel, C., Lathia, N., Picot-Clemente, R., Noulas, A.: Location recommendation with social media data. In: Social information access, pp. 624–653. Springer, Berlin (2018)
9. Albert, R., Barabási, A.-L.: Statistical mechanics of complex networks. Rev. Modern Phys. **74**(1), 47 (2002)
10. Jiang, J., Wen, S., Yu, S., Xiang, Y., Zhou, W.: Rumor source identification in social networks with time-varying topology. IEEE Trans. Dependable Secure Comput. **15**(1), 166–179 (2018)
11. Dou, W., Wang, X., Ribarsky, W., Zhou, M.: Event detection in social media data. In Proccedings of Proceedings of IEEE VisWeek Workshop Interactive Visual Text Analytics—Task Driven Analytics of Social Media Content, pp. 971–980 (2012)
12. Parikh, R., Karlapalem, K.: Et: Events from tweets. In: Proceedings of 22nd International Conference on World Wide Web, pp. 613–620. ACM, New York (2013)
13. Lee, R., Sumiya, K.: Measuring geographical regularities of crowd behaviors for Twitter-based geo-social event detection. In Proceedings of 2nd ACM SIGSPATIAL International Workshop on Location Based Social Network LBSN, pp. 1–10. ACM, NewYork (2010)
14. Motoi, S., Misu, T., Nakada, Y., Yazaki, T., Kobayashi, G., Matsumoto, T., Yagi, N.: Bayesian event detection for sport games with hidden Markov model. Pattern Anal. Appl. **15**(1), 59–72 (2012)
15. Washha, M., Qaroush, A., Mezghani, M., Sedes, F.: A topic-based hidden Markov model for real-time spam tweets filtering. Procedia Comput. Sci. **112**, 833–843 (2017)
16. Gama, J., Zliobaitė, I., Bifet, A., Pechenizkiy, M., Bouchachia, A.: A survey on concept drift adaptation. ACM Comput. Surv. **46**(4), 1–37 (2014)
17. Giatrakos, N., Deligiannakis, A., Garofalakis, M., Kotidis, Y.: Omnibus outlier detection in sensor networks using windowed locality sensitive hashing. Fut. Gener. Comput. Syst. (2018)
18. Barros, P.H., Cardoso-Pereira, I., Loureiro, A.A.F., Ramos, H.S.: Event detection in social media through phase transition of bigrams entropy. In: Proceedings of IEEE Symposium on Computers and Communictions (ISCC), pp. 1068–1073 (2018)
19. Li, R., Lei, K.H., Khadiwala, R., Chang, K.C.-C.: TEDAS: A Twitter based event detection and analysis system. In: Proceedings of IEEE 28th International Conference Data Engineering, pp. 1273–1276 (2012)
20. Sakaki, T., Okazaki, M., Matsuo, Y.: Earthquake shakes Twitter users: Real-time event detection by social sensors. In: Proceedings of 19th International Conference on World Wide Web WWW, pp. 851–860. ACM, New York (2010)
21. Giridhar, P., Amin, M.T., Abdelzaher, T., Wang, D., Kaplan, L., George, J., Ganti, R.: ClariSense+: an enhanced traffic anomaly explanation service using social network feeds. Pervas. Mobile Comput. **33**, 140–155 (2016)

22. D'Andrea, E., Ducange, P., Lazzerini, B., Marcelloni, F.: Real-time detection of traffic from Twitter stream analysis. IEEE Trans. Intell. Transp. Syst. **16**(4), 2269–2283 (2015)
23. Weng, J., Lee, B.-S.: Event detection in Twitter. Proc. ICWSM **11**, 401–408 (2011)
24. Rosso, O.A., Craig, H., Moscato, P.: Shakespeare and other English renaissance authors as characterized by information theory complexity quantifiers. Phys. A, Stat. Mech. Appl. **388**(6), 916–926 (2009)
25. Mathioudakis, M., Koudas, N.: TwitterMonitor: trend detection over the Twitter stream. In: Proceedings of International Conference on Management of Data (SIGMOD), pp. 1155–1158. ACM, NewYork (2010)
26. Deerwester, S., Duais, S., Furnas, G., Landauer, T., Harshman, R.: Indexing by latent semantics analysis. J. Amer. Soc. Inf. Sci. **41**(6), 391–407 (1990)
27. Li, C., Sun, A., Datta, A.: Twevent: segment-based event detection from tweets. In: Proceedings of 21st ACM International Conference on Information and Knowledge Management (CIKM), pp. 155–164 (2012)
28. Jarvis, R.A., Patrick, E.A.: Clustering using a similarity measure based on shared near neighbors. IEEE Trans. Comput. **C-22**(11), 1025–1034 (1973)
29. Dang, Q., Gao, F., Zhou, Y.: Early detection method for emerging topics based on dynamic Bayesian networks in micro-blogging networks. Expert Syst. Appl. **57**, 285–295 (2016)
30. Murphy, K.P., Russell, S.: Dynamic Bayesian networks: Representation, inference and learning. Ph.D. dissertation, Department of Electrical Engineering, University of California, Berkeley, CA, USA (2002)
31. Alsaedi, N., Burnap, P., Rana, O.: Can we predict a riot? Disruptive event detection using Twitter. ACM Trans. Internet Technol. **17**(2), 1–26 (2017)
32. Nguyen, D.T., Jung, J.E.: Real-time event detection for online behavioral analysis of big social data. Future Gener. Comput. Syst. **66**, 137–145 (2017)
33. Aiello, L.M., Petkos, G., Martin, C., Corney, D., Papadopoulos, S., Skraba, R., Goker, A., Kompatsiaris, I., Jaimes, A.: Sensing trending topics in Twitter. IEEE Trans. Multimedia **15**(6), 1268–1282 (2013)
34. Bhuvaneswari, A., Valliyammai, C.: Information entropy based event detection during disaster in cyber-social networks. J. Intell. Fuzzy Syst. **36**(5), 3981–3992 (2019)
35. Lin, J.: Divergence measures based on the Shannon entropy. IEEE Trans. Inf. Theory **37**(1), 145–151 (1991)
36. Guille, A., Favre, C.: Event detection, tracking, and visualization in Twitter: A mention-anomaly-based approach. Social Netw. Anal. Mining **5**(1), 18 (2015)
37. Benhardus, J., Kalita, J.: Streaming trend detection in Twitter. Int. J. Web Based Commun. **9**(1), 122 (2013)
38. Shetty, J., Adibi, J.: Discovering important nodes through graph entropy the case of enron email database. In: Proc. 3rd International Workshop Link Discovery LinkKDD, pp. 74–81 (2005)

Impact of Environmental Factors on COVID-19 Transmission Dynamics in Capital New Delhi Along with Tamil Nadu and Kerala States of India

Nishant Juneja, Sunidhi, Gurupreet Kaur, and Shubhpreet Kaur

1 Introduction

Today, world is suffering from the deadly viral disease COVID-19 which is causing a real threat to human life [2, 11]. The contagious disease was emerged in China and is spreading very quickly across the world [7]. Besides China, it quickly spread to further 24 countries, and by considering its alarming levels of spreading and severity, the WHO publicly announced coronavirus disease as pandemic disease on March 11, 2020. India too falls in trap of this deadly disease and is now at a critical phase in this deadlock war. Kerala was the first state of India to report a positive case for COVID-19 on January 30, 2020. With the passage of time, COVID-19 pandemic worsened in other states of country. The highest provinces in the COVID-19 affected are Maharashtra, Kerala, Tamil Nadu, New Delhi, Gujarat and Karnataka states of India [8]. It has been suggested that COVID-19 and other viruses, such as influenza and Ebola, had noticeable relationship with environmental indicators [1, 3–5, 9, 14, 17, 18]. Many geographical scientists and researchers elaborated in their research that the severity of coronavirus depends on geographical/climatological such factors mainly on temperature [6, 12, 13, 16, 19] and humidity [10, 15]. Their study reveals that increase and decrease of humidity along with temperature could affect transmission risk and thus leads to estimate the survival time of coronavirus. Therefore, it is realistic to analyze and elaborate the relation between environmental indicators (temperature and humidity) and new daily arrival of coronavirus cases.

Delhi is the national capital of India and is well known as the heart of the nation. The population of Delhi is around 3 crores, and this largest metropolitan city of India is well known for its vulnerability to climate change and global warming. It is famous for its enriched culture, heritage and historical monuments. The temperature variation in New Delhi between summer and winter is very much high with the

N. Juneja (✉) · Sunidhi · G. Kaur · S. Kaur
Department of Mathematics, Dev Samaj College for Women, Ferozepur, Punjab, India

M. Dua et al. (eds.), *Proceedings of the International Conference on Paradigms of Communication, Computing and Data Sciences*, Algorithms for Intelligent Systems,
https://doi.org/10.1007/978-981-16-5747-4_36

highest recorded temperature of 48.4 °C on May 26, 1998, in summers, and the lowest temperature in winters was minus 2.2 °C on January 11, 1967. Both of these temperatures are recorded at Indira Gandhi International Airport. The national capital reported its first case of coronavirus on March 2, 2020. At present, Delhi stood sixth in COVID-19 tally of infected cases in India with around 639,735 confirmed cases along with 10,911 deaths and 627,227 recoveries as reported on March 1, 2021. Delhi records its highest single day increase in COVID cases on November 11, 2020, with 8593 cases reported on the day. The similar situation occurs in Tamil Nadu state of India, where the state government announced coronavirus as a notified disease in the second week of March. In the first week of April, Government of Tamil Nadu selected some government hospitals as dedicated hospitals for the treatment of the patients suffering from this highly contagious disease. However, some private hospitals were also instructed to be fully prepared with all the necessary equipments along with trained teams of doctors to treat the COVID-19-infected patients. In spite of taking early combat actions by government, Tamil Nadu still stands on fifth position in COVID-19 tally of India with 852,016 cases of coronavirus as on March 1, 2021. The third state which is severely affected by this pandemic in last four months is Kerala. Kerala state of India reported its first COVID-19 case on January 30, 2020, in the district Thrissur, which was also the first case of India. The Government of Kerala takes every possible early step to control the rate of spread of this disease, and efforts of Kerala government were praised worldwide too. The rate of new arrival cases of COVID-19 in Kerala has been reduced to 0.25% as on April 30, 2020. But in the month of May, Kerala reported sudden increase in COVID-19 cases due to return of local citizens from other states of India. After that, Kerala reported its third increase in COVID-19 cases in the month of October. Kerala recorded its highest single day spike in cases with around 11,755 cases reported on October 10, 2020. This was the second-highest single day rise in India too after Maharashtra. However, the lowest fatality rate of 0.35% in Kerala as compared to 1.43% in India can be attributed to strict measures and early combat actions taken by the Kerala Government.

2 Methodology

The 90 days data set for daily confirmed cases of coronavirus is considered from September 30, 2020 to December 30, 2020, archive from the official Web site of Indian Government (https://www.covid19india.org/). The reason for choosing this time period is that there is sharp change in temperature in these states during these months. The data set of temperature was taken from the Weather Channel, IBM (https://weather.com/enIN/weather/monthly/l/New+Delhi+India+INXX0096:1:IN). SPSS software has been used for carrying out the statistical analysis. As the data obtained was not normally distributed, we calculate the Pearson's correlation coefficient to check the correlation between the environmental factors and infected cases of COVID-19 in national capital Delhi along with Tamil Nadu and Kerala states.

The purpose of the present research is to explore the relationship between temperatures (maximum, minimum and average), humidity and confirmed COVID-19 cases to offer constructive implications for policymakers and the public in New Delhi, Tamil Nadu and Kerala states of India. The prevalence of studies on the correlation between dynamics of coronavirus transmission and foremost environmental parameters (temperature and humidity) has been conducted for these three states.

3 Results and Discussions

In the statistical analysis of the data set, the pattern of COVID-19 cases in New Delhi has been analyzed with climate change parameters like temperature (maximum, minimum and average) and humidity. In Fig. 1a, the number of daily confirmed cases of COVID-19 has been plotted for 90 days (from 30/09/2020 to 30/12/2020). It can be seen that there is sudden surge in COVID-19 cases in Delhi in the second week of November with the highest number of case count 8593 on November 11, 2020. In the month of December, the number of cases starts decreasing with only 564 confirmed cases on December 28, 2020. Figure 1b, c shows the respective variation of maximum and minimum temperatures of the capital for the same period. A sharp variation in temperature has been noticed for this period with the highest maximum temperature of 36 °C in the first week of October and the lowest minimum temperature of 5 °C on December 30, 2020. Similarly, Fig. 1d shows the variation of average temperature with the highest value 30 °C on September 30, 2020. Figure 1e shows the variation of humidity with the highest value of 86% and the lowest being 31.9%.

Similar graphs have been plotted for Tamil Nadu and Kerala states of India as shown in Figs. 2 and 3.

Firstly, we will discuss the dynamics of COVID-19 cases in Delhi. It has been observed that daily confirmed COVID-19 cases are weakly correlated with maximum and minimum temperature with respective values 0.207 and 0.124. However, there is statistically significant correlation of confirmed cases with average temperature with correlation coefficient value 0.565. So we can say that there is sharp increase in COVID-19 cases with increase in average temperature. The calculated values of descriptive statistics and bivariate correlation coefficient for Delhi state of India are shown in Tables 1 and 2.

Now we see the statistical values of correlation coefficient in Tamil Nadu. It has been observed that there is strong correlation between COVID-19 cases and all other epidemiological factors with all correlation values above 0.6 as shown in Tables 3 and 4. It accounts for the sharp decrease in new confirmed cases in Tamil Nadu in November and December due to decrease in temperature.

Finally, we take the case of Kerala state of India. Here the values of correlation coefficients come out to be negative. So it has been observed that maximum, minimum and average temperatures are negatively correlated with COVID-19 case count in Kerala. The reason for sudden rise in confirmed cases in the month of November and December is due to fall in temperature (Tables 5 and 6).

Fig. 1 **a** Variation of COVID-19 cases in Delhi. **b** Variation of maximum temperature in Delhi. **c** Variation of minimum temperature in Delhi. **d** Variation of average temperature in Delhi. **e** Variation of humidity in Delhi

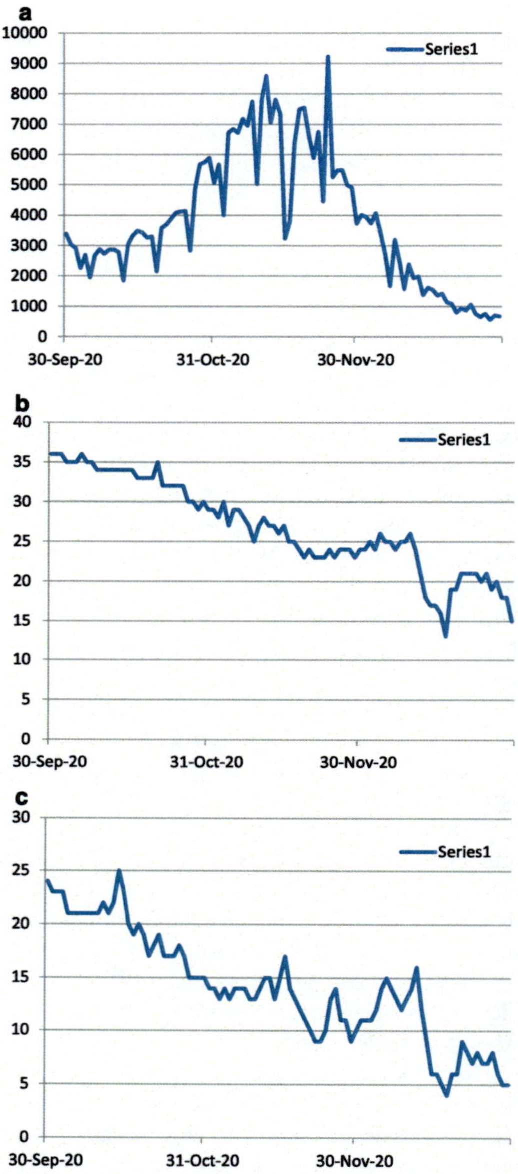

Fig. 1 (continued)

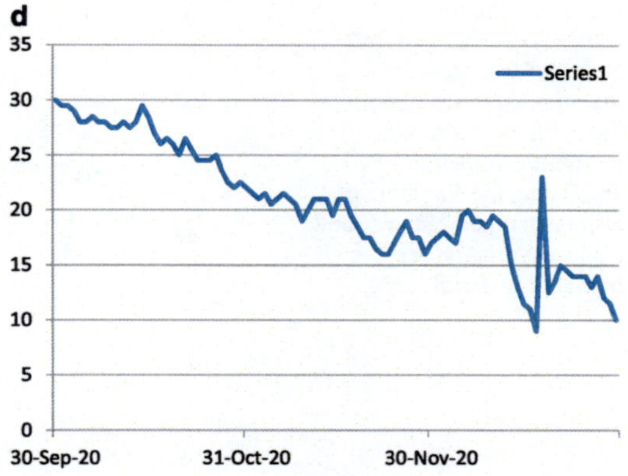

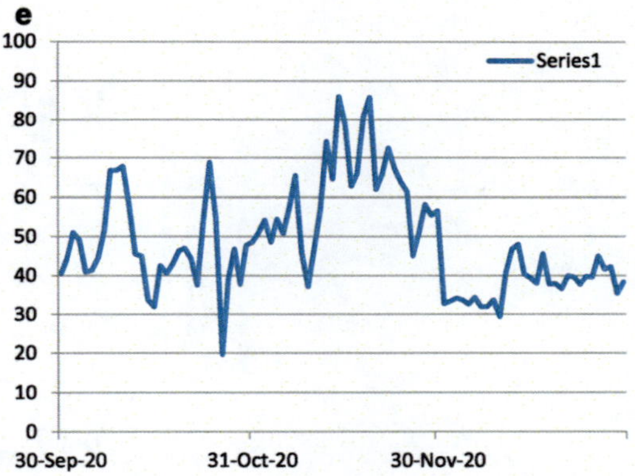

Finally, we study the impact of daily change in humidity in the three states on the dynamics of COVID-19. The correlation coefficient between COVID cases and humidity comes out to be non-significant in all the three states. So we can say that there is no significant relation between COVID-19 cases and humidity level.

4 Conclusion

The descriptive analysis and value of bivariate correlation coefficients have been found for all the three states. It has been found that in Tamil Nadu state of India, new arrival cases of COVID-19 increase with increase in temperature, whereas in Kerala, the cases decline significantly with increase in temperature. Humidity has no significant effect on daily confirmed cases in these two states. However, in Delhi, temperature is not significantly affecting the dynamics of COVID cases, whereas humidity is positively correlated with the new arrival of COVID cases. Previous

Fig. 2 **a** Variation of COVID-19 cases in Tamil Nadu. **b** Variation of maximum temperature in Tamil Nadu. **c** Variation of minimum temperature in Tamil Nadu. **d** Variation of average temperature in Tamil Nadu. **e** Variation of humidity in Tamil Nadu

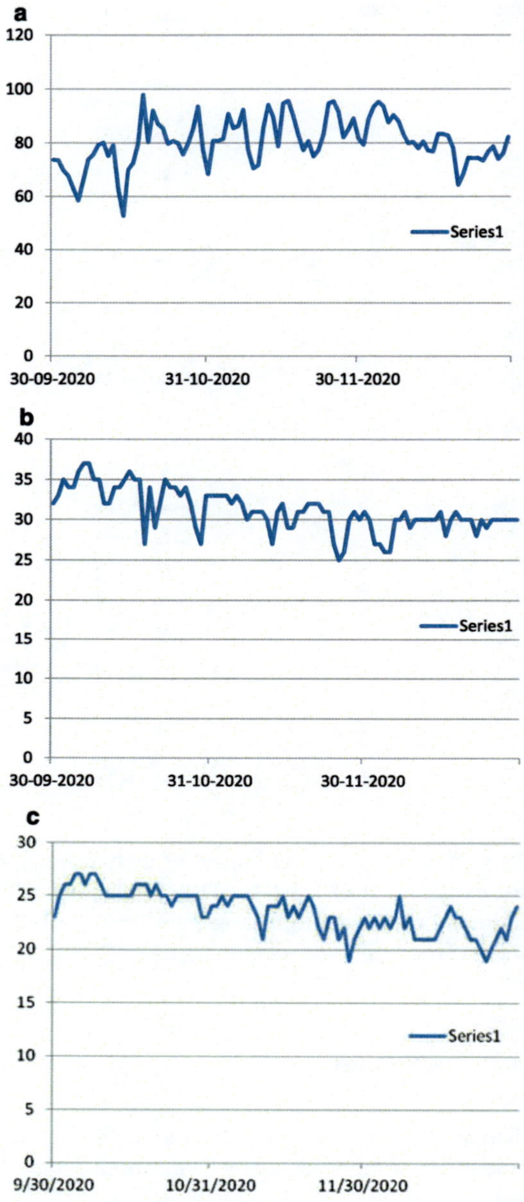

Fig. 2 (continued)

d

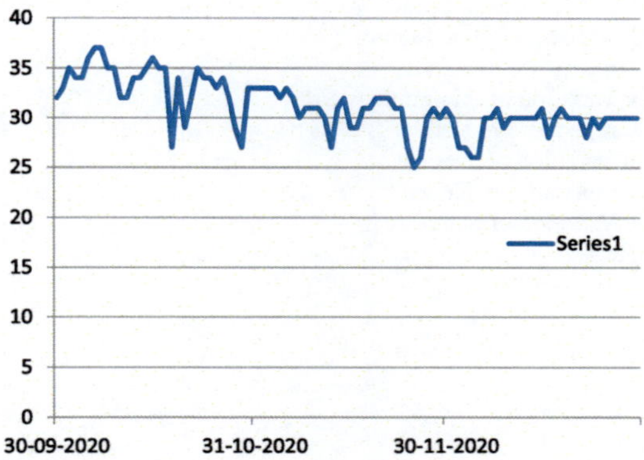

e

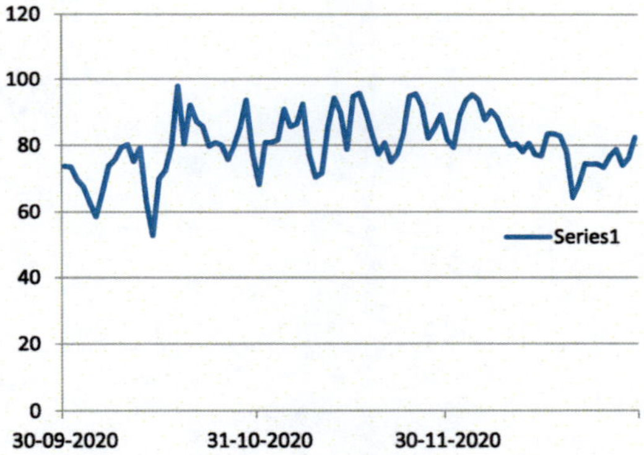

studies reveal that there exists an optimum temperature for this deadly virus, and elevated temperature will certainly decrease the viability of this virus. So the air temperature has been observed to be significantly affecting the COVID-19 transmission due to change in natural behavior of the virus at high temperatures. However, the present study reveals that the effect of environmental factors differs from region to region, and we cannot generalize it to the whole world. The study of Tamil Nadu and Kerala states of India indicates the temperature variation, and daily COVID-19 cases have highly significant relation, yet the precautionary measures like health hygiene, hand washing and social distancing should not be ignored and government policies should not wait for higher temperatures to defeat COVID-19.

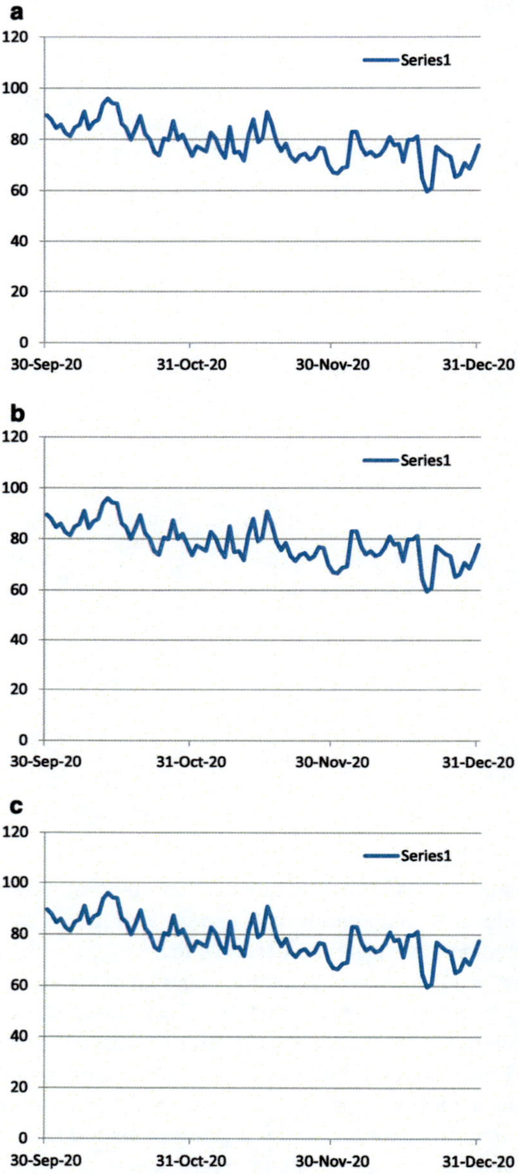

Fig. 3 **a** Variation of COVID cases in Kerala. **b** Variation of maximum temperature in Kerala. **c** Variation of minimum temperature in Kerala. **d** Variation of average temperature in Kerala. **e** Variation of humidity in Kerala

Fig. 3 (continued)

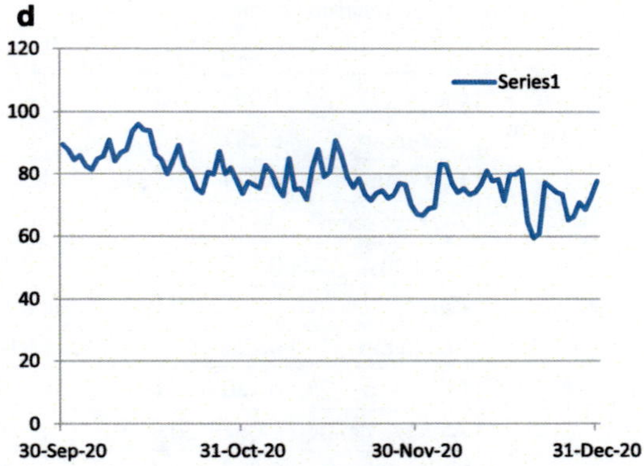

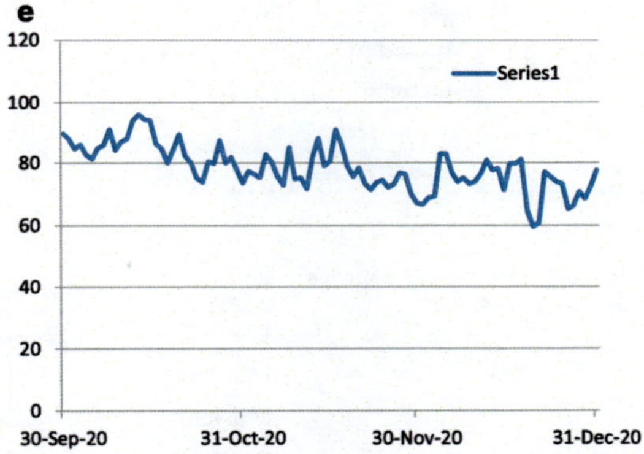

Fig. 3(e)

Table 1 Descriptive statistics for Delhi

	Mean	Std. deviation	*N*
Cases	3820.3261	2188.50323	92
Max temp	26.7826	5.80228	92
Min temp	14.0435	5.22023	92
Average temp	20.5163	5.35008	92
Humidity	48.4946	13.48544	92

Table 2 Bivariate correlation coefficients for Delhi

		Cases	Max temp	Min temp	Average temp	Humidity
Pearson correlation	Cases	1.000	0.207	0.124	0.146	0.565
	Max temp	0.207	1.000	0.931	0.965	0.084
	Min temp	0.124	0.931	1.000	0.949	0.074
	Average temp	0.146	0.965	0.949	1.000	0.067
	Humidity	0.565	0.084	0.074	0.067	1.000
Sig. (1-tailed)	Cases		0.024	0.120	0.082	0.000
	Max temp	0.024		0.000	0.000	0.212
	Min temp	0.120	0.000		0.000	0.242
	Average temp	0.082	0.000	0.000		0.261
	Humidity	0.000	0.212	0.242	0.261	
N	Cases	92	92	92	92	92
	Max temp	92	92	92	92	92
	Min temp	92	92	92	92	92
	Average temp	92	92	92	92	92
	Humidity	92	92	92	92	92

Table 3 Descriptive statistics for Tamil Nadu

	Mean	Std. Deviation	N
Cases	2447.8696	1469.93528	92
Max temp	31.1630	2.60716	92
Min temp	23.5326	1.89534	92
Average temp	27.3478	2.02150	92
Humidity	80.3967	8.90000	92

Table 4 Bivariate correlation coefficients for Tamil Nadu

		Cases	Max temp	Min temp	Average temp	Humidity
Pearson's correlation	Cases	1.000	0.691	0.746	0.795	−0.408
	Max temp	0.691	1.000	0.603	0.927	−0.635
	Min temp	0.746	0.603	1.000	0.857	−0.116
	Average temp	0.795	0.927	0.857	1.000	−0.464
	Humidity	−0.408	−0.635	−0.116	−0.464	1.000
Sig. (1-tailed)	Cases		0.000	0.000	0.000	0.000
	Max temp	0.000		0.000	0.000	0.000
	Min temp	0.000	0.000		0.000	0.136
	Average temp	0.000	0.000	0.000		0.000
	Humidity	0.000	0.000	0.136	0.000	
N	Cases	92	92	92	92	92
	Max temp	92	92	92	92	92
	Min temp	92	92	92	92	92
	Average temp	92	92	92	92	92
	Humidity	92	92	92	92	92

Table 5 Descriptive statistics for Kerala

	Mean	Std. Deviation	N
Cases	6167.7742	1749.60886	93
Max temp	33.1935	1.42385	93
Min temp	24.6559	0.94977	93
Average temp	28.9247	0.95822	93
Humidity	78.4548	7.42833	93

Table 6 Bivariate correlation coefficients for Kerala

		Cases	Max temp	Min temp	Average temp	Humidity
Pearson's correlation	Cases	1.000	−0.399	−0.060	−0.326	0.419
	Max temp	−0.399	1.000	0.275	0.879	−0.397
	Min temp	−0.060	0.275	1.000	0.700	0.016
	Average temp	−0.326	0.879	0.700	1.000	−0.287
	Humidity	0.419	−0.397	0.016	−0.287	1.000
Sig. (1-tailed)	Cases		0.000	0.285	0.001	0.000
	Max temp	0.000		0.004	0.000	0.000
	Min temp	0.285	0.004		0.000	0.440
	Average temp	0.001	0.000	0.000		0.003
	Humidity	0.000	0.000	0.440	0.003	
N	Cases	93	93	93	93	93
	Max temp	93	93	93	93	93
	Min temp	93	93	93	93	93
	Average temp	93	93	93	93	93
	Humidity	93	93	93	93	93

Acknowledgements The authors acknowledge Department of Biotechnology (DBT), New Delhi, and Dev Samaj College for Women, Ferozepur, Punjab, for providing research support.

References

1. Lowen, A.C., Steel, J.: Roles of humidity and temperature in shaping influenza seasonality. J. Virol. **88**(14), 7692–7695 (2014)
2. Huang, C., Wang, Y., Li, X., Ren, L., Zhao, J., Hu, Y., Zhang, L., Fan, G., Xu, J., Gu, X.: Clinical features of patients infected with 2019 novel Corona virus in Wuhan, China. Lancet **395**, 497–506 (2020)
3. Yip, C., Chang, W.L., Yeung, K.H., Yu, I.T.: Possible meteorological influence on the severe acute respiratory syndrome (SARS) community outbreak at Amoy Gardens, Hong Kong. J. Environ. Health **70**(3), 39–46 (2007)
4. Chu, C.M., Tian, S.F., Ren, G.F., Zhang, Y.M., Zhang, L.X., Liu, G.Q.: Occurrence of temperature-sensitive influenza A viruses in nature. J. Virol. **41**(2), 353–359 (1982)
5. Helm, D.: The environmental impacts of the corona virus. Environ. Resour. Econ. **76**, 21–38 (2020)
6. Prata, D.N., Rodrigues, W., Bermejo, P.H.: Temperature significantly changes COVID-19 transmission in (sub) tropical cities of Brazil. Sci Tot. Env. **729**(138), 862–868 (2020)
7. Benedetti, F., Pachetti, M., Marini, B., Ippodrino, R., Gallo, R.C., Ciccizzi, M., Zella, D.: Inverse correlation between average monthly high temperatures and COVID-19-related death rates in different geographical areas. J. Transl. Med. **18**(251), 2020 (2020)
8. https://www.covid19india.org/. Accessed 14 Nov 2007
9. Khan, I., Shah, D., Shah, S.S.: COVID-19 pandemic and its positive impacts on environment: an updated review. Int. J. Environ. Sci. Technol. **18**, 521–530 (2021)

10. Sharman, J., Kohm, M.: Absolute humidity modulates influenza survival, transmission and seasonality. Proc. Natl. Acad. Sci. USA **106**(9), 3243–3248 (2009)
11. Xie, J., Zhu, Y.: Association between ambient temperature and Covid-19 infections in 122 cities from China. Sci. Tot. Env. **724**, 201–205 (2020)
12. Chan, K.H., Malik Peiris, J.S., Lam, S.Y., Poon, L.L.M., Yuen, K.Y., Seto, W. H.: The effects of temperature and relative humidity on the viability of the SARS Corona virus. Adv. Virol. **2011**, 7, Article ID 734690 (2011). https://doi.org/10.1155/2011/734690
13. Casanova, L.M., Jeon, S., Rutala, W.A., Weber, D.J., Sobsey, M.D.: Effects of air temperature and relative humidity on corona virus survival on surfaces. Appl. Environ. Microbiol. **76**(9), 2712–3271 (2010)
14. Moriyama, M., Ichinohe, T.: High ambient temperature dampens adaptive immune responses to influenza a virus infection. Proc. Natl. Acad. Sci. U. S. A. **116**(8), 3118–3125 (2019)
15. Bashir, M.F., Ma, B., Komal, B.B., Bashir, M.A., Tan, D., Bashir, M.: Correlation between climate indicators and COVID-19 pandemic in New York, USA. Sci. Tot. Env. **728**(138), 835–838 (2020)
16. Arora, N.K., Mishra, J.: COVID-19 and importance of environmental sustainability. Environ. Sustain. **3**, 117–119 (2020)
17. Rupani, P.F., Nilashi, M., Abumalloh, R.A.: Corona virus pandemic (COVID-19) and its natural environmental impacts. Int. J. Environ. Sci. Technol. **17**, 4655–4666 (2020)
18. Thai, P.Q., Choisy, M., Duong, T.N., Thiem, V.D., Yen, N.T., Hien, N.T.: Seasonality of absolute humidity explains seasonality of influenza-like illness in Vietnam. Epidemics **13**, 65–73 (2015)
19. Tosepu, R., Gunawan, J., Effendy, D.S., Ahmad, L., Lestari, H., Bahar, H., Asfian, P.: Correlation between weather and Covid-19 pandemic in Jakarta, Indonesia. Sci. Tot. Env. **725**, 436–439 (2020)

Parallel Local Tridirectional Feature Extraction Using GPU

B. Ashwath Rao, Gopalakrishana N. Kini, Prakash K. Aithal, Konda Vaishnavi, and U. Nikhitha Kamath

1 Introduction

Local binary pattern (LBP) is one of the first and efficient local feature descriptors which efficiently captures the local structure of images. LBP efficiently describes characteristic textures of the surfaces. It is being used as a descriptor in many areas of computer vision and image processing areas, and it is most commonly used for texture discrimination like facial expressions, face, scene gesture and object recognition. It is an efficient method used for texture feature description. It is widely used in face detection and pattern recognition tasks. The LBP operator is an image operator, and it transforms an image into a vector of integer labels which describes small-scale look of the image. Pattern and strength are the two locally complementary aspects of a texture. Local binary pattern operator works in a 3×3 neighborhood. The pixels in the block are subtracted by its center pixel value to determine the sign and then multiplied by powers of two (Decimal) and then summed to obtain a label for the center pixel out of total 256 different labels.

LBP features are determined by concatenating all the signs obtained of the difference between neighbor and center pixel. The total number of neighbors around a radius of one is 8. The sign will be 1 if the neighbor intensity is greater than center pixel, and it will be 0 if the neighbor intensity is lesser than that of center pixel. The concatenation of signs can be done in the clockwise direction or in the anti-clockwise direction. For a matrix of 3×3, there will be eight signs. This vector of signs will enable to represent very fine-grained detail.

B. A. Rao · G. N. Kini · P. K. Aithal · K. Vaishnavi (✉) · U. N. Kamath
Department of Computer Science and Engineering, Manipal Institute of Technology, Manipal, Karnataka, India 576 104
e-mail: konda.vaishnavi@learner.manipal.edu

B. A. Rao
e-mail: ashwath.rao@manipal.edu
URL: https://www.manipal.edu

M. Dua et al. (eds.), *Proceedings of the International Conference on Paradigms of Communication, Computing and Data Sciences*, Algorithms for Intelligent Systems,
https://doi.org/10.1007/978-981-16-5747-4_37

LBP is based on the local pixel intensity for feature vector determination. Based on the appearance, the binary patterns are classified into uniform and non-uniform. Apart from this, patterns can be transformed into rotation invariant. Local binary pattern uses the difference of neighbor pixel and center pixel for binary pattern. Local binary pattern uses a single threshold value which is the center pixel intensity. The local ternary pattern(LTP) uses an interval rather than a single threshold value which is further converted into two binary patterns. The LBP is considered as the first-order derivative, and the second and more order derivative patterns are called as local derivative pattern (LDP).

Local binary pattern has many positive aspects and few negative aspects. The positive aspects include high discerning ability, simplicity in calculation, and invariance to intensity changes which may occur due to noise. The negative aspects include, it is variant to rotation, the time, space complexity of features increases with number of neighbors, and the amount of structural information captured is very limited.

The disadvantage of fixed neighborhood is that it fails to capture details corresponding to various scales. So in extended local binary pattern also called as circular LBP, we consider variable radius. This provides more number of patterns and also requires more computational effort. More the radius, the LBP will be smoother, and if we do not increase the sample points, the discriminative power of the feature will be less. If we put all features into a single histogram, all spatial information is lost. In applications like face detection and other pattern recognition problems, the spatial information is very important and has to be incorporated into the histogram. The idea is to divide the LBP image into grids and consider histogram of each cell separately. Finally, the histograms are concatenated to provide spatial information.

Local tridirectional pattern incorporates local intensity of pixels in three directions of the neighborhood. It is an extension of local binary pattern. Every pixel has some neighbors. Border pixels do not have eight neighbors, whereas all non-border pixels have eight neighbors. If we increase the radius to 2, there are 16 neighboring pixels. In local tridirectional pattern, mutual information of neighboring pixels is considered. Hence, this provides more information when compared with local binary pattern.

2 Litreture Review

It is a difficult problem to scan for relevant images based on visual properties or image quality and has gained significant attention from researchers in the last 20 years. The difference between low-level visual characteristics and higher-level semantic image understanding, also known as the problem of the semantic gap, is the bottleneck for further improving the performance of a content-based image retrieval system. One of the most common approaches in recent years is to shift the emphasis from the global content description of images to the local content description of regions toward solving this semantic gap problem [1].

The massive growth in digital data and huge amount of image and other multimedia data emphasize the need to invent thematic access methods that provide more than text-based retrieval systems by the lookup of database. To alleviate this research gap, a content-based image retrieval system (CBIRS) has been proposed by Siong and Zaki[2]. However, in the medical domain, these systems have drawbacks and require solution from researchers.

A one-dimensional micropattern determined by establishing relations between higher-order derivative of the pixel under consideration in four directions is the proposed local directional gradient pattern (LDGP). The feature descriptor defines the relation in four directions among the higher-order derivatives of the pixel under consideration to calculate the micropattern according to the local function. The descriptor's size is less which results in less feature extraction time and also time in matching. When tested on the AT&T, extended Yale B and CMU-PIE benchmark databases show that the descitor achieves identical accuracy to recent iteration methods with less extraction and matching time [3].

LBP for face recognition on the FRGC data set was implemented by Yang and Wang [4]. The results show that the hamming LBP is better than the original LBP, when there are variations in facial expression and visibility. A multi-scale LBP along with LDA was used for face recognition by Chan et al. [5]. The experimental results were better than recent approaches on the FERET and XM2VTS databases. By projecting multispectral LBP features obtained from local regions into the LDA subspace, the regional discriminative classification was achieved. They tested the efficiency on the FRGC and XM2VTS databases.

Since the global approaches require accurate normalization of pose, lighting and scale the component-based (local) approaches are better as shown by Heisele et al. [6]. To monitor the user interface on small smart devices, a head tracking system was proposed by Hannuksela [7]. Face and eye detection was performed by boosting LBP method. The system operated on resource poor handheld device in real time. LBP is also used in photo annotation method [8] in extracting facial features for face clustering and re-ranking. LBP can also be used in facial images as a technique in preprocessing. In minimizing lighting effects, LBP was used as a preprocessing step by Heusch et al.[9]. Brajovic [10] compared LBP to other techniques of preprocessing including histogram equalization.

3 Methodology

The local tridirectional pattern will consider the relationship based on different directions. In the proposed method, we have taken images of different size. Firstly, we determine the center pixel. Surrounding the center pixel are neighboring pixels in different distances. The closest neighbors will be the eight neighbors around the center pixel. The number of closest neighbors surrounding a pixel is limited in number, and hence, they give more salient information as they are very close to the center pixel. So, with these eight pixels, we create a pattern. We consider one neighborhood pixel at a time, and it is compared with the two most adjacent pixels and also with the center pixel. The two most adjacent pixels are either along vertical direction or horizontal direction of an image. We can consider it in a form of a matrix as shown in Fig. 1.

Consider the center pixel to be I_c and the eight neighborhood pixels be $I_1, I_2, \ldots I_8$. At first, we find out the difference of each of the neighborhood pixel with its two most nearest pixels and then calculate the difference between each neighborhood pixel with the center pixel. The difference can be named as D_1, D_2, D_3. Mathematically, it can be shown as follows:

$$D_1 = I_i I_{i-1},\ D_2 = I_i I_{i+1},\ D_3 = I_i I_c \forall i = 2, 3 \ldots 7 \tag{1}$$

$$D_1 = I_i I_8,\ D_2 = I_i I_{i+1},\ D_3 = I_i I_c for i = 1 \tag{2}$$

$$D_1 = I_i I_{i-1},\ D_2 = I_i I_1,\ D_3 = I_i I_c for i = 8 \tag{3}$$

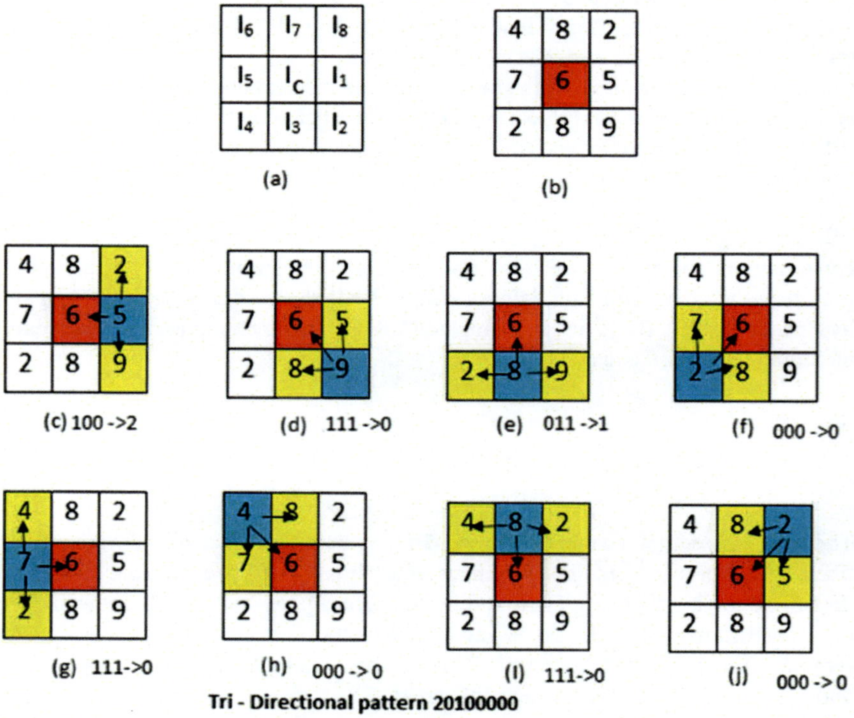

Fig. 1 Proposed method sample window example. **a** center and neighboring pixel notations. **b** Sample window, **c–j** Local tridirectional pattern and magnitude pattern calculation

The difference value will be 1 if the difference is greater than or equal to zero, otherwise the difference value is 0. After computing the values of D_1 , D_2 and D_3 for each of the neighborhood pixel, a label will be assigned based on the value of all the three differences.

The pattern number can take the values 0, 1 or 2 based on the values of the differences D_1 , D_2 and D_3. The pattern number will be 0 if all the values of Ds are 0 or if all the values of Ds are 1. Otherwise, the pattern value is calculated by the number of zeros present in the difference values D_1, D_2 and D_3. The pattern formation is demonstrated in Fig. 1

In the proposed method, we have taken DICOM images and computed tridirectional pattern for the images of different sizes using CUDA programming. The CUDA is for a graphical processing units from NVIDIA's software framework which provides direct access to the GPU's instruction set including parallel computational elements, for the execution of compute kernels. It works with programming languages C, C++, Fortran and others.

An example of pattern calculation is shown in Fig. 1. In window (a), the center pixel I_c and the neighborhood pixels $I_1, I_2, \ldots, I_8$ are shown. In window (b), the center pixel is marked in red color. In window (c), the first neighborhood pixel is marked in blue color, and the two most adjacent pixels are marked in yellow color. First, we compare blue pixel with yellow pixel and then with red pixel and assign a '0' or '1' using Eq. (3). Similarly, for the windows (d)–(i), compute the values using Eq. (3). For the eighth neighborhood pixel, calculate the value using Eq. (3). Finally, by combining all neighborhood pixel pattern values, the feature of local tridirectional pattern is obtained.

4 Result

The proposed method is tested for three DICOM images of varying size. Each DICOM image we have considered is an integral multiple of other. The time taken to obtain the tridirectional pattern is calculated. It is observed that the time taken for sequential execution was more compared to the time taken for parallel execution. The same is shown in Figs. 2 and 3. The parallel computing saves time, energy and money because many resources will be working together which in turn will help in reducing the time and potential costs. It is also impractical to solve time-consuming problems using serial computing. By using parallel programming, it unleashes a program's ability to execute multiple instructions at the same time which in turn increases the overall processing throughput. In sequential processing, since the entire computation is done by a single processor, the processor heats up quickly. It is also observed by varying the configuration parameters while launching kernel results in varying feature extraction time. In [11], we can observe that for the local binary pattern, the time taken for sequential execution is more when compared to the parallel execution. The same idea is being used for the tridirectional pattern. Also there has been no other research paper so far being done on the parallel computing of tridirectional pattern.

PARALLEL COMPUTATION		
Image Size	**Block Size**	**Time(ms)**
256x256	16x16	0.162686
	32x32	0.161937
	64x64	0.000822
	128x128	0.000941
	256x256	0.000804
512x512	16x16	0.611717
	32x32	0.581860
	64x64	0.000774
	128x128	0.000755
	256x256	0.000772
	512x512	0.000880
1024x1024	16x16	2.566435
	32x32	2.709298
	64x64	0.000835
	128x128	0.000762
	256x256	0.000827
	512x512	0.000794
	1024x1024	0.000806

Fig. 2 Parallel computation

Fig. 3 Time taken for sequential computation

SEQUENTIAL COMPUTATION	
Image Size	**Time(ms)**
256x256	10.304164
512X512	20.355789
1024x1024	150.124727

5 Conclusion

In this work, the feature vectors of DICOM images corresponding to local tridirectional pattern are extracted using parallel programming. The tridirectional pattern helps in extraction of the most local information and the magnitude pattern, which helps in creation of the informative feature vector. The magnitude pattern will help us provide information regarding intensity weight for each pixel. In future work, an optimized feature extraction technique will be tried. The optimization of feature extraction can be carried out using optimized memory usage in GPU for the data. Overall, we found parallel local tridirectional feature extraction using GPU results in significant speedup, thereby saving significant feature extraction time.

References

1. Elkapelli, S.S., Damahe, L.B.: A review: region of interest based image retrieval. In: 2016 Online International Conference on Green Engineering and Technologies (IC-GET), pp. 1–6, IEEE (2016)
2. Siong, L.C., Zaki, W.M.D.W., Hussain, A., Hamid, H.A.: Image retrieval system for medical applications. In: 2015 IEEE Symposium on Computer Applications & Industrial Electronics (ISCAIE), pp. 73–77. IEEE (2015)
3. Chakraborty, S., Singh, S.K., Chakraborty, P.: Local directional gradient pattern: a local descriptor for face recognition. Multimedia Tools Appl. **76**(1), 1201–1216 (2017)
4. Yang, H., Wang, Y.: A lbp-based face recognition method with hamming distance constraint. In: Fourth International Conference on Image and Graphics (ICIG 2007), pp. 645–649. IEEE (2007)
5. Chan, C.H., Kittler, J., Messer, K.: Multi-scale local binary pattern histograms for face recognition. In: International Conference on Biometrics, pp. 809–818. Springer (2007)
6. Heisele, B., Ho, P., Wu, J., Poggio, T.: Face recognition: component-based versus global approaches. Computer Vision Image Understanding **91**(1–2), 6–21 (2003)
7. Hannuksela, J., Sangi, P., Turtinen, M., Heikkilä, J.: Face tracking for spatially aware mobile user interfaces. In: International Conference on Image and Signal Processing, pp. 405–412. Springer (2008)
8. Cui, J., Wen, F., Xiao, R., Tian, Y., Tang, X.: Easyalbum: an interactive photo annotation system based on face clustering and re-ranking. In: Proceedings of the SIGCHI Conference on Human Factors in Computing Systems, pp. 367–376 (2007)
9. Heusch, G., Rodriguez, Y., Marcel, S.: Local binary patterns as an image preprocessing for face authentication. In: 7th International Conference on Automatic Face and Gesture Recognition (FGR06), 6pp. IEEE (2006)
10. Gross, R., Brajovic, V.: An image preprocessing algorithm for illumination invariant face recognition. In: International Conference on Audio-and Video-Based Biometric Person Authentication, pp. 10–18. Springer, Berlin (2003)
11. Badanidiyoor, A.R., Naravi, G.K.: θ (1) time complexity parallel local binary pattern feature extractor on a graphical processing unit. ICIC Express Lett. **13**(9), 867–874 (2019)

Digital Media and Global Pandemic

Dobrinka Peicheva, Dilyana Keranova, Valentina Milenkova, and Vladislava Lendzhova

1 Introduction

Over the last decade, communication via the Internet has grown tremendously at the expense of communication with traditional television and radio. This fact is especially actual for young age groups, but other age groups also increase their height. A series of studies show that social networks, which have positioned themselves on the Internet, constantly expand their users' number and take up an increasing share of people's time to communicate with media [1]. Highlighting television share in this process of permanent restructuring of media consumption is a significant scientific challenge. A particularly important challenge is highlighting its share during a pandemic, such as the COVID-19 pandemic, not only as a scientific fact but also in highlighting its role as a consolidating factor in extreme situations.

Despite widespread observations and publications to restructure media communication in favor of new media, television continued to account for a large proportion of 69% during the pandemic. Young people in Bulgaria also watch TV every day, and 23% of them spend on average more than 3 h a day [2].

The article aims at highlighting the role of traditional and new media in the context of COVID-19, emphasizing the degree of their consolidating role and trust in them. The sub-topics that emerge from the set goal and are discussed in the article are: Which types of digital media are perceived as reliable sources of information during the pandemic crisis and is there an urgent need for new digital literacy of people to facilitate the critical adoption of diverse and even contradictory information in the media, often offered with political purposes.

D. Peicheva (✉) · D. Keranova · V. Milenkova · V. Lendzhova
Department of Sociology, South-West University "Neofit Rilski", 2700 Blagoevgrad, 66 Iv. Mihailov Str., Blagoevgrad, Bulgaria

M. Dua et al. (eds.), *Proceedings of the International Conference on Paradigms of Communication, Computing and Data Sciences*, Algorithms for Intelligent Systems,
https://doi.org/10.1007/978-981-16-5747-4_38

2 Overview

There is no doubt that digital media offers the opportunity to develop new media formats for communication and expression. A series of existing publications in Bulgaria and abroad describe the advantages and manifestations of digital media: overcoming geographical boundaries, condensing time, meta media forms, convergence between media, opportunities to combine different media elements and even entire media, increased capacity, the duality of the media, increased capacity, transformation of the audience into a co-author and even into independent media units, etc.

The article starts from the mediatization direction in the media theory, in which the media are presented in the most general terms as determinants of each process or phenomenon in the societies, including as determinants for relevant information and management [3–5]. It is also based on the scientific achievements of the media-ecological theoretical direction, in which attention is paid to the especially important topics today, concerning the purity and quality of the presented media products, the reliability, trust in media and their sources related to their socializing role among adolescents. Given the growing role of media in educational processes for acquiring new knowledge in the pandemic environment and the resulting requirements for new digital media literacy associated with high skills for a critical view of media products. Within the two theoretical basic conceptualizations, the media's influence in the pandemic situation is linked to the preferences of consumers for individual media through which they are informed as to the media, as is well known, tell people what people should think [6]. Hence the importance of the media for setting priorities and their potential influence during the pandemic.

Moreover, the media contents often contain ideological, political, and value interpretations, which are understood subjectively. "Reflexivity becomes a starting point and a necessary condition for critical thinking of people" [7].

It is no coincidence that within the media determination of the processes and ecological dimensions of media products in the EU, a *European framework for digital competence of citizens and a European action plan in digital education*—2021–2027 has been developed, which set several priorities. This refers to the targeted application of digital technologies in various fields, including people's education for the permanent development of digital competencies and skills, to improve their analytical and prognostic skills, to improve digital learning [8].

The plan emphasizes that as a result of the lockdown caused by the COVID-19 pandemic, digital media is becoming a key factor in managing the pandemic crisis and a responsible factor in the effectiveness of any management.

3 Materials and Methods

The article is based on the results of a national sociological survey on "Media and COVID-19", conducted by the authors in the first year of the pandemic crisis. The method used was an individual direct online survey based on a quota principle and carried out on the principle of responded participants. The national survey was conducted in the period April 29–May 3, 2020 and received the answers of 906 respondents over the age of 18 from all over Bulgaria. The results were processed with the SPSS program.

In addition to the primary analysis of research results, the article also presents a secondary analysis of comparable results from similar questions of other surveys in Bulgaria and abroad, although conducted according to different methodologies. A comparison is also made with data from similar questions from a newly completed study by the authors' team related to digital literacy, conducted March 05–April 05, 2021.

4 Results

The first result was related to trust in the media as the main source of information. Confidence has been key to crisis management's success with the COVID-19 pandemic and the challenges facing the media's consolidating role during a crisis.

Our research turned out that people have the most incredible information trust in television (regardless of its public or private nature).

Television has proved to be a highly predominant source of information about people and the on-screen reality that people trust for the picture it creates of the global pandemic situation. To the question: "Whom do you trust as a source of information about the pandemic crisis?" more than 47 percent of the respondents in Bulgaria put the television in the first place, with significant differences over the other digital media. Other representative surveys in other countries also confirm similar facts [9]. A survey conducted by the world agency TVB in 2020 also ranked television first, 54%. With a significant difference, television was followed by government websites with 27% and social media with 12% [10, 11].

Our data are similar: Television is followed by 31, 7% by government sites and 15.9% by social networks.

Television ranks highest among other media in surveys from the autumn of 2020 and in the just-released results of our last survey conducted March 05–April 05, 2021, when it again has the highest relative share of 47%, as we saw from the first processed data.

Television regained its preference for it and proved to be a highly consolidating factor for dealing with the pandemic and awareness of the ongoing processes in Bulgaria and abroad. When asked whether they think that special broadcasts are enough to educate people on this particularly important topic for health, the answers are predominantly positive.

Almost half and the largest share of respondents (46, 3%) answered that the television programs they were mainly informed about are sufficient to get an idea and deal with COVID-19. Distrust and demand for information materials from other information sources turned out to be valid for a little over 20% of the respondents. The top-rated social network Facebook in Bulgaria has also revealed itself as a source of information about the pandemic, albeit with a relatively minor share. Even lower shares are searching for information from other social networks: Instagram, Twitter, TikTok, Snapchat, etc.

Television has regained its importance as a preferred primary media in crisis circumstances. According to the global agency Nielsen, in the first few weeks of the pandemic alone, TV viewing preferences are estimated at 60% in the USA [12]. The European Union Radio and Television data (2020) also reveal similar preferences.

The second significant result was related to the reliability of information and false news. The enhanced consolidating role of television in dealing with the pandemic and raising awareness of the country's processes and the world is also a reflection of the reliability of the information created and disseminated by it. Especially when compared to some new internet-based media and their uncontrolled information possibilities for spreading unverified and politically biased allegations. Feenstra et al. define information products as vital civic tools, including propaganda purposes and political populist abuses [13]. Bennett and Livingston attribute the heightened misinformation to the political polarization, fragmentation, and configuration of a very high media environment and the rise of populism [14, 15]. The constant saturation of the modern media ecosystem with traditional and new digital channels and platforms and an unlimited number of individual disseminators of information, including politicians, proved to be an environment for increased mistrust, misinformation, and populism during COVID-19.

This fact is confirmed by the data presented by the British research agency Statista [16], related to the trust in social networks for COVID-19. With 13% trust in the Statista survey and more than three times more distrust, the social network Facebook ranks last in trust. It turned out that YouTube has the highest confidence in the information about COVID-19 and, accordingly, with the lowest distrust—15% [16].

Social media is used more and more dynamically, not only to spread fake news but also to be defined as the largest generator in terms of scale and speed for their creation and distribution [17]. That is why the ability of people to deal with misinformation is extremely important.

The results of our study are encouraging. Despite the growing difficulties caused by the increasing digital opportunities for the emergence of counterfeit media products, just over 12% are people who say they cannot deal with this problem. However, the share of those already beginning to find it challenging to deal with some media products' credibility is growing. It turns out that about 80% of respondents believe that they do not have specialized skills for recognizing false information. The emergence and dissemination of specialized digital tools and videos to deal with, as well as the European project "Media Literacy for All," are precisely the expression of European concerns about "widespread misleading or outright misinformation" [18].

5 Discussion

We have reason to believe that television, regardless of its form of ownership—public or private, rehabilitates its leading role from the first years of its creation.

The special emphasis placed on many of the TV channels through the traditional columns and by creating new sections with news, the diversity of points of view and the presence of analysts with opposing points of view, intensified discussions on the topic, etc. proved to be relatively sufficient even for new phenomena such as COVID-19.

Simultaneously, attitudes toward pandemic management were accompanied by a denial of the pandemic's existence by some groups and neglecting management measures to reduce it. Our study gave us reason to believe that the reasons for differences in trust and management assessments of the crisis are mainly political and are related to the political orientation of the leading political leaders in our country. About half of the respondents in the survey support this thesis, and this is the highest share of the respondents' opinion.

The presence of political determination is typical in other countries, in the USA, for example, where people's opinions were almost divided in two during the same period [19].

We have reason to believe that the ability to read content from various digital media successfully is one of the relevant ways to eliminate political bias, especially during pandemics. Skills are proving critical to the overall social system, not just the media ecosystem. Regardless of the degree of preference or consolidating role of the media, we are convinced that this degree depends on people's presumptive attitudes to interpretations of facts and data about the ongoing processes in general and particularly on preconceived notions of interpretations of health policies. Dissemination of unverified information or skeptical statements during the crisis are the biggest ethical deficits for consolidation, and conversely, dissemination of verified and reliable information is the biggest positive for consolidation.

The opposition between individual leaders and groups can be tolerated in specific political situations, including playing a consolidating role, but during vital phenomena and health crises, it has destabilizing and destructive results.

The acquisition of constantly updated and complex digital skills and related skills for critical reading and assimilation of information from various media forms and content is a necessary and crucial tool for dealing with growing disinformation products and populist techniques.

6 Conclusion

Our survey revealed that, in general, most respondents in Bulgaria trust television for their relevant awareness and for dealing with the danger of the spread of COVID-19. During the pandemic, television regained its original role as a major source of information, including the population's mass consolidation.

The role of the media in consolidating people to deal with the global pandemic crisis COVID-19 as the main purpose of our study was proved.

Based on the direction of mediatization in media theory, it was found that the media and television, in particular, are a decisive factor in creating a mediatized image of the pandemic and promoting approaches to dealing with it. The media is revealed as a tool for the cohesion of groups and societies in the face of the challenges of one of the most important global processes in modern times—preserving human life and health.

The mediatization of the pandemic presupposed the mediatization of the possible tools and the mediatized solutions of people for coping.

The increase in trust in traditional media, particularly in television and public government websites during the crisis, has resulted in the environmental friendliness of the media's information products. The restructuring of people's trust in individual media, on the other hand, results in the reliability of the information and information products created and disseminated by them. Populist policies and the ambitions for certain political leaders for power further erode serious media ecosystem segments. The growing role of the media in education and the assimilation of behavioral patterns in general and for the pandemic situation poses serious challenges to the permanent improvement of people's digital media skills. The need for new media education and literacy for full inclusion in the mediatized democratic and cultural changes in the society is growing especially strong [20].

Acknowledgements The article is developed within the research project "Digital Media Literacy in the Context of the Knowledge Society: Status and Challenges" № KP-06-H25/4, funded by the National Science Fund.

References

1. Theoremus for Audience Profiling: Current data on the consumption of social networks in our country (2019). https://dialogical.team/потребление-на-социални-мрежи
2. Media consumption and the "influencers" of children and young people in Bulgaria. ESTAT UNICEF, Bulgaria (2020) https://www.unicef.org/bulgaria/доклади/медийното-потребление-и-инфлуенсърите-на-децата-и-на-младите-хора-в-България
3. Peicheva, D: Beginnings of a mediatization of modern society. Roman. Rev. J. Commun. **1**(2–3) (2006)
4. Hjarvard, S.: The mediatization of society. A theory of the media as agents of social and cultural change. Nordicom Rev. **29**(2), 105–134 (2008)

5. Lundby, K. (ed.): Mediatization: Concept, Changes. Consequences. Peter Lang, New York (2009)
6. McCombs, M., Shaw, D.: The agenda-setting function of the mass media. Public Opin. Quart. **73**, 176–187 (1972)
7. Peicheva, D., Milenkova, V.: Knowledge society and digital media literacy: foundations for social inclusion and realization in Bulgarian context. Qual. Life **1**, 50–74 (2017)
8. Digital Education Action Plan (2021–2027). Resetting education and training for the digital age. Education and Training, https://ec.europa.eu/education/education-in-the-eu/digital-education-action-plan_en (2020).
9. Hot in peak. Alpha-research with new data—Rumen Radev with record low confidence (https://pik.bg/горещо-в-пик-алфа-рисърч-с-нови-данни---румен-радев-срекордно-ниско-доверие-премиерът-бор-news927707.html) (2020, May11)
10. Johnson, J.: Trust in coronavirus news on social media in the UK in December 2020, by platform How much do you trust information/news about Coronavirus on social media. Statista (2020, December 9). https://www.statista.com/statistics/1112330/trust-in-coronavirus-news-on-social-media-in-the-uk/
11. 2020 Coronavirus Media Usage Study. TVB. https://www.tvb.org/Public/Rsearch/2020
12. CoronavirusMediaUsageStudy.aspxStaying put consumers forced indoors during crisis spend more time on media. Nielsen (2020) https://www.nielsen.com/us/en/insights/article/2020/staying-put-consumers-forced-indoors-during-crisis-spend-more-time-on-media/
13. Feenstra, R.A., Tormey, S., Casero-Ripollés, A., Keane, J.: La reconfiguración de la democracia. Granada, Comares (2016)
14. Bennett, W.L., Livingston, S.: The disinformation order: disruptive communication and the decline of democratic institutions. Eur. J. Commun. **33**, 122–139 (2018). https://doi.org/10.1177/0267323118760317
15. Waisbord, S.: Truth is what happens to news: journalism, fake news, and post-truth. J. Stud. **19**, 1866–1878 (2018). https://doi.org/10.1080/1461670X.2018.1492881
16. Statista. (2020, December 9) https://www.statista.com/statistics/1112330/trust-in-coronavirus-news-on-social-media-in-the-uk/
17. Fighting the Infodemic: The #CoronaVirusFacts Alliance (2020) https://www.poynter.org/coronavirusfactsalliance/
18. Tackling online disinformation: An European Approach. European Commission. Shaping Europe's digital future (2020). https://ec.europa.eu/digital-single-market/en/tackling-online-disinformation
19. Zhao, E., Wu, Q., Crimmins, E.M., et al.: Media trust and infection mitigating behaviours during the Covid 19 pandemic in the USA. BMJ Glob. Health (2020). https://doi.org/10.1136/bmjgh-2020-003323
20. Milenkova, V., Keranova, D., Peicheva, D.: Digital skills, new media and information literacy as a conditions of digitization. In: AHFEE 2019. Applied Human Factors and Ergonomics, pp. 65–72. Springer, Berlin (2019)

Data Sciences

Human Activities Analysis Using Machine Learning Approaches

Divya Gaur and Sanjay Kumar Dubey

1 Introduction

Human activity recognition (HAR) refers to identifying what an individual is doing by tracing their movements based on the accelerometer data recorded on smartphones or on some specific harnesses. The smartphone sensors mostly used are accelerometer, gyroscope, etc., and the movements usually include standard indoor activities like sitting, standing, going downstairs or upstairs, etc. In recent years, as the applications of human activities recognition, model has increased so there has been a rapid increase in research on human behavior recognition. Some of the applications are like surveillance system, anti-crime securities, health care, etc. Earlier recording sensor data used to be quite challenging and expensive as the collection of data required custom hardware, but now it has become quite easy as smartphones and other required tracking devices are easily affordable [4]. Despite the amount of research in this field, AR continues to be a challenging problem in unregulated smart environments. In the investigation of machine learning techniques for activity recognition (AR) issues, recently, there has been an improvement in the number as they were very successful in extracting and learning information from activity datasets [15]. This study proposes a human movement recognition model using logistic regression, decision tree, linear support vector classification and random forest classifier. Performances of all the classifiers have be analyzed. The data used for this study was collected using accelerometer and gyroscope sensors. The features were extracted from the dataset, and then the four machine learning approaches were applied for

D. Gaur (✉) · S. K. Dubey
Department of Computer Science and Engineering, Amity University, Sec-125, Noida Uttar Pradesh, India

S. K. Dubey
e-mail: skdubey1@amity.edu

M. Dua et al. (eds.), *Proceedings of the International Conference on Paradigms of Communication, Computing and Data Sciences*, Algorithms for Intelligent Systems,
https://doi.org/10.1007/978-981-16-5747-4_39

developing a human activity recognition model followed by comparison of accuracies of these algorithms. The paper is divided into four sections. Section 2 describes previous work analysis on HAR. Section 3 describes the techniques that has been used for performing this study. Section 4 includes the analysis and results that has been performed for developing the behavior recognition model. And finally, Sect. 5 presents conclusion and the future scope.

2 Literature Analysis

Various researchers have worked on human activity recognition in last years. Some of the research work has been listed in Table 1.

The state of the art of this paper is that from the above literature analysis, it has been found that logistic regression and random forest have achieved better accuracy, but in our study, we have found that the linear support vector classification has achieved better accuracy as compared to logistic regression and random forest classifier for human activity recognition.

3 Research Methodology

The dataset used for activity recognition was UCI HAR dataset. The data included readings of accelerometer and gyroscope captured from 30 volunteers or subjects performing six different activities. The study included activity recognition using four machine learning algorithms.

Logistic regression is applicable for binary variables not for continuous variables for example in order to predict whether the person is male (0) or female (1). This can be explained through one more example like for predicting whether the cancer in a patient is controllable or uncontrollable. In case of linear regression, a threshold level is set. So, if the patient is malignant, predicted continuous value is 0.4, and the threshold value is fixed to 0.6, then the cancer will be asserted as non-malignant. So, this case can led to highly serious consequences in real-world scenarios because of such kind of limitation logistic regression came into picture as it plays only with binary variables so will help in proper prediction [2].

Random forest as the name suggests is a set of large number of decision trees working in a group. In order for random forest giving good performance, firstly, the individual decision trees should have low correlation with each other, and secondly, some target indicators in the features should be there so that the model can be developed on those target features not on random guessing [16]. And other two machine algorithms used were decision tree and LinearSVC.

Table 1 Analysis of previous research on human activity recognition

Objective	Methods	Dataset	Analysis	Ref
Proposed a framework for recognizing complex human activities in real time	FR-DCNN (fast and robust deep convolutional neural network)	Data was collected by installing MATLAB application in smartphones (iPhone 6 s), and this data was then stored to cloud using Wi-Fi network	The experiment was conducted on 12 complex activities, and the FR-DCNN model achieved accuracy of 95.27% in 0.0029 s	[14]
Comparison of advantages and limitations of five algorithms	Convolutional neural network (CNN), long short-term memory (LSTM), bidirectional long short-term memory (BLSTM), multilayer perceptron (MLP) and support vector machine (SVM)	UCI and Pamap2 datasets	200 iterations were performed on each model for comparison. It was found that all the four algorithms, namely CNN, LSTM, BLSTM and MLP performance were recorded better on UCI dataset as compared to Pamap2 dataset. And CNN performed better than LSTM and BLSTM	[18]
Performed a systematic analysis about positions of sensors on body and acquisition of data for human activity recognition systems	LSTM	Collected from body-worn Inertial measurement units (IMU) sensors and Android mobile devices	The study identified that low sampling rate of data (~10 Hz) is sufficient for HAR and for sensor positioning, placing sensor one on the upper half and the other on the lower half can result in reasonable performance in identifying daily living activities	[5]

(continued)

Table 1 (continued)

Objective	Methods	Dataset	Analysis	Ref
Developed a model that automatically identified and classified the activity features	CNN-LSTM	UCI, WISDM and opportunity	The model achieved accuracy of 95.78, 95.85 and 92.63% on UCI HAR dataset, WISDM dataset and opportunity dataset, respectively	[19]
Developed a real-time human recognition system using data Augmentation and Two-Stage End-To-End CNN	Two-stage end-to-end (TSE)-CNN	Wearable sensor data	The study found that in comparison with neural network methods and other methods developed model achieved, improved accuracy in activity recognition and also reduced the computational complexity	[7]
Developed a human activity recognition model using deep learning approach	CNN, LSTM	iSPL dataset, UCI HAR dataset	The model achieved 99% accuracy on iSPL dataset and 92% on UCI HAR dataset	[12]
Developed a human activity recognition model using wearable sensors	CNN	Real-world human activity dataset	The motion signals from two sensors were collected, then these collected signals were converted to image sequences, and these image sequences were given as input to two CNNs and achieved recognition accuracy (F1-score) of 0.87	[10]

(continued)

Table 1 (continued)

Objective	Methods	Dataset	Analysis	Ref
Developed an activity recognition model for tracking activities of older people	Random forest	GOTOV dataset	A random forest activity recognition model was trained based on varying positions of sensors and the developed model which combined wrist and ankle accelerometer produced best accuracy > 80% for classification of 16 activities	[13]
Proposed a human activity recognition model on edge devices	RNN-LSTM	WISDM dataset	An accuracy of 99% was achieved by RNN-LSTM model on jogging and walking activities. For upstairs, activities minimum accuracy of 81% was achieved	[1]
Developed a new feature extraction algorithm for extraction of important features	Particle swarm optimization (PSO) and SVM algorithm	Motion sense dataset	The activities classification was performed using PSO and SVM algorithm and the accuracy achieved was 87.50%	[3]
Activity recognition performance comparison was done using logistic regression and support vector machine	Logistic regression (LR), support vector machine (SVM)	UCI HAR dataset	LR and SVM method achieved 100% accuracy in predicting the sitting activities with error rate less than 2% for standing and (up to) 4% for laying activities	[11]

(continued)

Table 1 (continued)

Objective	Methods	Dataset	Analysis	Ref
The study compared the machine learning and deep learning Techniques for recognition of human activities	Basic machine learning algorithms and deep learning techniques	[16]	The study analyzed that the deep learning is not always the best option for sensor-based and machine-based data. The performance highly depends on the topology and the parameters that are chosen	[2]
Presented a smartphone sensor-based approach for developing robust human activity recognition model	Kernel principal component analysis (KPCA), deep belief network (DBN)	UCI dataset	The proposed method was compared to the traditional SVM, and it was found that the proposed method outperforms the traditional SVM	[6]
Activity recognition in real time using CNN	CNN	WISDM and UCI datasets	The proposed method achieved good performance without requiring manual feature engineering and with low computational cost	[8]
Proposed an environment-independent activity recognition model	Deep learning		Conducted experiment on device-free four different activity recognition testbeds, namely Wi-Fi, ultrasound, visible light and 60 GHz mmWave and found that the proposed method was very effective	[9]

4 Analysis and Results

The data was gathered using sensors present in user's smartphones. The sensors used were accelerometer and gyroscope. The accelerometer is responsible for detection of directions and magnitude of proper acceleration. Six activities, namely walking, walking upstairs, walking downstairs, lying, sitting and standing, were performed by 30 volunteers. The activities were divided into stationary and dynamic activities. The stationary activity included sitting, standing and lying, whereas the dynamic activity included walking, walking upstairs and walking downstairs. Figure 1 depicts the graph for stationary and dynamic activities. Static activities will not be able to provide motion activities, whereas dynamic activities will be able to give significant motion information.

The readings of accelerometer were divided into two types gravity acceleration readings and body acceleration readings, respectively, which have x, y and z components. The measures of angular velocities having x, y and z components are gyroscope readings. The rate at which the device rotates about its axis (*x*-, *y*- and *z*-axis) is defined as angular velocity. The gyroscope's x, y and z attributes represent the angular velocity around *x*-, *y*- and *z*-axis, respectively.

The data was gathered by capturing linear acceleration of three axes from accelerometer and three-axial angular velocity from gyroscope. Firstly, by applying noise filters, the preprocessing of sensor signals was done, and sampling was done in fixed sliding windows of 2.56 s. Each window is having 128 readings. By calculating variables from time and frequency domains, feature vector was acquired from each window. Low-pass filter separated the acceleration signal into body acceleration signals and gravity acceleration signals. After all of this, the jerk signals were obtained by deriving the linear acceleration of body and angular velocity.

Some of the signals received from sensors were time-body acceleration signals around *x*-axis, *y*-axis and *z*-axis, time-gravity acceleration signals around *x*-axis, *y*-axis and *z*-axis, time-body acceleration jerk signals, time-body gyroscope signals,

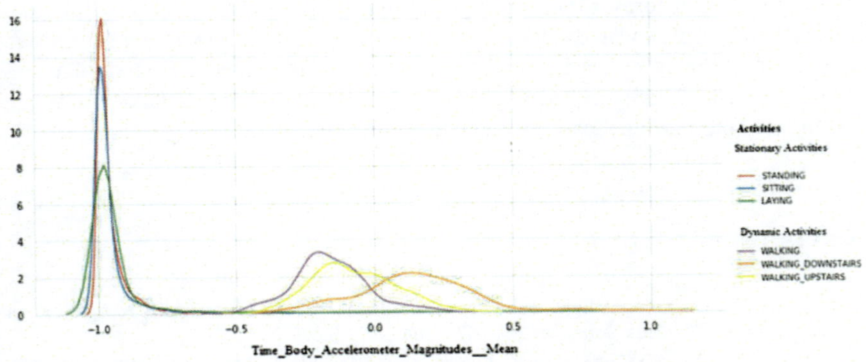

Fig. 1 Graph for depicting stationary and moving activities

Table 2 Results obtained for human activity recognition model

Machine learning algorithm	Accuracy (%)
Random forest classifier	91.26
Logistic regression	96.29
LinearSVC	96.35
Decision tree	83.44

time-body gyroscope jerk signals, etc. The training data and the test data were divided in the ratio of 70 and 30.

The study proposed human activity recognition using four machine learning algorithms. The algorithms used were logistic regression, random forest classifier, support vector classification in linear kernel, decision tree. The accuracies and error were formulated for each machine learning algorithms specified above. The accuracy achieved with logistic regression was 96.29%, and error obtained was 3.73%. The accuracy achieved by random forest classifier was 91.26% and error obtained 8.65%. An accuracy of 83.44% and error 13.54% were achieved from decision tree. The LinearSVC obtained accuracy of 96.35%. So, among logistic regression, decision tree, linear support vector classification and random forest classifier, LinearSVC outperforms among all four specified machine learning algorithms. Table 2. contains machine learning techniques along with their accuracies.

5 Conclusion and Future Scope

In this paper, human activity recognition model was proposed using four machine learning algorithms. These algorithms were logistic regression, random forest classifier, decision tree and linear support vector classification. Accuracies of these four algorithms were obtained, and it was observed that among these four machine learning algorithms, linear support vector classification achieved best accuracy with respect to random forest classifier, logistic regression and decision tree. The data for conducting study was captured using accelerometer and gyroscope sensors. A more lightweight and optimized human activity recognition model can be built using hybrid methods. The hybrid method can be formed by building a HAR model by combining both machine learning approaches and deep learning algorithms.

References

1. Agarwal, P., Alam, M.: A lightweight deep learning model for human activity recognition on edge devices. Procedia Computer Sci. **167**, 2364–2373 (2020)
2. Baldominos, A., Cervantes, A., Saez, Y., Isasi, P.: A comparison of machine learning and deep learning techniques for activity recognition using mobile devices. Sensors **19**(3), 521 (2019)

3. Batool, M., Jalal, A., Kim, K.: Sensors technologies for human activity analysis based on SVM optimized by PSO algorithm. In: 2019 International Conference on Applied and Engineering Mathematics (ICAEM), pp. 145–150. IEEE (2019, August)
4. Brownlee, J.: Deep Learning Model for Human Activity Recognition (2019)
5. Chung, S., Lim, J., Noh, K.J., Kim, G., Jeong, H.: Sensor data acquisition and multimodal sensor fusion for human activity recognition using deep learning. Sensors **19**(7), 1716 (2019)
6. Hassan, M.M., Uddin, M.Z., Mohamed, A., Almogren, A.: A robust human activity recognition system using smartphone sensors and deep learning. Futur. Gener. Comput. Syst. **81**, 307–313 (2018)
7. Huang, J., Lin, S., Wang, N., Dai, G., Xie, Y., Zhou, J.: Tse-cnn: A two-stage end-to-end cnn for human activity recognition. IEEE J. Biomed. Health Inform. **24**(1), 292–299 (2019)
8. Ignatov, A.: Real-time human activity recognition from accelerometer data using convolutional neural networks. Appl. Soft Comput. **62**, 915–922 (2018)
9. Jiang, W., Miao, C., Ma, F., Yao, S., Wang, Y., Yuan, Y., et al.: Towards environment independent device free human activity recognition. In: Proceedings of the 24th Annual International Conference on Mobile Computing and Networking, pp. 289–304 (2018)
10. Lawal, I.A., Bano, S.: Deep human activity recognition using wearable sensors. In: Proceedings of the 12th ACM International Conference on PErvasive Technologies Related to Assistive Environments, pp. 45–48 (2019, June)
11. Minarno, A. E., Kusuma, W. A., & Wibowo, H. Performance Comparisson Activity Recognition using Logistic Regression and Support Vector Machine. In *2020 3rd International Conference on Intelligent Autonomous Systems (ICoIAS)* (pp. 19–24). IEEE (2020, February)
12. Mutegeki, R., Han, D.S.: A CNN-LSTM approach to human activity recognition. In: 2020 International Conference on Artificial Intelligence in Information and Communication (ICAIIC), pp. 362–366. IEEE (2020, February)
13. Paraschiakos, S., Cachucho, R., Moed, M., van Heemst, D., Mooijaart, S., Slagboom, E.P., Beekman, M.: Activity recognition using wearable sensors for tracking the elderly. User Model. User-Adap. Inter. **30**(3), 567–605 (2020)
14. Qi, W., Su, H., Yang, C., Ferrigno, G., De Momi, E., Aliverti, A.: A fast and robust deep convolutional neural networks for complex human activity recognition using smartphone. Sensors **19**(17), 3731 (2019)
15. Ramasamy Ramamurthy, S., Roy, N.: Recent trends in machine learning for human activity recognition—a survey. Wiley Interdiscip. Rev. Data Mining Knowl. Discovery **8**(4), e1254 (2018)
16. Shoaib, M., Bosch, S., Incel, O.D., Scholten, H., Havinga, P.J.: Complex human activity recognition using smartphone and wrist-worn motion sensors. Sensors **16**(4), 426 (2016)
17. Swaminathan, S.: Logistic Regression—Detailed Overview (2018)
18. Wan, S., Qi, L., Xu, X., Tong, C., Gu, Z.: Deep learning models for real-time human activity recognition with smartphones. Mobile Networks Appl. **25**(2), 743–755 (2020)
19. Xia, K., Huang, J., Wang, H.: LSTM-CNN architecture for human activity recognition. IEEE Access **8**, 56855–56866 (2020)
20. Yiu, T.: Understanding Random Forest (2019)

An Approach to the Application of Ontologies in the Knowledge Management of Companies

Ihosvany Rodríguez González, Anié Bermudez Peña, and Nemury Silega Martínez

1 Introduction

Knowledge management (KM) is a process where organizations earn and produce value, due to the knowledge that their workers print, which is growing every day. We have selected these two definitions as a starting point for the development of the paper:

- Zapata [1] states that KM organizes, plans, and controls the knowledge chains that are created in the company, its interaction with the environment and its tasks, in order to promote essential competencies; which are the result obtained by the assessment of three main classes of competencies: Organizational, Technological and Personal.
- Senge [2] shows KM as a human and humanistic discipline, therefore, that does not focus much on the material, but is oriented mainly in relation to the human being as the main link of the organization.

The word ontology begins in philosophy; but in artificial intelligence (AI) it has several definitions. The strongest is that of Gruber [3] and refers to it as "an explicit and formal specification of a shared conceptualization". From this it is interpreted that ontologies define concepts and their relationships in a domain of discourse, and that this must be constituted in a readable, formal and usable way by computers.

To represent contents, characterizing and classifying the information through generalization of general and expert areas, it is significant to use ontologies.

I. R. González (✉)
Centro Nacional de Biopreparados, Bejucal, Mayabeque, Cuba
e-mail: ihosvany@biocen.cu

A. B. Peña · N. S. Martínez
Universidad de Las Ciencias Informáticas, Havana, Cuba
e-mail: nsilega@uci.cu

M. Dua et al. (eds.), *Proceedings of the International Conference on Paradigms of Communication, Computing and Data Sciences*, Algorithms for Intelligent Systems,
https://doi.org/10.1007/978-981-16-5747-4_40

The definition of ontology in AI is used to declare vocabularies that computers understand and that have the necessary precision to distinguish terms and can be better referenced.

Ontologies allow computers to catalog knowledge, and reuse it. From this we can deduce that organizations have advanced information and communication technologies, which implies that organizational knowledge can be categorized and widely disseminated, in addition to being managed with adequate feedback.

By adding meaning to the resources available on the web, through the semantic web (SW), it is possible to automate content, combine it, reason with it, and make reasonable assumptions to resolve situations more quickly and accurately.

To locate the bibliographic documents, several documentary sources were used, a bibliographic search of some of the electronic databases, books, theses, articles, high-impact journal, which are indexed by the search engine "academic google", using keywords and combinations between them. The records obtained ranged from 100 to 130, selecting the most relevant documents that appear in the references.

The techniques used in the research are: exploratory and descriptive. The theoretic methods: Analytic-Synthetic and Historical-Logical Analysis. As well as documentary analysis as an empirical method. Following, the most relevant data from the review are broken down by category.

Authors: Javier Salazar Argonza, Tim Berners-Lee, Thomas R. Gruber, Vitaliy Mezhuyev, Mónica Alarcón Quinapanta, Peter M. Senge, Laura Esther Zapata Cantú, Lan Yang, Ian Horrocks, Deborah L. McGuinness, Christopher J Mungall, Ian Horrocks, Mark Musen, Michael Vierhauser.

Databases: Medline, Pubmed, Springer, SciELO, Web of Science, Scopus, Doaj and Ebsco.

Countries: Great Britain, United States of America, Russia, China, Japan, Spain and Ukraine.

In this paper the following specific objectives are proposed:

1. Describe semantic web technologies that try to solve problems in the current web.
2. Analyze the benefits of ontologies in KM in organizations.
3. Expose domains in which the ontologies have been applied.

2 The Knowledge Management in Organizations

Next, the sections that give answers to the objectives are shown: KM in organizations, ontologies in SW, applications of ontologies in some areas of knowledge and the potentialities that these relationships have for the scientific community.

KM is based on explicit and tacit knowledge, the first one focuses on that which has been or can be articulated, encoded and stored in some type of medium and can be immediately transmitted to others. On the other hand, the second is much more complicated, it is part of our mental model, the result of our personal experience and involves intangible factors such as beliefs, values, intuition, points of view, etc., and therefore we cannot structure, store or to distribute. It is the most difficult to manage and at the same time the one that has the most possibilities of generating competitive advantage, since it is practically impossible to imitate, this type of knowledge has been and is currently highly valued by the most competitive companies and organizations.

The benefits of a good KM are related to avoiding the loss of personnel, which helps in the formation of a solid company and with trust among people, another benefit is the increase in production, because the employees will give the best of themselves for the company to help them, as it is necessary for the company to collaborate with the worker. The challenges that the KM has is to make the company see that people should not be ignored, nor their requests. Many times companies exercise KM when they motivate employees, with this a particular goal is given and the entire group of the company joins to achieve it [4].

The organization is considered as a living being, where the learning guidelines are similar to those of the human, which is why it is proposed that organizations, like people, manage their knowledge in four ways or style, taking into account the expressed approaches by Gómez [5], which are mentioned below:

- Active: organizations that learn from their own experience, without being risk-free. This is a special way of managing knowledge; organizations only rely on their own results.
- Reflective: organizations manage their information through systems, which will allow them to establish cause and effect relationships between the causes that originate them and the data.
- Theoretical: organizations manage knowledge based on their internal norms, their culture, their organizational principles, their vision and mission.
- Pragmatic: organizations only use knowledge that is useful to them. This knowledge provides usefulness in a short term, that is, it does not use or archive, and if necessary destroy irrelevant information.

Generally, those who have knowledge about the company teach the same to the other people in the company, so that the knowledge is public and everyone has the knowledge of that work. As not all information is important, it is in the hands of a company to say what knowledge they should have as primary.

The main process of KM systems is to obtain relevant sources of knowledge and use them in problem solving. Sources are divided into two categories: documents and formal rules. In order to carry out an efficient search for the knowledge in the second category, the content of the files is acquired from the declarations in a conditional condition-action, based on ontologies, using logical methods in their management.

Next, key knowledge guidelines are described, then it is discussed how ontologies can be applied to manage knowledge and the motivations for using them. In addition to how the KM group should be integrated into an organization.

2.1 Making the Intangible Tangible (Knowledge)

Organizations with a high development of technology are taking the model of intellectual capital, and the skills and capacities of their managers are the main assets. On the contrary, there are others where they do not admit the intellectual capital model, because they show business skills threatened by an uncertain future.

Alarcón [6] proposes the increase in the creation of the intellectual capital of an organization in six steps:

1. Missionary phase: begins with a few people seeing the underlying problem in order to convince the entire organization of the need for a fresh perspective.
2. Measurement: tries to indicate balance and taxonomy for a new model.
3. Direction: it is based on acting according to the new concepts from several levels.
4. Technology: focuses on the development of technology to increase transparency and information and communication systems, to distribute knowledge and the use of telecommunications and computing.
5. Capitalization: reaches the use of organizational technology and as intellectual property for the reaction of capital. In this phase it is key to reuse the main knowledge and the transformation of the company's structural capital.
6. Futurization: last phase, the systematic work of the invention is approached as the main competence of the organization in order to carry out an appropriate reform and development.

For Moran [7] employees who are working are not guaranteed their position anywhere, not even when the company is promoted. A company is not only measured with the assets it owns, not even with its products, intellectual capital is much more important, as it shows that a certain company is interested in creating information and knowledge.

In organizations, the human resource becomes more relevant every day, the key is in the minds of the people, who with their experience promote the growth of their knowledge; hence the importance that the different areas of an organization have the ability to learn in both directions and thus achieve collectively to solve the problems that arise, to identify new business opportunities for the growth of the company.

2.2 Key Knowledge Factors

Factors that can be related to knowledge, being intangible, play key roles in the organization, these are:

- Information management systems.
- Relations with clients.
- Treatment of the news.
- Internal information and communication tools.

- Data transmission.

In various organizational cultures, they correspond to dissimilar learning modes, which can be brakes to acquire or create knowledge.

2.3 *Knowledge Management Group in Organizations*

The knowledge management group is key to decision-making in the organization, through which the goals of the company can be achieved. The branches in which this internal communication function is applied is in data processing and human resources; especially the exchange of information throughout the administrative structure (collaborators, employees), as well as the knowledge between different levels of the organization. At present there is a growing role for information and knowledge, and it can lead the organization to establish and project strategies to improve KM.

3 Semantic Web and Ontologies

Berners-Lee [8], creator of the term SW. He defines it as the extension of web 2.0 where information already has a clear meaning, offering collaboration between humans and computers. The W3C in the last decade, has been building technologies and standards that support the accelerated growth of the SW. See Fig. 1.

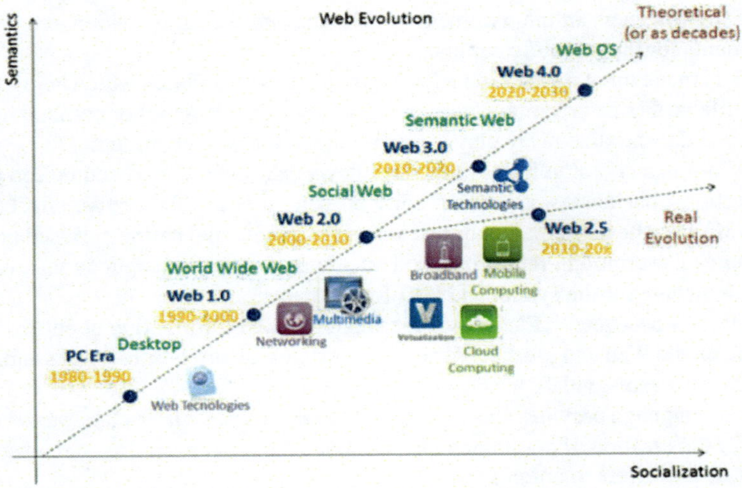

Fig. 1 Evolution and prediction of the web

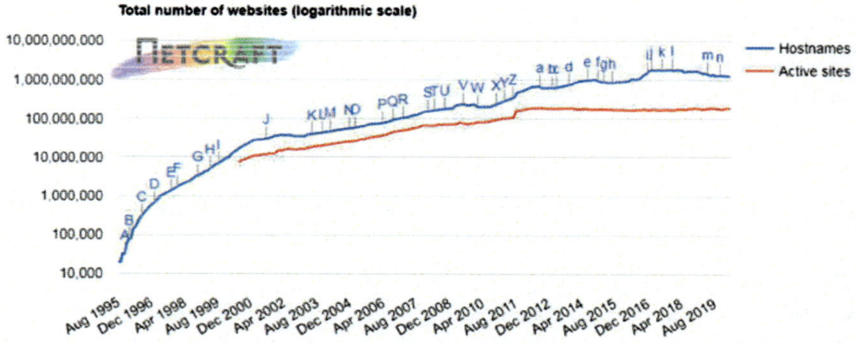

Fig. 2 Number of existing websites until 2019. *Source* [9]

The volume of information on the web continues to grow exponentially, there are more than 1 billion websites today. See Fig. 2.

The emergence of the SW is focused on the user, to develop and search capacity, increase the possibilities of improving their queries and deal with more robust applications to express knowledge. It allows computers to manage knowledge, until now reserved only for people and makes use of artificial intelligence.

Ontologies in SW play an important role in empowering computers with the ability to operate and organize information based on a semantic evaluation of its content, so that AI can understand the meaning of this information. content.

SW has a strong tendency to develop with advances in AI (rule-based expert systems, neural networks and other branches), and ontologies as an essential component of SW have been applied more to fields such as genetics, medicine, biology, education, among other areas of knowledge. The adoption of ontologies contributes to the improvement in the treatment of relevant, organized, classified and updated documents for the scientific community.

SW is made up of a series of technologies as shown in Fig. 3, which help to solve many life problems in various areas of knowledge and each tool or combinations of these in a decentralized way play a crucial role in the web evolution.

SW offers a common framework that allows data to be shared and reused across applications, businesses, and community borders. It is a collaborative effort led by the W3C with the participation of researchers and industrial partners. Based on RDF (Resource Description Framework) and integrates a variety of applications using the XML language for the syntax and URI for the names.

RDF is a data model that provides simple detailed information about resources located on the Web and used, for example: in cataloging directories, books and photo collections, events, video, music, etc.

OWL language provides specific terminologies or vocabularies interpreted by the Web for the creation of ontologies and facilitates the interoperability and integration of data OWL offers to ontologies:

- Be distributed through various systems.

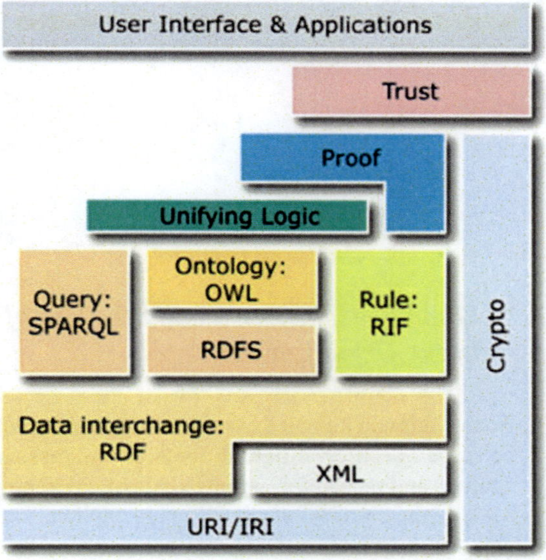

Fig. 3 Scheme of the SW and its technologies, by the W3C

- Scalable.
- Compatible with Web standards.

Ontologies are in charge of defining the terms that are used to formalize the information in a field or domain, using the necessary tools.

To create the ontological model, use Protégé [10] among others; most popular software tool that employs a user interface that allows the creation of an ontology with their respective identities, properties, rules and instances in an integrated way.

Descriptive logic (DL) is a family of languages that personalize knowledge with the vocabulary of a specific domain in a well-structured and understood way, one of these languages is OWL-DL, which is used in the tool Protects. It also makes use of a domain assertion set (ABOX) and a domain terminology (TBOX).

On the other hand, reasoners are known who work with ABOX and TBOX to verify if the TBOX is true, or to check the slots that are included in others. Inference rules are always established for searching by intelligent agents.

Ontologies need a logical and formal language to be expressed. In artificial intelligence, many languages have been created for this purpose, some based on predicate logic and others based on frames (attributes and class taxonomies), which have more expressive power, but less power of inference; there are even languages more oriented to reasoning. These languages have been used to develop other languages applicable to the Web. For this reason, Protégé is used, which implements the OWL language for modeling ontologies based on Frames. It is used for the development of Ontologies and Systems based on knowledge through a user interface that facilitates the creation of frame structures with classes, slots and instances in an integrated way.

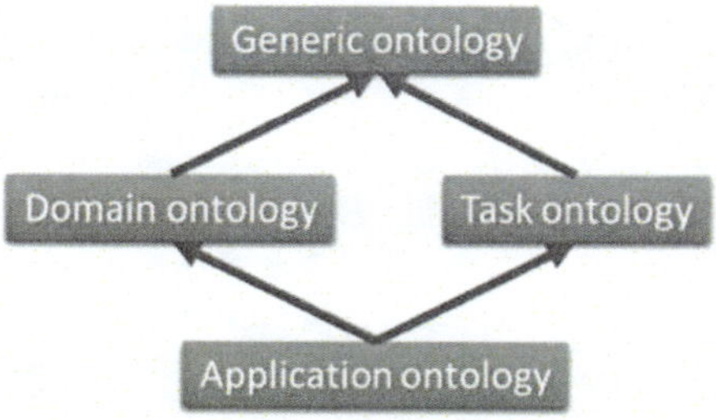

Fig. 4 Types of ontology

3.1 *Some Reasons that Justify the Use of Ontologies in the Organizations*

- Enable domain knowledge reuse, to determine which aspects of an ontology can be reused between different domains and tasks. Ontology libraries that can be adapted and reused for various kinds of environments and problems.
- Making the premises about a domain explicit. The availability of explicit ontologies for information resources are basic in the mediation-based approach to information integration.
- Separating the operational knowledge from the domain knowledge, we can differentiate which ontology to develop according to its classification: of domains, of tasks or of applications.
- Formally analyze the knowledge of the domain, graphically describe a domain that is more understandable by humans who are not close to this domain and this helps considerably the good management of information and KM.
- To share an understanding about the semantics of the information between software and human agents for a better KM.
- Ontologies presume a vital resource for many systems that have relationships between their elements and a degree of reasoning for learning.

Guarino [11] catalogs the ontologies according to their level of dependency and the relationship with a specific activity, See Fig. 4.

- Generic ontologies: they describe basic and general concepts.
- Domain ontologies: they represent a general domain terminology.
- Basic task ontologies: they represent a specific activity.
- Ontologies application: they represent concepts that depend on a specific task and/or domain.

3.2 *Methodologies for Designing Ontologies*

Next, methodologies for designing ontologies and the fundamental steps of one in particular are listed:

- Cyc
- Uschold and King
- Grüninger and Fox
- KACTUS
- METHONTOLOGY
- SENSUS
- On to Knowledge
- DILIGENT
- Upon
- Finish
- No and
- Stuart
- Neon

At present Neon is one of the most applied methodologies according to academic google, it is based on nine stages, which can be combined among them for the construction of ontologies and ontology networks, with special attention on the reuse and reengineering of knowledge resources (ontological and non-ontological):

1. Treatment of ontology networks from specification to implementation.
2. Reuse and reengineering of non-ontological resources.
3. Reuse of ontological resources.
4. Reuse and reengineering of ontological resources
5. Reuse and mix of ontological resources.
6. Reuse, mix and reengineering of ontological resources.
7. Reuse of ontological design patterns
8. Restructuring of ontological resources.
9. Location of ontological resources.

Figure 5 shows more clearly a graph with the relationships between the stages.

3.3 Applications with Ontologies

Cantor in 2007 classified the use cases of ontologies to support knowledge management as:

- Web Portal [13].
- Rule to optimize search on the web [14].
- Multimedia repository [15].
- Corporate Websites—administration [16].
- Taxonomic organization of data and documents [17].
- Allocation between Corporate Sectors [18].
- Administration of restrictions [19].
- Expression and interests of users [20].
- Mapping between websites [21].

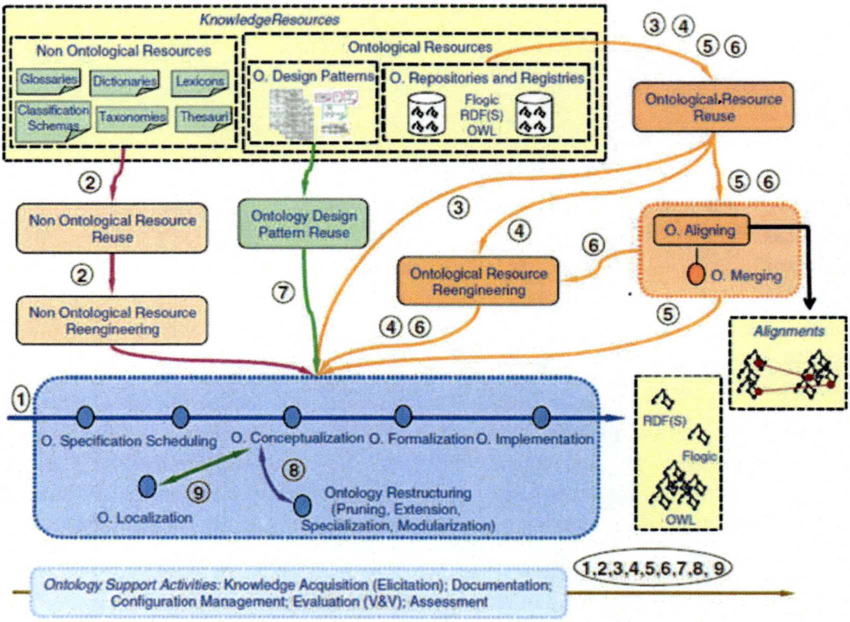

Fig. 5 Stages for the construction of ontologies and ontology networks according to the NEON methodology. *Source* [12]

- Discover Web Services and their composition [22], among other.

The review carried out revealed the widespread application of ontologies in different domains. Below are examples of its application in the domains of public health and related areas such as in the field of biomedicine, domain of digital libraries and domain of software engineering. In these domains, the benefits generated by the adoption of ontologies to support their processes have been widely documented.

3.4 Ontologies Applied in the Management of Knowledge in Public Health

Building an effective large-scale epidemiological system to model the immunity of a given population to an infection depends on the integration of data not only from biology and medicine, but also from statistics, sociology and geography.

The system should incorporate data from the entire society on the rates of occurrence of this infection, mode of transmission, birth rates, vaccination rates, family structures, age distribution and other relevant demographic factors [23], and also patient-specific data on clinical manifestations of disease, diagnoses and treatments received. The formation of silos is a difficulty both in research and in public health prevention.

Ontologies are an effective tool for sharing data [24]. But for ontologies to be effective, it is important that they are designed in a coordinated way; otherwise, the ontologies themselves will lead to the creation of a new type of silo.

One of the most successful and widely adopted approaches to coordinated ontology development is that of the Open Biomedical Ontology Foundation, where the principles of ontologies have been modeled on the practices of genetic ontology [24], which has served as a model for a series of life science ontologies that follow in his footsteps. Examples of projects are:

- Ontologies using relational terms and expressions taken from existing ontologies and the Relationship Ontology.
- The infectious disease ontology provides coverage for aspects of infectious diseases on a biological scale (gene, cell, organ, organism, population), disciplinary perspective (clinical, biological, epidemiological, etc.), relevant type of organism (host, pathogen, vector, reservoir), including biomedical research, clinical care, and public health.
- The Consortium of gene ontology, produces a vocabulary that applies to all organisms. provides three structured networks of defined terms to describe the attributes of gene products.
- Sequence Ontology is a part of the gene ontology project and its goal is to develop an ontology suitable for describing biological sequences.
- The Biomedical Research ontology, integrated for the description of biological and clinical research. The common formal language used is the Web Ontology Language.
- The Plant Ontology Consortium aims to share structured controlled vocabularies (ontologies) that describe plant structures and growth/development stages.
- Phenoscape: Ontology of teleost fish, uses annotations that combine terms from an anatomy ontology, an accompanying taxonomic ontology, and quality terms from the PATO ontology of phenotype qualities.

This is one of the most advanced sciences in the creation of ontologies due to the impact it has on the health of human beings and plants as a vital element of our diet, together with other sciences with greater interaction.

3.5 Ontology Applied in the Management of Knowledge in Digital Libraries

The ontologies that are mostly used in the domain of bibliographic data and the institutions that handle the largest volumes of this data were identified [25]. Next, the columns that represent the ontologies and the rows the institutions are shown and the relationship that exists is appreciated. See Table 1.

Table 1 Use of ontologies by institutions

	DC	FOAF	SKOS	FRBR	ISBD	RDA	MADS	FRAD	FRSAD	BIBO	OWLT	ORE	EDM
LIBRIS	X	X	X	X									
BNE	X		X	X	X	X	X	X	X				
GNB	X	X			X					X			
BNB	X		X		X	X				X	X		
BNF	X	X	X	X		X							
Europeana	X	X										X	X
OU		X								X			

It can be verified by the results of Table 1 that the most widely used ontologies worldwide are: DC, FOAF and SKOS.

Dublin Core: It is mostly used for the representation of bibliographic metadata. Currently protected by the Dublin Core Metadata Initiative (DCMI). It stands out for its simplicity and generality. This metadata schema has been used in other fields of application as well.

FOAF: used for information about people and their tasks. It uses the RDF-based format, facilitating semantic interoperability between systems.

SKOS: used to represent Knowledge Organization Systems such as: thesauri, taxonomies and classification schemes in the context of the SW. uses RDF, facilitating semantic interoperability between systems.

3.6 Ontology Applied in Knowledge Management in Software Engineering

Much research has been done on improving software engineering (SE) through the use of ontologies in the last two decades.

Ontologies can not only make explicit the nature and structure of engineering systems and their components, but can also help different stakeholders to better understand the complexities inherent in large engineering systems and their socio-technical environments. Despite the continuing focus on merging ontologies in systems engineering, there is no comprehensive and systematic interpretation of how ontologies influence and support computer systems, or which ontologies are established to best represent domain knowledge.

In recent years, there have been several reviews that focus on different software engineering topics, such as future challenges oriented to services and model-based computer engineering tools [26].

To address emerging challenges, several reviews have synthesized approaches to architecture, representation of knowledge, quality attributes, system integration [27] and requirements engineering [28], but there is still a manifest research gap in the literature on the state of the art of ontologies in this domain. Also, much literature has reported on the challenges and problems faced in this engineering.

First, current knowledge of the SE domain shows a poorly structured representation, which is due to the high level of fragmentation in the discourse of this domain.

Secondly, it is of heuristic origin, which implies that the performance of this engineering is highly dependent on personal experiences, this results in a lack of consistent evaluation of systems engineering standards, both in the form of manuals and meta-models. In particular, the standards are limited to human-readable descriptions and are not computer-interpretable [29]. Therefore, they are shared as text documents that are not the format of choice for semantic representations. Furthermore, the vocabulary used to build the meta-models for these systems is not the most common in the software community, and researchers, in turn, argue that these meta-models need improvements in terms of their semantics.

Systems science and engineering need a foundational, universal, general, necessary, and sufficiently well-defined ontology to support the concepts and terms it uses to be accurate and unambiguous. Research is being done to improve in this regard and that this branch of science has ontologies that contribute more to Software Engineering.

4 Discussion

As a result of the review carried out, it was possible to verify the application of SW technologies to various domains. Specifically, applications in public health, digital libraries, as well as in software engineering were described. In all cases there are potential benefits of ontology-based approaches. However, it is difficult to find works with empirical evidence that demonstrate the real impact of these approaches. This finding shows the need to carry out studies that provide this evidence in order to really quantify the impact of ontologies in the different domains. For example, Gassen et al. [30] perform an experiment to evaluate the impact of ontologies on business process modeling. This study yields interesting results and can serve as a guide for conducting similar studies in other settings.

On the other hand, it was found that most of the works have an academic orientation and it is difficult to find works that describe real applications in various industries. Therefore, it is desirable to develop work for this purpose. Likewise, the attention paid to the integration of ontologies, both from the same domain and from different domains that share concepts, is insufficient. For example, Infectious Disease Ontology (IDO) is an ontology that describes infectious diseases [31], there are various ontologies that also describe these diseases, instead of reusing the ontological resources that are included in IDO. In general, a low reuse of ontological resources is evidenced, which prevents the benefits of ontologies from being fully exploited.

5 Conclusions

The SW and its tools have become a new area with excellent results. The progress between professionals and scientists in various areas is appreciated, as they collaborate to a great extent with access to cognitive processes, reasoning and collections of structured information.

Ontologies are the tools with which you can model and organize definitions or concepts, meanings and their relationships of domains or subdomains not explored openly on the web, and increase visibility. They are a fundamental link for the representation and management of knowledge.

Organizations must come together to advance in innovation and teamwork issues, creating efficient web services that offer profitable, dynamic and self-managed sources and means of information. Especially pay attention to tacit knowledge.

A group of applications that use ontologies are mentioned and evaluations are made in some more advanced domains on the subject.

All these aspects mentioned above contribute to proper knowledge management and improve decision-making.

Knowledge management is a very important technique to facilitate the work of an organization, so it is necessary to have a method to detect its sources and location; one way to do it efficiently is by using intelligent agents through ontologies.

References

1. Zapata, L.E.: Los determinantes de la generación y la transferencia del conocimiento en pequeñas y medianas empresas del sector de las tecnologías de la información de Barcelona. Autonomous University of Barcelona (2005)
2. Senge, P.M.: La quinta disciplina 9Ed: Como impulsar el aprendizaje en la organización inteligente. Ediciones Granica S.A. (2012)
3. Gruber, T.R.: A translation approach to portable ontology specifications. Knowl. Acquis. **5**(2), 199–221 (1993)
4. Ardila, J., Rojas, A., Rodríguez, L.: Gestión del conocimiento. Tecnología Investigación y Academia **6**(2), 46–51 (2018)
5. Gómez, W.A.: Estilos de aprendizaje y aprendizaje significativo de los estudiantes de la Facultad de Ciencias Contables de la Universidad Privada San Andrés. Lima (2018)
6. Alarcón, M., Gómez, S.D., García, J.F., Barral, O.P.: Estudio y análisis del capital intelectual como herramienta de gestión para la toma de decisiones. Revista Digital del Instituto Internacional de Costos **10**, 49–65 (2012)
7. Moran, F.J.: Inteligencia emocional y rendimiento académico en estudiantes de la Universidad Estatal del Sur, Facultad de Auditoria y Contabilidad del Noveno Semestre de Jipijapa-Manabi-Ecuador. Revista FADMI: Administración y Tecnología **2**(2), 17–21 (2018)
8. Berners-Lee, T., Hendler, J., Lassila, O.: The semantic web. Sci. Am. **284**(5), 34–43 (2001)
9. Netcraft: Web Server Survey (2020). https://news.netcraft.com/archives/2020/03/20/march-2020-web-server-survey.html
10. Monroy, M.E., Arciniegas, J.L., Rodríguez, J.C.: Modelo Ontológico para Contextos de uso de Herramientas de Ingeniería Inversa. Información tecnológica **27**(4), 165–174 (2016)
11. Guarino, N.: Formal ontology and information systems. In: Proceedings of FOIS'98, 6–8 June 1998, Trento, Italy. IOS Press, Amsterdam, pp. 3–15 (1998)
12. Suárez-Figueroa, M.C.: NeOn methodology for building ontology networks: Specification, scheduling and reuse. PhD Thesis. Computing (2010)
13. Gorga, N.: Projecte de Millora de la visibilitat del Portal Jurídic de Catalunya al web (2019)
14. Quintero, B., Piraquive, F., Ruiz, H.: Ingeniería ontológica aplicada a la gestión de interesados de un proyecto. Revista vínculos **16**(1) (2019)
15. Moreno, C., Hernández, J., Bravo, D., Gaona, P., Rodríguez, F.: Desarrollo de un modelo Ontológico para la Indexación Semántica de Contenidos Digitales. Letras con Ciencia Tecno-Lógica, pp. 75–87 (2018)
16. Vasquez, M.G.: Editor gráfico de ontologías para los lenguajes semánticos RDF, RDF (S) y OWL, como extensión del framework Ontoconcept. Ph.D. Thesis. Universidad Tecnológica de Pereira. Facultad de Ingenierías Eléctrica (2016)

17. Fernández, A., López, M.J., Prevot, Y.: Modelo de sistema de organización del conocimiento basado en ontologías. Revista Cubana de Información en Ciencias de la Salud, vol. 26, no. 4 (2015)
18. Di Luca, M.A.: Modelo basado en ontología para implementar web Semántica que apoye la gestión de la información y el conocimiento (2019)
19. Peñaherrera, B., Bárcenas, G.R.: Caracterización de la Gestión basada en Tecnologías Semánticas para la Administración de Redes Informáticas. Ciencias de la Ingeniería y Aplicadas **2**(1), 29–45 (2018)
20. González, G.: Ontología del perfil de usuario para personalización de sistemas de u-learning universitarios (2019)
21. Dévora, G., Rayas, F., del Real, A., Alvarado, C.V.: Identificando hallazgos de mejora en Pymes de TI utilizando un modelo ontológico para CMMI-DEV v1.3. Revista electrónica de Computación, Informática, Biomédica y Electrónica **2**(3) (2017)
22. Rodríguez, C.: Incorporación de semántica a los procesos de gestación y aprobación de programas de postgrado en la UCLV. Ph.D. Thesis. Universidad Central Marta Abreu de Las Villas (2016)
23. Pesquit, C.: The epidemiology ontology: an ontology for the semantic annotation of epidemiological resources. J. Biomed. Semant. **5**(4), 1–7 (2014)
24. The Gene Ontology Consortium: The gene ontology resource: 20 years and still going strong. Nucleic Acids Res. **47**(D1), pp. 333–338 (2019)
25. Hidalgo, Y.: Marco de trabajo basado en los datos enlazados para la interoperabilidad semántica en el protocolo OAI-PMH. Ph.D. Thesis. Universidad de las Ciencias Informáticas, Cuba (2020)
26. Rashid, M., Anwar, M.W., Khan, A.M.: Toward the tools selection in model based system engineering for embedded systems—a systematic literature review. J. Syst. Softw. **106**, 150–163 (2015)
27. Vargas, I.G., Gottardi, T., Braga, T.: Approaches for integration in system of systems: a systematic review. In: 2016 IEEE/ACM 4th International Workshop on Software Engineering for Systems-of-Systems (SESoS), pp. 32–38 (2016)
28. Vierhauser, M., Rabiser, R., Grünbacher, P.: Requirements monitoring frameworks: a systematic review. Inf. Softw. Technol. **80**, 89–109 (2016)
29. Yang, L., Cormican, K., Yu, M.: Towards a methodology for systems engineering ontology development—an ontology for system life cycle processes. In: 2017 IEEE International Systems Engineering Symposium (ISSE), pp. 1–7 (2017)
30. Gassen, J.B., Mendling, J., Bouzeghoub, A., Thom, L.H., de Oliveira, P.M.: An experiment on an ontology-based support approach for process modeling. Inf. Softw. Technol. **83**, 94–115 (2017)
31. Cowell, L.: Infectious diseases ontology. In: Infectious Disease Informatics, pp. 373–395 (2020)

Analysis of Long-Term Rainfall Trends Over Punjab State Derived from CHIRPS Data in the Google Earth Engine Platform

Harpinder Singh, Aarti Kochhar, P. K. Litoria, and Brijendra Pateriya

1 Introduction

Dynamics of rainfall have changed across the world over the past few years. World has faced extreme variability in climate due to variation of the occurrence, spatial extent, duration and the intensity of precipitation. Secondly, rain is a major component in water cycle and has significant dependence of many ecosystems, agriculture and freshwater availability on it. To minimize uncertainties and create better management strategies in dependent fields and climate variation, it is important to measure and monitor rainfall. Although rain gauge is a common method of measuring rainfall, many developing countries do not have sufficient density of on-site rain gauge stations [1]. So, compared to this traditional method, it is better to adopt methods including remotely sensed data.

Products such as Climate Hazards Group InfraRed Precipitation with Station data (CHIRPS) have provided access to areas uncovered by gauged stations to monitor the variability of rainfall with high spatial resolution. The CHIRPS data process includes three main mechanisms, i.e. the climate hazards group monthly precipitation climatology (CHP_{clim}), the satellite-only CHIRP and the blending of station data that produces the CHIRPS. More specifically, CHIRPS is a reanalysis product that combines cold cloud duration (CCD) with CHP_{Clim}, tropical rainfall measuring mission's (TRMM) 3B42 product and precipitation observations from various national or regional meteorological stations [2]. However, there are other similar products available, but the significant difference of this precipitation database with other databases is its high resolution of 0.05° that is less than resolution of 0.5° or 0.25° available from majority of products.

Along with the asset of higher resolution of CHIRPS data set for monitoring and observation-related tasks, quantum of data follows. Running complex statistical

H. Singh (✉) · A. Kochhar · P. K. Litoria · B. Pateriya
Punjab Remote Sensing Centre, Ludhiana, Punjab, India

M. Dua et al. (eds.), *Proceedings of the International Conference on Paradigms of Communication, Computing and Data Sciences*, Algorithms for Intelligent Systems,
https://doi.org/10.1007/978-981-16-5747-4_41

operations on high-resolution large data from satellite sensors turns out to be a challenge. With desktop computing resources, it is difficult to download, process, collaborate, manage and explore big data from sensors related to remote sensing. In order to provide a special system specifications that can address problems related to big geospatial data analytics, cloud computing lends a helping hand [3, 4]. Platforms based on cloud computing can provide on-demand resources or services related to computing like server operating system, storage, processing, analytics, software, etc., over Internet. Earth on Amazon Web Services (AWS), Azure-AI for earth and Google Earth Engine (GEE) are few common cloud computing platforms for geospatial big data analytics.

Being most widespread, GEE has extended across wide range of applications related to remote sensing. GEE has a diversified range of applications that includes studies related to water (including rivers, lakes, oceans, etc.), natural disasters, nuclear non-proliferation, wetlands, forests, climate change, urban, soil, land use/land cover, archaeology, habitat mapping, crop mapping, vegetation and agricultural monitoring. GEE supports applications related to data processing also like cloud detection, radiometric correction and mosaic image generation [5].

The objective of this study is to analyse the long-term rainfall trends over Punjab state for the time period 1987–2020 using satellite-based CHIRPS version 2.0 final data, in the GEE cloud platform. This time series data analysis can be used for identifying rainfall trends and also for seasonal drought monitoring.

2 Literature Survey

Rivera et al. [6] validated 30 years of data from CHIRPS on Central Andes of Argentina region. Ground stations are highly scarce in this region. Depending on the rainy season, area was divided in two zones. Seasonal, annual and spatial variabilities were investigated by authors. While one of the zones showed good results, other seemed to overestimate it. Pearson's correlation coefficient, mean absolute error, Nash–Sutcliffe efficiency and per cent bias were calculated for performance evaluation. Region with summer precipitation performed better than winter precipitation region. Similarly, Paca et al. [7] studied the region of Amazon River Basin using CHIRPS precipitation product. Density of on-site stations was less in the region, so authors explored the capabilities of mentioned satellite product. Variability of precipitation in Amazon Rainforest can affect world's climate, so even clusters within the basin were also studied. Over the time, many other authors have also proved the significance of such products for wide coverage and better resolution in precipitation analysis [8].

Other than implementing CHIRPS data for studying areas with less rain gauge stations, authors have tried it over areas with dense network of on-site stations. Katsanos et al. [9] studied the area of Cyprus and investigated monthly and annual rainfall. Although the satellite data from CHIRPS was in correlation with station data, results seemed to be overestimated over the last decade. Correlation of CHIRPS with other related data set like TRMM and E-OBS (daily observational precipitation, temperature and sea-level pressure data set from Europe) was also computed. CHIRPS seemed to correlate better with TRMM data than E-OBS for the studied area. TRMM correlated better because CHIRPS itself contain TRMM data and E-OBS correlated weakly because of limited incorporation of station data. Paredes-Trejo et al. [10] validate the CHIRPS-based satellite precipitation estimates in Northeast Brazil. The CHIRPS version 2.0 data was compared to the ground stations data for the period 1981–2013. Various metrics were generated, and the results show that CHIRPS data correlates well with ground observations for all stations ($r = 0.94$). But the data tends to overestimate low and underestimate high rainfall values. Creation of hydrological models is very important for water resource management. CHIRPS data has been used as an input to the hydrological models in [11] for the study undertaken in Nzoia Basin in Western Kenya over the temporal range 1990–2000. The stream flow model results were compared with in situ rainfall gauge station data, the Climate Forecast System Reanalysis (CFSR) data set and the CHIRPS data set. Simulated stream flow estimates were poor with rainfall gauge station data but improved significantly with the CHIRPS and CFSR data sets.

Venkatesh et al. [12] compared 16 products from four different categories of gauge-based, gauge-adjusted, reanalysis and satellite-based data sets. Authors selected a tropical hilly terrain of Tungbhadra from Karnataka, India. Other than CHIRPS, Asian Precipitation Highly Resolved Observational Data Integration Towards Evaluation (APHRODITE), National Centre for Environmental Prediction-Climate Forecast System Reanalysis (NCEP-CFSR), TRMM also performed well in detecting rainfall. Prakash [13] assesses the performance of latest versions of four multi-satellite precipitation products: CHIRPS, Multi-Source Weighted-Ensemble Precipitation (MSWEP), SM2RAIN-Climate Change Initiative (SM2RAIN-CCI), TRMM Multi-satellite Precipitation Analysis (TMPA) across India using gauge-based observations for the period of 1998–2015 at monthly scale. According to the results, CHIRPS and TMPA are comparable to gauge-based precipitation estimates at all-India and sub-regional scales followed by MSWEP estimates. The study highlights that CHIRPS data set could be used for long-term precipitation analyses with rather higher confidence. Another study used CHIRPS data for drought hazard mapping and trend analysis in the Bundelkhand region of Uttar Pradesh [14]. The objective was to identify district-wide drought and its trend characterization from 1981 to 2018. Standardized precipitation index (SPI) was computed from the CHIRPS data. The main results indicated the average of nine severe drought events occurred in all the districts in the last 38 years. Also, the most intense drought was recorded for the Jalaun District in 1983–1985.

Guo et al. [15] demonstrates the use of CHIRPS for drought monitoring in the Lower Mekong Basin (LMB) where there is a shortage of ground observations. SPI has been computed at various timescales from January 1981 to July 2016. According to the results, CHIRPS could easily identify the drought characteristics at various timescales, and the best performance is achieved at three month scale. The results also show that the study area experienced four severe droughts during the last three decades. According to the rainfall analysis by Divya and Shetty [16], the accuracy of CHIRPS data was very high mainly in the low-lying areas of Kerala (India), i.e. at the coastal areas, and it was found to be decreasing when it approaches towards the Western Ghats. Monitoring of agricultural drought in semi-arid ecosystem of Peninsular India through indices derived from time series CHIRPS and MODIS data sets has been done by Sandeep et al. [17]. Thirty years CHIRPS data along with various indices generated from the MODIS datasets was analysed for the drought analysis. The authors have found the CHIRPS data to be very useful.

Even with availability of CHIRPS data, it is difficult to process and analyse it in scenario of limited computing resources. GEE provides a decent cloud computing platform. GEE is most prevalent from other available cloud computing resources. It provides prompt platform with high computational capabilities. GEE can handle processing and analytics of planetary scale geospatial big data. GEE platform can aid spectral, temporal and spatial analyses. GEE hosts competencies for simple to complex statistical algorithms, image processing techniques for a single image or batch of images and machine learning operations [18]. Combining big data analytics potential of GEE with data available from CHIRPS can help in an easy and reliable analysis. Torres Batlo and Marti-Cardona studied Bolivia region for spatio-temporal analysis from 1981–2018. CHIRPS satellite product was used. GEE cloud platform was used for processing of data. It was compared with on-site data for seasonal and regional data. Both data were in agreement with each other with correlation coefficient above 0.8. Authors represented zone wise and month wise increasing and decreasing trends [19]. Banerjee et al. [20] studied long-term spatial patterns and trends over years using GEE. Monthly, annual and seasonal rainfall was analysed for complex topography of Himalayan region of Bhilangana River Basin in Uttarakhand, India. For the period of 1983 to 2008, data from India Meteorological Department (IMD) was also integrated with remotely sensed data of CHIRPS and Precipitation Estimation from Remotely Sensed Information Using Artificial Neural Networks-Climate Data Record (PERSIANN-CDR). For 2009 to 2018, CHIRPS and PERSIANN-CDR products were used without on-site integration. Along with mean absolute error (MAE), root mean square error (RMSE), mean–variance between in situ and satellite rainfall values (Bias), multiplicative bias, relative bias and correlation coefficient was calculated using GEE platform to evaluate the performance and accuracy. Authors concluded the good correlation, performance and accuracy of CHIRPS data set with ground observed data, i.e. it confirmed the decreased annual and seasonal rainfall.

Based on the literature survey and to the best of author's knowledge, CHIRPS data has not been utilized for the rainfall analysis of Punjab state. Also, there are very few studies in India which have processed CHIRPS data in the GEE cloud platform for rainfall analysis.

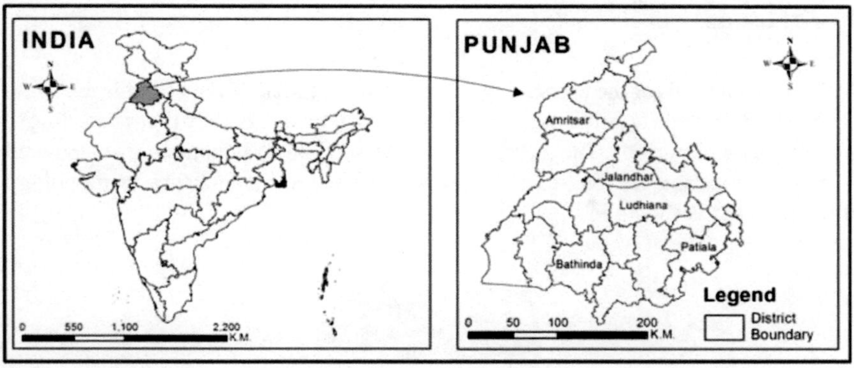

Fig. 1 Study area

3 Study Area and Datasets

3.1 Study Area

The study area is the whole Punjab state of India. Its geographical area is 50,362 km^2 (Fig. 1), and it lies between 29° 33′ and 32° 31′ N latitude and 73° 53′ and 76° 55′ E longitude. The state experiences three distinct seasons, the hot season from April to mid-June, the rainy season from June to September and the winter season extending from October to March. The Punjab state is intensively cultivated, and 86.77% of its area is under agriculture. The cropping pattern of Punjab shows predominance of wheat, rice and cotton.

3.2 Datasets

CHIRPS is a 30+ year quasi-global rainfall data set. It integrates 0.05° resolution satellite imagery with in situ station data to create a gridded time series rainfall data set. This time series data is used for trend analysis and seasonal drought monitoring. This data is available from 1981 to the near present. CHIRPS pentad (version 2.0 final) data has been used in this research. There are six pentads in each calendar month. Each of first five pentads in a month has five days. The last pentad contains all the days from the 26th to the end of the month. The units used in CHIRPS pentad are total mm/pentad. CHIRPS data of the time periods 1987 to 2019 has been used to derive the long-term monthly averages, and 2020 data has been used as current data set.

4 Methods

GEE has been used for the processing and analysis of CHIRPS data for this research. The methods include filtering the images based on the study area and then according to the required time periods. The monthly and seasonal rainfall deviations are calculated based on comparing the long-term averages with the current data. The methodology of the research is shown in Fig. 2.

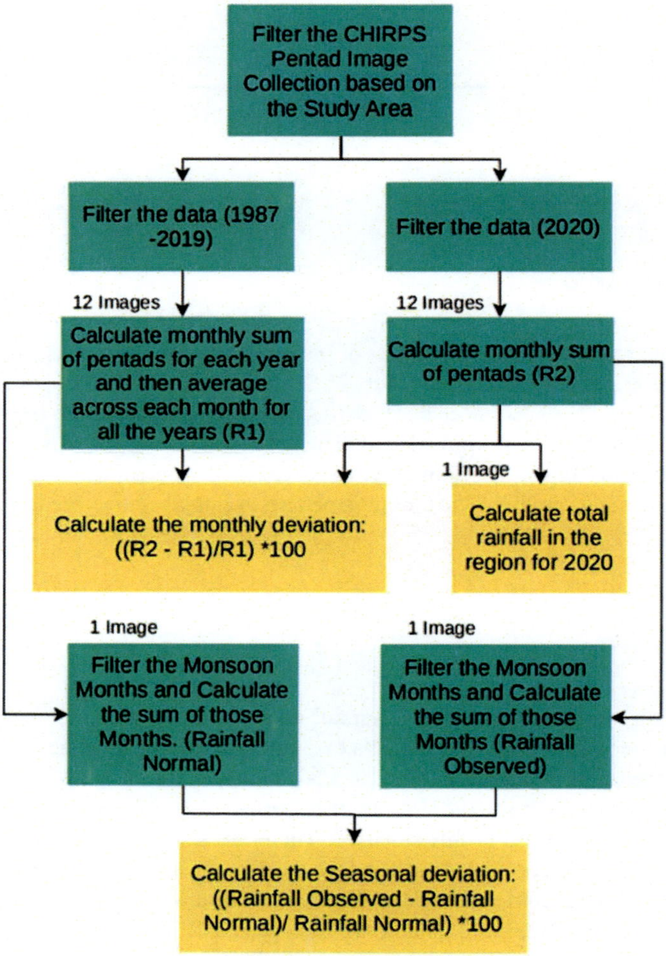

Fig. 2 Research methodology

4.1 Filtering the Image Collection

The CHIRPS images were first filtered based on the study area, i.e. Punjab state. From this set of images, two sets were created. One was the images of the time period 1987–2019 and other of 2020.

4.2 Long-Term Averages

Long-term averages were calculated from the data set of 1987–2019. First a total of all the pentads were done for all the months in the above time period. Then an average is taken across each month for all the years.

4.3 Current Rainfall (2020)

The current rainfall was calculated from the data set of 2020. Total of all the pentads was done for all the months in the above time period.

4.4 Monthly and Seasonal Deviations

Monthly deviations (%) were calculated by comparing the long-term monthly averages with the rainfall in the month of the year 2020. Seasonal deviations (%) were calculated by comparing the sum of the rainfall (in the monsoon season, i.e. June, July, August and September) of the time period 1987–2019 with that of 2020.

5 Results and Discussions

The results of the research work are divided into various sections as follows.

5.1 Total Rainfall in Punjab (2020)

According to the CHIRPS data analysis, the total rainfall in Punjab for the year 2020 is 777.25 mm. Figure 3 shows the map generated in the GEE. The darker blue areas in the north (Pathankot, Gurdaspur and north of Hoshiarpur) have received more

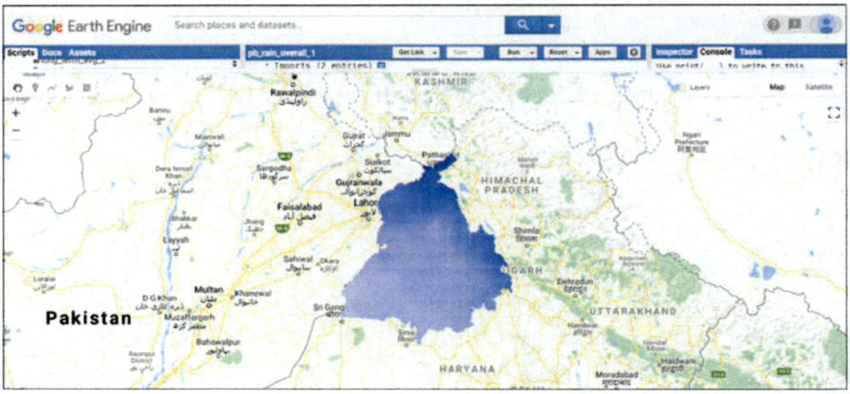

Fig. 3 Total rainfall in Punjab (2020)

rainfall than the lighter areas in the south-western part of Punjab (Fazilka, Muktsar, Bathinda, Mansa).

5.2 *Long-Term Monthly Mean Rainfall Over Punjab*

Figure 4 shows the long-term monthly mean rainfall over Punjab (1987–2019). According to the graph, the months of July and August experience the highest average rainfall.

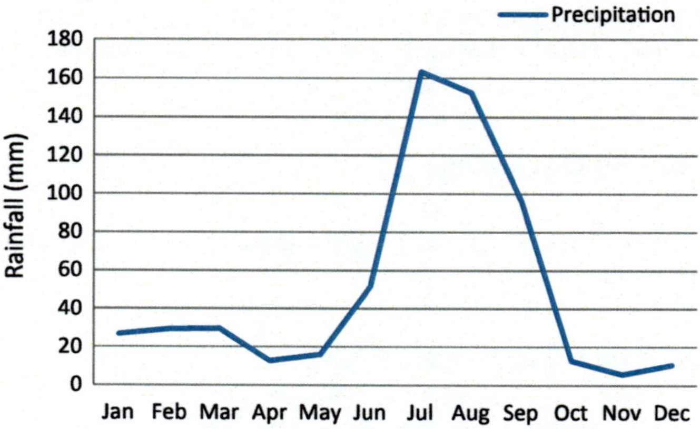

Fig. 4 Long-term monthly mean rainfall over Punjab (1987–2019)

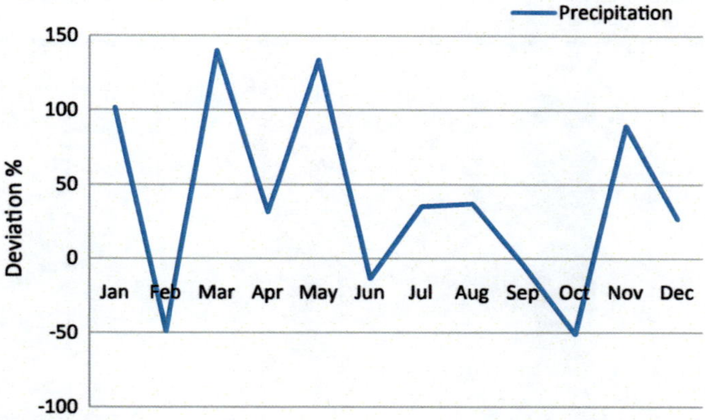

Fig. 5 Rainfall deviation over long-term mean

5.3 Rainfall Deviation Over Long-Term Mean

Figure 5 shows the rainfall deviation over long-term mean for Punjab. The deviation has been calculated by comparing the current year (2020) rainfall with the long-term mean of 1987–2019. According to the graph, the months of March and May 2020 have experienced more rainfall than the long-term mean, while the months of February and October have received less rainfall.

5.4 Seasonal (Monsoon) Rainfall Deviation for Punjab

Figure 6 shows the seasonal rainfall deviation. The deviation has been calculated for the monsoon season, i.e. June, July, August and September. The rainfall in the monsoon season of 2020 has been compared with the monsoon season of 1987–2019. According to the analysis, almost whole of Punjab has received more than the average long-term rainfall. The central and the north-eastern parts of Punjab (Jalandhar, Kapurthala, Ludhiana, Fatehgarh Sahib and Southern parts of Hoshiarpur districts, etc.) have received less rainfall as compared to the other districts. There are few dark red pixels (areas around the Ludhiana city, southern parts of Jalandhar and northern part of Fatehgarh Sahib Districts) in the central part of Punjab which show negative deviation, i.e. less than average rainfall in the year 2020. (Legend: towards red colour shows negative deviation and towards blue colour shows positive deviation).

Figure 7 shows the seasonal deviation by the districts in Punjab. The pixel values (deviation) are averaged across the district boundary. After calculating the average, there is no district with negative deviation.

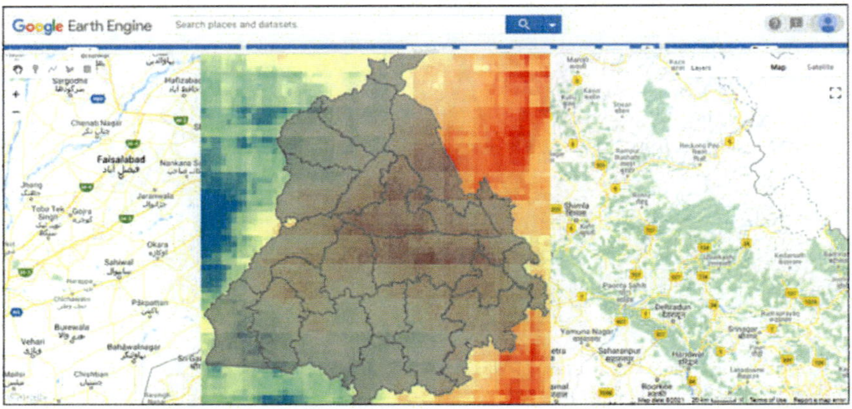

Fig. 6 Seasonal deviation (Legend: towards red colour shows negative deviation and towards green/blue colour shows positive deviation)

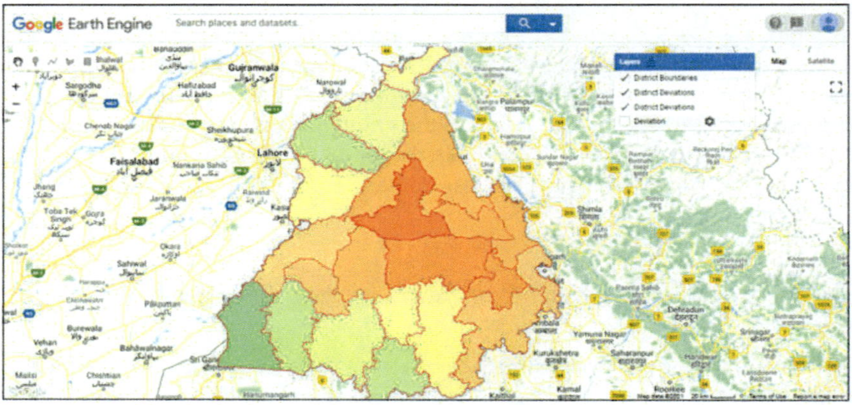

Fig. 7 Seasonal deviation by districts (Legend: towards red colour shows negative deviation and towards green/blue colour shows positive deviation)

The report [21] brings out observed rainfall variability and trends over the Punjab state as an impact of climate change based on recent 30 years of data (1981–2018). Ground station data has been analysed to generate the maps and statistics. Rainfall pattern of monsoon months, south-west monsoon season and annual of the state and its districts as well as extreme rainfall event of different intensity of stations are analysed in this report. According to the report subsection "Tend in district rainfall", during the whole south-west monsoon season Hoshiarpur, Jalandhar, ShahidBhagat Singh Nagar, Ferozepur, Fazilka, Patiala and Fatehgarh Sahib Districts have shown significant decreasing trend.

The results of our research also corroborate with the results in the report as our research (seasonal deviation) also shows a decreasing trend (for the year 2020) in Jalandhar, ShahidBhagat Singh Nagar, Hoshiarpur and Fatehgarh Sahib Districts. From the report, we can see that based on the 30 year data, Punjab gets maximum rainfall in July followed by August which confirms our results, i.e. long-term monthly mean rainfall over Punjab (1987–2019).

6 Conclusion and Future Work

This study demonstrates an effective way of utilizing cloud-based earth observation tool for analysing rainfall over large study areas. Usage of freely available CHIRPS pentad data set and openly available analysis tool ensures the extension of work by other researchers.

Future enhancement in the study can be done with the integration and comparison with other machine learning classification techniques in addition to the variety of remotely sensed data sets. Such big data analysis involving processing of huge EO data sets on local computers requires a lot of resources (computation, storage, specialized software, time, etc.). This study amply demonstrates the use of GEE for big geospatial data analytics on the cloud using only a simple computer and network connectivity.

References

1. World Meteorological Organization: Guide to hydrological practices. Secretariat of the World Meteorological Organization (1994)
2. Funk, C., Peterson, P., Landsfeld, M., Pedreros, D., Verdin, J., Shukla, S., Husak, G., Rowland, J., Harrison, L., Hoell, A., Michaelsen, J.: The climate hazards infrared precipitation with stations—a new environmental record for monitoring extremes. Scientific Data **2**(1), 1–21 (2015)
3. Chi, M., Plaza, A., Benediktsson, J.A., Sun, Z., Shen, J., Zhu, Y: Big data for remote sensing: challenges and opportunities. Proc. IEEE **104**(11), 2207–2219 (2016)
4. Ma, Y., Wu, H., Wang, L., Huang, B., Ranjan, R., Zomaya, A., Jie, W: Remote sensing big data computing: challenges and opportunities. Future Gener. Comput. Syst. **51**, 47–60 (2015)
5. Tamiminia, H., Salehi, B., Mahdianpari, M., Quackenbush, L., Adeli, S., Brisco, B: Google Earth engine for geo-big data applications: a meta-analysis and systematic review. ISPRS J. Photogrammetry Remote Sens. **164**, 152–170 (2020)
6. Rivera, J.A., Marianetti, G., Hinrichs, S: Validation of CHIRPS precipitation dataset along the Central Andes of Argentina. Atmos. Res. **213**, 437–449 (2018)
7. Paca, V.H.D.M., Espinoza-Dávalos, G.E., Moreira, D.M., Comair, G: Variability of trends in precipitation across the amazon river basin determined from the CHIRPS precipitation product and from station records. Water **12**(5), 1244 (2020)
8. Dai, Q., Bray, M., Zhuo, L., Islam, T., Han, D: A scheme for rain gauge network design based on remotely sensed rainfall measurements. J. Hydrometeorology **18**(2), 363–379 (2017)
9. Katsanos, D., Retalis, A., Michaelides, S: Validation of a high-resolution precipitation database (CHIRPS) over Cyprus for a 30-year period. Atmos. Res. **169**, 459–464 (2016)

10. Paredes-Trejo, F.J., Barbosa, H.A., Kumar, T.L: Validating CHIRPS-based satellite precipitation estimates in Northeast Brazil. J. Arid Environ. **139**, 26–40 (2017)
11. Le, A.M., Pricope, N.G: Increasing the accuracy of runoff and streamflow simulation in the Nzoia Basin, Western Kenya, through the incorporation of satellite-derived CHIRPS data. Water **9**(2), 114 (2017)
12. Venkatesh, K., Krakauer, N.Y., Sharifi, E., Ramesh, H: Evaluating the performance of secondary precipitation products through statistical and hydrological modeling in a mountainous tropical basin of India. Adv. Meteorol. **2020** (2020)
13. Prakash, S.: Performance assessment of CHIRPS, MSWEP, SM2RAIN-CCI, and TMPA precipitation products across India. J. Hydrol. **571**, 50–59 (2019)
14. Pandey, V., Srivastava, P.K., Singh, S.K., Petropoulos, G.P., Mall, R.K: Drought identification and trend analysis using long-term CHIRPS satellite precipitation product in Bundelkhand, India. Sustainability **13**(3), 1042 (2021)
15. Guo, H., Bao, A., Liu, T., Ndayisaba, F., He, D., Kurban, A., De Maeyer, P: Meteorological drought analysis in the Lower Mekong Basin using satellite-based long-term CHIRPS product. Sustainability **9**(6), 901 (2017)
16. Divya, P., Shetty, A: Evaluation of CHIRPS satellite rainfall datasets over Kerala, India. Trends Civ. Eng. Challenges Sustain. 655–664 (2021)
17. Sandeep, P., Reddy, G.O., Jegankumar, R., Kumar, K.A: Monitoring of agricultural drought in semi-arid ecosystem of Peninsular India through indices derived from time-series CHIRPS and MODIS datasets. Ecol. Ind. **121**, 107033 (2021)
18. Amani, M., Ghorbanian, A., Ahmadi, S.A., Kakooei, M., Moghimi, A., Mirmazloumi, S.M., Moghaddam, S.H.A., Mahdavi, S., Ghahremanloo, M., Parsian, S., Wu, Q: Google Earth engine cloud computing platform for remote sensing big data applications: a comprehensive review. IEEE J. Sel. Top. Appl. Earth Observations Remote Sens. **13** (2020)
19. Torres-Batlló, J., Martí-Cardona, B.: Precipitation trends over the southern Andean Altiplano from 1981 to 2018. J. Hydrol. **590**, 125485 (2020)
20. Banerjee, A., Chen, R., E Meadows, M., Singh, R.B., Mal, S., Sengupta, D: An analysis of long-term rainfall trends and variability in the Uttarakhand Himalaya using Google Earth engine. Remote Sens. **12**(4), 709 (2020)
21. Guhathakurta, P., Pednekar, R.A., Khedikar, S., Menon, P., Prasad, A.K., Sangwan, N.: Observed rainfall variability and changes over Punjab State. Climate Research and Service, India Meteorological Department, Ministry of Earth Sciences, Pune. Met Monograph No.:ESSO/IMD/HS/Rainfall Variability/21(2020)/45, January 2020. Available Online at: https://imdpune.gov.in/hydrology/rainfall%20variability%20page/punjab_final.pdf, Accessed on: 22-Feb-2021

Weed Classification from Paddy Crops Using Convolutional Neural Network

J. Dhakshayani, Sanket S. Kulkarni, Ansuman Mahapatra, B. Surendiran, and Malaya Kumar Nath

1 Introduction

Paddy crop is one of the vital and significant crops in India. Its productivity is affected by various factors, such as stress, nutrition deficiency, pests, and weeds. Among them, excessive use of fertilizers and high salinity have huge impact and reduce the yield by 50%. Weeds are one of the key factors that decrease the paddy yield. The common approach for the prevention of weed affecting the crop is by spraying herbicides. Its excessive use leads to ecosystem imbalance. Weeds intensify the pest and disease problem, by serving as an alternate host that reduces the efficiency of harvesting, increases water contamination, and results in poor quality of paddy.

The process of identification and classification of paddy and weed depends on experts' knowledge to recognize, which is very difficult and erroneous that leads to incorrect identification [1]. So, weeds have to be detected in advance for increasing the productivity. In earlier days, stresses are identified by visual observations. This method is time-consuming, requires experts' knowledge, and is also destructive in nature. Some of the images of paddy crops along with the various kind of weeds are shown in Fig. 1. Figure 1a, b shows the healthy paddy crops and low fertile nutrient paddy crops, respectively. Figure 1c–f represents the different kind of weeds. Large amount of similarity can be noticed between the paddy crops and the weed types. So, identification is difficult by visual observation. In order to overcome such

J. Dhakshayani (✉) · S. S. Kulkarni · A. Mahapatra · B. Surendiran · M. K. Nath
National Institute of Technology Puducherry, Karaikal, India

A. Mahapatra
e-mail: ansuman.mahapatra@nitpy.ac.in

B. Surendiran
e-mail: surendiran@nitpy.ac.in

M. K. Nath
e-mail: malaya.nath@nitpy.ac.in

M. Dua et al. (eds.), *Proceedings of the International Conference on Paradigms of Communication, Computing and Data Sciences*, Algorithms for Intelligent Systems,
https://doi.org/10.1007/978-981-16-5747-4_42

Fig. 1 Paddy crops and various weeds grew in paddy field: **a** healthy paddy crop, **b** low fertile nutrient deficiency paddy crop, **c** weed plant type-1, **d** weed type-2, **e** weed type-3, **f** weed type-4. (Images have been taken from Kaggle Web site https://www.kaggle.com/archfx/paddyimages)

difficulties, many ICT-based techniques have been developed by the researchers. Few of them uses machine learning and deep learning methods for classification. But, due to advances in information and technology, deep learning provides better performance compared to the others. Some of the techniques used in agriculture field for classification of stresses, diseases, weeds, etc., from the paddy crops are discussed in below subsection.

Fig. 1 (continued)

1.1 Literature Review

Many techniques were proposed in order to control these weed growth without affecting the yield. These techniques should be highly judicious for distinguishing the weeds from the crops for meaningful functionality. Several machine learning techniques have been introduced for classification of plant stress [2–5]. The problem with these approaches was lack of proper features. CNN has been used as a great training model for solving many real-time problems including visualization and identification of patterns from the data. Researchers have yielded the deep neural network for various agricultural activities, as it has currently dominated the era of agriculture. Some of them are stress management [6], counting of fruits [7], pest and weed control [8], etc.

Few research articles have been found in the literature regarding the application of deep learning in paddy weed classification. A segmentation method cascaded encoder–decoder network (CED-Net) [9] was proposed by Khan et al. for differentiating the weeds from crops. This encoder–decoder network architecture consists of two stages, and each has two models for training. The first two models were involved for weed prediction, and the next two models were considered for crop prediction. The experimental study was carried out with four different open access dataset such as rice seeding & weed [10], BoniRob [11], carrot crop and weed [12], and paddy–millet [13]. For all the databases, both qualitative and quantitative analyses were made in order to prove the performance of the network. This method has 1/5.74, 1/5.77, 1/3.04, and 1/3 24 times lesser parameters compared to other models such as U-Net, SegNet, fully convolutional network (FCN-8s), and DeepLabv3.

Kamath et al. [14] developed a multi-classifier system (MCS) that merges the output of two or more different classifiers to provide the final result, for paddy crop and weed classification. The classifiers used here are support vector machine (SVM) and random forest. This method attains an accuracy of 91.36% for weed management. This accuracy is low for efficient weed classification.

Anami et al. [15] have proposed the early access of stresses in paddy crops at booting growth stage in order to avoid the qualitative and quantitative loss of agricultural yield. Here, authors mainly focus on designing a deep convolutional neural network (DCNN) for automatic identification and classification of various abiotic and biotic paddy crop stresses from the field images. This uses VGG-16 model for automatic categorization of stressed from five different types of paddy crop images captured during the booting growth stage. The trained models achieved an accuracy of 92.89% on the experimental dataset.

Konstantinos [16] has proposed convolutional neural network to detect and diagnose plant diseases from various leaf images. This model is trained using an open database that consists of 87,848 images, containing 25 different classes. This architecture achieved a success rate of 99.53% in identifying the disease from the plant leaves. This helps in providing the early stages of the disease and may be extended further to support and integrate the methods to classify plant diseases. This method was experimented to give the best results during training after suitable number of iterations. A particular annealing schedule was followed by the learning rate, starting from 0.01 and descending every 20 epochs by 1/2 or 1/5, alternately, down to 0.0001. The model comparison was made based on their performance with respect to the set of tests conducted (all models achieved 100 percent accuracy on the training set).

The different methods/techniques are used for increasing the yield in agriculture and are summarized in Table 1. The performance of all the methods is limited to very less accuracy. There is a scope for improvement of performance. In the literature, researchers have used some pretrained models for classification of diseases in medical images [17].

1.2 Motivation

Analysis results from Table 1 motivate to develop a crop classification technique using pretrained deep learning model VGG-16 as researchers have not performed weed classification from paddy field for improving paddy yield. The work proposed in this paper uses the Kaggle database for classification of weeds from the paddy plants. Present work demonstrates the difference in patterns of healthy crops, low fertile nutrient crops (mainly influenced by sunlight, nutrient content, etc.), and various weed types. Appearance of different crops depends on the illumination conditions. These factors are different to machine, while it looks similar to the human observer. This leads to more error and less accuracy. These limitations are addressed in the proposed method with the help of deep learning paradigm. In this paper, the strength of VGG-16 pretrained model is used for paddy crop classification. A study was

Table 1 Description of related work for paddy crop classification

Method	Description	Database	Accuracy (%)	Remark
Cascaded encoder–decoder network (CED-Net), 2020 [9]	Semantic segmentation method for detecting and extracting the exact location of crops and weeds	1. Rice seedling and weed dataset 2. BoniRob dataset 3. Carrot crop versus weed dataset 4. Paddy–millet dataset	95	Number of parameters and inference time can be improved
Multiple classifier systems, 2020 [14]	Multiple classification using support vector machines and random forest for classifying paddy crop and weeds	Image in this database is captured by canon power-shot SD3500 IS and Sony Cybershot DSC-W220	91.36	Accuracy is low for efficient weed classification
Deep convolutional neural network (VGG-16), 2020 [6]	Automatic recognition and classification of stresses in paddy crop	Authors have used their own database that consists of 30,000 images having 11 different stress classes and five various paddy crops	92.89	Applicable only for paddy crop type and not for other conditions
Transfer learning, (VGG-16), 2020 [20]	Used for predicting diseases from paddy plant	Images in this database consist of three disease classes (Hispa, brown spot, and leaf blast) and one healthy class	90	Accuracy of the model is low, and images have low resolution
Convolutional neural network, 2017 [16]	Various CNN models such as AlexNet, AlexNetOWTBn, GoogLeNet, Overfeat, VGG for detection of plant diseases	It uses plant village dataset	99.53	Uncontrolled acquisition of pesticides causing effect to environment
Support vector machine and K nearest neighbor, 2016 [4]	Identify disease affected portion of paddy plant using AdaBoost classifier and disease recognition using SVM and KNN	Publically available paddy plant leaf image dataset from Kaggle	93.33	Used classifiers provide comparatively better accuracy and can be improved with other classifiers

(continued)

Table 1 (continued)

Method	Description	Database	Accuracy (%)	Remark
Bayes' and SVM classifier, 2012 [5]	Helps in classifying the leaf brown spot and the leaf blast diseases from rice plant	The images present in the databases are captured by Nikon Coolpix P4 digital camera in different parts of Midnapur	79.5 (Bayes' classifier) 68.1 (SVM)	Misclassification occurs due to shadow effect and color distortion of aging leaves

carried out with different pretrained models, and selection of this depends on the application of interest. VGG-16 model has been preferred in this work as it is used in the literature for varying classification purpose and found to be good architecture for benchmarking on a particular classification task.

The rest part of the paper is organized as follows. Section 2 discusses the proposed methodology for classification of weeds from the paddy crops. Experimental outcomes are presented in Sect. 3. Finally, Sect. 4 summarizes the work and presents a roadmap for future directives.

2 Proposed Methodology

The proposed methodology for classification of various types of weeds from paddy crops is mentioned in Fig. 2. The important blocks are input image, pretrained VGG-16 model for classification and performance evaluation.

Input image block: The input images are taken from publicly available Kaggle Web site [18]. This consists of 3923 images, having dimension of 1280 × 720 in RGB format. The database has six classes (healthy, low fertile nutrient crop, weed type-1, weed type-2, weed type-3, and weed type-4). The number of images in different classes is mentioned in Table 2.

Classification block: The images are preprocessed before feeding to the VGG-16 network for classification [19]. The model is implemented using a Keras higher-level Python library that runs as a backend over a TensorFlow open-source deep learning platform. The input image dimension is 224 × 224 pixels with depth 3, and images are passed through stack of convolutional neural networks and convolutional filters of size 3 × 3 × 3. The activation function ReLU has been used for hidden layers, and softmax has been applied for final prediction in order to confirm the probability that lies between 0 and 1. The network uses different epoch with batch size 128. The network has been optimized using the three different optimization algorithms with categorical cross-entropy logarithmic loss function. The model receives the paddy images of size 224 × 224 × 3 pixels as input. It has a sequence of two

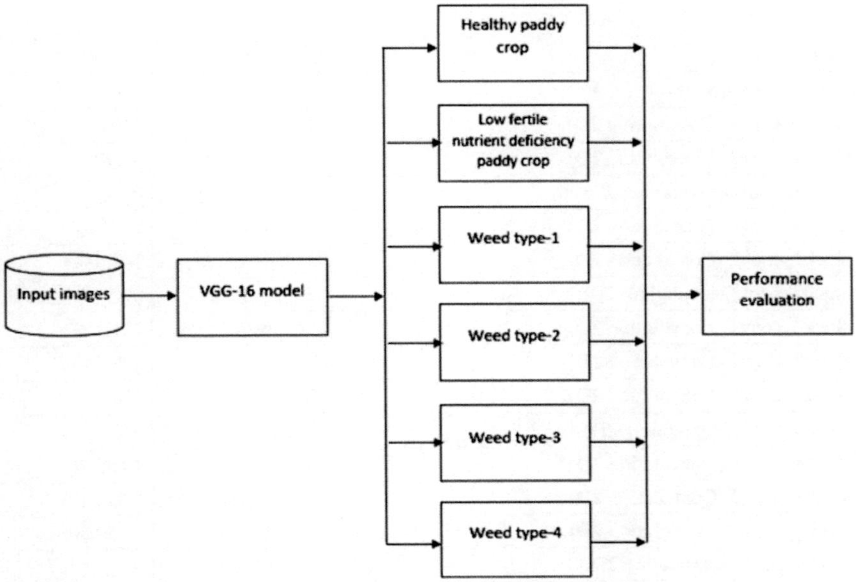

Fig. 2 Proposed methodology

Table 2 Description of input images

Classes	Number of images	Number of training images (70%)	Number of validating images (30%)
Healthy paddy crop	676	473	203
Low fertile nutrient deficiency paddy crop	898	629	269
Weed type-1	482	337	145
Weed type-2	521	365	156
Weed type-3	991	694	297
Weed type-4	355	248	107
Total images	**3923**	**2746**	**1177**

convolutional and pooling layers as feature extractors, followed by a fully connected layer to interpret the features and an output layer. The information about the layers is mentioned in Table 3.

The output layer has six neurons, which correspond to the number of paddy classes that the input images are needed for classification. The softmax function is used in VGG-16 which is a more generalized version of sigmoid function. The softmax function would squeeze the outputs for each class ranging between 0 and 1 and would also divide by the sum of the outputs. This essentially gives the probability of the input being in a class. Table 3 gives the description about layers, their output shape,

Table 3 Schematic structure of VGG-16 model

Layer name (type)	Output shape	Parameters
input_1 (Input layer)	[224, 224, 3]	0
block1_conv1 (Convolution 2D)	[224, 224, 64]	1792
block1_conv2 (Convolution 2D)	[224, 224, 64]	36,928
block1_pool (Maxpooling 2D)	[112, 112, 64]	0
block2_conv1 (Convolution 2D)	[112, 112, 128]	73,856
block2_conv2 (Convolution 2D)	[112, 112, 128]	147,584
block2_pool (Maxpooling 2D)	[56, 56, 128]	0
block3_conv1 (Convolution 2D)	[56, 56, 256]	295,168
block3_conv2 (Convolution 2D)	[56, 56, 256]	590,080
block3_conv3 (Convolution 2D)	[56, 56, 256]	590,080
block3_pool (Maxpooling 2D)	[28, 28, 256]	0
block4_conv1 (Convolution 2D)	[28, 28, 512]	1,180,160
block4_conv2 (Convolution 2D)	[28, 28, 512]	2,359,808
block4_conv3 (Convolution 2D)	[28, 28, 512]	2,359,808
block4_pool (Maxpooling 2D)	[14, 14, 512]	0
block5_conv1 (Convolution 2D)	[14, 14, 512]	2,359,808
block5_conv2 (Convolution 2D)	[14, 14, 512]	2,359,808
block5_conv3 (Convolution 2D)	[14, 14, 512]	2,359,808
block5_pool (Maxpooling 2D)	[7, 7, 512]	0
flatten (Flatten)	[25088]	0
fc1 (Fully connected)	[4096]	102,764,544
fc2 (Fully connected)	[4096]	16,781,312
dense (Dense)	[6]	24,582
Total number of parameters: 134,285,126 Total number of trainable parameters: 24,582 Total number of non-trainable parameters: 134,260,544		

and the parameters for training VGG-16 model. For training model, the total number of trainable and non-trainable parameters is 24,582 and 134,260,544, respectively.

Performance evaluation block: The performance of the classified output has been computed by accuracy and loss function. The accuracy is represented by the equation given in Eq. 1

$$\text{Accuracy} = \frac{\#\hat{X}}{\#X} \tag{1}$$

where $\#\widehat{X}$ denotes the number of correct predictions, $\#X$ denotes the total number of predictions. Cross-entropy function is used for calculating the loss. These results are computed for various hyperparameters, such as batch size, number of epochs, and optimizers. The experiment is conducted with different learning rates such as 0.01, 0.001, and 0.0001 on different optimizers such as Adam, SGD, Adadelta. The accuracy and validation values are noted for a batch size of 128.

Initially, the experiment is carried out with a learning rate of 0.01 on each optimizer with increasing step of epochs 10, 20, 30, up to 140. It is observed that the value remains constant when it reaches the saturation point. The next set of experiments are conducted based on the previous result (considering the maximum accuracy), by changing the learning rate to 0.001 and 0.0001 for the same set of optimizers. This provides an extensive observation about the classification accuracy with respect to the applied dataset and helps to find better optimizer for classification of weed from the paddy crops.

3 Results and Discussion

The proposed method has been applied to the database obtained from Kaggle Web site consisting of 3923 paddy crop images along with different types of weed. Out of the total images, 70% (2746 images) are used for training, and the rest 30% (1177 images) are used for validation. The images are present in a RGB color format. The detailed description of data for various classes is mentioned in Table 2. The database is mainly divided into six classes, which include healthy paddy crops, low fertile nutrient deficiency, weed type-1, weed type-2, weed type-3, and weed type-4. The input images are resized to the input dimension of VGG-16 (i.e., $224 \times 244 \times 3$) for classification. The main task is classification of weed from paddy crop using pretrained VGG-16 model.

The classification uses transfer learning approach [20], which uses pretrained deep learning model. Images of different classes are given as input to the pretrained VGG-16 model. Both softmax and fully connected layers are included in the architecture with a batch size of 128. The performance measures have been calculated for different learning rates (0.01, 0.001, and 0.0001) with various epochs (10, 20,…, 100, 120, 140) and class mode as categorical. The performance measure is also calculated for various optimizers such as Adam, SGD, and Adadelta. For each optimizer, the best learning rate along with the best epoch has been decided based on their accuracy and loss function. The detailed performance measures along with their variations for different optimizers, learning rate, and epoch have been tabulated.

From Tables 4, 5, and 6, it is found that Adam optimizer provides comparatively better result for 0.0001 learning rate with 100 epochs. An accuracy of 99.85% has been noticed with a loss of 0.0164 for Adam optimizer. For SGD optimizer, the learning rate of 0.001 with 100 epochs provides the accuracy of 98.99% and loss of 0.0251. From table, it is seen that for Adadelta optimizer, the accuracy and loss have been found to be 93.73% and 0.0063, respectively, for 0.0001 learning rate with 120 epochs.

Table 4 Performance measure for different optimizers and epochs for the learning rate 0.01

Optimizer	Learning rate	Epochs	Training performance		Validation performance		Training time/image (seconds)	Validation time/image (seconds)
			Accuracy	Loss	Accuracy	Loss		
Adam	0.01	10	0.8999	0.4646	0.8887	0.4682	0.0546	0.1274
		20	0.9552	0.2056	0.9516	0.2100	0.1092	0.2548
		30	0.9760	0.1144	0.9550	0.1313	0.1638	0.3823
		40	0.9865	0.0717	0.9720	0.0851	0.2184	0.5097
		50	0.9916	0.0457	0.9856	0.0619	0.2731	0.6372
		60	0.9960	0.0290	0.9941	0.0333	0.3277	0.7646
		70	0.9971	0.0187	0.9952	0.0298	0.3823	0.8920
		80	0.9974	0.0178	0.9954	0.0261	0.4369	1.0195
		90	0.9980	0.0123	0.9953	0.0243	0.4916	1.1469
		100	0.9983	0.0093	0.9956	0.0160	0.5462	1.2744
		120	**0.9983**	0.0076	**0.9962**	0.0115	0.6554	1.5293
		140	0.9952	0.0235	0.9963	0.0451	0.7647	1.7841
SGD		80	0.9546	0.3493	0.9115	0.4366	0.4369	1.0195
		90	0.9530	0.2529	0.9329	0.2190	0.4916	1.1469
		100	0.9612	0.1848	0.9526	0.1818	0.5462	1.2744
		120	**0.9686**	0.1522	**0.9588**	0.1484	0.6554	1.5293
Adadelta		80	0.9134	1.6317	0.8710	1.6587	0.4369	1.0195
		90	0.9201	1.5164	0.8875	1.5364	0.4916	1.1469
		100	0.9219	1.4012	0.9042	1.4210	0.5462	1.2744
		120	**0.9264**	1.2863	**0.9057**	1.3061	0.6554	1.5293

Table 5 Performance measure for different optimizers and epochs for the learning rate 0.001

Optimizer	Learning rate	Epochs	Training performance		Validation performance		Training time/image (seconds)	Validation time/image (seconds)
			Accuracy	Loss	Accuracy	Loss		
Adam	0.001	80	0.9834	0.0949	0.9773	0.0952	0.4952	1.1554
		90	0.9859	0.0420	0.9811	0.0469	0.5571	1.2999
		100	0.9953	0.0677	0.9922	0.0865	0.6190	1.4443
		120	**0.9969**	0.0315	**0.9914**	0.0390	0.7428	1.7332
SGD		80	0.9799	0.0209	0.9685	0.0316	0.4952	1.1554
		90	0.9867	0.0213	0.9769	0.0316	0.5571	1.2999
		100	**0.9899**	0.0251	**0.9835**	0.0315	0.6190	1.4443
		120	0.9856	0.0248	0.9768	0.0310	0.7428	1.7332
Adadelta		80	0.9056	0.0173	0.8978	0.0165	0.4952	1.1554
		90	0.9160	0.0168	0.9106	0.0163	0.5571	1.2999
		100	0.9189	0.0165	0.9148	0.0161	0.6190	1.4443
		120	**0.9197**	0.0193	**0.9139**	0.0158	0.7428	1.7332

Table 6 Performance measure for different optimizers and epochs for the learning rate 0.0001

Optimizer	Learning rate	Epochs	Training performance		Validation performance		Training time/image (seconds)	Validation time/image (seconds)
			Accuracy	Loss	Accuracy	Loss		
Adam	0.0001	80	0.9894	0.0597	0.9801	0.0622	0.5243	1.2234
		90	0.9940	0.0279	0.9895	0.0296	0.5899	1.3763
		100	**0.9985**	0.0164	**0.9961**	0.0195	0.6554	1.5293
		120	0.9982	0.0075	0.9951	0.0094	0.7865	1.8351
SGD		80	0.9756	0.0057	0.9569	0.0096	0.5243	1.2234
		90	0.9794	0.0068	0.9644	0.0096	0.5899	1.3763
		100	0.9811	0.0068	0.9717	0.0096	0.6554	1.5293
		120	**0.9854**	0.0068	**0.9758**	0.0095	0.7865	1.8351
Adadelta		80	0.9284	0.0073	0.9012	0.0065	0.5243	1.2234
		90	0.9310	0.0068	0.9133	0.0063	0.5899	1.3763
		100	0.9356	0.0065	0.9232	0.0061	0.6554	1.5293
		120	**0.9373**	0.0063	**0.9241**	0.0058	0.7865	1.8351

Table 7 Performance comparison with two different sizes of training set

Optimizer	Learning rate	Epochs	Set 1: Training set 70% and validation set 30%				Set 2: Training set 80% and validation set 20%			
			Training performance		Validation performance		Training performance		Validation performance	
			Accuracy	Loss	Accuracy	Loss	Accuracy	Loss	Accuracy	Loss
Adam	0.01	120	0.9983	0.0076	0.9962	0.0115	0.9987	0.0065	0.9966	0.0178
	0.001	120	0.9969	0.0315	0.9914	0.0390	0.9973	0.0287	0.9908	0.0377
	0.0001	100	0.9985	0.9985	0.9961	0.0622	0.9989	0.0115	0.9965	0.0195
SGD	0.01	120	0.9686	0.1522	0.9588	0.1484	0.9691	0.0983	0.9904	0.1142
	0.001	100	0.9899	0.0251	0.9835	0.0315	0.9887	0.0127	0.9852	0.0215
	0.0001	120	0.9854	0.0068	0.9758	0.0095	0.9853	0.0063	0.9743	0.0088
Adadelta	0.01	120	0.9264	1.2863	0.9057	1.3061	0.9298	1.1432	0.9078	1.3213
	0.001	120	0.9197	0.0193	0.9139	0.0158	0.9203	0.0211	0.9167	0.0198
	0.0001	120	0.9373	0.0063	0.9241	0.0058	0.9399	0.0045	0.9267	0.0179

Table 7 shows the variation in terms of accuracy and loss when the number of training is increased from 70 to 80% and the rest 20% are used for validation. The performance accuracy does not increase much crops.

4 Conclusion

In the literature, researchers have not performed weed classification from paddy field for improving paddy yield. So, in this work, a pretrained VGG-16 model has been used for classification of different weed types grew in the paddy field. The performance of weed type classification from paddy crop by pretrained network has been computed for different optimizers, various learning rates, and epochs. It is found that Adam optimizer provides a better accuracy of 99.85% with 100 epochs and the learning rate of 0.0001. VGG-16 can perform better classification of weed from paddy crop.

In this experiment, less inaccuracies are noticed, and it may be due to the limited amount of database used. This can be further improved by considering large database with more number of classes. Since the current focus of interest is classification of one type of crop and various categories of weed type, the training can also be performed on different crop categories and also by varying different environmental conditions (such as stable/proper sunlight and different climatic conditions).

References

1. Ali, M.M., Al-Ani, A., Eamus, D., Tan, K.Y.: Leaf nitrogen determination using non-destructive techniques-a review. J. Plant Nutr. **40**(7), 928–953 (2017)
2. Sun, Y., Tong, C., He, S., Wang, K., Chen, L.: Identification of nitrogen, phosphorus, and potassium deficiencies based on temporal dynamics of leaf morphology and color. Sustainability **10**(3), 762 (2018)
3. Chen, L., Lin, L., Cai, G., Sun, Y., Huang, T., Wang, K., Deng, J.: Identification of nitrogen, phosphorus, and potassium deficiencies in rice based on static scanning technology and hierarchical identification method. PLoS ONE **9**, 1–17 (2014)
4. Jagan, K., Balasubramanian, M., Palanivel, S.: Detection and recognition of diseases from paddy plant leaf images. Int. J. Comput. Appl. **144**(12), 34–41 (2016)
5. Phadikar, S.: Classification of rice leaf diseases based on morphological changes. Int. J. Inf. Electron. Eng. **2**(3), 460–463 (2012)
6. Anami, B.S., Malvade, N.N., Palaiah, S.: Deep learning approach for recognition and classification of yield affecting paddy crop stresses using field images. Artif. Intell. Agric. **4**, 12–20 (2020)
7. Rahnemoonfar, M., Sheppard, C.: Deep count: fruit counting based on deep simulated learning. Sensors **17**(4), 1–12 (2017)
8. Cheng, X., Zhang, Y., Chen, Y., Wu, Y., Yue, Y.: Pest identification via deep residual learning in complex background. Comput. Electron. Agric. **141**, 351–356 (2017)
9. Khan, A., Ilyas, T., Umraiz, M., Mannan, Z.I., Kim, H.: CED-NET: crops and weeds segmentation for smart farming using a small cascaded encoder-decoder architecture. Electronics **9**(10), 1–16 (2020)

10. Ma, X., Deng, X., Qi, L., Jiang, Y., Li, H., Wang, Y., Xing, X.: Fully convolutional network for rice seedling and weed image segmentation at the seedling stage in paddy fields. PLoS ONE **14**(4), 1–13 (2019)
11. Chebrolu, N., Lottes, P., Schaefer, A., Winterhalter, W., Burgard, W., Stachniss, C.: Agricultural robot dataset for plant classification, localization and mapping on sugar beet fields. Int. J. Robot. Res. **36**(10), 1045–1052 (2017)
12. Haug, S., Ostermann, J..: A crop weed field image dataset for the evaluation of computer vision based precision agriculture task. In: Agapito, L., et al. (eds.) Springer International Publishing Switzerland 2015 ECCV 2014 Workshops, LNCS 8928, pp. 105–116. Springer, Switzerland (2015)
13. Adhikari, S.P., Yang, H., Kim, H.: Learning semantic graphics using convolutional encoder–decoder network for autonomous weeding in paddy. Front. Plant Sci. **10**, 1–12 (2019)
14. Kamath, R., Balachandra, M., Prabhu, S.: Paddy crop and weed discrimination: a multiple classifier system approach. Int. J. Agronomy **3**(4), 1–14 (2020)
15. Anami, B.S., Malvade, N.N., Palaiah, S.: Classification of yield affecting biotic and abiotic paddy crop stresses using field images. Inf. Process. Agric. **7**, 272–285 (2020)
16. Ferentinos, K.P.: Deep learning models for plant disease detection and diagnosis. Comput. Electron. Agric. **145**, 311–318 (2017)
17. Nath, M.K., Kanhe, A., Mishra. M.: A novel deep learning approach for classification of covid-19 images. In: International Conference on Computing Communication and Automation (ICCCA), pp. 752–757. IEEE Xplore, Greater Noida, UP, India, (2020)
18. https://www.kaggle.com/archfx/paddyimages
19. Wang, G., Sun, Y., Wang, J.: Automatic image-based plant disease severity estimation using deep learning. Comput. Intell. Neurosci., 1–8 (2017)
20. Rautaray, S.S., Pandey, M., Gourisaria, M.K., Sharma, R.: Paddy crop disease prediction—a transfer learning technique. Int. J. Recent Technol. Eng. **8**(6), 1490–1495 (2020)

Data Analytics: The Challenges and the Latest Trends to Flourish in the Post-COVID-19

T. R. Mahesh, V. Vivek, C. Saravanan, and K. Vinay Kumar

1 Introduction

While developments in business technology have mostly been noticeable through cloud computing and smart technologies, advances have also been important in the analysis of business data [1]. Trends in corporate data recording, encoding, extracting, and analyzing have improved drastically over the years, and future developments are even showing greater capacity to develop and improve market analytics. Business intelligence systems and analytical technologies for data mining are now best fused with transactional systems that create a closed loop between research and activities [1]. Several new trends have characterized the real changes people are experiencing in the business analytics world. The purpose of this paper is to examine emerging patterns in the study of industry.

Business intelligence developments have led to advances in many fields of business analytics, such as multi-polar analytics, cloud analytics, analytical technology, fluid analytics, ecosystems of analytics, and privacy systems of data. Multi-polar analytics are a method of data processing and analysis where researchers capture, store, and analyze large-scale distributed data via a combined analytical model [2]. This data analytics technology decreases the chances of departmental corruption

T. R. Mahesh (✉) · V. Vivek · C. Saravanan
Faculty of Engineering and Technology, JAIN (Deemed-to-be University), Bangalore, India
e-mail: t.mahesh@jainuniversity.ac.in

V. Vivek
e-mail: v.vullikanti@jainuniversity.ac.in

C. Saravanan
e-mail: c.saravanan@jainuniversity.ac.in

K. Vinay Kumar
Kakatiya Institute of Technology and Science, Warangal (KITSW), Warangal, India
e-mail: kvk.cse@kitsw.ac.in

M. Dua et al. (eds.), *Proceedings of the International Conference on Paradigms of Communication, Computing and Data Sciences*, Algorithms for Intelligent Systems,
https://doi.org/10.1007/978-981-16-5747-4_43

by data thefts. Cloud analytic technology is another prevalent and renowned technology that constantly seems to reshape the field of business analytics [2]. The management of cloud computing technology has improved. Some of the prominent cloud technologies include Amazon Corporation's Red-shift hosted BI data warehouse, Google Corporation's Big Query data analytics system, and the International Business Machines (IBM) Corporation's Blue-mix cloud platform [3]. The cloud computing approaches used in business analytics have improved protection of data, reliability of data, and compactness of data.

Wide and highly secure analytical cloud technologies have improved data storage, data coding, database management, and data retrieval from trusted global internet and software companies [4]. The use of fluid analytics is another impressive technique of data handling that analysts have considered in emerging technologies as an imperative instrument for information systems. The developments of fluid analytics, popularly known as big data lakes, have allowed corporate organizations to manage large chunks of data, provide high-level data descriptions on these data, and produce independent views of the data recorded [4].

Other advanced technologies that have allowed businesses to build broad business data networks are sophisticated predictive analytics. Sophisticated predictive analytics help analysts minimize large data, work with large numbers of records of data, and gain rapid insights and alert about possible losses of data through linked data networks [5]. In the field of business analytics, another new technology is the use of methodology of dynamic event management (CEP). Complex event processing is business analytical innovations and methodologies that enable analysts to analyze data streams for pattern detection from live sources and as business indicators that support decision making for entrepreneurs [5].

The Apache Spark and the Spark SQL business analytic software are some of the excellent predictive analytic tools. The increased use of the Hadoop Structured Query Language (SQL), which is one among cloud data analytics that involves a framework as well as many collection of tools of data used for huge amount of data processing [5]. The incorporation of a quicker, better, and improved procedure, SQL servers on Hadoop have allowed the processing, coding, and storage of data. Data analytical tools that maintain SQL-like querying processes allow business owners with SQL server expertise to improve proper data storage and handling [2]. Deep learning is another business analytical advancement which, through its advanced analytical techniques, continually promises many advances in the business analytics paradigm [2]. A range of machine learning techniques derived from neural networking systems are defined by Deep Learning Technology.

Data collection happens when business analytics technology is able to expose previously invisible processes to allow stored transaction data to be effectively monitored and optimized [6]. While the method is a traditional approach that has undergone a sequence of modifications and altered its original architecture and structures with minor adjustments, new data patterns have re-emerged to embrace high processing speeds. Owing to the development and availability of real-time operational analytics systems, the criteria for this approach have increased. Furthermore, the rising prices in the data collection processes have made the data collection models marketable.

The global spread of the COVID-19 pandemic has created an enormous amount of data that can be used to enhance our understanding of big data management science, as well as demonstrating the need among academics, practitioners, and policymakers for a better and deeper understanding of a variety of analytical tools that can be used to better predict and respond to disasters [7–9]. The year 2020 has been full of unforeseen difficulties. Having said that, it also acted as a rare opportunity on many fronts to exploit technology. The industry has gone through numerous digital touch points, from implementing it in various sectors such as retail, e-commerce and others, to adopting it to ensure the protection of workers at work from home scenarios, and enhancing customer experiences[10–12]. In order to bring about improvements to suit the changing market scenario, the implementation of data, analytics, AI, cyber security, and other emerging technology has seen exponential growth. The next sections provide the challenges in data analytics and the newest trends which will flourish in the post-COVID-19 [12, 13].

2 Challenges in Data Analytics

Tim McGuire, the McKinsey director, said in 2013: Analytics will definitely define the dissimilarities between losers as well as winners who going forward," and he was definitely not wrong. Since then, we have seen businesses leverage knowledge resources across the board to execute smarter decisions, drive transparency, boost their financial health, and keep a closer eye on their performance.

The data analytics market is projected to expand at an annual growth rate of 30 percent by 2023, hitting $77.6 billion in annual spending, according to a recent report. Such figures demonstrate just how important data capabilities have become, indicating a future where it would be simply impossible to accept digital business without data. Efficient analytics have become such a determinative factor that it is now clear that it will succeed for those who master it. The path toward that aim, however, is not possible without barriers. The most common data analytics challenges and how these companies confidently confront them are discussed below.

A powerful database can remove any problems with usability. Registered workers would be able to access or edit information securely from anywhere, highlighting organizational improvements and allowing high-speed decision-making.

2.1 Pressure Coming from the Top

The top executives expect more performance by risk managers since risk management is very common in organizations. They demand higher returns on all kinds of data and a large number of studies. Risk managers can go above and beyond standards with a robust analysis framework and produce the desired analysis easily.

2.2 Confusion or Anxiety

Users, though they realize the advantages of automation, they may sense confused or sometimes anxious about transforming from traditional methods of data analysis. No one likes to change especially when they are already comfortable and very familiar with the kind things that are done.

2.3 Navigating Constraints on the Budget

Due to rapid growth in technology trends, budding analytical tools are deployed in the same phase which makes challenging for the enterprise to keep up. Top two major challenges faced by most of the analytical tools are to be in front line on data analytical platform and continues testing test bed resources. Leaders of data analytics need to work in the moment, but still think about the future. This means producing data-driven projects with business results while also creating an efficient data system for tomorrow. It allows them to take responsibility in creating a transparent and systematic plan to balance these needs. Businesses, on the other hand, often lack the necessary data and analytics organizational structure to rely on. Many organizations' data and analytics departments come under IT and are granted very little in terms of headcount and budget, as they are usually seen as cost centers, making it difficult to justify large investments in analytics tools and skills. To work around budget limits, these organizations' administrators must first consider their particular organizational requirements before implementing their own data management strategy. This way, they do not have to spend millions of dollars on complicated data storage infrastructure just to discover that they only need a fraction of that shortage of abilities.

Owing to a lack of talent, some companies struggle with research. In those without structured risk departments, this is particularly true. Employees may not have proper experience or capacity to conduct analysis of data in-depth. This problem is diminished using two ways: by discussing analytical capabilities in the recruiting process and providing an easy to use analysis framework. The first approach ensures that skills are available; however, the second approach will simplify everyone's research process. This form of method, regardless of skill level, can be used by anyone.

2.4 Lack of Support

Nevertheless, organizations also lack the right organizational framework for data and analytics to lean on. The data and analytics community falls under IT in many businesses and is given very little in terms of headcount and budget, as they are usually seen as cost centers, making it hard to justify high spending on analytics tools and skills. Managers in these groups should first understand the basic organizational requirements to develop their particular data management plan in order to overcome budget limitations. They stop spending millions of dollars in complex data storage infrastructure in this way, only to find out that they need even less than that. In addition, analytics leaders can secure a budget for tools and skill sets by evaluating a system's ROI and highlighting short-term and long-term benefits.

2.5 Data Analytics for Scaling

As businesses, and the amount of data they collect, expand over time, it can become increasingly difficult to manage data analytics. The process of gathering information, working on projects, and producing reports can easily go awry without a plan in place. To avoid inaccuracies, it is important to introduce a framework that can evolve with the company and adapt to the rapid pace of change. Increased collaboration, especially among IT, line-of-business, and analytics groups, can drive the success of big data and analytics. That is why risk managers should look at scalable tools that provide a 360-degree view of data and exploit the capabilities of advanced processing and analysis. Some of the most important considerations to consider are automated processing and sorting, simple sharing and retrieval, real-time collaboration, and the ability to condense complex datasets into a single type of study.

2.6 Poor Data Quality

Data is the lifeblood of an organization, so decisions would invariably be adversely affected if they are not of good quality. One of the most common challenges when it comes to streamlining analytics is to provide access to large quantities of data from multiple sources, often with distinct formats and consistency. It begins with the input: as we say, garbage in, garbage out. Companies are at risk of making uninformed business decisions, achieving inaccurate contact with consumers, and not meeting regulatory expectations. This technique is incredibly inefficient and duplicative. Companies should instead invest in centralized systems and data cleaning automation to solve data quality problems. These tools allow the input of data automatically with field quality controls, leaving little space for errors of humans and also, businesses guarantee that a sudden shift in one region does not cause any changes by utilizing simple device integrations.

3 Trends in Data Analytics from 2021 Onwards

Businesses now realize that they have to expect the unexpected, be prepared in advance to navigate. Considering this, in the advanced analytics industry in 2021, there are big developments that will flourish.

Cost-saving initiatives, since they are costly, sluggish, and moment-in-time driven, can lead to less dependency on consulting firms so optimization is essential: 2020 was really disruption year of the workforce, and definitely 2021 and further are going to be the recovery years and pushing automation efficiencies and efficient utilization of internal capital. It is not only about gathering information, but also about taking the information and placing it into effect. Due to the need to continuously adapt to evolving market trends, direct-to-consumer shopping the dramatic change to e-commerce, and frequent changing consumer buying habits would increase the focus on data as well as analytics as against to reading one-time reports.

Organizations require to access as well as use more information channels, namely centers, and other contact points for consumers: In performing so, corporations would be more forced to build a single origin of truth. Of course, this is, unmet need and when it comes to their implementation of advanced analytics, companies must have the right mentality to be successful. This will also drive further collaborations in the data ecosystem.

Virtualization as well as specialization of platforms for data as well as analytics can take on greater significance: The demand for analytics is established well, and it has matured with generic platforms that crunch information and generate visualizations. Companies will now, however, demand a degree of domain experience and understanding of how actually data as well as analytics can serve particular use cases.

It will blur the boundaries between IT and other agencies: Areas such as data governance, open data systems, data integration and use in various parts of organization will allow users of business to execute tasks historically booked for IT teams, and data generated by units of business will be fed into platforms controlled by IT. This also ensures that data systems can become more seamless and easier to deploy, combined with a scarcity of analytics experts and data scientists, so that all areas of an enterprise can take advantage of it.

For data storytelling, No-code interactive dashboards: No-code and low-code systems have become a widely debated technology that has drawn the interest of technology communities and large investments in startups that provide such solutions. It is not just startups, however, that are operating on these instruments, Apple-owned

Claris also strives to deliver "powerful technology accessible to everyone." Companies that supply such instruments destroy two birds with one stone. With each interaction, product teams and developers are liberated from the repetitive work of manually writing code, as it paves the way for "non-technical" individuals to contribute to such projects. The importance of these types of tools for companies is difficult to overestimate, especially when it comes to interactive data storytelling dashboards. Data analysts are able to run pattern analysis and almost immediately enjoy personalized visualization without coding. We see from our own experiences that more than 75% of requests are based on no-code analytics and visualization, a trend driven by companies searching for ways to embrace a fast-paced workflow without recruiting additional experts. Interestingly, even though there are data story apps that provide detailed explanations behind the changes in data, businesses are still likely to use dashboards that only control and view data points.

Quicker and faster exchange of data: Sharing is affectionate. Sharing also implies teamwork and growth in the case of data. However, data protection issues have raised government concerns, which have resulted in several laws on privacy. This makes organizations legally obligated to use the information of users securely under the danger of fines, litigation, and bans on the web. As a result, it turns the heads of business owners and lets investors reach into their wallets when data can be exchanged easily, smoothly, and without risks. The need for fast and simple data sharing solutions is clearly demonstrated by well-funded startups focused on file transfers. Dataflow constraints should not be the only element that determines data processing creation. The problem of data monopolies also needs to be discussed where a small number of corporations control the vast majority of internet data. Working together to build data infrastructures that allow organizations as well as individual users to have full control over their data, including secure ways to share it, is essential for market players and governments.

Moving more rapidly toward 3D visualization: When the number of sources of knowledge and advanced analytics software increases, so does the sophistication of organization processes. The need to avoid these patterns from being a handicap paves the way for the rapid implementation of solutions for 3D visualization. For businesses that process geospatial data and combine classical analytics with location monitoring, this tool has already proven to be particularly successful. The 3D rendering market is expected to rise to $6 billion by the year 2025, considering the interest in 3D visualizations in both outdoor and indoor areas. Companies, government agencies, and utility companies are looking more closely at smart solutions to try to integrate all layers, including LiDAR (light detection and range) data. Digital twins, virtual replicas built using data technology, are one of these. They are used to enhance environmental efficiency and productivity, and to investigate places without compromising the well-being of employees. The issue with conventional 2D presentation of data is the limitation of how much data it holds and how much of it can actually be used. Many observations, concealed under the layers of maps, bars, and pies, remain untapped. Showing data in three dimensions, however, can lead to the disclosure of geospatial knowledge that can help companies gain insights and discover new business opportunities. It helps leaders and experts not only to look at "what is" but also to investigate and anticipate the responses to the questions of "what if" and "what will."

It is not possible to ignore or understate the influence of COVID: From 2021 onwards, companies will start raising questions about what patterns they see are linked to COVID, whether data or insight anomalies are to be allocated to either short or else long term or how to supervise the company in the future. In order to improve accuracy, predictive analytics would have to consider this and exploit data which is continually refreshed and linked to as many sources of data possible.

4 Conclusion

In conclusion, we assume that, while staying hopeful, the problems facing the world in 2020 will soon become history. These incidents, however, will probably speed up the ongoing trends that we have described. Working remotely, as well as collaborating with long-distance teams, would increase the importance of secure and effortless data sharing, 3D geospatial information visualization, and the popularity of solutions that allow a greater number of experts to use data analytics tools without coding. Looking at the previous year, 2021 looks like an opportunity to expand into newer arenas for tech trends. In the coming year, the highlights will be smart computers, hybrid cloud, increased NLP adoption, and an increased emphasis on data science and AI overall. Pragmatic AI, containerization of analytics and AI, algorithmic separation, augmented data management, differential privacy, quantum analytics, among others are some of the other developments that could see a rise in the coming years. Considering these patterns, it can be said that since the pandemic, knowledge is gradually becoming a vital part of organizations. Finally, this methodology corner article aimed to include a summary of the newest developments that will thrive in the post-COVID period, as well as how they can be used to analyze contemporary organizational and management issues emerging from the global pandemic triggered by COVID19, as well as other grand challenges of our time.

References

1. Kohavi, R., Rothleder, N., Simoudis, E.: Emerging trends in business analytics. Commun. ACM **45**(8), 45–48 (2002)
2. Mitchell, R.: 8 big trends in big data analytics (2014). Retrieved from http://www.computerworld.com/article/2690856/big-data/8-big-trends-in-big-dataanalytics.html
3. Hard, H.: Organizational Applications of Business Intelligence Management: Emerging Trends: Emerging Trends. IGI Global, New York (2012)
4. Hota, J.: Workforce analytics approach: an emerging trend of workforce management. AIMS Int. J. **7**(3), 167–179 (2013)
5. Bihani, P., Patil, S.: A comparative study of data analysis techniques. Int. J. Emerg. Trends Technol. Comput. Sci. **3**(2), 95–101 (2014)
6. Loshin, D.: Business Intelligence: The Savvy Manager's Guide. Newnes, New York (2012)

7. Kambatla, K., Kollias, G., Kumar, V., Grama, A.: Trends in big data analytics. J. Parallel Distrib. Comput. **74**(7), 2561–2573 (2014). https://doi.org/10.1016/j.jpdc.2014.01.003
8. Faroukhi, A.Z., El Alaoui, I., Gahi, Y., Amine, A.: Big data monetization throughout Big Data Value Chain: a comprehensive review. J. Big Data (2020). https://doi.org/10.1186/s40537-019-0281-5,7,1
9. Lavalle, A., Teruel, M.A., Maté, A., Trujillo, J.: Improving sustainability of smart cities through visualization techniques for big data from IoT devices. Sustainability (2020). https://doi.org/10.3390/su12145595,12,14,(5595)
10. Lavalle, A., Maté, A., Trujillo, J.: Requirements-driven visualizations for big data analytics: a model-driven approach. Conceptual Modeling (2019). https://doi.org/10.1007/978-3-030-33223-5_8,(78-92)
11. Han, W., Jochum, M.: 2017 IEEE International Geoscience and Remote Sensing Symposium (IGARSS), pp. 384-387 (2017). https://doi.org/10.1109/IGARSS.2017.8126976
12. Indriasari, E., Soeparno, H., Gaol, F.L., Matsuo, T.: 2019 8th International Congress on Advanced Applied Informatics (IIAI-AAI), pp. 877–883 (2019). https://doi.org/10.1109/IIAI-AAI.2019.00178
13. Lavalle, A., Mate, A, Trujillo, J., Rizzi, S.: 2019 IEEE 27th International Requirements Engineering Conference (RE), pp. 109–119 (2019). https://doi.org/10.1109/RE.2019.00022

Sentiment Analysis of Tweets in Social Media Over Covid-19 Span

S. Uma Maheswari and S. S. Dhenakaran

1 Introduction

In the history of Corona in 2003, at Guangdong in China, the first SARS-CoV, a Severe Acute Respiratory Syndrome Corona Virus was detected. This virus infected 8098 people across 26 countries in the world, with a 9% of mortality rate [1–4]. In 2012, Saudi Arabian Nationals detected a Corona Virus called Middle East Respiratory Syndrome Corona Virus (MERS-CoV). The World Health Organization (WHO) reported more than 2428 people infected by (MERS-CoV) and the death rate was 838 [5]. In December 2019, WHO announced the first Covid19 in Hubei Province of Wuhan in China. The International Committee on Taxonomy of Viruses (ICTV) named this virus as SARS-CoV-2 [1–4, 6, 7]. The augmented growth of Covid-19 initially infected people in the seafood market of China. Hundreds of millions of people are killed by this Novel Corona virus. Many countries across the world have taken many steps to control the outbreak of this virus. Now, Covid-19 is a little bit managed and spreading is controlled by maintaining 'Social Distancing', masking, and regular hand washing at certain intervals of time. Social distancing restricted people to have physical contact and nation-wide lockdown has decreased requirements for daily living. The less contact with the outside world raised the need for social networks rises which has connected friends, shopping centers, family circles, and the education system. Also working from home, business meetings, office meetings, video conferences, education meetings, webinar programs, educational conferences, online classes, online exams, virtual presentation, everything conducted via Social Networks like Google Meet, Zoom Meet, WhatsApp, WeChat, Instagram, Twitter, Facebook, and YouTube, etc.

S. Uma Maheswari (✉) · S. S. Dhenakaran
Department of Computer Science, Alagappa University, Karaikudi, India

M. Dua et al. (eds.), *Proceedings of the International Conference on Paradigms of Communication, Computing and Data Sciences*, Algorithms for Intelligent Systems,
https://doi.org/10.1007/978-981-16-5747-4_44

Also the government announced the causes and situations of Covid-19 through mass media and social media. Sharing some misinformation of Covid-19 through social media also has increased the panic situation [8–10]. So special attention is needed to post information especially of this Covid-19 outbreak on social media Networks because social media has given support to all walks of our life like office work and business, work from home, online classes, and online exams, etc. Though the social network gives support, on the other hand, controversial information is also shared for the social network influence during this Coivd-19 period. Some set of people have a good opinion about social media, online classes, online exams, and doing official work from home. Some other people blame about the work from home and online learning because of network signal problem, internet speed problems. Network and internet speed are the key factors for online learning and work from home. Again, a set of people think that online classes virtual presentations are not much interactive, and online exams are not held honestly or genuinely. They think that students are simply copying and submitting the class works, assignments, and exams. Though office work and business activities are smoothly happened due to objectives and need, sincere studies, writing exams and listening to online meetings, etc., have an impact on social media because monitoring and controlling are difficult in these cases.

Another set of people have gossip about the treatment of Corona positive cases. The Corona positive cases are admitted to the Health Care Center but treatment was just by giving paracetamol medicine for 5 days and sent patients back to home. Because of this nature of Covid-19 treatment failed to take seriousness about Covid-19 disease, some of the people in the village interior are even not believed the Corona Virus. Also, such kinds of people do not have seriousness and have not followed the responsibilities of Government rules and regulations to decrease the Corona Virus. The people of this category never give importance to wearing masks and never maintained social distancing. This problematic and most seriously considerable situation and environment such as online education and work from home, treatment on Covid-19 has motivated to analyze the people opinion about work from home, online education, and treatment of Covid-19 positive cases. Consequently, social media has influenced all walks of life to complete necessary tasks. This will helps the people to know about the social media influencing doing a good job or not, whether it is useful or not and whether has to be improved or not on these mentioned areas. Also, the Government and organizations also can know the public supports for social media and activities taken to manage the Covid-19 pandemic situation. And if necessary can give more attention to manage the situation, treatment on Covid-19 positive cases, online classes, and work from home.

The remaining of this article is organized as follows: Sect. 2 presented the related works of sentiment analysis on Covid-19, Sect. 3 discusses the proposed work of sentiment analysis on Covid-19 situation, classification, and prediction, Sect. 4 illustrates the experimental results, and finally, Sect. 5 summarizes the conclusion of the proposed work.

2 Literature Review

The increase of Corona death cases and rapid panic growth of infection has accelerated the governments, doctors, and researchers to give more attention to control the proliferation and diagnosis. Literature reviews can give a road map to understand the Covid-19 pandemic, key factors of Covid-19, most considerable situation, and necessity of the distinct researches in this Covid-19 outbreak.

In this work, the authors have collected two types of tweets considering Covid-19 rampant problems. Nearly, 23,000 re-tweeted tweets of Covid-19 are gathered from Jan 01, 2019 to March 23, 2020. The observation of tweets is stated that the maximum number of comments is neutral and negative. Then, tweets of Covid-19 from December 2019 to May 2020 are gathered, and its analysis expressed positive and neutral comments. It is seen that the deep learning classifier model for sentiment analysis has produced 81% accuracy. The sentiment analysis of Gaussian membership function based fuzzy logic technique [11–14] has produced 79% accuracy.

Government, political leaders, and celebrities can move forward and help the followers to make sure, while posting information about Covid-19 through social media is not fake and informative. While sharing information through social media people believe the information that they are seeing is true. Everybody should have the responsibilities before sharing any information and make sure that sharing information should not be rumors/fake [15–18]. Social media helped a lot in this period for quick information sharing, say, an infographic system developed for airway management for patients with confirmed Corona positive and suspected cases [19]. It was shared via social media such as Twitter and WeChat. The dissemination of information has caused anxiety, panic, stress, and depression in public [20]. Social media has to follow some important and useful protocols for the followers and users to share the information. Important healthy and diagnostic information can be shared rapidly but the dissemination of information must be restricted. The fear of Covid-19 and its consequences lead to some more diseases in public. Twitter dataset for 7 days Corona Virus is scraped via python beautiful soap package. Then, tweets are analyzed for positive or negative opinions. Likewise, tweets of online classes and work from home (WFH) are scraped and analyzed by deep learning methods called LSTM and ANN for sentiment classification [21–23].

Sentiment analysis on Covid-19 pandemic has five phases since tweets are preprocessed into a structured format by removing unwanted words and symbols. Tokenization, stemming, and lemmatization have done by tweets cleaning, words converted into vectors and most frequent words are selected by feature selection, algorithm modeled by a set of rules and equation and algorithm evaluated by performance evaluation methods [24].

3 Proposed Method

The proposed sentiment analysis on online classes, work from home, and treatment of Corona positive case has five phases. Tweets collecting, preprocessing, feature selection, building the sentiment classification model and finally evaluating the classification model.

3.1 Tweets Collecting

Tweets of online education, work from home, and treatment of Corona Virus positive case are collected in real-time using the Python Tweepy package. Then, tweets are stored in the Big Data Analysis tool Hadoop Distributed File System (HDFS).

3.2 Preprocessing

Collected tweets are usually in an unformatted and unstructured format. To make tweets meaningful, formatted, and structured format tweets preprocessed using natural language processing methods such as removing unwanted symbols, removing stop words, tokenization, stemming, and lemmatizing processes (Fig. 1).

3.3 Feature Selection

The features are important terms in the tweets that make comments meaningful and carry information to readers. The feature selection sometimes called the keywords (important words) of tweets is done from the preprocessed tweets, and features list or dictionary words are created using natural language processing. Intensifier, modifier, negation words, emoji, slang words, POS-Tag are considered as features in this proposed work. A word dictionary is created with a combination of positive and negative sentiment-oriented words with corresponding polarity scores. Intensifier words list is created with !, #, ?, happyyyy (a word with a sequence of repeated characters), AWESOME (Capitalized Word) [24, 25]. These kinds of symbols and words include words giving more attention in the sentiment analysis. But, the negation word list is not good, not excellent. The slang word dictionary is created like OMG with their proper expansion. Modifier word list and conjunction word list are created. The emoji dictionary is created with emoji unicode with their proper polarity score. The POS-tagging words, part of speech words such as noun, verb, adverb, and adjective are extracted from each tweet. Re-tweet count and likes count of the tweets also selected as the feature in this work.

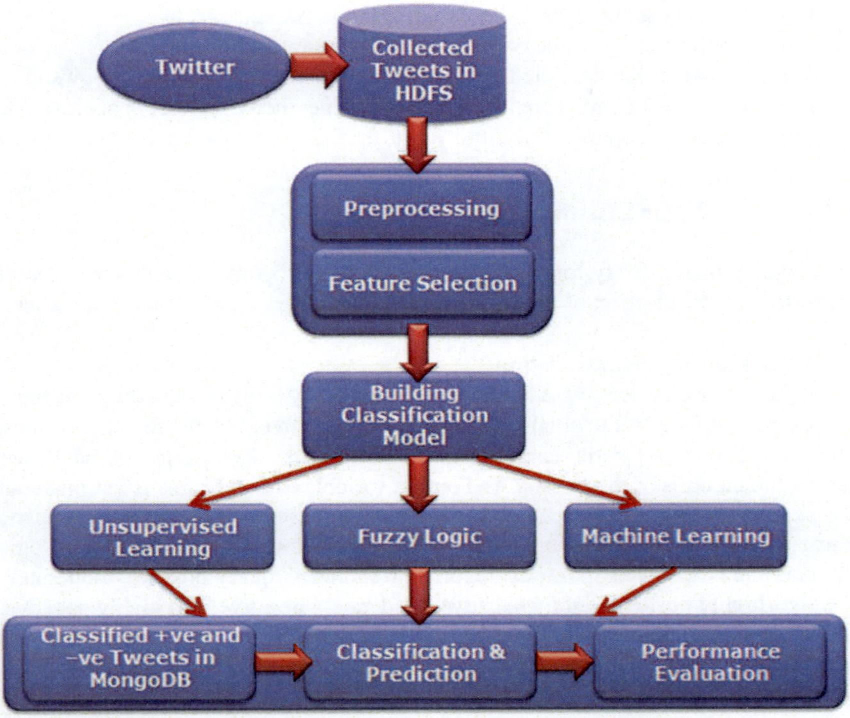

Fig. 1 Architecture of the proposed model

3.4 Building Classification Model

For better understanding, the comments of customers/people for the tweets (data sets) considered for experimentation, and the tweets are processed by the following methods.

3.4.1 Unsupervised Learning of Classification Tweets

Sentiment score is calculated for each feature of the tweet by polarity score provided with word list and dictionary words. Then, all sentiment scores are aggregated to find the total sentiment score of each tweet. The sentiment of tweets is classified based on the rules stated below.

If the sentiment score of the tweet is >0, then the tweet is positive.
If the sentiment score of the tweet is ≤0, then the tweet is negative.
Classified tweets are predicted based on certain predefined rules as follows.
If the tweet's sentiment score is ≥ threshold value, then the tweet is positive.
If the sentiment score is < threshold value, then the tweets are negative.

3.4.2 Fuzzy Logic Classification of Tweets

In this experiment, fuzzy logic is embedded in this proposed model. The rules of fuzzy logic, classification of tweets are done with the sentiment score of all tweets.

Fuzzification and Defuzzification
In the fuzzy logic system, fuzzification is performed by transforming the crisp input values into the fuzzy set to obtain the fuzzified values. To obtain the fuzzified values, input and output linguistic variables are identified first. Here, input variables are the sentiment class of the tweets. And output variables are public satisfaction about the particular category of tweets. For classification based on fuzzy logic, linguistic terms must be taken for the input and output linguistic variables. Here, fuzzy logic has classified the tweets into six categories. It includes lightly positive, moderately positive, highly positive, lightly negative, moderately negative, and highly negative. In the defuzzification process, the need for information to the public, the tweet is labeled as highly supportive, moderately supportive, and lightly supportive. For fuzzification, IF–THEN rules are applied.

3.4.3 Machine Learning Techniques of Classification Tweets

In this experiment, existing emerging machine learning techniques such as Naïve Bayes, support vector machine, decision tree, logistic regression, and random forest methods are applied. Initially, the sentiment score for all tweets is calculated, and then the dataset is split into training set and testing set. The preprocessing tweets are further divided into training and testing data sets. For training, only 80% of the data are used and for testing 20% of the data used. All classified and predicted results are stored in the big data analysis tool MongoDB. The results have tweets with sentiment score and polarity stating tweet either positive or negative.

3.5 *Performance Evaluation*

The proposed classification model is evaluated by the confusion matrix and the associated parameters namely accuracy, precision, recall, and F1-score.

Actual values	Predicted values	
	Positive (1)	Negative (−1)
Positive (1)	True positive (TP)	False positive (FP)
Negative (−1)	False negative (FN)	True negative (TN)

TP positive class is predicted as positive
TN negative class is predicted as negative
FP positive class is predicted as negative
FN negative class is predicted as positive
Classification accuracy rate = (TP + FN) / (TP + TN + FP + FN)
Precision = TP / (TP + FP)
Recall = TP / (TP + FN)
F1-score = 2 * (Precision * Recall) / (Precision + Recall)

The proposed sentiment analysis is implemented with three different data sets viz online classes-online education, tweets on work from home, and tweets on treatment of Covid-19 positive cases. The results of proposed unsupervised learning, fuzzy logic are compared to existing machine learning techniques.

4 Results

In this proposed sentiment analysis, three different experimental analysis done for using tweets about online class and online education, tweets about work from home, and tweets about treatment on Covid-19 positive cases.

4.1 Experiment I: Unsupervised Learning

This experiment analysis is done on tweets of medical treatment for Covid-19 positive cases, work from home, and online education. In Table 1, the number of tweets 3537 of Covid-19 treatment cases shown. The preprocessing phase has generated only 3528 number of tweets. Further, this method has classified 2085 number of positive tweets, 1443 number of negative tweets. Similarly, 2161 tweets of work from home are collected. The preprocessing has generated only 1528 number of tweets. In this case, only 936 tweets are found positive, and 592 tweets are indicated negative. It states that the opinion about the treatment for Covid-19 positive cases is supported by the public. And for WFH tweets, the number of positive tweets is greater than the number of negative tweets. These results state that the opinion about the medical treatment, and work from home is liked and highly supported by the public.

Table 1 Experimental analysis of unsupervised learning

Tweets	Total tweets	Preprocessed tweets	No. of positive tweets	No. of negative tweets
Treatment of Covid-19	3537	3528	2085	1443
	6463	4548	2954	1594
	4394	3504	2450	1054
	6064	4468	3047	1421
	5751	4114	2754	1360
WFH	2161	1528	936	592
	1920	1313	834	479
	1761	1226	761	465
	1832	1303	771	532
	1833	1356	787	569
Online education	3441	2434	1208	1226
	4411	3106	1766	1340
	5743	3819	2954	865
	5396	3500	2608	892
	5311	3527	2634	893

In the case of online education, only 3441 number of tweets obtained. The preprocessing has produced 2434 number of tweets. By classification, only 1208 number of tweets are positive and 1226 tweets negative which shows that the number of negative tweets is greater than the number of positive tweets. Classification of online education using the small number of tweets shows negative opinions from the public. For do more clarification, more data about online education have been collected, analyzed, and classified.

Totaly for five days, real-time tweets on online education, work from home and treatment of Covid-19 have been collected, preprocessed, and classified based on the user-defined approach, fuzzy logic, and machine learning approaches. In Table 1, total tweets represents the collected noumber of tweets, preprocessed tweets represent the tweets after cleaned, no. of positive and no. of negative represent the number of positive and negative tweets after classified based on the user-defined classification model. All classification report shows that positive tweets gretater than negative tweets. Which implies that treatment of Covid-19, online education, and work from home are supported by public.

Table 2 shows the metrics values of tweets, and these values are above 90%. The accuracy, precision, recall, F1-score, and ROC-AUC curve are measures qualifying the proposed model. From the report, the classification of tweets by the proposed model has performed well (Figs. 2, 3, 4, 5, 6 and 7).

Table 2 Performance report for unsupervised learning

Tweets	Pre-processed tweets	Accuracy (%)	Precision (%)	Recall (%)	F1-score (%)
Treatment of Covid-19	3528	95.49	100	92.37	96.04
	4548	96.61	100	94.79	97.32
	3504	93.04	100	90.04	94.76
	4468	91.94	100	88.19	93.72
	4114	90.64	100	86.02	92.48
WFH	1528	89.66	100	83.12	90.78
	1313	89.19	100	82.97	90.69
	1226	90.13	100	84.1	91.36
	1303	90.71	100	84.31	91.48
	1356	90.86	100	84.24	91.45
Online education	2434	94.99	100	89.9	94.68
	3106	94.85	100	90.94	95.26
	3819	95.16	100	93.74	96.77
	3500	94.89	100	93.14	96.45
	3527	94.89	100	93.13	96.44

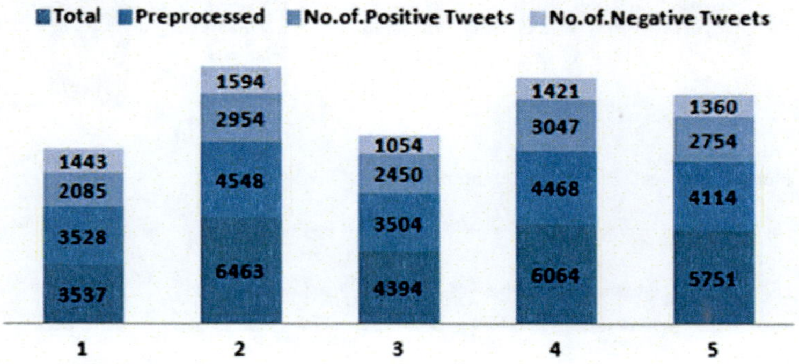

Fig. 2 Number of tweets on treatment of Covid-19

4.2 *Experiment II: Fuzzy Logic*

Table 3 shows data set of tweets used for fuzzy logic classification. For the medical category, 6463 tweets are collected and preprocessing has displayed only 4499 tweets. The classification of the Fuzzy principle has brought out 2905 tweets positive and 1594 tweets negative. For the WFH case of data sets, 1920 tweets are collected, only 1301 tweets are obtained by preprocessing. The classification has shown 821 positive tweets and the remaining 480 are negative tweets. For Online Education cases, 4411 tweets are gathered, and preprocessing has 3117 tweets. In this case,

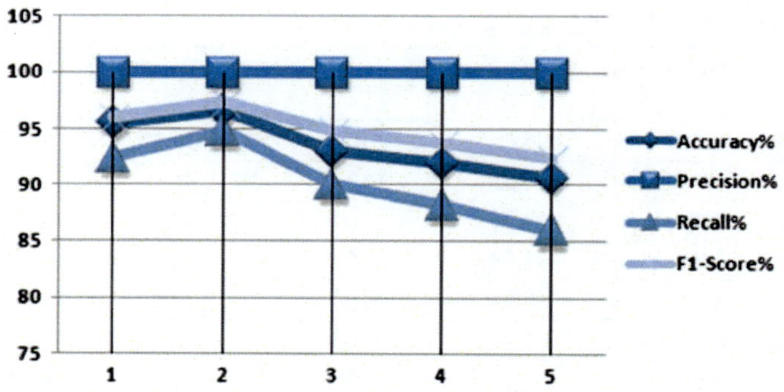

Fig. 3 Tweets performance analysis on treatment of Covid-19

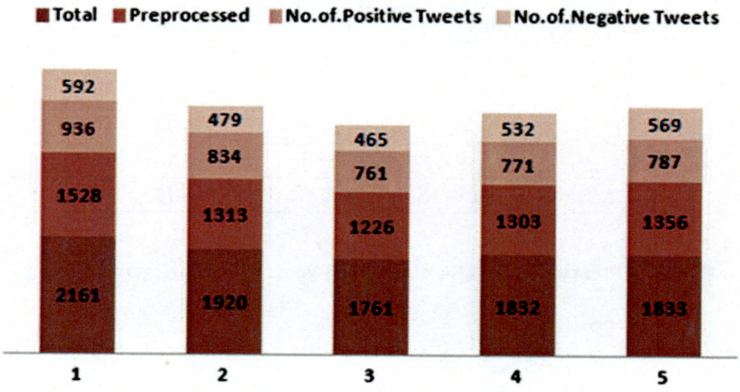

Fig. 4 Number of tweets on work from home

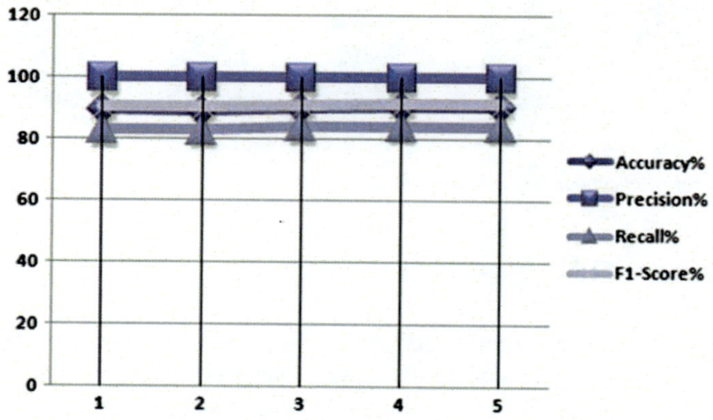

Fig. 5 Tweets performance analysis on work from home

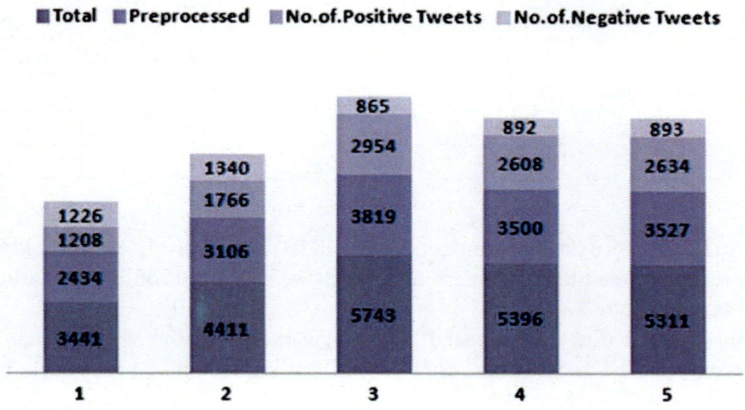

Fig. 6 Number of tweets on online education

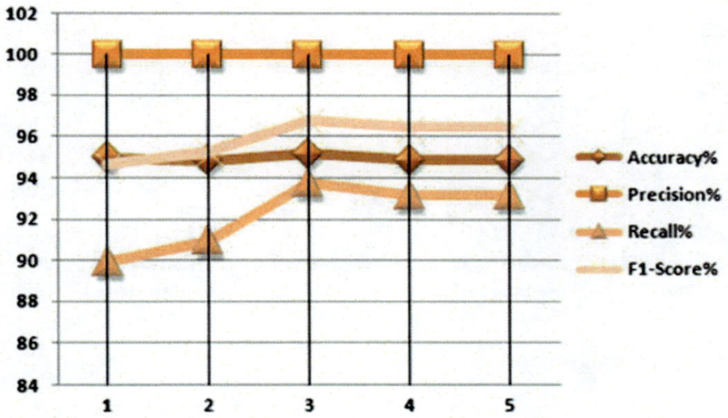

Fig. 7 Tweets performance analysis an online education

Table 3 Fuzzy logic based experimental analysis

Tweets	Total tweets	Preprocessed tweets	No. of. positive tweets	No. of. negative tweets
Treatment of Covid-19	6463	4499	2905	1594
WFH	1920	1301	821	480
Online education	4411	3117	1767	1350

Table 4 Performance analysis report for fuzzy logic

Tweets	Accuracy (%)	Precision (%)	Recall (%)	F1-score (%)
Treatment of Covid-19	96.58	100	94.7	97.28
WFH	89.09	100	82.7	90.53
Online education	94.83	100	90.89	95.23

only 1767 tweets are positive, and remaining 1350 are negatively commended. The results recommended that treatment for Covid-19, WFH, and online education are highly supported.

Table 4 shows that the proposed fuzzy logic model produced more than 90% of accuracy, and F1-score. This result proves that implemented fuzzy logic model performed well (Figs. 8 and 9).

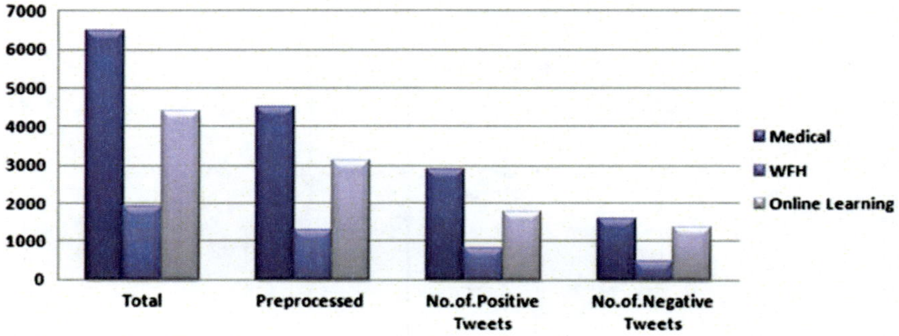

Fig. 8 Fuzzy representation of tweets

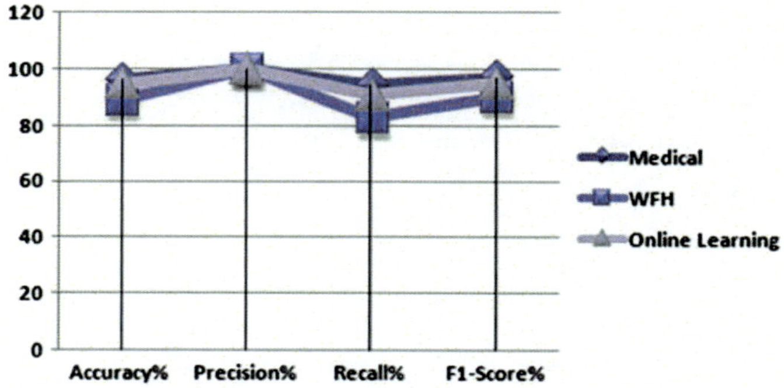

Fig. 9 Performance of fuzzy representaion

4.3 Experiment III: Machine Learning Techniques

In experiment III, five existing machine learning techniques are used to analyze the tweets of the social networks. The experiments are conducted on the same three data sets viz, Covid-19 medical treatment, WFH, and online education. The tweets classification by these methods is identified by accuracy, precision, and F1 metrics. The metric values of existing methods are nearer to the metric values of proposed unsupervised learning and fuzzy logic approaches. Hence, the performance of the proposed work is better (Table 5; Figs. 10, 11, 12 and 13).

5 Conclusion

This work has been implemented on tweets of Covid-19 medical treatment, WFH, and online education during Covid-19 pandemic period. It is attempted to observe the opinion of the public on social media relating to the above implementation. An unsupervised machine learning model is applied for sentiment classification. The classification is satisfied by metric values namely accuracy and F1-score. The results of the work can help the Government and Organization to understand the public opinion of the above working policies to encourage further and suggestions to improve working policy. From this experimentation, it is understood that online education though social media is not much appreciated by the public, necessary steps to be taken to remove obstacles in online education and work from home and Covid-19 medical treatment are well-supported by the public.

Table 5 Performance analysis of machine learning techniques

Tweets	Techniques	Accuracy (%)	Precision (%)	Recall (%)	F1-score (%)	ROC
Treatment of Covid-19	Naïve Bayes	95.62	97.11	96.13	96.62	99.19
	Support vector machine	97.37	97.5	98.48	97.99	99.77
	Decision tree	96.5	98.12	96.46	97.28	95.23
	Logistic regression	96.61	95.92	98.99	97.43	99.53
	Random forest	96.94	95.79	99.66	97.69	99.54
WFH	Naïve Bayes	84.73	80.49	100	89.19	92.91
	Support vector machine	86.64	84.21	96.97	90.14	95.11
	Decision tree	81.68	83.43	88.48	85.88	75.78
	Logistic regression	80.92	76.74	100	86.84	94.87
	Random forest	85.88	82.99	97.58	89.69	93.31
Online education	Naïve Bayes	96.17	93.7	100	96.75	99.71
	Support vector machine	96.97	95.68	99.16	97.39	99.86
	Decision tree	96.65	96.93	97.2	97.06	95.96
	Logistic regression	93.14	89.25	100	94.32	99.86
	Random forest	96.81	94.69	100	97.28	99.61

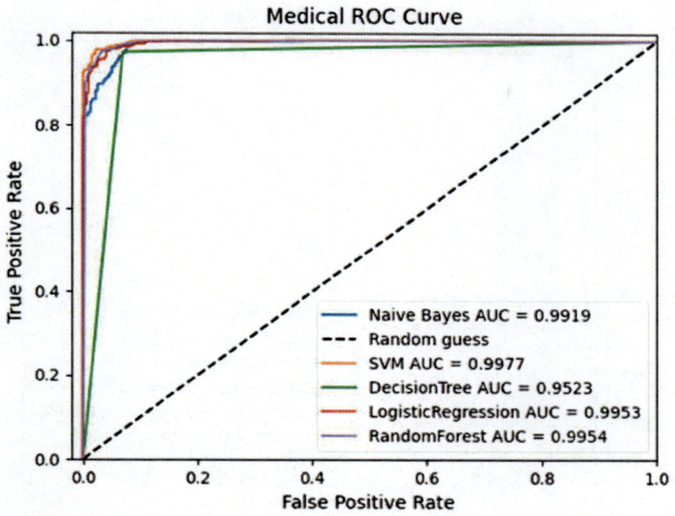

Fig. 10 ROC-AUC-curve for treatment of Covid-19

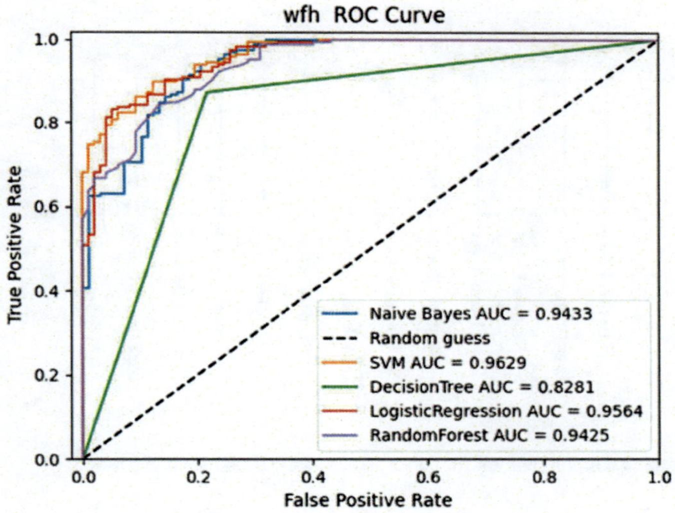

Fig. 11 ROC-AUC-curve for work from home

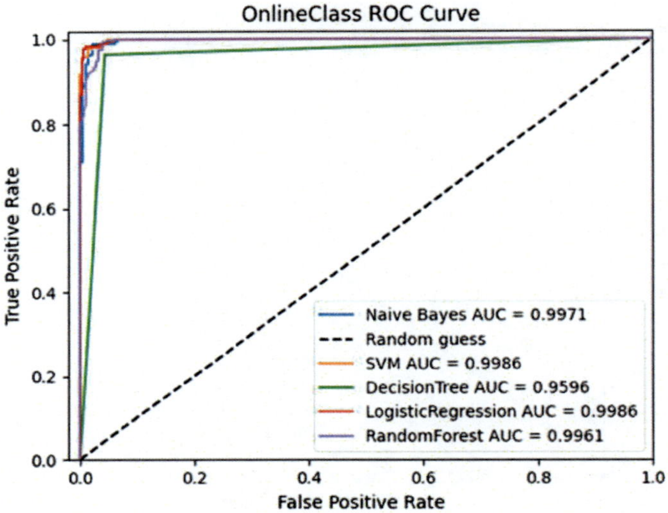

Fig. 12 ROC-AUC-curve for online eductaion

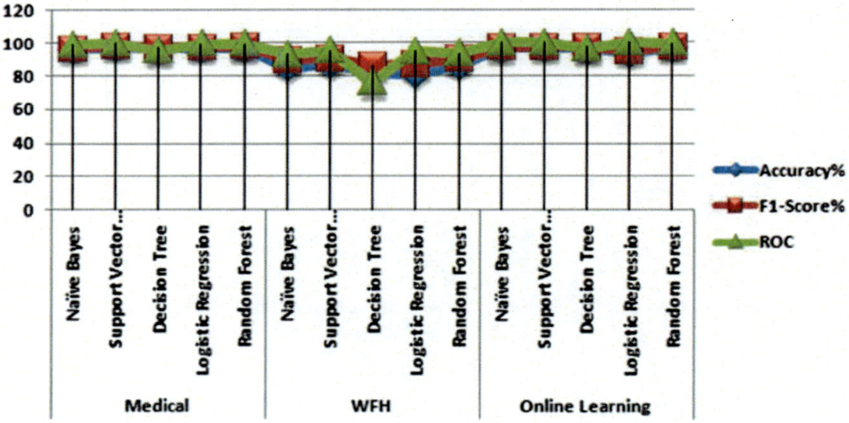

Fig. 13 Graphical representation for performance analysis on machine learning

Acknowledgements This Proposed Research Work "**Sentiment Analysis of Social Media Over Covid-19 Span**", has been done with the financial support of RUSA—Phase 2.0 grant sanctioned vide Letter No. F. 24-51/2014-U, Policy (TNMulti-Gen), Dept. of Edn. Govt. of India, Dt. 09.10.2018.

References

1. Shereen, M.A., Khan, S., Kazmi, A., Bashir, N., Siddique, R.: COVID-19 infection: origin, transmission, and characteristics of human coronaviruses. J. Adv. Res. **24**, 91–98 (2020)
2. Cui, J., Li, F., Shi, Z.-L.: Origin and evolution of pathogenic corona viruses. Nat. Rev. Microbiol **17**(3), 181–192 (2019)
3. Lai, C.-C., Shih, T.-P., Ko, W.-C., Tang, H.-J., Hsueh, P.-R.: Severe acute respiratory syndrome coronavirus 2 (SARS-CoV-2) and corona virus disease-2019 (COVID-19): the epidemic and the challenges. Int. J. Antimicrob. Agents **55**, 105924 (2020)
4. World Health Organization: Laboratory Testing for Coronavirus Disease 2019 (COVID-19) in Suspected Human Cases: İnterim Guidance, 2 Mar 2020. World Health Organization (2020)
5. Rahman, A., Sarkar, A.: Risk factors for fatal middle east respiratory syndrome coronavirus infections in Saudi Arabia: analysis of the WHO line list, 2013–2018. Am. J. Publ. Health **109**(9), 1288–1293 (2019)
6. Ogunleye, O.O., et al.: Response to the Novel Corona Virus (COVID-19) Pandemic Across Africa: Successes, Challenges, and Implications for the Future Frontiers in Pharmacology, vol. 11, Sept 2020. Article 1205
7. Hussain, W.: Role of social media in COVID-19 pandemic. Int. J. Front. Sci. **4**(2), 59–60
8. Kietzmann, J.H., Hermkens, K., McCarthy, I.P., Silvestre, B.S.: Social media? Get serious! understanding the functional building blocks of social media. Bus. Horiz. **54**(3), 241–251 (2011). https://doi.org/10.1016/j.bushor.2011.01.005
9. Perrin, A.: Social media usage. Pew Research Center. **8**, 52–68 (2015)
10. Cinelli, M., Quattrociocchi, W., Galeazzi, A., Valensise, C.M., Brugnoli, E., Schmidt, A.L., Zola, P., Zollo, F., Scala, A.: The covid-19 social media infodemic. arXiv preprint arXiv: 2003.05004. 10 Mar 2020
11. Sentiment analysis of COVID-19 tweets by deep learning classifiers—a study to show how popularity is affecting accuracy in social media. Appl. Soft Comput. J. **97**, 106754 (2020)
12. Hameed, I.A.: Using Gaussian membership functions for improving the reliability and robustness of students' evaluation systems. Exp. Syst. Appl. **38**(6), 7135–7142 (2011)
13. Ali, S.: Introductory chapter: which membership function is appropriate in fuzzy system? In: Fuzzy Logic Based in Optimization Methods and Control Systems and its Applications. IntechOpen. https://doi.org/10.5772/intechopen.79552
14. Mamdani, E.H., Assilian, S.: An experiment in linguistic synthesis with a fuzzy logic controller. Int. J. Man-Mach. Stud. **7**(1), 1–13 (1975)
15. Chen, E., Lerman, K., Ferrara, E.: Covid-19: the first public corona virus twitter dataset. arXiv preprint arXiv: 2003.07372. 16 Mar 2020
16. Li, C., Chen, L.J., Chen, X., Zhang, M., Pang, C.P., Chen, H.: Retrospective analysis of the possibility of predicting the COVID-19 outbreak from Internet searches and social media data, China, 2020. Eurosurveillance **25**(10), 2000199 (2020). https://doi.org/10.2807/1560-917.ES.2020.25.10.2000199
17. González-Padilla, D.A., Tortolero-Blanco, L.: Social media influence in the COVID-19 pandemic. Int. Braz. J. **46**(Suppl 1), 120–124 (2020)
18. Chan, A.K.M., Nickson, C.P., Rudolph, J.W., Lee, A., Joynt, G.M.: Social media for rapid knowledge dissemination: early experience from the COVID-19 pandemic. Anaesthesia. 2020. Epub ahead of print
19. Ahorsu, D.K., Lin, C.Y., Imani, V., Saffari, M., Griffiths, M.D., Pakpour, A.H.: The fear of COVID-19 scale: development and ınitial validation. Int. J. Ment. Health Addict., 1–9 (2020) (Table 1. WHO report about confirmed cases of nCoV-19 on 20 Jan 2020, 1–9)
20. Mansoor, M., Gurumurthy, K., Anantharam, R.U., Badri Prasad, V.R.: Global sentiment analysis of COVID-19 tweets over time. arXiv: 2010.14234v2 [cs.CL] 10 Nov 2020
21. Taspinar: GitHub repository, https://github.com/taspinar/twitterscraper
22. Covid19 Kaggle Dataset https://www.kaggle.com/imdevskp/coronavirus-report
23. Priya Iyer, K.B., Kumaresh, S.: Twitter sentiment analysis on coronavirus outbreak using machine learning algorithms. Eur. J. Mol. Clin. Med. **07**(03) (2020). ISSN 2515-8260

24. Uma Maheswari, S., Dhenakaran, S.S.: Opinion mining on ıntegrated social networks and E-commerce blog. IETE J. Res., 1–9 (2021)
25. Uma Maheswari, S., Dhenakaran, S.S.: Opinion exploration of tweets and amazon reviews. Int. J. Sci. Res. (IJSTR), 1–9 (2020)

S. Uma Maheswari is presently pursuing research in the Depart. of Computer Science, Alagappa University, Karaikudi-630003, Tamil Nadu, India. Her areas of interest are Opinion Mining, Big Data Analytics and Sentiment Analysis.

Dr. S. S. Dhenakaran , Professor, is currently working in the Dept. of Computer Science, Alagappa University, India. He has rich experience in teaching and research with dual PG degrees viz. Mathematics and Computing and Doctoral degree in computer Science. His main areas of research are Information Security, Data Mining and Mathematical Algorithms.

Effective Prediction of COVID-19 Using Supervised Machine Learning with Ensemble Modeling

Alka Kumari and Ashok Kumar Mehta

1 Introduction

The novel Coronavirus disease was firstly found in China in November 2019. In December 2019, World Health Organization (WHO) has received the first report of Coronavirus disease. In the medical domain, the novel word refers to a virus or bacterial strain that is not previously known. Coronavirus is a new disease which is caused by the SARS-CoV2. The virus of novel coronavirus has created a global threat on 11th February 2020 and then it was named COVID-19 (Corona VIrus Disease-2019). World Health Organization (WHO) has declared this outbreak as a pandemic and also mentioned that whenever a healthy person comes in contact with the COVID-19 infected persons the viruses being transmitted through the respiratory tract. This fatal virus has spread across the whole world like a fire in different places. It affected millions of people and also killed thousands of population worldwide. That's why it is very important to classify the infected people so that they can take precautions in the initial stage. The medical domain has vast data and it requires real-time collection and processing of medical data. This infectious disease may give rise to a serious issue to public health and its well-being. The symptoms of Covid-19 can be seen between 2 and 14 days in infected persons. The usual symptoms of Covid-19 are cold, fever, cough, running nose which are very serious in older adults who is having any kind of chronic lung disease. The person who is affected with Diabetes, Heart Disease, Asthma, and Gastrointestinal may have more chances to get infected with COVID-19 when they come in contact with infected persons. Due to lack of diagnosis systems, resources in the medical domain may exposes the patients and healthcare members to the infections. The use of different machine learning

A. Kumari (✉) · A. K. Mehta
Department of Computer Applications, NIT Jamshedpur, Jamshedpur, India

A. K. Mehta
e-mail: akmehta.ca@nitjsr.ac.in

M. Dua et al. (eds.), *Proceedings of the International Conference on Paradigms of Communication, Computing and Data Sciences*, Algorithms for Intelligent Systems,
https://doi.org/10.1007/978-981-16-5747-4_45

algorithms may be helpful in prediction and diagnosis of COVID-19 infections in early stage.

Machine learning methods are used successfully in many problems. In the COVID-19 Pandemic, much research work has been done using machine learning methods. Some of them are used to identify and track this infectious disease in early stages and also estimating the patient's health conditions with early detection of COVID-19 infection. Machine learning provides us various important concepts which can be used to develop objective-based algorithmic, automated, and complex techniques for disease analysis. Machine learning has the potential of diagnosing and detecting the disease in earlier stages based on several symptoms.

In this work, we have followed 3 steps process to handle the collected COVID-19 data: (a) The first phase is used to collect the separate data from different resources. (b) The second phase is used to preprocess the collected data to fix the data related issues by discarding and avoiding the invalid and missing values so that data can be analyzed and can extract useful or hidden information from the past experiences. (c) In this step, the performance evaluation model is used to get accurate values using new rules and methods. According to this work, different supervised machine learning algorithms were used for the development of the predictive models for the COVID-19 disease which included Linear Discriminant Analysis, Logistic regression, Naive Bayesian, Support Vector Machine, K-Nearest Neighbor algorithms along with ensemble learning methods to decrease the variance and bias value or improve predictions.

In this paper, our main motive is to develop a model which is beneficial in predicting the COVID-19 in early stage which may reduce the burden of healthcare industry and also save the human lives. The rest part of this paper is organized as follows: Sect. 2 describes the concept of various machine learning approach in prediction of COVID-19 disease. Section 3 describes the overview of proposed work. Section 4 describes the experimental setup to predict COVID-19. Section 5 describes the results and discussions followed by conclusion and references.

2 Literature Survey

Regression Analysis of COVID-19 using machine learning Algorithms by Ekta et al. [1] has proposed a system in which Polynomial Regression Algorithm is compared with Support Vector Machine Algorithm, which shows an accuracy of approximately 93%. Predictive data mining models for novel coronavirus (COVID-19) infected patients' recovery by Muhammad [2] has proposed a model with Decision Tree, which was analyzed as the most efficient with the highest percentage of accuracy of 99.85%, followed by Random Forest with 99.60% accuracy, then Support Vector Machine with 98.85% accuracy, then KNN with 98.06% accuracy, then Naïve Bayesian with 97.52% accuracy and Linear Regression with 97.49% accuracy. The developed models were very helpful in healthcare for the prediction of COVID-19. COVID-19 Future Forecasting Using Supervised Machine Learning

Models by Rustam [3] has proposed a machine learning-based prediction system for predicting the risk of COVID-19 outbreak globally. The use of LR and LASSO Regression also performed well for forecasting death rates and confirm cases due to COVID-19. The results of these two models were used to depicted the death rates will increase in the upcoming days, and recoveries rate will be slowed down. Prediction of Covid-19 pandemic based on Regression by Mandayam [4] has proposed a model in which Support Vector Regression (SVR) compared to Linear Regression (LR) with time-series data, the Linear Regression algorithm performs better since the data set used is also linear and the SVR cannot handle large linear datasets very well. Explainable Machine Learning for Early Assessment of COVID-19 Risk Prediction in Emergency Departments by Casiraghi [5] has demonstrated a model where best results are obtained when they imputed the missing data with missForest, where the univariate imputation order is based on the increasing amount of missing values, and also selected the most discriminative features by combining Boruta and permutation-based feature selection through internal cross-validation, and trained RFs on the selected features and on that basis prediction of COVID-19 risk can be done in early stages. Covid Prediction from X-ray Images by Haritha et al. [6] has proposed trained the CNN with VGG model using X-ray images to predict the novel covid-19 disease. We obtained an accuracy of 99% in detecting covid-19 virus and also 98% for other different classes that included pneumonia along with covid from chest X-rays. A comparative study on the prediction Model of COVID-19 by Zhou et al. [7] has proposed a model ARIMA model, Logistic model, SIR model, and improved SEIR model for experimental verification, and further changes have been done to the relevant models or parameters. Finally, by comparing the prediction results with the real data, we found the improved SEIR model has the best effect of fitting the real data and predicting the short-term trend with the highest accuracy. Cardiovascular disease detection using a new ensemble classifier by Esfahani and Ghazanfari [8] has proposed a model where the ability of a new data mining technique was used for early diagnosis of heart disease. These data mining techniques used a fusion strategy in which three classifiers such as neural network, Rough Set, and Naïve Bayesian have been combined by a weighted majority vote. Early Prediction of Parkinson's Disease (PD) Using Ensemble Classifiers by Anisha and Arulanand [9] has proposed a model for the prediction of Parkinson's disease with the help of an Ensemble classifier. The experimental results showed that standard scalar preprocessing provided better results than min–max scalar preprocessing, Principal Component Analysis (PCA) performs better than Linear Discriminant Analysis (LDA). Hence, PCA has been used for further prediction analysis, grid search hyperparameters tuning process provides optimal parameters than random search. An accuracy of 94% is obtained in the proposed work by boosting ensemble classifiers which are higher than the accuracy obtained by other existing methods. Diabetes Prediction Using Ensembling of Different Machine Learning Classifiers by Hasan and Alam [10] has proposed a model where the predictions of diabetes have been developed using an ensemble model, in which the preprocessing of the datasets played a very important role in robust and precise prediction. The quality of the dataset was improved by the proposed preprocessing concept, where outlier rejection and filling missing values was a core concern for the analysis of the result which can be accurate and efficient.

3 Proposed Methodology

The proposed methodology consists of various steps as shown in Fig. 1. (a) Firstly, the data collection needs to be performed then (b) the relevant data is needed to be refined. (c) In Step 3, the extracted data needs to be preprocessed. After preprocessing of the dataset (d) the feature extraction or selection process will be done. (e) Then, finally the machine learning algorithms are used to process which will provide us the results of the disease along with ensemble classifier methods.

3.1 Data Collection

According to WHO, the COVID-19 pandemic was declared as a health emergency all over the world. We have collected our dataset from an open-source data repository Kaggle, in which 5434 patients data were stored with different attribute values. COVID-19 dataset consists of about 22 attributes.

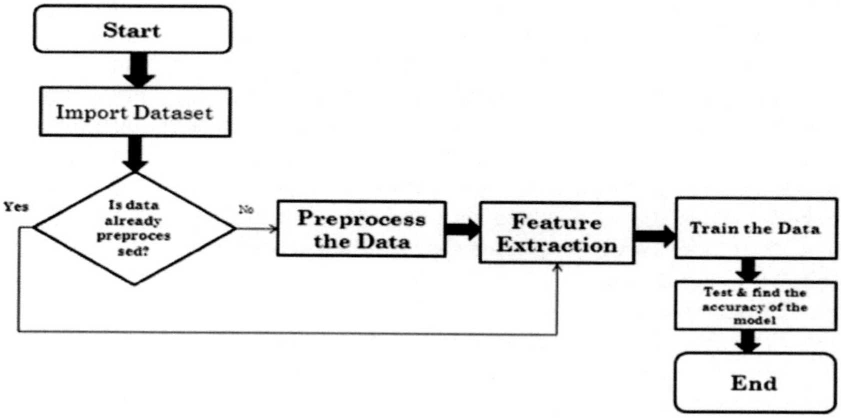

Fig. 1 Proposed methodology

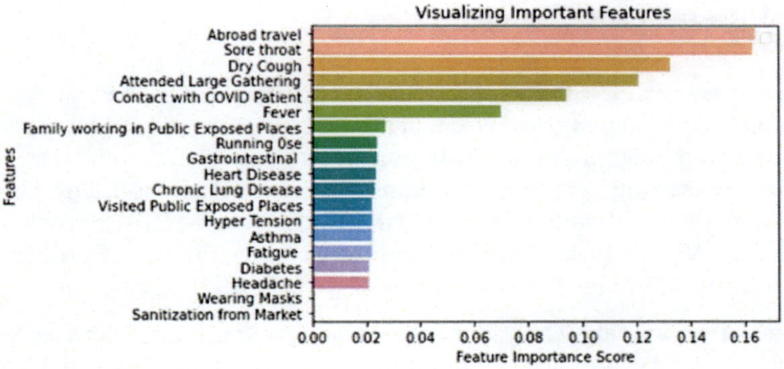

Fig. 2 Visualization of important features

3.2 *Finding Relevant Data*

As our work is regarding personal and medical information of the patients so we need to refine the related values of the given dataset. In our dataset total 5434 patients, details needed and their value was also considered as 0 and 1. The results are categories as affected or not affected with COVID-19.

3.3 *Preprocessing of the Dataset*

The dataset which we have collected is in the instructed form so we need to be preprocess the data. So, in this phase we have followed various steps to remove the unnecessary and missing values from the collected dataset. So that classification can be achieved with better accuracy.

3.4 *Feature Extraction*

Once the preprocessing has been done, we need to extract the important features from the dataset so that we can achieve a result that can be more accurate. Here Fig. 2 depicts the most important features of the COVID-19 dataset.

3.5 Machine Learning Algorithms

As we can say that classification is used to classify the given dataset into two categories such as affected or not affected. In this work, we have put various supervised machine learning algorithms for classifying the attribute values into two categories. The machine learning algorithm like Linear Discriminant Analysis (LDA), Logistics Regression (LR), *K*-Nearest Neighbor (KNN), Naive Bayesian (NB), Support Vector Machine (SVM), AdaBoost, Bagging, Gradient, Boosting, Classifier, Random Forest were used for performing the classification of the COVID-19 disease.

- **Linear Discriminant Analysis (LDA)**—This algorithm is also known as Normal Discriminant Analysis and Discriminant Function Analysis. This analysis comes under the concept of the supervised learning algorithm of machine learning. It is a dimensionality reduction technique which can be used for segregating the differences in 2 or more groups, i.e., separating two or more classes. Linear Discriminant Analysis is used to separate two groups or classes efficiently. These Classes may have more than one feature. So, considering only one feature to classify them may create an overlapping of features. That's why we needed to increase the no. of features for proper classification. In LDA, we analyzed both the axes (X and Y) to projects the data onto a new axis. That is why we need to maximize the separation of 2 categories that means reducing 2-dimensional spaces into only 1-dimensional space.
- **Logistic Regression (LR)**—It is a predictive model which is used to predict the results of dependent variables based on independent variables. This algorithm is used to conduct analysis when we have the dependent variable as binary. This regression is used to analyze the relationship between one dependent binary variable with more than one independent binary variable. This algorithm calculates the membership probability of the class.
- **Support Vector Machine (SVM)**—One more algorithm of the supervised machine learning algorithm is a support vector machine that uses a classification algorithm for the two different group classification problems. It is known as a discriminative classifier formally designed by a boundary line or separating line i.e., hyperplane. The main task of SVM is to create or draw a decision boundary that is used to put a new data category or class. So the best decision boundary is considered as a hyperplane. SVM always uses the extreme points that help us to create the hyperplane. These extreme points are called as support vectors. So this concept is termed SVM.
- **K-Nearest Neighbor (KNN)**—KNN is used to considered for classification problems. So in the case of COVID-19 detection case, KNN will play a very important role. It is one of the simplest ML of which comes under the concept of supervised learning technique. KNN is used to consider the similarity between the case data value and we need to put the new data point into that category which is closest to the available categories. It analyses all available data points and classify a new data point based on its similarities which means when we consider new data point then

that data point can be easily classified with a good suite category by considering the concept of KNN.

- **Naive Bayesian Classifier (NBC)**—The concept of naïve Bayesian algorithm arises from the concept of Baye's theorem, which is used for solving the classification algorithm. It predicts the results based on the probability of a data point. In naïve Bayesian algorithm, the occurrence of an exact feature is totally independent of the occurrence of the other features, where Bayes law determines the probability of a hypothesis with prior knowledge.

3.6 Ensemble Classifier Methods

Ensemble methods are used to create different types of models, and then after that, these models are combined to produce the improved accurate results. As we can say, ensemble methods always provide the more accurate and efficient result of a problem than an individual would do. That's why machine learning algorithms always consider ensemble classifiers. The ensemble classifier uses multiple diverse kinds of models that are already created to predict an accurate outcome. The result which we have got from the given dataset either by using different types of modeling algorithms or by using different kinds of training datasets is more accurate as compared with previous work. The main focus is to aggregate the prediction of each base model and produce results of the final prediction for the hidden patterns.

- **AdaBoost**: The AdaBoost stands for Adaptive Boosting. It is considered adaptive because the succeeding weak learners are adjusted in just a way to favor those attributes that are not accurately classified by the previous classification techniques. The COVID-19 dataset, which is processed by the different classification algorithms, can be used to enhance the results.
- **Bagging**: The word Bagging in ensemble classifier created by the term Bootstrap Aggregating. It is used to enhance the performance and precision of machine learning algorithms. Thus, accurate results are used in statistical classification by avoiding the over-fitting of the model.
- **Gradient Boosting Classifier (GBC)**: This algorithm creates trees from the training dataset samples. The 19 attributes have been selected in feature extraction, with some values are represented as input values. Gradient Boosting Classifier is used for minimizing the co-relations between the trees. Henceforth, in each step, subsamples of the training data are chosen randomly without replacement from the full training dataset. These subsample data, which was randomly chosen used to fit the base learner.
- **Random Forest**: One of the most important ensemble learning algorithm which is used to improve the performance of the model. It is used in two phases: (i) Create random forest by combining the different decision trees (ii) Predict each tree which is created in the first phase. In our dataset, the random forest is used to develop a model which is providing a model that is more efficient and accurate.

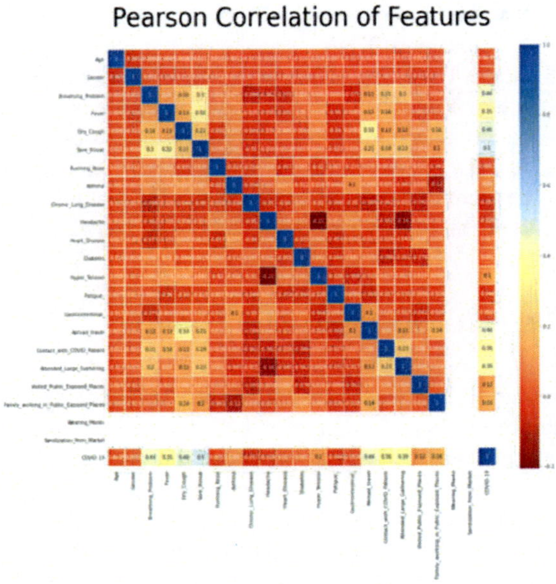

Fig. 3 Pearson's correlation coefficient

3.7 Experimental Setup

In this work, the supervised machine learning algorithms used with python programming language in which we have used Scikit learn tool for performing classification for different machine learning algorithms. Various libraries have been used for improving the accuracy of all the machine learning algorithms with the window operating system. All the useful python libraries were installed and used in Google Colaboratory Notebook for the data and correlation analysis in the given dataset.

Predictive Model for COVID-19 Infection.

As we can see in Fig. 3, the Pearson's correlation coefficient is used to analyze the association between the dependant and independent data of the dataset. Pearson's correlation coefficient was used to draw the strong relationship between the attribute values which were collected in a dataset.

4 Results and Discussion

The early prediction of COVID-19 may lessen the load of the healthcare systems. In this work, Logistic Regression, Linear Discriminant Analysis, KNN, Naïve Bayesian and Support Vector Machine were used for early detection of COVID-19 infection in patients. The performance and accuracy of all the classification models were analyzed based on certain parameters along with ensemble classifier methods. The

Table 1 Comparative analysis of machine learning algorithms

Machine learning algorithms	Accuracy scores (%)
Logistic regression	96.99
Linear discriminant analysis	95.83
K-nearest neighbor	93.07
Naïve Bayesian	84.21
Support vector machines	80.24

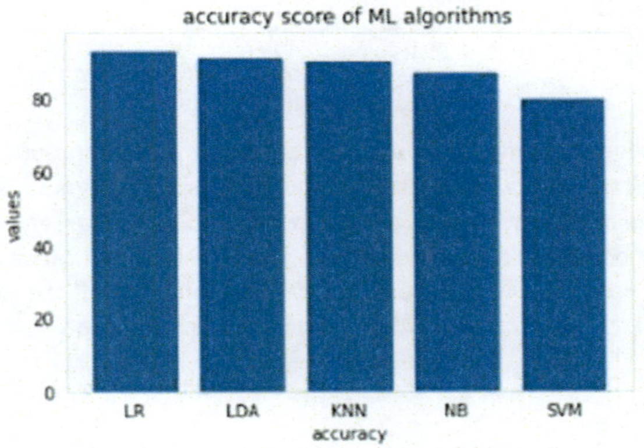

Fig. 4 Accuracy score of machine learning algorithms

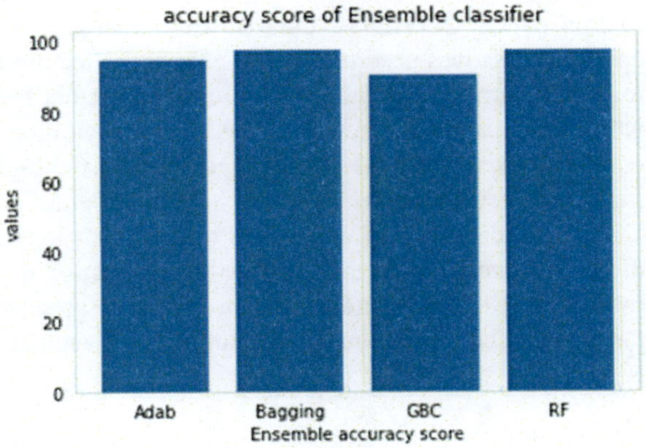

Fig. 5 Accuracy score of ensemble classifier

model which was developed with the help of Logistic Regression concept is the leading model among all the classification models in the terms of accuracy with 96.99% when compared with other models such as linear discriminant analysis, KNN, Naive Bayes, and SVM which have 95.83%, 93.07%, 84.21%, and 80.24% accuracy, respectively, as shown in Table 1 (Fig. 4). As we can see in Fig. 5, with the help of ensemble classifier methods random forest algorithm has a 97.88% accuracy

Table 2 Comparative analysis of ensemble classifier methods

Ensemble classifier methods	Accuracy scores (%)
Random forest	97.88
Bagging	97.74
AdaBoost	95.98
Gradient-based classifier	90.59

Table 3 Confusion matrix for random forest algorithm

FAR	FDR	Recall	Precision	Accuracy
1.00	0.098	1.00	0.90	0.95

score as compared with other ensemble classifier model such as AdaBoost, Bagging, Gradient Boosting Classifier, which has accuracy scores of 97.74%, 97.78%, 96.59%, respectively, as we can see in our Table 2. Thus, the supervised machine learning models can be considered the COVID-19 infection cases.

As we can see from Table 3 the values of all the factors of confusion matrix of random forest algorithm is analyzed with all the values.

5 Conclusion

In this paper, COVID-19 dataset was collected from the Kaggle website. The collected data was preprocessed to extract the feature and then different supervised machine learning algorithms were used to predict the accurate results. The result value which was collected from the machine learning algorithms was feed into the ensemble classifier model which provided us the highest accuracy score with the random forest algorithm. With the help of accurate result values, the prediction of COVID-19 disease will be an easy task to improve the socio-economic factor. By the use of two or more than two and hybridization of supervised machine learning models can be used to forecast the more distant spread of the virus in future. In future, the data can be farther analyzed, and the more accurate results can be achieved.

References

1. Ekta, G., Ritika, J., Alankrit, G., Uma, T.: Regression analysis of COVID-19 using machine learning algorithms. In: Proceedings of the International Conference on Smart Electronics and Communication (ICOSEC 2020). IEEE Xplore Part Number: CFP20V90-ART. ISBN: 978-1-7281-5461-9
2. Muhammad, L.J., et al.: Predictive data mining models for novel coronavirus (COVID-19) infected patients' recovery. SN Comput. Sci. **1**(4), 1–7 (2020). Lagzian, S., et al.: Robust

watermarking scheme based on RDWT-SVD: embedding data in all sub bands. 978-1-4244-9834-5/11. ©2011 IEEE
3. Rustam, F., et al.: COVID-19 future forecasting using supervised machine learning models. IEEE Access **8**, 101489–101499 (2020). https://doi.org/10.1109/ACCESS.2020.2997311
4. Mandayam, A.U., Rakshith, A.C., Siddesha, S., Niranjan, S.K.: Prediction of Covid-19 pandemic based on regression. In: 2020 Fifth International Conference on Research in Computational Intelligence and Communication Networks (ICRCICN), pp. 1–5. Bangalore, India (2020). https://doi.org/10.1109/ICRCICN50933.2020.9296175
5. Casiraghi, E., et al.: Explainable machine learning for early assessment of COVID-19 risk prediction in emergency departments. IEEE Access **8**, 196299–196325 (2020). https://doi.org/10.1109/ACCESS.2020.3034032
6. Haritha, D., Praneeth, C., Pranathi, M.K.: Covid prediction from X-ray images. In: 2020 5th International Conference on Computing, Communication and Security (ICCCS), pp. 1–5. Patna, India (2020) https://doi.org/10.1109/ICCCS49678.2020.9276795
7. Zhou, Q., Tao, W., Jiang, Y., Cui, B.: A comparative study on the prediction model of COVID-19. In: 2020 IEEE 9th Joint International Information Technology and Artificial Intelligence Conference (ITAIC), pp. 1348–1352. Chongqing, China (2020). https://doi.org/10.1109/ITAIC49862.2020.9338466
8. Esfahani, H.A., Ghazanfari, M.: Cardiovascular disease detection using a new ensemble classifier. In: 2017 IEEE 4th International Conference on Knowledge-Based Engineering and Innovation (KBEI), pp. 1011–1014. Tehran, Iran (2017). https://doi.org/10.1109/KBEI.2017.8324946
9. Anisha, C.D., Arulanand, N.: Early prediction of Parkinson's disease (PD) using ensemble classifiers. In: 2020 International Conference on Innovative Trends in Information Technology (ICITIIT), pp. 1–6. Kottayam, India (2020). https://doi.org/10.1109/ICITIIT49094.2020.9071562
10. Hasan, M.K., Alam, M.A., Das, D., Hossain, E., Hasan, M.: Diabetes prediction using ensembling of different machine learning classifiers. IEEE Access **8**, 76516–76531 (2020). https://doi.org/10.1109/ACCESS.2020.2989857

Comparative Analysis by Transfer Learning of Pre-trained Models for Detection of COVID-19 Using Chest X-ray Images

Divyanshu Malik, Anjum, and Rahul Katarya

1 Introduction

According to WHO (World Health Organization), the (COVID-19) Coronavirus disease is an infectious disease related to a family of RNA (Ribonucleic Acid) viruses which cause illness and disease such as respiratory tract infections, with mild varieties being the common cold and lethal/severe varieties being SARS (Severe acute respiratory syndrome), MERS (Middle East respiratory syndrome), and COVID-19. It is a contagious and deadly disease that spreads between mammals and birds [1]. The COVID-19 is the seventh type of Coronavirus due to which humans have been affected, and in the month of Dec. 2019, due to a pneumonia outbreak reported in Wuhan, China, it has been named COVID-19. Since the Coronavirus is 96 percent similar to a Coronavirus found in bats, that is why it has been suspected to have originated from bats [2, 3]. The most common signs and symptoms of being affected by the COVID-19 are respiratory problems, headaches, coughing, loss of smell or taste, fever, the severe stage of COVID causes constant pain/pressure in chest/pneumonia, organ failure, and death. People are infectious, i.e., they spread the virus to other people for up to ten days in moderate cases, whereas up to 20 days in severe cases after the symptom onset. Infected people can transmit the Coronavirus to other people for up to two days before they themselves have any symptoms, as can asymptomatic people, i.e., who do not show any kind of symptoms [4–6]. The RT-PCR (Reverse Transcription Polymerase Chain Reaction) test is the standard way to analyze and test the Coronavirus on any suspicious cases. Coronavirus presence in a patient can also be tested from Chest CT scan images showing pneumonia, risk factors, and some mix of symptoms [7, 8]. A lot of research is being done across the globe in

D. Malik · Anjum (✉) · R. Katarya
Delhi Technological University, New Delhi, Delhi 110042, India

R. Katarya
e-mail: rahulkatarya@dtu.ac.in

M. Dua et al. (eds.), *Proceedings of the International Conference on Paradigms of Communication, Computing and Data Sciences*, Algorithms for Intelligent Systems,
https://doi.org/10.1007/978-981-16-5747-4_46

order to handle and control the epidemic. Studies in the field of machine learning and deep learning are being done and have proposed different models and research work to predict the COVID-19.

In this paper, we have done comparative analysis by using transfer learning techniques performed on famous pre-trained models so that they can classify COVID-19 utilizing Chest X-ray images. Section 2 explains the literature survey of our work. The framework/architecture used in the paper is discussed in Sect. 3. Section. 4, the experiments performed, the results, visualizations, dataset, etc., have been discussed in detail. The paper is discussed in Sects. 5 and later concluded in Sect. 6.

2 Related Work

In order to control the spreading of the Novel Coronavirus 2019, i.e., COVID-19 [9], testing a lot of suspicious patients is required. Then proper treatment of vaccine and medicine is also needed with adequate care and quarantine. The RT- PCR (Real-Time Reverse Transcription- Polymerase Chain reaction) is considered as the standard way of testing for the Novel Coronavirus, but it has some significant number of wrong outcomes. All the medical professionals are anticipating effective techniques to fight the Coronavirus. Becky McCall [10] discusses the use of artificial intelligence to protect the workers engaged in health and frontline services and to control the spread of COVID-19. Presumed asymptomatic carrier transmission of COVID-19 related to transmission of the Coronavirus from one person to another are discussed in [11].

In [12], the pathological findings of COVID-19 associated with acute respiratory distress syndrome have been discussed, and it explains the pathological presence in lungs, heart, and liver tissue in a severely affected pneumonia patient caused by the COVID-19. This paper [13] addresses the diagnostic, clinical, epidemiological, and therapeutic aspects as well as the perspectives on vaccines and prevention steps already implemented globally to fight the pandemic. Chen [14] finds that there is little information on pregnancy-related presenting symptoms and outcomes. This study gives clinical data of transmission of the Coronavirus in pregnant patients. Shan [15] has made use of deep learning by employing "VB- net" named neural network for automatic quantification and segmentation of infected areas of a lung utilizing CT scan image of the chest. The study shows the importance of doing Chest CT scans and deep learning for screening Coronavirus in patients. Gozes [16] developed an AI-based automation tool for the analysis of CT images in order to detect, quantify, and track COVID. Wang [17] uses a DL algorithm that makes use of Chest CT images in order to screen patients for Coronavirus. Wang [18] has developed a way to classify abnormal respiratory patterns using a physical device that classifies a COVID patient by their breathing pattern as a person with COVID has more rapid respiration.

Rajpurkar [19] has developed a DL model called cheXNet, which is a 121 layered CNN that uses Chest X-ray images to detect Coronavirus in suspicious cases. Xu [20] discussed that RT-PCR method for detecting viral RNA from nasopharyngeal swabs has a low accuracy score. The study proposes a DL model to distinguish

Coronavirus-19 pneumonia from IAVP by screening CT scans. Yoon [21] reports the computed tomography (CT) and chest radiography images of COVID-19 pneumonia in Korea. Singh [22] shows through the study the importance of age in assessing the country-specific impact of social distancing. Recently Roy [23] makes use of deep learning techniques in order to analyze the lung ultrasonography images of COVID-19 patients. Singh [24] has used differential evolution—based CNN in order to classify Coronavirus patients from their chest CT images. In the study [25], Harrison X. Bai assesses different radiologists' performance in differentiating viral pneumonia from COVID-19 pneumonia using the dataset as chest CT images. Apostolopoulos [26] and Ayan [27, 28] proposed state-of-the-art CNNs for automatic Coronavirus disease detection. Harsono [29] implemented deep learning, computer-aided diagnosis (CAD), and artificial intelligence to detect lung cancer. Hemdan [30] developed a deep learning classifier framework, i.e., COVIDX-Net, in order to detect COVID-19 using CXR images. Asif Iqbal Khan [31] developed another deep neural network called CoroNet for diagnosis and detection of Coronavirus using CXR images. Similar studies are done by Oh [32], Shi [33], Wang [34], Wang [35], Xue [36].

3 Proposed Work

In the following section, comparative analysis using TL (Transfer learning) on pre-trained DL models is discussed. The pre-trained models used by us are VGG16, InceptionResnetV2, Resnet50, InceptionV2, MobileNetV1, MobileNetV2, Alexnet, Densenet, Xception, NasNet, and simple CNN. The comparative analysis has been done in the proposed work by using transfer learning on pre-trained deep learning models. Transfer learning (TL) is a technique used in ML which works by storing some knowledge or training experience gained while solving some problem and applying that accumulated knowledge to a different but similar situation. A large number of datasets are used from training in transfer learning in order to classify the Coronavirus efficiently. In Fig. 1, the design of the proposed framework of this paper is shown. Firstly, the dataset of the CXRs is pre-processed in order to get good quality images.

The *CLAHE* (Contrast Limited adaptive histogram equalization) has been performed on our dataset in order to increase contrast, remove noise as these techniques improve images' quality. *CLAHE* is a different type of adaptive histogram equalization in which the limitation has been done in contrast amplification so as to reduce the noise amplification problem. Using histogram equalization, we can enhance the quality of our images with any loss of initial information. OpenCV library has been used to resize the photos in our dataset. Figure 2 shows that the images after image enhancement techniques have been applied to them.

After pre-processing all the CXR images, we perform TL on them. We take different types/varieties of pre-trained models and train each one, one by one and find out the different classification accuracy of those models. We take our dataset

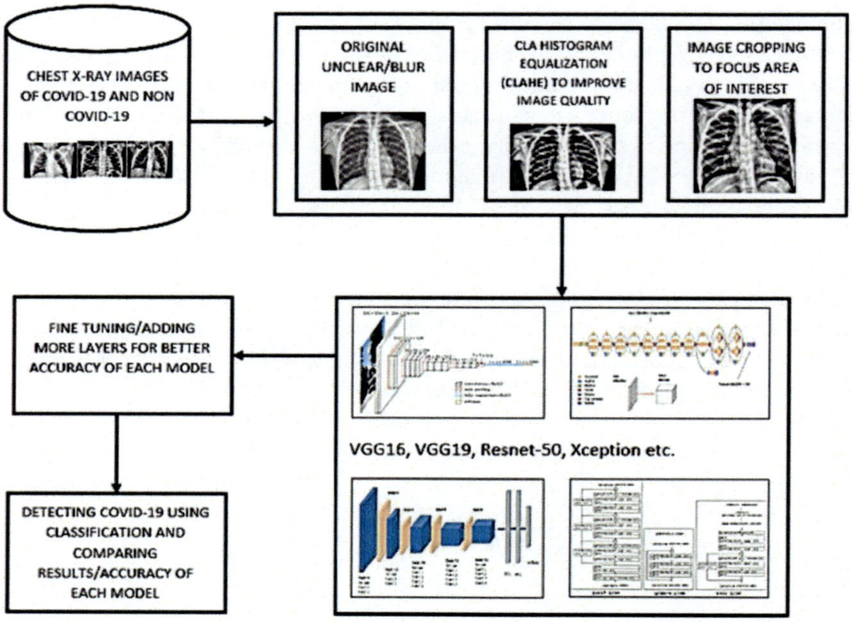

Fig. 1 COVID-19 detection system framework

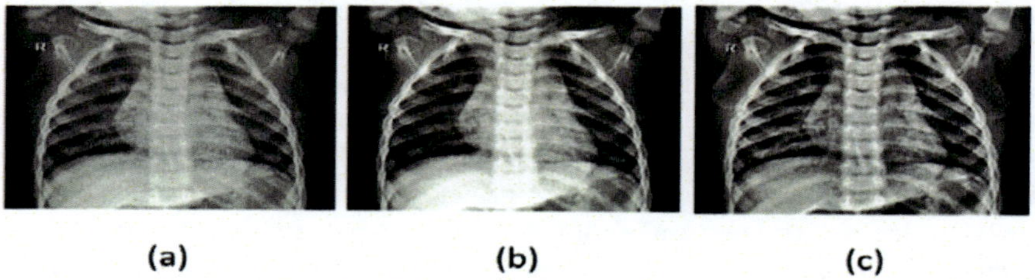

Fig. 2 **a** Normal image, **b** histogram equalized image, **c** CLAHE image

and train pre-trained models like MobileNetV1, MobileNetV2, Resnet-50, Xception, InceptionV2, Densenet, NasNet, VGG16, InceptionResnetV2, etc., one by one on our Chest X-ray dataset. We also fine-tune models that have significantly less accuracy by either making changes in the overall structure of the neural network or by adding a greater number of neural net layers or by fine- tuning parameters of those models like batch size, epochs, learning rate, etc. After training the different models, we calculate each model's accuracy and analyze the results using graphs/plots.

4 Experiments and Results

4.1 Experiments

The experiment that has been executed in this paper is that all pre-trained models were trained on our Chest X-ray dataset one by one. The models were trained in Spyder (Python IDE). Firstly, the models were downloaded from their respective databases available on the Internet. Some models were downloaded by directly calling libraries such as Keras.applications, whereas some models were downloaded from other libraries, and the leftover models were built by using concepts of deep neural networks and convolutional neural networks by studying their respective architectures on the Internet and making use of libraries such as Keras.layers, Keras.models, Keras, pre-processing, etc. Models such as Vgg16, Resnet50, Xception, MobileNetV1, etc., were directly downloaded, whereas simple CNN, Alexnet, etc., were built first with the Internet's help.

After every model was correctly built, the already pre-processed images were fed to each model one by one and trained. Each image's image size was taken as (224, 224), whereas the number of epochs being 25 and the batch size used during training being 32. After training the model completely, we plotted the accuracy and loss graphs, and the use of Matplotlib.pyplot library was made to plot the respective graphs. After plotting, we predicted and tested the accuracy; some models have excellent accuracy, whereas some have average accuracy in predicting COVID-19.

4.2 Dataset

The dataset of Normal and Viral pneumonia CXR images are downloaded from Kaggle [37–39] with a total of 6213 images, as described in Table 1. The following Table shows the distribution of COVID and Non-COVID images in our dataset. The images are collected from three sources on Kaggle and merged to form our dataset. The number of normal Chest X-rays is 1583, whereas that of COVID-19 CXRs is 4630. The images are of size 1024 × 1024, which are then pre-processed using the OpenCV library and other measures. The final size of the images is after being pre-processed, becomes 224 × 224.

Table 1 Distribution of the normal and COVID-19 images in our dataset

Disease name	Number of images
Normal	1583
COVID-19	4630

4.3 Results

The accuracy and loss are evaluated for all the discussed pre-trained models; also rate of their change per epoch has been visualized in graphs for every model. After plotting all the accuracy and loss graphs, the comparative analysis has been done in Table 2. Since the training accuracy and validation accuracy increases epoch by epoch, whereas the training loss and validation loss decrease epoch by epoch for all the models, this shows that our models are learning more epoch by epoch and becoming more accurate.

Since the VGG16 is only 16 layers deep, the time taken for training it on our dataset is less, whereas resnet50 is 50 layers deep and inceptionV3 being 48 layers deep, therefore, taking more time to train. Spyder (Python IDE) has been used in order to train our models and perform all the experiments. From Fig. 3, we observe that our trained model VGG16 classifies the given data of images correctly. In the following figure, we can clearly observe that the right-side image has some increased

Table 2 Comparative analysis of accuracy and loss of all models

Model's name	Validation accuracy	Training accuracy	Validation loss	Training loss
Vgg16	0.93	0.99	0.6	0.1
Inceptionresnetv2	0.60	0.93	8.0	0.8
Resnet50	0.64	0.98	6.0	0.0
Inceptionv2	0.68	0.95	0.65	0.20
Mobilenetv1	0.65	0.97	8.0	0.9
Mobilenetv2	0.64	0.98	2.0	0.8
Alexnet	0.63	0.84	0.65	0.2
Densenet	0.63	0.76	0.0	0.0
Xception	0.50	0.98	2.5	0.1
Nasnet	0.84	0.95	2.0	0.8
Basic CNN	0.865	0.975	0.8	0.05

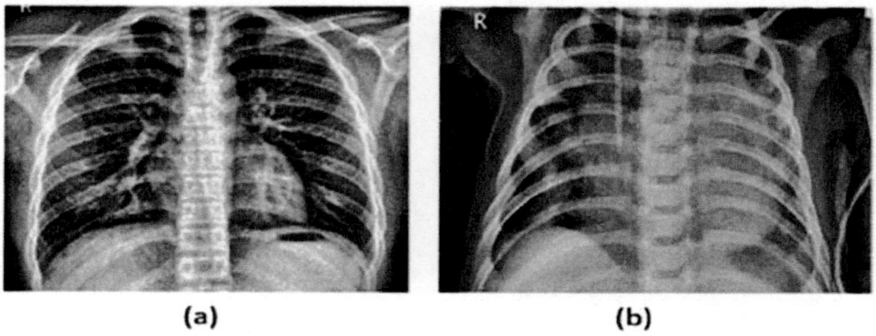

Fig. 3 **a** Normal lungs, **b** COVID-19 patient lungs

patchy opacity or blurriness at the bottom right side of the lungs. Whereas the left side image seems to be clearer/transparent and more defined, this shows that our VGG16 model has classified both images correctly as the right side being COVID patient lungs and the left side being normal lungs.

The loss and accuracy graphs have been visualized in the following section. In the section, we can also observe that during the training and testing process of our experiment, there is a steady decrease in the (Training + Validation) loss values and a steady increase in the (Training + Validation) accuracy values, which shows that our model is getting more efficient and effective.

5 Discussion

The WHO has recommended the RT-PCR tests for all people suspected of the Coronavirus disease, but due to the limitations of architecture, finance, shortage of testing equipment like testing kits, trained medical staff, and a vast populace this recommended procedure has not been able to be followed by many countries. The work done in this paper can help aid the difficulty of testing using standard ways, as the deep learning models and TL techniques can give a different and fast alternative in the diagnosis process so that the Coronavirus epidemic can be, if not stopped completely, be at least limited and slowed down. This work's primary purpose is to save researchers from the massive task of training different models and finding better accuracy results, and merely seeing which deep learning model best suits their needs. Even Doctors and medical professionals can use these pre-trained models and take help in diagnoses of COVID-19.

6 Conclusion and Future Work

The comparative analysis in this research paper has given us the accuracy results of famous deep learning models in the classification of COVID-19 using Chest X-rays. The most accurate deep learning model, in order to classify Coronavirus, is VGG16, showing an accuracy of 93%, followed by a simple CNN (convolutional neural network) with an accuracy of 86.5% and NasNet with an accuracy of 84%, and most of the others in the range of 65–75% accuracies. The paper results are good enough and consistent for usage other than the study related to COVID-19. We hope the results and output of this paper can help serve as a little step for developing a more accurate and precise model for Coronavirus detection using CXR images. In the future, as more people get affected by this disease, a lot more data can be acquired so that a better model can be made using a considerable amount of data, which will improve research on this topic.

References

1. WHO EMRO I About COVID-19 I COVID-19 I Health topics. http://www.emro.who.int/health-topics/corona-virus/about-covid-19.html
2. Rathore, J.S., Ghosh, C.: Severe acute respiratory syndrome coronavirus-2 (SARS-CoV-2), a newly emerged pathogen: an overview. Pathog. Dis. **78**(6), 42. https://doi.org/10.1093/femspd/ftaa042
3. Report of the WHO-China joint mission. https://www.who.int/docs/default-source/coronaviruse/who-china-joint-mission-on-covid-19-final-report.pdf
4. Coronavirus disease (COVID-19). https://www.who.int/emergencies/diseases/novel-coronavirus-2019/question-and-answers-hub/q-a-detail/coronavirus-disease-covid-19
5. Oran, D.P., Topol, E.J.: The proportion of SARS-CoV-2 infections that are asymptomatic. Ann. Intern. Med. (2021). https://doi.org/10.7326/m20-6976
6. Transmission of COVID-19. https://www.ecdc.europa.eu/en/covid-19/latest-evidence/transmission
7. Ai, T., et al.: Correlation of chest CT and RT-PCR testing for coronavirus disease 2019 (COVID-19) in China: a report of 1014 cases. Radiology **296**(2), E32–E40 (2020). https://doi.org/10.1148/radiol.2020200642
8. Li, C., Zhao, C., Bao, J., Tang, B., Wang, Y., Gu, B.: Laboratory diagnosis of coronavirus disease-2019 (COVID-19). Clin. Chim. Acta **510**, 35–46 (2020). https://doi.org/10.1016/j.cca.2020.06.045
9. Huang, L., et al.: Serial Quantitative Chest CT Assessment of COVID-19: Deep-Learning Approach
10. McCall, B.: COVID-19 and artificial intelligence: protecting health-care workers and curbing the spread. Lancet Digit. Health **2**(4), e166–e167 (2020). https://doi.org/10.1016/s2589-7500(20)30054-6
11. Bai, Y., et al.: Presumed asymptomatic carrier transmission of COVID-19. JAMA J. Am. Med. Assoc. **323**(14), 1406–1407 (2020). https://doi.org/10.1001/jama.2020.2565
12. Xu, Z., et al.: Pathological findings of COVID-19 associated with acute respiratory distress syndrome. Lancet Respir. Med. **8**(4), 420–422 (2020). https://doi.org/10.1016/S2213-2600(20)30076-X
13. Dhama, K., et al.: Coronavirus Disease 2019–COVID-19 (2020) [Online]. Available: http://cmr.asm.org/
14. Arentz, M., et al.: Characteristics and outcomes of 21 critically ill patients with COVID-19 in Washington State. JAMA J. Am. Med. Assoc. **323**(16). American Medical Association, pp. 1612–1614, 28 Apr 2020. https://doi.org/10.1001/jama.2020.4326
15. Shan, F., et al.: Lung infection quantification of COVID-19 in CT images with deep learning author list
16. Gozes, O., et al.: Title: Rapid AI Development Cycle for the Coronavirus (COVID-19) Pandemic: Initial Results for Automated Detection and Patient Monitoring Using Deep Learning CT Image Analysis Authors
17. Wang, S., et al.: A deep learning algorithm using CT images to screen for Corona Virus Disease (COVID-19) (2020). https://doi.org/10.1101/2020.02.14.20023028
18. Wang, Y., Hu, M., Li, Q., Zhang, X.-P., Zhai, G., Yao, N.: Abnormal respiratory patterns classifier may contribute to large-scale screening of people infected with COVID-19 in an accurate and unobtrusive manner, Feb 2020 [Online]. Available: http://arxiv.org/abs/2002.05534
19. Rajpurkar, P., et al.: CheXNet: Radiologist-Level Pneumonia Detection on Chest X-Rays with Deep Learning, Nov 2017 [Online]. Available: http://arxiv.org/abs/1711.05225
20. Xu, X., et al.; A deep learning system to screen novel coronavirus disease 2019 pneumonia. Engineering (2020). https://doi.org/10.1016/j.eng.2020.04.010
21. Yoon, S.H., et al.: Chest radiographic and CT findings of the 2019 novel coronavirus disease (Covid-19): analysis of nine patients treated in korea. Korean J. Radiol. **21**(4), 498–504 (2020). https://doi.org/10.3348/kjr.2020.0132

22. Singh, R., Adhikari, R.: Age-structured impact of social distancing on the COVID-19 epidemic in India, Mar 2020 [Online]. Available: http://arxiv.org/abs/2003.12055
23. Roy, S., et al.: Deep learning for classification and localization of COVID-19 markers in point-of-care lung ultrasound. IEEE Trans. Med. Imaging **39**(8), 2676–2687 (2020). https://doi.org/10.1109/TMI.2020.2994459
24. Singh, D., Kumar, V., Vaishali, Kaur, M.: Classification of COVID-19 patients from chest CT images using multi-objective differential evolution–based convolutional neural networks. Eur. J. Clin. Microbiol. Infect. Dis. **39**(7), 1379–1389 (2020). https://doi.org/10.1007/s10096-020-03901-z
25. Bai, H.X., et al.: Performance of radiologists in differentiating COVID-19 from viral pneumonia on chest CT
26. Apostolopoulos, I.D., Mpesiana, T.A.: Covid-19: automatic detection from X-ray images utilizing transfer learning with convolutional neural networks. Phys. Eng. Sci. Med. **43**(2), 635–640 (2020). https://doi.org/10.1007/s13246-020-00865-4
27. İstanbul AREL Üniversitesi, IEEE Engineering in Medicine and Biology Society, Institute of Electrical and Electronics Engineers. Turkey Section, and Institute of Electrical and Electronics Engineers: Scientific Meeting on Electrical-Electronics, Computer and Biomedical Engineering, 24–26 Apr 2019. Istanbul AREL University, Kemal Gözükara Campus, Prof. Dr. Aziz Sancar Conference Hall = Uluslararası Katılımlı Elektrik-Elektronik, Bilgisayar, Biyomedikal Mühendislikleri Bilimsel Toplantısı : 24–26 Nisan 2019 . T. C. İstanbul AREL Üniversitesi, Kemal Gözükara Yerlekeşi, Prof. Dr. Aziz Sancar Amfisi
28. Asnaoui, K.E., Chawki, Y., Idri, A.: Automated Methods for Detection and Classification Pneumonia based on X-Ray Images Using Deep Learning
29. Harsono, I.W., Liawatimena, S., Cenggoro, T.W.: Lung Nodule Texture Detection and Classification Using 3D CNN (2019)
30. Hemdan, E.E.D., Shouman, M.A., Karar, M.E.: COVIDX-Net: A Framework of Deep Learning Classifiers to Diagnose COVID-19 in X-Ray Images
31. Khan, A.I., Shah, J.L., Bhat, M.M.: CoroNet: a deep neural network for detection and diagnosis of COVID-19 from chest X-ray images. Comput. Methods Programs Biomed. **196** (2020). https://doi.org/10.1016/j.cmpb.2020.105581
32. Oh, Y., Park, S., Ye, J.C.: Deep learning COVID-19 features on CXR using limited training data sets. IEEE Trans. Med. Imaging **39**(8), 2688–2700 (2020). https://doi.org/10.1109/TMI.2020.2993291
33. Shi, F., et al.: Large-Scale Screening of COVID-19 from Community Acquired Pneumonia using Infection Size-Aware Classification
34. Wang, S., et al.: A fully automatic deep learning system for COVID-19 diagnostic and prognostic analysis. Eur. Respir. J. **56**(2) (2020). https://doi.org/10.1183/13993003.00775-2020
35. Wang, X., Peng, Y., Lu, L., Lu, Z., Bagheri, M., Summers, R.M.: ChestX-ray8: Hospital-Scale Chest X-ray Database and Benchmarks on Weakly-Supervised Classification and Localization of Common Thorax Diseases, May 2017. https://doi.org/10.1109/CVPR.2017.369
36. Xue, Z., et al.: Chest X-ray image view classification. In: Proceedings—IEEE Symposium on Computer-Based Medical Systems, vol. 2015, pp. 66–71, 2015. https://doi.org/10.1109/CBMS.2015.49
37. COVID-19 chest X-ray image dataset | Kaggle. https://www.kaggle.com/alifrahman/covid19-chest-xray-image-dataset
38. COVID-19 chest xray | Kaggle. https://www.kaggle.com/bachrr/covid-chest-xray
39. Chest X-ray images (pneumonia) | Kaggle. https://www.kaggle.com/paultimothymooney/chest-xray-pneumonia

A Novel Hybrid Method for Melanoma Classification from Skin Images

Duggani Keerthana and Malaya Kumar Nath

1 Introduction

Skin is a complex structure made up of different types of tissues, which may undergo cancerous conversions during their development process [1]. Skin cancer is a common cancer type in humans which mainly occurs because of ultraviolet (UV) rays exposure on the skin. The factors responsible for skin cancer are lifestyle, environment and genetic problems [2]. There are three different types of skin cancer namely, basal cell carcinoma, melanoma and squamous cell carcinoma. Basal cell carcinoma are slow growing and are mainly found on the face and back of hands. Squamous cell carcinoma are fast growing and mainly arise in upper level epidermis. Among the different types of skin cancer melanoma is very dangerous as it develops in melanocytes, which produce the pigment melanin [3, 4].

The accuracy obtained for detecting the melanoma using clinical diagnosis approach is 65–80% [5]. By using dermoscopic images diagnostic accuracy of skin lesion can be improved by 49% [6]. Dermoscopy is a non-invasive skin imaging technique used for obtaining magnified image of a particular region of skin [7, 8]. Identification of melanoma through naked eye is very difficult. Due to this several computerized analysis technique of dermoscopic images became an important research area [9, 10].

Most prevalent techniques used by clinical specialists for melanoma diagnosis are seven-point checklist and ABCD rule [1]. Seven-point checklist [11] identifies seven dermoscopic criteria related to melanoma namely, pigment network, blue whiteish vile, vascular structures, pigmentation, streaks, dots and globules and regression structures. These are obtained based on visual similarities. In ABCD rule [12] lesion is analyzed based on asymmetry, border, color and diameter. According to the ABCD rule, the two halves of the area of the image differ in shape, the edges of the image area are irregular, color is uneven and the diameter is greater than 6 mm [13, 14].

Automatic and accurate segmentation is important for diagnosing skin cancer. With the help of semantic segmentation, the objects or features within the image are

D. Keerthana (✉) · M. K. Nath
Department of ECE, NIT Puducherry, Karaikal, India
e-mail: malaya.nath@nitpy.ac.in

M. Dua et al. (eds.), *Proceedings of the International Conference on Paradigms of Communication, Computing and Data Sciences*, Algorithms for Intelligent Systems,
https://doi.org/10.1007/978-981-16-5747-4_47

segmented and classified. Deep learning is mainly used for semantic segmentation in various fields [15]. CNNs have attained exceptional results in computer vision tasks and are used for semantic segmentation. CNNs are neural networks which learn relationship between input objects and class labels, they are used mainly for image recognition and classification [16]. Among various CNN based architectures AlexNet, VGG16, ResNet, DenseNet, U-Net etc. are mainly used by the researchers [17, 18]. Various methods using CNN architectures are discussed below.

Seeja and Suresh proposed a method for melanoma segmentation and classification, which uses U-Net architecture for segmentation and VGG-16 architecture for the classification process. The method is tested on ISBI 2016 dataset. The method achieved an accuracy of 83.18%, sensitivity of 95.53% and specificity of 96.22% [19].

Agilandeeswari et al. proposed a method for skin lesion detection using texture based segmentation and classification by CNN. The steps involved in this method are image acquisition, image pre-processing, segmentation using ostu thresholding, feature extraction using GLCM and image classification using ResNet-50 pre-trained network. The method is tested on PH2 dataset and achieved an accuracy of 96% [20].

Suganya proposed a method for skin lesion classification using support vector machine classifier and k-means clustering technique. The steps involved in this method are hair removal, segmentation using k-means clustering, feature extraction, feature selection and classification using support vector machine. The dataset used for evaluation is taken from dermweb. The method obtained an accuracy of 96.8%, sensitivity of 95.4% and specificity of 89.3% [21].

Baghersalimi et al. [22] has proposed a method called DermoNet used for the segmentation of skin cancer which is made of encoder and decoder architecture. The method is tested over ISBI 2017, ISBI 2016 and PH2 datasets. Authors reported an average Jaccard coefficient of 78.3%, 82.5% and 85.3% on PH2, ISBI 2016 and ISBI 2017 datasets, respectively [22].

Yuan and his co. [23] have proposed a method for skin cancer segmentation, which consists of 19 layers divided into convolution, maxpooling, dropout, deconvolution and upsampling layers. The method is tested on ISBI 2016 and PH2 dataset. The method obtained a Jaccard index of 84.7% and Dice index of 93.8%, respectively [23].

In this paper, hybrid CNN architecture is proposed for segmentation and classification of skin cancer using deep learning approach. The model uses U-Net architecture of 26 layers for segmentation and DenseNet architecture for classification. The U-Net architecture is symmetric and it consists of contraction, bottleneck and expansion paths. The location information obtained from the contraction path is combined with the contextual information obtained in the expansion path to provide a good segmentation map. In DenseNet, each layer receives input from all the previous layers and makes it computational intensive. It also performs binary classification for early melanoma detection. The rest part of the paper is categorized as follows, Sect. 2 deals with the methodology and Sect. 3 deals with the results and discussion followed by conclusion described in Sect. 4.

2 Methodology

The proposed hybrid model for melanoma skin cancer classification is shown in Fig. 1. It has three main steps: pre-processing, segmentation and classification. Pre-processing is used to avoid issues related to low contrast images [24]. The pre-processing process involves cropping the input image to obtain the required aspect ratio. Then the cropped image is resized to 552 × 552. The secondary objective is to convert the image into NumPy format (.npy) for easy access in the segmentation process. Mean subtraction is implemented for centring the RGB values from input data around zero value along every image dimension. Image normalization [25] is applied by dividing every dimension of input images with its standard deviation and the values are normalized to 0 and 1. The image is further converted into greyscale for reducing image dimension.

The pre-processed images are given to the segmentation network which is the second step in the proposed model which is carried out using U-Net architecture. Ronnerberger et al. [26] proposed U-Net architecture which is made up of contraction, bottleneck and expansion path as shown in Fig. 2. The contraction path contains contraction blocks which applies two 3×3 convolution operation and a 2 × 2 max-pooling for the input. After each block in the contraction path the feature maps are doubled so that it can learn effectively even the complex structures. The intersection between the contraction and expansion path is the bottleneck. It has two convolution layer of dimension 3 × 3 and followed by an upsampling layer of 2 × 2. The expansion path consists of expansion blocks made of two 3 × 3 convolution layers and a 2 × 2 upsampling layer and the feature maps are reduced to half after each block for maintaining the symmetry. Each time the input is attached to the corresponding contraction block by feature maps for guaranteeing that the features learned are only used for reconstruction. The number of expansion and contraction blocks are equal and it is passed through another 3 × 3 convolution layer to get the desired number of segments. All the convolution layers uses ReLU activation function.

The proposed model uses U-Net network which is trained from scratch. It has 26 layers divided into contraction, bottleneck and expansion paths. In contraction path, each block is made up of two convolution layers and a pooling layer. The bottleneck is the intersection of contraction and expansion path which is made of two convolution layers. In the expansion path, each block is made of one upsampling layer and two convolution layers. ReLU activation function is used for all the convolution layers. The initial weights are considered randomly and values are optimized by backpropagation algorithm. The output is obtained in NumPy array and further converted to image (JPG) having dimension 64 × 80. In this case, Dice coefficient loss function is used for training.

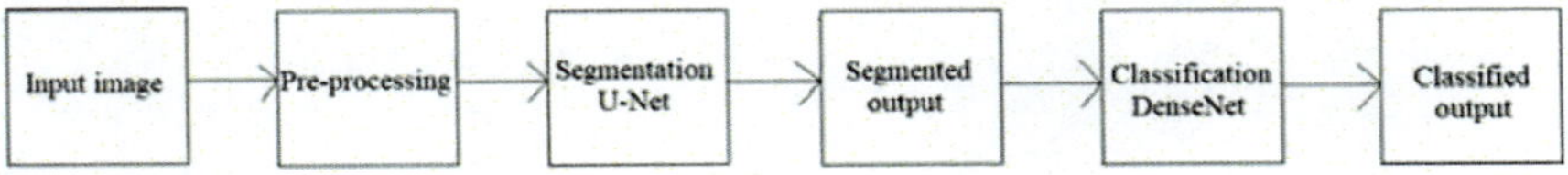

Fig. 1 Block diagram of the proposed model

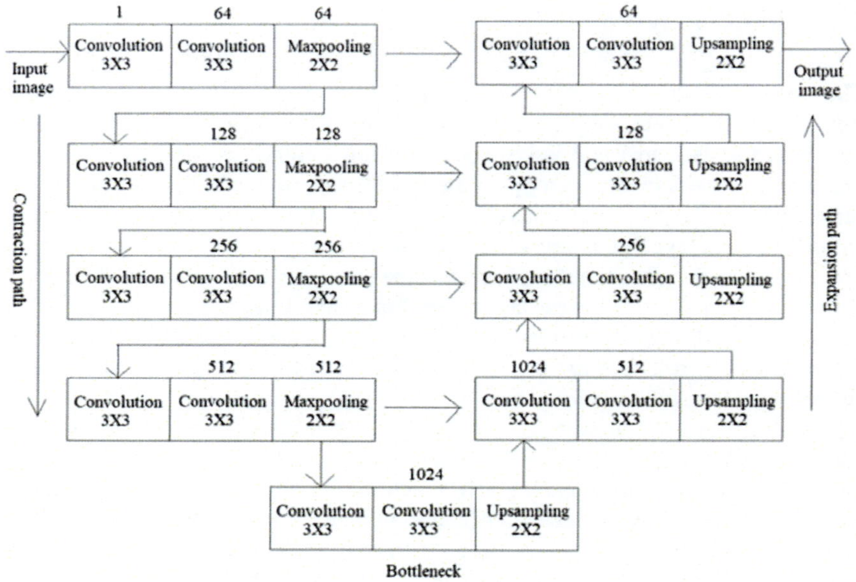

Fig. 2 U-Net architecture [26].

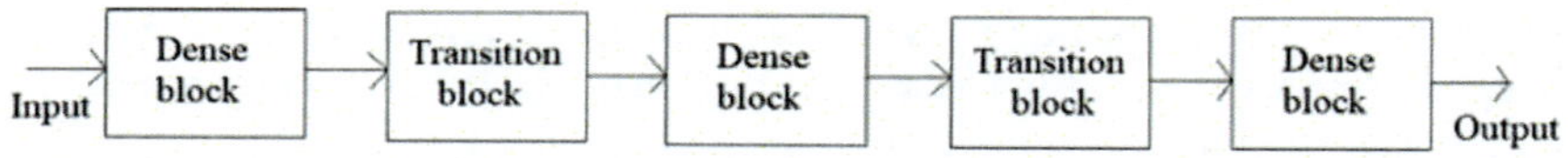

Fig. 3 DenseNet architecture [27]

The segmented image is provided as input to the classification process. It uses DenseNet architecture for classification shown in Fig. 3. This architecture is advantageous as it requires fewer parameters than other architectures. The layers in the DenseNet are very narrow and only add a few new feature-maps. The input image and gradients obtained from loss function are provided to each dense block. The output of the dense block is given to other blocks after performing complex operations which includes convolution, pooling, activation and batch normalization. Last layer represents the output layer and performs binary classification to determine whether it is benign or malignant. The proposed model uses DenseNet containing three dense blocks. The layers between two dense blocks are called as transition layers and it changes the size of the feature maps using convolution and pooling operation. The convolution layer uses ReLU activation function. Sigmoid activation function is used in the output layer [27, 28].

3 Results and Discussion

The proposed method for melanoma skin cancer classification is experimented on freely available database obtained from ISBI 2016 challenge [22]. The details about the database is mentioned in Sect. 3.1. The quality of the method was qualitatively evaluated by accuracy and Dice coefficient. The qualitative or performance measures are elaborated in Sect. 3.2.

3.1 Dataset

ISBI 2016 dataset is a subset of International skin imaging collaboration (ISIC) dataset. ISBI 2016 dataset contains 1279 skin lesion dermoscopic images. All the images present in the dataset are in 8-bit RGB format. Image resolution varies between 1022 × 767 and 4288 × 2848. 70% of images are used for training and 30% are used for testing. Ground truth is available for all the images [22, 23].

3.2 Performance Measures

Dice coefficient is used for evaluating the segmentation process and accuracy is used for evaluating the classification process. Dice coefficient is calculated by the comparision of the pixel-wise agreement between segmented image and ground truth. Accuracy is calculated by the probability of correct predictions made for all the samples [29]. Detailed representation of the performance measures are represented in Table 1.

3.3 Results

The images are first pre-processed as stated above and then the segmentation process begins. The loss function used in segmentation process is Dice coefficient (Table 2). Inorder to acquire more information from the training samples and for increasing the accuracy, the ISBI 2016 dataset is augmented with the help of transformations

Table 1 Various performance measures used for measuring the performance of the model

Name	Representation	Range
Accuracy [30]	$\frac{TP+TN}{TP+FP+FN+TN}$	[0–100]
Dice coefficient [22]	$\frac{2*TP}{2*TP+FN+FP}$	[0–100]

TP true positive, FP false positive, TN true negative, FN false negative

Table 2 Dice coefficient for various optimizers and learning rates using 70% of training images and 30% testing images

Optimizers	Learning rate		
	1e−5	1e−4	1e−3
Adam	0.0016	0.0018	0.0027
RMSprop	0.0016	0.0024	0.0028
Adamax	0.0016	0.0018	0.0026
SGD	0.0016	0.0016	0.0016
Adagrad	0.0016	0.0018	0.0028
Adadelta	0.0016	0.0016	0.0016

Table 3 Optimizers versus accuracy using 70% of training images and 30% testing images for batch size 24 and epoch 7

Optimizers	Accuracy
Adam	84.2
RMSprop	85.5
Adamax	78.9
SGD	65.7
Adagrad	55.2
Adadelta	47.5

such as size rescaling, rotating the image with 40°, zooming the image, horizontally shifting the image and horizontally flipping the image. Data augmentation prevents overfitting problem.

The data is trained through the learning algorithm that applies the stochastic gradient descent (SGD) and after learning the weights. A prediction algorithm is used to classify the testing data. During training, some parameters can be changed for obtaining better performance they are batch size, epoch and optimizer. ReduceLR function is used for adjusting the learning rate. Table 3 gives us the information about different optimizers used and the highest accuracy attained for the respective optimizer. Among the various optimizers used for testing, RMSprop optimizer has obtained highest accuracy of 85.5%. Table 4 provides the information regarding the batch size and accuracy for the segmented data using RMSprop optimizer. Table 5 provides the information about various epoches and accuracy attained on segmented images using RMSprop optimizer. Comparision of various architectures present in the literature using ISBI 2016 dataset is shown in Table 6. Some of output images are shown in Fig. 4.

The minimum Dice co-efficient loss obtained is 0.0016 and the Dice coefficient is 0.9984. The optimum epoch, batch size are obtained as 7 and 24 for segmented data. The highest accuracy obtained after performing 7 epoch on the segmented images is about 85.5% and further increase in number of epochs does not affect the accuracy.

Table 4 Batch size versus accuracy using 70% of training images and 30% testing images for epoch 7 and rmsprop optimizer

Batch size	Accuracy
5	80.2
10	80.2
15	82.9
20	81.5
24	85.5
25	76.3
30	77.6
32	75
35	77.6

Table 5 Epoch versus accuracy using 70% of training images and 30% testing images for batch size 24 and rmsprop optimizer

Epoch	Accuracy
5	76.3
6	82.8
7	85.5
8	80.2
10	80.2

Table 6 Comparision of various architectures

Method	Architecture	Datasets	Performance (%)
Yuan et al. [23], 2017	Fully convolution network	ISBI 2016	JC—84.7
Baghersalimi et al. [22], 2019	Dermonet	ISBI 2016	JC—82.5
Seeja and Suresh [19], 2019	U-Net and VGG-16	ISBI 2016	ACC—83.18, SEN—95.53 and SPE—96.22%

4 Conclusion

This paper discusses the skin cancer classification by using a novel method which consists of segmentation of cancer tissues by U-Net followed by the classification by DenseNet. U-Net consists of 26 layers having three stages namely, contraction, bottleneck and expansion. DenseNet consists of three dense blocks and transition blocks. The proposed method has been tested on ISBI 2016 dataset which consists of 1279 images. From the dataset 70% of images are used for training the network and rest 30% of images are used for testing. The method has obtained a Dice coefficient of 99.84% for segmentation and an accuracy of 85.5% for classification. In this case,

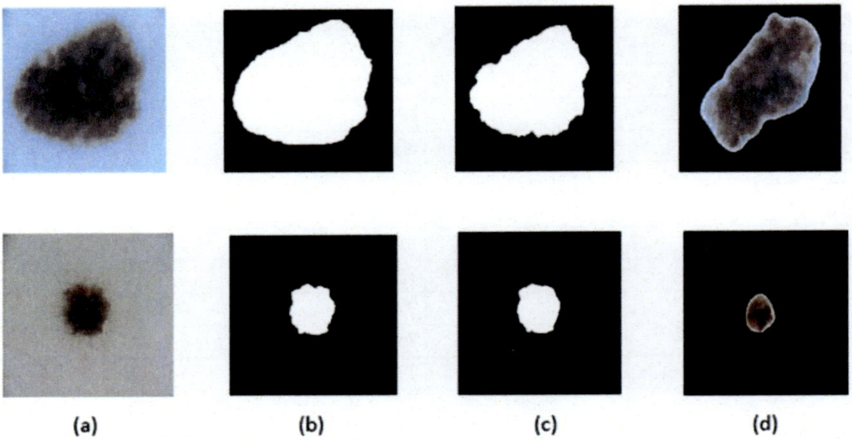

Fig. 4 Segmentation process: **a** input image **b** ground truth **c** binary mask **d** segmented output

RMSprop optimizer is used with an epoch of 7 and batch size of 24. This performance may be further improved with some pre-processing and post-processing steps.

References

1. Pham, H., Koay, C.Y., Chakraborty, T., Gupta, S., Tan, B.L., Wu, H., Vardhan, A., Nguyen, Q., Palaparthi, N.R., Nguyen, B., Chua, M.: Lesion segmentation and automated melanoma detection using deep convolutional neural networks and xgboost. In: International Conference on System Science and Engineering (ICSSE), pp. 142–147, 20 July 2019
2. Younis, H., Bhatti, M.H., Azeem, M.: Classification of skin cancer dermoscopy images using transfer learning. In: 2019 15th International Conference on Emerging Technologies (ICET), pp. 1–4 (2019)
3. Hosny, K., Kassem, M.A., Foaud, M.M.: Classification of skin lesions using transfer learning and augmentation with alex-net. PLoS ONE, 21 May 2019
4. Trovitch, P.., Gupte, A.., Ciftci, K..: Early detection and treatment of skin cancer. Turk. J. Cancer **32**(4), 129–137 (2002)
5. Argenziano, G., Soyer, H.P.: Dermoscopy of pigmented skin lesions—a valuable tool for early diagnosis of melanoma. Lancet Oncol. **2**, 443–449 (2001)
6. Kittler, H., Pehamberger, H., Wolff, K., Binder, M.: Diagnostic accuracy of dermoscopy. Lancet Oncol. **3**, 159–165 (2002)
7. Binder, M., Schwarz, M., Wrinkler, A.: Epiluminescence microscopy: a useful tool for diagnosis of pigmented skin lesions for formally trained dermatologist. Arch. Dermatol. **131**, 286–291 (1995)
8. Seeja, R., Suresh, A.: Deep learning based skin lesion segmentation and classification of melanoma using support vector machine. Asian Pacific J. Cancer Prev. **20**, 1555–1561 (2019)
9. Mahbod, A., Schaefer, G., Wang, C., Ecker, R., Ellinge, I.: Skin lesion classification using hybrid deep neural networks. In: 2019 IEEE International Conference on Acoustics, Speech and Signal Processing (ICASSP), pp. 1229–1233. Brighton, United Kingdom (2019)
10. Ozturk, Ş., Ozkaya, U.: Skin lesion segmentation with improved convolutional neural network. J. Digit. Imag. (2020)

11. Kawahara, J., Daneshvar, S.: Seven-point checklist and skin lesion classification using multitask multimodal neural nets. IEEE J. Biomed. Health Inform. **23**(4), 538–546 (2018)
12. Nachbar, F., Stolz, W., Merkle, T., Cognetta, A., Vogt, T., Landthaler, M., Bilek, P., Falco, B., Plewig, G.: The abcd rule of dermatoscopy. High prospective value in the diagnosis of doubtful melanocytic skin lesions. J. Am. Acad. Dermatol. **30**(4), 551–559 (1994)
13. Damian, A., Ponomaryov, V., Sadovnychiy, S., Fernandez, C.: Melanoma and nevus skin lesion classification using handcraft and deep learning feature fusion via mutual information measures. Entropy **22**(4) (2020)
14. Keerthana, D., Nath, M.K.: A technical review report on deep learning approach for skin cancer detection and segmentation. Data Anal. Manag. 87–99 (2021)
15. Kar, M.K., Ravichandran, G., Elangovan, P., Nath, M.K.: Analysis of diagnostic features from fundus image using multiscale wavelet decomposition. ICIC Exp. Lett. Part B: Appl. **10**, 175–184 (2019)
16. Brinker, T., Hekler, A., Utikal, J., Grabe, N., Schadendorf, D., Klode, J., Berking, C., Steeb, T., Enk, A., Kalle, V.: Skin cancer classification using convolutional neural networks: systematic review. J. Med. Internet Res. **20**(10) (2018)
17. Lameski, J., Jovanov, A., Zdravevski, E., Lameski, P., Gievska, S.: Skin lesion segmentation with deep learning. In: IEEE EUROCON 2019—18th International Conference on Smart Technologies, Novi Sad, Serbia, pp. 1–5, 1 July 2019
18. Harangi, B.: Skin lesion classification with ensembles of deep convolutional neural networks. J. Biomed. Inf. **86** (2018)
19. Seeja, R.D., Suresh, A.: Melanoma segmentation and classification using deep learning. Int. J. Innov. Technol. Explor. Eng. (IJITEE) **8**, 2667–2672 (2019)
20. Agilandeeswari, L., Sagar, M.T., Keerthana, N.: Skin lesion detection using texture based segmentation and classification by convolutional neural networks. Int. J. Innov. Technol. Explor. Eng. (IJITEE) **9**, 2117–2120 (2019)
21. Suganya, R.: An automated computer aided diagnosis of skin lesions detection and classification for dermoscopy images. In: 2016 International Conference on Recent Trends in Information Technology (ICRTIT), Sept 2016
22. Baghersalimi, S., Bozorgtabar, B., Schmid-Saugeon, P., Ekenel, H.K., Thiran, J.-P.: Dermonet: densely linked convolutional neural network for efficient skin lesion segmentation. EURASIP J. Image Video Process. **71** (2019)
23. Yuan, Y., Chao, M., Lo, Y.-C.: Automatic skin lesion segmention using deep fully convolutional networks with jaccard distance. IEEE Trans. Med. Imag. **36**, 1876–1886 (2017)
24. Sushma, M., Nath, M.K., Lokeshwari, R., Premalatha, T., Santhini, J.: Wavelet-narm based sparse representation for bio medical images. Int J Image Graph Sig Process **3**, 38–44 (2015)
25. Kocioleka, M., Strzeleckia, M., Obuchowiczb, R.: Does image normalization and intensity resolution impact texture classification? Comput. Med. Imag. Graph **81** (2020)
26. Ronneberger, O., Fischer, P., Brox, T.: U-net: convolutional networks for biomedical image segmentation. In: International Conference on Medical Image Computing and Computer-Assisted Intervention, pp. 234–241, 18 Nov 2015
27. Huang, G., Liu, Z., Weinberger, K.: Densely connected convolutional networks. In: IEEE Conference on Computer Vision and Pattern Recognition (CVPR), pp. 2261–2269, 26 July 2017
28. Krizhevsky, A., Sutskever, I., Hinton, G.E.: Imagenet classification with deep convolutional neural networks. Neural Inf. Process. Syst. **25**, 1097–1105 (2012)
29. Unver, H..M.., Ayan, E..: Skin lesion segmentation in dermoscopic images with combination of yolo and grabcut algorithm. Diagnostics **9**(72), 97–114 (2019)
30. Hosny, K.M., Kassem, M.A., Foaud, M.M.: Skin cancer classification using deep learning and transfer learning. In: 9th Cairo International Biomedical Engineering Conference (CIBEC), pp. 90–93, 22 Dec 2018

Deep Learning for Satellite Image Reconstruction

Jaya Saxena, Anubha Jain, and P. Radha Krishna

1 Introduction

Recent advancement in deep learning neural networks has emerged as a popular tool for image generation, classification, and restoration [1, 2].

The outstanding results are due to the large training dataset and the ability to learn practical picture priors from it. However, since such a training dataset is not always accessible, this is a bottleneck in their execution. This sparks an idea for research in which the distorted image, as a handcrafted prior, serves as an input to the model, eliminating the need for additional data.

Clouds are one of the most important factors that lead to missing information in optical remote sensing images, limiting their use for Earth observation. Clouds make it difficult to track trees, land surfaces, and water bodies, among other things. The availability of remote sensing imagery can be improved significantly by removing cloud cover from satellite images.

In this work, we utilize randomly initialized neural network as a handcrafted prior and applied to satellite images reconstruction obtaining promising results. The study is also used for cloud removal because cloud is the most important and prominent atmospheric interferer in remote sensing imagery. The analysis employs an encoder–decoder "hourglass" architecture with varying hyperparameters. The MSE, PSNR,

J. Saxena (✉)
NRSC, ISRO, Hyderabad, India
e-mail: jayasaxena@nrsc.gov.in

J. Saxena · A. Jain
Department of Computer Science & Information Technology, IIS University, Jaipur, India
e-mail: anubha.jain@iisuniv.ac.in

P. R. Krishna
Department of Computer Science, National Institute of Technology, Warangal, India
e-mail: prkrishna@nitw.ac.in

M. Dua et al. (eds.), *Proceedings of the International Conference on Paradigms of Communication, Computing and Data Sciences*, Algorithms for Intelligent Systems,
https://doi.org/10.1007/978-981-16-5747-4_48

SSIM, and image hash evaluation metrics were used to determine the findings, and they were found to be promising. Doing this also removes the dependency on huge pretraining datasets, in contrast with most of the other deep learning algorithms.

The proposed model has a significant benefit over other deep learning models such as CNN, RNN, SpaGans, and others in that it is not dependent on pretraining datasets. It can be difficult to get near real-time good cloud-free images of the interest area to train the model, particularly during the rainy season, and this becomes a bottleneck.

The organization of the paper is as follows; Sect. 2 briefs about the related works in this area, Sect. 3 covers the methodology, architecture adopted, and operational requirements. Section 4 contains experiments. Conclusion is presented in Sect. 5, followed by References.

2 Related Works

For image recognition, classification, segmentation, and reconstruction, extensive research and application work is being done in the field of neural networks, including convolutional, completely convolutional, recurrent, residual, generative adversarial networks, and so on [3, 5]. Many algorithms are explored depending upon the requirements which started after the seminal work of Juergen Schmidhuber [2] which marks the beginning of a new era in this field. Image inpainting technique for irregular holes using partial convolutions was investigated by Liu et al. [4]. Deep recurrent neural networks were used by van den Oord et al. [6] to develop generative models for natural images and tested pixel RNNs with up to 12 LSTM layers. For cloud removal from optical satellite imagery, Jordi Inglada and Sebastien Garrigues [7] used simple linear interpolation with temporal images. For thick cloud removal from MODIS data, Ren et al. [8] used overlapping region detection, matching point pair extraction, and image rectification. For cloud removal from optical satellite imagery, Inglada and Garrigues [7] used simple linear interpolation with temporal images. For thick cloud removal from MODIS data, Ren et al. [8] used overlapping region detection, matching point pair extraction, and image rectification. Traditional approaches like wavelet transformations, Savitzky–Golay filters, ECDR algorithm, etc. used for cloud removal in satellite imagery were also explored. Deep learning algorithms were also tested, and it was discovered that they are more effective and yield better performance [14–21].

3 Methodology

The research uses four monochromatic RGB images from the Sentinel 2 satellite with different spectral features in the red, green, and blue bands. Masks covering between 15 and 35% of the image area are produced for masks of various shapes and sizes. To cover the picture region, these masks are overlaid on satellite images. Due to the masks, the covered satellite area portrays holes that can occur due to cloud presence, line loss, pixel losses, band misregistration, and other factors. These blurred satellite images are now being restored with the aid of handcrafted priors. In all three spectral bands, the reconstructed image is evaluated. A similar analysis was also conducted on a real clouded image, with the blurred image being reconstructed to produce a cloud-free output image. The output is evaluated using a temporal image of the same position with the same spectral characteristics.

3.1 Software

Python 3.7 is used in this project. From the available library, the necessary image reconstruction tools are selected and checked, and only these tools are kept in suitable folders for efficient working.

3.2 Architecture

With a variety of dependent variables and hyperparameters, the model employs an encoder–decoder "hourglass" architecture (possibly with skip-connections) [9–11]. The best results come from fine-tuning the relationship between dependent variables and hyperparameters, such as learning rate, batch size, momentum, and number of epochs. In our model, we have used 2000 epochs per image, which can be increased or decreased as per the time accuracy trade-off need in particular case.

The details of depending variables and hyperparameters in the architecture are:

z ∈ R32 X W X H ~ U(0,1/10)	nu = nd = [16, 32, 64, 128, 128, 128].
kd = [3, 3, 3, 3, 3, 3]	Satellite Image.
ku = [5, 5, 5, 5, 5, 5]	ns = [0, 0, 0, 0, 0, 0]
	σp = 0.
num iter = 5000L	R = 0.1
	upsampling = nearest.

3.3 Evaluation Metrics

The research employs mean squared error (MSE), peak signal-to-noise ratio (PSNR), structural similarity (SSIM), and image hash [12, 13]. The MSE is a metric for assessing an estimator's quality; it is always positive, and values closer to zero are better. The PSNR ratio is used to compare the quality of an original and a compressed image. The better the quality of the compressed or reconstructed image, the higher the PSNR. The SSIM algorithm is used to determine how close two images are. Traditional methods such as peak signal-to-noise ratio (PSNR) and mean squared error are being replaced by SSIM (MSE).

4 Experiments

Dataset: Four monochromatic RGB images with bounding box coordinates as (79.76, 23.00) (81.00, 23.00) (79.76, 22.24) (81.01, 22.25) from Sentinel 2 satellite with 10 m spatial resolution and spectral resolution between 458 and 899 nm along with four different shapes and size masks are considered for the study, as shown in Fig. 1:

Further observations are recorded on these four sample images having shape (3, 256, 256) with different percentage and types of black masks as shown in Table 1. These masks are applied bitwise. Mask and image combinations M-i. M: Mask, i: Image.

We observe from above results that structural symmetry index (SSIM) values are better for red and green band rather than blue band. Now considering the general trend observed here for each image.

M1 < M3 < M4 < M2
M1 < M4 < M3 < M2
M3 < M4 < M1 < M2
M3 < M2 < < M1 < M4

According to above information, we see that masks other than M2, which is having a horizontal rectangular mask, perform better than M2.

Now considering the least and maximum obtained MSE we have following results, as shown in Figs. 2 and 3.

4.1 Cloud Removal

The study was also applied to detect the presence of cloud in the image and then for cloud removal and reconstruction of features beneath cloud on the land or water body. The similar study was earlier done using traditionally followed approaches of cloud removal like image fusion, wavelet transformations, filters, ECDR technique, etc.

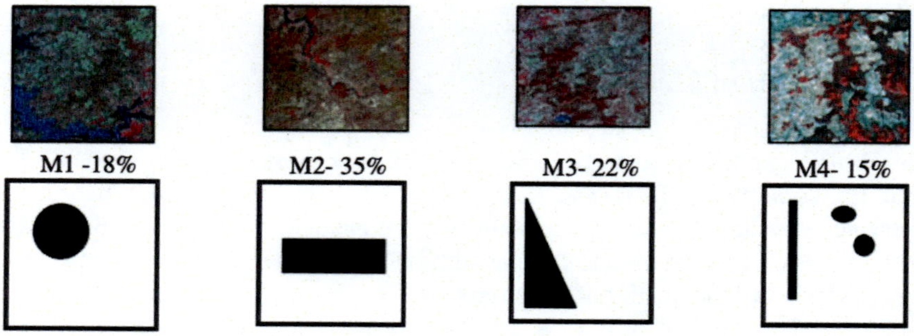

Fig. 1 Mask shapes, as applied on images. From left to right: M1, M2, M3, and M4

Table 1 Loss function (MSE) after 2000 iterations SSIM of individual bands on applying masks over images

Mask-image	MSE ($\times 10^{-4}$)	SSIM		
		Red	Green	Blue
M1-1	3.310	0.893	0.900	0.811
M2-1	5.703	0.806	0.814	0.734
M3-1	3.669	0.852	0.864	0.781
M4-1	3.852	0.900	0.910	0.819
M1-2	2.742	0.904	0.916	0.854
M2-2	3.658	0.847	0.845	0.785
M3-2	3.494	0.863	0.866	0.803
M4-2	3.353	0.922	0.924	0.859
M1-3	3.382	0.918	0.915	0.881
M2-3	3.901	0.843	0.837	0.812
M3-3	2.479	0.871	0.870	0.837
M4-3	2.933	0.917	0.9171	0.882
M1-4	18.406	0.846	0.832	0.830
M2-4	15.588	0.785	0.781	0.774
M3-4	8.527	0.799	0.786	0.787
M4-4	19.170	0.861	0.848	0.850

Fig. 2 Least overall MSE → M3-3, 0.0002.479

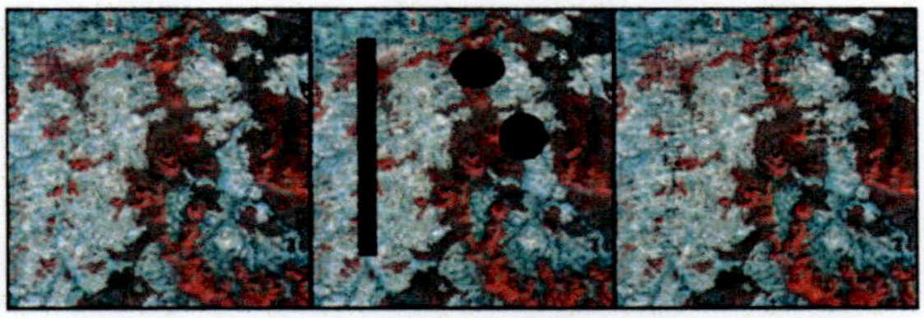

Fig. 3 Max overall MSE → M4-45, 0019.170

The proposed model was also compared over other deep learning models like CNN, RNN, SpaGans, etc., and among all found to be very effective, efficient, adaptive and can be applied in all conditions. Its non-dependency on pretraining datasets proved to be its major advantage. Sometimes, it becomes really difficult to get near real-time good cloud-free images of the interest area, especially in the rainy season to train the model and this becomes a bottleneck. Also, this model efficiently removes thick and thin clouds unlike SpaGans, which can only remove haze and thin clouds.

The algorithm is applied to the cloudy satellite image in the following steps:

Step 1: #Read the image with cloud using cv2.imread("filepath").
Step 2:# Calculating mask using threshold values for white color.

gray = cv2.cvtColor(image, cv2.COLOR_BGR2GRAY)
blurred = cv2.GaussianBlur(gray, (11, 11), 0)

 # Calculate threshold the image to reveal light regions in the blurred image.

thresh = cv2.threshold(blurred, 110, 255, cv2.THRESH_BINARY) [1]
cv2_imshow(image)
cv2_imshow(gray)
cv2_imshow(thresh)

Step 3:# Inverting the mask

mask=np.invert (thresh)
cv2_imshow(mask)) (Please refer Fig. 6, shown above)

Step 4
#Applying mask on image using bitwise and

output = cv2.bitwise_and(image, image, mask = mask)
cv2_imshow(output)

(Please refer Fig. 4).

For evaluation of above image, we would now have some metric evaluation:

MSE of reconstructed portion—7.155447383411229e-05
MSE of whole image—3.314674387401269
PSNR of whole image—67.00879454985356. Please refer Fig. 5.

Image hash

Calculating hash difference values of images is process which uses the fingerprint technique to store value of an image as a hash value of 16 bits. After calculating hash value of reconstructed image, another image retrieved of the same location is

Fig. 4 Left-side vertical three images: image, gray scale, and masked by threshold values. Right-side two images: inverted mask and mask applied on cloudy image

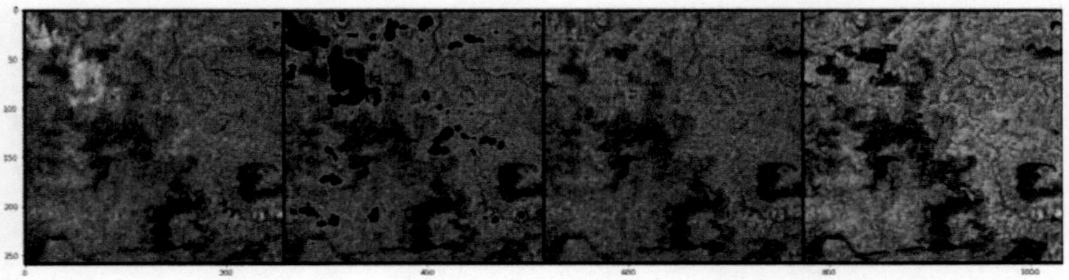

Fig. 5 Left to right; original cloudy image, masked image, reconstructed and temporal image

generated and then their difference is calculated in order to find differences between two images. We have calculated average, perception, and difference hash values.

Average hash 10
Perceptual hash 32
Difference hash 13.

Another ways of evaluating the reconstruction are by using the values of only reconstructed pixels rather than the whole image. Work has started in this direction and can be progressed further. The mask is a rectangular mask thus can be easily cropped using available libraries. We find the coordinates of the mask in whole image, using following code:

```
img = Image.open('/gdrive/My Drive/Reconstruct/data/mask/m7.png').convert(''1')
pixels = img.load()
xlist = []
ylist = []
for y in range(img.size[1])
    for x in range(img.size[0])
        if pixels[x, y] = = 0
            xlist.append(x)
                ylist.append(y)

# Four corners of the black square
xleft = min (xlist)
xright = max(xlist)
ytop = min(ylist)
ybot = max(ylist)
print(xleft,xright,ytop,ybot)
```

Next, we will crop the original and output image using the bounding box coordinates obtained from the above code. Further applying the evaluation metrics we got:

MSE of whole image 0.1341

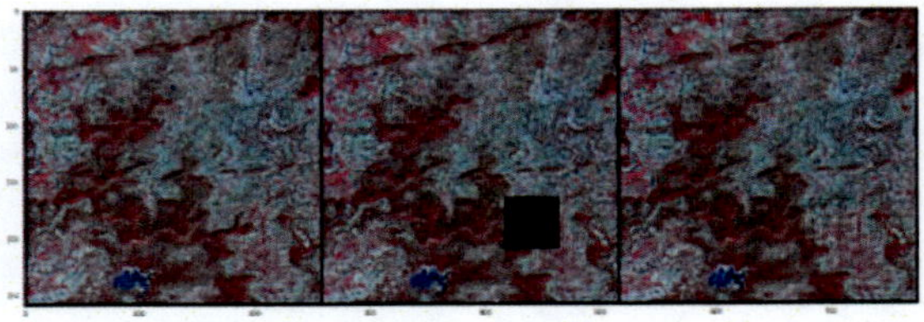

Fig. 6 Original, masked, and output images after Iteration 05000 loss 0.00014157035911921412

MSE of masked region 0.0186
PSNR of whole region 80.9369
PSNR of masked region 69.3652
hash1 = ffffd4f5e1031f1e
hash2 = ff1c1e643717203f
Hash1-hash2 = 27 (Please refer Fig. 6).

Here are some constraints that need further study to generate better results. Firstly, we are able to apply only the rectangular mask hence the application area is getting very limited. Secondly, although the MSE value is quiet acceptable, the other values are not suitable hence some hyperparameter and parameter tuning is required which can be done easily using the provided architectures, like setting different skip depth, changing the learning rate, number of iterations (more the number of iterations more smooth the output will be), etc.

5 Conclusions

The RGB monochromatic satellite images are riddled with masks of different shapes that obscure an area of 15–35%. Reconstructed images have a minimum overall MSE of 0.000247 when using a triangular mask and a maximum overall MSE of 0.00191 when using rectangular and circular masks. We also discovered that the red and green bands have higher structural symmetry index (SSIM) values than the blue and purple bands. The algorithm was then used to remove clouds from the image, yielding precise reconstructed cloud-free satellite images. The values for the average hash, perceptual hash, and difference hash were all satisfactory. Furthermore, rather than evaluating the entire image, the reconstructed satellite image was evaluated using values from only the reconstructed pixels. MSE 0.000141 and PSNR 80 are the best results obtained.

Given that blue band is not used by many sensors, the design may be useful for predicting better structural symmetry and applying deep learning to image reconstruction. In light of the above findings, we assume that using the relationship between dependent variables and hyperparameters, we can develop this model even further. Changes in variables such as the number of iterations skip depth and so on can affect the results. The skip depth used in this case is 6. With a skip depth of 8, the results can differ. Alternative architectures, such as ResNet and UNet, may also be investigated.

References

1. Malladi, R. M. V., Nizami, A., Mahakali, M.S., Krishna, B.G.: Cloud masking technique for high-resolution satellite data: an artificial neural network classifier using spectral & textural context. J. Indian Soc. Remote Sens. **47**(4), 661–670 (2019)
2. Schmidhuber, J.: Deep learning in neural networks: an overview. Neural Netw. **61**, 85–117 (2015)
3. Simard, P., Bottou, L., Haffner, P., LeCun, Y.: Boxlets: a fast convolution algorithm for neural networks and signal processing. In: Advances in Neural Information Processing Systems (NIPS 1998), vol. 11. MIT Press (1999)
4. Liu, G., Reda, F.A., Shih, K.J., Wang, T.-C., Tao, A., Catanzaro, B.: image inpainting for irregular holes using partial convolutions. In: Computer Vision—{ECCV} 2018—15th European Conference, Munich, Germany, September 8–14, 2018, Proceedings, Part XI, pp. 89–105
5. Cheng, J.Y., Chen, F., Alley, M.T., Pauly, J.M., Vasanawala, S.S.: Highly scalable image reconstruction using deep neural networks with band pass filtering. Comput. Res. Repository, abs/1805.03300 (2018)
6. van den Oord, A., Kalchbrenner, N., Kavukcuoglu, K.: Pixel recurrent neural networks. In: Proceedings of the 33nd International Conference on Machine Learning, ICML, 2016, New York City, NY, USA, 19–24 June 2016, pp. 1747–1756
7. Inglada, J., Garrigues, S.: Land-cover maps from partially cloudy multi-temporal image series: optimal temporal sampling and cloud removal. In: IEEE International Geoscience & Remote Sensing Symposium, IGARSS 2010, 25–30 July 2010, Honolulu, Hawaii, USA, Proceedings
8. Ren, R., Guo, S., Gu, L., Wang, H.: Automatic thick cloud removal for MODIS remote sensing imagery. In: 2009 International Conference on Information Engineering and Computer Science, Wuhan, pp. 1–4 (2009)
9. Li, E.Y.: Human pose estimation with stacked hourglass network and TensorFlow, Towards DataScience, Mar 2020
10. Newell, A., Yang, K., Deng, J.: Stacked hourglass networks for human pose estimation. In: European Conference on Computer Vision (ECCV) (2016)
11. Babu, S.C.: A 2019 guide to Human Pose Estimation with Deep Learning, Nanonets
12. Rajkumar, S., Malathi, G.: A Comparative analysis on image quality assessment for real time satellite images. Indian J. Sci. Technol. **9**(34) (2016). https://doi.org/10.17485/ijst/2016/v9i34/96766
13. Silva, E.A., Panettaa, K., Agaianb, S.S.: Quantifying image similarity using measure of enhancement by entropy
14. Schmidt, U., Roth, S.: Shrinkage fields for effective image restoration. In: IEEE Conference on Computer Vision and Pattern Recognition (CVPR), June 2014
15. Ilya, S., Martens, I., Hinton, G.E.: Generating text with recurrent neural networks. In: Proceedings of the 28th International Conference on Machine Learning (2011)
16. Hornik, K., Stinchcombe, M., White, H.: Multilayer feed forward networks are universal approximators. Neural Netw. **2**, 359–366 (1989)
17. Bengio, Y.: Greedy layer-wise training of deep networks. In: Advances in Neural Information Processing Systems (2007)
18. Lowe, D.G.: Distinctive image features from scale-invariant key points. Int. J. Comput. Vis. **60**(2), 91–110
19. Mahendran, A., Vedaldi, A.: Understanding deep Image representations by inverting them. In: IEEE 2015 IEEE Conference on Computer Vision and Pattern Recognition (CVPR) - Boston, MA, USA (7 June 2015–12 June 2015)]. IEEE (2015)
20. Saxena, J., Jain, A., RadhaKrishna, P.: A review on applicability of deep learning for remote sensing applications. Solid State Technol. **63**(06), 3823–3832 (2020)
21. Chawan, A.C., Kakade, V.K., Jadhav, J.K.: Automatic detection of flood using remote sensing images. J. Inf. Technol. **2**(01), 11–26 (2020)

ANN and M5P Approaches with Statistical Evaluations to Predict Compressive Strength of SCC Containing Silicas

Pranjal Kumar Pandey and Yogesh Aggarwal

1 Introduction

Infrastructural growth plays a pivotal role in growth of any country economies, and concrete occupies the core of construction industry. With time, concrete has evolved from conventional portland cement concrete to a durable, high strength self-compacting concrete (SCC). Back in 1980s when Japan construction industry was suffering from durability issues and lack of skilled labour, Prof. Hajjim Okamura came up with the concept of self-compacting concrete [1]. As the name suggest, it is a special type of high-performance concrete with properties like "flow ability", "filling ability", "passing ability" and "segregation resistance". Therefore, it avoids bleeding and segregation and maintains its stability at the same time [2, 3]. SCC is amongst most practical concrete in construction industry. It sets under its own weight without use of any external vibrators and can easily occupies the confined spaces between reinforcements. It finds huge application in prefabricated industries.

Global warming is rising at an alarming rate, and cement industry contributes 7–9% of total global CO_2 causing a negative impact on environment. Thus, researchers across the globe are working towards eco-friendly concrete, as in cement production fossil fuels are burnt and huge amount of CO_2 is emitted in atmosphere. Thus, the concept of blended cement with supplementary cementitious materials (SCMs) are in play. Commonly used SCMs are fly ash [4], silica fume (Si) [5–8], nanosilica fume (nSi) [9–11], micro-silica fume (mSi) [12, 13], perlite powder (PP), ground-granulated blast-furnace slag (GGBFS) [14], viscosity modifying admixtures (VMA) and new generation of super plasticizers (SP). The use of fly ash in SCC reduces the amount of superplasticizer required to obtain slump flow as in that of ordinary

P. K. Pandey (✉) · Y. Aggarwal
Department of Civil Engineering, NIT Kurukshetra, Kurukshetra, Haryana 136119, India

Y. Aggarwal
e-mail: yogeshkumar@nitkkr.ac.in

M. Dua et al. (eds.), *Proceedings of the International Conference on Paradigms of Communication, Computing and Data Sciences*, Algorithms for Intelligent Systems,
https://doi.org/10.1007/978-981-16-5747-4_49

Portland cement concrete. Fly ash also improves the rheological properties of SCC and reduces the cracks in concrete as heat of hydration is lowered.

Silica fume, a pozzolanic material, is used to overcome low early compressive strength of SCC. It is a by-product of powder industry, thus incorporating silica as SCMs gives boost to the environmental cause. Not only silica fume reduces the water absorption and porosity, but also it increases the electrical resistivity and highly effective in controlling the chloride ion penetration. Recent uses of nanotechnology in construction industry are revolutionary as it modifies the material to nano- and micro-level to enhance its properties. In this series, according to various studies, the use of nanosilica (nSi) is quite outstanding. With use of nSi, permeability decreases and compressive strength increases because it not only acts as a filling agent with C–S–H gel for voids but also forms a nanostructured C–S–H gel material, thereby improving the micro-structure.

Within last decade, researchers have explored multiple techniques of modelling the properties of material including computational modelling, statistical techniques and various tools like artificial neural network (ANN) [15], M5P model tree in predicting 28-day compressive strength of concrete. These machine learning tools have ability to learn from input output relation of any complex problem, thus eliminating use of any specific equation form. Feed-forward network or multilayer perceptron mostly uses ANN model for various applications. Padmini et al. [16] have successfully modelled neuro-fuzzy technique to find bearing capacity of shallow foundation. ANN finds its application in structural engineering. [17]

The objective of this paper is to create the model and compare the potential of ANN and M5P model technique in estimating the 28-day compressive strength of SCC incorporating silica using data obtained from literature. To build the model, a total of 99 data are collected. These data are arranged in form of eight input parameters, and machine learning tools are used to predict the strength as output parameter. It was observed that in spite of intricate data these tools can be effectively used to predict the compressive strength and help in decision-making.

2 Modelling

2.1 Artificial Neural Network (ANN)

ANN is the computing system designed to simulate the way how the human brain processes and analyses. It finds its application in prediction problem in construction engineering. It is a multilayer feed-forward network consisting of input layer, hidden layer and output layer programmed with eight input parameters and compressive strength as output. Based on lowest average square error criteria, maximum number of hidden layers and neurons are calculated by trial and error.

Optimum number of epochs are chosen during the training that gives minimum MAE and RMSE values and higher R^2 value. With optimum architecture designed,

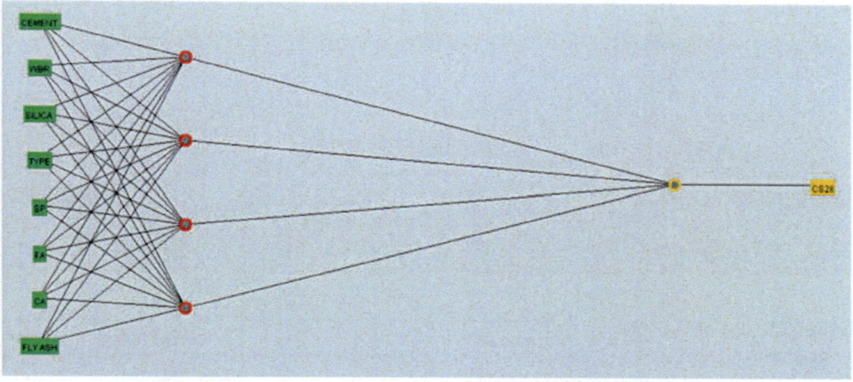

Fig. 1 Network structure of neural network model

total data set was divided in to training and testing data sets. Training consists of calculating outputs, comparing the measured outputs and adjusting the weight of each node to minimize the error. Multiple coefficient of determination R^2 value compares model accuracy with basic benchmark model wherein prediction is samples mean. $R^2 = 1$ results in perfect fit model. To have a non-complicated network with the better inference capability, the back propagation neural network-based modelling algorithm was done which required setting up of different learning parameters and optimum number of nodes in the hidden layer. Further for selected hidden layers and nodes, suitable value of parameters like learning rate and momentum is also required. A neural network model with one hidden layer and four nodes is shown in Fig. 1.

2.2 *M5P Tree Model (M5P)*

M5P tree was first introduced as genetic algorithm learner for regression problems. This tree algorithm divides different areas of data into multiple different spaces and fits multivariate linear regression model on each sub-location and linear regression features on the terminal nodes. [18] The M5P tree approach can handle high-dimensional functions rather than discrete segment. Nonlinear relationship of data sets is approximated by the linear model component developed. Tree division criteria on each node are presented with error estimation information. The default value variance of the class entering the node measures the error. Any function of that node is evaluated by the attribute that maximizes the expected error reduction. The M5P error is determined by the standard deviation of the class values at the node. This division creates large tree like structure that results in overfitting. Secondly, the tree is trimmed and replaced by linear regression functions (Fig. 2).

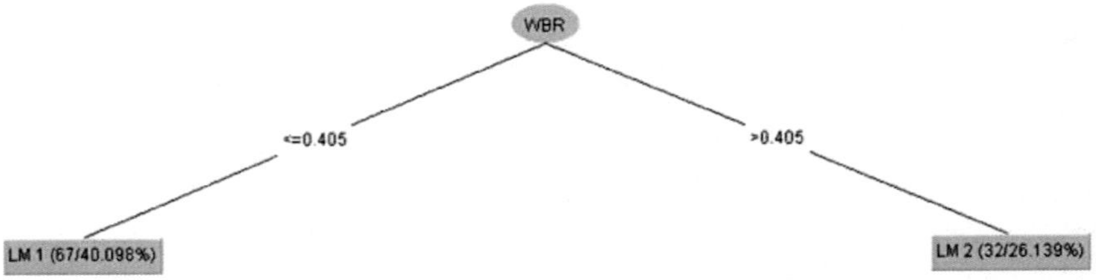

Fig. 2 M5P model tree

Table 1 Range of variables for database

Parameters	Database range
Cement (kg/m^3)	135–570
Fine aggregate (kg/m^3)	720–1166
Coarse aggregate (kg/m^3)	595–940
Fly ash (kg/m^3)	0–420
Silica	
• Silica fume (kg/m^3)	0–85
• Micro-silica (kg/m^3)	0–52.5
• Nanosilica (kg/m^3)	0–11.5
Superplasticizer (%)	0.25–2
Water (kg/m^3)	180–285

3 Results and Discussion

3.1 Data Set

The data set used for training purpose consists of seven input variables and 28-day compressive strength as output. Table 1 gives the ranges of the data obtained from literature.

3.2 Analysis

The model is accepted or rejected based on its ability to predict the compressive strength based upon the data it was trained on. In this paper, a tenfold cross-validation was performed on the data. Cross-validation is the method in which complete data set is divided in parts and one part is used for testing the trained model created from remaining parts. The correlation coefficient, root mean square error (RSME) and mean absolute error (MAE) are used to analyse the performance of computing approach in predicting the strength.

Table 2 Summary of network parameters

Network parameters	ANN
No. of hidden layers	1
Number of hidden neurons	8
Learning rate	0.3
Momentum	0.2
Iterations	500

$$\text{RMSE} = \sqrt{\frac{1}{N}\sum_{n=1}^{N}(\text{ACS} - \text{PCS})^2}$$

$$R^2 = 1 - \frac{\Sigma\left(x - x'\right)^2}{\sum(x - \overline{x})^2}$$

where "ACS—actual compressive strength, PCS—predicted compressive strength, x is the actual value, x' is the predicted value and $(\overline{x})$ is the mean of x values".

Statistical approach for nonlinear relationship is complex and circuitous. In comparison, modelling process is more direct as no mathematical relationship between input and output variables is to be set up. Summary of network parameters incorporated into the model for predicting strength through ANN is given in Table 2.

In training, irrelevant input variable is provided with low weightage. To determine the significance of various inputs by portioning, the connection weight is determined by method proposed by Garson [19]. The maximum weightage was observed for cementitious material (cement + fly ash + silica) in both the models. Thus, input parameters can be varied to achieve the desired output compressive strength.

Table 3 provides the correlation coefficient (R^2), mean absolute error (MAE) and RMSE values for both the models. Table 4 gives the values of actual strength and predicted strength for ANN and M5P model and values for error in the prediction. To compare the model performance, a graph is plotted between actual and predicted 28-day compressive strength. The graph shows that most of the points are plotted within +20% and −20% range which suggest that these AI techniques can be used to predict the strength of self-compacting concrete. Figures 3 and 4 show the plot between actual and predicted strength through ANN and M5P model tree. Figure 5 shows the plot between the predicted strength through ANN and M5P technique.

Table 3 Summary of coefficients for models for 28-day strength

Model	Correlation coefficient (R^2)	Mean absolute error (MAE)	Root mean square error (RMSE)
ANN	0.8948	5.7115	7.5915
M5P	0.9132	4.8407	6.4841

Table 4 Actual strength and predicted strength in (MPa) for ANN and M5P model and values for error in the prediction

S. No.	Actual 28-day compressive strength	Predicted 28-day compressive strength through ANN	Error	Predicted 28-day compressive strength through M5P	Error
1	47.87	51.077	3.207	50.882	3.012
2	75.2	69.445	−5.755	67.142	−8.058
3	57.84	56.347	−1.493	54.662	−3.178
4	35.7	34.459	−1.241	42.799	7.099
5	44.5	52.644	8.144	52.166	7.666
6	61.2	50.374	−10.826	57.838	−3.362
7	63.4	54.6	−8.8	60.115	−3.285
8	34.29	30.386	−3.904	24.81	−9.48
9	21.63	27.62	5.99	29.484	7.854
10	57.84	57.034	−0.806	54.352	−3.488
11	68.65	57.329	−11.321	59.219	−9.431
12	47.7	43.522	−4.178	44.723	−2.977
13	64.31	56.382	−7.928	58.716	−5.594
14	41.49	41.926	0.436	45.3	3.81
15	50	55.859	5.859	60.75	10.75
16	44.2	41.266	−2.934	44.156	−0.044
17	55.49	51.657	−3.833	49.623	−5.867
18	66.4	61.051	−5.349	62.391	−4.009
19	31.7	25.754	−5.946	35.308	3.608
20	42.49	52.374	9.884	49.822	7.332
21	53.52	47.354	−6.166	44.614	−8.906
22	31.61	32.096	0.486	30.186	−1.424
23	60.3	57.242	−3.058	54.472	−5.828
24	45.88	48.048	2.168	45.341	−0.539
25	62.82	57.865	−4.955	57.023	−5.797
26	40.89	47.456	6.566	44.704	3.814
27	58	54.796	−3.204	55.458	−2.542
28	91.6	95.726	4.126	86.757	−4.843
29	57.4	54.266	−3.134	51.928	−5.472
30	85.2	87.48	2.28	74.079	−11.121
31	44.96	52.177	7.217	48.154	3.194
32	29.24	35.749	6.509	28.844	−0.396
33	54.85	61.298	6.448	54.672	−0.178

(continued)

Table 4 (continued)

S. No.	Actual 28-day compressive strength	Predicted 28-day compressive strength through ANN	Error	Predicted 28-day compressive strength through M5P	Error
34	51.1	58.204	7.104	50.798	−0.302
35	45.28	52.744	7.464	45.489	0.209
36	32.03	33.694	1.664	27.353	−4.677
37	75.4	74.772	−0.628	63.819	−11.581
38	95.3	106.597	11.297	103.563	8.263
39	42.9	45.854	2.954	44.443	1.543
40	100.5	104.882	4.382	105.831	5.331
41	45.13	46.261	1.131	51.888	6.758
42	36.02	35.799	−0.221	44.551	8.531
43	46.15	42.07	−4.08	51.05	4.9
44	52.6	40.679	−11.921	47.133	−5.467
45	48.03	43.334	−4.696	50.059	2.029
46	67.05	48.455	−18.595	55.483	−11.567
47	41.9	41.053	−0.847	46.248	4.348
48	49.07	48.509	−0.561	55.626	6.556
49	54.3	40.672	−13.628	48.821	−5.479
50	48.3	46.749	−1.551	52.1	3.8
51	25.7	53.575	27.875	57.29	31.59
52	32.68	31.395	−1.285	29.836	−2.844
53	64.7	61.594	−3.106	59.946	−4.754
54	39.2	42.884	3.684	45.661	6.461
55	65.78	57.38	−8.4	58.225	−7.555
56	27.42	26.349	−1.071	26.702	−0.718
57	43.17	47.801	4.631	44.853	1.683
58	33.1	14.967	−18.133	29.553	−3.547
59	54.31	52.355	−1.955	47.83	−6.48
60	44.9	55.759	10.859	48.877	3.977
61	33.64	27.769	−5.871	29.406	−4.234
62	57.33	57.814	0.484	58.246	0.916
63	60	55.002	−4.998	53.228	−6.772
64	55.29	55.062	−0.228	56.223	0.933
65	72.4	67.122	−5.278	72.331	−0.069
66	43.2	39.039	−4.161	36.74	−6.46

(continued)

Table 4 (continued)

S. No.	Actual 28-day compressive strength	Predicted 28-day compressive strength through ANN	Error	Predicted 28-day compressive strength through M5P	Error
67	58.23	57.632	−0.598	57.999	−0.231
68	52.51	49.302	−3.208	50.522	−1.988
69	47.45	42.229	−5.221	44.752	−2.698
70	45.6	40.668	−4.932	48.836	3.236
71	52.44	79.102	26.662	68.103	15.663
72	68.4	67.648	−0.752	58.772	−9.628
73	29.89	32.962	3.072	28.874	−1.016
74	28.2	37.891	9.691	40.191	11.991
75	55.37	61.683	6.313	64.32	8.95
76	50.78	64.741	13.961	61.017	10.237
77	42.18	49.107	6.927	44.021	1.841
78	39.8	42.841	3.041	39.871	0.071
79	54.05	64.802	10.752	60.14	6.09
80	51.5	57.295	5.795	50.968	−0.532
81	56.64	62.166	5.526	51.699	−4.941
82	39.54	42.366	2.826	38.329	−1.211
83	75.3	87.771	12.471	70.646	−4.654
84	25.38	33.511	8.131	28.239	2.859
85	62.26	64.278	2.018	60.136	−2.124
86	83	80.48	−2.52	80.444	−2.556
87	47.67	56.876	9.206	48.863	1.193
88	49.1	60.599	11.499	48.866	−0.234
89	53.7	64.873	11.173	56.247	2.547
90	79	87.386	8.386	71.957	−7.043
91	28.92	28.528	−0.392	29.527	0.607
92	58.43	61.411	2.981	59.409	0.979
93	60.89	53.566	−7.324	51.658	−9.232
94	59.2	58.774	−0.426	51.414	−7.786
95	45.9	47.224	1.324	46.228	0.328
96	73.6	69.914	−3.686	69.59	−4.01
97	36.65	28.241	−8.409	29.59	−7.06
98	56.47	57.872	1.402	55.75	−0.72
99	85.3	91.218	5.918	84.629	−0.671

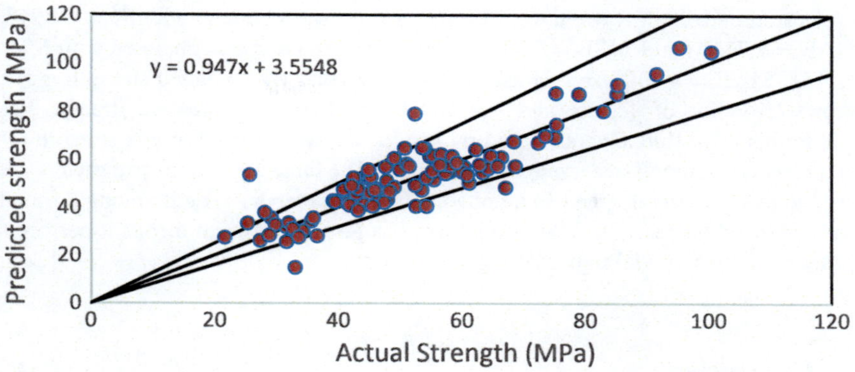

Fig. 3 Actual versus predicted value of 28-day strength for ANN

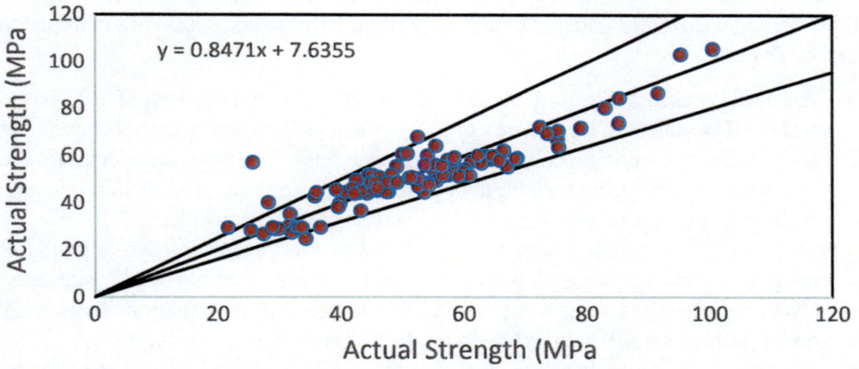

Fig. 4 Actual versus predicted value of 28-day strength for M5P

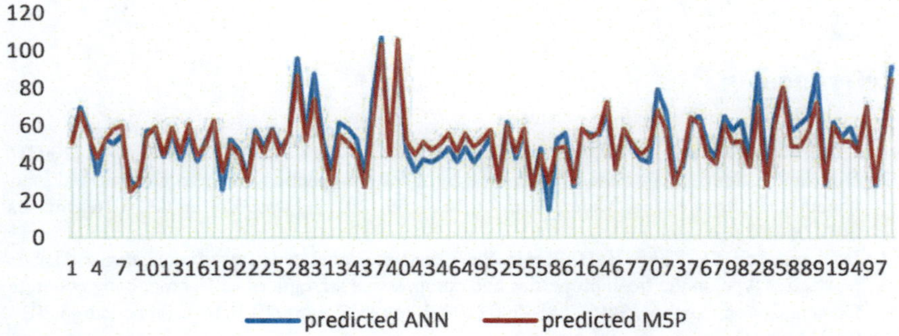

Fig. 5 Predicted value of 28-day strength for ANN and M5P

As summarized in above table, the ANN model was obtained with $R^2 = 0.8948$, MAE = 5.7115 and RMSE = 7.5915, and M5P model tree was obtained with $R^2 = 0.9132$, MAE = 4.8407 and RMSE = 6.4841. Actual v/s predicted strength graph shows error line of + 20% and −20%. The predicted compressive strength lies within this error line. Overall based on results, the M5P model tree is suitable for prediction of compressive strength of this data. The three coefficient parameters are used to evaluate the degree of compliance of compressive strength amongst actual and predicted values. The M5P model tree has less error which means it performs better than ANN model in predicting the strength.

4 Conclusion

Based on data obtained from literatures and simulation of the compressive strength of SCC using 99 different data set under eight input parameters, following conclusion can be drawn.

- This study presents the application to predict compressive strength of SCC based on several parameters. Design mix of SC is different from conventional concrete as it requires more workability and contains more fines. The study demonstrates the feasibility of computing technique in developing a nonlinear interaction between various parameters in solving complex civil engineering problems.
- ANN and M5P model tree were used to predict the compressive strength. The M5P model predicted the strength with coefficient of correlation = 0.9132; similarly, ANN predicted the strength with coefficient of correlation = 0.8948. Thus, M5P model performs better than ANN in predicting compressive strength.
- Based on training and testing data set, the ANN and M5P model predict the compressive strength with minimum deviation from actual strength.

References

1. Okamura, H., Ouchi, M.: Self-compacting concrete. J. Adv. Concr. Technol. **1**, 5–15 (2003)
2. Dinakar, P., Sethy, K.P., Sahoo, U.C.: Design of self-compacting concrete with ground granulated blast furnace slag. Mater. Des. **43**, 161–169 (2013). https://doi.org/10.1016/j.matdes.2012.06.049
3. Boukendakdji, O., Kadri, E.-H., Kenai, S.: Effects of granulated blast furnace slag and superplasticizer type on the fresh properties and compressive strength of self-compacting concrete. Cement Concr. Compos. **34**(4), 583–590 (2012). https://doi.org/10.1016/j.cemconcomp.2011.08.013
4. GuraJawahar, J., Sashidhar, C., RamanaRedddy, I.V., Annie Peter, J.: Design of costeffective M 25 grade of self compacting concrete. Mater. Des. **46**, 687–692 (2013)
5. Wongkeo, W., Thongsanitgarn, P., Chaipanich, A.: Compressive strength and drying shrinkage of fly ash-bottom ash-silica fume multi-blended cement mortars. Mater Des **36**, 655–662 (2012)

6. Yazici, H.: The effect of silica fume and high-volume Class C fly ash on mechanical properties, chloride penetration and freeze–thaw resistance of self-compacting concrete. Constr. Build. Mater. **22**, 456–462 (2008)
7. Borges, C., Santos Silva, A., Veiga, R.: Durability of ancient lime mortars in humid environment. Constr. Build. Mater. **66**, 606–620 (2014)
8. Kawashima, S., Hou, P., Corr, D.J., Shah, S.P.: Modification of cement-based materials with nanoparticles. Cem. Concr. Compos. **36**, 8–15 (2013)
9. León, N., Massana, J., Alonso, F., Moragues, A., Sánchez-Espinosa, E.: Effect of nano-Si_2O and nano-Al_2O_3 on cement mortars for use in agriculture and livestock production. Biosyst. Eng. **123**, 1–11 (2014)
10. Güneyisi, E., Gesoglu, M., Al-Goody, A., Ipek, S.: Fresh and rheological behavior of nano-silica and fly ash blended self-compacting concrete. Constr. Build. Mater. **95**, 29–44 (2015)
11. Jalal, M., Pouladkhan, A., Harandi, O.F., Jafari, D.: Comparative study on effects of Class F fly ash, nano silica and silica fume on properties of high performance self compacting concrete. Constr. Build. Mater. **94**, 90–104 (2015)
12. Nazari, A.: The effects of SiO_2 nanoparticles on physical and mechanical properties of high strength compacting concrete. Compos. Part B: Eng. **42**(3), 570–578 (2011)
13. Jalal, M., Mansouri, E., Sharifipour, M., Pouladkhan, A.R.: Mechanical, rheological, durability and microstructural properties of high performance selfcompacting concrete containing SiO_2 micro and nanoparticles. Mater. Des. **34**, 389–400 (2012)
14. Kavitha, S., Felix Kala, T.: Evaluation of strength behavior of self-compacting concrete using alccofine and GGBS as partial replacement of cement. Indian J. Sci. Technol. **9**(22) (2016)
15. Siddique, R., Aggarwal, P., Aggarwal, Y.: Prediction of compressive strength of self-compacting concrete containing bottom ash using artificial neural networks
16. Padmini, D., Ilamparuthi, K., Sudhir, K.P.: Ultimate bearing capacity prediction of shallow foundations on cohesionless soils using neurofuzzy models
17. Rogers, J.L.: Simulating structural analysis with neural network. J Comput Civil Eng **8**(2), 252–265 (1994)
18. Mohammed, A., Rafiq, S., Sihag, P., Kurda, R., Mahmood, W., Ghafor, K., Sarwar, W.: ANN, M5P-tree and nonlinear regression approaches with statistical evaluations to predict the compressive strength of cement-based mortar modified with fly ash
19. Garson, G.D.: Interpreting neural-network connection weights. AI Expert. **6**(7), 47–51 (1991)

Ensemble of Deep Learning Approach for the Feature Selection from High-Dimensional Microarray Data

Nabendu Bhui

1 Introduction

Modern medical science allows researchers to explore thousands of features and protein functions in a particular sample [1]. Generally, a continuous change in any normal cell of a body causes damage in DNA (De-oxyribonucleic Acid), which is the main cause of forming any disease [2]. Thanks to the biomedical scientists for discovering the microarray technology, which contains the gene expression values. More discussion about microarray technology is discussed in Sect. 3.1. Microarray data consist of thousands of features with a minimum number of samples. Due to high dimensionality as well as redundant and irrelevant features, it is very critical to diagnose and predict a particular disease. The classification of high-dimensional microarray data is very complex and for that reason, we need fast and efficient algorithms [3]. Health is everything for any living being, if it is good then everything is well. Generally, any feature selection or dimension reduction techniques generate a new subset without reducing classification accuracy. There are different types of microarray datasets in medical care, and they have different sizes, different values, and samples. So, it becomes very troublesome to select the minimum number of features from high-dimensional data. To overcome these situations and put our health, disease-free, many feature selection, and classification algorithms are proposed. In addition, in this paper, I have proposed an ensemble of the Deep Learning approach for the feature selection from high-dimensional microarray data. In two stages, dimension is reduced and a new reduced dataset is generated. In the first stage, a simple autoencoder with multiple hidden layers, and in the second stage, a folded autoencoder is used to reduce the dimension of features. After that all the reduced datasets are merged and top-*n* features are selected with the help of the T-Score value. Finally, I classified the proposed model with the help of SVM and MLP classifiers by using a total of five microarray datasets, namely, CNS, Colon, Leukaemia, Ovarian, and Prostate.

N. Bhui (✉)
Department of Computer Science and Engineering, National Institute of Technology Warangal, Warangal, Telangana 506004, India

M. Dua et al. (eds.), *Proceedings of the International Conference on Paradigms of Communication, Computing and Data Sciences*, Algorithms for Intelligent Systems,
https://doi.org/10.1007/978-981-16-5747-4_50

The remaining part of this paper is arranged as follows. The second part (Sect. 2) gives a discussion about related works, a brief overview of all the basic concepts are presented in the third part (Sect. 3), fourth part (Sect. 4) gives the details about the proposed model, and in the fifth part, simulation analysis is explained (Sect. 5).

2 Related Works

In previous research, many researchers introduced different types of techniques to select the gene or features from the microarray data. In [4], the authors introduced two reduction methods: attribute selection and principal component analysis for the exploration of gene expression. They used a dimensionality reduction technique in the microarray technology to increase the feature prediction quality in less time. And a combination of minimum redundancy maximum relevance (mRMR) and consistency-based subset evaluation technique is used to improve the better classification. W. H. Chan et al. proposed an improved SVM with T-test for the identification of informative genes, which improves the feature selection performance [5]. J. Lv et al. used the mRMR filter method for the pre-selection of data and after that they have introduced two rules: higher-fewer rule for sorting according to accuracy with a comparison of the number of selected features. Forcibly Decrease Rule: Reducing the NSF then create new individuals and find the probability vector [6]. In [7], authors selected top-ranked genes with the help of the feature selection method and a metaheuristic approach for the selection of top n genes. Top n genes are ranked by the mRMR selection criteria. J. Xu et al. proposed a deep flexible neural forest (DFNForest) network model, where fisher ratio and the neighbor rough set are combined. This method can change any multiclass problem into a binary class problem in each forest. Here, the fisher ratio is used as a filtering technique, and the neighbor rough set is used to avoid the loss in information [8]. D. Zhang et al. combined the Principal Component Analysis and autoencoder neural network for identification, selection, and extraction of feature expressions. The AdaBoost algorithm is created for the final prediction of cancer [9]. In [10], the Convolutional Neural Network is introduced for the selection and classification of the feature or genes. Large relative movements are happened in between large and small features due to the pooling layer of the convolutional neural network. Finally, they get a softmax probabilistic result and classify that result.

N. Almugren and H. Alshamlan explained in their survey paper that there is a curse of dimensionality problem in microarray data because of a high number of features with respect to the samples [11]. They described that we have to avoid these problems and for that reason, we need hybrid approaches like a filter, wrapper, and embedded methods. And also parameter setting is a more important part of any model or program because it will affect classification accuracy. In [12], the authors proposed a hybrid framework for the selection and classification of gene samples of microarray data. For a primary gene selection, they used the multiple fusion filter technique and then combined the Genetic Algorithm with Tabu Search and SVM for selecting the relevant genes. In [13], the kernelized fuzzy rough set is combined with semi-supervised SVM to predict the cancer biomarker from the microarray data, which improved the classification result. In [14], the authors proposed an approach for the selection of the minimum number of features which is based on the Quantum Inspired Genetic Algorithm, and it gives higher accuracy with minimum features. This newly generated algorithm result is compared with the Genetic Algorithm by using the SVM classifier. W. Ke et al. introduced score based criteria fusion method by combining the symmetrical uncertainty (SU) and ReliefF for the feature selection. SU is mainly used for measuring the relevance after avoiding the redundant features and relief is used to distinguish the similar distanced weight features among all the features [15]. Y. Wang et al. proposed a correlation-based feature selection technique, which is combined with the Decision Tree, Naive Bayes, and Support Vector Machine. This combination of different classification techniques and the feature selection technique is capable of selecting the relevant genes with confidence [16].

Inspired by these, a model is proposed to select the minimum feature subset after reducing the dimension. Based on the ensemble of the Deep Learning approach and T-Score, I have selected the minimum number of features.

3 Brief Overview of Concepts

An overview of Microarray Technology, Simple Autoencoder and Folded Autoencoder Technique is discussed in the following.

3.1 Microarray Technology

In microarray (a silicon chip) thousands of gene samples (features) are represented at the same place, and microarray technology is the way to find the difference between normal or abnormal tissues. Generally, different genetic samples are expressed in a microarray, which is high dimensional. It consists of more and more features with the least number of samples.

Description of Microarray Datasets: I have used publically available five Microarray Datasets in this paper (https://www.ncbi.nlm.nih.gov, http://csse.szu.edu.cn/staff/zhuzx/Datasets.html, https://web.stanford.edu/~hastie/CASI_files/DATA/leukemia.html and http://www.biolab.si/supp/bi-cancer/projections/) [2]. A small description of datasets are described below.

1. **CNS**: The CNS (Central Nervous System) contains 7129 features, 21 samples from a normal person and 39 samples from cancer patients.
2. **Colon**: The Colon cancer dataset contains 2000 features, 22 samples from a normal person and 40 samples from an abnormal patient.

3. **Leukaemia**: The Leukaemia cancer dataset contains 7129 features, 25 samples from AML (Acute Myeloid Leukaemia) patients, and 47 samples from ALL (Acute Lymphoblastic Leukaemia) patients.
4. **Ovarian**: The Ovarian cancer dataset contains 15,154 features, 91 samples from a normal person and 162 samples from an abnormal patient/cancer patient.
5. **Prostate**: The Prostate cancer dataset contains 12,533 features, 50 samples from a normal person, and 52 samples from an abnormal patient.

3.2 Simple Autoencoder and Folded Autoencoder

Autoencoder is a deep learning technique that is used to reduce the dimension of high-dimensional data. It can retrieve the original input data from the reduced data. By using the encoder process, it reduces the data and by using the decoder process it reconstructs the data, the output of encoder and the input of decoder part is called as encoded representation. Hidden layers are used to compress and construct the input dimension. Encoded Representation gives the newly generated feature subset [17] (Fig. 1).

Folded autoencoder (FA) is the autoencoder with a fewer number of hidden layers. FA uses $(N-1)/2$ hidden layers and works approx similar to unfolded autoencoder, which uses N hidden layers. FA reduces the computing cost. In FA, feature dimension is reduced by the power of two in each layer [18] (Fig. 2).

Fig. 1 Simple autoencoder with multiple hidden layers

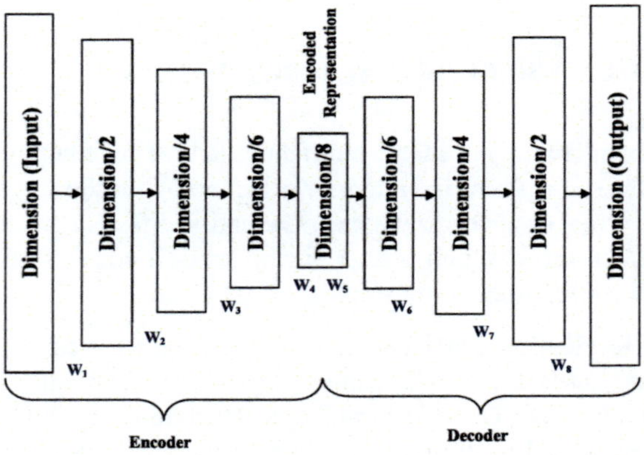

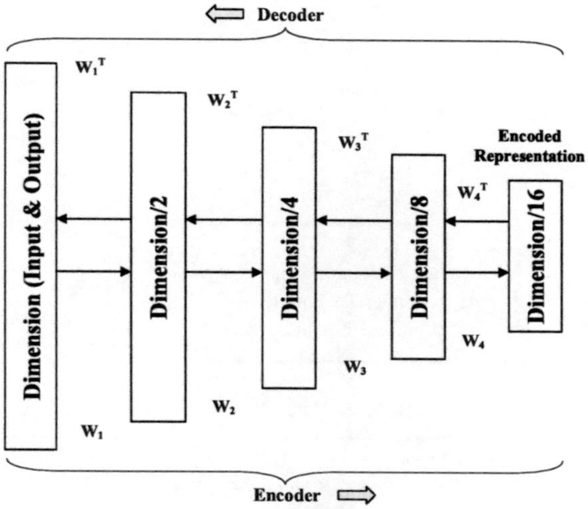

Fig. 2 Folded autoencoder

4 Proposed Method

Microarray data consist of high-dimensional data, so at first, the reductions of data and after that selection of features are the main focus. The proposed method flow chart is shown in Fig. 3.

Original raw data is normalized and then divided into training-testing. After that dimension is reduced by using a simple autoencoder and folded autoencoder, both techniques are applied to training data. From these two types of reduced data, I have selected top-n features on the basis of T-Score value.

4.1 *Proposed Approach*

Original raw data are normalized by using the min-max normalization approach (Eq. 1). It improves efficiency by arranging all the data between 0 and 1. Then, dataset is divided into training and testing parts (80:20, 70:30 and 60:40). Now, simple autoencoder and folded autoencoder both used the training set and reduced the dimension.

$$\hat{X}[:, j] = \frac{X[:, j] - min(X[:, j])}{max(X[:, j]) - min(X[:, j])} \tag{1}$$

where, 'X' is the total dataset and '$X[:, j]$' describes feature 'j' [19].

I have used seven hidden layers in the autoencoder part, where encoded representation is (Original_Dimension/8), input–output dimension is same and each hidden layer dimension is reduced by half in the encoder part, increased by double in decoder part. Similarly, in the FA part, I have used a total of four hidden layers,

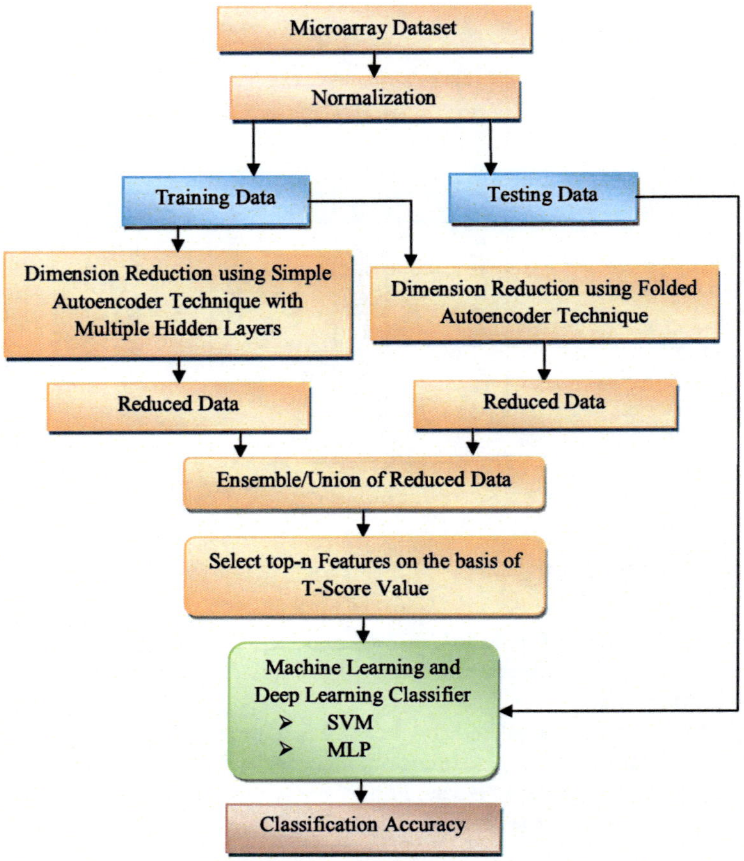

Fig. 3 Flow chart of proposed approach

where encoded representation is (Original_Dimension/16), input–output dimension (and layer) is the same and each hidden layer dimension is reduced by the power of two in the encoder part, increased by the power of two in decoder part.

$$T\text{-Score} = \frac{\mu_x - \mu_y}{\sqrt{\frac{\sigma_x}{S_x} + \frac{\sigma_y}{S_y}}} \tag{2}$$

where, 'μ' is the mean, 'σ' is the standard deviation of the sample and 'S' is the sample of two class (x and y) level [20].

After that I assembled the reduced dimensional data. Now, I have selected top-n features on the basis of T-score value (Eq. 2). Finally, Machine Learning and Deep Learning Classifier algorithms (SVM and MLP) are applied for classification purposes, where the model is tested.

4.2 Parameter Selection

All the parameters are based on trial and error; I have selected parameter values that give us better accuracy. In the neural network, weights are initialized by using a random initialization process. At the time of dimensionality reduction, network parameters are: in the autoencoder input layer, dimension is the same as the output layer dimension with seven hidden layers; in FA, input/output layer is the same but at the time of decoder operation weights are transposed. The learning rate is 0.001 and the training epoch is 500 for both the dimension reduction model.

5 Simulation Analysis and Discussion

I have performed the proposed method by using Jupyter Notebook and python 3.7 versions. The system environment is a 4 GB RAM Intel i5 2.40 GHz processor and Ubuntu operating system. To validate the performance of our approach, ML and DL classifier algorithms are applied. Finally, the results are compared with the original raw dataset and evaluated the model classification accuracy, sensitivity, and specificity.

5.1 Result for Selected Five Datasets

The results of our experiment for each microarray dataset are shown in Table 1, where I created a comparison table of the total microarray dataset and reduced microarray dataset. Also, the average (80:20, 70:30 and 60:40) of classification accuracy, sensitivity, and specificity is shown in this table. After applying the proposed method, approximate 6–7% features are selected and for that reason, classification is getting faster.

6 Conclusion

It is an inescapable fact that feature selection is a significant task in microarray technology, especially in high-dimensional feature/data-driven technology. A microarray consists of more and more features with fewer samples. So, we have to select the least number of features and important features. For that reason, in this paper, an ensemble of deep learning approaches has occurred with T-Score for feature selection. Dimensionality is reduced by using autoencoder and folded autoencoder technique and a new subset is generated, after that important features are selected by using the T-Score. Finally, the evaluation process is happened by using machine learning

Table 1 Simulation results of different classifiers for microarray datasets

Dataset	Classifier	Test	Classification without ensemble approach				Classification after applying ensemble approach			
			$Test_1$ (80 : 20)	$Test_2$ (70 : 30)	$Test_3$ (60 : 40)	Average	$Test_1$ (80 : 20)	$Test_2$ (70 : 30)	$Test_3$ (60 : 40)	Average
CNS	SVM	Sensitivity	0.7388	0.7319	0.7247	0.7318	0.9432	0.9390	0.9352	0.9391
		Specificity	0.7102	0.7057	0.7011	0.7057	0.9134	0.9089	0.9026	0.9083
		Accuracy	0.7222	0.7177	0.7121	0.7173	0.9337	0.9282	0.9222	0.9280
	MLP	Sensitivity	0.8910	0.8886	0.8832	0.8876	0.9806	0.9761	0.9713	0.9760
		Specificity	0.8433	0.8391	0.8344	0.8389	0.9504	0.9466	0.9398	0.9456
		Accuracy	0.8777	0.8721	0.8652	0.8717	0.9688	0.9622	0.9571	0.9627
Colon	SVM	Sensitivity	0.7454	0.7426	0.7394	0.7425	0.9253	0.9201	0.9167	0.9207
		Specificity	0.7231	0.7198	0.7166	0.7198	0.9023	0.8961	0.8914	0.8966
		Accuracy	0.7368	0.7311	0.7281	0.7320	0.9136	0.9092	0.9047	0.9092
	MLP	Sensitivity	0.8711	0.8670	0.8639	0.8673	0.9623	0.9566	0.9495	0.9561
		Specificity	0.8328	0.8290	0.8237	0.8285	0.9254	0.9197	0.9120	0.9190
		Accuracy	0.8421	0.8388	0.8354	0.8388	0.9411	0.9367	0.9301	0.9359
Leukaemia	SVM	Sensitivity	0.8592	0.8505	0.8424	0.8507	0.9542	0.9493	0.9422	0.9486
		Specificity	0.8331	0.8271	0.8201	0.8268	0.9206	0.9166	0.9111	0.9161
		Accuracy	0.8477	0.8412	0.8377	0.8422	0.9386	0.9333	0.9275	0.9331
	MLP	Sensitivity	0.8833	0.8781	0.8728	0.8781	0.9711	0.9654	0.9572	0.9646
		Specificity	0.8402	0.8376	0.8335	0.8371	0.9318	0.9255	0.9183	0.9252
		Accuracy	0.8518	0.8478	0.8422	0.8473	0.9506	0.9467	0.9399	0.9457
Ovarian	SVM	Sensitivity	0.8763	0.8688	0.8612	0.8688	0.9657	0.9604	0.9560	0.9607
		Specificity	0.8537	0.8481	0.8426	0.8481	0.9332	0.9280	0.9221	0.9278
		Accuracy	0.8642	0.8575	0.8491	0.8569	0.9492	0.9433	0.9387	0.9437
	MLP	Sensitivity	0.7209	0.7156	0.7113	0.7159	0.9228	0.9176	0.9123	0.9176
		Specificity	0.6659	0.6610	0.6561	0.6610	0.8930	0.8881	0.8829	0.8880
		Accuracy	0.6842	0.6805	0.6757	0.6801	0.9066	0.9013	0.8971	0.9016
Prostate	SVM	Sensitivity	0.6553	0.6505	0.6461	0.6506	0.8687	0.8622	0.8573	0.8627
		Specificity	0.6299	0.6223	0.6178	0.6233	0.8372	0.8320	0.8267	0.8319
		Accuracy	0.6483	0.6430	0.6377	0.6430	0.8531	0.8477	0.8402	0.8470
	MLP	Sensitivity	0.9052	0.9007	0.8967	0.9008	0.9894	0.9833	0.9764	0.9830
		Specificity	0.8536	0.8482	0.8441	0.8486	0.9503	0.9443	0.9382	0.9442
		Accuracy	0.8838	0.8771	0.8720	0.8776	0.9702	0.9652	0.9584	0.9646

and deep learning classifiers, and I have observed that reduced new data gives better classification result and alleviate computing time. In future, my plan is to update the feature selection technique where I want to combine evolutionary algorithm and deep learning for feature selection.

References

1. Wahid, A., Khan, D.M., Iqbal, N., Khan, S.A., Ali, A., Khan, M., Khan, Z.: Feature selection and classification for gene expression data using novel correlation based overlapping score method via chou's 5-steps rule. Chemometr. Intell. Lab. Syst. **199**, 103958 (2020)
2. Ghosh, M., Begum, S., Sarkar, R., Chakraborty, D., Maulik, U.: Recursive memetic algorithm for gene selection in microarray data. Exp. Syst. Appl. **116**, 172–185 (2019)
3. Kilicarslan, S., Adem, K., Celik, M.: Diagnosis and classification of cancer using hybrid model based on relieff and convolutional neural network. Med. Hypotheses **137**, 109577 (2020)
4. De Souza, J.T., De Francisco, A.C., De Macedo, D.C.: Dimensionality reduction in gene expression data sets. IEEE Access **7**, 61136–61144 (2019)
5. Chan, W.H., Mohamad, M.S., Deris, S., Zaki, N., Kasim, S., Omatu, S., Corchado, J.M., Al Ashwal, H.: Identification of informative genes and pathways using an improved penalized support vector machine with a weighting scheme. Comput. Biol. Med. **77**, 102-115 (2016)
6. Lv, J., Peng, Q., Chen, X., Sun, Z.: A multi-objective heuristic algorithm for gene expression microarray data classification. Exp. Syst. Appl. **59**, 13–19 (2016)
7. Mohamed, N.S., Zainudin, S., Othman, Z.A.: Metaheuristic approach for an enhanced MRMR filter method for classification using drug response microarray data. Exp. Syst. Appl. **90**, 224–231 (2017)
8. Xu, J., Wu, P., Chen, Y., Meng, Q., Dawood, H., Khan, M.M.: A novel deep flexible neural forest model for classification of cancer subtypes based on gene expression data. IEEE Access **7**, 22086–22095 (2019)
9. Zhang, D., Zou, L., Zhou, X., He, F.: Integrating feature selection and feature extraction methods with deep learning to predict clinical outcome of breast cancer. IEEE Access **6**, 28936–28944 (2018)
10. Zeebaree, D.Q., Haron, H., Abdulazeez, A.M.: Gene selection and classification of microarray data using convolutional neural network. In: 2018 International Conference on Advanced Science and Engineering (ICOASE), pp. 145–150. IEEE (2018)
11. Almugren, N., Alshamlan, H.: A survey on hybrid feature selection methods in microarray gene expression data for cancer classification. IEEE Access **7**, 78533–78548 (2019)
12. Bonilla-Huerta, E., Hernandez-Montiel, A., Morales-Caporal, R., Arjona-Liopez, M.: Hybrid framework using multiple-filters and an embedded approach for an efficient selection and classification of microarray data. IEEE/ACM Trans. Comput. Biol. Bioinform. (TCBB) **13**(1), 12–26 (2016)
13. Chakraborty, D., Maulik, U.: Identifying cancer biomarkers from microarray data using feature selection and semisupervised learning. IEEE J. Transl. Eng. Health Med. **2**, 1–11 (2014)
14. Ram, P.K., Bhui, N., Kuila, P.: Gene selection from high dimensionality of data based on quantum inspired genetic algorithm. In: 2020 11th International Conference on Computing, Communication and Networking Technologies (ICCCNT), pp. 1–5. IEEE (2020)
15. Ke, W., Wu, C., Wu, Y., Xiong, N.N.: A new filter feature selection based on criteria fusion for gene microarray data. IEEE Access **6**, 61065–61076 (2018)
16. Wang, Y., Tetko, I.V., Hall, M.A., Frank, E., Facius, A., Mayer, K.F., Mewes, H.W.: Gene selection from microarray data for cancer classification—a machine learning approach. Comput. Biol. Chem. **29**(1), 37–46 (2005)

17. Mallick, P.K., Ryu, S.H., Satapathy, S.K., Mishra, S., Nguyen, G.N., Tiwari, P.: Brain MRI image classification for cancer detection using deep wavelet autoencoder-based deep neural network. IEEE Access **7**, 46278–46287 (2019)
18. Wang, J., He, H., Prokhorov, D.V.: A folded neural network autoencoder for dimensionality reduction. Proc. Comput. Sci. **13**, 120–127 (2012)
19. Bhui, N., Ram, P.K., Kuila, P.: Feature selection from microarray data based on deep learning approach. In: 2020 11th International Conference on Computing, Communication and Networking Technologies (ICCCNT), pp. 1–5. IEEE (2020)
20. Ram, P.K., Kuila, P.: Feature selection from microarray data: genetic algorithm based approach. J. Inf. Optim. Sci. **40**(8), 1599–1610 (2019)

A Comparison Study of Abstractive and Extractive Methods for Text Summarization

Shashank Bhargav, Abhinav Choudhury, Shruti Kaushik, Ravindra Shukla, and Varun Dutt

1 Introduction

With the growing availability of online documents, the importance of automated text summarization has been increasing [1]. It is almost impossible to talk about summarization without first mentioning a number of types of summaries and summary tasks recognized by the information sciences and computational linguistic communities [2]. Automatic text summarization is useful in several ways compared to human-generated summaries; automatic summaries are less biased, provide more efficient indexing, reduce reading time, and are key to question–answering systems [1]. Automatic summarizers have been used by commercial abstracting providers to efficiently increase their throughout by generating a large number of summaries in a short time [3]. Overall, machine-driven text summarization provides promising applicability in many tasks associated with NLP, such as text summarization, question answering, an overview of legal documents, a summary of news, and generation of headlines [1].

A better approach for automatic text summarization could be through abstractive and extractive models [3]. Prior research has investigated multiple abstractive and extractive models for text summarization [4]. These approaches include sequence-to-sequence models, TextRanking, and K-nearest neighbor (KNN) algorithms [4]. For example, Tjandra et al. [6] applied multiscale convolution with several attention vectors to the decoder state to improve the sequence-to-sequence attention model [19]. Liu et al. proposed BERTSUM, a variant of the BERT model, for extractive summarization with a baseline non-pretrained transformer model with similar

S. Bhargav · A. Choudhury · V. Dutt (✉)
Indian Institute of Technology Mandi, Kamand 175005, India
e-mail: varun@iitmandi.com

S. Kaushik · R. Shukla
RxDataScience, Inc., Raleigh, NC 277709, USA

M. Dua et al. (eds.), *Proceedings of the International Conference on Paradigms of Communication, Computing and Data Sciences*, Algorithms for Intelligent Systems,
https://doi.org/10.1007/978-981-16-5747-4_51

BERT architecture [19]. Fang et al. [13] implemented a sentence scoring method for extractive summarization. The novel approach combined the word- and graph-based ranking models [13]. Jo et al. [14] implemented two text summarization approaches, KNN, and support vector machines (SVM). Lastly, Moratanch et al. [16] conducted a comparison study between different abstractive and extractive approaches. Results revealed that abstractive summarization methods produce an information-rich and less redundant summary [13].

The primary drawback of automated summary systems is their emphasis on the text's parameters while neglecting the importance of the text. However, in our research, we have mainly focused on the important features contained in the sentences and proposed abstractive and extractive methods that use the semantics of the text as features [15]. Although prior literature has proposed multiple abstractive and extractive summarization models, a comprehensive evaluation of these models lacks literature. Moreover, the investigation and benchmarking of abstractive and extractive text summarization models on the text summaries are yet to be performed. As a result, the purpose of this study is to fill in these gaps in the literature by formulating a detailed assessment of the output of various abstractive and extractive text summarization models. We approach text summarization as an NLP problem and use different abstractive and extractive text summarization models to capture the text's semantics. The data used in this research is obtained from Amazon fine food reviews dataset [14] and consists of the multiple Amazon customer reviews and their summaries.

2 Background and Related Work

Prior research has investigated different abstractive methods for text summarization. For example, Miao et al. [6] proposed a strategy for encoding the convolutional source and then used a context-sensitive attentional feed-forward neural network to produce the abstractive summary. Similarly, Nallapati et al. [7] developed a sequence-to-sequence model based on abstractive text summarization. They applied an attentional recurrent neural network (RNN) encoder–decoder, and a powerful hybrid pointer–generator network model was used to copy words by pointing in the source text. Lastly, to prevent repetition, coverage of track summary was exploited in the model training phase. Similarly, multiple extractive models for text summarization have also been proposed in the literature.

Various early works like Mohd et al. [6] used single document extractive summarization like ranking algorithms for sentence scoring to generate extractive summaries from the text. An automated summarizer is used in this experiment to catch the semantics of the summaries using the semantic model. To determine the document's successful description, Gong et al. [12]. The relationship between topics and sentences was determined using latent semantic analysis. In this experiment, evaluation on two summaries is done by comparing the summary output with the manual summary generated by human evaluators. Similarly, Wong et al. [8] used machine learning models like support vector machine (SVM) and Naive Bayes to generate the extractive summaries by features, relevance, and topic. In this experiment, results are obtained by measuring the relevant features and content word relatedness from the sentences.

Neto et al. [9], to retrieve the text's original text log, used statistical and linguistic features. In this experiment, the author captures the document's semantics of text and fundamental features for summarizing the document. Furthermore, Rabiener et al. [10] implemented the hidden Markov chain methods to summarize the text.

In this experiment, different sizes of overlapping and hierarchical windows were used to generate from each sentence for evaluation of the text. Khali al-hind [8] used a KNN and Naive Bayes for feature selection to generate extractive summaries from the text. The results are of the extractive models. However, it showed less accuracy in document classification tasks. Moreover, benchmarking was limited in these studies, and to the best of the author's knowledge, no comparative research of extractive and abstractive was performed. This research aims to overcome the literature mentioned above gaps by performing a comprehensive evaluation of different abstractive and extractive text summarization models. Lastly, the abstractive and extractive results were evaluated using other performance metrics.

3 Data

3.1 Data Description

In this research, we used the Amazon fine food review dataset [14] for our analyses. The data consisted of 568,454 customer reviews and their summaries. Lastly, we consider only 100 reviews from the dataset for training the model.

3.2 Data Preprocessing

Firstly, the data was preprocessed by converting all words to lowercase characters and removing HTML tags, parenthesis, punctuations, special characters, and stop-words.

4 Evaluation Metrics

Abstractive models (sequence-to-sequence decoder with attention) and extractive models (TextRank, KNN, and BERT) were tested on an Amazon fine food reviews dataset. And the following metrics were used to evaluate the performance of the models: cosine similarity, Rouge-1 percentage, tf-idf vectorizer, count vectorizer, and soft cosine similarity cosine [17].

Count vectorizer: Count vectorizer function transforms the entire text into a vector on the basis of the word count of each word that appears in the entire text. Here, the individual feature name is represented by any vector (term) in a document and a matrix represents its occurrence.

Rouge-1: Rouge-1 calculates the overlap between each word present in the model summary and actual summary [17]. It is a percentage metrics that are defined as follows described as:

$$\text{Rogue} - 1 = \frac{x}{y} \tag{1}$$

where x represented the difference between the model summary and the actual summary and y represents the total number of words in the reference summary.

Tf-idf: tf-idf is a weighting matrix which is used to evaluate each word's importance in the document collection [17]. It is represented as follows:

$$\text{tf-idf}(t, d, D) = tf(t, d)idf(t, d) \tag{2}$$

where t represents the terms; d represents each document; D represents the collection of documents. Term frequency: $tf(t. d)$ evaluates the number of times the terms t appears in document d. Inverse document frequency:

idf(t, d) measures how much information the attribute value contains. It is represented as follows:

$$\text{Idf}(t, d) = \log\log\log\log \frac{|D| + 1}{1 + |d \in D : t \in d|} \tag{3}$$

Soft cosine similarity: The soft cosine similarity can be described as terms similar in meaning should be treated as similar [17]. The mathematical equation of soft cosine similarity can be defined as:

$$\text{soft\_cosine}(a, b) = \frac{\sum_{i,j}^{N} s_{i,j} a_i b_j}{\sqrt{\sum_{i,j}^{N} s_{i,j} a_i a_j} \sqrt{\sum_{i,j}^{N} s_{i,j} b_i b_j}} \tag{4}$$

where s_{ij} = similarity (feature$i$, feature) If there is no similarity between features ($s_{ii} = 1$, $s_{ij} = 0$ for $i \neq j$), the given equation is equivalent to the conventional cosine similarity formula.

5 Methodology

5.1 K-Nearest Neighbor

KNN can be defined as a nonparametric supervised learning algorithm used for both classification and regression tasks and predicts new points in the cluster. There are no training steps for the data points as all the data points will be considered during prediction. It uses the similarity between the features to predict new points in the cluster.

$$d(q, p) = \sqrt{\sum_{i=1}^{n} qi - pi} \quad (5)$$

The above formula takes in n number of dimensions or features. The test data point is assumed to belong to the same class as its k-nearest neighbors.

5.2 TextRank Algorithm

TextRank algorithm can be defined as an unsupervised extractive text summarization technique similar to the PageRank algorithm. In TextRank, all the texts in the document are combined. The text from the document is then split into sentences. Next, the word embedding of each sentence and sentence words are generated, and the similarities between the word embeddings are calculated and stored in the matrix. The similarity matrix is converted into the graph with similarity scores as edges and sentences as vertices. From this procedure, top-ranked sentences are selected.

5.3 Sequence-to-Sequence Decoder with Attention

There are two parts of a standard sequence-to-sequence model: an encoder and a decoder. The two components are two separate models of neural networks fused into one giant network. This representation is then forwarded to a network of decoders that produces its sequence representing the output. Attention was first used in natural language processing (NLP) for machine translation tasks [18]. As the attention mechanism belongs to the neural architecture, it can highlight relevant features from the text. In contrast, NLP works on the sequence of textual elements, which enable it to apply directly to raw input data [18]. The main characteristic behind attention is to measure a weight distribution on the input data, assigning more important elements to higher values. [18], whereas context vector *ct* is defined as the sum of hidden states of the input sequence for weighted alignment score.

$$C_t = \sum_{i=1}^{n} \alpha_{t,i}, h_i \tag{6}$$

$$\alpha_{t,i} = \text{align}(y_t, x_i) \tag{7}$$

$$= \frac{\exp(\text{score}(s_t - 1, h_i)}{\sum_{i=1}^{n} \exp(\text{score}(s_t - 1, h_i))} \tag{8}$$

5.4 Bidirectional Encoder Representations from Transformers (BERT)

BERT is a machine learning model built on transformers that use a method of attention to learning semantic relations between words. The transformer contains two mechanisms: an encoder that reads the text input and a decoder that outputs a prediction. Firstly, the input text is translated into tokens and fed into the BERT model, where a [MASK] token replaces 15 percent of the words in each series. The model subsequently aims to identify the original meaning of the masked words in the sequence, based on the other non-masked words' context, during pretraining. Using labeled data from downstream operations, the parameters are fine-tuned. Despite the fact that they are all initialized with the same pretrained parameters, each downstream mechanism has its own fine-tuned model. The downstream tasks' named data is used to fine-tune all parameters. And if they are all initialized, each downstream mission has its own fine-tuned model.

6 Model Training

We had tested our models on the Amazon fine food dataset. We randomly extracted 100 reviews and their summaries from the dataset and fed the reviews as input to the different models. Only 100 reviews were used to decrease the training time of the model. The summaries were used as class labels for checking the accuracy of our models. Firstly, we cleaned the reviews by removing punctuations and stop words and then separated them into tokens. These tokens were then fed into the BERT model. We used the BERT model from huggingface's Python pretrained-BERT library. Based on these tokens, the pretrained BERT models output a $N \times E$ matrix where N is represented as sentences and E is represented as an embedding dimension. These embeddings were then clustered using the K-nearest neighbor approach, and the embeddings that were closest to the centroid were selected as candidate summary sentences.

In the KNN approach, we generate word2vec embeddings for each word. During training, each word is classified as either being part of the summary or not. Next, during the testing phase, each word is classified based on majority voting by its K nearest neighbors. This process was repeated for every word in the test set.

In the BERT model, we first clean the reviews by removing punctuations, stops words, and special characters from the review test. Similarly, we clean the review summary and add two special tokens, first and last, at the initial and end of the cleaned summary. Next, all the reviews' length was analyzed to set the reviews and summaries' maximum length. Now we convert each summary into a set of tokens using a tokenizer. The vocabulary is built by a tokenizer, which transforms a word set to an integer sequence. A sequence-to-sequence model is then applied to the tokens. An encoder and a decoder make up a sequence-to-sequence system. The encoder consisted of three stacked LSTM, and the decoder consisted of a single LSTM layer. After that, the attention layer receives the encoder and decoder's output. The alignment score (e_{ij}) is calculated by the attention layers depending on how well the word in the input is associated with the target word using a score function. The attention layer's output is then concatenated with the decoder output and fed to a time-distributed dense layer. We used the sparse categorical cross-entropy as the loss function for our model.

TextRank is an extractive text summarization technique. It takes as input the summaries and splits the whole summary into individual sentences. Then it uses the pretrained GloVe word embedding to generate a vector representation of each sentence. Lastly, it calculates the similarities between each sentence embedding and stores it in a matrix. For sentence rank calculation, the similarity matrix is then transformed into a graph, where the sentences are vertices, and similarity scores are edges.

7 Results

Tables 1 and 2 contain the results of all the models for different metrics. As shown in Table 1, the best results on the training data were obtained by KNN on the Rouge-1 metrics, while BERT performed the best on count vectorizer, tfidf vectorizer, and soft cosine similarity.

Table 1 Training results of text summarization from similarity measures

Algorithms	Rouge-1	Count vectorizer	Tf-idf vectorizer	Soft cosine similarity
TextRank algorithm	0.028	0.809	0.781	0.935
BERT algorithm	0.002	0.912	0.901	0.989
Attention mechanism	0.029	0.431	0.332	0.527
KNN algorithm	0.050	0.501	0.381	0.607

Table 2 Test results of text summarization from similarity measures

Algorithms	Rouge-1	Count vectorizer	Tf-idf vectorizer	Soft cosine similarity
TextRank algorithm	0.001	0.728	0.665	0.853
BERT algorithm	0.020	0.840	0.796	0.928
Attention mechanism	0.095	0.413	0.577	0.808
KNN algorithm	0.071	0.756	0.636	0.792

As shown in Table 2, attention mechanism performed better for Rogue-1 compared to count vectorizer, tfidf vectorizer, and soft cosine similarity. Similarly, BERT performed better for count vectorizer compared to Rogue-1, tf-idf vectorizer, and soft cosine similarity. Furthermore, BERT performed better for tf-idf compared to Rogue-1, tfidf vectorizer, and soft cosine similarity. Next, BERT performed better for soft cosine similarity compared to rouge-1, tf-idf vectorizer, and count vectorizer.

8 Discussion and Conclusion

The need for text summarization is growing rapidly worldwide due to the large amount of textual data surfacing digitally [5]. Therefore, there is a need for the natural language models to take out the relevant and important information from the long text and keep the main information. Text summarization techniques have become the most necessary solution in textual data surfacing in all the fields, from e-commerce to financial and Internet-based firms [8]. All firms get huge amounts of text data and work out the most important information from these data, which helps the business. This research aimed to develop a comparative study of abstractive summarization (sequence-to-sequence decoder with attention) and extractive summarization techniques (BERT, KNN, and TextRank algorithms). The different models were tested on Amazon's fine food dataset to summarize the customer's reviews [14]. Results revealed that the text summarization could be carried out using both abstractive methods (sequence-to-sequence with attention) and extractive methods (BERT, KNN, and TextRank algorithm). Furthermore, the BERT model performed the best both on the training data and test data across all the comparison metrics except the Rouge-1 score.

First, we found that the KNN model gave a higher Rouge-1 score on the training data while the sequence-to-sequence decoder with the attention model gave a higher Rouge-1 score on test data compared to other algorithms. These results can be explained by the fact that the KNN nonparametric approach does not make any underlying assumptions about the distribution of data. The KNN approach works by finding the distances between a query and all the examples in the data, selecting the specified number of examples (K) closest to the query. The KNN approach then votes for the most frequent label. The words extracted by KNN can represent each class clustering well and show high quality for semantic expressions. While in the case of the attention mechanism, we compute a context vector for each time step by computing the attention weights every time. This process makes the attention mechanism model face difficulties in dealing with long sentences and makes the model unable to perform better on the text.

Second, the BERT model performed better than the sequence-to-sequence decoder with attention, KNN, and TextRank models on all the comparison metrics except the Rouge-1 score. This finding agrees with the prior literature [6]. The BERT model has been found to perform better than other text summarization models [19]. It may also be that the default pretraining in the BERT model enables this model to represent the most important points from the text. The BERT model uses a powerful transformer architecture that enables it to get the best results in summarization. As a pretrained BERT model is trained on a large dataset, no further training is required.

There are multiple implications of this work at the industry levels. For example, if the summary produced is concise and fluent, then that could help the businesses determine the customers' needs. In extractive methods, the critical task is to find the text's key phrases for inclusion in summary [11]. In abstractive methods, the sentences paraphrase sections of the source text [8]. This paraphrasing could reduce the reading time and assist in researching the information that can fit in a particular area. Furthermore, the models built can be implemented in the production environment at minimal cost across several domains, including digital marketing agencies, financial research, legal contracts, question answering bots, and e-commerce.

Future research could expand on this work by comparing various BERT implementations for summarization tasks, such as OpenAI's GPT-2 model and sequence-to-sequence with side dependencies model [6]. We can also increase the training data size, which could help the models generalize better and aid in better text summarization. Next, it may also be possible to evaluate text summarization models on bilingual evaluation understudy (BLEU) score, which could help find better text quality. Implementing the transformer model in the sequence-to-sequence model and bidirectional LSTM models for improved output test results could be other objectives in addition to enhancing the text quality [19]. We want to try some of these ideas in our research program concerning text summarization.

References

1. Aksenov, D., Julián, M., Peter, B., Robert, S., Leonhard, H., Georg, R.: Abstractive Text Summarization based on Language Model Conditioning and Locality Modeling. arXiv arXiv: 2003.13027 (2020).
2. Teufel, S., Moens, M:. Summarizing scientific articles: experiments with relevance and rhetorical status. Comput. Linguist., 409–445 (2002)
3. Gambhir, M., Gupta, V.: Recent automatic text summarization techniques: a survey. Artif. Intell. Rev. **47**(1), 1–66 (2017)
4. Suanmali L, Mohammed, S., Naomie, S.: Sentence features fusion for text summarization using fuzzy logic. In: 2009 Ninth International Conference on Hybrid Intelligent Systems, vol. 1, pp. 142–146 (2009)

5. Mudasir, M., Jan, R., Shah, M.: Text document summarization using word embedding. Expert Syst. Appl. **143**, 112958 (2020)
6. Andros, T., Sakti, S., Nakamura, S.: Multi-scale alignment and contextual history for attention mechanism in sequence-to-sequence model. In: 2018 IEEE Spoken Language Technology Workshop (SLT), pp. 648–655 (2018)
7. Rafael, F., Cabral, L., Lins, R.D., Silva, G.B, Freitas, F., Cavalcanti, D.C., Lima, R., Steven, J.S., Luciano, F.: Assessing sentence scoring techniques for extractive text summarization. Expert Syst. Appl. **40**(14), 5755–5764 (2013)
8. Wong, F., Mingli, K., Wenjie, L.: Extractive summarization using supervised and semi-supervised learning. In: Proceedings of the 22nd International Conference on Computational Linguistics, pp. 985–992, (Coling) (2008)
9. Neto, L.J., Freitas, A., Kaestner, C.A.: Automatic text summarization using a machine learning approach. In: Brazilian Symposium on Artificial Intelligence, pp. 205–215. Springer, Berlin, Heidelber (2002)
10. Rabiner, L., Juang, B.: An introduction to hidden Markov models. IEE ASSP magazine **3**(1), 4–16 (1986)
11. Conroy, J.M., Dianne, P.O.: Text summarization via hidden Markov models. In: Proceedings of the 24th Annual International ACM SIGIR Conference on Research and development in information retrieval, pp. 406–407 (2001)
12. Gong., Y, Liu X.: Generic text summarization using relevance measure and latent semantic analysis. In: Proceedings of the 24th Annual International ACM SIGIR Conference on Research and Development in Information Retrieval, pp. 19–25 (2001)
13. Fang, C., Mu, D., Deng, Z., Wu, Z.: Word-sentence co-ranking for automatic extractive text summarization. Expert Syst. Appl., 189–195 (2017)
14. Bhati, V., Jayveer, K.: Survey for Amazon fine food reviews. Int. Res. J. Eng. Technol. (IRJET) **6**(4) (2019)
15. Taeho, J.: K nearest neighbor for text summarization using feature similarity. In: 2017 International Conference on Communication, Control, Computing and Electronics Engineering (ICCCCEE), pp. 1–5. IEEE (2017)
16. Moratanch, N., Chitrakala, S.: A survey on extractive text summarization. In: 2017 International Conference on Computer, Communication and Signal Processing (ICCCSP), pp. 1–6 (2017)
17. Baoli, L., Han, L.: Distance weighted cosine similarity measure for text classification. In International Conference on Intelligent Data Engineering and Automated Learning, pp. 611–618 (2013)
18. Dichao, H.: An introductory survey on attention mechanisms in NLP problems. In: Proceedings of SAI Intelligent Systems Conference, pp. 432–448. Springer, Cham (2019)
19. Liu, Y.: Fine-tune BERT for extractive summarization. arXiv preprint arXiv, 1903.10318 (2019)

An Efficient Deep Neural Network-Based Framework for Building an Automatic Attendance System

Rahul Thakur, Harshit Singh, Charanpreet Singh Narula, and Harsh

1 Introduction

Teachers have a limited amount of time for lectures, and a lot of their time is taken away through attendance. In order to save time and use it for more valuable interactions between teachers and students, an automated attendance system [1] using deep neural networks [2] can be used to tackle the manual class attendance problem. The proposed system aims to achieve real-time detection of faces of each student at an efficient and accurate rate, using state-of-the-art object detection and face recognition algorithms.

In deep learning models, data is passed through multiple layers to allow the algorithm to learn patterns in the data. Deep learning models are just function approximations, fit for adapting to all kinds of data. Some of the most widely used applications in deep learning are face detection and face recognition. Face recognition is a well-researched field; its application is widely seen in face unlock systems of mobile phones. While tackling this attendance issue, we can separate the whole system into an object detection and an object classification problem where the classes are all the individual students. The issue articulation for an article recognition structure can be built as an issue of picture confinement, which means the neural network finds the

R. Thakur · H. Singh · C. S. Narula (✉) · Harsh
Department of Electronics and Communication Engineering, Delhi Technological University, New Delhi, India
e-mail: charanpreetsinghnarula_2k17ec50@dtu.ac.in

R. Thakur
e-mail: rahulthakur@dtu.ac.in

H. Singh
e-mail: harshitsingh_2k17ec66@dtu.ac.in

Harsh
e-mail: harsh_2k17ec64@dtu.ac.in

M. Dua et al. (eds.), *Proceedings of the International Conference on Paradigms of Communication, Computing and Data Sciences*, Algorithms for Intelligent Systems,
https://doi.org/10.1007/978-981-16-5747-4_52

coordinates of the image at which the object to be identified lies, and furthermore, the certainty with which the object is associated with a specific class.

The advancement in the development of object detection has reached a state where 155 instances can be detected every second in real time through the YOLOv4 [3] network. Through this system, the lecturer's task of attendance is taken away via a programmed facial recognition system. Procedures utilizing deep learning structures which give quick and exact outcomes and furthermore conquer the downsides of the current strategies.

2 Literature Survey

There are numerous methodologies for taking care of the object detection issue and largely can be classified into two fundamental strategies: branched networks [4] and single flow network. Branched network is neural network designed to play out the distinctive sub-assignments needed to achieve the task. With regard to object detection organizations, one expanded sub-network chips away at the issue of area proposition, implying that the organization attempts to realize the most probable location at which the desired object is situated. The proceeding branch of the network labels the located object from the known categories.

One of the past techniques for automated attendance used the eigenface method in principal component analysis. The framework of the eigenfaces [5] method involves extracting the features of the face and representing the face as a linear combination called 'eigenfaces' obtained from the feature extraction process. Recognition is done by prediction of the face into the space formed by the eigenfaces. But the limitation of this system is that a proper centered face is required for the training and testing. Further, the eigenface method is also sensitive to lighting, shadows and scale of face in the image.

R-CNN [6] can also be used for precise face detection. It involves extracting the features from input, and then the extracted features from regions are classified by support vector machines, and these features are represented in low-dimensional space proficient of capturing the necessary information, but it cannot be implemented in real time. Fast R-CNN [7] overcomes the drawbacks of R-CNN, it identifies the region, and ROI pooling is done on the extracted features to ensure that the regions are of same size. Finally, the regions are feeded to fully connected layers which classify and return the bounding boxes. Both R-CNN and Fast R-CNN use the selective search algorithm, and this algorithm is slow and time-consuming.

Faster R-CNN [8] overcomes the problem of R-CNN and Fast R-CNN by replacing the selective search algorithm with a regional proposal network. First step involves extracting the features from the input source using ConvNet, then using RPN on extracted features to get the object proposals. Then finally, the proposals are classified and predict the bounding box for the input.

Another popular model for automated attendance is by using YOLOv3 [9]; it is a state-of-the-art algorithm for object detection across a variety of categories. It uses a

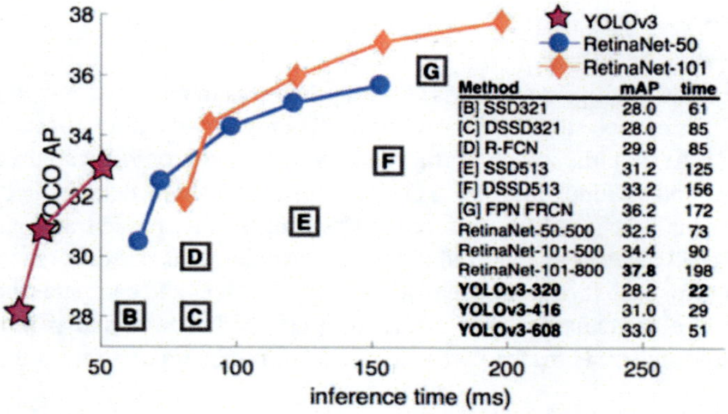

Fig. 1 Performance of YOLOv3 with COCO

modified version of the Resnet53 model called Darknet53 for feature extraction. This model improved a lot in its accuracy compared to its previous version YOLOv2 [10]. In addition, it added the ability to detect small objects which is ideal for attendance in the classroom as cameras will be placed far away from students. By adding residual blocks, skip connections and upsampling are a staple for any state-of-the-art model. YOLOv3-416 shows higher accuracy and speed as compared to SSD321 (see Fig. 1).

The existing system for automated attendance system uses Single-Shot Multi-Box (SSD) [11] detector for face detection and VGGFace2 [12] network for multi-class face recognition. SSD uses multi-scale feature and default box as it uses lower-resolution images which increase its speed. It can achieve 41.2% mAP at 59 FPS on COCO test-dev2015. Its predictions are defined as positive and negative matches and only use positive matches in calculating the cost. The network had good performance because it used multiple feature maps to get better prediction results per class and changing resolutions. But in the proposed system, we replaced the SSD model with the revamped YOLOv4 including numerous methods to improve speed and accuracy in the system and accuracy in the system.

Proposed system uses the You Only Look Once (YOLO) [13] model. The latest version of the object detector, YOLOv4, contains a base network and an assisting auxiliary network. The feature maps are created by the main base network. The feature maps are used by the auxiliary network to predict bounding boxes of three particular sizes. So, for multi-scale prediction, the auxiliary network consists of additional convolution layers for small-sized objects which was an issue encountered in the previous version of YOLO. Additionally, the locale proposition-based decoupled models are moderate and struggle accomplishing combination to a close ideal arrangement. Branched networks although highly accurate like the R-CNN family but cannot be used in real time as they are very slow and inefficient. To resolve this issue of the branch networks, single-layer networks are used like the YOLO and SSD which provide precision without compromising on the speed. The proposed system gives quick and precise outcomes when multiple objects are to be recognized.

3 Methodology

The goal of this intelligent system is to verify the faces in the classroom with the pre captured faces of each student in the database. There are two main steps of the system (see Fig. 2). At first, the image of the students in the classroom is captured through the camera and is input into the YOLOv4 network, which detects and localizes all the student's faces in the classroom. These faces are cropped and passed into the VGGFace2 network specialized in face verification. This network creates unique embeddings for each face using each face's unique features and compares the cropped faces with the pre-existing faces in the university's database, and if a match has occurred that student is marked present, and this process repeats for all the cropped faces.

It is quite possible that it is the most remarkable object detector almost on the verge of calling it as completely real time. It can achieve accuracy, at the level similar to Region-CNN or its improved Fast Region-CNN which are state of the art in terms of accuracy but lack in speed. Most networks either compromise on the accuracy or the speed. We can see that YOLOv4 is multiple times quicker than EfficientDet [14] (see Fig. 3), with similar generally speaking execution. Furthermore, average precision and frames per second (FPS) raised by 10% and 12% contrasted with YOLOv3.

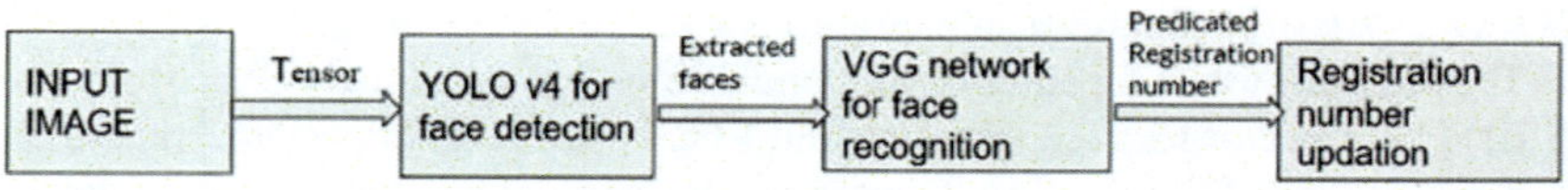

Fig. 2 Workflow of automatic attendance system

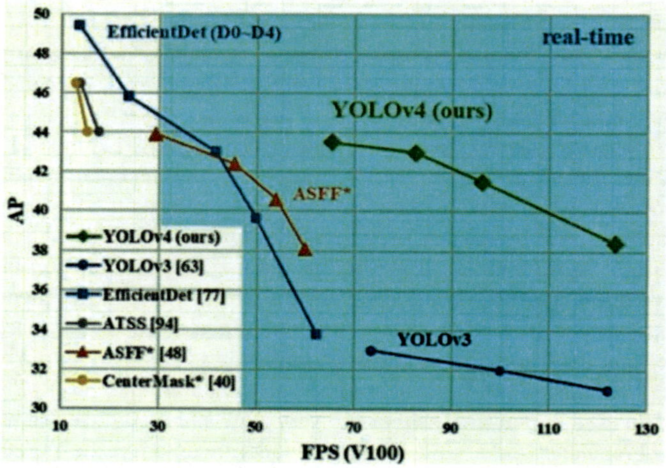

Fig. 3 Speed and accuracy of YOLOv4

The Bag of Freebies (BoF) improves the precision of the finder, without expanding the induction time. They just increment the preparation cost. Then again, the Bag of Specials (BoS) increment the induction cost, and they altogether improve the precision of the final bounding box. YOLOv4 has a mean average precision (mAP) estimation of 43.5% on the COCO dataset alongside an ongoing speed of 65 FPS on the Tesla V100, beating the quickest detectors regarding both speed and precision.

4 Results and Analysis

For the results (see Fig. 4), we can easily verify four of the main faces from the classroom, i.e., Caleb, Dustin, Finn and Will. YOLOv4 which gives a faster frames per second (FPS) speed in detection and very high accuracy. The network detects occluded, side faces, cut-faces and blurred faces well. This network was trained on

Fig. 4 Faces detected from classroom

Google Colab on 6000 images of human faces and validated on 1500 images. The dataset was created using Google's Open Images Dataset V6 detection of human faces and using random images from categories in which group photos were used [15]. After the face detection part, the faces of each student were cropped and saved as a new image in a new folder. Further, the four protagonist faces were verified if they were present in the classroom using VGGFace2 face recognition algorithm. The embeddings of the faces were compared using Cosine metric [16] for calculating the distance between the two compared Embeddings. The range is being from 0 to 1, where 0 being completely identical and 1 being no match. A threshold of 0.5 was kept, and below are examples of Caleb and Will (see Figs. 5 and 6).

We can see that the face detection is providing good results but the face recognition shows one false positive (see Table 1). It is not easy to differentiate the face of Will and Finn even for the human eye. The network fails to differentiate between them (see Figs. 6 and 8).

On comparing the dataset with its counterparts on a standard dataset such as WIDERFACE [17], it can be seen that even without a huge dataset of 6000 images and lower training time spanning through 4000 iterations, batch size of 64 and subdivision of 16 and learning rate 0.001, good precision is observed at an IOU threshold of 0.50 or 50% as shown in (see Table 2). Note that the FPS is higher for the YOLOv4 custom model since it has been inferred on a high-end GPU provided by Google

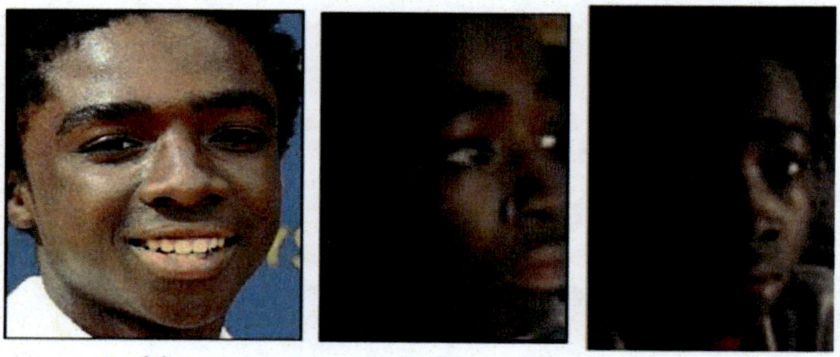

(a) (b) (c)

Fig. 5 **a–c** Caleb

Fig. 6 Will

Table 1 Face recognition results

Images	Distance between face embeddings	Face detection accuracy (%)
Figure 5a, b (Caleb)	0.38	96
Figure 5a, c (Caleb)	0.38s	99
Figures 6 and 7 (Will)	0.33	84
Figures 6 8 (Finn)—Mismatch	**0.26**	89

Fig. 7 Will

Fig. 8 Fin

Table 2 Overall analysis of different models on WIDERFACE dataset

Model	Inference time of the entire dataset(s)	FPS	No of images	mAP @ (0.5:0.95) (mean average precision)	GPU
YOLOv4 (custom)	99	30.3	3000	81.4% (mAP@0.5)	High-end GPU (Tesla K80)
YOLOv4	1329	12.1	16,067	–	Low-end single GPU
Tiny-YOLOv4	533	30.1	16,067	54%	Low-end single GPU
SwiftFace	407	39.5	16,067	51%	Low-end single GPU

Colab (Tesla K80). On lower-end edge devices such as Raspberry Pi and Nvidia Jetson, a striped down version of YOLOv4 which are the tiny-YOLOv4 [18] and SwiftFace [19] would be more efficient though the accuracy will drop down.

5 Conclusion

Deep learning has advanced to a level where we can substitute the manual attendance with automatic camera-based detection. Models are being updated frequently with higher speed and accuracy. Most models either compromise with speed or accuracy. With the development of YOLOv4, the updated version from previous YOLOv3 speed and accuracy have further increased by 10% and 12% respectively, becoming the current state of the art. When the captured faces are propagated to the VGGFace2 network, one of the most robust facial data recognizers, it is possible to recognize the faces in a classroom by matching pre-existing facial data in the database. Each student with their unique roll no. can be used to update the attendance of that student in the university's database when connected to a server. Overall, the proposed model performs well with a mAP of 81.4% without needing much training time and large datasets, but we found one false positive case in the recognition part. It was also noted that if limited computation is available but speed is a concern and accuracy can be compromised, then we have lighter models such as the tinyYOLOv4 and the SwiftFace. The limitation of this work is that accuracy is compromised since training time, and dataset is small. A higher accuracy can be achieved if trained on the WIDERFACE dataset. Also, the system cannot differentiate between facially very similar faces which are difficult even for the human eye. Further, the hardware edge device required for this is still expensive and does not reach the low-cost category.

6 Future Scope

In future works, a more advanced face recognition network can be used to improve upon the recognition results and be robust enough to recognize masked faces. We can also improve the accuracy of the model by using a large standard dataset and optimize the model for a lower-end hardware by using scaled down version of YOLOv4 models such as SwiftFace and tiny-YOLOv4. Working on lower-end development boards will help in developing a low-cost system ready to communicate attendance automatically to the institution's database.

Acknowledgements This project was supported by Delhi Technological University, New Delhi, India.

References

1. Athanesious, J.J., Adithya, S., Bhardwaj, C.A., Lamba, J.S., Vaidehi, A.V.: Deep learning based automated attendance system. Procedia Comput. Sci. **165**, 307–31 (2019). ISSN 1877-0509, https://doi.org/10.1016/j.procs.2020.01.04
2. Benuwa, B., Zhan, Y., Ghansah, B., Wornyo, D., Banaseka, F.: A review of deep machine learning. Int. J. Eng. Res. Afr. **24**, 124–136 (2016). https://doi.org/10.4028/www.scientific.net/JERA.24.124
3. Bochkovskiy, A., Wang, C.-Y., Liao, H.-y.: YOLOv4: Optimal Speed and Accuracy of Object Detection (2020)
4. Alrahhal, M., Bazi, Y., Abdullah, T., Mekhalfi, M., Alhichri, H., Zuair, M.: Learning a multi-branch neural network from multiple sources for knowledge adaptation in remote sensing imagery. Remote Sensing **10**, 1890 (2018). https://doi.org/10.3390/rs10121890
5. Tamimi, A., Al-Allaf, O., Alia, M.: Eigen faces and principal component analysis for face recognition systems: a comparative study. Int. J. Comput. Technol. **14**, 5650–5660 (2015). https://doi.org/10.24297/ijct.v14i4.1967
6. Girshick, R.B., et al.: Rich feature hierarchies for accurate object detection and semantic segmentation. In: 2014 IEEE Conference on Computer Vision and Pattern Recognition, pp. 580–587 (2014)
7. Fast R-CNN: 2015 IEEE International Conference on Computer Vision (ICCV), Santiago, pp. 1440–1448 (2015). https://doi.org/10.1109/ICCV.2015.169
8. Sun, X., Wu, P., Hoi, S.C.H.: Face detection using deep learning: an improved faster RCNN approach. Neurocomputing **299**, 42–50 (2018). ISSN0925-2312. https://doi.org/10.1016/j.neucom.2018.03.030. (https://www.sciencedirect.com/science/article/pii/S0925231218303229)
9. Redmon, J., Farhadi, A.: YOLOv3: An Incremental Improvement (2018)
10. Redmon, J., Farhadi, A.: YOLO9000: better, faster, stronger. In: 2017 IEEE Conference on Computer Vision and Pattern Recognition (CVPR), 6517–6525 (2017)
11. Liu, W., Anguelov, D., Erhan, D., Szegedy, C., Reed, S., Fu, C.-Y., Berg, A.: SSD: Single Shot MultiBox Detector, vol. 9905, pp. 21–37 (2016). https://doi.org/10.1007/978-3-319-46448-0_2
12. Cao, Q., Shen, L., Xie, W., Parkhi, O., Zisserman, A.: VGGFace2: A Dataset for Recognising Faces across Pose and Age, 67–74 (2018). https://doi.org/10.1109/FG.2018.00020
13. Ahmad, T., Ma, Y., Yahya, M., Ahmad, B., Nazir, S., Haq, A., Ali, R.: Object Detection through Modified YOLO Neural Network. Scientific Programming (2020). https://doi.org/10.1155/2020/8403262
14. Tan, M., Pang, R., Le, Q.V.: EfficientDet: scalable and efficient object detection. In: 2020 IEEE/CVF Conference on Computer Vision and Pattern Recognition (CVPR), Seattle, WA, USA, 2020, pp. 10778–10787. https://doi.org/10.1109/CVPR42600.2020.01079
15. Open Images Dataset V6. https://storage.googleapis.com/openimages/web/visualizer/index.html?set=train&type=detection&c=%2Fm%2F0dzct
16. Nguyen, H., Bai, L.: Cosine similarity metric learning for face verification. ACCV **6493**, 709–720 (2010). https://doi.org/10.1007/978-3-642-19309-5_55
17. Yang, S., Luo, P., Loy, C.C., Tang, X.: Wider face: A face detection benchmark, Apr 2016. [Online]. Available: http://shuoyang1213.me/WIDERFACE/
18. Jiang, Z., et al.: Real-time object detection method based on improved YOLOv4-tiny. ArXiv abs/2011.04244 (2020): n. pag
19. Ramos, L., Morales, B.: SwiftFace: Real-Time Face Detection. ArXiv abs/2009.13743 (2020): n. pag
20. Redmon, J., Divvala, S., Girshick, R., Farhadi, A.: You only look once: unified, real-time object detection. In Proceedings of the IEEE Conference on Computer Vision and Pattern Recognition, pp. 779–788 (2016)
21. Stranger Things, *Imdb*, 1234 (2016)

Multi-fruit Classification Using a New FruitNet-11 Based on Deep Convolutional Neural Network

Raghavendra and Satishkumar Mallappa

1 Introduction

In India, fruit consumption is increased rapidly in the time of pandemic of COVID-19. India is popular for its variety of fruit production. According to the statistical records published by Statista Research Department, October 16, 2020, the overall production of fruit was approximately more than 98 million metric tons in the fiscal year 2019. Such fruits include bananas, mangos, papayas, and many other fruits. In the time of the COVID-19 outbreak, many doctors have suggested fruits for all in their diets to increase the immunity in the body to fight against the novel coronavirus. Now it is time to choose the right and good fruits for good health, by seeing the color, shape, and texture of the fruits is made easy to customers to choose the fruit. It is difficult in the automation to select the fruits by their appearance for the fruits vendors. The classification of fruits is a challenging task [1–3]. In this pipeline, this proposed system is useful in the classification of fruits. Considering the tremendous success of CNN, it is attracted many researchers to utilize its features to develop a robust system for any classification problem. Hence, we have designed an algorithm named FruitNet-11 based on deep convolutional neural network (DCNN) modalities.

Raghavendra
Department of Computer Science, University of Horticultural Sciences, Bagalkot, Karnataka, India

S. Mallappa (✉)
Department of Computer Science, Gulbarga University, Kalaburagi, Karnataka, India

M. Dua et al. (eds.), *Proceedings of the International Conference on Paradigms of Communication, Computing and Data Sciences*, Algorithms for Intelligent Systems,
https://doi.org/10.1007/978-981-16-5747-4_53

2 Literature Review

In the literature, many papers are found in connection with fruit classification. The classification of fruits using wavelet transformation is mentioned in [4], and they have used apple and banana fruits for the classification and obtained encouraging recognition accuracy. The work mentioned in [5], is based on the color and GLCM features using KNN and SVM classifiers, they have extracted features from four types of fruits. The geometrical features-based fruit classification is given in [6], and they have used multiple classification methods to yield the highest recognition accuracy, and from the multiclass SVM, they got the maximum of 87.5 and 91.67% recognition accuracy. By using the machine vision, the fruit classification review is proposed in [7], and they have employed many feature extraction methods viz, HOG, SURF, LBP, and for the classification, they have taken KNN, ANN, CNN, and SVM classifiers. The date fruit classification is given in [8], and they have focused on the date fruit only they have extracted the color, shape, size, and texture features from the date images, and they have achieved 99% accuracy from the ANN classifier. In [9], they have discussed some approaches-related approaches for the fruit detection and classification of the disease of fruits. The model of a neural network is mentioned in [10], based on the color of the surface of fruits they have classified the fruits and got the 92% recognition accuracy. The long type watermelon classification is proposed in [11], and they have studied the fruit shapes for the classification. The following Fig. 1 shows the schematic diagram of the proposed method.

Fig. 1 Sample images of fruits. (i) Apple, (ii) Banana, (iii) Cherry, (iv) Grapes, (v) Grapes Blue, (vi) Lemon, (vii) Mango, (viii) Orange, (ix) Papaya, (x) Pomgrenete, (xi) Watermelon

Table 1 Details of dataset used for the experiment

S. No.	Fruit name	No. of images
1	Apple	1000
2	Banana	1000
3	Cherry	1000
4	Grape	1000
5	Grape blue	1312
6	Lemon	656
7	Mango	1000
8	Orange	639
9	Papaya	656
10	Pomegranate	656
11	Watermelon	632
Total		9551

3 Dataset Details

The standard dataset fruits-360 [12, 13] is considered for the proposed experiment. From that dataset, the 11 different types of fruits viz., Apple, Banana, Cherry, Grape, Grape blue, Lemon, Mango, Orange, Papaya, Pomegranate, and Watermelon. Following Table 1 shows the details of the dataset.

4 Proposed Method

For the problem of fruit classification, it is necessary to develop an model, which have the ability to classify given set of fruit images. In this process, the proposed method has given the efficient and reliable result.

Following Fig. 2 shows the block diagram of the proposed model.

The following Table 2 shows the details of layers with parameters used to design a FruitNet-11 model.

The layers mentioned in the above table have been taken to design the proposed model. In the selection of layers, the input parameters were fixed empirically. By considering few layers, the proposed model was developed.

The proposed model has a total of 20 layers (1 input and 1 output layer with 18 hidden layers) viz., convolution2d, Batchnormalization, Relu, Maxpooling, Group-convolution, Softmax, Fullyconnected, and Classificationoutput layers. The first and most important layer is the convolution layer, and it uses an input image and applies the 2×2 convolution mask, and it is sliding over the input image. From the convolution layer, the meaningful features (border, corner, blobs, etc.) are extracted. Next the batch normalization layer this layer normalizes the output values, which are given

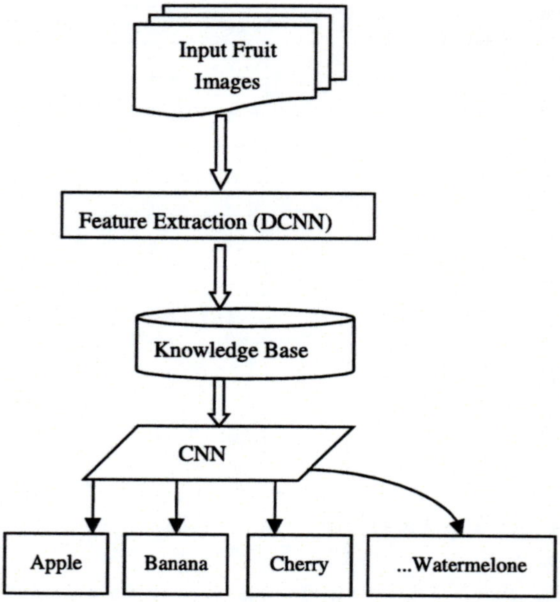

Fig. 2 Block diagram of the proposed method

by the previous layer. It standardizes the input values of the network. After this layer, the Relu layer has been utilized, and this layer takes many inputs from other nodes and produced one output to the next node. It normalizes the result between 0 and 1. Next, the maxpooling layer is considered, and this layer performs downsampling of the spatial size and removing the redundant spatial information. In this section, the 2×2 sized mask is applied on the feature map and takes the max value from the mask, and makes another normalized feature map. To recognize the larger patterns, the fully connected layer is used, and it combines all the features learned in previous layers all over the feature map. The feature vector of the previous layer is transformed into a value between 0 and 1, and the value may be negative, positive, or greater than one this is done by the softmax layer. The final layer is the classification output layer gives the output of the proposed model. For multiclass classification problems with mutually exclusive classes, a classification layer computes the cross-entropy loss.

5 Experimental Results and Discussion

The proposed experiment is carried out on a dataset containing 9551 fruit images belongs to 11 classes. The FruitNet-11 model obtained efficient results the DCNN features proved their robustness.

Following Table 3 experimental results were obtained from the proposed method (Table 4 and Fig. 3).

Table 2 Layers used in the proposed FruitNET-11 model

S. No.	Layer name	Type	Activations	Learnables
1	Imageinput (100 × 100 × 3) images with zerocenter	Image input	100 × 100 × 3	–
2	Conv_1 8 3 × 3 with stride[1 1] 'same' padding	Convolution	100 × 100 × 8	Weights 3 × 3 × 3 × 8 Bias 1 × 1 × 8
3	Batchnorm_1 With 8 channels	Batch normalization	100 × 100 × 8	Offset 1 × 1 × 8 Scale 1 × 1 × 8
4	relu_1	ReLU	100 × 100 × 8	–
5	Maxpool_1 2 × 2 maxpooling with stride [2 2], [0 0 0 0] padding	Maxpooling	50 × 50 × 8	
6	Conv_2 16 3 × 3 × 8 with stride[1 1] 'same' padding	Convolution	50 × 50 × 16	Weights 3 × 3 × 8 × 16 Bias 1 × 1 × 16
7	Batchnorm_2 With 16 channels	Batch normalization	50 × 50 × 8	Offset 1 × 1 × 16 Scale 1 × 1 × 16
8	relu_2	ReLU	50 × 50 × 16	
9	Maxpool_2 2 × 2 maxpooling with stride [2 2], [0 0 0 0] padding	Maxpooling	25 × 25 × 16	
10	grouped_conv 16 groups of 32 3 × 3 × 1 convolutions, [2 2] stride and padding 'same'	Grouped convolution	13 × 13 × 512	Offset 3 × 3 × 1 × 32 × 16 Scale 1 × 1 × 32 × 16
11	Batchnorm_3 With 512 channels	Batch normalization	13 × 13 × 512	Offset 1 × 1 × 512 Scale 1 × 1 × 512
12	relu_3	ReLU	13 × 13 × 512	
13	Maxpool_3 2 × 2 maxpooling with stride [2 2], [0 0 0 0] padding	Maxpooling	6 × 6 × 512	
14	Conv_3 643 × 3 × 512 with stride[1 1] 'same' padding	Convolution	50 × 50 × 16	Weights 3 × 3 × 8 × 16 Bias 1 × 1 × 16
15	Batchnorm_4 With 64 channels	Batch normalization	6 × 6 × 64	Offset 1 × 1 × 64 Scale 1 × 1 × 64

(continued)

Table 2 (continued)

S. No.	Layer name	Type	Activations	Learnables
16	relu_4	ReLU	6 × 6 × 64	
17	Maxpool_3 2 × 2 maxpooling with stride [2 2], [0 0 0 0] padding	Maxpooling	3 × 3 × 64	
18	fc 11 fully connected layer	Fully connected	1 × 1 × 11	
19	softmax	Softmax	1 × 1 × 11	
20	classoutput	Classification output		

Table 3 Average recognition accuracy obtained from FruitNet-11 using DL

Method	Recognition accuracy (%)
FruitNet-11	96.15%

Table 4 Confusion matrix obtained from DL method on a validation set

Fruits	A	B	C	G	GB	L	M	O	P	PM	W	Rec. Acc. (%)
A	**299**	0	0	0	0	0	0	0	0	0	0	100
B	0	**300**	0	0	0	0	0	0	0	0	0	100
C	0	0	**269**	0	0	0	0	0	0	0	0	90
G	0	0	0	**211**	0	0	0	0	0	0	0	70
GB	0	0	16	87	**394**	0	0	0	0	0	0	100
L	0	0	0	0	0	**197**	0	0	0	0	0	100
M	0	0	0	0	0	0	**300**	0	0	0	0	100
O	0	0	0	0	0	0	0	**192**	0	0	0	100
P	0	0	0	0	0	0	0	0	**197**	0	0	100
PM	1	0	0	2	0	0	0	0	0	**193**	0	98
W	0	0	0	0	0	0	0	0	0	0	**190**	100
Average recognition												**96.15**

Note A apple, *B* banana, *C* cherry, *G* grape, *GB* grape_blue, *L* lemon, *M* mango, *O* orange, *P* papaya, *PM* pomgranate, *W* watermelon

5.1 Comparative Analysis

To show the robustness of the proposed model, we have considered the popular pre-trained CNN model; alexnet. We have applied our fruit dataset to both models viz., alexnet and proposed FruitNet-11 model with the same input parameters. Following Table 5 shows the comparative analysis between alexnet and FruitNet-11 models.

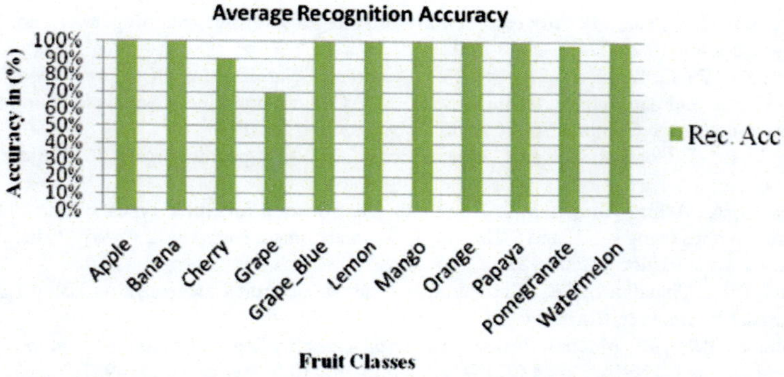

Fig. 3 Average recognition accuracy of 11 types of fruit images using FruitNet-11

Table 5 Comparative analysis of proposed model with other popular model

Model	Epochs	Learning rate	Mini batch size	Recognition accuracy (%)
Alexnet	25	0.0003	128	87.36
Proposed (FruitNet-11)	25	0.0003	128	**96.15**

From the above table, it is clear that our proposed model has obtained the highest recognition accuracy over the alexnet. In both the model same dataset has used, and some important parameters are also set same. In the experimental setup, the proposed FruitNet-11 model has outperformed over the alexnet model.

6 Conclusion and Future Work

This proposed model FruitNet-11 can classify different fruit qualities based on texture, color, shape, and size using the DCNN method. The new model has been developed based on the DCNN method. The model has obtained a significantly high recognition accuracy of 96.15%. The model consists of 20 layers. This FruitNet-11 model is compared with another popular method; alexnet and the highest recognition result is obtained from the proposed model. Hence, it shows the superiority of the proposed model. In future work, we put effort to increase the dataset size and make an attempt to design a new method that can identify the internal defect in the fruits. Capturing the internal defect by passing rays belongs to different bands.

References

1. Nayak, M.A.M.: Fruit Recognition using Image Processing, vol. 7, no. 08, pp. 1–6 (2019)

2. Mureşan, H., Oltean, M.: Fruit recognition from images using deep learning. arXiv, no. (2017). https://doi.org/10.2478/ausi-2018-0002
3. Mercol, J.P., Gambini, J., Santos, J.M.: Automatic classification of oranges using image processing and data mining techniques. In: XIV Congr. Argentino Ciencias la Comput. XIV Argentine Congr. Comput. Sci. (CACIC 2008), pp. 1–12 (2008)
4. Pl, C.: Int. J. Comput. Sci. Eng. Open Access (2019). https://doi.org/10.26438/ijcse/v7si5.131135
5. Nosseir, A., Ashraf Ahmed, S.E.: Automatic classification for fruits' types and identification of rotten ones using k-NN and SVM. Int. J. Online Biomed. Eng. **15**(3), 47–61 (2019). https://doi.org/10.3991/ijoe.v15i03.9832
6. Patel, C.C., Chaudhari, V.K.: Comparative Analysis of Fruit Categorization Using Different Classifiers. Springer, Singapore
7. Naik, S., Patel, B.: Machine vision based fruit classification and grading—a review. Int. J. Comput. Appl. **170**(9), 22–34 (2017). https://doi.org/10.5120/ijca2017914937
8. Haidar, A., Dong, H., Mavridis, N.: Image-based date fruit classification. Int. Congr. Ultra Mod. Telecommun. Control Syst. Work. 357–363 (2012). https://doi.org/10.1109/ICUMT.2012.6459693
9. Barot, Z.R., Limbad, N.: An approach for detection and classification of fruit disease: a survey. Int. J. Sci. Res. **4**(12), 838–842 (2015). https://doi.org/10.21275/v4i12.8121502
10. Hambali, H.A., Abdullah, S.L.S., Jamil, N., Harun, H.: Fruit classification using neural network model. J. Telecommun. Electron. Comput. Eng. **9**(1–2), 43–46 (2017)
11. Sadrnia, H., Rajabipour, A., Jafary, A., Javadi, A., Mostofi, Y.: Classification and analysis of fruit shapes in long type watermelon using image processing. Int. J. Agric. Biol **1**(9), 68–70 (2007)
12. Sakib, S., Ashrafi, Z.: Implementation of Fruits Recognition Classifier using Convolutional Neural Network Algorithm for Observation of Accuracies for Various Hidden Layers, pp. 8–11 (1980)
13. Dubey, S.R., Jalal, A.S.: Application of image processing in fruit and vegetable analysis: a review **24**(4), 405–424 (2015). https://doi.org/10.1515/jisys-2014-0079

Breast Cancer Prediction Using Intuitionistic Fuzzy Set with Analytical Hierarchy Process with Delphi Method

S. Rajaprakash, R. Jaichandaran, and S. Muthuselvan

1 Introduction

Breast cancer growth has now overwhelmed cellular breakdown in the lungs as the world's generally normally analyzed disease, as indicated by insights delivered by the International Agency for Research on Cancer (IARC) in December 2020.

So, on World Cancer Day, WHO will have the first of a progression of discussions to set up another worldwide bosom malignancy activity, which will dispatch later in 2021. This community-oriented exertion between WHO, IARC, the International Atomic Energy Agency, and other multi-sectorial accomplices will diminish passing from bosom malignancy by advancing bosom wellbeing, improving ideal disease identification and guaranteeing admittance to quality consideration.

WHO and the malignancy local area are reacting with restored desperation to deliver bosom disease and to react to the developing malignant growth trouble around the world that is stressing people, networks, and wellbeing frameworks?

In the previous twenty years, the general number of individuals determined to have malignancy almost multiplied, from an expected 10 million of every 2000 to 19.3 million out of 2020. Today, one out of 5 individuals worldwide will create malignant growth during their lifetime. Projections recommend that the quantity of individuals being determined to have malignancy will expand even further in the coming years and will be almost half higher in 2040 than in 2020. The number of deaths from cancer has also increased, from 6.2 million in 2000 to 10 million in 2020. More than one out of every six deaths are due to cancer.

While changes in way of life, for example, unfortunate weight control plans, deficient active work, utilization of tobacco, and destructive utilization of liquor have all added to the expanding malignant growth trouble, and a huge extent can likewise

S. Rajaprakash (✉) · R. Jaichandaran · S. Muthuselvan
Aarupadai Veedu Institute of Technology, Vinayaka Mission Research Foundation, Salem, Tamil Nadu, India

M. Dua et al. (eds.), *Proceedings of the International Conference on Paradigms of Communication, Computing and Data Sciences*, Algorithms for Intelligent Systems,
https://doi.org/10.1007/978-981-16-5747-4_54

be ascribed to expanding life span, as the danger of creating disease increments with age. This fortifies the need to put resources into both malignant growth anticipation and disease control, zeroing in on significant tumors like bosom, cervical, and youth diseases.

2 Intuitionistic Fuzzy Set

Intuitionistic fuzzy set (IFS) based on the two functions, like membership and non-membership functions. A be the subset of the universal set *U*, and it is defined by $A^* = \{\langle x, 0 \le \mu_A(x) + \nu_A(x) \le 1\rangle \,|x \in U\}$ where $0 \le \mu_A(x)+\nu_A(x) \le 1$. The functions membership and non-membership are $\mu_A(x){:}U \to [0, 1]$ and $\nu_A(x){:}\ U \to [0, 1]$. $\pi(x) = 1 - \mu_A(x) + \nu_A(x)$ represent the degree of indeterminacy or hesitation degree. Where $\pi(x) : U \to [0, 1]$.

- Triangular Intuitionistic Fuzzy Number

A triangular intuitionistic fuzzy number *A*(*x*) is an intuitionistic fuzzy set in *R* with membership function and non-membership function as given below.

$$\mu_A(x) = \begin{cases} \frac{x-a_1}{a_2-a_1} & \text{for } a_1 \le x \le a_2 \\ \frac{a_3-x}{a_3-a_2} & \text{for } a_2 \le x \le a_3 \\ 0 & \text{otherwise} \end{cases} \tag{1}$$

$$\nu_A(x) = \begin{cases} \frac{(a_2-x)}{a_2-a_1'} & \text{for } a_1' \le x \le a_2 \\ \frac{x-a_2}{\beta'} & \text{for } a_2 \le x \le a_3' \\ 1 & \text{otherwise} \end{cases} \tag{2}$$

3 Past Work

The notion of intuitionistic fuzzy sets (IFS) as a generalization of fuzzy sets was introduced by Atanassov in the year of 1986 [1]. The scholar has added in the definition of fuzzy set new component which determines the degree of non-membership. In the year 1996, Burillo et al. [2, 3] have learned around two hypotheses in intuitionistic fuzzy set, the hypotheses which permits to build an intuitionistic fuzzy set from two fuzzy sets. The creators likewise talked about the fuzzy sets utilized in the development from the intuitionistic fuzzy sets. The new meaning of distance between two intuitionistic fuzzy sets from separated from that Atanassov concentrated by Szmidt et al. [4]. Intuitionistic fuzzy pair, intuitionistic fuzzy couple, and intuitionistic fuzzy worth have been concentrated by Atanassov et al. [4] in 2013. Intuitionistic Fuzzy Delphi Method was utilized as anticipating instrument dependent on master's proposal. They utilized three-sided fuzzy number and total cycle dependent on the assessment of the master proposed by Kumar et al. [5] in 2012. The thought of intuitionistic fuzzy *t*-standard and *t*-conorm is found in 2002 by Deschrijver et al. [6] intuitionistic fuzzy analytic hierarchy process was carried out in positioning the four options in human resources assessment. The novel thought was a combination of two-sided assessment of intuitionistic fuzzy sets with pair-wise examination of scientific analytic hierarchy process by Abdullah et al. [7]. In 2007, Xu proposed the intuitionistic fuzzy three-sided standards. Utilizing the intuitionistic fuzzy sets as an instrument finding of sickness D dependent on the manifestations S was concentrated by Szmidt et al. [8, 9] in 2001. In this work, the creator utilized arrangement of side effects as a data set [10]. A non-probabilistic sort entropy measure for intuitionistic fuzzy set was proposed by Eulalia et al. (2000), and in this work, the measure is reliable with customary fuzzy sets. Utilizing the intuitionistic fuzzy rationale, an intuitionistic fuzzy framework was created to control the warmer fans. In this work, the speed of the radiator fan is determined utilizing intuitionistic fuzzy principles applied in an induction motor utilizing defuzzification strategy by Akram et al. [4, 11]. In the year 2013, Rajaprakash et al. [12] discussed about fuzzy analytical hierarchy process over the stress factors of Indian school students. In the year 2014, S. Rajaprakash et al. [13] have been study about ranking the customer satisfaction factors in the automobile sectors, using intuitionistic fuzzy analytical hierarchy process. In the year 2015, Rajaprakash et al. [14] study above the customer requirement key factors is ranked in many levels in the automobile sector using the intuitionistic fuzzy analytical hierarchy process. In the year 2009, Sadiq et al. [15] proposed IF-AHP methodology was implemented with an illustrative example to particular best drilling fluid (mud) for drilling operations under multiple environmental criteria. In 2016, business and balance scorecard important are ranked to increase the production in an organization using intuitionistic fuzzy analytical hierarchy process by Rajaprakash et al. [16] Fuzzy logic is the extension of set theory replacing the binary truth values. It was described by Zadeh L. A., at the University of California, Berkeley [17, 18]. The analysis of breast cancer includes a few degrees of vulnerability which shows itself in an unexpected way, contingent upon the patient and general climate and force [19]. Intuitionistic fuzzy set and rough set in combination with statistical feature extraction techniques over the cancer diagnosis with medical images by Chowdhary [2]. In the year 2016, assigns the computerized execution of a model, in view of an intuitionistic fuzzy histogram and possibility fuzzy c-mean bunching calculation for early bosom malignant growth identification. Bunching assumes a vital part in division section by Chowdhary [20].

4 Methodology

In this work on IF-AHP, triangular membership function is used. It is a combination of intuitionistic fuzzy set and AHP. The intuitionistic fuzzy set theory gives very good output over the vagueness compared to the FAHP.

The following steps involved in IF-AHP with fuzzy Delphi method.

Step 1: Fuzzy Delphi Method

- Questions are framed based on the expert's suggestion on the relevant area.
- Take the opinion from the experts for the questionnaire. Convert the value into membership and non-membership values.
- Convert the linguistic value convert the value into membership and non-membership value.
- Mean of the linguistic value calculated, the variation between experts' value and mean of each value calculated.
- The difference is high then question is sent again to expert for revaluation. It will be continuing until two successive are very less else expert satisfied.

Step 2: Comparison Matrix

Comparison table or matrix are framed with the final value of Delphi method.

Step 3: Relation preference of Intuitionistic fuzzy set

To check the consistency utilizing the intuitionistic inclination connection as per Xu et al. [19], the steady stretch fuzzy intuitionistic relations are follows.

$R = (M_{ik})_{n\times n}$ with $M_{ik} = (\mu_{ik}, \nu_{ik})$ $(i, k = 1, 2, 3, \ldots, n)$ is multiplicative consistent if

$$\mu_{ik} = \begin{cases} 0, & \text{if } (\mu_{it}, \mu_{tk}) \in \{(0,1), (1,0)\} \\ \frac{\mu_{it}\mu_{tk}}{\mu_{it}, \mu_{tk} + (1-\mu_{it})(1-\mu_{tk})}, & \text{otherwise} \end{cases} \tag{3}$$

$$\nu_{ik} = \begin{cases} 0, & \text{if } (\nu_{it}, \nu_{tk}) \in \{(0,1), (1,0)\} \\ \frac{\nu_{it}\nu_{tk}}{\nu_{it}, \nu_{tk} + (1-\nu_{it})(1-\nu_{tk})}, & \text{otherwise} \end{cases}$$

Xia and Xu have proved the fuzzy preference relation.

Theorem: [*Xe*] In the fuzzy preference relation, the following statement are equivalent:

$$b_{ik} = \frac{\sqrt[n]{\prod_{s=1}^{n} b_{is}b_{sk}}}{\sqrt[n]{\prod_{s=1}^{n} b_{is}b_{sk}} + \sqrt[n]{\prod_{s=1}^{n} (1-b_{is})(1-b_{sk})}} \quad i, k = 1, 2, 3, \ldots, n \tag{4}$$

Based on the above result and theorem of zeshuri Xu et al. [19] in the "intuitionistic fuzzy analytic hierarchy process" developed two algorithms. But used only one algorithm. Based on that algorithm, we can have the following formula.

For $k > i + 1$, and let $\overline{M}_{ik} = \left(\overline{\mu}_{ik}, \overline{\nu}_{tk}\right)$ where

$$\overline{\mu}_{ik} = \frac{\sqrt[k-i-1]{\prod_{t=i+1}^{k-1} \mu_{it}\mu_{tk}}}{\sqrt[k-i-1]{\prod_{t=i+1}^{k-1} \mu_{it}\mu_{tk}} + \sqrt[k-i-1]{\prod_{t=i+1}^{k-1} (1-\mu_{it})(1-\mu_{tk})}} \quad k > i+1 \tag{5}$$

$$\overline{v}_{ik} = \frac{\sqrt[k-i-1]{\prod_{t=i+1}^{k-1} v_{it}v_{tk}}}{\sqrt[k-i-1]{\prod_{t=i+1}^{k-1} v_{it}v_{tk}} + \sqrt[k-i-1]{\prod_{t=i+1}^{k-1} (1-v_{it})(1-v_{tk})}} \quad k > i+1 \tag{6}$$

And for $k = i + 1$ let $\overline{M}_{ik} = M_{ik}$ and $k < i$ let $\overline{M}_{ik} = \left(\overline{v}_{ki}, \overline{\mu}_{ki}\right)$.

Multiplicative consistent intuitionistic relation is obtained from the above Equations, the lower triangular elements of the matrix.

Step 4: Consistence Check

The distance between intuitionistic relations Xu [19] is calculated by the following formula.

$$d\left(\overline{M}, M\right) = \frac{1}{2(n-1)(n-2)} \sum_{i=1}^{n} \sum_{k=1}^{n} \left(|\overline{\mu}_{ik} - \mu_{ik}| + ||\overline{v}_{ik} - v_{ik}| + ||\overline{\pi}_{ik} - \pi_{ik}|\right) \tag{7}$$

$$d\left(\overline{M}, M\right) < \tau \tag{8}$$

If Eq. (8) less than or equal to threshold value, then the relation matrix is consistent in the intuitionistic relation. Here τ is the threshold for the consistency. If Eq. (8) is not satisfied, then intuitionistic preference relation is not consistent then go to the step 1.

Step 5: Calculation of weight

The priority of the intuitionistic preference relation (zeshuri Xu) is calculated by using the following.

$$w_i = \frac{\sum_{k=1}^{n} M_{ik}^{1}}{\sum_{i=1}^{n} \sum_{k=1}^{n} M_{ik}^{1}} = \frac{\sum_{k=1}^{n} [\mu_{ik}, 1 - v_{ik}]}{\sum_{i=1}^{n} \sum_{k=1}^{n} [\mu_{ik}, 1 - v_{ik}]}$$

$$w_i = \left(\frac{\sum_{k=1}^{n} [\mu_{ik}]}{\sum_{i=1}^{n} \sum_{k=1}^{n} [1 - v_{ik}]}, 1 - \frac{\sum_{k=1}^{n} [1 - v_{ik}]}{\sum_{i=1}^{n} \sum_{k=1}^{n} [\mu_{ik}]} \right) \tag{9}$$

According to the Szmidt and Kacprzyk [10, 15], a function in mathematical form

$$\rho(\alpha) = 0.5(1 + \pi_{\alpha})(1 + \mu_{\alpha}) \tag{10}$$

Step 6: Preference Ranking

After calculating the weight of Eq. (10), the value $\rho(\alpha)$ is calculated by using the above Eq. (10). Then, based on the $\rho(\alpha)$ value, one can easily order the preference (Rank) attributes.

5 Implementation

Breast cancer has eleven attributes $A1, A2, A3, A4, A5, A6, A7, A8, A9, A10, A11$ which represent radius, texture, perimeter, area, smoothness, compactness, concavity, concave points, symmetry, fractal dimension, and age of the patients based on the doctors and pathologist suggestion. The initial Intuitionistic fuzzy comparison matrix is given (Table 1).

To check the consistency, generate the intuitionistic preference relation matrix using Eq. (1). Matrix $(\overline{M})$ is framed (Table 2).

Difference between intuitionistic fuzzy values using the above two matrices by Eq. (5). The $d(\overline{M}, M) = 0.0935$. Which is less than $\tau = 0.1$. Therefore, the above matrix is consistent. Weight of the attributes are calculated using (7) (Table 3).

Using Eq. (8), the weight $\rho(\alpha)$ is calculated and ranking the attribute based on the calculated values (Table 4).

6 Comparison with Existing Work

In the existing work (Xu, Z., Liao, H) with our proposed work IF-AHP with fuzzy Delphi method, we can observe the following.

- The deviation is very less in the proposed system compare with existing system. So, we can observe that proposed work is consistent.

In the existing system, the output of data are inconsistent; but in the proposed system, the output data are consistent, and it is observed from Fig. 1.

7 Result

In the present work, five experts are accepted from different hospitals in Tamil Nadu. Using the online survey questionnaires are designed and sent to the experts to generate

Table 1 Comparison matrix-I

M	*A*1		*A*2		*A*3		*A*4		*A*5	
	μ	ν	μ	ν	μ	ν	μ	ν	μ	ν
*A*1	0.5	0.5	0.5	0.5	0.5	0.5	0.4	0.5	0.4	0.5
*A*2	0.5	0.5	0.5	0.5	0.5	0.4	0.5	0.5	0.4	0.5
*A*3	0.5	0.5	0.4	0.5	0.5	0.5	0.4	0.5	0.4	0.5
*A*4	0.5	0.4	0.5	0.5	0.5	0.4	0.5	0.5	0.4	0.5
*A*5	0.5	0.4	0.5	0.4	0.5	0.4	0.5	0.4	0.5	0.5
*A*6	0.6	0.4	0.5	0.5	0.5	0.4	0.5	0.5	0.5	0.5
*A*7	0.5	0.4	0.5	0.4	0.5	0.5	0.5	0.5	0.5	0.5
*A*8	0.7	0.2	0.7	0.3	0.7	0.3	0.6	0.3	0.5	0.4
*A*9	0.8	0.2	0.8	0.1	0.8	0.2	0.7	0.3	0.5	0.4
*A*10	0.5	0.5	0.5	0.5	0.5	0.4	0.5	0.5	0.5	0.5
*A*11	0.5	0.4	0.5	0.5	0.6	0.4	0.5	0.5	0.5	0.5

*A*6		*A*7		*A*8		*A*9		*A*10		*A*11	
μ	ν	μ	ν	μ	ν	μ	ν	μ	ν	μ	ν
0.4	0.6	0.4	0.5	0.2	0.7	0.2	0.8	0.5	0.5	0.4	0.5
0.5	0.5	0.4	0.5	0.3	0.7	0.1	0.8	0.5	0.5	0.5	0.5
0.4	0.5	0.5	0.5	0.3	0.7	0.2	0.8	0.4	0.5	0.4	0.6
0.4	0.5	0.5	0.5	0.3	0.7	0.3	0.7	0.5	0.5	0.5	0.5
0.5	0.5	0.5	0.5	0.4	0.5	0.4	0.5	0.5	0.5	0.5	0.5
0.5	0.5	0.5	0.5	0.4	0.5	0.3	0.7	0.4	0.5	0.5	0.5
0.5	0.4	0.5	0.5	0.5	0.4	0.4	0.6	0.5	0.5	0.2	0.8
0.7	0.3	0.5	0.4	0.5	0.5	0.4	0.6	0.4	0.6	0.4	0.5
0.7	0.3	0.6	0.4	0.6	0.4	0.5	0.5	0.5	0.5	0.5	0.5
0.5	0.4	0.5	0.5	0.6	0.4	0.5	0.5	0.5	0.5	0.5	0.5
0.5	0.5	0.8	0.2	0.5	0.4	0.5	0.5	0.5	0.5	0.5	0.5

the data for the problem. Fuzzy analytical hierarchy process with intuitionistic fuzzy set has been implemented over the attributes of breast cancers. Based on Table 5 and output of the above work for predict the breast cancer mainly radius, perimeter, texture, area, smoothness, fractal dimension, concavity, concavity points, age and symmetry. From the fiure-1, we can easily observe that proposed system is consistent with existing model.

Table 2 Consistence matrix

$\overline{M}$	A1		A2		A3		A4		A5	
	μ	ν	μ	ν	μ	ν	μ	ν	μ	ν
A1	0.500	0.500	0.500	0.500	0.500	0.449	0.449	0.449	0.308	0.500
A2	0.500	0.500	0.500	0.500	0.500	0.400	0.500	0.449	0.352	0.500
A3	0.449	0.500	0.400	0.500	0.500	0.500	0.400	0.500	0.500	0.500
A4	0.449	0.449	0.449	0.500	0.500	0.400	0.500	0.500	0.400	0.500
A5	0.500	0.308	0.500	0.352	0.500	0.500	0.500	0.400	0.500	0.500
A6	0.500	0.308	0.449	0.352	0.500	0.400	0.449	0.500	0.500	0.500
A7	0.634	0.386	0.449	0.449	0.400	0.500	0.500	0.400	0.500	0.500
A8	0.814	0.077	0.609	0.160	0.555	0.225	0.449	0.308	0.500	0.500
A9	0.552	0.006	0.897	0.048	0.778	0.135	0.778	0.135	0.696	0.263
A10	0.821	0.057	0.707	0.171	0.741	0.160	0.741	0.160	0.551	0.308
A11	0.853	0.026	0.859	0.046	0.824	0.104	0.824	0.104	0.667	0.214

A6		A7		A8		A9		A10		A11	
μ	ν	μ	ν	μ	ν	μ	ν	μ	ν	μ	ν
0.308	0.500	0.386	0.634	0.077	0.814	0.006	0.552	0.057	0.821	0.026	0.853
0.352	0.449	0.449	0.449	0.160	0.609	0.048	0.897	0.171	0.707	0.046	0.859
0.400	0.500	0.400	0.500	0.225	0.555	0.135	0.778	0.160	0.741	0.104	0.824
0.449	0.500	0.400	0.500	0.308	0.449	0.135	0.778	0.160	0.741	0.104	0.824
0.500	0.500	0.500	0.500	0.500	0.500	0.263	0.696	0.308	0.551	0.214	0.667
0.500	0.500	0.500	0.500	0.500	0.449	0.352	0.600	0.304	0.652	0.151	0.753
0.500	0.400	0.500	0.500	0.500	0.400	0.449	0.500	0.449	0.551	0.400	0.500
0.449	0.500	0.500	0.400	0.500	0.500	0.400	0.600	0.449	0.551	0.400	0.600
0.600	0.352	0.500	0.449	0.600	0.400	0.500	0.500	0.500	0.500	0.500	0.500
0.652	0.304	0.551	0.449	0.551	0.449	0.500	0.500	0.500	0.500	0.500	0.500
0.753	0.151	0.500	0.400	0.600	0.400	0.500	0.500	0.500	0.500	0.500	0.500

8 Conclusion

In this work, a study of intuitionistic fuzzy analytical hierarchy process with fuzzy Delphi method over breast cancer factors has been attempted. The main important critical issue to diagnosis of a breast cancer disease in the medical field. The major problem is to identify the factors and to correlate them with existing information. In this work, we are taken a breast cancer factors that are identified and ranked them which factors will affect more for the patients and which factors will affect less comparing to the other factors. Further, this work can be extended into many levels of attributes to predict the breast cancers.

Table 3 Weight calculation

Attributes	μ	ν
$W(A1)$	0.06015609	0.8725673
$W(A2)$	0.06425087	0.8668095
$W(A3)$	0.06173075	0.8709366
$W(A4)$	0.06037432	0.8647277
$W(A5)$	0.07134361	0.8549978
$W(A6)$	0.06678494	0.8523731
$W(A7)$	0.07140146	0.8541532
$W(A8)$	0.07271441	0.8355862
$W(A9)$	0.08648703	0.8129741
$W(A10)$	0.16943978	0.819528
$W(A11)$	0.17567961	0.8046788

Table 4 Ranking

Attribute	Proposed work
Radius	0.93628368
Perimeter	0.93340473
Texture	0.93546856
Compactness	0.93236356
Area	0.92749856
Smoothness	0.92618625
Fractal dimension	0.9270756
Concavity	0.9176531
Concave points	0.9064654
Age	0.9097640
Symmetry	0.9023393

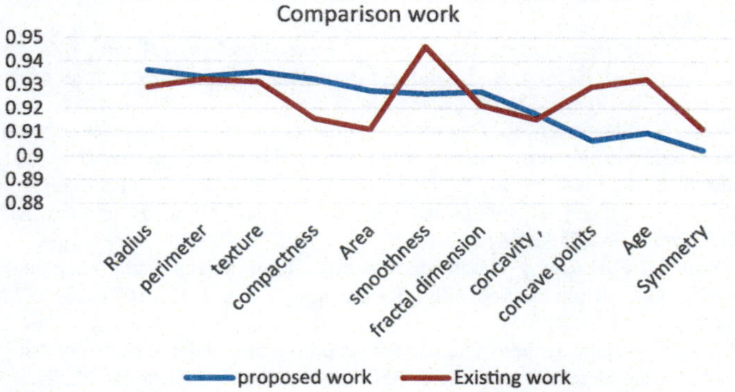

Fig. 1 Comparison graph

Table 5 Comparison matrix

Attribute	Proposed work	Existing work
Radius	0.93628368	0.929169
Perimeter	0.93340473	0.932383
Texture	0.93546856	0.931283
Compactness	0.93236356	0.91565
Area	0.92749856	0.911122
Smoothness	0.92618625	0.936354
Fractal dimension	0.9270756	0.921354
Concavity	0.9176531	0.915313
Concave points	0.9064654	0.929169
Age	0.9097640	0.932383
Symmetry	0.9023393	0.931283

References

1. Atanassov, K.T.: Intuitionistic fuzzy sets. Fuzzy Sets Syst. **20**(1), 87–96 (1986)
2. Chowdhary, C.L.: A hybrid scheme for breast cancer detection using intuitionistic fuzzy rough set technique. Int. J. Healthc. Inf. Syst. Inf. **11**(2) (2016)
3. Bustince, B.H.: Construction theorems for intuitionistic fuzzy sets **84**(3), 271–281 (1996)
4. Atanassov, K., Szmidt, E., Kacprzyk, J.: On intuitionistic fuzzy pairs. Notes Intuitionistic Fuzzy Sets **19**(3), 1–13 (2013)
5. Tapan Kumar Roy, A.G.: Intuitionistic fuzzy delphi method: more realistic and interactive forecasting tool. Notes on Intuitionistic Fuzzy Sets **18**(50), 37–50 (2012)
6. Deschrijver, G., Cornelis, C., Kerre, E.: On the representation of intuitionistic fuzzy tnormsand t-conorms. Notes Intuitionistic Fuzzy Sets **8**(3), 1–10 (2002)
7. Abdullah, L., Jaafar, S., Imran: Intuitionistic fuzzy analytic hierarchy process approach in ranking of human capital indicator. J. Appl. Sci. **3**(1), 423–429 (2013)
8. Szmidt, E., Kacprzyk, J.: Intuitionistic fuzzy sets in some medical applications. In: Fifth International Conference on IFSs, Soa, pp. 58–64 (2001)
9. Szmidt, E., Kacprzyk, J.: Distances between intuitionistic fuzzy sets. Fuzzy Sets Syst. **114**(3), 505–518 (2000)
10. Szmidt, E.: Entropy for intuitionistic fuzzy sets. Fuzzy Sets Syst. **118**, 467–477 (2001)
11. Akram, M., Shahzad, S., Butt, A., Khaliq, A.: Intuitionistic fuzzy logic control for Heater fans. Math. Comput. Sci. **7**(3), 367–378 (2013)
12. Rajaprakash, S., Ponnusamy, R.: Determining students expectation in present education system using fuzzy analytic hierarchy process. MIKE 2013, LNAI 8284, pp. 553–566 (2013)
13. Rajaprakash, S., Ponnusamy, R., Pandurangan, J.: Determining the customer satisfaction in automobile sector using the intuitionistic fuzzy analytical hierarchy process. MIKE 2014, LNAI 8891, pp. 239–255 (2014)
14. Rajaprakash, S., Ponnusamy, R., Pandurangan, J.: Intuitionistic fuzzy analytical hierarchy process with fuzzy delphi method. Glob. J. Pure Appl. Math. **11**(3), 1677–1697 (2015). ISSN 0973-1768
15. Sadiq, R., Tesfamariam, S.: Environmental decision-making under uncertainty using intuitionistic fuzzy analytic hierarchy process (IF-AHP). In: Stochastic Environmental Search and Risk Assessment, pp. 75–91. Springer, Berlin (2009)
16. Rajaprakash, S., Ponnusamy, R.: Ranking business scorecard factor using intuitionistic fuzzy analytical hierarchy process with fuzzy delphi method in automobile sector. MIKE 2015, LNAI 9468, pp. 1–12 (2015)

17. Zadeh, L.A.: Fuzzy sets. Information control **8**, 35–40 (1965)
18. Zadeh, L.A.: Biological application of the theory of fuzzy sets and systems. In: Proceedings International Symposium Biocybernetics of the Central Nervous System, Little Brown and Co., Boston, pp. 199–212 (1969)
19. Xu, Z.: Intuitionistic preference relations and their application in group decision making. Inf. Sci. **177**(11), 2363–2379 (2007)
20. Chowdhary, C.L.: Breast cancer detection using intuitionistic fuzzy histogram hyperbolization and possibilitic fuzzy c-mean clustering algorithms with texture feature based classification on mammography images. In: AICTC '16: Proceedings of the International Conference on Advances in Information Communication (2016)

Content-Based Medical Image Retrieval Using Pretrained Inception V3 Model

B. Ashwath Rao, Gopalakrishana N. Kini, and Joshua Nostas

1 Introduction

During the last period, there have been incredible technological advances in the field of imaging, improving the ways of taking, storing, and searching images. These advances have also affected the medical field. More and better medical images are being obtained, increasing the number of medical image databases in hospitals. With this increase, the search for a particular image becomes more difficult and costly, so it is necessary to improve the methods of image search and retrieval. The first approaches in the field of image retrieval were based on text search methods [1], but this approach had many limitations as it relied on manually entering a description of the image. This task is costly and time-consuming, and this information can be affected by the criteria and perspective of the person who fills it in, this bias is much more evident in the field of medicine since it creates a dependence on the interpretation of the physician who performs this task.

A method of image search based on image characteristics, such as shape, colors, textures, etc., rather than text [2], was then approached. This approach is called content-based image retrieval (CBIR) for non-medical images. The techniques proposed in CBIR have been used in different fields of study and work, as well as in different projects. The medical field is not the exception; in the medical field, the CBIR field is known as content-based medical image retrieval (CBMIR). Techniques and methods used in CBMIR must be more precise because medical images have subtle differences that cannot be considered irrelevant.

In recent times, the evolution and increased use of deep learning and convolutional neural networks present an opportunity to discover and adopt new techniques that can be used to solve the problems that CBIR and CBMIR present. Machine learning and deep learning models are increasing day by day, but training a model from scratch

B. Ashwath Rao (✉) · G. N. Kini
Manipal Institute of Technology, Manipal Academy of Higher Education, Manipal, India
URL: http://www.manipal.edu

J. Nostas
Departamento de Ciencias Exactas e Ingenierias, Universidad Católica Boliviana "San Pablo", Cochabamba, Bolivia

M. Dua et al. (eds.), *Proceedings of the International Conference on Paradigms of Communication, Computing and Data Sciences*, Algorithms for Intelligent Systems,
https://doi.org/10.1007/978-981-16-5747-4_55

requires a lot of time and a huge amount of data. To solve this problem, there is a technique called transfer learning, which consists of using pre-trained models with a huge amount of data, and taking advantage of their weights to classify with the data, we have available, regardless of the amount of data or if the classes are different.

This paper proposes to use the Inception V3 model with the weights of the ImageNet database, adapting the last layers of the model to use a database with medical images with six different classes.

This paper presents the following sections: Sect. 2 contains related work, Sect. 3 contains the proposed methodology, Sect. 4 presents the experimental results, and finally, Sect. 5 presents the conclusion and future work.

2 Related Work

2.1 Content-Based Image Retrieval

Content-based image retrieval is a cutting-edge research field of computer vision that addresses the problem of searching for images in large databases [1]. The initial work on image retrieval was based on the textual annotation of images, recent methods in this field support full retrieval of visual content, and properties of images [2]. A classic content-based image retrieval architecture is shown in Fig. 1.

This architecture presents two phases, one online phase, and one offline phase. At first, in the offline phase, features of every single image in the database are extracted and saved in a features database. This phase is generally time-consuming and needs a lot of images to fill a general features database. On the other hand, in the online phase, the user interacts with the system. The user first inputs a query image to the system. Similar to the offline phase, the features of the image are extracted, and they are subjected to a similarity measure function, which calculates the distance between features of the query image and the features stored in the features database.

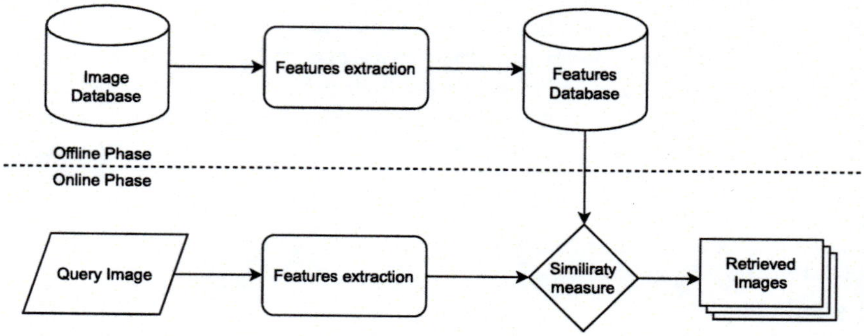

Fig. 1 Classic CBIR architecture

The system retrieves a set of images from the database which are the most similar images to the query image based on the similarity function [3].

One of the most important steps in the image retrieval task is feature extraction of both images stored in the database and the query image. Features extracted from images are primitive visual features such as color, texture, shape, and among others.

In this area, many studies have been and continue to be carried out to improve existing techniques or to use new techniques to improve results. Different methods such as Grey level co-occurrence matrix (GLCM), Discrete wavelet transform (DWT), Gabor transform, Curvelet, and Local binary pattern (LBP) were tested to extract the texture from the images, comparing the results of each method [4]. On the other hand, there are some studies that explore the information security side, using encryption techniques and convolutional neural networks [5] to perform the CBIR task in a way where the image data is not compromised. This area is quite large, and there is still much more to study and test.

2.2 *Content-Based Medical Image Retrieval*

The task of content-based image retrieval is important and can be applied to different areas of study and use similar methods to perform it, but the area of medicine requires much more precise methods and techniques because in medical imaging every detail of the image is very important and cannot be ignored. For this reason, there is a content-based medical image retrieval, an area of study that seeks to apply and improve CBIR techniques to apply them to medical images [1, 2, 6].

There are some studies that propose different techniques and methods to perform the CBIMR task were proposed the use of clustering method using dictionary learning to group similar images [3]. Another proposal is to consider the fused and context-sensitive similarity of the image by fusing the semantic and visual similarities between the query image and each image in the database as their pairwise similarities and then form a graph where the nodes represent an image and the edges represent the pairwise similarity. Then, using the shortest path algorithm, the most similar images are obtained [7]. And so, there are different approaches to tackle this challenging and ambiguous task [8–10].

Another popular method is the use of deep learning, due to the great advances in this area, the use of deep learning for the CBIMR task is an idea that many share. The use of parallel and Siamese convolutional neural networks was proposed to take advantage of the benefits they present [11, 12]. The use of convolutional neural networks with transfer learning was also proposed [13, 14]. Another approach is the proposal of new models and architectures of convolutional neural networks, some more specific for the data used, others with different convolutions convolutional filters, some deeper and others simpler [15–18].

3 Methodology

This paper proposes a classification-based framework to retrieve similar medical images from a database with medical images belonging to different classes. The representation of the proposed framework is shown in Fig. 2. This paper proposes to use a trained deep convolutional neural network, with different final layers to fit with the data. The neural network used is Inception V3 with weights of the ImageNet dataset. First we trained final layers with a randomly chosen 90% of the total images in the database and it was tested with the last 10% of images. Once the model was trained, each image in the dataset was fed to the model, and image features from the last fully connected layers are extracted to create a features database. With a features database in place, a query image is fed forward to the trained model. Its features are extracted and similarity measure between features of the query image and the features from the features database are calculated to retrieve the most similar images.

3.1 *Dataset*

This paper was used an open dataset named "Medical MNIST" [19] available in the Kaggle platform. The dataset has 58,954 images belonging to 6 classes: Abdomen CT with 10,000 images, Breast MRI with 8954, Chest X-ray with 10,000, Chest CT with 10,000, Hand X-ray with 10,000, and Head CT with 10,000. Images were originally taken from other datasets and processed into this format. The data presented is balanced between the classes.

The data were randomly split into a training and test set using 90% for the training set and 10% for the test set. In total, there are 53,058 images in the training set and

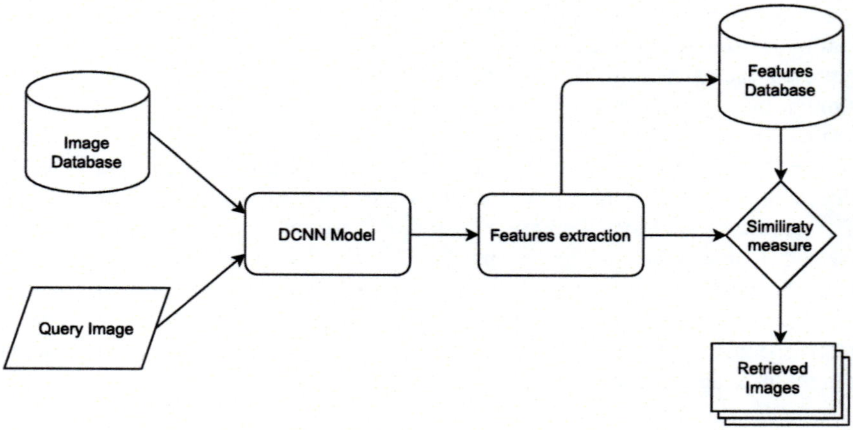

Fig. 2 Proposed CBIR architecture using a DCNN model

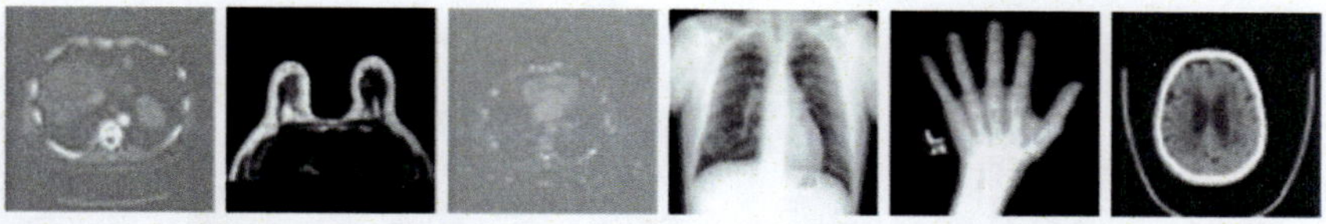

(a) Abdomen CT (b) Breast MRI (c) Chest CT (d) ChestX-Ray (e) Hand X-Ray (f) Head CT

Fig. 3 Images of examples of each class in the data set

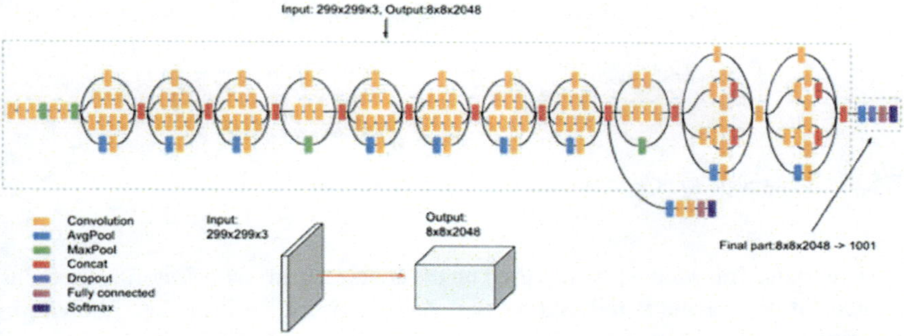

Fig. 4 Inception V3 model architecture

5896 images in the test set. All images were resized to 75×75 to use the Inception V3 model. Numerical labels were assigned to classes in the dataset for supervised learning. Figure 3 shows example images belonging to each of the classes.

3.2 *D-CNN Model*

Inception V3 The model used in this paper is the Inception V3, available in Keras [3], using ImageNet weights, changing last layers to fit the data, and using images with size: $75 \times 75 \times 3$.

Inception V2 and Inception V3 were presented in the same paper [20]. This actualization was proposed some upgrades to increase the accuracy and reduce the computational complexity. The first proposed upgrade was: factorize convolutions with a large filter size. To do this, the authors proposed three different inceptions modules. The first one replaces one 5×5 filter with two 3×3 filters, this factorization reduces the number of parameters by 28%. This technique is called: 'Factorization Into Smaller Convolutions'. The second module replaces a single 3×3 filter with a 1×3 filter followed by a 3×1 filter reduce the number of parameters by 38%. This technique is called: 'Factorization Into Asymmetric Convolutions'. The last module was proposed for promoting high-dimensional representations with expanded filter

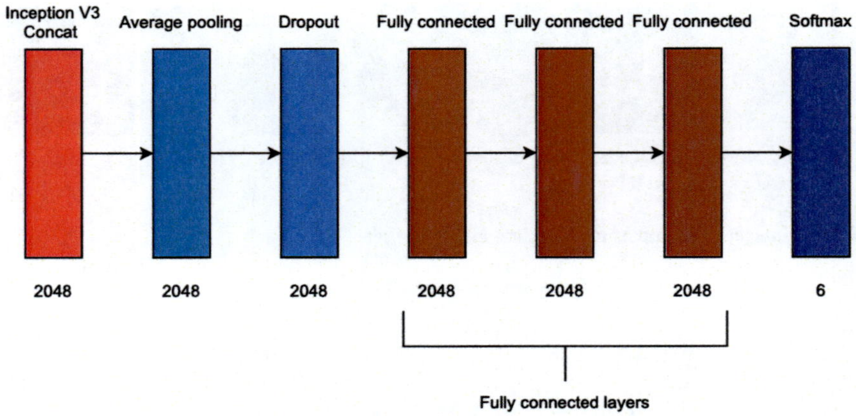

Fig. 5 Last layers to work with our data and extract features from it

bank outputs. This module is only used on the coarsest grid. With factoring modules, number of parameters is reduced for the whole network, and it is less likely to be overfitting, and the network can go deeper.

Another proposed upgrade was the utility of Auxiliary Classifiers, the authors proposed that using auxiliary classifiers in early training not result in improved convergence, but near the end of the training, the network with the auxiliary classifier starts to overtake the accuracy of the network without any auxiliary branch and reaches a slightly higher plateau [20].

Another one is efficient grid size reduction. The authors proposed another variant to reduce the computational cost even further, while removing the representational bottleneck, by using two parallel strides 2 blocks. One is a pooling layer and the other is a convolution layer, both are concatenated and their results on another inception module that reduces the grid-size while expands the filter banks [20].

The last proposed upgrade was the: Model regularization via label smoothing, where they propose a mechanism to regularize the classifier layer by estimating the marginalized effect of label-dropout during training [20].

The model architecture is presented in Fig. 4. It presents 42 layers deep, using 10 Inception modules, 3 using the factorization into smaller Convolutions technique, 5 using factorization into asymmetric convolutions technique and the last 2 using the last proposed module.

Last layers architecture To use the Inception V3 model with the data, the last layers were changed as shown in Fig. 5. The first layer connected to Inception V3 'Concat' layer is an Average pooling layer, then a dropout layer, then three fully connected layers, which gave us the features to fill the features database. And at the end, a Softmax layer with six outputs, which produces probability distributions for each class label.

The proposed model accepts images with exactly three channels and with and shape not smaller than 75. Inception V3 model output has a shape of 2048, and then this result is fed to a global average pooling layer and a dropout layer with a 0.5 rate. The output of the dropout layer is fed to three fully connected layers, each one with 2048 neurons. And at the end, a Softmax layer with 2048 inputs and 6 outputs.

Training details All images in the database were re-shaped to (75 × 75) before training because this is the minimum size allowed to use with the Inception V3 model. The model was trained with Adam optimizer with a learning rate of 0.001 with a maximum of 18 epochs and a batch size of 32. Categorical cross-entropy was used as a loss function.

The Adam optimization algorithm is an extension to stochastic gradient descent and was used in place of classical stochastic gradient descent because it updates the network weights iterative based on training data. In the original paper where the algorithm was introduced, the authors listed the major benefits of using this algorithm. These benefits include straightforward implementation, computational efficiency, little memory requirements, and among others [21].

3.3 *Feature Extraction*

To continue the retrieval task. Once the deep convolutional neural network is successfully optimized and trained for classifying multi-class medical images, the following step is to extract the features of all images in the dataset to create the features database as it is required for the retrieval task.

Therefore, each image in the dataset is fed forward to the trained model and extracts image features from the last three fully connected layers. These features are saved associated with each image, his label, and the fully connected layer where came from.

3.4 *Similarity Measure*

To finish the retrieval task is needed to extract image features from a query image and compare them with the features in the feature database and retrieve the most similar images. For this, the query image is fed forward to the trained model and are extracts image features from the last fully connected layers. To compare query image features with our features database, we use the Euclidean distance showed in Equation, where a is query image features and b are featured in the features database.

$$d(a,b) = \sqrt{\sum_{i=1}^{n}(a_i - b_i)^2} \tag{1}$$

When distances are calculated between the features of the query image and the features of images in the feature database, most similar or shortest distance images are retrieved.

4 Results

In this work, it has been used Keras, an API written in Python running on top of the machine learning platform TensorFlow for developing and training the framework. The simulations have been performed in Kaggle notebooks available on the Kaggle platform. Notebooks run in a remote computational environment. Has 4 CPU cores and RAM of size 16 GB. The proposed framework has been evaluated in terms of classification and retrieval because these are the two most important phases of the framework.

4.1 Classification Performance

Classification performance has been measured in terms of average precision (AP), average recall (AR), accuracy, and F1 measure. The formulas to calculate this measure terms are presented in Eqs. (2), (3), (4) and (5), respectively. Where TP is for true positive and denote the count of images from class i and is correctly classified, TN is for true negative and denotes the count of images that are correctly classified as not belonging to class i, FP is for false-positive and denotes the number of images not from class i but miss-classified as class i, FN is for false negative and denotes the images that are from class i but are miss-classified and n represents the total number of classes [17].

$$\text{Average Precision} = \frac{1}{n}\sum_{i=1}^{n}\frac{\text{TP}_i}{\text{TP}_i + \text{FP}_i} \tag{2}$$

$$\text{Average Recall} = \frac{1}{n}\sum_{i=1}^{n}\frac{\text{TP}_i}{\text{TP}_i + \text{FN}_i} \tag{3}$$

$$\text{Accuracy} = \frac{1}{n}\sum_{i=1}^{n}\frac{\text{TP}_i + \text{TN}_i}{\text{TP}_i + \text{TN}_i + \text{FP}_i + \text{FN}_i} \tag{4}$$

$$\text{F1 measure} = 2 * \frac{\text{AP} * \text{AR}}{\text{AP} + \text{AR}} \tag{5}$$

The results given by the presented model are shown in Table 1 which shows 100% in all the measure terms. The confusion matrix presented in Table 2 shows an average

Table 1 Results of the model in classification task

Measure metric	Results (%)
Average precision	100
Average recall	100
Accuracy	100
F1 measure	100

Table 2 Confusion matrix

	Abdomen CT	Breast MRI	Chest X-Ray	Chest CT	Hand X-Ray	Head
Abdomen CT	1	0	0	0.002	0	0
Breast MRI	0	1	0	0	0	0
Chest X-Ray	0	0	1	0	0.001	0
Chest CT	0.001	0	0	1	0	0
Hand X-Ray	0	0	0	0	1	0
Head CT	0	0	0	0	0	1

accuracy of 100% in almost all classes, the Abdomen CT class shows an average accuracy of 99%.

Obtaining a high average accuracy in the classification phase not only demonstrates that the classification method is effective but is also beneficial to the medical image retrieval framework, as the features obtained from the images will be more reliable for each class.

4.2 *Retrieval Performance*

Retrieval performance has been measured in terms of precision. The mathematical expression for precision is presented in Eq. (6) [17]. A total of 20 retrieved image performance were analyzed using two strategies, the first one using class prediction, to reduce the size of the dataset where the images should be retrieved, because the result of the classification stage is high enough, it is very unlikely to obtain a wrong class prediction of the query image. The second strategy does not use class prediction but compares the query image with all the images in the dataset.

$$\text{Precision} = \frac{\text{Number of relevant images retrieved}}{\text{Number of retrieved images}} \tag{6}$$

Image features used in the retrieval task were obtained from the last three fully connected layers of the neural network. Analysis of these representations has been

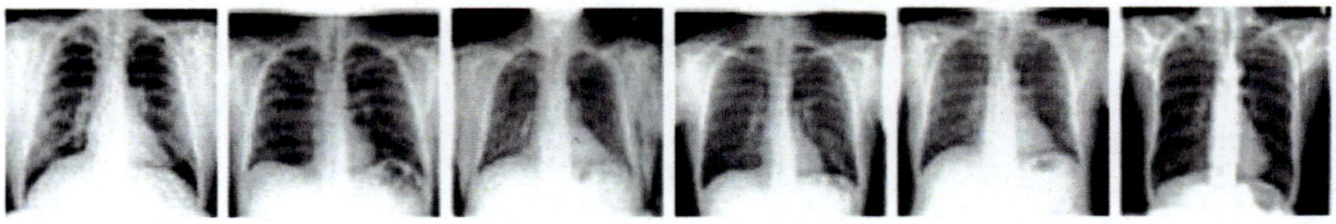

(a) Query image (b) (c) (d) (e) (f)

Fig. 6 Retrieval results without using class prediction

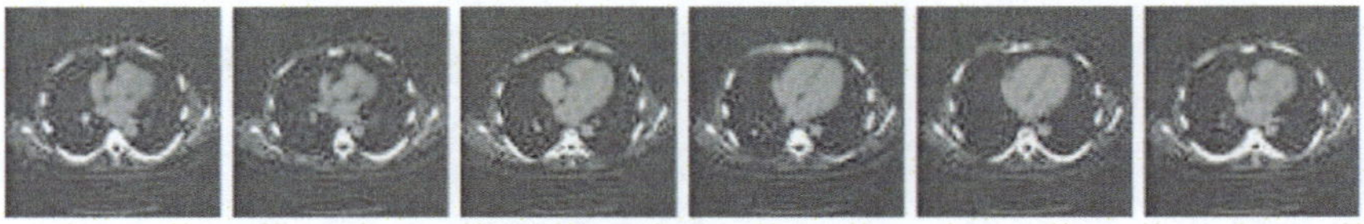

(a) Query image (b) (c) (d) (e) (f)

Fig. 7 Retrieval results using class prediction

performed in terms of retrieval precision without using the predicted class labels. The results of this analysis show that the representation of the first fully connected layers has better results than the second and third fully connected layers.

Retrieval results using the first fully connected layer without class prediction are shown in Fig. 6. The first image in the figure is the image to be searched, and the following images are the first 5 images retrieved by the framework. Figure 7 shows retrieval results with class prediction with same format as Fig. 6.

The framework achieved an average accuracy of 0.97 without class prediction and 0.99 with class prediction in the task of retrieving similar images, and a small improvement in accuracy can be seen between both strategies. When using the predicted class, the dataset where the images must be compared is smaller since all the images belonging to that class are filtered, this in terms of computation and performance is beneficial when you have a very large dataset with many different classes.

5 Conclusion

This paper proposes a deep learning framework for content-based medical image retrieval using the Inception V3 model pre-trained with the ImageNet dataset for the classification phase. Two strategies were used for the medical image retrieval task. The first one is to obtain the prediction of the class of the query image to reduce the size of the dataset to be retrieved. The second is to not obtain the class prediction and perform the retrieval task on the entire dataset, as proposed in the paper "Medical image retrieval using deep convolutional neural network" [17]. The solution proposed in this paper reduces the semantic gap by learning the characteristics of the images directly from them. It proposed to use, transfer learning, with the Inception V3 model, and the weights of the ImageNet dataset. The last layers of the model were replaced and trained for six classes of medical images with an average classification accuracy of 99.98%. Image features were extracted from the last 3 fully connected layers of the trained model to create the image feature dataset for the retrieval task. To test the performance of the presented framework in the task of medical image retrieval, the precision measure was used. The proposed framework achieves a mean average precision of 0.99 multi-modal image data with class prediction and 0.97 for multi-modal image data without class prediction.

As future work, it is considered to expand the experiments by testing the proposed framework with different medical image data sets and also use different pre-trained models with the same medical image datasets. To validate the use of transfer learning as an option in the medical image retrieval task [22, 23].

References

1. Deserno, T.M.: Biomedical Image Processing. Aachen, Germany. Springer, Berlin, Heidelberg (2011)
2. Cai, W., Kim, J., Feng, D.D.: Biomedical Information Technology. Academic Press Inc., San Diego (2007)
3. Srinivas, M., Naidu, R.R., Sastry, C.S., Mohan, C.K.: Content based medical image retrieval using dictionary learning. Neurocomputing **168**, 880–895 (2015). ISSN 0925-2312. https://doi.org/10.1016/j.neucom.2015.05.036
4. Dhingra, S., Bansal, P.: Experimental analogy of different texture feature extraction techniques in image retrieval systems. Multimed. Tools Appl. **79**, 27391–27406 (2020). https://doi.org/10.1007/s11042-020-09317-3
5. Ahmed, K.W., Al Aziz, M.M., Sadat, M.N., Alhadidi, D., Mohammed, N.: Nearest neighbour search over encrypted data using intel SGX. J. Inform. Secur. Appl. **54**, 102579 (2020). ISSN 2214-2126. https://doi.org/10.1016/j.jisa.2020.102579
6. Müller, H.: Medical image retrieval: applications and resources. In: Proceedings of the 2020 International Conference on Multimedia Retrieval (ICMR'20). Association for Computing Machinery, New York, NY, USA, pp. 2–3 (2020). https://doi.org/10.1145/3372278.3390668
7. Ma, L., Liu, X., Gao, Y., Zhao, Y., Zhao, X., Zhou, C.: A new method of content based medical image retrieval and its applications to CT imaging sign retrieval. J. Biomed. Inform. **66**, 148–158. (2017). ISSN 1532-0464. https://doi.org/10.1016/j.jbi.2017.01.002
8. Shamna, P., Govindan, V.K., Nazeer, K.A.A.: Content based medical image retrieval using topic and location model. J. Biomed. Inf. **91**, 103112 (2019). ISSN 1532-0464. https://doi.org/10.1016/j.jbi.2019.103112
9. Wei, C.-H., Li, C.-T., Wilson, R.: A general framework for content-based medical image retrieval with its application to mammograms. In: Proceedings of the SPIE 5748, Medical Imaging 2005: PACS and Imaging Informatics, 15 Apr 2005. https://doi.org/10.1117/12.594929
10. Park, S.C., Sukthankar, R., Mummert, L., Satyanarayanan, M., Zheng, B.: Optimization of reference library used in content-based medical image retrieval scheme. Med. Phys. **34**(11), 4331–4339 (2007). https://doi.org/10.1118/1.2795826
11. Haripriya, P., Porkodi, R.: Parallel deep convolutional neural network for content based medical image retrieval. J. Ambient Intell. Human Comput. **12**, 781–795 (2021). https://doi.org/10.1007/s12652-020-02077-w

12. Chung, Y.-A., Weng, W.-H.: Learning Deep Representations of Medical Images Using Siamese CNNs with Application to Content-Based Image Retrieval, arXiv:1711.08490 (2017)
13. Khatami, A., Babaie, M., Tizhoosh, H.R., Khosravi, A., Nguyen, T., Nahavandi, S.: A sequential search-space shrinking using CNN transfer learning and a Radon projection pool for medical image retrieval. Exp. Syst. Appl. **100**, 224–233 (2018). ISSN 0957-4174. https://doi.org/10.1016/j.eswa.2018.01.056
14. Owais, M., Arsalan, M., Choi, J., Park, K.R.: Effective Diagnosis and Treatment through Content-Based Medical Image Retrieval (CBMIR) by Using Artificial Intelligence. Journal of Clinical Medicine. **8**(4), 462 (2019). https://doi.org/10.3390/jcm8040462
15. Roth, H.R., Lee, C.T., Shin, H.-C., Seff, A., Kim, L., Yao, J., Lu, L., Summers, R.M., Anatomy-specific classification of medical images using deep convolutional nets. In: 2015 IEEE 12th International Symposium on Biomedical Imaging (ISBI). IEEE, 2015. ISSN 9781479923748. https://doi.org/10.1109/ISBI.2015.7163826
16. Judah, E.S., Sklan, A.J. Plassard, D.F., Landman, B.A.: Toward content-based image retrieval with deep convolutional neural networks. In: Proceedings of the SPIE 9417, Medical Imaging 2015: Biomedical Applications in Molecular, Structural, and Functional Imaging, 94172C, 19 Mar 2015. https://doi.org/10.1117/12.2081551
17. Qayyum, A., Anwar, S.M., Awais, M., Majid, M.: Medical image retrieval using deep convolutional neural network. Neurocomputing **266**, 8–20 (2017). ISSN 0925-2312. https://doi.org/10.1016/j.neucom.2017.05.025
18. Öztürk, Ş.: Stacked auto-encoder based tagging with deep features for content-based medical image retrieval. Exp. Syst. Appl. **161**, 113693 (2020). ISSN 0957-4174. https://doi.org/10.1016/j.eswa.2020.113693
19. Apolanco3225: Medical MNIST classification. GitHub repository 2017. Available: https://github.com/apolanco3225/Medical-MNIST-Classification
20. Szegedy, C., Vanhoucke, V., Ioffe, S., Shlens, J., Wojna, Z.: Rethinking the inception architecture for computer vision. arXiv:1512.00567 (2015)
21. Kingma, D.P., Ba, J.: Adam: A Method for Stochastic Optimization. CoRR abs/1412.6980: n. pag (2015)
22. Chollet, F., et al.: Keras [Online]. Available: https://keras.io (2015)
23. Zhang, F., Song, Y., Cai, W., Hauptmann, A.G., Liu, S., Pujol, S., Kikinis, R., Fulham, M.J., Feng, D.D., Chen, M.: Dictionary pruning with visual word significance for medical image retrieval. Neurocomputing **177**, 75–88 (2016). ISSN 0925-2312. https://doi.org/10.1016/j.neucom.2015.11.008

Iris—Palmprint Multimodal Biometric Recognition Using Improved Textural Representation

Neeru Bala, Anil Kumar, and Rashmi Gupta

1 Introduction

The expanding interest for robust and secure authentication frameworks currently utilized in numerous arenas is a clear proof that biometric recognition approaches deserve more considerations. Biometric system refers to an automated process of authenticating an individual precisely depending upon physiological characteristics (iris, palmprint, face, ear, palm-vein, finger-knuckle-print, DNA) as well as behavioral characteristics (keystrokes, gait, signature, and hand-writing) [1]. These characteristics cannot be disremembered as passwords or PINs and cannot be misplaced like cards. Biometric seeks it applications in wide range of arenas such as: border control, investigation department, personal gadgets, business, online purchase or sale, buildings [2]. Unimodal biometric frameworks rely on one trait for analysis and authentication. But unimodal biometric frameworks have several shortcomings such as proneness to fraudulent attacks, sensitive to noise, non-universality, dearth of uniqueness, intra-class discrepancies. To resolve these snags of unimodal biometric frameworks more than one trait can be utilized in the same authentication procedure, and these types of biometric frameworks are termed as multimodal biometric frameworks [3]. The integration of evidences acquired through different sources is names information fusion. The information from different sources can be integrated at many levels such as at sample level, at feature level, at score level, at rank level, at decision level and at hybrid level [4]. In sensor level fusion, the samples acquisition is done using different sensors, and the samples are integrated at image level so as to boost the authentication rate. Various feature descriptors are employed to extract the features of different traits and then a combined feature vector is obtained. But

N. Bala (✉) · A. Kumar
Amity School of Engineering and Technology, Amity University, Gurugram, India

R. Gupta
Netaji Subhash University of Technology, East Campus, Delhi 110031, India

M. Dua et al. (eds.), *Proceedings of the International Conference on Paradigms of Communication, Computing and Data Sciences*, Algorithms for Intelligent Systems,
https://doi.org/10.1007/978-981-16-5747-4_56

feature level fusion suffers from various issues such as: compatibility between feature vectors to be concatenated and huge dimensionality. In score level fusion, matching scores of individual trait are obtained by matchers. Matching module compares the template already stored in the database and test template and generates scores in terms of genuine and imposter scores. Then, the scores obtained from different traits are fused to obtain the final matching score. In rank level fusion, the identity of an individual is established by integrating the ranks produced by the classifiers for each trait. In decision level fusion, the decisions generated by classifiers of different traits are integrated using operations such AND, OR to establish a final decision about the identity of a person. In hybrid level fusion, two or more fusion strategies may be applied to integrate the traits.

There are two stages in the authentication process of biometric recognition framework, registration and acknowledgment [5]. In registration stage, the traits of all the users are acquired and kept in the database. In acknowledgment stage, the sample templates are captured from the users, and then these are matched with the templates already kept in the database. Furthermore, a biometric recognition framework can be operated into two manners, identification manner and verification manner. In identification manner, there is no sued identity or the test template is matched with all the templates kept in the database. In this mode, the identity of an individual is determined if the test template matches with any of the templates kept in the database otherwise not. For example, if identification of a criminal is to be done then biometric framework must be operated in identification manner. In verification mode, there is a sued identity so the test template is matched with the corresponding one template kept in the database. For example, in the offices to mark the attendance of officer's biometric framework must be operated in verification manner.

In last two decades, numerous multimodal biometric frameworks have been proposed. Some of the recent works includes: Authors of [6] have proposed fusion of face and speech Signal at decision level. In [7], decision level fusion of fingerprint and iris has been suggested. Rank level fusion of face, fingerprint and iris has been proposed in [8]. Authors of [9] proposed hybrid level of fusion, i.e., feature level and score level fusion, to fuse face, left and right ocular sections. An approach to produce cancelable multimodal biometrics by integrating iris, and fingerprint was presented in [10]. A multimodal biometric framework to fuse fingerprint and face at feature level was proposed in [11]. In [12] integration of face and fingerprint at feature level has been presented. Authors of [13] combined iris and face at feature level. Another technique to fuse features of finger dorsal surface, major knuckle-prints and minor knuckle-prints has been proposed in [14]. Integration of features of face and iris have been suggested in [15].

A conventional multimodal biometric framework consists of pre-processing of images, feature extracting module, matching module and decision module [16]. The feature descriptor to extract the features of biometric traits plays a vital role in the performance of a multimodal biometric framework. This paper proposes a multimodal framework based on iris and palmprint, which utilizes improved Xor-Sum Code [17] to extract the features of iris and palmprint.

The organization of the rest of paper is as follows: Sect. 2 refers to the proposed multimodal biometric system. Section 3 presents the results of the experiment performed to validate the performance of the proposed approach and Sect. 4 concludes the article.

2 Proposed Approach

In this work, improved Xor-sum code (IXSC) [17] is applied to develop a multimodal biometric framework fusing iris and palmprint to achieve augmented authentication performance. In this proposed work, we have selected iris and palmprint as the traits to be fused due to following reasons: Iris is the most precise (appreciably lesser FAR and FRR), adaptable (can be employed effectively in both giant and tiny applications), permanent (consistent texture throughout life of an individual), accessible and contactless trait [18]. Another chosen universal trait is palmprint, which is a multi-textural unique pattern of the interior surface of the region between wrist and the roots of fingers of a humanoid hand. It has higher user acceptance and discerning capability. The superlative vital property of palmprint is that it has multiple-textural region of interest. The details of proposed approach are portrayed in this section.

2.1 *Pre-processing of Traits*

Pre-processing of images is significant step in any biometric recognition framework. The goal of this step is to acquire substantial information existing in the images and also to lessen the noise or irrelevant data existent in the image. Before feature extraction each eye image from the IITD database has been segmented and normalized to rectangular strip of 64×512 pixels and in order to extract ROI of palm images index-middle and small-ring finger's endpoints [19] are taken as reference positions so that images can be distinguished in terms of rotation, scale and transformation. The square shaped area is extracted taking the line fragment between these two reference positions as one of the axis. The extracted ROI is normalized using histogram equalization to 128×128 pixels. Examples of pre-processed images of both the traits are presented in Fig. 1.

2.2 *Feature Extraction Using Improved X-OR Sum Code*

In this work, features of the traits are extracted through followings stages [17]:

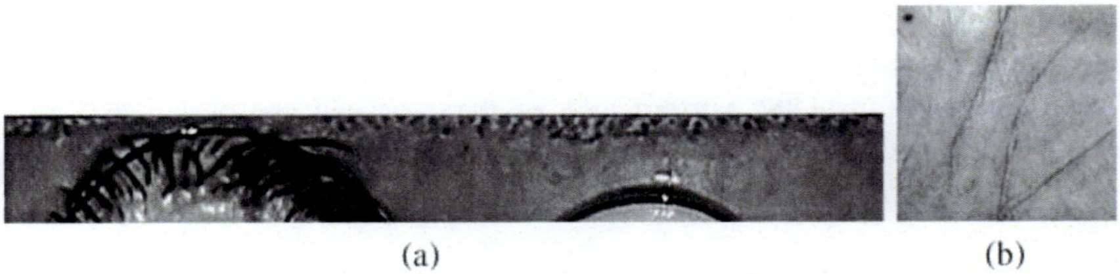

Fig. 1 Example images of a normalized and segmented iris, b extracted palmprint ROI

- Decomposition using Haar Wavelets: To augment the recognition accuracy and speed of performance of the system, the dimensions of templates are reduced using Haar wavelet at two levels, which splits the template into four parts at each level and out of these four parts only one part with greatest energy and key information is selected for further processing [20].
- Curvature Gabor Filtering: Gabor filter is proficient in evaluating any image using textural information existing in it. In curvature gabor filter (CGF), there is an extra curvature parameter which differentiates it from the traditional Gabor filter [21]. Mathematically, it can be epitomized as shown in Eq. (1):

$$\xi(x, y, \sigma, v, \varphi, c) = \frac{1}{\sqrt{2\pi\sigma^2}} \exp\left\{\frac{x^2 + y^2}{2\sigma^2}\right\} * \exp\left\{2\pi i\left(v(x\cos(\phi) + y\sin(\phi)) + c\sqrt{x^2 + y^2}\right)\right\} \quad (1)$$

where (x, y) epitomizes the kernel size, σ is Standard deviation, v is the frequency of the sinusoidal function, Phase Offset $\phi = n * pi/N$ where N is the number of orientations, n varies from 0, 1, …, $N - 1$. Example images of CGF filtered images of both the traits are presented in Fig. 2.

- X-OR sum coding for each orientation: In this stage, convolution of image is performed with both real and imaginary parts of the Curvature Gabor Filter. The results of convolutions are converted into binary form taking 0 as threshold. The X-OR operator is employed to merge the binarized real and imaginary parts of

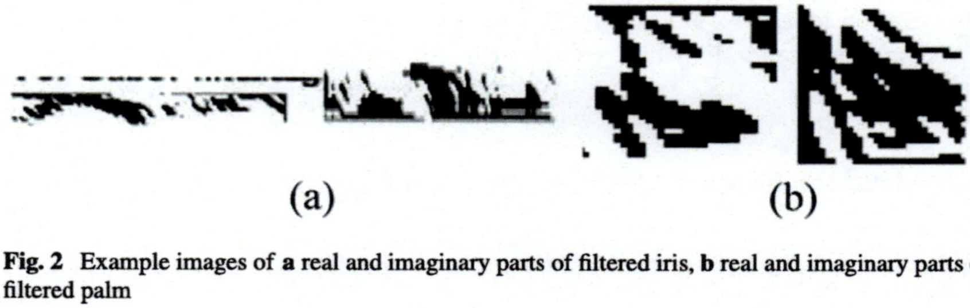

Fig. 2 Example images of a real and imaginary parts of filtered iris, b real and imaginary parts of filtered palm

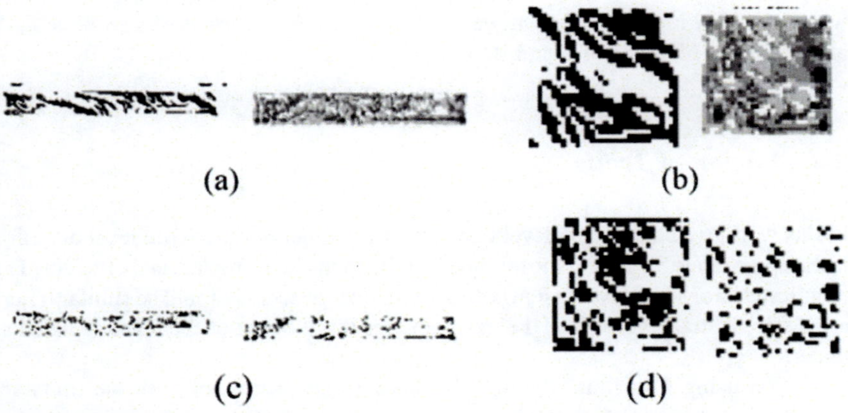

Fig. 3 Example images of **a** X-OR and X-OR sum of iris, **b** X-OR and X-OR sum of palm, **c** Bit 1 and Bit 2 of X-OR sum of iris, **d** Bit 1 and Bit 2 of X-OR sum of palm

images for every orientations. And then for all the orientations, the outputs of X-OR operator are summed together to obtain the texture feature vector. The outputs of X-OR and X-OR Sum operators are presented in Fig. 3.

- Encoding: The encoding of the X-OR summed output to obtain compact code is done using the following equation [20]:

$$XSC(b) = \begin{pmatrix} 1 \; if \, b \leq S < b + \left(\frac{N+1}{2}\right) \\ 0 \qquad \text{otherwise} \end{pmatrix} \tag{2}$$

where $b = 1, 2, \ldots \lceil \frac{N+1}{2} \rceil$ epitomizes the number of bits necessary to encode every element of IXSC and operator $\lceil \rceil$ provides the subsequent greater integer.

2.3 *Classification of Traits*

In this work, hamming distance is used for classification of both the traits. Hamming distance is the most convenient approach to match the binary images. It works by computing the similarity between feature vectors of two images from same trait. If while matching there are identical images it will produce the score as the genuine score else it will produce the imposter score. Mathematically, it can be epitomized as under [20]:

$$HD(A, B) = \frac{\sum_W \sum_L \sum_{i=1}^{b} A_i \oplus B_i}{b * W * L} \tag{3}$$

where A and B represents feature vectors that are to be matched of size $W \times L$, A_i and B_i are the ith plane of A and B.

2.4 Fusion of Traits

In this work, we chose score level fusion of iris and palmprint. Score level of fusion is the appropriate for fusing the biometric traits as this the simplest and efficient. The requirement for score level fusion is that the number of scores should be similar. There are different strategies to fuse the traits at score level. Some of them are depicted as under:

Fusion using Min-Rule Strategy: In this strategy, scores of both the traits are compared and the minimum of all these final scores are chosen as final scores. Mathematically, it can be depicted as:

$$S_{\text{Final}} = \text{MIN}\{I_L, I_R, P_L, P_R\} \tag{4}$$

Fusion using Max-Rule Strategy: In this strategy, scores of both the traits are compared and the maximum of all these final scores are chosen as final scores. Mathematically, it can be depicted as:

$$S_{\text{Final}} = \text{MAX}\{I_L, I_R, P_L, P_R\} \tag{5}$$

Fusion using Sum-Rule: Under this rule, the scores of all the traits are summed together to obtain final scores. Mathematically, it can be depicted as:

$$S_{\text{Final}} = \text{Average}(I_L, I_R, P_L, P_R) \tag{6}$$

Fusion using Product-Rule: Fusion using product rule is the most popular approach in the literature. Under this rule, the scores of all the traits are multiplied together to obtain final scores. Mathematically, it can be depicted as:

$$S_{\text{Final}} = \sqrt{I_L * I_R * P_L * P_R} \tag{7}$$

In this proposed approach, fusion using sum and product rules are employed for fusion iris and palmprint images. The schematic diagram of proposed approach is presented in Fig. 4.

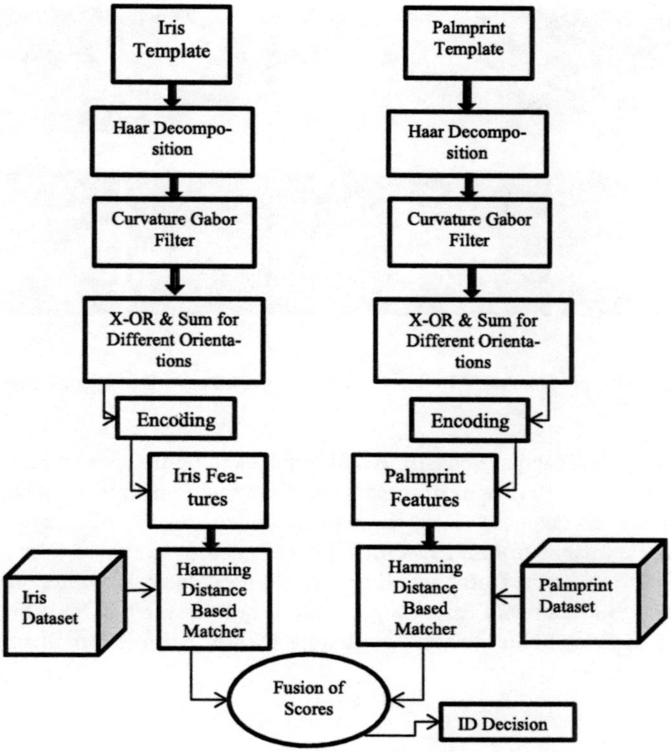

Fig. 4 Schematic diagram of proposed multimodal biometric framework

3 Experimental Results

The objective of this paper is to develop an effective multimodal framework. In this paper, we began with assessing the performance of individual unimodal framework of iris and palmprint, then present a multimodal biometric framework fusing iris and palmprint at score-level fusion. This work is validated using IITD v1.0 iris database and PolyU-II palmprint database. The details of these databases and evaluation metrics are given in the subsequent sections. Figure 4 presents the schematic diagram of proposed work.

3.1 Database Information

IIT Delhi v1.0 Iris Database: This database was developed by students and personnel at IIT Delhi, acquired with CMOS camera, JIRIS and JPC1000 in indoor

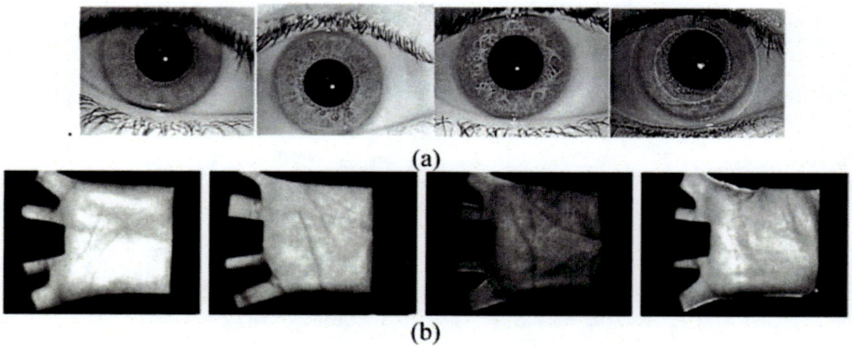

Fig. 5 Sample Images from **a** IIT Delhi v1.0 iris database, **b** PolyU-II palmprint database

environment. This database consist of 2240 images in bitmap format acquired from 224 volunteers, i.e., 10 images from each volunteer including images of both left and right eyes. In this database, each image has a dimension of 320 × 240. Figure 5a shows sample images from IIT Delhi v1.0 Iris Database.

PolyU-II Palmprint Database: It is a widely used database comprising of 7752 images from 386 palms. These images were acquired using CCD camera in two sessions with different illumination conditions. Figure 5b shows sample images from PolyU-II Palmprint Database.

3.2 Evaluation Metrics

To evaluate the proposed multimodal framework, various metrics are used. False acceptance rate (FAR) arises when a framework faultily admits a person, whom it should have prohibited, False Rejection Rate (FRR) occurs when a framework faultily discards a person, whom it should have admitted. Equal error rate (EER) is demarcated as the point on receiver operating characteristic curve where the FAR and FRR possess identical values. Decidability index (DI) is the measure of normalized separation between mean distribution of genuine and imposter scores. And at a particular threshold the percentage of truly accepted users is termed as genuine acceptance rate (GAR). The tiniest values of FAR, FAR, EER and greatest values of GAR and DI govern the better performance of the framework. Mathematically, these can be epitomized as under in Eq. 8 [17]:

$$\text{FAR} = \frac{\text{falsely\_accepted}}{\text{length of imposter score}}$$

$$\text{FRR} = \frac{\text{falsely\_rejected}}{\text{length\_genuine score}}$$

$$\text{EER} = \frac{\text{FAR} + \text{FRR}}{2} \tag{8}$$

$$\text{GAR} = 1 - \text{FRR}$$

$$\text{DI} = \frac{|\mu_g| - |\mu_i|}{\sqrt{\sigma_g^2 + \sigma_i^2}}$$

3.3 Performance Assessments of Unimodal Biometric Frameworks

Iris Recognition Framework: For experimentation, each eye image from the IITD database has been segmented and normalized to rectangular strip of 64 × 512 pixels. To discern the superlatively, appropriate value of curvature parameter '*c*', the experimentation has been done by varying its values. The performance metrics for five different value of 'c' are acquired and presented in Table 1. Augmentation in all the evaluation metrics due to inclusion of an additional curvature parameter 'c' is ostensible from the Table 1. From table, we can conclude that the optimal value of '*c*' is 0.060 which is highlighted as well using bold letters, as it results in greatest DI, AUC, GAR and least EER.

Palmprint Recognition Framework: In this work for experimentation, each palmprint image from PoluU-II database has been segmented and normalized to square shaped 128 × 128 pixels. The experiments have been performed using five different values of curvature parameter '*c*'. Performance metrics thus obtained using different values of '*c*' are presented in Table 2.

Table 1 Performance of iris recognition framework

Curvature parameter '*c*'	DI	AUC	EER (%)	GAR (%)
0.020	2.5620	0.9917	2.44	96.69
0.040	2.6903	0.9929	2.26	97.10
0.060	**2.7437**	**0.9948**	**2.18**	**97.40**
0.080	2.6642	0.9956	2.51	96.50
0.100	2.6376	0.9955	2.33	96.87

Table 2 Performance of palmprint recognition framework

Curvature parameter '*c*'	DI	AUC	EER (%)	GAR (%)
0.020	2.7631	0.9924	2.31	96.98
0.040	2.8912	0.9936	2.18	97.32
0.060	**2.9473**	**0.9955**	**2.10**	**97.63**
0.080	2.8636	0.9963	2.43	96.54
0.100	2.5417	0.9961	2.26	96.90

3.4 Performance Assessments of Multimodal Biometric Framework

The goal of score level fusion is to achieve augmented authentication performance than that can be achieved using unimodal framework. The scores acquired from iris and palmprint recognition frameworks are fused using product rule and sum rule. The requirement for score level fusion is that the number of scores should be similar. As in IITD database, there are 2240 eye images so from PolyU-II palmprint database also we have selected 2240 palm images. The performance metrics of proposed multimodal biometric framework for five different value of 'c' are acquired and presented in Table 3. Augmentation in all the evaluation metrics due to inclusion of an additional curvature parameter '*c*' is ostensible from the Table 3. Different fusion schemes can be employed to fuse the scores obtained from different traits such as (MIN) minimum rule, (MAX) maximum rule, (SUM) sum rule and (PROD) product rule. In this work, we have fused iris and palmprint images using SUM and PROD rules. From table, it can be seen apparently that although both the strategies employed for fusion results in improved evaluation metrics but PROD strategy outperforms in terms of all the evaluation metrics.

Table 3 Performance of proposed multimodal recognition framework

Curvature parameter '*c*'	Fusion using PROD strategy				Fusion using SUM strategy			
	DI	AUC	EER (%)	GAR (%)	DI	AUC	EER (%)	GAR (%)
0.02	4.4343	0.9973	1.41	98.79	3.9782	0.9946	1.92	98.09
0.04	4.6007	0.9984	1.28	98.96	4.1447	0.9957	1.79	98.26
0.06	**4.7102**	**0.9989**	**1.01**	**99.02**	**4.2542**	**0.9962**	**1.52**	**98.32**
0.08	4.547	0.9992	1.3	98.73	4.091	0.9965	1.81	98.03
0.1	4.1985	0.999	1.23	98.88	3.7425	0.9963	1.74	98.18

Table 4 Performance of proposed multimodal recognition framework

Technique used	DI	AUC	EER (%)	GAR (%)
LBP	3.6782	0.9902	1.81	98.43
Proposed approach	**4.7102**	**0.9989**	**1.01**	**99.02**

3.5 Comparative Analysis

The proposed multimodal biometric framework has employed improve X-OR sum code (IXSC) approach for feature extraction. This section compares the proposed multimodal framework with local binary patterns (LBPs) [22]. Based on multimodal framework, it is ostensible from Table 4 that proposed approach outperforms the LBP based approach in terms of all the evaluation metrics.

Notably, LBP is a textural feature descriptor which analyses the pixel intensities in the local neighborhood of a central pixel and generates decimal codes ranging in [0, 255] for each center pixel [23].

4 Conclusions

This paper offers a vigorous multimodal biometric framework based upon iris and palmprint. Curvature gabor filter is applied to extract the distinctive features of iris and palmprint. These traits are fused at score level using four fusion rules, i.e., min rule, max rule, product rule and average rule. Rigorous experimentations have been performed on two benchmark databases such as IITD v1.0 iris database and polyU-II palmprint database. The experimental results have revealed that proposed iris-palmprint multimodal framework augments the recognition performance of unimodal iris or palmprint framework achieving the EER of as low as 1.01%. The proposed approach also outperforms the popular LBP approach significantly. In future, deep learning based algorithms to extract the highly distinctive and expedient features can be explored.

References

1. Jain, A.K., Ross, A., Prabhakar, S.: An introduction to biometric recognition. IEEE Trans. Circ. Syst. Video Technol. **14**(1), 4–20 (2004)
2. Jain, A.K., Ross, A., Pankanti, S.: Biometrics: a tool for information security. IEEE Trans Inf Forensics Secur **1**(2), 125–143 (2006)
3. Ross, A., Jain, A.K.: Multimodal biometrics: an overview. In: 2004 12th European Signal Processing Conference, Vienna, Austria, pp. 1221–1224 (2004)
4. Srivastava, R.: A score level fusion approach for multimodal biometric fusion. Int. J. Sci. Technol. Res. **9**(01), 4241–4245 (2020)

5. Ma, Y., Huang, Z., Wang, X., Huang, K.: An overview of multimodal biometrics using the face and ear. Math. Prob. Eng. **2020**(Article ID 6802905), 17 (2020)
6. Ali, Z., Hossain, M.S., Muhammad, G., Ullah, I., Abachi, H., Alamri, A.: Edge-centric multimodal authentication system using encrypted biometric templates. Futur. Gener. Comput. Syst. **85**, 76–87 (2018)
7. Mohamed Elhoseny, A.E.H., Elkhateb, A., Sahlol, A.: Multimodal biometric personal identification and verification. Adv. Soft Comput. Mach. Learn. Image Process. Stud. Comput. Intell. **730**, 249–276 (2018)
8. Gunasekaran, K., Raja, J., Pitchai, R.: Deep multimodal biometric recognition using contourlet derivative weighted rank fusion with human face, fingerprint and iris images. Automatika **60**(3), 253–265 (2019)
9. Kondapi, L., Rattani, A., Derakhshani, R.: Cross-illumination evaluation of hand crafted and deep features for fusion of selfie face and ocular biometrics. In: 2019 International Symposium on Technologies for Homeland Security (HST), Woburn, MA, USA, pp. 1–4 (2019)
10. Gupta, K., Singh, G., Kapil, S.: Novel approach for multimodal feature fusion to generate cancelable biometric. Vis. Comput. 0123456789 (2020)
11. Kalingani, S., Ekka, B.K.: Multimodal verification system using face and retina features. Int. Res. J. Mod. Eng. Technol. Sci. **2**(06), 251–259 (2020)
12. Singh, L.K., Khanna, M., Garg, H.: Multimodal biometric based on fusion of ridge features with minutiae features and face features. Int. J. Inf. Syst. Model. Des. (IJISMD) **11**(1), 37–57 (2020)
13. Mahmoud, R.O., Selim, M.M., Muhi, O.A.: Fusion time reduction of a feature level based multimodal biometric authentication system. Int. J. Sociotechnol. Knowl. Dev. (IJSKD) **12**(1), 67–83 (2020)
14. Attia, A., Akhtar, Z., Chahir, Y.: Feature-level fusion of major and minor dorsal finger knuckle patterns for person authentication. Signal, Image Video Process (2020)
15. Talreja, V., Member, S., Valenti, M.C., Nasrabadi, N.M.: Deep hashing for secure multimodal biometrics **16**, 1306–1321 (2021)
16. Jain, A.K., Nandakumar, K., Ross, A.: 50 years of biometric research: accomplishments, challenges, and opportunities, **79**, 80–105 (2016)
17. Vyas, R., Kanumuri, T., Sheoran, G.: Iris recognition using 2-D gabor filter and XOR-SUM Code. In: 1st India International Conference on Information Processing (IICIP), pp. 0–4 (2016)
18. Daugman, J., Downing, C.: Radial correlations in iris patterns, and mutual information within IrisCodes. IET Biometrics **8**(3), 185–189 (2019)
19. Naderi, H., Soleimani, B.H., Matwin, S.: Manifold learning of overcomplete feature spaces in a multimodal biometric recognition system of iris and palmprint. In: Proceedings of 2017 14th Conference on Computer Robot Vision, CRV 2017, vol. 2018, pp. 191–196 (2018)
20. Tamrakar, D., Khanna, P.: Palmprint verification with XOR-SUM Code. Signal, Image Video Process. **9**(3), 535–542 (2015)
21. Adamović, S., et al.: An efficient novel approach for iris recognition based on stylometric features and machine learning techniques. Futur. Gener. Comput. Syst. **107**, 144–157 (2020)
22. Attallah, B., Brik, Y., Chahir, Y., Djerioui, M.: Fusing palmprint, finger-knuckle-print for bi-modal recognition system based on LBP and BSIF, **44**, 89–103 (2019)
23. Ojala, T., Pietikainen, M., Maenpaa, T.: Multiresolution gray-scale and rotation invariant texture classification with local binary patterns. IEEE Trans. Pattern Anal. Mach. Intell. **24**(7), 971–987 (2002)

Enhanced Human Identification Technique Using Deep Learning

Shashi Shreya and Kakali Chatterjee

1 Introduction

The identification system is a new demand in the modern world with a high crime rate. If we talk about India, the National Crime Records Bureau (NCRB) which is under the Ministry of Home Affairs (India) (MHA) reported that cognizable crimes are about 51,56,172 comprising Indian Penal Code crimes of about 3,225,701 and Special Local Laws crimes of about 1,930,471 are registered nationwide as of 2019. It also states the annual increase in the registration of cases by 1.6%, still very few cases are being solved by law enforcement due to lack of evidence. So to speed up the increase in solved cases, identification systems are valuable. For identification, various physical, as well as behavioral characteristics of a human, are taken into consideration. One of the physical characteristics is latent fingerprint. Latent fingerprints are the unintentional evidence left by the criminal in the crime area that never lies [1].

From past 50 years, latent fingerprints are sensitive data to be presented for final verdict of the court [2–5]. For making it proper evidence, latent fingerprint goes through different phases because it contains noise as it is lifted from different surface depending upon the situation such as glass, wood, paper, and metal. These surfaces significantly vary in texture, pattern, or color. Not only the surface-based issues but also pressure-based issue like the pressure of the hand on the surface are responsible for the quality of latent fingerprint image. Therefore, it is a challenging

S. Shreya (✉) · K. Chatterjee
Computer Science and Engineering, National Institute of Technology Patna, Patna, Bihar 800005, India
e-mail: shashis.phd19.cs@nitp.ac.in

K. Chatterjee
e-mail: kakali@nitp.ac.in

M. Dua et al. (eds.), *Proceedings of the International Conference on Paradigms of Communication, Computing and Data Sciences*, Algorithms for Intelligent Systems,
https://doi.org/10.1007/978-981-16-5747-4_57

task to model/predict the latent fingerprint. Before feature extraction, these fingerprints are preprocessed. The preprocessing includes noise removal, reason of interest (ROI) segmentation, and enhancement. After preprocessing, the human identification system, proceed with feature extraction and matching to get a final decision.

Leaving behind the manual process, automated fingerprint recognition systems (AFRS) are in practice. These systems can do all the steps automatically. It identifies based on the fingerprint features of three different levels like ridge orientation (level-1), minutiae (level-2) and pores (level-3).

This paper is taking level-3 as its main focus. Level-3 features are pores, dots, incipient ridges, and ridge edge shapes. The mention of pores in this paper is actually the sweat pores that present on the ridge of a fingerprint from our birth. These pores, also have some features like the very small size and round in shape and are densely present in a small section, for example, more than 9 pores can be found within a cm of the ridge. So pores are considered as a new add-on parameter apart from minutiae and ridge orientation for human identification processes using fingerprint. Pores play a heroic role in the case of a small portion of the fingerprint image, as sufficient pores can be collected for identification. And this is the reason that level-3 should be helpful in identification. The major contribution of the paper is-

- Evaluating deep learning method to extract the level-3 feature of latent fingerprint and correctly matching with their labels using the many-to-one concept.
- Comparing it with other pore detection and extraction methods like Gabor filter, inversion, and convolution neural network (CNN).

The organization of the paper is as follows, Sect. 2 describes the literature survey; Sect. 3 discusses the proposed framework; Sect. 4 provides a detailed description of results and performance analysis; Sect. 5 concludes the paper with future work discussion.

2 Literature Survey

This section describes the survey performed on level-3 features of the latent fingerprint. The features that are included in level-3 are dimensional features (pores, dots, incipient ridges, and ridge edge shapes) of fingerprint ridges. There are various methods for pore extraction that uses a static isotropic pore model to detected pores. In every passing time, the pore extraction method is also changing. The literature survey is summarized in Table 1. After the literature survey, a human identification system based on level-3 feature extraction is proposed.

Table 1 Literature survey

Paper	Year	Description	Limitations
[6]	1994	Automated system for fingerprint authentication using pores and ridges structure	Image is binarized and then skeletonize Binarization is a process of putting vales of 0 and 1 in the image. After that, skeletonization is performed. This skeleton image ridge is then tracked by some method and then pores are extracted. This whole process is time-taking as well as expensive, noise sensitivity, and strictly working on high definition images or high-quality images
[7]	2004	Extraction of level-2 and level-3 features for fragmentary fingerprints	Binarization and skeletonization
[8]	2004	Study of the distinctiveness of level-2 and level-3 features in fragmentary fingerprint comparison	Binarization and skeletonization
[9]	2005	A novel approach to fingerprint pore extraction	It uses modified two-dimensional Gaussian function. This model nature is isotropic, but when it comes to open pores of rolled or fingerprint, it is not isotropic
[10]	2006	Pores and ridges: Fingerprint matching using level 3 features	Jain's pore model is introduced by Jain et al. that uses Mexican hat wavelet transform for pore extraction. It observes the facts that the area of the pore have maximum negative frequency due to abrupt change in intensity values from white to black at the pores. But this model is limited to some specific database. It is useless for other data-sets
[11]	2008	Rotationally invariant statistics for examining the evidence from the pores in fingerprints	DoG (Difference of Gaussian) pore model is introduced by Parsons et al. This pore detection is done with the help of a band-pass filter. It considers circle looking feature as pore. This model is also anisotropic
[12]	2010	Adaptive fingerprint pore modeling and extraction	It is a dynamic anisotropic pore model It uses orientation along with scale parameters This model also have limitation like it cannot adopt difference in shape and size for several person

(continued)

Table 1 (continued)

Paper	Year	Description	Limitations
[13]	2016	Toward touchless pore fingerprint biometrics: a neural approach	It is a theory of extracting a level-3 feature from the image of the fingerprint. The used image is self-captured and are touchless acquiesced
[14]	2017	Deeppore: fingerprint pore extraction using deep convolutional neural networks	It is a pore extraction method is proposed using deep ConvNet (CNN) that has a method that shows pores coordinates of any fingerprint images. But CNN is not suitable because pores in latent fingerprints are very small and sensitive to image normalization techniques
[15]	2018	A novel pore extraction method for heterogeneous fingerprint images using convolutional neural networks	It is a pore extraction method is proposed using ConvNet (CNN). Again pores in latent fingerprints are very small and sensitive to image normalization techniques
[16]	2019	End-to-end pore extraction and matching in latent fingerprints: Going beyond minutiae	Have very marginal improvement

3 Proposed Framework

The identification system works in two stages: enrollment and identification stages as shown in Fig. 1.

3.1 Enrollment Stage

- **Image Enhancement**: Enhancement is the process that makes an image better for feature extraction. A latent fingerprint is the type of prints that get deposited on surfaces and remain invisible through the naked eye. During the development of the latent fingerprint through chemical treatment, various unwanted things (background noise) also come along with the region of interest (ROI). The enhancement process helps to remove these background noise and provide a clear image in which reliable friction ridges is available. It helps in further processing. Two basic techniques, (i) short-time Fourier transform (STFT) [17], and (ii) cartoon-texture decomposition [18] are used in this paper for the process of enhancement.
- **Feature Extraction**: The detection accuracy is not satisfactory of abovementioned proposal of literature survey due to the limited approximation capability of static isotropic models to various types of pores. CNN, from a machine

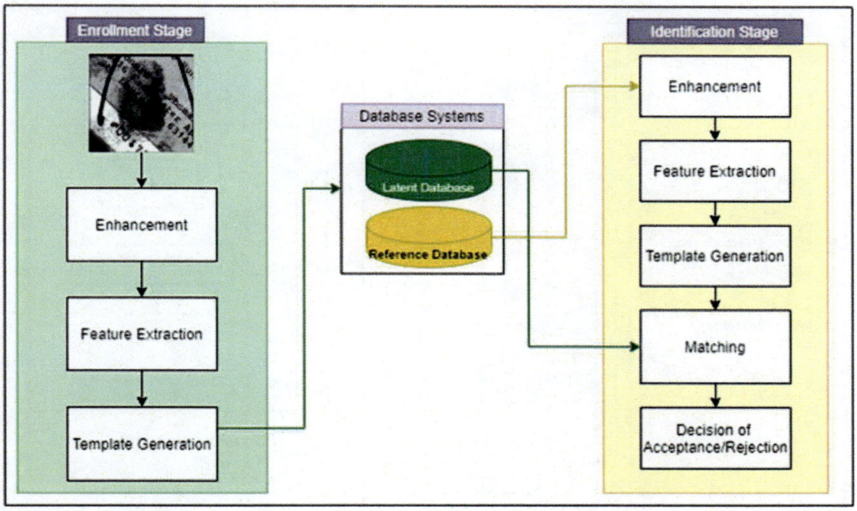

Fig. 1 Identification framework

learning perspective, is working well for the pore extraction but to yield the best result, DeepPore-based CNN is an efficient way. The advantage of D-CNN (deep convolution neural network) is its receptive field, whose size is bigger. This helps in analyzing more information present over the latent fingerprint image along with precisely detected pore. In brief, CNN with depth x and filter size (3×3) pixels, have receptive field of size $(2x + 1) \times (2x + 1)$. Based on this thought, a framework is designed and shown in Fig. 2.

Here, pore extraction is done in three steps described as follows:

Step-1: Pore Detection Network working: To detect pores, a network is designed as C_{DP} which takes image I of $m \times n$. The architecture of the CNN of this paper is based on deep CNN. It is influenced by the CNN visual geometry group (VGG) where training is performed in a supervised learning manner. In this paper, DeepPore-based CNN architecture performs feature extraction in k-layers. Each layer has 64 filters of size (3×3). This architecture is more described in having k learnable layer and $(k - 1)$ nonlinearity function. This is because each layer, from first till $(k - 1)$ performs three different works, i.e., (1) convolution, (2) batch normalization, and (3) rectified linear units (ReLU) function. The last layer, i.e., kth layer is restricted to do the only convolution. C_{DP}, gives the output as pore intensity map of the input latent fingerprint I. The size of the output is same as the input image because zeros are padded before each layer of convolution work.

Training: Aim of C_{DP} training is basically to yield a pore intensity map. The map has enhanced pores only because the background is reduced. The information (pore) is provided in the map using soft labels, which describes low-value labels to the

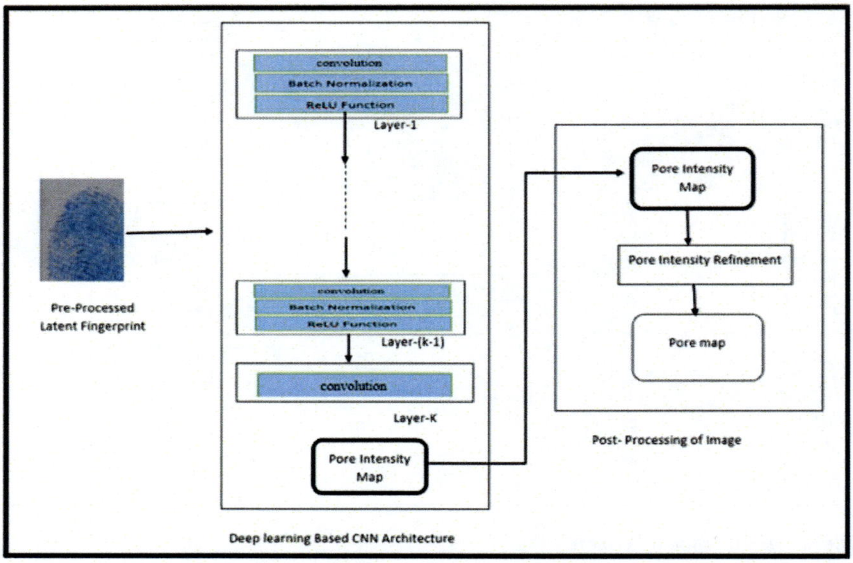

Fig. 2 Proposed framework for feature extraction

maximum Euclidean distance to the pores and high-value label to the minimum distance to the pore. Mathematically, let us suppose a and b denotes the latent image and labels of its pore intensity map. The training dataset is given as:

$$a_i, b_i^n \; i = 1$$

where $n =$ number of batch. Hence, if the model 'l' is given, then it is trained to predict $\hat{b} = l(a)$. Here, $\hat{b}$ is the estimate defined in the pore intensity map.

Step-2: Pore Intensity Refinement: The intensity map of the pore is a mountain-like structure whose peaks are pore pixels. This intensity of the peak varies mostly due to its associated ridge thickness and its type, i.e., close or open. So it is not possible to apply the binary threshold concept over the whole map. Here, the concept of local maxima is used for the pore estimation. It states that the estimation of local maxima is performed only when the intensity of the pore is equal to or larger than the pore pixel.

Step-3: Estimation of Pore Coordinates After the pore estimation, the final pore map gets generated by assigning 1 to the pore pixel coordinate and 0 to the others.

Template Generation: Template of latent image is finally generated, say T^{let} and given by,

$$T^{\text{let}} = \{p = (x, y)\}_{i=1}^{t} \tag{1}$$

Here, p is pore on the latent image with labeled coordinate (x, y), and 't' is the total number of pores extracted from the image.

3.2 *Identification Stage*

The identification stage identifies the claimed person through various processes. Half of the process (up-to pore extraction and template generation) is the same as the enrollment stage. And then, two more processes of matching and decision making are also included in this identification stage. Matching is done using the bipartite graph technique introduced in [16] and accordingly decision is made. The decision is also dependent on a threshold value because matching is done in a one-to-many fashion. The decision is accepted when the matching score is within the threshold value otherwise rejected.

4 Implementation Results and Performance

Features of the latent fingerprint are very sensitive and personal data of a person. A person cannot deny his presence if his fingerprints are collected and claimed as suspected during any criminal investigation. Therefore, many researchers are focusing on improving performance and accuracy. This paper is also contributing to the increase in performance. Data-sets of [15, 16, 19] are not available currently, so for this paper, latent fingerprint dataset is created in our own laboratory. For this, not only the forensic procedures, a document scanner is also used. The dataset comprises 30 labeled latent fingerprints. The characteristic of these images is: (1) 320×240 *pixels*, (2) 1200 DPI, which is considered as high resolution. A total of 12,085 pores are extracted, and coordinates are estimated with the method proposed in the paper.

For C_{DP} training, each one of the latent fingerprint images is partitioned in an overlapped sub-block of size (80×80) pixel. A total of 86,240 sub-blocks are generated to be used in the experiment. Data augmentation using flips and rotation counter-clockwise is also practiced to gather enough training dataset. The receptive field by the sub-block is accommodated of size (21×21) with the learnable size is $(d = 10)$. As mentioned above, the activation function used is ReLU. Around 0.01 rate is set for base learning and five batches are used. The experiment is trained for 100 epochs. Adaptive learning rates are computed through the adaptive moment estimation (ADAM) in order to optimize network learning. The threshold for the pore distance is taken as five and pore intensity refinement sliding window is 11×11. For detecting pore, the threshold taken is 0.25, where the sliding window and pore distance threshold are adjusted according to the better performance.

The proposed method is compared with other existing models such as the inversion method, Gabor filtering and CNN for pore extraction. For testing the proposed deep

Table 2 Average performance metrics in percentage (standard deviation)

	Gabor Filter	Inversion	CNN (*CD*)	C_{DP}
R_{Tr}	33.8 (28.8)	34.8 (13.0)	52.5 (31.3)	78.15 (10.2)
R_{fls}	70.8 (10.8)	79.3 (7.9)	25.0 (10.2)	12.65 (5.5)

Table 3 Average performance metrics in percentage (standard deviation)

	Gabor Filter	Inversion	CNN (*CD*)	C_{DP}
R_{Tr}	45.2 (18.1)	45.0 (25.4)	63.6 (15.0)	92.09 (4.63)
R_{fls}	41.9 (20.3)	55.0 (27.1)	17.5 (10.6)	9.64 (4.15)

learning-based approach, N-fold ($N = 5$) validation strategy is followed to estimate the coordinates of pore, i.e., folds for training is 3, validation is 1, and testing is 1. For accuracy measurement of pore extraction two concepts: True detection rate R_{tr} and R_{fls}. R_{tr} is the ratio of true detected pores to all pores, whereas R_{fls} is the ratio of false detected pores to all pores. The true pore is considered only when the Euclidean distance between coordinates of the true pore to detected pore is less than pixel P. Here, P is set through $\frac{\mathrm{avg}}{2}$, where 'avg' is the average ridge width of the specific latent fingerprint image. The average performance metrics are shown in Table 2.

As this paper is all about latent fingerprint, it is not necessary that we can get the whole image due to the presence of noise. So for this, a subset of the latent fingerprint is taken that contains only 20 images of 100×100 pixels size. And again, the whole process is repeated. The results are shown in Table 3.

4.1 Limitations of the Study

The limitation of the paper is the clear background of the image. If the work was performed on the original latent dataset, the results would be different and more precise. But as mentioned above, the dataset is created, not collected from the forensic database. It may become more complex through any random entry.

5 Conclusion and Future Research

The existing pore extraction method depends upon the individual, area, pressure, and type of pore. This method is an adaptive method that is difficult in nature. So to overcome the problem, this paper has presented a pore extraction framework based on deep learning. From Tables 1 and 2, it is seen that the DeepPore method performs better than others by a sufficient margin.

In the future, various other methods can be applied to experiment more. Also, a higher resolution image can be taken into consideration. Issues mentioned in Sect. 4.1 can be reduced.

References

1. Anush, S., Mayank, V., Richa, S.: Latent fingerprint matching: a survey. IEEE access. IEEE Biometric Conpendium RFIC Vitual J. **2**, 982–1004 (2014)
2. Li, J., Feng, J., Jay Kuo, C.-C.: Deep convolutional neural network for latent fingerprint enhancement. Signal Process. Image Commun. **60**, 52–63 (2018)
3. Cao, K., Jain, A.K.: Automated latent fingerprint recognition. IEEE Trans. Pattern Anal. Mach. Intell. **41**(4), 788–800 (2018)
4. Cao, K., Jain, A.K.: Latent fingerprint recognition: role of texture template. In: 2018 IEEE 9th International Conference on Biometrics Theory, Applications and Systems (BTAS). IEEE (2018)
5. Manickam, A., et al.: Score level based latent fingerprint enhancement and matching using SIFT feature. Multim. Tools Appl. **78**(3), 3065–3085 (2019)
6. Stosz, J.D., Alyea, L.A.: Automated system for fingerprint authentication using pores and ridge structure. In: Automatic Systems for the Identification and Inspection of Humans, vol. 2277. International Society for Optics and Photonics (1994)
7. Kryszczuk, K., Drygajlo, A., Morier, P.: Extraction of level 2 and level 3 features for fragmentary fingerprints. In: Proceedings of 2nd COST275 Workshop, vol. 8388 (2004)
8. Kryszczuk, K.M., Morier, P., Drygajlo, A.: Study of the distinctiveness of level 2 and level 3 features in fragmentary fingerprint comparison. In: International Workshop on Biometric Authentication, pp. 124–133. Springer, Berlin (2004)
9. Ray, M., Meenen, P., Adhami, R.: A novel approach to fingerprint pore extraction. In: Proceedings of the Thirty-Seventh Southeastern Symposium on System Theory, SSST'05, pp. 282–286. IEEE (2005)
10. Jain, A., Chen, Y., Demirkus, M.: Pores and ridges: fingerprint matching using level 3 features. In: 18th International Conference on Pattern Recognition (ICPR'06), vol. 4. IEEE (2006)
11. Parsons, N.R., et al.: Rotationally invariant statistics for examining the evidence from the pores in fingerprints. Law, Prob. Risk **7**(1), 1–14 (2008)
12. Zhao, Q., Zhang, D., Zhang, L., Luo, N.: Adaptive fingerprint pore modeling and extraction. Pattern Recogn. **43**(8), 2833–2844 (2010)
13. Genovese, A., Munoz, E., Piuri, V., Scotti, F., Sforza, G.: Towards touchless pore fingerprint biometrics: a neural approach. In: 2016 IEEE Congress on Evolutionary Computation (CEC), pp. 4265–4272. IEEE (2016)
14. Jang, H.-U., Kim, D., Mun, S.-M., Choi, S., Lee, H.-K.: Deeppore: fingerprint pore extraction using deep convolutional neural networks. IEEE Signal Process. Lett. **24**(12), 1808–1812 (2017)
15. Labati, R.D., Genovese, A., Munoz, E., Piuri, V., Scotti, F.: A novel pore extraction method for heterogeneous fingerprint images using convolutional neural networks. Pattern Recogn. Lett. **113**, 58–66 (2018)
16. Nguyen,D-L., Jain, A.K.: End-to-End Pore Extraction and Matching In Latent Fingerprints: Going Beyond Minutiae. arXiv preprint arXiv:1905.11472 (2019)
17. Chikkerur, S., Cartwright, A.N., Govindaraju, V.: Fingerprint enhancement using stft analysis. Pattern Recogn. **40**(1), 198–211 (2007)
18. Buades, A., et al.: Cartoon+ texture image decomposition. Image Process. Line **1**, 200–207 (2011)
19. Zhao, Q., Feng, J., Jain, A.K.: Latent fingerprint matching: utility of level 3 features. MSU Tech. Rep. **8**, 1–30 (2010)

Hyperparameter Tuning of Dense Neural Network for ECG Signal Classification

S. Clement Virgeniya and E. Ramaraj

1 Introduction

In the growing era of big data, classification plays a supreme role. In a classification problem, data with labels (i.e., training data) is used to assign a label to new data (i.e., testing data) using a set of rules of training data [1]. Although there are many classification algorithms, depending upon the problem, the algorithm varies.

Deep learning, one such algorithm is indeed an essential advancing system in today's science and technology. The successful implementation of deep learning algorithms highly depends on hyperparameters. It is not quite easy to tune the hyperparameters. It is all based on trial and error. Certain models are very sensitive in setting hyperparameters. Tuning the model with optimal value is the current issue in the PTB-XL dataset [2, 3] for ECG classification. Moreover, hyperparameters like learning rate, momentum, batch size, weight decay, etc., each have a specific role in training a model.

In the proposed work, data from PTB-XL is currently used for ECG signal classification. Initially, data is preprocessed and converted into multiple labels. Since the dataset consists of five different labels, the target is a multilabel classification problem. Label powerset technique is used to handle multiple labels in the dataset. The dataset is trained using multilayer perceptron (MLP); the results are observed and then the dataset is trained using DNN with and without hyperparameters mentioned above. It is validated with k fold cross validation using the grid search technique. At the storage level, we use MongoDB.

Tinu et al. [4] in his work underwent hyperparameter optimization using Bayesian optimization. The author divided the data into several chunks and used transfer learning to get further information from the data. The author also evaluated it using two machine learning algorithms and proved that the proposed method used less

S. Clement Virgeniya (✉) · E. Ramaraj
Department of Computer Science, Alagappa University, Karaikudi, India

M. Dua et al. (eds.), *Proceedings of the International Conference on Paradigms of Communication, Computing and Data Sciences*, Algorithms for Intelligent Systems,
https://doi.org/10.1007/978-981-16-5747-4_58

computational time compared to other state of art techniques. Rafael et al. [5] proposed hyperparameter optimization for the decision tree algorithm. The author used four different algorithms using 102 datasets to evaluate how sensitive is decision tree algorithm. Although the authors achieved low average improvement, in most of the datasets, the results are satisfactory. James et al. [6] introduced hyperopt, a reusable engine for hyperparameter optimization. It contains inbuilt optimization algorithms. Hyperopt parallelizes sequential optimization and updates the results asynchronously. It is moreover a hyperparameter tool. Aaron et al. [7] used a new Bayesian optimization method using the entropy search variant. The authors achieved good performance in terms of computation. Nowadays many automatic hyperparameter optimizations have resolved the previous issues, and in some cases, they even surpassed them. One of the toughest jobs of automatic hyperparameter optimization is that it requires high computational resources [8]. Matthias et al. [9] "discussed most prominent Hyperparameter optimization techniques like Bayesian and Black box function optimization methods." Since these methods are computationally costly, the author discussed how to solve these issues and discussed its alternatives. Zhang et al. [10] proposed a modified sequential number theoretical optimization algorithm. The author showed promising improvements in tuning machine leaning algorithms using this modified optimization algorithm with sequential uniform designs.

Authors contribution to the proposed work include data preprocessing, training data with MLP for comparison, training data with DNN with and without hyperparameters, using grid search cross validation technique and plotting the results in terms of the learning rate, activation function, dropouts, momentum, weights, epochs, batch size, optimizers, etc. In the rest of the paper, Sect. 2 describes the Preliminaries needed. Section 3 describes the Methodology. Section 4 displays the Experimental Results and Sect. 5 ends up with a Conclusion.

2 Preliminaries

This section discusses the preliminaries used in training the dataset for classification.

2.1 Stochastic Gradient Decent (SGD)

Gradient descent is a repetitive algorithm whose goal is to find a minimal point in the function as it reaches down the slope. Here for each iteration, the entire dataset is used until minima are reached. This algorithm is difficult in case the dataset size is big. "SGD compared to gradient descent is a repetitive method that inconstantly selects one data point from the whole dataset at each iteration to reduce the computations enormously" [11]. For every iteration "*i*" in a range, SGD is given in Eq. (1).

$$\theta_j = \theta_j - \alpha(\hat{y}^i - y^i)x_j^i \tag{1}$$

The stochastic algorithm does not always remember the previous repetitions since it optimizes the expected risk directly as the samples drawn are random.

2.2 Root Mean Square Propagation (RMSprop)

RMSprop usually moves the squared gradients for every weight. It then divides the gradient by taking the square root of the mean square. It is the same as gradient descent but shows slight variations in the update of parameters [12, 13]. The learning rate adapted for each parameter is given in Eq. (4).

$$c_j^1 \; and \; r_j^1, \forall \mathbf{j} \in \{1, \ldots, \mathrm{s}_1\} \; and \; \forall 1 \in \{1, \ldots, \mathrm{L}\} \tag{2}$$

$$E\left[g^2\right]_t = 0.9E\left[g^2\right]_{t-1} + 0.1g_t^2 \tag{3}$$

$$\theta_{t+1} = \theta_t - \frac{\eta}{\sqrt{E\left[g^2\right]_t + \epsilon}} gt \tag{4}$$

where η is the learning rate.

2.3 Adam

Adaptive moment estimation (Adam) is an optimization algorithm where the learning rate is maintained for each parameter and individually adjusted as learning unfolds. "Adam is a combination of RMSprop and Stochastic Gradient Descent with momentum. Adam uses squared gradients to scale the learning rate like RMSprop and it takes advantage of momentum by using moving average of the gradient instead of the gradient itself like SGD with momentum" [14, 15].

The preliminaries described above are optimizers that optimize the entire neural network. Therefore, the best optimizer is chosen through trial and error for performance enhancement.

3 Methodology

The various steps starting from preprocessing to hyperparameter tuning are discussed below in the fore coming section.

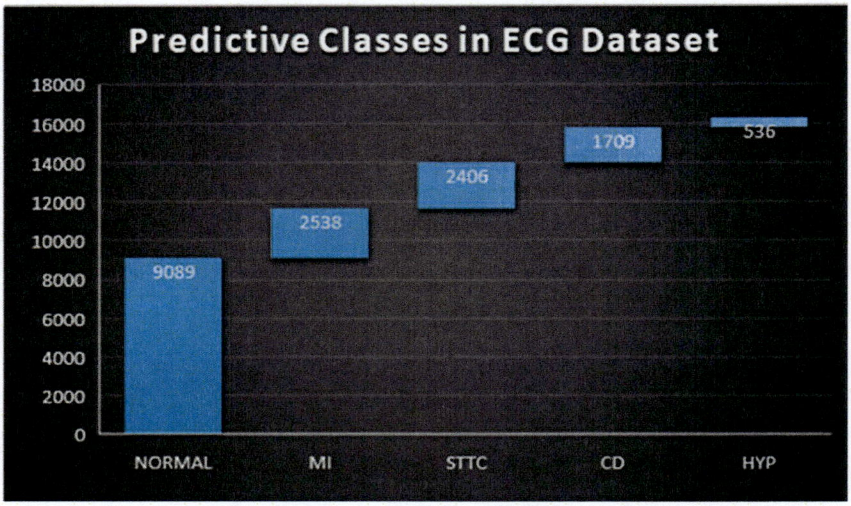

Fig. 1 Count of classes in ECG dataset

3.1 Data Preprocessing

In PTB-XL dataset, there are 21,837 records. Null values are removed and columns like height and weight are reduced to BMI (body mass index). The dataset consists of five different predictive outcomes. These outcomes are divided into different labels namely normal ECG (NORM), conduction disturbance (CD), myocardial infarction (MI), hypertrophy (HYP), and ST/T change (STTC). This belongs to the multilabel classification problem.

Null values in the dataset are imputed using KNN imputation. There are also null values in the target variable. These values are dropped, and the dataset is shaped to 16,278 records. Since the target is a multilabel classification problem, label powerset is used to handle multilabel. "Here one multi-class classifier is trained on all each of the five labels" [16]. The count of different classes is given in Fig. 1.

3.2 Multilayer Perceptron (MLP)

Initially, the dataset is trained using MLP. Since MLP undergoes only feedforward, there are no adjustments of weights and it consists of only one hidden layer. The dataset used in the proposed work does not yield bearable accuracy using MLP. So, more than two hidden layers are needed to increase the accuracy. The steps involved in the proposed work are shown in Fig. 2.

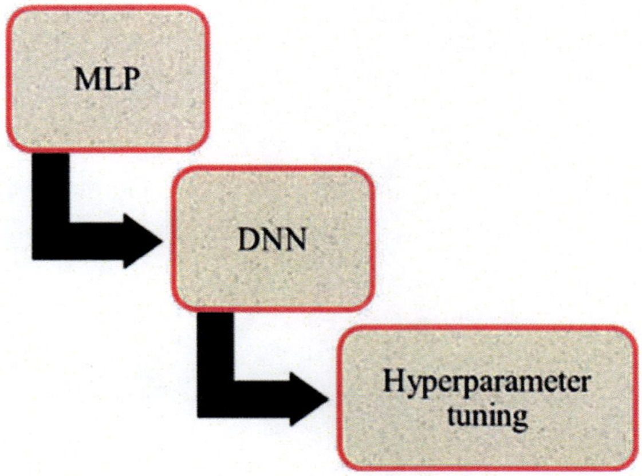

Fig. 2 Steps involved in the proposed work

3.3 Dense Neural Network (DNN)

DNN is an artificial neural network with a feedforward pass. It consists of multiple hidden layers between input and output layers. Every hidden unit '*j*' contains a nonlinear activation function $f(.)$ which maps all inputs from lower layer x_j to scalar state y_j which is later fed into upper layer [17]. 'b_j' in the Eq. (5) represents bias of unit '*j*', '*i*' is the index of the lower layer, and w_{ij} is the weights of unit '*i*', and '*j*'.

$$x_j = b_j + \sum_i y_i w_{ij} \tag{5}$$

3.4 Hyperparameter Tuning (HT)

Tuning hyperparameters is one way to increase the accuracy of the model. In our dataset, an increase from 52 to 79% is achieved via Tuning. HT largely affects the deep learning model. It is chosen only through trial and error. HT or optimization problem is given in Eq. (6)

$$H = H_1 * H_2 * \ldots H_k \tag{6}$$

where HT tuning is given as $a \in A, H, and\ D \in D$ is given in Eq. (7)

$$h^* \arg_{h \in H} \max f(a, D, h) \tag{7}$$

HT is usually of two types, manual and automatic [18]. Automatic HT usually optimizes the process. There are several methods a HT gets tuned. In our work, we

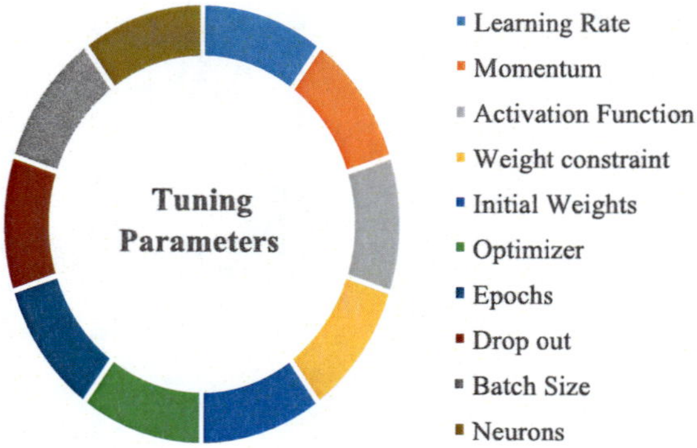

Fig. 3 Different parameters used in tuning a DNN

trained the model using grid search. Usually, a grid of possible values is created. Each repetition tries a different combination of hyperparameters in a specific order. It fits the model and records the model performance and, finally, returns the best model with the best hyperparameters. The tool used for HT optimization is Scikit–learn. The different parameters used in training the model are given in Fig. 3.

In this section, we discussed MLP and DNN with and without tuning. How to increase the accuracy from a simple MLP layer to a more complicated DNN layer by using the best optimization technique. Finally, the parameters that suit them best are given as input to train the dataset using DNN.

4 Experimental Results

The experiments were carried out using PTB-XL dataset. Initially, the dataset is preprocessed and reduced to 16,278 records. Since it is a multilabel classification problem, a label powerset is used to handle the target data. Afterward, the dataset is trained using MLP and DNN (with and without optimization) and the results are compared and tabulated. The different tuning parameters and their best values using Hyperparameter optimization of DNN are tabulated in Table 1. The DNN classifier is optimized using grid search cross validation and performs best with 0.001 learning rate, Relu activation function, RMSprop optimizer, with a batch size of 100 and momentum of 0.5. It is also noted that the dropout rate is null here. Depending upon the dataset, the values for tuning parameters vary. This is based on several trials and errors. Other attributes like epochs, neurons, dropouts, and weight constraints are also taken into account and given due importance at every stage.

After undergoing many observations, our model showed increasing accuracy at all levels. First of all, the lowest accuracy of 52% is achieved, and this value is achieved in multilayer perceptron. Later, DNN without using any optimizers is tested. It showed a slight increase from 52 to 55%. Although this is not good accuracy to consider,

Table 1 Parameters that suits best for the DNN model with the best accuracy

Tuning parameters	Grid search
Learning Rate	0.001
Momentum	0.5
Activation	Relu
Initial weights	he_uniform
Optimizer	RMSprop
Drop out	0.0
Neurons	64
Batch size	100
Weight constraint	1
Epochs	10
Accuracy	79

Table 2 Model and accuracy

Model	Accuracy (in %)
MLP	52
DNN (before optimization)	55
DNN (after optimization)	79

later training the samples with different optimizers proved to give an outstanding result.

DNN after undergoing hyperparameter optimization shows the best results with an increase of 27%. Tuning usually takes some time to execute. But choosing the right set of parameters is an art. Table 2 shows the results of different models with their accuracy.

5 Conclusion

In this paper, we have investigated MLP, DNN, and tuning DNN with different hyperparameters to improve the accuracy of the model. The overall accuracy of the model is increased by tuning the model with different parameters and with different values. Tuned DNN models were compared with MLP and DNN without tuning. It is observed that an increase of nearly 27% is achieved using DNN with grid search cross validation technique. The dataset from PTB-Xl is used for tuning the hyperparameters. It is indeed a good increase in terms of the accuracy of the model stated. Results showed that DNN while tuning achieved an accuracy of 79% from 52% using MLP and 55% using DNN without tuning.

Acknowledgements This Paper has been written with the financial support of RUSA—Phase 2.0 grant sanctioned vide Letter No. F.24-51/2014-U, Policy (TNMulti-Gen), Dept. of Edn. Govt. of India, Dt. 09.10.2018.

References

1. Suthaharan, S.: Machine Learning Models and Algorithms for Big Data Classification, vol. 36 (2016)
2. Goldberger Ary, L., Amaral Luis, A.N., Leon Glass, H.J.M., Ivanov Plamen, Ch., Mark Roger, G., et al.: PhysioBank, PhysioToolkit, PhysioNet. Circ. vol. 101, pp. e215–e220 (2000)
3. Wagner, P., et al.: PTB-XL, a large publicly available electrocardiography dataset. Sci. Data **7**(1) (2020). https://doi.org/10.1038/s41597-020-0495-6
4. Joy, T.T., Rana, S., Gupta, S., Venkatesh, S.: Hyperparameter tuning for big data using Bayesian optimisation. In: Proceedings of International Conference on Pattern Recognition, vol. 0, pp. 2574–2579 (2016). https://doi.org/10.1109/icpr.2016.7900023
5. Mantovani, R.G., Horvath, T., Cerri, R., Vanschoren, J., De Carvalho, A.C.P.L.F.: Hyperparameter tuning of a decision tree induction algorithm. In: Proceedings of 2016 5th Brazilian Conference on Intelligence Systems BRACIS 2016, pp. 37–42 (2017). https://doi.org/10.1109/bracis.2016.018
6. Bergstra, J., Bardenet, R., Kégl, B., Bengio, Y.: Implementations of algorithms for hyperparameter optimization. In: NIPS Work. Bayesian Optimus, p. 29 (2011) [Online]. Available: https://papers.nips.cc/paper/4443-algorithms-for-hyper-parameter-optimization.pdf
7. Klein, A., Bartels, S., Falkner, S., Hennig, P., Hutter, F.: Towards efficient bayesian optimization for big data. Adv. Neural Inf. Process. Syst. 1–5 (2015) [Online]. Available: file:///Files/89/892ef754-0c81-42db-9c7c-e2ba70787acf.pdf
8. Domhan, T., Springenberg, J.T., Hutter, F.: Speeding up automatic hyperparameter optimization of deep neural networks by extrapolation of learning curves. International Joint Conferences on Artificial Intelligence, vol. 2015, pp. 3460–3468 (2015)
9. Feurer, M., Hutter, F.: Hyperparameter Optimization (2019)
10. Zhang, A., Yang, Z.: Hyperparameter tuning methods in automated machine learning. Sci. Sin. Math. **50**(5), 695–710 (2020). https://doi.org/10.1360/n012019-00092
11. Bottou, L.: 18 Stochastic Gradient Descent Tricks, pp. 421–422 (2012)
12. Kurbiel, T., Khaleghian, S.: Training of Deep Neural Networks based on Distance Measures using RMSProp. arXiv, pp. 1–6 (2017)
13. Tieleman, T., Hinton, G.: Lecture 6.5-rmsprop: divide the gradient by a running average of its recent magnitude. COURSERA Neural Netw. Mach. Learn. **4**(2), 26–31 (2012)
14. Bushaev, V., Bushaev, V.: Adam—Latest Trends in Deep Learning Optimization. https://towardsdatascience.com/adam-latest-trends-in-deep-learning-optimization-6be9a291375c
15. Bock, S., Goppold, J., Weiß, M.: An improvement of the convergence proof of the ADAM-Optimizer, arXiv, pp. 1–5 (2018)
16. Jain, S.: Solving Multi-Label Classification problems (2017). https://www.analyticsvidhya.com/blog/2017/08/introduction-to-multi-label-classification/
17. Yao Qian, F.K.S., Fan, Y., Hu, W.: On the Training Aspects of Deep Neural Network (DNN) for Parametric TTS Synthesis, vol. 63, pp. 3829–3833 (2014). https://doi.org/10.1109/icassp.2014.6854318
18. ES, S.: Hyperparameter Tuning in Python: A Complete Guide 2021 (2021). https://neptune.ai/blog/hyperparameter-tuning-in-python-a-complete-guide-2020

Detection of Lung Malignancy Using SqueezeNet-Fc Deep Learning Classification Technique

Vinod Kumar and Brijesh Bakariya

1 Introduction

According to the World Health Organization, in 2017, 819.570 or 9.30% of mortality were observed to induce lung disease in India. Indians get the second-highest maturity level mortality rate in lung disease in the world of 89.36 with 100.000 rises. The most prevalent type of cancer is in males relative to females with such a 4.5:1 proportion at 93 and 23 nations, it is the lead cause of mortality. The number of deaths in males and females probably due to lung cancer in 2018 was 1,184,947 and 576,060 [1]. During the first quarter of the generation, a growing epidemic of lung cancer is expected to increase because mortality and morbidity of lung cancer and survivorship throughout the world are growing. The growth of signs during the first acute evaluation in certain lung cancer patients has been delayed significantly. But many sufferers are diagnosed with the disease often at advanced stages, so most of the deaths are caused a year after their prognosis [2].

CT images yet very complex analyses are held by radiologists. As severe non-transmissible diseases like cancer occurred across both sexes caused by lack of hygiene, insufficient facilities, and highly developed standard of living, the situation started changing drastically [3]. Living beings mortality in 2018, especially in lung cancer, is supposed to drop to 9.6, 1.76 million, respectively [4]. Lung cancer is diagrammatically diagnosed with its stages like (i) IA1, IA2, IA3, IB, (ii) IIA, IIB (iii) IIIA, IIIB, IIIC (iv) IVA, IVB related to primary tumor cell location in the image [5]. A combination of T, N, and M is available for them as a whole level of 0, I, II, III, IV as benign or malign.

V. Kumar (✉) · B. Bakariya
I. K. Gujral Punjab Technical University, Kapurthala, Punjab, India

B. Bakariya
e-mail: dr.brijeshbakariya@ptu.ac.in

M. Dua et al. (eds.), *Proceedings of the International Conference on Paradigms of Communication, Computing and Data Sciences*, Algorithms for Intelligent Systems,
https://doi.org/10.1007/978-981-16-5747-4_59

CT is now the most common trend in the clinical diagnosis of cancer. A pulmonary nodule is a concentrated pulmonary abnormality usually seeing it as a circular of widths 3–30 mm in digital images. As any examination offers the most detailed knowledge on the tumor, a specific patent of lung nodules is quite critical [6]. The deep neural network is a part of computer vision and has been extensively used for positive outcomes for medical image analysis and is expected to receive three hundred million dollars in the medical image segment by 2021 [7]. Yet accessibility and usability are still the two biggest challenges of deep learning. It is directly extracted to neural nets incorporate while a deep network can itself delete the functionalities of many layers [8]. Deep learning provides a few advantages: (a) it can be extracted directly from a test phase. (b) The function's selection is simple. (c) Just same scope, the procedure of learning can involve important aspects in the extraction, selection of features, and classification. The recent study of deep chest diagnostic tests is an effective area of study. Quite a few deep learning networks are currently being seen including ResNet, GoogleNet, AlexNet, DenseNet, and SqueezeNet with different classifiers like SVM, LDA, GNB, and KNN.

It is sometimes difficult to classify as well as divide a nodule from their unrevealed role and reduce the severity. It is a long-term medical issue with such a smaller true positive rate. With the advent of innovative imaging technology, the amount of image files has exploded. Health professionals will often not manually cut the lesion with CT images because anyone can create thousands of CT scan images of a lung. It facilitates awareness and also the success of the monitoring process by healthcare professionals. It performed multiple threshold analyses to separate the nodules and that can be easily identified medically formation. Lung CT images, therefore, make it hard for oncology and thoracic initiatives to differentiate between accurate lung nodules. In this way, the false-positive rate rises.

Work in progress: The machine mainly enhances precision, decreases diagnostic imaging time, identifying tumors, and also its positioning. This has carried out various studies to test its most common and effective form of lung cancer. It tests and increases the sensitivity, accuracy, and precision of lung nodules. Its analysis seeks to propose and discuss the ways lung cancer has been diagnosed.

Motivation: The main objectives of the approach are high efficiency, minimized diagnostic duration, and the detection of nodules and extracts its features. There were some findings in which the total impact of lung cancer was identified. The sensitivity, precision, and well-being of all nodules have been assessed. These investigations continue to improve and improve lung cancer identification.

2 Methodology

Many functionalities used to identify and measure the features are learned. The conventional model employs craft features, but again the standard neural network-based model supports healthcare professionals, as they can be able to understand the machine via their data (Fig. 1).

2.1 Image Dataset Description

Cancerous CT images are trailed by radiologists, oncologists, respectively. It would seem that the researcher has the challenge to categorize a CT tumor. Numerous images acquired are being identified by radiologists. It is all just widespread. Therefore, 1010 is perceived to an observant incident in LIDC-IDRI, with that kind of a large number of scanners for 244,527 images. Inability to emphasize the final view, technologists and third radiologists simultaneously investigated the characteristics. Again for XML-annotated data review its most nodes in three categories, like ≥ 3, nodule < 3, and non-nodule ≥ 3. The estimate of the cancer nodule is just above 3 and below 3. Another of the four radiologists reported such intertitles and described all such nodules within the third.

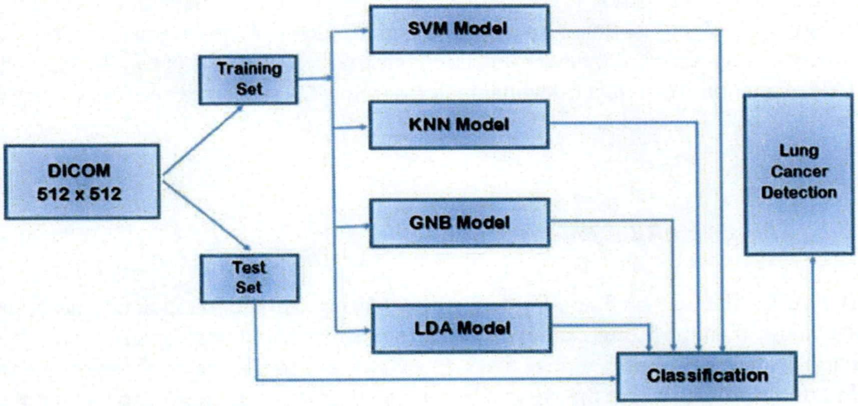

Fig. 1 The classification process diagram of the lung cancer detection system

2.2 Image Preprocessing

De-noising is critical for ensuring gray pixel density is widely supported by Gaussian noise, Poisson noise, and irregularities. Some techniques like a dictionary, flatter, and transform-based strategies would remove noisy pixels from scanning and low-dose images of CT [8]. Effective preprocessing to allow efficient separation of the lung nodule. It processed the lung CT image to Hounsfield Unit, which is binaries through the threshold, assigned the linked portion of the lung, isolated the left lobe and right lobe of the lung with dilation and erosion, determined the convex hull, dilated and integrated with left and right mask and filled the mask with luminance [9]. Pattern adjustment may be used in separating lung nodules. CNN's architecture such SqueezeNet usually requires input images with the same extent 227 × 227 image size, and therefore, it transformed the image through RGB to grayscale and implemented it dilation, segmentation, and erosion to get the mask of that image [10]. Even so, its comprehensive implementation about the R channel then skipped the B or G channel rather than conversion gray-scale to RGB image. It showed the performance of CNN was higher on unsmoothed images than it is on the smoothed image [11]. Even as lung volumes are much too massive that are of various sizes, going to feed the information in a deep neural network is problematic, and therefore, a resizing is required. CT images have been preprocessed and resized from 512 × 512 and 227 × 227 for 2D CT images based on maximum nodule size. Gaussian scale area filters or Gaber filters can also enhance the image [12]. Due to the difference in node size, interposition is also performed on preprocessed nodule images of a CT image by applying isotropic voxel and Gaussian filters to the original images and binary coding, threshold, binary reflection, obstruction, and so on [13]. It can apply field extraction. To extract triangular features from a CT scan, it can combine several methods of CT scanning.

2.3 Image Enhancement

A median filter, a 3 × 3 non-linear feature, is being chronic to efficiently decorate the image by minimizing natural distractions whereas a low-pass filtration has been chronic after putting off high-frequency components in conformity with limited photo interruptions and except the noise. The y is standard dislodgement with the aid of a Gaussian distribution. x differs beside its odd straight dosage then y differs through a foremost axis.

A Gabor filter, its almost strong technology because of analysis regarding texture, is a planar band-pass filter. Further elevated here is the region around the nodule. An equation indicates the announcements concerning the 2D Gabor filter. Next, α is the wavelength; so in course of the inclination, β out of grade in imitation of helm slices; μ is the segment of off-set. It describes the Gaussian and Gabor filter.

$$g(x, y) = \frac{1}{2\pi\sigma^2} \exp - \left(\frac{x^2 + y^2}{2\sigma^2} \right) \text{ 2D Gaussian Filter}$$

$$g(x, y) = \exp\left(\frac{x'^2 + \gamma'^2}{2\sigma^2} \right) \cos\left(2\pi \frac{x'}{\lambda} + \psi \right) \text{ 2D Gabor Filter}$$

2.4 Image Segmentation

Generally, in lung images, a calculation is being utilized for Otsu and Watershed thresholds. Seed areas are grownup in such a way to segment them regionally. The article is utilizing a seed size of 800 for extra handling. The region that belongs to a pulmonary is described again. To recognize the nodule and its context from the processed image a watershed segmentation approach has been applied, with the means illustrated: (a) Image read and change image in grayscale (b) The angle size is controlled by the segmentation interaction (c) Label all components residing in the image. (d) Elaborate image background markings. (e) Now the processed image is converted into its features that belong to that image (c, d).

When this type of nodule is uneven, it manually maybe with the advent of deep convolutions neural networks such as CNN faster region or a completely converting neural network extricate a bounding box from its nodule. Nodule segmentation may be streamlined with different tactics such as boundary assessment, prototype, structure, and classification techniques mostly on basis of a region [14]. The regional methodology contains growth in the region, thresholds, and the technique of cluster analysis. The process threshold will include the entropy and recursive strategy. A standard approach to extracting the region of interest was the random field for Markov Gibbs. To use other filters, candidate nodes can also be segmented and their sensitivities enhance [15]. Its most lung noodles are allocated unequally on the radiologist's analysis.

The IDRI-LIDC dataset often represents a useful DCM file and its accurate label to promote the implementation of object detection. The container and the primary area of expertise have been used in the field [16]. Although the type of the node is uneven, a discrete node, like the fastest CNN or entire curving neural networks, is properly or peripherally extracted by using automated node distribution. Node distribution is being implemented through different methods, including boundaries analysis, region structures, feature, and probability concepts [17]. Regional strategies including regional growth, thresholds, and classification approaches. Which include the repeated thresholding that optimizing inter-class variability, and the entropy approach. Full confusion networks merged with a trained randomized field are being used to convey the lung object. It applied dual masks, thresholds, slices, filtering, and closing processes to distribute the illustration of the organs [18]. Also, separated the vessel as well as the bronchial region to use a Hess-based 3D streak filter, separated

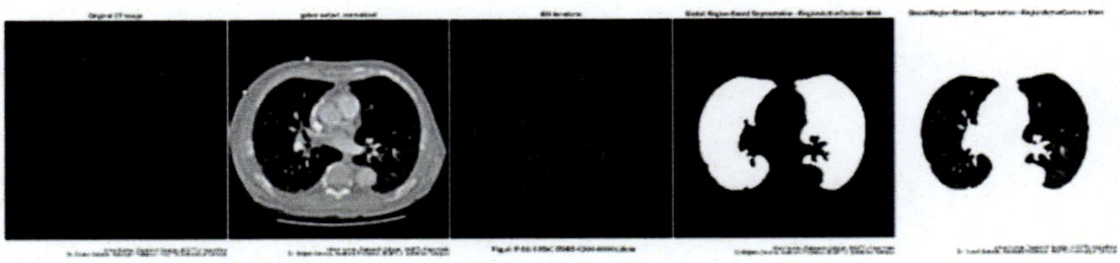

Fig. 2 Process of nodule segmentation

the separated regions using thickness, length, and roundness scales, again, and separated the lungs to use a threshold factor. This also decreased the false-positive effect using some advanced loop filters. The graph-cutting techniques and five various characteristics of the nodule, along with standard deviation, depth, flexibility, mean intensity, abilities, and vector high concentration, and the SVM, were obtained from time reduction images [19]. Decreased the use of false positives in this, shows the accuracy of the classification in comparison with the feature extracted from tumor sites or holes gained from sliding windows, whenever the tumor is downscale to size. Combining features got from just a multi-display of CNN, only in case of posterior opaqueness, gives excellent distribution findings. A few bad patterns seem to be difficult to accomplish and take a glance such as nodules. A few negatives patterns have always been difficult to label that are like nodules. While using complicated passive extraction, one solves such a problem. Because so many of the false positives in the nodules' detection are mostly caused by air passes within the lungs, so the segmentation algorithm is necessary. Feature removal and precision estimation are just the simplified characteristics of a tumor node that is neither benign nor malignant [20]. The next stage involves classifying the nodes with classification parameters. A filtered image has been used as a basic layer I_0 for something like a specific lung image. Figure 2 indicates the segmentation of the nodule.

$$I_\mathrm{d} = I - I_0. \text{ The normalized } I \text{ is computed as } I \leftarrow \frac{I_\mathrm{d} - \mu_\mathrm{d}}{\sigma_\mathrm{d}}$$

The optimal cross-section resolution has been changed from 512 to 512 pixels to 227 × 227 pixels, making it easier to understand the dynamics of its structure to be processed, a redesigned image with a better location. Use the feature to check the process performance [21]. These types of trained diagnostics have identified the cancerous goals of lung cancer by SqueezeNet.

2.5 Feature Extraction

It segments the pulmonary nodules all through its extraction processes. Extraction of its characteristic is a nodule component and the image is commonly used; a sub-feature with fine estimating its results and is chosen as the remaining collection at final. It used to be a necessary step for reworking information to go through environment functionality efficiently way [22, 23]. Every phase extracted special facets with unique features of the remote region. It investigates the dimensions of lung tumors and sorts as in lungs with three kinds by using eradication, like area, perimeter, and eccentricity [24].

$$\text{Area: } A = \sum ap, \text{ Perimeter: } P = \frac{\pi}{N}\sum_{\alpha}^{\pi} I_\alpha d_\text{L},$$

$$\text{Eccentricity: } E = \frac{\text{Major Axis Length}}{\text{Minor Axis Length}}$$

$$\text{Compactness: } C = \frac{(\text{Perimeter})^2}{(4\pi \text{ Area})}, \text{ Solidity: } S = \frac{\text{Area}}{\text{Convex Area}},$$

$$\text{Roundness or Circulatory: } R = \frac{(4\pi \text{ Area})}{(\text{Convex Perimeter})^2}$$

The segmentation is characterized, to be specific extraction from images concerning the area over its interest [25]. A numerical morphology, twofold into the lung area is an amiable benign strategy. From that point onward, it prepared over the gifted scope of the methodology through twofold images [26]. As among even parts of the images, the morphological starting along with disk intensity being driven overseas in impersonation of getting mindful of bothersome particles adjacent to an image [27]. Subsequently, a show line method is noticed regarding a surveyed image. The pulmonary mask has been done by ensuring each opening and hole concerning a lung [28].

3 Classification

The population is split before it is clarified, through preparing the sets of training and testing. The ordinary SqueezeNet result shows malignant growth; notwithstanding, as exhibited by the Table, there are 9113, 6610 cancer disease cases, and 2137 cases in the information assortment; however, 2103 have no malignancy: (1) AI models have been utilized to perceive a knob of malignancy implemented by the supervised classifier. A marked gathering of information results is utilized by the classifiers. Harmful or benevolent labeling of images is vital during the characterization of classification.

3.1 CNN Based and Traditional Based System

The structure historically uses frameworks such as HOG, GLCM, SURF, LBP, and several modern technologies. Such five steps of the conventional technique as illustrated in Fig. 1 are (a) image collection, (b) preprocessing, (c) nodule recognition, (d) feature extraction, and (e) classification. Some explanations of such approaches have recently been proposed using crafts features. To separate the nodule regions into patterns and produce figures, we have to use the HOG descriptor. It is engaged GLCM to verify the structural isolation, any attribute estimate, and eliminates image redundant information. This analysis refers to the interaction between center and each adjacent pixel using LBP feature extraction. About enhancing the performance of its technique, just use evolutionary algorithms to use as feature representation.

The article proposes a data-driven CNN-based approach to identify lung nodules and predict tumor growth. Many others have suggested major advances in nodule identification, which may be based totally on the data-driven paradigm, because of its deep learning techniques. The system focuses on learning machine functionality through CNNs including such RPN, DPN, etc., will obtain image heights, that increases its performance, and it has been effective in getting detected and making maximum improvements in medical support to identify such as head, breast, and lungs cancer.

3.2 State-of-The-Art CNN Studies

Variations focused on interactions, process side effects, and operational activities in various layers are discerned in those technologies. The latest advanced efficiency gains still show that AlexNet, VGGNet, GoogLeNet, and DenseNet are certainly is among the most famous frameworks. Google Net and ResNet are established for the analysis, and the VGG network is only a general framework. Interconnection is a closely packed design set, like SqueezeNet, for example. Figure 1 represented the lung cancer detection system with its classification process.

3.3 Architecture of SqueezeNet

A neural network is a collection of strategies, including different views of the machine but many fields, such as radiology. It contains a range of critical components such as convolution, cluster, even completely connected layers. It eventually optimizes that to adjust and improve features via regional identification backpropagation iterations. The uniformity of concepts, advantages, and disadvantages of a fully convolutional deep net is critical to increasing the productivity of a radiologist and, without a question, good result outcomes. Justify the CNN architecture of SqueezeNet as illustrated in Fig. 2. SqueezeNet starts with an autonomic convolution layer (conv1), preceded by eight fire modules (fire2-9) (conv10). The number of filters per fire module is progressively increased from its start to the end of a network. With just a phase of 2 SqueezeNet carries out max pools as 1, fire 4, fire 8, and conv10 layers; those are all late-stage pools according to strategy 3.

A network of recently qualified image classifications may use images to retrieve successful and recognizable characteristics and transfer them to unfamiliar jobs. This part of the image repository for a volume of conceptual identify challenges could be used in pretrained nodes. More than a million images were acquired in the network, so its images are being categorized into 1000 object classes, including certain clocks, ground coffee, colored pencils, and so many more. Figure 3 shows the SqueezeNet architecture used for lung malignancy analysis.

Deep learning is generally applied rapidly and broadly in a pretrained network as a framework. SqueezeNet is a deep-seated neural network and a pretrained network model which can be accessed through it. SqueezeNet is a deep neural network and even a pretrained network model which is capable of collecting upwards of one million objects from its image repository and training. The size of the image input is 227 × 227. The network pretrained image is subdivided into thousands of categories of items, including keys, cursor, and styles. Figure 4 demonstrates the basic steps of SqueezeNet. Then, Internet even trained to view several images in their entirety.

In the papers referred to here, there is a detailed explanation of the reasoning for these decisions as: (i) In the input data, a 1-pixel null-padding edge is applied to 3 × 3 expanding modulus filtering to make sure the activations from 1 × 1 and 3 × 3 filters have about the same width and length. (ii) The ReLU is implemented for squeeze triggering and expanding layers. (iii) A fire 9 modules with even a 50% ratio is added to drop-out. (iv) Mention that SqueezeNet lacks completely interconnected

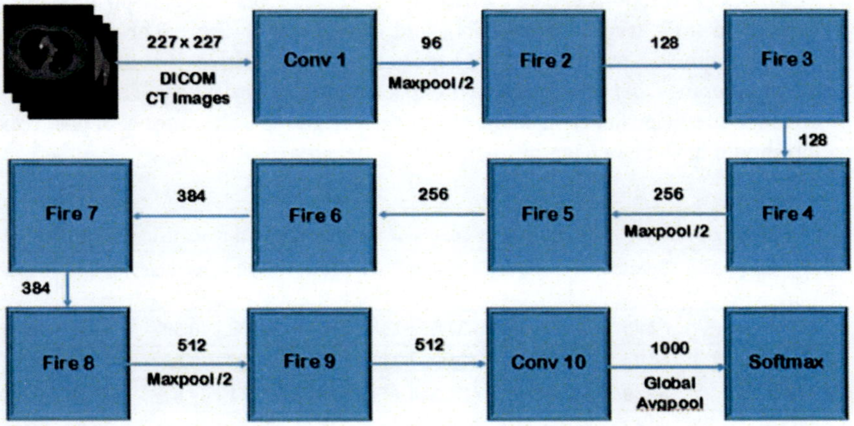

Fig. 3 SqueezeNet architecture for analysis of lung malignancy

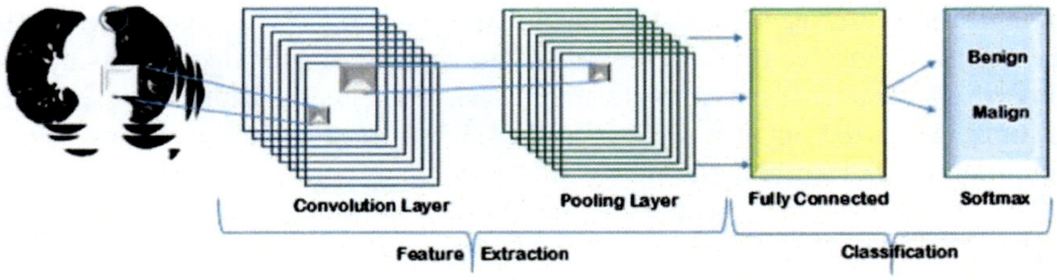

Fig. 4 Basic steps of classification using SqueezeNet

layers but was influenced by NiN architecture. (v) It starts with an even learning rate of 0.04 when training with SqueezeNet and reduces the learning rate linearly during training [29]. To solve that, use an expanding layer through two different overall layers: the 1 × 1 filter layer and the 3 × 3 filter layer. So together like a remaining pixel, concatenate the results of such layers. It was similar statistically to one layer which comprises all 1 × 1 and 3 × 3 filters.

4 Experiment and Results

Layers

A quantitative quantity of images is used to differentiate 50% training and testing, and training from 60 to 40% testing and the next 70% testing, 30% testing, and 20% testing with 80% training, 90% training, and 10% testing. There were all differences in the classification of the figures in the uncertainty matrix for deep CNN networks (Table 1).

The model predictions can be graded in four separate classes: true positive (TP), false positive (FP), true negative (TN), and false negative (FN). The findings are genuinely optimistic, and the outlook is projected to be positive. It is the number of very bad outcomes that false positive is expected to be positive [29]. The number of positive and predicted negative outcomes are false negatives. The number of positives, which should have been negative, is a genuine negative (Table 2).

Table 1 Total number of 9113 test images considered in the proposed work using GNB network

TTR (%)	Sensitivity	Specificity	Precision	FPR	FNR	FDR	FOR	Accuracy	*F*1 score	MCC
50–50	76.58	62.45	73.11	37.55	23.41	26.88	33.33	70.52	74.81	39.40
60–40	75.09	64.74	73.93	35.26	24.90	26.07	33.87	70.65	74.51	39.94
70–30	73.16	65.49	73.93	34.51	26.84	26.06	35.41	69.87	73.55	38.58
80–20	66.14	69.47	74.34	30.53	33.85	25.66	39.44	67.56	70.00	35.23
90–10	66.67	68.08	73.68	31.91	33.33	26.31	39.62	67.27	70.00	34.40

Table 2 Total number of 9113 test images considered in the proposed work using KNN network

TTR (%)	Sensitivity	Specificity	Precision	FPR	FNR	FDR	FOR	Accuracy	*F*1 score	MCC
50–50	96.83	88.60	91.89	11.39	03.16	08.10	04.54	93.30	94.29	86.39
60–40	94.07	93.15	94.82	06.84	05.92	05.17	07.81	93.67	94.44	87.12
70–30	94.21	97.18	97.81	02.81	05.78	02.18	07.38	95.48	95.97	90.91
80–20	96.06	95.78	96.82	04.21	03.93	03.17	05.20	95.94	96.42	91.73
90–10	93.65	1.00	1.00	0.00	06.34	0.00	07.84	96.36	96.72	92.90

In this paper, the author listed the dataset of training and tests: the SqueezeNet is trained at 50% and then TP is 60%, and TN 46%. Both FP and FN values are 1 and 3%, which means 60% of the total malignancy show and 46% convey benignity. Even measurements are made of $60/(60 + 1) = 98.36\%$ and $46/(46 + 3) = 93.88\%$ positive and negative predictive values, for LDA at 90–10 TTR, respectively, as per Table 3.

The next is that TP is 59% and TN 47%. The meaning for both FP and FN is 0 and 4%, which means 59% of overall malignancy shows and 47% shows benignity. Also, as per table 2, the calculation of the positive and negative predictive values for KNN at 90–10 TTR is $59/(59 + 0) = 100\%$, and $47/(47 + 4) = 92.16\%$, respectively.

Furthermore, according to Table 4, TP is 61%, while TN is 38%. The values of both FP and FN are 9 and 2%, which means 70% are healthy. The figure indicates

Table 3 Total number of 9113 test images considered in the proposed work using LDA network

TTR (%)	Sensitivity	Specificity	Precision	FPR	FNR	FDR	FOR	Accuracy	*F*1 score	MCC
50–50	91.77	91.98	93.85	08.01	08.22	06.14	10.65	91.86	92.80	83.47
60–40	95.65	89.47	92.36	10.52	04.34	07.63	06.07	93.00	93.98	85.70
70–30	96.84	90.84	93.40	09.15	03.15	06.59	04.44	94.27	95.09	88.31
80–20	93.70	88.42	91.53	11.57	06.29	08.46	08.69	91.44	92.60	82.48
90–10	95.24	97.87	98.36	02.12	04.76	01.64	06.12	96.36	96.77	92.67

Table 4 Total number of 9113 test images considered in the proposed work using the SVM network

TTR (%)	Sensitivity	Specificity	Precision	FPR	FNR	FDR	FOR	Accuracy	*F*1 score	MCC
50–50	89.87	90.71	92.81	09.28	10.12	07.18	12.95	90.23	91.32	80.22
60–40	92.88	91.57	93.62	08.42	07.11	06.37	09.37	92.32	93.25	84.36
70–30	91.57	88.02	91.09	11.97	08.42	08.90	11.34	90.06	91.34	79.68
80–20	91.33	91.57	93.54	08.42	08.66	06.45	11.22	91.44	92.43	82.62
90–10	96.82	80.85	87.14	19.14	03.17	12.85	05.00	90.00	91.73	79.88

40% malignancy. Even calculation of the positive and negative predictive values for SVM at 90–10 TTR is 61/(61 + 9) = 87.14%, and 38/(38 + 2) = 95.0%, respectively.

At a training set of 80%, only seven malignant images and seven benign images were found, respectively, in TP and TN, of 71 and 29%. Again, accuracy is 100% up.

This implies, relative to other classifiers, the maximum classification time taken by KNN. The other LDA, GNB, and SVM have about the same average time, described in Tables 1, 2, 3 and 4. The analysis indicates the highest accuracy at the full training set ratio when the classifier is trained at several training sets of 50–90 percentiles.

On a test dataset of 60%, otherwise, TP and TN are both 67.8 and 30.5%, FP is nil, while FN is 1% and 40 images are malignant. Sensitivity and characteristic: 97.6% and 100%, respectively.

In train–test ratio 60–40, the SVM and GNB models give the maximum accuracy while 90–10 are given by KNN and LDA. SVM is better than others in general. The accuracy of the KNN and LDA has improved somewhat.

When it is taught in the model's dataset at 70%, then TP and TN are 66.7%, and FP at 30.6%, while FN is 1%, which means 24 images are malignant and 11 benign. Sensitivity and specialty: 96% and 100%, respectively.

Just eight malicious images and four benign images were found on a 90% training sample, 66.7%, and 33.3%, respectively. Also, precision is 100% higher. The DCM images are malignant in an average period of 18.19 s and are 97.12 less accurate, with an average of 10.94 s, with a 100% degree of accuracy in the same number of DCM images. In general, as shown in Figs. 5, 6, 7 and 8, that takes too much time.

Figures 9 and 10 indicate a small improvement in detection as compared with the falsification rate at different training percentiles of 40–90. The identification of the malignancy is nearly constant. Figures 11 and 12, which shows that all rates initially display variations of up to 60% of the training period in which the saturation occurs, then describes the erroneous rate along with a detection rate from forty to nineties. Similarly, that takes longer than it.

Tables 1, 2, 3 and 4 show the best achievements of profound learning technics and shows, in comparison with other literary approaches, that the system adopted increases classification accuracy. The author explains that it has achieved the highest

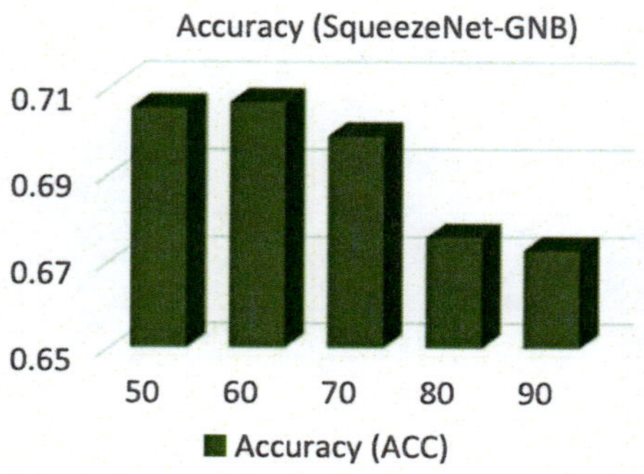

Fig. 5 Accuracy SqueezeNet with GNB

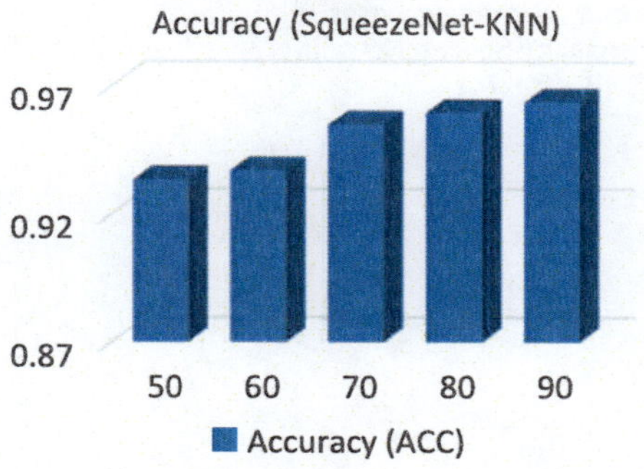

Fig. 6 Accuracy SqueezeNet with KNN

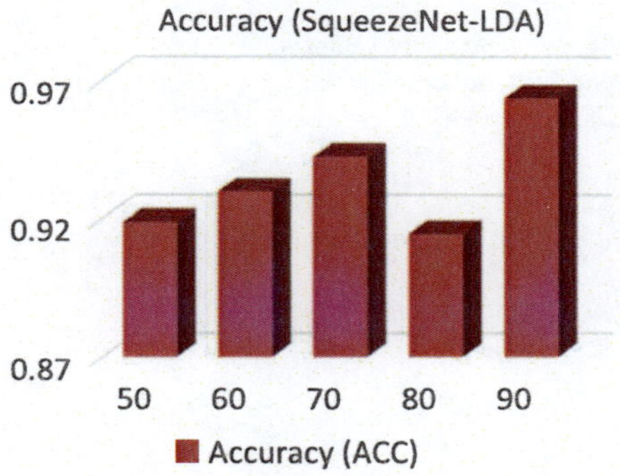

Fig. 7 Accuracy SqueezeNet with LDA

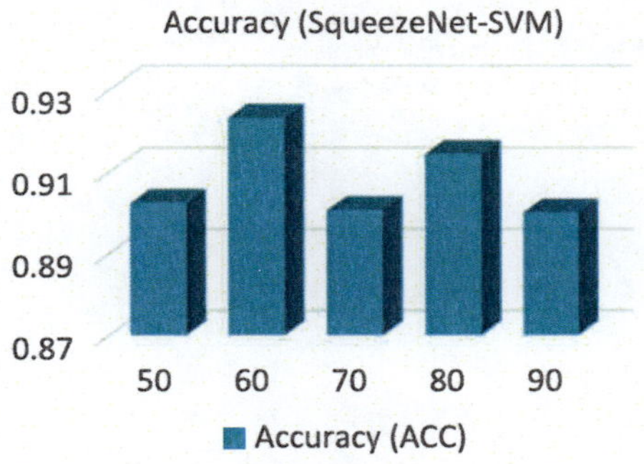

Fig. 8 Accuracy SqueezeNet with SVM

results with much shorter processing time and a multi-class SVM classification, contrasting to it. As well as the best results obtained by the use of in-depth teaching techniques, the extraction time is shown in Fig. 18, which AlexNet takes, is better and demonstrates that the method adopted improves the classification accuracy, as Fig. 20 is shown here in comparison to other methods. The author explains that in a considerably shorter period and an SVM multi-classification AlexNet provided the highest results in comparison to GoogleNet. Figure 19, therefore, demonstrates that

Fig. 9 Sensitivity versus specificity SqueezeNet with GNB

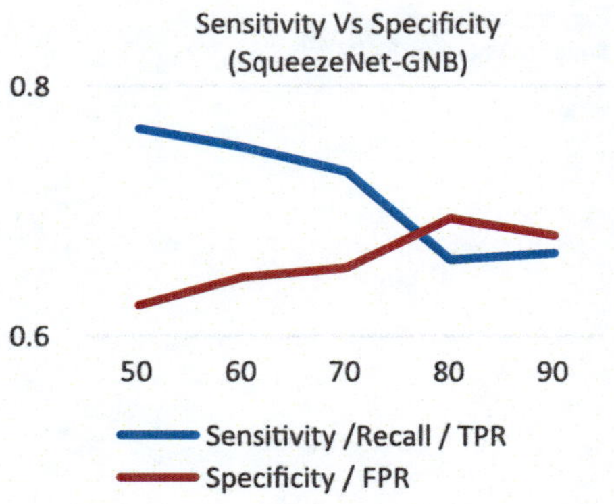

Fig. 10 Sensitivity versus specificity SqueezeNet with KNN

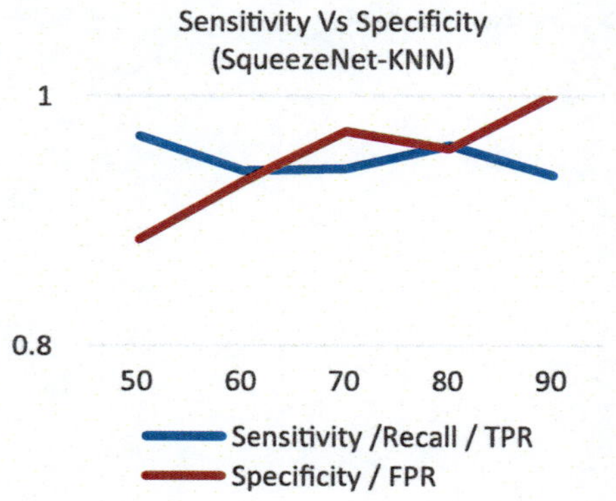

Fig. 11 Sensitivity versus specificity SqueezeNet with LDA

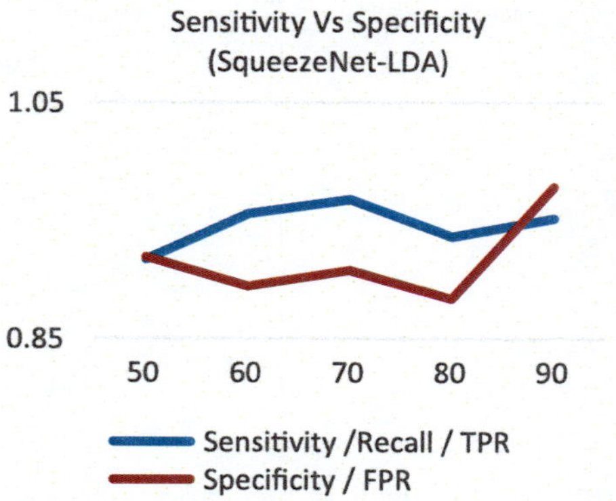

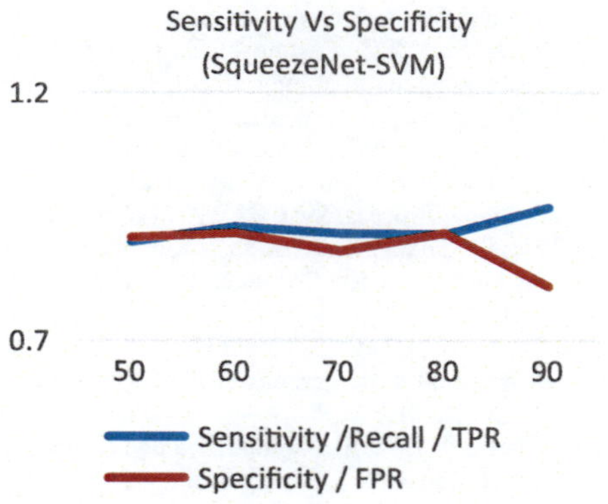

Fig. 12 Sensitivity versus specificity SqueezeNet with SVM

the time taken to demonstrate its usefulness in this regard is much greater than in the other techniques.

5 Conclusion

The author proposes SqueezeNet deep neural networks in this article. Experiments have been conducted with pretrained CNN on LIDC processing DICOM datasets. The characteristics should not be extracted based on form or texture. The highest resolution images are automatically extracted. The importance of the initial network layers should in the future also be taken into account and the required number of starting layers tested for greater efficiency in the proposed network tests. In DCM images, all training processes, including preprocessing, lung separation, and removal, were effectively carried out. Efficiency, sensitivity, specificity 100% at zero false discovery rate were measured. Further analysis into such strategies on the segmentation is needed. The SVM and GNB models have the highest precision on the train–test ration 60–40, while the KNN and LDA model achieves 90–10. In general, SVM is better than others. KNN and LDA's accuracy were somewhat improved. The proposed diagnostic system promises to give elite medical professionals a reliable and timely diagnostic impression.

References

1. Key Statistics for Lung Cancer retrieved on 3rd December 2019. https://www.cancer.org/cancer/nonsmallcelllungcancer/about/keystatistics.html (2019)
2. Detterbeck, F.C.: The eighth edition TNM stage classification for lung cancer: what does it mean on the main street? J. Thoracic Cardiovasc. Surg. **155**(1), 356–359 (2018)
3. Razzak, M.I., Naz, S., Zaib, A.: Deep learning for medical image processing: over-view, challenges and the future. In: Dey, N., Ashour, A., Borra, S. (eds.) Classification in Bio Apps.

Lecture Notes in Computational Vision and Biomechanics, vol. 26. Springer, Berlin (2018)
4. Detterbeck, F.C., Postmus, P.E., Tanoue, L.T.: The stage classification of lung cancer diagnosis and management of lung cancer, 3rd ed: American College of chest physicians evidence-based clinical practice guidelines. Chest **143**(5), e191S-e210S (2013)
5. De Carvalho Filho, A.O., Silva, A.C., de Paiva, A.C., Nunes, R.A., Gattass, M.: Lung-nodule classification based on computed tomography using taxonomic diversity indexes and an SVM. J. Signal Process. Syst. **87**, 179–196. https://doi.org/10.1007/s11265-016-1134-5 (2016)
6. He, K., Zhang, X., Ren, S., Sun, J.: Deep residual learning for image recognition. In: Proceedings of the IEEE Conference on Computer Vision and Pattern Recognition, pp. 770–778 (2016)
7. Wang, W., Chen, G., Chen, H., Anh Dinh, T.T., Jinyang Gao, B.C., Ooi, K.-L.T., et al.: Deep learning at scale and ease. ACM Trans. Multim. Comput. Commun. Appl. (TOMM) **12**(4), 1–25 (2016)
8. Cheng, J.-Z., et al.: Computer-aided diagnosis with deep learning architecture: applications to breast lesions in US images and pulmonary nodules in CT scans. Sci. Rep. **6**, 24454. https://doi.org/10.1038/srep24454 (2016)
9. Liao, F., Liang, M., Li, Z., Hu, X., Song, S.: Evaluate the malignancy of pulmonary nodules using the 3d deep leaky noisy-or network. arXiv preprint arXiv:1711.08324 (2017)
10. Nagao, M., et al.: Detection of abnormal candidate regions on temporal subtraction images based on DCNN. In: 2017 17th International Conference on Control, Automation, and Systems (ICCAS), Jeju, pp. 1444–1448 (2017)
11. Sathyan, H., Panicker, J.V.: Lung nodule classification using deep ConvNets on CT image. In: 2018 9th International Conference on Computing, Communication, and Networking Technologies (ICCCNT) (2018)
12. Fan, L., Xia, Z., Zhang, X., Feng, X.: Lung nodule detection based on 3D convolutional neural networks. In: 2017 International Conference on the Frontiers and Advances in Data Science (FADS) (2017)
13. Paul, R., Hawkins, S.H., Hall, L.O., Gold of, D.B., Gillies, R.J.: Combining deep neural network and traditional image features to improve survival prediction accuracy for lung cancer patients from diagnostic CT. In: 2016 IEEE International Conference on Systems, Man, and Cybernetics (2016)
14. Setio, A.A.A., Jacobs, C., Gelderblom, J., van Ginneken, B.: Automatic detection of large pulmonary solid nodules in thoracic CT images. Med. Phys. **42**(10), 5642–5653 (2015)
15. Kumar, D., Wong, A., Clausi, D.A.: Lung nodule classification using deep features in CT images. In: 2015 12th Conference on Computer and Robot Vision, pp 133–138 (2015)
16. Wang, S., Liu, Z., Chen, X., Zhu, Y., Zhou, H., Tang, Z., Wei, W., Dong, D., Wang, M., Tian, J.: Unsupervised deep learning features for lung cancer overall survival analysis. In: 2018 40th Annual International Conference of the IEEE Engineering in Medicine and Biology Society (EMBC) (2018)
17. Wei, L., Cao, P., Zhao, D., Wang, J.: Pulmonary nodule classification with deep convolutional neural networks on computed tomography images. In: Computational and Mathematical Methods in Medicine, pp. 1–7 (2016). https://doi.org/10.1155/2016/6215085
18. Ding, J., Li, A., Hu, Z., Wang, L.: Accurate pulmonary nodule detection in computed tomography images using deep convolutional neural networks. In: MICCAI (2017)
19. Shafe, A., Soliman, A., Ghazal, M., Taher, F., Dunlap, N., Wang, B., van Berkel, V., Gimel'farb, G., Elmaghraby, A., El-Baz, A.: A novel autoencoder-based diagnostic system for early assessment of lung cancer. In: 2018 25th IEEE International Conference on Image Processing (ICIP) (2018)
20. Kockelkorn, T.J.P., Rikxoort, M., Grutters, C., et al.: Interactive lung segmentation in CT scans with severe abnormalities. In: IEEE International Symposium on Biomedical Imaging: From Nano to Macro, vol. 14, pp. 564–567 (2010)
21. Meng, Y., Yi, P., Guo, X., Gu, W., Liu, X., Wang, W., Zhu, T.: Detection for pulmonary nodules using RGB channel superposition method in the deep learning framework. In: 2018 Third International Conference on Security of Smart Cities, Industrial Control System, and Communications (SSIC) (2018)

22. Lakshmanaprabu, S.K., Mohanty, S.N., Shankar, K., Arun Kumar, N., Ramirez, G.: Optimal deep learning model for classification of lung cancer on CT images. Future Gener. Comput. Syst. **92**, 374–382 (2018). ISSN: 0167-739X
23. Xie, Y., Zhang, J., Xia, Y., Fulham, M., Zhang, Y.: Fusing texture, shape and deep model-learned information at decision level for automated classification of lung nodules on chest CT. Data Inf. Fusion **42**, 102–110 (2018). ISSN: 1566-2535
24. Cao, P., Liu, X., Zhang, J., Li, W., Zhao, D., Huang, M., et al.: A _ 2, 1 norm regularized multi-kernel learning for the false-positive reduction in Lung nodule CAD. Comput. Methods Program. Biomed. **140**, 211–231 (2017)
25. Singh, G.A.P., Gupta, P.K.: Performance analysis of various machine learning-based approaches for detection and classification of lung cancer in humans. Neural Comput. Appl. **31**(10), 6863–6877 (2018)
26. Sun, W., Zheng, B., Qian, W.: Automatic feature learning using multichannel ROI based on deep structured algorithms for computerized lung cancer diagnosis. Comput. Biol. Med. **89**(1), 530–539 (2017)
27. Kim, B., Sung, Y.S., Suk, H.: Deep feature learning for pulmonary nodule classification in a lung CT. In: 2016 4th International Winter Conference on Brain-Computer Interface (BCI), Yongpyong, pp. 1–3 (2016)
28. Sun, W., Zheng, B., Qian, W.: Computer-aided lung cancer diagnosis with deep learning algorithms. In: Proceedings of SPIE 9785, Medical Imaging 2016: Computer-Aided Diagnosis, pp. 97850Z (2016). https://doi.org/10.1117/12.2216307
29. Kumar, V., Bakariya, B.: Classification of malignant lung cancer using deep learning. J. Med. Eng. Technol. **45**(2), 85–93 (2021). https://doi.org/10.1080/03091902.2020.1853837

Deep Learning Models for Early Detection of Pneumonia Using Chest X-Ray Images

Avnish Panwar and Siddharth Gupta

1 Introduction

Pneumonia causes life-threatening lung disease that many times damage one or both the lungs. In a severe case, the air sacs are blocked with fluid or pus (purulent material) that causes cough and difficulty in breathing. According to the World Health Organization report, 15% of all deaths of children under the age group of 5 years is due to pneumonia [1]. Also in 2017, the total number of children who lost their lives due to pneumonia are 808,694 [2]. Pneumonia is caused by bacteria, fungus, or viruses. The early symptoms are: cough (greenish/yellow mucus), fever, shaking chills, and sometimes shortness of breath. The early diagnosis of pneumonia is necessary to save the lives of many people. In these days, several techniques such as chest X-ray [3], magnetic resonance imaging (MRI) [4], ultrasound imaging (Ultrasound) [5], computed tomography (CT) [6], and many others are used by the radiologist for detection of pneumonia. However, among the aforementioned techniques, the preferred one is chest X-ray images due to its very low cost it can be afforded by anyone, and also the chest X-ray images in comparison with other radio techniques generate very few radiations [7]. Several parameters such as lung texture, patchy shadow, shadow density, and inflammation site are proved to be efficient in the diagnosis of pneumonia. Sometimes, the X-ray images are similar, and the appearance of pneumonia in X-rays is often low contrast due to which the expert doctors may misdiagnose the disease. Therefore, the accurate and early detection of pneumonia manually are one of the very big challenges for radiologists and doctors.

Doctors benefit greatly from the progress of technology in the field of medical science these days. Several convolution neural network (CNN) deep learning [8]

A. Panwar
Graphic Era Hill University, Dehradun, Uttarakhand, India

S. Gupta (✉)
Graphic Era Deemed to be University, Dehradun, Uttarakhand, India

M. Dua et al. (eds.), *Proceedings of the International Conference on Paradigms of Communication, Computing and Data Sciences*, Algorithms for Intelligent Systems,
https://doi.org/10.1007/978-981-16-5747-4_60

models are used to diagnose pneumonia infection in the chest X-ray scans. Deep learning especially the CNN models proves to be very efficient in medical images. Lots of work are carried out by many researchers for diagnosis of chest X-ray image to detect the pneumonia infection. Wang et al. [9] have used a new chest X-ray database namely "chest X-ray8". This dataset is formed by collecting the X-ray labels of 32,717 patients. Multiple labels of eight different diseases were collected. Total 108,948 front view X-ray images are formed. The authors formed a machine human-annotated model that has the capability of handling at least ten thousand images of patients. Very promising results are obtained using CNN models. Gu et al. in [10], have used a novel computer-aided diagnosis (CAD) system for the classification of bacterial and viral pneumonia from chest radiography images. The dataset used is the open Japanese Society of Radiological Technology (JSRI) that comprises 241 images and Montgomery Country, Md (MC, 138 images) dataset. DCNN is used for the feature extraction, and SVM classifier is used for the classification.

2 Background of CNN Models and Classifiers

2.1 Convolution Neural Network (CNN)

As described in previous studies, CNN models are popular due to the better performance for image classification. In these days, the advance deep learning models [11] used in computer vision make use of CNN. Several layers such as the convolution layer, pooling layer, and ReLU layer along with the filters help in extracting the spatial and temporal features from the image. The weight sharing techniques present in these layers help in reducing computation efforts [12].

CNN is described as a feed-forward artificial neural network (ANN). The architecture of CNN's is very simple and comprises three major building blocks. The first layer called a convolutional layer which is the building block of CNN. This layer comprises filters (kernels) that are used for capturing the spatial and temporal dependencies in an image. The second layer, known as the pooling layer, is responsible for reducing the size of the presentation generated by the previous layer. With the help of the pooling layer, the overall computation efforts is reduced. Max-pooling is commonly used. The third layer is known as a fully connected layer that helps in the equipment of the network with classification capabilities [12]. The overall architecture may be extracted from Fig. 1.

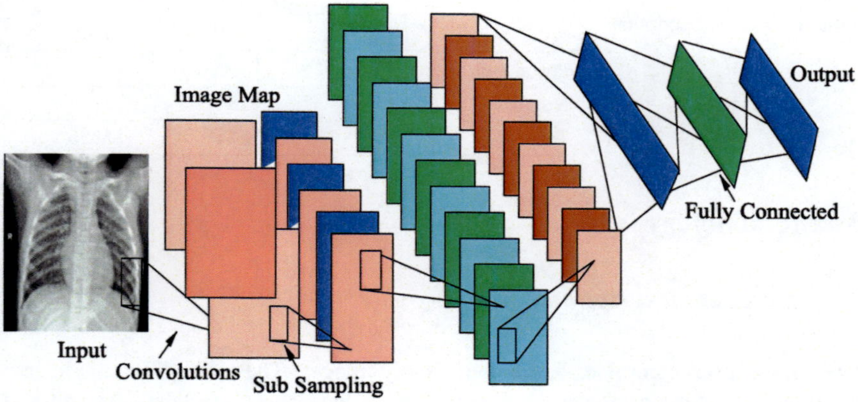

Fig. 1 Convolutional neural network (CNN) architecture

2.2 Model Description

In this study to achieve the state-of-the-art classification results, the pre-trained model VGG [13] is used. This model is already trained on the ImageNet dataset. After training, the model is used on the chest X-ray images.

2.2.1 VGG16 and VGG19 [13]

VGG16 and VGG19 are CNN architectures that come into consideration due to their simple structure. These models comprise 3×3 small convolutional filters that are stacked on each other with the increasing depth and stride size of 1. This design helps the network to achieve high accuracy for image recognition applications. The difference between VGG16 and VGG19 is in the depth of several layers: convolution layer, pooling layer, and fully connected layer. The VGG16 model comprises 16 layers in the base model, and VGG19 model consists of 19 layers in the base model [13].

2.2.2 Inception V3

The model of Inception v3 is based on the family of Inception models but with few important changes and improvements. The basic focus in this model is on less computational cost that is why the large-sized filters have been replaced with the low-sized filters, producing lesser parameters. An auxiliary classifier (small CNN) has been added in between the layers to regularize the loss. Along with batch normalization, label smoothing has been added to stop the model from over-fitting [14].

Table 1 Dataset description

	Training	Testing	Total
Pneumonia	3883	390	
Normal	1349	234	
Total	5232	624	5856

3 Methodology

3.1 Dataset Description

The dataset used comprises 5,856 chest X-ray images. These images are split into training and testing sets. The images present in the dataset are labeled as normal and pneumonia infection. Further, the testing set comprises 234 normal images and 390 pneumonia chest X-ray images, whereas in the training set 1349 images are normal chest X-ray images and 3883 images are of patients infected from pneumonia. These images were taken from Guangzhou Women and Children's Medical Center, Guangzhou. The images are of small children whose age group range from 1 to 5 years [15] (Table 1).

3.2 Dataset Pre-processing

The collected dataset consists of varying images. The shape and size of every image are different. For the image classification, all the input images should have the same resolution. To solve this issue, several data pre-processing techniques such as image resizing and image cropping are used. After pre-processing the images, the resolution of all the extracted images is the same, as shown in Fig. 2.

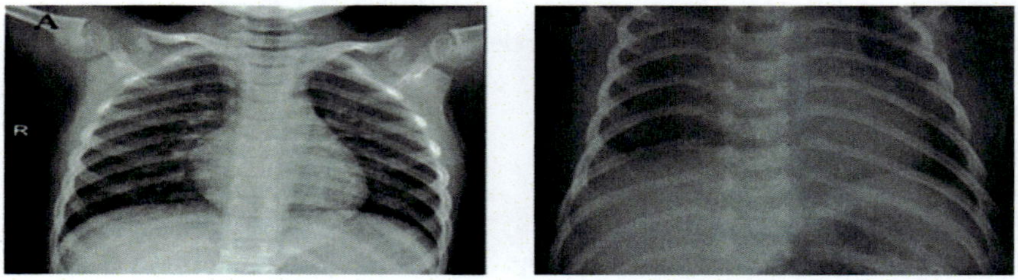

Fig. 2 Dataset images. **a** Normal chest X-ray image, **b** pneumonia chest X-ray image

3.3 Model and Classifiers Used

In this work, the dataset images are pre-processed and fed to several deep learning CNN models such as VGG16, VGG19, and Inception V3. These models are used for the purpose of feature extraction. Once the features are extracted, these images are passed to several machine learning classifiers such as k-Nearest Neighbors (kNN) [16], Tree [17], Support Vector Machine (SVM) [18], Random Forest (RF) [19], Neural Network (NN) [20], Naïve Bayes [21], Logistic Regression (LR) [22], AdaBoost for the classification of the chest X-ray images as normal or pneumonia infected.

3.4 Evaluation Parameters

To calculate the performance of various models and classifiers, several parameters are considered such as area under the ROC curve (AUC) [23], accuracy [24], F1 score [24], precision [24], and recall [24]. The formulas for calculating these parameters can be extracted from Eqs. (1)–(4).

$$\text{Accuracy (Acc)} = \text{TP} + \text{TN}/\text{TP} + \text{FP} + \text{FN} + \text{TN} \tag{1}$$

$$F1 \text{ score} = 2 * (\text{Precision} * \text{recall} /\text{Precision} + \text{recall}) \tag{2}$$

$$\text{Precision(Pre)} = \text{TP}/\text{TP} + \text{FP} \tag{3}$$

$$\text{Recall} = \text{TP}/\text{TP} + \text{FN} \tag{4}$$

where TP stands for True Positive, TN stands for True Negative, FP stands for False Positive, FN stands for False Negative.

4 Results

The prime aim of this study is to diagnose the chest X-ray images constituting the pneumonia infection. For this purpose, the training and testing sets containing the normal and pneumonia chest, X-ray images were prepared. Tables 2, 3, and 4 show the results for various deep learning CNN models along with machine learning classifiers for the diagnosis of pneumonia chest X-ray images. Table 2 shows the Inception v3 model along with several ML classifiers. It can be seen that the Inception v3 model along with the neural network gives an accuracy of 94.6%.

Table 3 shows the VGG19 model along with several ML classifiers.

Table 2 Inception v3 model along with ML classifiers

Model	AUC	Accuracy	$F1$ score	Precision	Recall
kNN	0.958	0.917	0.917	0.919	0.917
Tree	0.794	0.829	0.829	0.830	0.829
SVM	0.982	0.942	0.942	0.942	0.942
RF	0.946	0.886	0.886	0.886	0.886
NN	0.985	**0.946**	0.946	0.946	0.946
Naive	0.947	0.899	0.900	0.904	0.899
LR	0.986	0.944	0.944	0.944	0.944
AdaBoost	0.805	0.812	0.813	0.815	0.812

Table 3 VGG19 model along with several ML classifiers

Model	AUC	Accuracy	$F1$ score	Precision	Recall
kNN	0.947	0.888	0.885	0.892	0.888
Tree	0.758	0.798	0.799	0.800	0.798
SVM	0.951	0.878	0.879	0.879	0.878
RF	0.962	0.886	0.886	0.886	0.886
NN	0.960	0.910	0.910	0.910	0.911
Naive	0.914	0.861	0.862	0.865	0.861
LR	0.974	**0.915**	0.915	0.916	0.914
AdaBoost	0.791	0.809	0.808	0.808	0.809

Table 4 VGG16 model along with several ML classifiers

Model	AUC	Accuracy	$F1$ score	Precision	Recall
kNN	0.939	0.883	0.880	0.887	0.883
Tree	0.747	0.814	0.814	0.814	0.814
SVM	0.954	0.878	0.878	0.878	0.878
RF	0.949	0.885	0.884	0.884	0.885
NN	0.949	0.897	0.898	0.899	0.897
Naive	0.884	0.835	0.836	0.841	0.835
LR	0.977	**0.933**	0.933	0.933	0.933
AdaBoost	0.806	0.816	0.816	0.817	0.816

Table 4 shows the VGG16 model along with several ML classifiers.

To check the performance of any model, area under the ROC curve (AUC)/receiver operating curve (ROC) may be considered. Figure 3 shows the ROC curve for the VGG16, VGG19, and Inception V3 model along with the several classifiers.

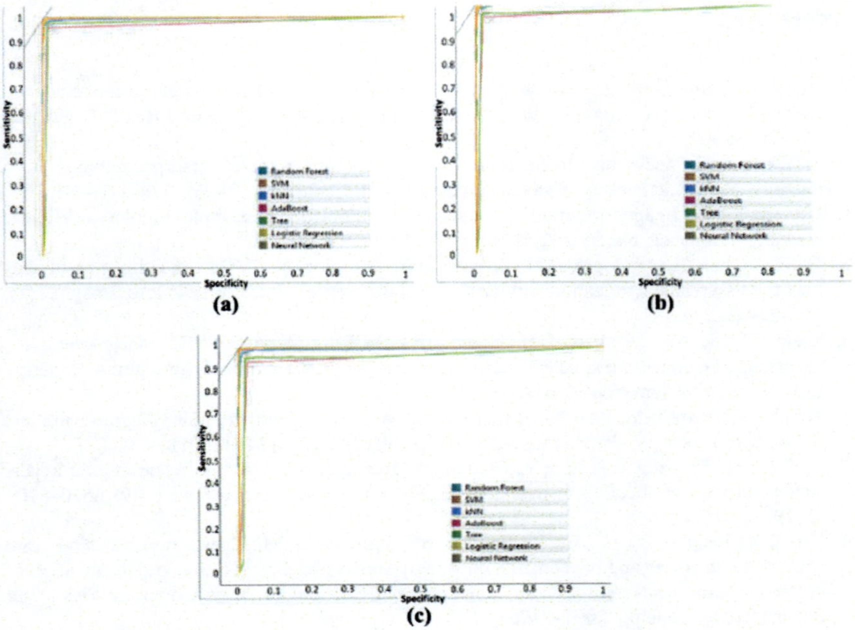

Fig. 3 ROC curve. **a** VGG16 model, **b** VGG19 model, **c** inception V3 model

5 Conclusion

Covid-19 is a life-threatening disease caused by coronavirus or SARS-CoV-2 virus. The WHO declared it as a "Worldwide Pandemic" due to the large number of casualties caused by Covid-19. There is no registered vaccine available to cope up with this virus. The only possible solution to prevent the spread of Covid-19 is by maintaining the social distance, covering the face with a mask, regularly washing hands, and avoiding mass gatherings. Therefore, early diagnosis of Covid-19 patients is utmost. In this work, chest X-ray images are used pre-processed and fed to the various deep learning CNN models such as VGG16, VGG19, and Inception V3 model for the feature extraction. Once the features are extracted, the images are passed to several machine learning classifiers for the classification of images as Covid-19, normal, or pneumonia infection. The result shows that 94.6% accuracy is given by Inception V3 along with neural network which is the highest in comparison with other models. In future, we will try to enhance the number of images in the dataset to obtain high accuracies. Also, we will try to implement the same dataset on other state-of-art deep learning CNN models to obtain much better results.

References

1. Rudan, I., Tomaskovic, L., Boschi-Pinto, C., Campbell, H.: Global estimate of the incidence of clinical pneumonia among children under five years of age. Bull. World Health Organ. **82**, 895–903 (2004)
2. WHO report for pneumonia in 2017: who.int/news-room/fact-sheets/detail/pneumonia
3. Gregory, N.W.: Elements of X-ray diffraction. J. Am. Chem. Soc. **79**(7), 1773–1774 (1957)
4. Lauterbur, P.C.: Image formation by induced local interactions: examples employing nuclear magnetic resonance. Nature **242**(5394), 190–191 (1973)
5. Garcìa-Garcìa, H.M., Gogas, B.D., Serruys, P.W., Bruining, N.: IVUS-based imaging modalities for tissue characterization: similarities and differences. Int. J. Cardiovasc. Imag. **27**(2), 215–224 (2011)
6. Naime, J. de M., Mattoso, L.H.C., da Silva, W.T.L., Cruvinel, P.E., Martin-Neto, L., Crestana, S.: Conceitos e aplicações da instrumentação para o avanço da agricultura. Embrapa Instrumentação-Livro científico (ALICE) (2014)
7. Ter Haar Romeny, B.M, Van Ginneken, B., Viergever, M.A.: Computer-aided diagnosis in chest radiography: a survey. IEEE Trans. Med. Imag. **20**(12), 1228–1241 (2001)
8. Li, S., Song, W, Fang, L., Chen, Y., Ghamisi, P., Benediktsson, J.A.: Deep learning for hyperspectral image classification: an overview. IEEE Trans. Geosci. Remote Sens. **57**(9), 6690–6709 (2019)
9. Wang, X., Peng, Y., Lu, L., Lu, Z., Bagheri, M., Summers, R.M.: Chestx-ray8: hospital-scale chest x-ray database and benchmarks on weakly-supervised classification and localization of common thorax diseases. In: Proceedings of the IEEE Conference on Computer Vision and Pattern Recognition, pp. 2097–2106 (2017)
10. Gu, X., Pan, L., Liang, H., Yang, R.: Classification of bacterial and viral childhood pneumonia using deep learning in chest radiography. In: Proceedings of the 3rd International Conference on Multimedia and Image Processing, pp. 88–93 (2018)
11. Al-Saffar, A.A.M., Tao, H., Talab, M.A.: Review of deep convolution neural network in image classification. In: 2017 International Conference on Radar, Antenna, Microwave, Electronics, and Telecommunications (ICRAMET), pp. 26–31. IEEE (2017)
12. Guo, T., Dong, J., Li, H., Gao, Y.: Simple convolutional neural network on image classification. In: 2017 IEEE 2nd International Conference on Big Data Analysis (ICBDA), pp. 721–724. IEEE (2017)
13. Simonyan, K., Zisserman, A.: Very deep convolutional networks for large-scale image recognition. arXiv preprint arXiv:1409.1556 (2014)
14. Xia, X., Xu, C., Nan, B.: Inception-v3 for flower classification. In: 2017 2nd International Conference on Image, Vision and Computing (ICIVC), pp. 783–787. IEEE (2017)
15. Kermany, D., Zhang, K., Goldbaum, M.: Large dataset of labeled optical coherence tomography (oct) and chest x-ray images. Mendeley Data, **3**, 10–17632 (2018)
16. Altman, N.S.: An introduction to kernel and nearest-neighbor nonparametric regression. Am. Stat. **46**(3), 175–185 (1992)
17. Tsipouras, M.G., Tsouros, D.C., Smyrlis, P.N., Giannakeas, N., Tzallas, A.T.: Random forests with stochastic induction of decision trees. In: 2018 IEEE 30th International Conference on Tools with Artificial Intelligence (ICTAI), pp. 527–531. IEEE (2018)
18. Li, X., Wang, L., Sung, E.: Multilabel SVM active learning for image classification. In: 2004 International Conference on Image Processing, 2004. ICIP'04., vol. 4, pp. 2207–2210. IEEE (2004)
19. Denisko, D., Hoffman, M.M.: Classification and interaction in random forests. Proc. Natl. Acad. Sci. **115**(8), 1690–1692 (2018)
20. Jürgen, S.: Deep learning in neural networks: an overview. Neural Netw. **61**, 85–117 (2015)
21. McCann, S., Lowe, D.G.: Local naive bayes nearest neighbor for image classification. In: 2012 IEEE Conference on Computer Vision and Pattern Recognition, pp. 3650–3656. IEEE (2012)
22. Gasso, G.: Logistic Regression (2019)

23. Gupta, S., Panwar, A., Goel, S., Mittal, A., Nijhawan, R., Singh, A.K.: Classification of Lesions in retinal fundus images for diabetic retinopathy using transfer learning. In: 2019 International Conference on Information Technology (ICIT), Bhubaneswar, India, pp. 342–347 (2019). https://doi.org/10.1109/ICIT48102.2019.00067
24. Panwar, A., Semwal, G., Goel, S., Gupta, S.: Stratification of the Lesions in color fundus images of diabetic retinopathy patients using deep learning models and machine learning classifiers. In: 26th annual International Conference on Advanced Computing and Communications (ADCOM 2020), Silchar, Assam, India (2020). (in press)

Machine Learning for Human Activity Detection Using Wearable Healthcare Device

K. Sornalakshmi, Revathi Venkataramanan, and R. Pradeepa

1 Introduction

The COVID-19 pandemic has profoundly changed working life. Many organizations have implemented work from home options to prevent from new infections [1]. As per the Statista report [2], due to the COVID-19 outbreak, around 1.5 billion children were out of school, many of them spend more time on screens as part of online learning or entertainment. The Lenovo Research report [3] reveals that 91% of Indian respondents agree that they have increased laptop usage during the pandemic, which is even higher than the global average of 85%. In this new normal working environment where there are no specific working hours and no routine, people lack physical activities and engage themselves by sitting in front of a screen for long periods. It has been shown that extended spells of sitting in the same place or being "sedentary" negatively impact our well-being. Extended use of digital platforms has been shown to lead to visual impairment effects, musculoskeletal disorders, etc. Lack of frequent computer breaks was identified as a significant factor among the many potential causes of injury [4].

The Internet of Things (IoT) plays an eminent role in recognizing human activity and assisting users to achieve their wellness goals [5]. IoT enables the number of medical devices that sense and diagnosis to be connected [6]. A healthcare strategy to

K. Sornalakshmi (✉) · R. Venkataramanan · R. Pradeepa
School of Computing, SRM Institute of Science and Technology, Kattankulathur, Chennai 603203, India
e-mail: sornak@srmist.edu.in

R. Venkataramanan
e-mail: revathin@srmist.edu.in

R. Pradeepa
e-mail: pr4365@srmist.edu.in

M. Dua et al. (eds.), *Proceedings of the International Conference on Paradigms of Communication, Computing and Data Sciences*, Algorithms for Intelligent Systems,
https://doi.org/10.1007/978-981-16-5747-4_61

deliver personalized testing and activity monitoring with portable devices has been implemented to address a sedentary lifestyle.

The IoT architecture consists of four layers: sensor layer, network layer, data processing layer, and application layer [7]. The sensor layer is responsible for sensing and acquiring the data from the sensors. In the healthcare scenario, this data includes human activity data, nutritional and medicine intake, etc. The network layer comprises gateway devices that facilitate communication between the sensors and the rest of the system. It collects the information gathered by sensors, converting the sensor data into easily transferable formats to a back-end destination. The third layer is the data processing layer or a cloud-based system that stores and processes data for analysis and decision making. The final layer is implemented through a dedicated application at the device end. It is responsible for delivering application-specific services. The following Fig. 1 shows the architecture of the IoT system.

A wearable system is necessary to monitor users' activity data in real time and to provide recommendations based on the sensed data [8]. The wearable can be fitness trackers or other human activity recognition systems integrated into their manufacturer's mobile apps to offer personalized feedback about the users' activities. Wearable sensors such as accelerometers and gyroscopes shown in Fig. 2 provide triaxial acceleration and orientation to identify human activity. The main challenge in recognizing human activity is classifying sequences of sensed data recorded by wearable or smart phones into known well-defined movements.

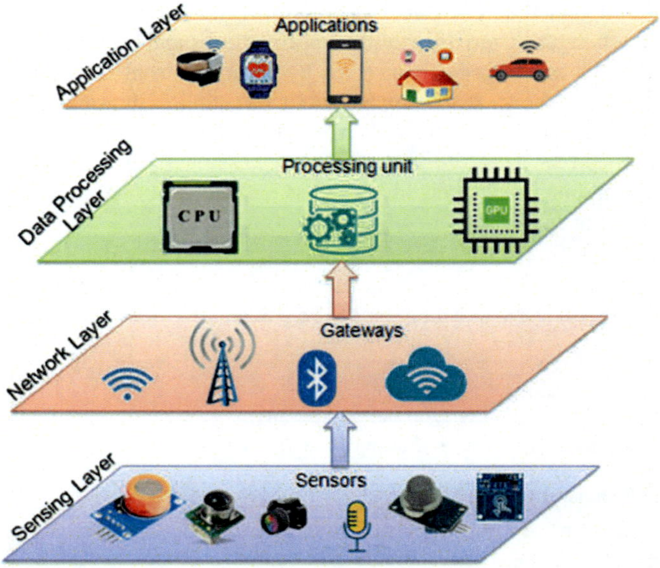

Fig. 1 IoT architecture

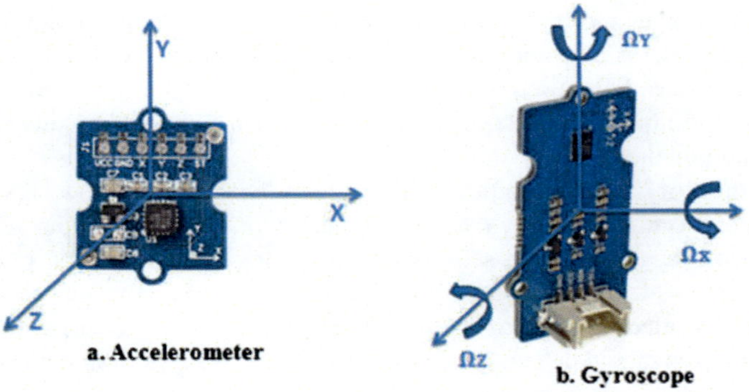

Fig. 2 Accelerometer and gyroscope sensors

Machine learning methods have proved to be more effective than conventional statistical techniques in extracting and recognizing corresponding activity from sensed data [9].

The application for monitoring human activity requires methods for recognizing human activity that provide a high level of accuracy and recall of the human activity classification. A number of previous studies have already mentioned some of the ways in which machine learning can be used to classify human activities. The most familiar methods, including the k-nearest neighbor (KNN), artificial neural networks (ANNs) and support vector machine (SVM), have already been tested [10]. This paper compares the performance of machine learning methods: multi-class logistic regression (LR), linear discriminant analysis (LDA), k-nearest neighbors (KNN), classification and regression tree (CART), and support vector mechanism (SVM) to find the most efficient.

The rest of the paper is structured in the following way. Section 2 discusses the related literature survey. The implementation and evaluation of machine learning methods are presented in Sect. 3. Section 4 gives the result analysis of machine learning methods. Finally, the conclusion is presented in Sect. 5.

2 Related Works

The method proposed in [6] used an Apple watch to gather accelerometer and gyroscope data. From the collected data, nine features have been extracted using the statistical formulas and the basic machine learning models have been combined with ensemble approaches such as stacking and majority voting to detect the sitting activity. The stacking-based model has shown 93.57% accuracy in detecting the sitting activity.

Similarly, in [11], accelerometer and gyroscope sensors from smartphones have been used to recognize human activities such as sitting, standing, walking, climbing up, climbing down, and laying. Different classification methods including decision trees, support vector machines (SVM), k-nearest neighbors (KNN), and ensemble classification have been used and evaluated in the recognition of human activity, and support vector machine classification has shown improved results than other classification methods.

The authors of [12] have used a wearable named "Bioharness 3 model K113045" to measure pulse rate, respiratory rate, skin temperature, and acceleration. The device must be worn at the chest to sense these physiological parameters. For the rule tree classifiers, C4.5 and Naive Bayes were implemented and tested, and the C4.5 algorithm was selected based on the size of the rule set. This classification achieved 95.83\% success rate in detecting activities such as walking, sitting, and jogging.

Smart belt has been designed to recognize human postures such as sit, stand, squat, walk, and jump in [8]. The triaxial angular rate sensors and triaxial acceleration sensors have been integrated in the smart belt. The support vector machine classification with particle swarm optimization has been applied to identify human postures, and the experimental results showed that SVM achieved 92.3% overall classification accuracy in recognizing human postures.

To identify various human activities, accelerometer-based smart watch data used in the work presented in [13]. An artificial neural network (ANN) is constructed to classify human activity. The performance of the ANN has been compared with the other machine learning models such as decision tree (DT), random forest (RF), and support vector machine (SVM) where ANN outperforms other models in terms of accuracy.

3 Smart Belt-Based Sitting Detection Model

The sitting detection using a smart belt involves four steps: data collection, training the models, activity classification, and sitting duration detection. The following architectural model can be used to utilize the procedures within a system for detection. The procedure starts with collecting data from wearable, training various models for the dataset, classification of different users' activities, and ends with the detection of sitting duration, as designated in Fig. 3.

The summary of three stages of the proposed model is presented as follows.

3.1 Data Collection

The raw data for this study is collected from thirty volunteers. The participants are aged between 14 and 35 and are asked to wear a smart belt a whole day to identify four various activities: sitting, standing, lying, and walking. A smart belt is used to sense human activity, the corresponding application gathers triaxial accelerometer

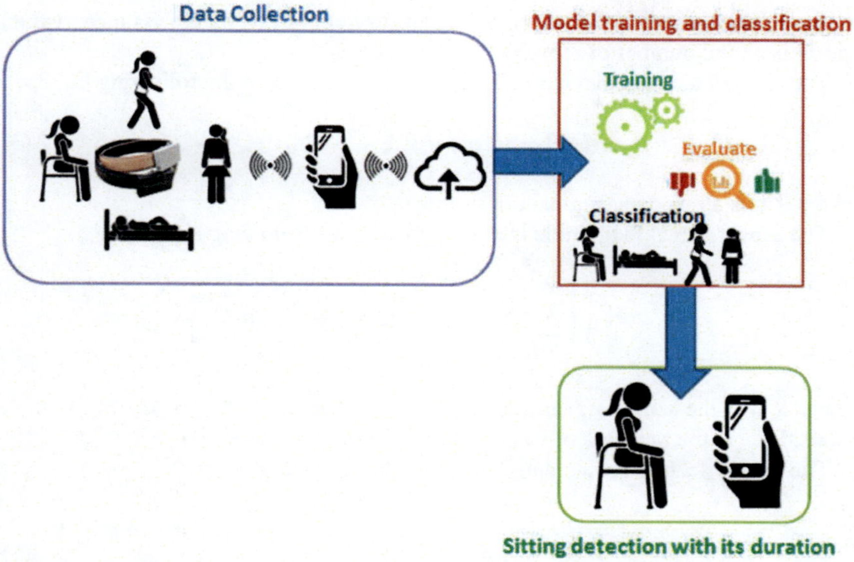

Fig. 3 Proposed system architecture

and gyroscope readings from the sensors in the smart belt, and the sample rate is 30 Hz.

Feature Generation. Classification is not possible with the data collected from the accelerometer and gyroscope sensors directly. Data pre-processing has been done in the form of feature extraction from sensors' raw data. The data window size is imagined with a particular data fragment ($\boldsymbol{Rd_t}$) within ($\boldsymbol{Rd_i}$) $\boldsymbol{ith}$ data item provided by the sensors.

Proper feature sets lead to an effective machine learning model. In order to generate the features, the following statistical calculations are applied.

The mean value is calculated based on the following Eq. 1.

$$\mu = \frac{1}{N}\sum_{i=1}^{N} \text{Rd}_i \tag{1}$$

where μ is the mean value, Rd_i is an ith item of raw data received from the accelerometer and gyroscope sensors, and N is a total number of samples.

The feature standard deviation is generated with the following Eq. 2.

$$\sigma = \sqrt{\frac{1}{N}\sum_{i=1}^{N} (\text{Rd}_i - \mu)^2} \tag{2}$$

where σ is the standard deviation, Rd_i is an ith item of raw data, μ is mean value, and N is a total number of samples.

The median absolute deviation (MAD) is calculated with the following Eq. 3.

$$\text{MAD} = \text{median}(|\text{Rd}_i - \mu(\text{Rd}_t)|) \tag{3}$$

where Rd_t is a raw data fragment in data window.

The sensor signal magnitude is evaluated with the following Eq. 4.

$$\vec{S}_{xyz} = \frac{1}{3}\left\{\sum_{i=1}^{N}|\text{Rd}_{xi}| + \sum_{i=1}^{N}|\text{Rd}_{yi}| + \sum_{i=1}^{N}|\text{Rd}_{zi}|\right\} \tag{4}$$

where $\vec{S}_{xyz}$ is the sensor signal magnitude of xyz directions, Rd_{xi} is the raw data in x-axis, Rd_{yi} is the raw data in y-axis, Rd_{zi} is the raw data in z-axis.

The energy feature is calculated based on the following Eq. 5.

$$\xi = \frac{1}{N}\sum_{i=1}^{N}\text{Rd}_i^2 \tag{5}$$

where ξ is energy, Rd_i is an ith item of raw data, and N is a total number of samples.

The correlation between axes is generated with the help of following Eqs. 6 and 7.

$$\rho_{xy} = \frac{\sum_{i=1}^{N}(\text{Rd}_{xi} - \mu(\text{Rd}_x))\sum_{i=1}^{N}(\text{Rd}_{yi} - \mu(\text{Rd}_y))}{\sqrt{\sum_{i=1}^{N}(\text{Rd}_{xi} - \mu(\text{Rd}_x))^2\sum_{i=1}^{N}(\text{Rd}_{yi} - \mu(\text{Rd}_y))^2}} \tag{6}$$

$$\rho_{yz} = \frac{\sum_{i=1}^{N}(\text{Rd}_{yi} - \mu(\text{Rd}_y))\sum_{i=1}^{N}(\text{Rd}_{zi} - \mu(\text{Rd}_z))}{\sqrt{\sum_{i=1}^{N}(\text{Rd}_{yi} - \mu(\text{Rd}_y))^2\sum_{i=1}^{N}(\text{Rd}_{zi} - \mu(\text{Rd}_z))^2}} \tag{7}$$

where ρ_{xy} is a correlation between x and y axes, ρ_{yz} is a correlation between y and z axes, Rd_{xi} is an ith item of the x-axis, Rd_{yi} is an ith item of the y-axis, Rd_{zi} is an ith item of the z-axis.

Features for the classification model are generated with the help of sensor data using the above equations.

3.2 Classification Model and Sitting Detection

This section introduces the classification methods used to classify human activity and the siting detection, including k-nearest neighbors (KNN), classification and regression tree (CART), and support vector mechanism (SVM). Mainly focused on the linear discriminant analysis (LDA) and the multi-class logistic regression (LR) model, we proposed to classify various human activities by analyzing the generated features.

Classifiers. The proposed method focuses on implementing and evaluating multi-class logistic regression and the linear discriminant analysis classifiers to classify the various human activities.

Multi-class Logistic Regression. Multi-class logistic regression is used to classify the multiple possible target outcomes of human activities such as standing, walking, sleeping, and sitting. This multi-class logistic classifier uses a softmax function, to compute the probability of the target y being in each potential class $c \in C$, $p(y = c|x)$. The input to the softmax function is $\vec{v} = \left[\overrightarrow{v_1}, \overrightarrow{v_2}, \ldots, \overrightarrow{v_m}\right]$ of m arbitrary values, and these values are mapped to a probability distribution, with each value in the range (0, 1).

For a $\vec{v}$ of dimensionality m, the softmax is defined as following Eq. 8.

$$\text{softmax}\left(\overrightarrow{v_i}\right) = \frac{\exp\left(\overrightarrow{v_i}\right)}{\sum_{j=1}^{m} \exp\left(\overrightarrow{v_j}\right)} 1 \leq i \leq m \tag{8}$$

The softmax of an input vector $\vec{v} = \left[\overrightarrow{v_1}, \overrightarrow{v_2}, \ldots, \overrightarrow{v_m}\right]$ is thus a vector itself shown in Eq. 9.

$$\text{softmax}(\vec{v}) = \left[\frac{\exp\left(\overrightarrow{v_1}\right)}{\sum_{i=1}^{m} \exp\left(\overrightarrow{v_i}\right)}, \frac{\exp\left(\overrightarrow{v_2}\right)}{\sum_{i=1}^{m} \exp\left(\overrightarrow{v_i}\right)}, \ldots, \frac{\exp\left(\overrightarrow{v_m}\right)}{\sum_{i=1}^{m} \exp\left(\overrightarrow{v_i}\right)}\right] \tag{9}$$

The denominator $\frac{\exp\left(\overrightarrow{v_m}\right)}{\sum_{i=1}^{m} \exp\left(\overrightarrow{v_i}\right)}$ is used to normalize all the values into probabilities represented in Eq. 10. To produce separate weight vectors (and bias) for each of the M classes,

$$p(y = c|x) = \frac{\exp(w_c \cdot x + b_c)}{\sum_{j=1}^{m} \exp\left(w_j \cdot x + b_j\right)} \tag{10}$$

Linear Discriminant Analysis. Linear discriminant analysis is a classification technique that is used to find a data projection matrix W using the following criterion Eq. 11.

$$\text{LDA} = \underset{W}{\arg\min}\, T_r\left(\left(W^T S_b W\right)^{-1}\left(W^T S_w W\right)\right) \tag{11}$$

where S_w and S_b are the between-class and within-class scatter matrices given by the following Eqs. 12 and 13.

$$S_{\mathrm{w}} = \sum_{k=1}^{K} \sum_{i,l_i=k} (x_i - m_k)^{\mathrm{T}} \tag{12}$$

$$S_{\mathrm{b}} = \sum_{k=1}^{K} (m_k - m)(m_k - m)^{\mathrm{T}} \tag{13}$$

where m_k is the mean vector of class k, m is the total mean vector. W is obtained by solving the generalized eigenvalue decomposition problem $S_{\mathrm{w}} v = \lambda S_{\mathrm{b}} v$.

Sitting Detection. Based on classification and timestamp, the overall sitting time of a person is calculated. Whenever the proposed LR classifier classifies the target as sitting, the duration is noted till the next target. The durations are accumulated for the entire day to identify the overall time based on the following Eq. 14.

$$O_{\mathrm{st}} = \sum_{t=1}^{n} \mathrm{St}_t \tag{14}$$

where O_{st} is overall sitting time, St_t detected sitting time of t_{th} time, and n is the total duration of the evaluation.

4 Result Analysis

This section presents the experimental results for the sitting detection with various classification methods and their comparisons.

The performance of various machine learning models is compared, and the best model is selected based on results observed from the cross-validation report shown in Fig. 4.

We compared various classification models such as LR, LDA, KNN, CART, and SVM. Based on the above cross-validation, multi-class logistic regression (LR) and linear discriminant analysis (LDA) are selected for next level analysis. Different metrics are used to analyze and compare LR and LDA models.

Confusion Matrix. Confusion matrix is a best and easy reference guide used to show the performance of the selected models. The confusion matrix presented in Table 1 describes the common format. In this confusion matrix, the predicted values are mentioned as positive and negative, and the actual values are mentioned as true and false.

Figure 5 shows the confusion matrix of the LR model. LR model confusion matrix diagonals clearly show the detection performance of the model. The normalized value of one is present almost for all the classes. Based on that, the accuracy score is an average of 0.98.

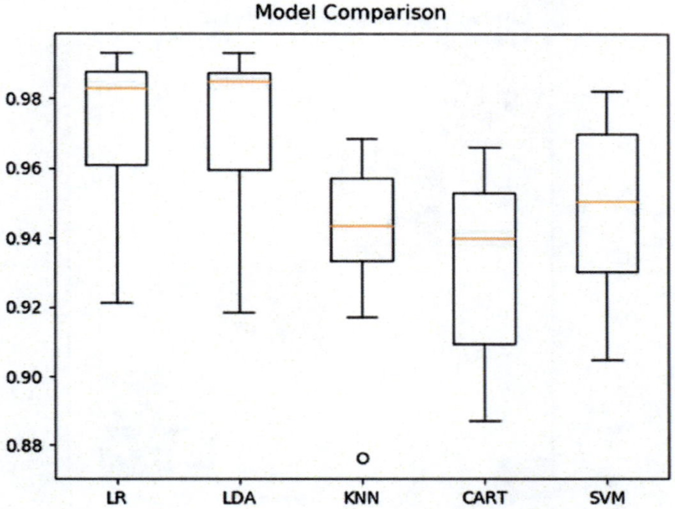

Fig. 4 Model comparison

Table 1 Confusion matrix

	Positive (P)	Negative (N)
True (T)	TP	TN
False (F)	FP	FN

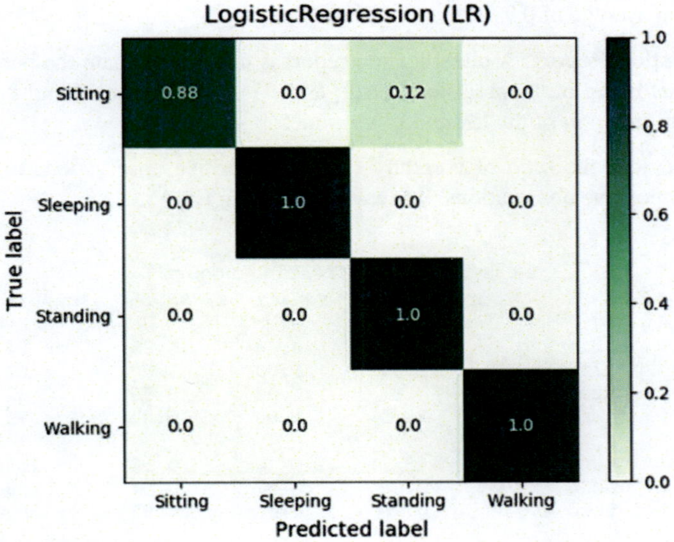

Fig. 5 LR model confusion matrix

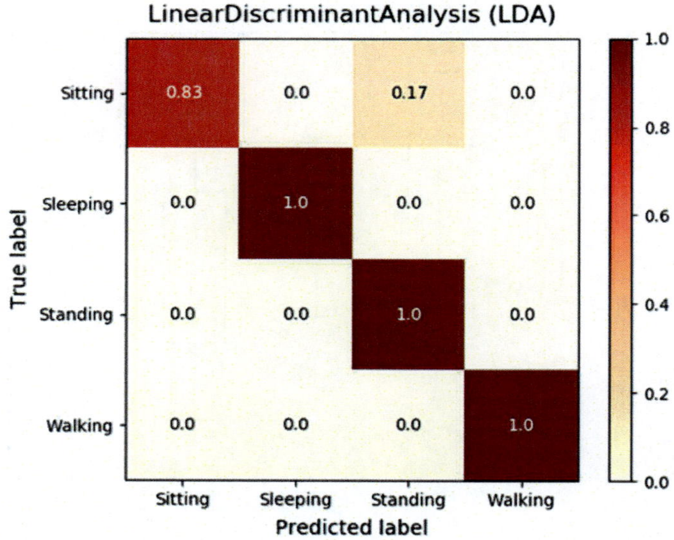

Fig. 6 LDA model confusion matrix

Figure 6 shows the confusion matrix of the LDA model. LDA model confusion matrix diagonals clearly show the detection performance of the model. The normalized value of one is present almost for all the classes. Based on that, the accuracy score is an average of 0.97.

Classification Report. A classification report is used to evaluate the performance of a model based on its precision, recall, and $F1$ score. Figures 7 and 8 show the classification report of the LR and LDA models.

Precision. It is the ratio of correctly classified positive observations to the total classified positive observations. It is calculated using Eq. 15.

LR model Classification Report

	precision	recall	f1-score	support
Sitting	1.00	0.88	0.93	24
Sleeping	1.00	1.00	1.00	24
Standing	0.91	1.00	0.96	32
Walking	1.00	1.00	1.00	79
accuracy			0.98	159
macro avg	0.98	0.97	0.97	159
weighted avg	0.98	0.98	0.98	159

Fig. 7 LR classification report

LDA model Classification Report

	precision	recall	f1-score	support
Sitting	1.00	0.83	0.91	24
Sleeping	1.00	1.00	1.00	24
Standing	0.89	1.00	0.94	32
Walking	1.00	1.00	1.00	79
accuracy			0.97	159
macro avg	0.97	0.96	0.96	159
weighted avg	0.98	0.97	0.97	159

Fig. 8 LDA classification report

$$\text{Precision(Pr)} = \frac{\text{TP}}{(\text{TP} + \text{FP})} \tag{15}$$

Recall. It is the ratio of correctly classified positive observations to all observations in type as represented, and it is calculated using the Eq. 16.

$$\text{Recall(Re)} = \frac{\text{TP}}{(\text{TP} + \text{FN})} \tag{16}$$

F1 score. It is calculated based on the weighted average of precision and recall. It is calculated using Eq. 17.

$$F1 - \text{Score} = \frac{2 * \text{Pr} * \text{Re}}{(\text{Pr} + \text{Re})} \tag{17}$$

From this LR and LDA classification report, the accuracy score of 98% accuracy is achieved in the LR model, and the accuracy score of 97% accuracy is achieved in the LDA model.

Receiver Operating Curve (ROC). It facilitates in visualizing a model by plotting the true-positive rate against the false-positive rate of the model. It is represented in the form of Eq. 18. The AUC of the ROC gives an approximation of the performance of the model [14]. In multi-class classification, the class detection for each case is frequently made based on a continuous random variable R. Given a threshold parameter t,

$$\text{Classification} = \begin{cases} \text{if } R > t, & f_1(x) \text{ Positive} \\ \text{Otherwise}, & f_0(x) \text{ Negative} \end{cases} \tag{18}$$

$$\text{TP}_{\text{rate}}(t) = \int_t^{\infty} f_1(x)\text{d}x$$

$$\text{FP}_{\text{rate}}(t) = \int_t^{\infty} f_0(x)\text{d}x$$

Area under the curve (AUC). AUC is identical of the probability that ranks a classifier with a random positive case than a random negative. This can be calculated using Eq. 19.

$$\mathrm{TP}_{\mathrm{rate}}(t) : t \to y(x)$$

$$\mathrm{FP}_{\mathrm{rate}}(t) : t \to x$$

$$A = \int_{-\infty}^{\infty} \int_{-\infty}^{\infty} I(t' > t) f_1(t') f_0(t) \mathrm{d}t' \mathrm{d}t = P(R_1 > R_0) \tag{19}$$

where R_1 is the score of positive and R_0 is the score of negative.

The ROC graph for LR model and LDA model between the classes is in Figs. 9 and 10. From the above ROC curve, multi-class logistic regression (LR) achieves 98% accuracy and performs better than the LDA model.

All the above metrics show the performance of the LR model is better than the other models. We implemented the multi-class LR model to detect the sitting time of humans for a day. Repeat process for continues 30 days, and average human activity is represented in following Fig. 11.

The pie chart shows that the detected average sitting time of a human is more than 60% based on our intelligent model. Our healthcare application will send activity information every hour to care for a human to avoid long-time sitting issues.

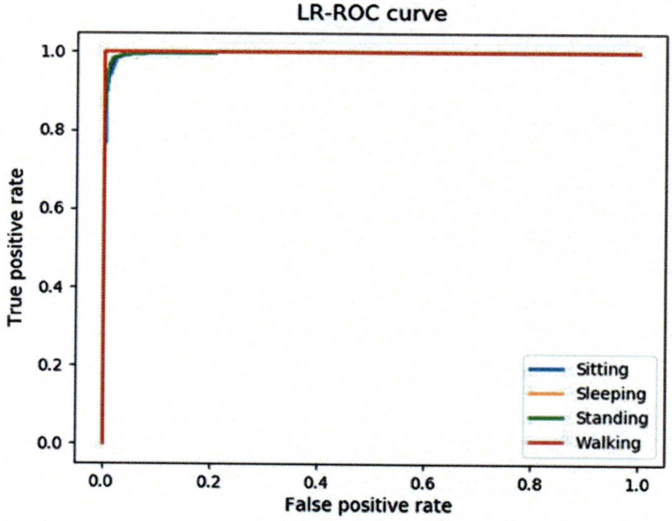

Fig. 9 LR ROC

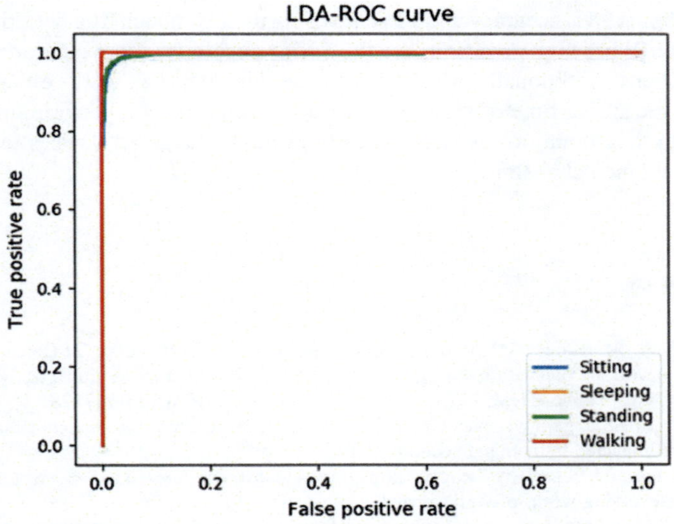

Fig. 10 LDA ROC

Fig. 11 Human activity distribution

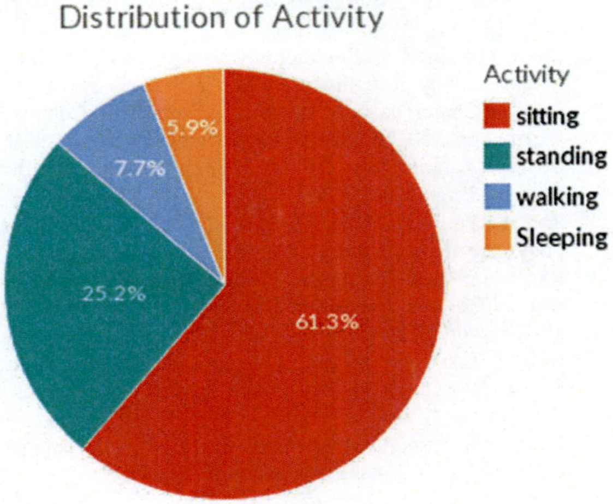

5 Conclusion

In this proposed work, a smart belt, an emerging IoT solution to detect sitting time, has been used. It has an in-built accelerometer and gyroscope sensors in the center position. The smart belt is a promising wearable device compared with other smart device positions. For sitting detection, we evaluated various machine learning methods to train and classify the data retrieved from the smart belt. Our experiment results showed almost 98% accuracy and F1 score using the LR model. The detection using this smart belt has high precision, and the device position is more comfortable than other solutions. Additionally, we confirmed that this smart belt works efficiently for detecting the sitting time of the user. For future work, we will perform this experiment with long duration for many users to examine the smart devices to provide physical and mental health.

References

1. Moretti, A., Menna, F., Aulicino, M., Paoletta, M., Liguori, S., Iolascon, G.: Characterization of home working population during COVID-19 emergency: a cross-sectional analysis. Int. J. Environ. Res. Public Health 17 (2020). https://doi.org/10.3390/ijerph17176284
2. COVID-19's Staggering Impact On Global Education. https://www.statista.com/chart/21224/learners-impacted-by-national-school-closures/
3. Global research report by Lenovo. https://news.lenovo.com/press-kits/lenovos-technology-and-the-evolving-world-of-work-report/
4. Wang, S.C., Chern, J.Y.: Impact of intermittent stretching exercise animation on prolonged-sitting computer users' attention and work performance. In: International Conference on Human-Computer Interaction, pp. 484–488. Springer, Cham (2015)
5. Islam, S.R., Kwak, D., Kabir, M.H., Hossain, M., Kwak, K.: The internet of things for health care: a comprehensive survey. IEEE Access. **3**, 678–708 (2015). https://doi.org/10.1109/ACCESS.2015.2437951
6. Mekruksavanich, S., Hnoohom, N., Jitpattanakul, A.: Smartwatch-based sitting detection with human activity recognition for office workers syndrome. In: International ECTI Northern Section Conference on Electrical, Electronics, Computer and Telecommunications Engineering, pp. 160–164. IEEE (2018). https://doi.org/10.1109/ECTI-NCON.2018.8378302
7. Sethi, P., Sarangi, S.R.: Internet of things: architectures, protocols, and applications. J. Electr. Comput. Eng. (2017)
8. Zhu, Y., Yu, J., Hu, F., Li, Z., Ling, Z.: Human activity recognition via smart-belt in wireless body area networks. Int. J. Distrib. Sens. Netw. **15** (2019). 1550147719849357
9. Ramasamy Ramamurthy, S., Roy, N.: Recent trends in machine learning for human activity recognition—a survey. Wiley Interdisc. Rev. Data Min. Knowl. Disc. **8** (2018). https://doi.org/10.1002/widm.1254
10. Bustoni, I.A., Hidayatulloh, I., Ningtyas, A.M.: Classification methods performance on human activity recognition. J. Phys. Conf. Series 1456 (2020). https://doi.org/10.1088/1742-6596/1456/1/012027
11. Bulbul, E., Cetin, A., Dogru, I.A.: Human activity recognition using smartphones. In: 2018 2nd International Symposium on Multidisciplinary Studies and Innovative Technologies, pp. 1–6. IEEE (2018). https://doi.org/10.1109/ISMSIT.2018.8567275
12. Castro, D., Coral, W., Rodriguez, C., Cabra, J., Colorado, J.: Wearable-based human activity recognition using an iot approach. J. Sens. Actuator Netw. 6 (2017). https://doi.org/10.3390/jsan6040028
13. Kwon, M.C., Choi, S.: Recognition of daily human activity using an artificial neural network and smartwatch. Wirel. Commun. Mob. Comput. (2018). https://doi.org/10.1155/2018/2618045
14. Narkhede, S.: Understanding AUC–ROC Curve Towards Data Science (2018). https://towardsdatascience.com/understanding-auc-roc-curve-68b2303cc9c5

Fault Tolerance Analysis in Neural Networks Using Dropouts

Farhana Kausar, P. Aishwarya, and Gopal Krishna Shyam

1 Introduction

One of the leading and state-of-the-art models in machine learning is neural network models and deep models. In several different domains, they have been implemented. The ones with several layers that significantly increase their number of parameters are the most effective deep neural models. A large number of training samples are required to train such models, which are not always available. Overfitting, which is the problem discussed, is one of the fundamental problems in neural networks. This issue also arises when the training of large models is carried out using a few training samples. Several methods have been suggested [1, 2] to avoid overfitting and to improve efficiency of generalization, such as data augmentation, early stopping, sharing criteria, unsupervised learning, dropouts, normalization of batches, etc.

The building blocks of any machine learning architecture are neural networks. They consist of one layer of data, one or more hidden layers, and a layer of output. It might become too dependent on the dataset we are using when we train our neural network (or model) by updating each of its weights. Therefore, it will not provide satisfactory results when this model has to make a prediction or classification. This is referred to as overfitting. Through a real-world example, we might understand this problem: If a math student reads just one chapter of a book and then takes a test on the entire syllabus, he will probably fail, this analogy of neural network where training

F. Kausar (✉) · P. Aishwarya
Atria IT, Visvesvaraya Technological University, Bangalore, India
e-mail: farhanakausar@atria.edu

P. Aishwarya
e-mail: aishwarya.p@atria.edu

G. K. Shyam
REVA University, Bangalore, India
e-mail: gopalkrishna@revainstitutions.org

M. Dua et al. (eds.), *Proceedings of the International Conference on Paradigms of Communication, Computing and Data Sciences*, Algorithms for Intelligent Systems,
https://doi.org/10.1007/978-981-16-5747-4_62

datasets may lead to overfitting. In its simplest form, feature detectors are deleted with probability $q = 1 - p = 0.5$ during training, at each example presentation, and the rest weights are trained by backpropagation. All weights are divided into every case. The weights are divided by two during the prediction. The primary motivation for the algorithm is to prevent co-adapting or over adaptation of feature sensors by requiring neurons to become flexibly populated instead of individual units. The dropouts are reported on multiple benchmark datasets to achieve state-of-the-art performance. It is also found that for one logistics unit, there is a kind of drop off. "Geometric averaging" inferred something over the possible subset of subnetworks. Likewise, the opinion in multilayer networks that dumping the neurons can also be economical.

In a neural network, the topological relationship of its neurons determines its architecture. The parameters in the model of the neural network are the "weights" on the connections between various neurons. The best way describes the activation of each neuron from its incoming signals as a function of the weighted sum $a_j = g\left(\sum_i w_i a_i\right)$, w_i is the weight on the connection from neuron i to neuron j, and $g(.)$ is the activation function. The activation function is of the form of a logistic function $g(x) = \frac{1}{1+e^{-x}}$. The modern topology of the neural network is mostly constructed with interconnected neurons forming into layers. A collection of neurons directly forwarding a vector of input features is typically the first layer. The set of output prediction is computed by neurons in the last layer. Typically, there is one or more concealed in between, the layers. Initially, the weights are set using random values and are updated by learning algorithms to boost neural network performance. A simple technique for avoiding overfitting in feedforward neural networks is given by the original proposed dropout process, implemented in 2012. Each neuron is omitted from the network, with probability p, during each training iteration. The complete network is used once trained, while neuron outputs the likelihood that the neuron has been omitted is multiplied by p. Now that no neurons are dropped, this compensates for the bigger size of the network and can be interpreted as averaging over the potential networks during training. For each sheet, the probability will vary; $p = 0.2$ is suggested for the input layer and $p = 0.5$ for the hidden layer. Neurons are not dropped in the output layer. This technique is generally simply known as dropout, but we call it a regular dropout for the purposes of this article to differentiate it from other dropout methods (Fig. 1).

2 Literature Survey

The author [1] specifies in dropout methods; the most popular research path has been improving dropout for regularization. It is widely agreed that a wide range of neural network models can be regularized by standard dropouts, but there is space to achieve either faster convergence in training or better final results. As dropout decreases the neuron exposure to each training sample, which can slow down training, the former concern is significant. Techniques such as rapid dropouts that reduce this impact are useful as neural networks become larger and more computationally expensive to train.

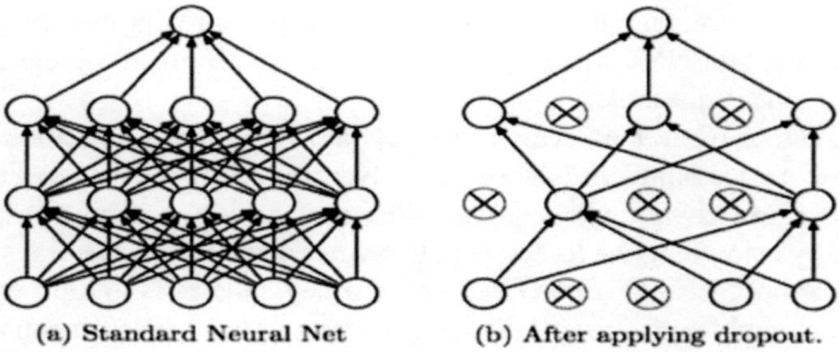

Fig. 1 An example of dropout. **a** is standard neural network and **b** neurons dropped by probability = 0.5

It is also an ongoing issue to develop how dropouts impact the efficiency of qualified networks. In a more intelligent or logically justified way than normal dropouts, attempting to drop neurons showed promise. Both, in practice, the development of convolution and recurrent neural networks and the development of advanced techniques that work better than normal dropouts have prompted specific neural network forms. As new types of neural networks and neural network layers continue to be created, specialized dropout opportunities continue to be built or improved methodologies. In [3], dropout is a relatively new neural network training algorithm that depends on training on stochastically "dropping out" neurons to keep the neurons from dropping out co-adaptation of detectors with features. In order to evaluate, we implement the general formalism for the study of dropout on either units or links, with arbitrary probability values, by the averaging and regularizing properties of dropout in both linear and nonlinear networks. Three recursive equations define the average properties of dropout for deep neural networks, including the approach of expectations by normalized weighted geometric means. We approximate, limit, and validate the results with simulations for these approximations. We also demonstrate how dropout results, among other results, are stochastically descending on a regularized error function. For, applications focused on convolutional neural networks (CNNs) Where proper regularization is greatly required, and they have become ubiquitous. To stop overfitting large neural network models, dropout has in practice, and it has been commonly used as an important regularization method. Many recent studies, however, show that the traditional dropout is counterproductive or even harmful to CNN training. We are, in this document, to revisit this problem and in an effort to analyze different dropout variants for CNNs to strengthen current dropout-based regularization techniques. Drop-Conv2d offers a structurally more suitable dropout version; for deep CNNs, a more powerful and productive regularization. For these dropouts, it is possible to easily incorporate variants into the building blocks of CNNs and implemented on current frameworks for deep learning. Wide-ranging experiments compare the current building blocks and the proposed ones, and benchmark datasets like CIFAR, SVHN and ImageNet are compared. With preparation for dropouts, the results suggest that our building blocks are improving over the considerably state-of-the-art CNNs, which can be traced to the better impact of regularization and implicit model ensemble.

For [4], author focuses on deep neural nets are very efficient machine learning systems with a wide number of parameters. Overfitting, however, is a serious issue for such networks. Large networks are often slow to use, making it difficult to solve them by combining the forecasts in test times of many large neural networks. Dropout is a tool for solving this dilemma. The key concept is to drop units from the neural randomly during preparation and network (along with their connections). This avoids too much co-adapting by units. During training, dropout samples of various "thinned" networks from an exponential number are tested, it is easy to estimate the combined impact of predictions of all these thinned networks with a single unthinned network with smaller weights. This greatly decreases overfitting and provides substantial advantages over other methods of regularization. We demonstrate that dropout increases neural network efficiency in supervised learning by obtaining state-of-the-art findings on several benchmark datasets in vision, speech recognition, text classification, and computational biology tasks.

In [5], the focus is on the reasons, and the intuitions behind the design of the supervised learning algorithm are to prevent overfitting. Overlap usually happens if there are too many parameters to change in a model of learning. A common approach to alleviate this problem is to include a penalty term in an objective feature called regularization in order to stop network parameters from being too large. Another suggested technique known as dropout is applied to avoid co-adaptation by randomly dropping certain units out during the training phase on each training data. In this article, the MNIST dataset examines an overview of various regularization methods of L2 and dropout in a single hidden neural layer network. In our experiment, the two methods of regularization apply to the same hidden layer neural network with different scales complexity of the network. For complex networks that involve a large number of secret neurons, the findings suggest that dropping out is more successful than L2-norm. The findings of this study help design neural networks with sufficient regularization choices.

We know well that adding noise to the neural network input data during training can lead to significant improvements in generalization performance under certain circumstances. Earlier work has shown that such noise training equals a regularization of the error function in which an additional term is applied. The regularization concept, however, which contains second error function derivatives, is not restricted below and may therefore lead to problems when used directly within an algorithm error reduction based. In this paper, we show that the regularization term can be reduced to a given positive form with regard to network training. The regularization term is part of the class of generalized Tikhonov regulator for a sum of quadrants error function. A realistic alternative to training is the direct minimization of the regularized error function with noise.

3 Dropouts

Dropout is a regularization strategy to avoid complex co-adaptations to the training data to minimize excessive fitness of artificial nerve networks. It is an effective way of using neural networks for the model averaging. Dropout has modified the idea of gathering all weights to learn a fraction of the network weights in each training (Fig. 2).

First of all, it is instructive to analyze some dropping properties of neural networks since these can be described most commonly as an acyclic graph in a multilayer feedforward network for unit i of layer h can be expressed as:

$$S_i^h(I) = \sum_{l<h} \sum_j w_i^{hl} j S_j^l \text{ with } S_j^0 = I_j \tag{1}$$

where the variable w denote the weights l number of layers and I input vector.

Dropouts can be expressed in the form

$$S_i^h(I) = \sum_{l<h} \sum_j w_i^{hl} j \delta_j^l S_j^l \text{ with } S_j^0 = I_j \tag{2}$$

where δ_j^l is Bernoulli variable, with $P(\delta_j^l = 1) = p_j^l$. The activity in unit is independent of weights and each other. Similarly, dropouts to connections can be applied as,

$$S_j^h(I) = \sum_{l<h} \sum_j \delta_{ij}^{hl} w_{ij}^{hl} S_j^l \text{ with } S_j^0 = I_j \tag{3}$$

For model estimation, we minimize the loss function. For this network, consider the least square loss function

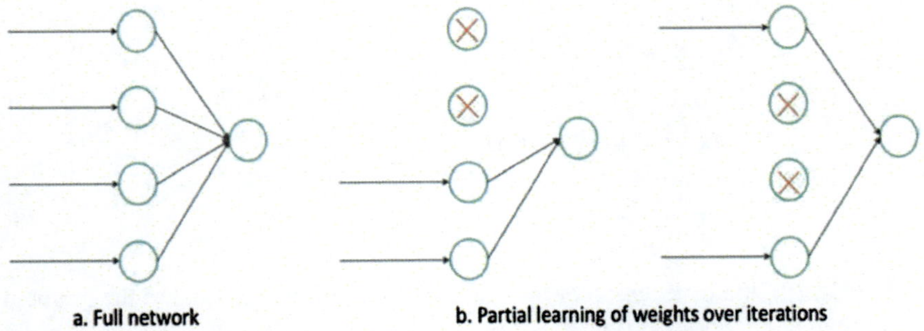

Fig. 2 Illustration of learning a part in network in each iteration

$$E_N = \frac{1}{2}\left(t - \sum_{i=1}^{n} w_i^t I_i\right)^2 \tag{4}$$

$$E_D = \frac{1}{2}\left(t - \sum_{i=1}^{n} \delta_i w_i^t I_i\right)^2 \tag{5}$$

Equation shows the loss for regular network, Eq. (5) is for a dropout network, and the dropout rate is δ, where $\delta \sim$ Bernoulli(p). This means δ is equal to 1 with probability p and 0 otherwise.

The backpropagation for the network training uses a gradient descent approach for the gradient of the dropout network is

$$\frac{\partial E_D}{\partial w_i} = -t\delta_i I_i + w_i \delta_i^2 I_i^2 + \sum_{j=1, j\neq 1}^{n} w_j \delta_i \delta_j I_i I_j \tag{6}$$

So, we apply a relationship between the gradient and the gradient of the regular network. Suppose we make $w' = p * w$, in Eq. (4). Therefore,

$$E_N = \frac{1}{2}\left(t - \sum_{i=1}^{n} p_i w_i I_i\right)^2 \tag{7}$$

Taking the derivate of Eq. (7), we find

$$\frac{\partial E^N}{\partial E_i} = -t p_i I_i + w_i p_i^2 I_i^2 + \sum_{j=1, j\neq 1}^{n} w_j p_i p_j I_i I_j \tag{8}$$

If we find the gradient of the dropout network, we get

$$\begin{aligned} E\left[\frac{\partial E^N}{\partial E_i}\right] &= -t p_i I_i + w_i p_i^2 I_i^2 + w_i \mathrm{Var}(\delta_i) I_i^2 + \sum_{j=1, j\neq i}^{n} w_i p_i p_j I_i I_j \\ &= \frac{\delta E_n}{\delta w_i} + w_i \mathrm{Var}(\delta_i) I_i^2 \\ &= \frac{\delta E_n}{\delta w_i} + w_i p_i (1 - p_i) I_i^2 \end{aligned} \tag{9}$$

For Eq. (9), the expectation of the gradient with dropout is equal to the gradient of the regular network E_N if $w' = p * w$.

Table 1 Comparison of dropout with uniform and Gaussian distribution

S. No.	Dropout at hidden and output layers	Error rate for uniform distribution	Error rate for Gaussian distribution
1	Execution time	4.990	2.1886
2	Dropout of 1 connection	4.5403	4.1905
3	Dropout of 2 connection	3.5670	2.5679
4	Dropout of 3 nodes	3.1897	1.2543
5	Dropout of 5 nodes	3.0234	1.1987
6	Dropout of 6 nodes	2.1712	0.5273

4 Experimental Setup

To see how dropouts work, the network model was built with MNSIT datasets with sigmoid and ReLU activation function. The setup included dropouts both at connection and neurons as well with variation of different learning rate. The setup showed that the neural network was able to improve the generalization performance that did not use dropout.

Table 1 shows the performance of uniform distribution and Gaussian distribution with dropouts applied on different connection and nodes. The setup considers network size and learning rate and momentum. Dropping units reduces a neural network's capability, which is expected. In assumption, if n is the number of hidden units in any layer and p is the probability of retaining a unit, then instead of n hidden units, only pn units will be present after dropouts. Furthermore, this set of pn units will be unique each time and the units are not permitted to overlap and create as many coadaptations required. As opposed to normal stochastic gradient descent, dropout adds a large amount of noise in the gradients. As a result, several gradients appear to cancel each other out. To compensate for this, a dropout net can normally use 10–100 times the learning rate that a normal neural net will use. Using a high momentum is another way to reduce the influence of noise. Although standard nets usually use momentum values of 0.9, we found that values of 0.95 to 0.99 function much better with dropout. Learning is greatly accelerated when a high learning rate and/or momentum is used.

5 Result

We have applied dropouts on different layers, on both connection and node. The dropouts were also examined with the building blocks of different levels like connection and node. The different levels of dropouts are regularized with different training levels of neural network and also evaluate effectiveness of dropout operations. We can further improvise the by applying effective dropout training and Bayesian inference (Figs. 3 and 4).

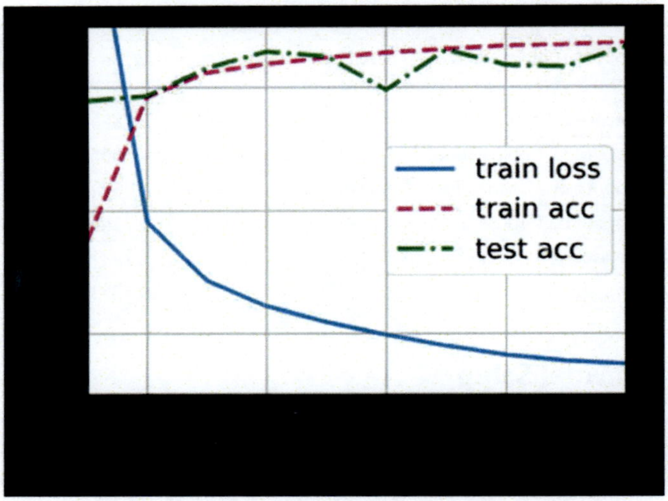

Fig. 3 Train versus test accuracy with training loss

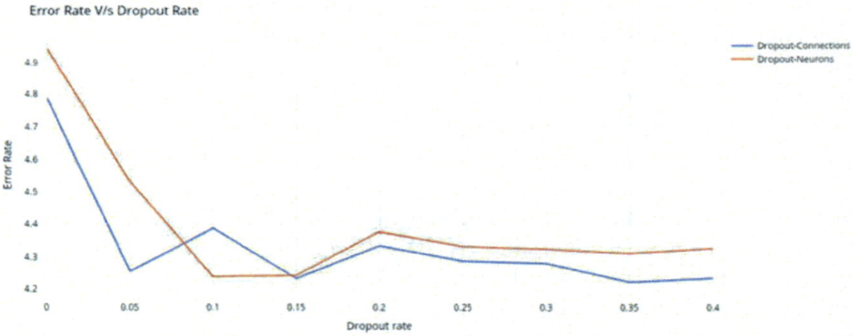

Fig. 4 Dropout at connection and neuron

6 Conclusion

In this paper, there are two levels of dropout training which includes dropout training for drop-connection and dropout-neuron. Dropout aims at producing robust units that are not dependent on other units' activation information. As a result, it seeks to generate units with activities with small dropout variance through dropout subnetworks. This partial variance minimization is achieved by keeping the weights low and by using a slick encoding, which in turn enhances the accuracy and consistency of the dropout approximation. So, to some extent, dropout leads to broad but energy-efficient networks through the use of small weights, and small coding that is probably biologically important is well known. However, a more thorough analysis shows an elegant mathematical network underlying this. It is doubtful if this mathematical structure is the only result of chance and makes one believe that dropout is more than a smart hack and can be an important concept for a certain amount of time, especially in AI and machine learning.

References

1. Labach, A., Salehinejad, H., Valaee, S.: Survey of Dropouts Methods for Deep Neural Networks. arXiv:1904:13310v1 (2019)
2. Matsuoka, K.: Noise injection into inputs in back-propagation learning. IEEE Trans. Syst. Man Cybern. **22**(3), 436–440 (1992)
3. Baldi, P., Sadowski, P.: Understanding Dropouts. University of California
4. Srivastava, N., Hinton, G., Krizhensky, A., Sutskever, I., Salakhutdinov, R.: Dropout: a simple way to prevent neural network from overfitting. J. Mach. Learn. Res. (JMLR) (2014)
5. Phaisangittiasagul, E.: An analysis of the regularization between L_2 and dropout in the single hidden layer neural network. In: International Conference on Intelligent Systems, Modelling and Simulation (2016)

Visual Secret Share Creation with Grayscale Image Converted to RGB Images Using Zigzag Scanning Algorithm

M. Karolin and T. Meyyappan

1 Introduction

Today, more and more digital materials are being transmitted and exchanged over the Internet. Image security is becoming very important for the transmission of images over the Internet or the wireless network. Cryptography includes a set of techniques to keep transmitted or stored data confidential. Visual cryptography can be classified into three separate schemes of symmetric key cryptography, asymmetric key cryptography and secret sharing. Conventional symmetric and asymmetric cryptography modifies a message given to a random string using a secret key or a public key [1, 2]. Visual cryptography is a new cryptographic technique that makes it possible to encrypt visual information in such a way that the decryption processes. The proposed technique includes the image variance composed of the grayscale image block and the white and color image [3, 4]. The objective of the article is to provide an introduction to various technologies like OpenCV, Java and MATLAB. These technologies convert an image into a grayscale to a color image [5]. The visual cryptography of the color image into a grayscale image including prominent features is a complicated process. To convert the color image into a grayscale image, the new algorithm approximates RGB, decreases and adds luminance and chrominance [6]. This article is organized as following: Sect. 1: Introduction; Sect. 2: Description of related work performed by various researchers; Section 3: Discuss how the zigzag scanning algorithm can be extended to the grayscale image converted to RGB color image encryption and suggested decryption method of proposed method; Section 4 performance analysis of result and discussion; and Sect. 5: finally conclusion of the work.

M. Karolin (✉) · T. Meyyappan
Department of Computer Science, Alagappa University, Karaikudi, India

T. Meyyappan
e-mail: meyyappant@alagappauniversity.ac.in

M. Dua et al. (eds.), *Proceedings of the International Conference on Paradigms of Communication, Computing and Data Sciences*, Algorithms for Intelligent Systems,
https://doi.org/10.1007/978-981-16-5747-4_63

2 Related Work

Apurva et al. suggested that adding color to the grayscale picture is not possible. Image splitting will provide a more efficient color addition. For converting grayscale image into color image the following steps might apply; that is image selection, image splitting, pattern recognition, changing color a grayscale image, joining split color image and and also applying image difference method [7, 8].

Preetha et al. suggested that encrypting the shared visual images generated by cryptography using public key encryption. The RSA algorithm is used to provide double security for secret text. Thus, a shared secret is not present in their real form for any modification by adversaries who try to generate fake actions. The system provides more secure and robust secret sharing against a number of attacks and writings, texts, images and printed materials over the public network [9, 10].

Manika Sharma et al. proposed a cryptographic technique for the color image using color error diffusion technique using by the XOR operation. Shares are expanded by using this random number. The key generated by the decryption process is sent securely to the network using the RSA algorithm [11, 12].

Karolin and Meyyappan proposed the images are transmitted after applying the visual cryptographic methods. The attacker cannot appreciate the fuzzy image and thus the data communication becomes secure. This technique overcomes the disadvantage of complex computations required in classical cryptography. Visual cryptography can also be applied to RGB color images by converting them to binary images in black and white [13, 14].

3 Proposed Methodology

In this article, the proposed method of visual cryptography is converted to color images. The original visual cryptographic image values are RGB pixel values and separate share values. Encrypting and decrypting the images using the zigzag scanning algorithm. The image encryption process is the reverse of the decryption process. The proposed method is performance analysis of a calculation of the radio noise peak signal value (PSNR) and the average square error (MSE). This work is implemented within the MATLAB encoding. Figure 1 shows an encrypted and decrypted secret image sharing creation and shared in stacked images. Phase 1 is representing the secret image. Phase 2 represents the creating separate units. Phase 3 represents the share 1 and share 2 encryption and decryption using the zigzag scanning algorithm. Phase 4 represents the exit of the stacked image.

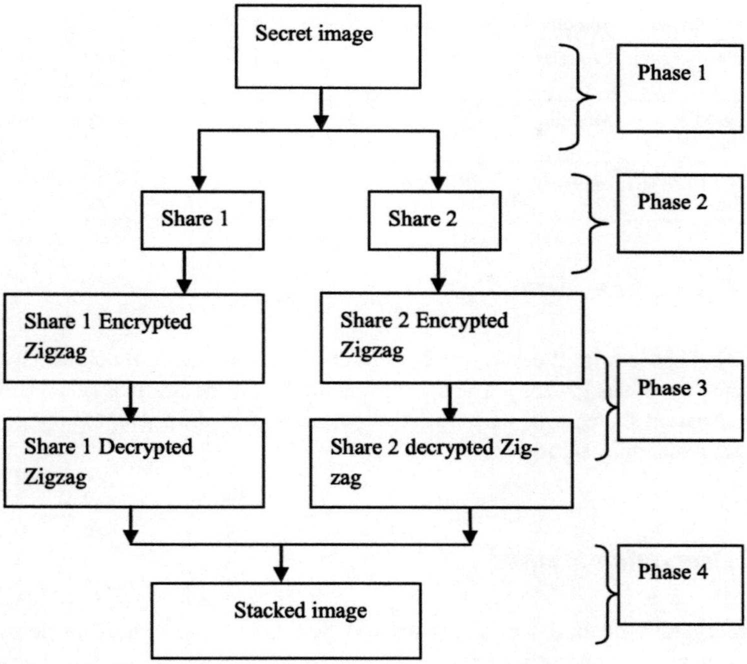

Fig. 1 Proposed methodology

3.1 Block Diagram

See Fig. 1.

3.2 Zigzag Scanning Algorithm

Visual cryptography zigzag scanning algorithm is 1, 2, 9, 17, 10, 3, …, 64, secret image matrix size is MXN, and vector size is $1 \times M * N$. Visual cryptography zigzag scanning algorithm is used for compare the image matrix at vector size. The zigzag scanning algorithm is created as 2 out of 2 shares and the techniques of creating visual secrecy sharing. Visual cryptography inverse zigzag scanning algorithm is (1, 2, 9, 17, 10, 3, …, 64) was 2 out of 3 visual secret sharing schemes and combine with the stacked image bit AND operation.

Table 1 shows the existing AES algorithm in the pepper binary image and compares the proposed zigzag scanning algorithm to the pepper color image.

Table 1 Comparison of complexity and security

Algorithm	Complexity	Security	Generated no of share	Original image to stacked image
Existing AES algorithm	Medium	Increase	2	Grayscale image
Proposed zigzag algorithm	Medium	Increase	2	Color image

3.3 Encryption Method

The global process for the encryption method is the original RGB color images and the input key values (0–255). The encryption method is the original shared image 1 and transparent shared image 2 to the stacked image, then the encrypted image using the zigzag scanning algorithm.

3.4 Decryption Method

The encryption method works against the decryption methods. The decryption method is put in encryption because the encrypted image shared 1 and shared 2 are transparent to the stacked image and then decrypted process is using by zigzag algorithm.

3.5 Grayscale Image Converted to Color Image

Converting a grayscale image into a binary image is exactly two colors usually black and white. Grayscale images also have two levels (0, 1). In Grayscale image, all the pixels from the input image with a luminance higher than the level with the value 1 (White) and all other pixels with the value 0 (Black) BW = im2bw (RGB, level). A color image is a digital image that includes information about the color per pixel. A color image has three values or channels per pixel, and they measure the intensity and light of smoothness.

```
RGB = imread ('peppers.png');
Imshow (RGB).
   Figure
   I = rgb2gray (RGB);
   Imshow (I);
```

4 Result and Discussion

Experimental results are gotten from the binary image converted into RGB color image. The color image is recovered, and the time taken to encrypt shares is lower than for existing systems. In this document, the result and discussion are included in the original algorithm of the zigzag image using methods of encryption and decrypting images, and then stacked the image on the secret image or the original image. The results of some experiments are presented in the following pictures of the pepper grayscale images are converted into color images.

Figure 2 shows that the pepper grayscale image. In that the original image and created image are share 1 and share 2. Visual cryptography zigzag scanning algorithm is used for this share 1 and share 2 encryption and decryption process to the stacked image.

Figure 3 shows that the visual cryptography pepper grayscale image. It is converted into the RGB pepper color image using by color conversion method. Share 1 and share 2 are encrypted and decrypted to the stacked image using by zigzag scanning algorithm (Chart 1).

Table 2 shows the existing method is as follows the picture quality evaluation, Floyd and Steinberg dithering algorithm, adaptive order dithering and color error diffusion using XOR as compared to the proposed zigzag scanning algorithm (Chart 2).

Table 3 shows that the pepper grayscale image AES existing method is converted to a proposed RGB pepper color image value as PSNR 62.94 and MSE value 106.7 with calculated the image quality and NPCR 44.44 and UACI value is 13.88 calculated the image security analysis and image quality measurement.

Fig. 2 Pepper grayscale image

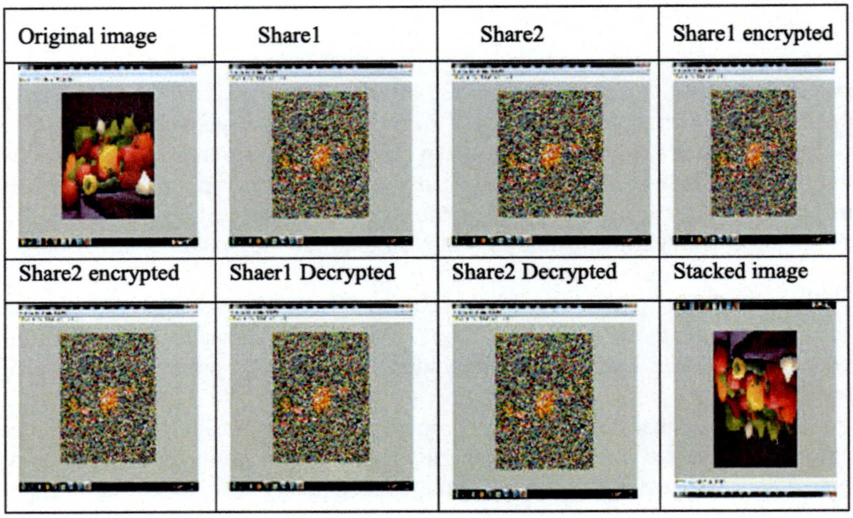

Fig. 3 Converted into pepper color image

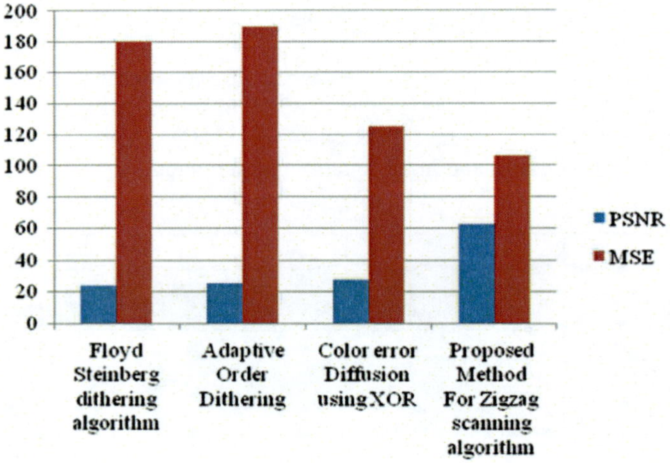

Chart 1 PSNR and MSE values

Table 2 Performance of various methods for PSNR and MSE values

Picture quality evolution	Floyd–Steinberg dithering algorithm	Adaptive order dithering	Color error diffusion using XOR	Proposed method for zigzag
PSNR	24	25.25	27.17	62.94
MSE	180.90	190	125	106.7

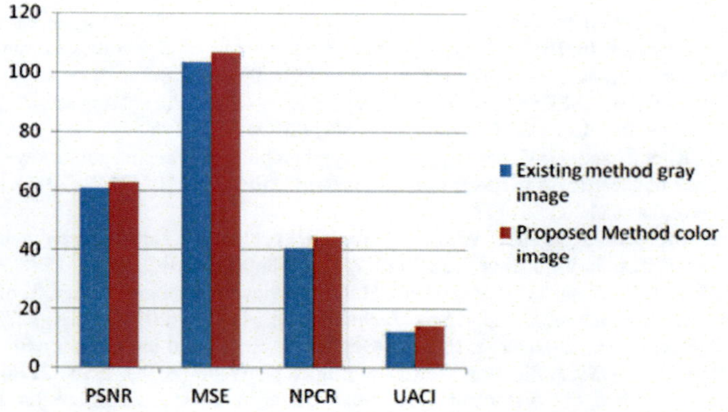

Chart 2 Pepper grayscale image comparison into color image

Table 3 Performance analysis of existing method compared to proposed method

Pepper	PSNR	MSE	NPCR	UACI
Grayscale image	60.94	103.36	40.32	12.08
Color image	62.94	106.7	44.44	13.88

5 Conclusion

In this article, it is assumed that the original image is produced from three color channel images R, G and B. Therefore, the quality of the original image depends on the image quality of the R, G and B channels. In the visual cryptography, secret sharing scheme images are converted with the help of zigzag scanning algorithm. To add more security to the secret sharing scheme of the image, the zigzag scanning algorithm is used to share1 and share 2 creation and then share 1 share 2 encrypted and decrypted to stacked images. In this paper, the analysis of the proposed binary image approach is converted to the RGB color image. When performing the PSNR and MSE values, calculate the original image and the decrypted color images. Image security analyzes the NPCR and UACI values for image measurement. In visual cryptography, converted RGB image from grayscale image is highly secured. Its image quality and security are comparatively high.

References

1. Naor, M., Shamir, A.: Visual cryptography. Adv. Cryptol. Eurocrypt'94 (1995)
2. Naor, M., Shamir, A.: Visual cryptography. Adv. Cryptol. Euro crypt 94", Lect. Notes Comput. Sci. **950**, 1–12 (1995). https://doi.org/10.1007/BFb0053419, Print ISBN: 978-3-540-60176-0. Online ISBN: 978-3-540-44717-7

3. Koga, H., Yamamoto, H.: Proposal of a "lattice based visual secret sharing scheme for color and gray-scale images. In: IEICE Transactions on Fundamentals of Electronics Communications and Computer Sciences, pp. 1262–1269 (1998). ISSN: 0916-8508
4. Shetty, S., Abraham, M.P.: A secure visual cryptography scheme for sharing secret image using RSA. J. Innov. Res. Comput. Commun. Eng. **3**(4) (2015)
5. Kaur, K., Khemchandani, V.: Securing visual cryptographic shares using public key encryption. In: Institute of Electrical and Electronics Engineers (IEEE), pp. 1108–1113 (2013). https://doi.org/10.1109/IAdCC.203.6514382
6. Kondra, A.K., RatnaKumari, U.V.: An improved (8, 8) color visual cryptography scheme using floyd error diffusion. Int. J. Eng. Res. Appl. **2**(5), 2248–9622 (2012). ISSN: 2248-9622
7. Parandekar, A.B., Dhande, S., Vhywhare, H.R.: A review on changing image from grayscale to color. Int. J. Adv. Res. Comput. Eng. Technol. **3**(1), 143–146 (2014). ISSN: 2278-1323
8. Taki El_Deen, A.E., El-Badawy, E.-S.A., Gobrans, S.N.: Digital image encryption based on RSA algorithm. IOSR J. Electron. Commun. Eng. **9**(1), 69–73 (2014). ISSN: 2278-8735
9. Preetha, M., Nithya, M.: A study and performance analysis of RSA algorithm. Int. J. Comput. Sci. Mobile Comput. **2**(6), 126–139 (2013). ISSN: 2320-088X
10. Chepuri, S.: An RGB image encryption using RSA algorithm. Int. J. Current Trend Eng. Res. **3**(3), 1–7 (2017). eISSN: 2455-1392
11. Sharma, M., Saraswat, R.: Secure visual cryptography technique for color images using RSA algorithm. Int. J. Eng. Innov. Technol. **2**(10), 183–186 (2013). ISSN: 2277-3754
12. Asha Bhadran, R.: An improved visual cryptography scheme for color images. Int. Res. J. Eng. Technol. **02**(05), 1391–1395 (2015). ISSN (e): 2395-0056, ISSN (p): 2395-0072
13. Karolin, M, Meyyappan, T.: RGB based secret sharing scheme in color visual cryptography. Int. J. Adv. Res. Comput. Commun. Eng. **4**(7), 151–155 (2015). ISSN(online): 2278-1021, ISSN(Print): 2319-5940. https://doi.org/10.17148/IJARCCE.2015.4734
14. Chouksey, P., Rai, R.: Secret Sharing based Visual Cryptography Scheme for color preservation using RGB Color Space, vol. 5, No. 5 (2015)

Performance Optimization of Short Video Using Convolutional Neural Network for IOT Applications

Sneha Venkateshalu and Santosh Deshpande

1 Introduction

Internet of Things (IoT) is an Internet community connecting every edge devices and processing them which becomes a very major thing because everyday communication accounts are informative. While the Internet of things-related models are capable of understanding the respective tool of technology with methods to achieve the end user requirements, the present models are proposed to solve real-time issues by developing the adaptation of powerful IoT edge devices. But presently, we should look after and take care of the exact power of the devices and contribute models with respective to their configurations. In order to carry out this systematically, convolutional neural network (CNN) and deep learning methods processed the computer vision problems such as language processing and image segmentation with multi-class labels. They extended traditional artificial neural network (ANN) by increasing depth features with additional constraints to the previous layers. This pre-trained model has different data sets with respective of functionalities and models that can attain with the boosting performance of these networks. The focused work is about understanding, analyzing and modifying the respective model with available functions like activation functions, about layers modifications using mathematical functions and models applied to them for achieving the performance of the CNN model which reduces space and time which then saves the power consumption of IoT interconnected devices.

CNN task consists in extracting features from the images as follows:

- The grayscale image is an input layer to the neural network with output of multiple labels using hidden layers containing convoluted layers with activation functions

S. Venkateshalu · S. Deshpande (✉)
Department of Computer Network Engineering, Visvesvaraya Technological University, Belagavi, India
e-mail: sld@vtu.ac.in

M. Dua et al. (eds.), *Proceedings of the International Conference on Paradigms of Communication, Computing and Data Sciences*, Algorithms for Intelligent Systems,
https://doi.org/10.1007/978-981-16-5747-4_64

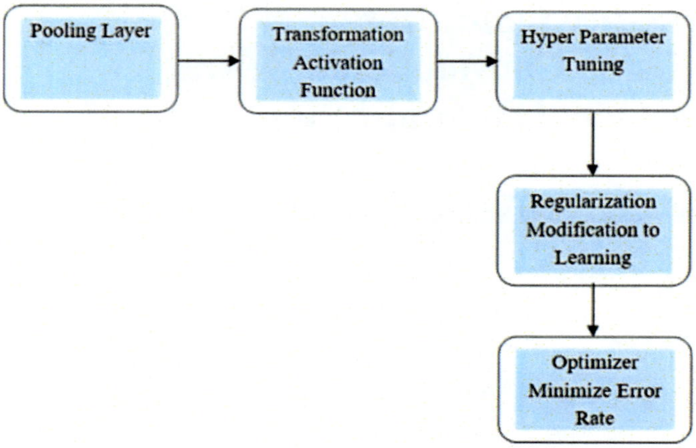

Fig. 1 Flow of CNN video analytics

as output signals accordingly with the pooling layers and lastly fully connected layer (as shown in Fig. 1).

ANN containing convolution and pooling layers with various neurons does not have the capability of extracting depth features from the input data image. Similarly, these layers cannot perform classification for this reason required for connecting about the full neural network. CNN task processes the images into an easier form without affecting features. Training data contains grayscale images to the convoluted layer for extracting features. The layers consisting of distinct weights are used to extract features from the input data.

Kernel shifts across the image, and the process is called stride. Stride value with 1 is about covering the entire image. The CNN learns the weights by its own. The convolution operation extracts depth features from the image and can always add more than one convolution layer to build the neural network, first convolution layer is responsible for capturing gradients whereas the second layer captures the edges, and adding layers depend on the complexity of the image. Padding with the values depends on filter and maintains the original dimension of an image.

ReLU is an activation function applied to increase the nonlinearity of the network without affecting convolution layers of the neural network, making faster the training of the data. ReLU produces linearity for positive values and making zero for negative values. It takes less time to train and run the computations due to its simple math. When the large data is the input, then its linearity does not saturate, and the convergence is faster with the sparse network. In sparse network instead of firing all every time, distinct roles are activated by the different signals. The model with finer predictive power and less noise impact is due to its sparsity producing in concise manner.

The different types of nonlinear activation functions other than the ReLU are sigmoid function which is used for the output required for the predicting type probability model existing between 0 and 1. But this type of function stuck at training data set in neural network is stuck due to its logistic function. The more generalized function is softmax function used for multi-class classification. Tanh overcomes sigmoid function ranging the values from − 1 to 1, used for the classification of between two classes. Feed forward networks use both the types of tanh and logistic sigmoid activation function. Deep learning methods mostly used the ReLU activation function, and there are few more varieties of ReLU functions, namely leaky ReLU which solves the dying of the network model of the ReLU function. The leaky function increases the range of ReLU with a 0.01 value. If the value a is not 0.01, then this type of ReLU is called as randomized ReLU. The leaky and randomized ReLU functions increase or decrease the function as well as with their derivatives and are called as monotonic function.

Bilinear is a calculation made by linear interpolation between the values of the four neighbor points and is the point value in the new image. Real-world data is complex, and to solve complex problems, solutions are needed with fewer weight parameters, preventing model from complex. With respective of complexity, weight parameters are added to the loss function and because of some parameters are positive and some other with negative. So in order to obtain the result, a best model is required to add all the squares of the parameters to loss function setting parameters to 0. The sum of the squares is multiplied with another smaller number referred to weight decay. The weights updated during training process are called as learning rate and are a configuration hyper-parameter that can be externally modified to the model, and whose value is not estimated from the original data controls how much to change the model in response to the estimated error each time the model weights are updated.

Convolutional neural networks are trained by using algorithm called stochastic gradient descent. This optimization algorithm is used to estimate the gradient error for the current state of the model using training data set examples, and then weights are updated using the algorithm of back-propagation error, referred to as back-propagation. Similarly, the learning rate is also a configurable hyper-parameter used during the training in neural networks which has a small positive value, within the range of 0 and 1.

Model with high bias creates under-fitting problem and can be solved using regularization. Bias is an additional parameter in neural network used in order to adjust the output with weighted sum of the inputs to the neuron. Regularization is the technique which forces the weights of a model to be small that further penalizes the loss function and finally helps to simplify the model.

1.1 Survey

Video frame interpolation proposed using adaptive convolution with fully connected CNN estimates kernels of adaptive convolutions for every pixel from the two input consecutive frames. The pixels of the interpolated frames are estimated for a separate kernel. The pixel-wise estimation of each convolutions is applied to generate the new interpolated frame [1]. Traditional methods of frame interpolation involves motion estimation and resampling methods. The kernels account these methods, and they process locally the input frames with encoding technique along the coefficient for the frame pixel generation. The video frame interpolation proposed in neural network is trained from end–to-end layers by using features of occlusion and brightness to alter and to increase the quality of video frame interpolation [2]. Optical flow-based technique is made unable to handle these challenges, is dealt with frame interpolation for the reduction in optical flow errors with queues of system in a multithreaded environment and provided a quick solution to enhance the frame rate of the system [3]. Further, with better results, kernel-based methods acquired the edge-based kernels of estimation in neural network. Two-0dimensional convolution kernels computations for each pixel made the technique slower and also expensive, failing to employ on high-resolution video [4].

Convolutional neural networks make use of video frame interpolation with the purpose of function minimizing the pixel difference generated one along with ground-truth image correspondingly. CNN removes optical flow method steps and process as a convolution of video frame interpolation. CNN has the best compatibility features compared to other neural networks which deals with single, two-dimensional feature extraction. CNN has its achievements on its architectural model built with important categories based on extracting spatial features. It has the capability of extracting depth, height, width and multipath features using filters boosting the channel and mapping features of a neural network model with ability of extracting multiple features and makes it to learn from the data representations very well in timely. Architecture of neural network made improvements with its model using the activation functions with error functions enabling how well is the model working. Performance optimization of CNN can be done using the parameter optimizers of regularization techniques, hyper-parameter tuning of the learning weights [5]. This further improves the hardware implementations of the technology used. Thus, architecture is provided to be adapted on the edge devices applications including for detecting objects with multiple label extractions, image classification of video with image segmentations, processing about recognizing the language and speech in a real-time video.

The features of activation function discuss vanishing gradient problem, zero centered, computational expense and differentiable. Neural networks use the gradient algorithms to train the network and optimize the model with necessary parameters. Target of this algorithm is to reduce the errors of every epoch using back-propagation method. Gradients tend to vanish because of the depth features in network, shifting value to zero. So to overcome this problem, activation functions are used to shift the gradient toward zero. Calculating functions multiple times at fully connected layer or at each layer makes it computationally inexpensive to calculate [5]. Gradient descent process of layers in neural network needs to be least differentiable or differentiable in the parts of the network.

Neural network does feature extraction or any other processing of image with respective to video analyzing is done in time series. The ReLU activation function provides this feature that returns the model with precision analysis in time process of the video frames in a network. This activation function is simple to calculate, so it does not require much computation for mathematical analysis. The advantage of using the function is its sparsity nature which makes the sparse matrix. Sparsity provides the model with quick good prediction power with less overfitting where neurons process only the required features of the data. ReLU overcomes the vanishing gradient problem and has better convergence property for larger input of the model. The processing of intermediate frame between the consecutive frames is said to be video frame interpolation in which for further video synthesis, activation function applied is for the better convergence of video frames at the output layer [6]. At the decline of ReLU, it creates a problem lasting of neurons when all the negative values are made zero. LReLU with $y = 0.1x$, when $x < 0$, alternatively making all together the negative values with small slope. LReLU fixes ReLU decline issue, making training fast possible with mean activation. The function with balanced property makes neuron to learn quickly, and this has been applied for estimating quality of intermediate interpolated frames of a video [7]. The phase network calculates the decay process of the intermediate frame with the network decoder. The neural network model is proposed to collaborate phase-based approach to produce an intermediate image with input of two neighboring frames. ReLU gets the altered feature maps and values predicted from the previous layers to produce nonlinearity changing the output state of neuron [8]. To overcome the drawback of ReLU nonlinearity function, the PReLU has been introduced, is a new abstraction of ReLU activation function, in which this activation function learns the parameters adaptively of the rectifiers and improves accuracy. The PReLU attempts to fix the problem of completely saturating ReLU by providing a slight negative slope toward when the input values are negative. When the unit is not active, the LReLU function allows even small nonzero gradients. The PReLU solves leaking of coefficients and further making the parameters adaptively learn with other neural network parameters [6, 9]. Bilinear interpolation is a method that produces using with continuous function [10] and efficient on many platforms which are available in hardware, making it practically possible for real-time applications.

1.2 Recent Work

Frame interpolation to achieve real-time frame rate

Frame interpolation task is implemented on real-time short video data applications in order to add the number of frames per second. Video streaming adapts frame interpolation technique for streaming lower frame rate video that is interpolated to create more frames and produce smoothly with high-frame rate video. Adding intermediate frames called as interpolated frames between the two consecutive frames, which is called as frame interpolation, requires processing, where interpolated frames may contain the noise while adding in between in real-time implementation. So we require deep learning. Frame interpolation increases additively the smoothness of motions.

Frame rate and image quality are one of the two major key factors that affect service quality.

To achieve real-time frame rate, one of the techniques called deep learning is used to enhance the quality of interpolated frames. Deep convolution model using bilinear interpolation which is also meant for bilinear filtering with ReLU activation function can better perform enhancing the quality of interpolated frame task. The linear interpolation on 2D network is about interpolating functions of two points and is used when one coordinate is a function of the other. Linear interpolation is the convex combination of 2D points. The bilinear term refers to performing interpolation process with the coordination of using linear interpolation first in one direction and later then again in the other direction. The new image is calculated using the four nearest neighbor points, and the point value is referred to as bilinear interpolation, where recently, the bilinear interpolation is used for the skin lesion segmentation.

The image coordinates, $f(p_{10}, p_{20})$, $f(p_{11}, p_{21})$, $f(p_{12}, p_{22})$ and $f(p_{13}, p_{23})$, are the nearest points of $g(p_1, p_2)$.

The interpolated image with the intensity values of the $g(n1, n2)$ is calculated as follows:

$$g(p_1, p_2) = B_0 + B_1 p_1 + B_2 p_2 + B_3 p_1 p_2 \tag{1}$$

This is the equation representing coordinates (p_1, p_2) of bilinear function. Here, f refers to an intensity value with given $f(p_{10}, p_{20})$, $f(p_{11}, p_{21})$, $f(p_{12}, p_{22})$ and $f(p_{13}, p_{23})$, and these are the pixel locations applying with g specifying intensity value of the interpolated image using matrix in (p_1, p_2) coordinates being applied before the interpolation and B specified as the bilinear weights.

$$\begin{bmatrix} B_0 \\ B_1 \\ B_2 \\ B_3 \end{bmatrix} = \begin{bmatrix} 1 & p_{10} & p_{20} & p_{10}p_{20} \\ 1 & p_{11} & p_{21} & p_{11}p_{21} \\ 1 & p_{12} & p_{22} & p_{12}p_{22} \\ 1 & p_{13} & p_{23} & p_{13}p_{23} \end{bmatrix}^{-1} \begin{bmatrix} f(p_{11}, p_{21}) \\ f(p_{12}, p_{22}) \\ f(p_{13}, p_{23}) \\ f(p_{10}, p_{20}) \end{bmatrix} \tag{2}$$

The result with $g(p_1, p_2)$ is defined as a bilinear interpolation. The assigned values of $g(p_1, p_2)$ and when making least squares fit into it, and the process is made with optimal averaging producing smoother results. ReLU with nonlinear nature is made to process the back-propagation of the multiple layers of neurons along with errors. ReLU is defined using function $f(x) = \max(0, x)$ and where $f(x)$ is represented as

$$f(x) = \begin{cases} x, & \text{if } x \geq 0 \\ 0, & \text{if } x < 0 \end{cases} \tag{3}$$

ReLU activation function is performed as shown in the below image where negative numbers are replaced with 0 (Fig. 2).

ReLU activation function is simple to calculate, so it does not require much computation for mathematical analysis. The advantage of using the function is its

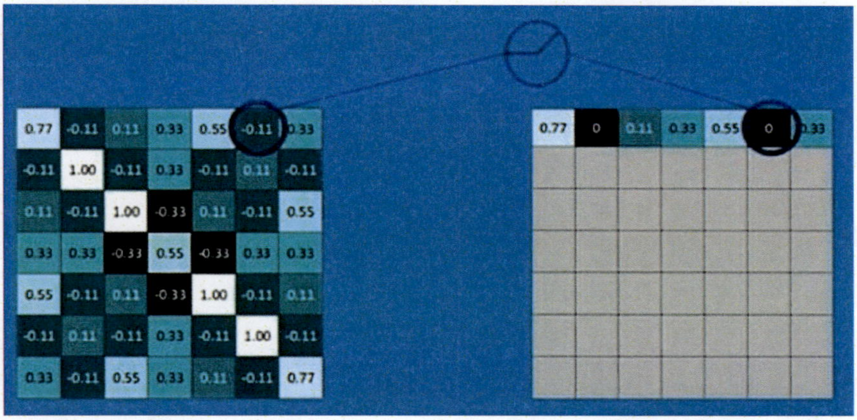

Fig. 2 Process of ReLU activation function

sparsity nature which makes the sparse matrix. Sparsity provides where the neurons process only the required features of the data. ReLU overcomes the vanishing gradient problem and has better convergence property for larger input of the model. Many gradients can be stopped training, causing a weight unstable that gets deactivated data points in neural network, resulting in ReLU with dead neurons in a model.

To overcome this activation function drawback, PReLU solved the problem of getting stopped the training model when nothing is there to learn for the model further, which is generalization function of ReLU activation function. PReLU function learns the parameters adaptively of the rectifiers and improves accuracy.

$$f(x_i) = \begin{cases} x_i, & \text{if } x_i > 0 \\ a_i x_i, & \text{if } x_i \leq 0 \end{cases} \tag{4}$$

x_i is the input of the nonlinear activation function f on the ith channel. To control the negative part of the slope, a_i is the coefficient been used and i specifies activation varying on distinct channels. The a_i is the adaptive learnable parameter figuring out itself in the neural network. The end–to-end activations applied will lead to more specialized when required.

Optimization is carried out with PReLU while training using the back-propagation method and is used to optimize at the same time with other layers of network. The a_i updating has been done with applying chain rule, for one layer the gradient of a_i is

$$\frac{\partial \varepsilon}{\partial a_i} = \sum_{x_i} \frac{\partial \varepsilon}{\partial f(x_i)} \frac{\partial f(x_i)}{\partial a_i} \tag{5}$$

where ε is the objective function. The $\frac{\partial \varepsilon}{\partial f(x_i)}$ is the term gradient from the deeper layer of neural network. The activation function gradient is specified by

$$\frac{\partial f(x_i)}{\partial a_i} = \begin{cases} 0, \text{ if } x_i > 0 \\ x_i, \text{ if } x_i \leq 0 \end{cases} \tag{6}$$

The feature map positions run over using the summation $\sum x_i$ for sharing channel the variant of the gradient a are

$$\frac{\partial \varepsilon}{\partial a} = \sum_i \sum_{x_i} \frac{\partial \varepsilon}{\partial f(x_i)} \frac{\partial f(x_i)}{\partial a_i} \tag{7}$$

where $\sum$i is applied over all the channels of the layer in neural network. Overall, the PReLU activation function solves leakage of coefficients, and the parameters adaptively learn the other neural network parameters. Frame interpolation convolves the two frames to interpolate the pixel, and this is meant as bilinear interpolation producing smoother results along with parametric ReLU activation function.

2 Statistical Analysis

Peak Signal-to-Noise Ratio (PSNR)

- PSNR is frequently used to estimate the efficiency of filters, compressors, etc., and used to define PSNR value of correlation to original image and interpolated frame with neural network model.
- PSNR is the measure of ratio between a signals maximum power of an original image and power of corrupted signals noise of interpolated frame.
- To estimate PSNR, PSNR is a log scale ratio related to the MSE between interpolated frame and its original frame and is the measure of quality of reconstructed image that have been compressed.

$$\text{PSNR} = 10 \log 10\big((L-1)^2/\text{MSE}\big) = 20 \log 10(L - 1/\text{RMSE}) \tag{8}$$

where L is the maximum number of intensity levels (the minimum intensity level supposed to be 0) in an image, MSE is the mean square error used to analyze the performance between the interpolated frame and the original frame, and RMSE is the root mean square error between a tested interpolated image and an "original image."

$$\text{RMSE} = \sqrt{\text{MSE}} = \sqrt{\frac{1}{NM} \sum_{i=0}^{N-1} \sum_{j=0}^{M-1} (A(i,j) - B(i,j))^2} \tag{9}$$

- PSNR penalizes the noise created to real-time video frame during frame interpolation.

2.1 Hyper-Parameter Tuning Technique to Attain Low Latency

A model hyper-parameter is a configurable value that is externally tuned and whose value is not possible to estimate from data. To attain low latency, weight decay-type hyper-parameter tuning technique is carried out in deep learning. Weight decay value is used as optimal hyper-parameter in a gradient descent optimization method of deep learning algorithms. The main idea of this technique is to adjust the weights "ws" to minimize the loss or cost.

The algorithm used is gradient descent to perform the product of gradient by a scalar said to as the learning rate to determine the next point. This parameter conveys how far it is to move the weights in the direction toward of the gradient. The wsd and bsd are used as gradients to update parameters ws and bs for these algorithm as follows:

$$\text{ws} = \text{ws}-\text{learning rate} * \text{wsd} \tag{10}$$

$$\text{bs} = \text{bs}-\text{learning rate} * \text{bsd} \tag{11}$$

If smaller is the learning rate, time taken is more to minimize loss function tiny. If it has high weight, changes are bigger that it makes optimizer overshoot the minimum making the loss worse. Hence, the objective aims to find the minimum loss with optimal learning rate. Initializing weights of the model may lead to rest sometime in the local minimum; in order to avoid this, weight vectors are initialized with values from a random distribution, using wssd and bsd independently for each epoch, exponentially weighted averages of wsd and bsd are computed as

$$\text{Vwsd} = \beta * \text{Vwsd} + (1-\beta) * \text{wsd} \tag{12}$$

$$\text{Vbsd} = \beta * \text{Vbsd} + (1-\beta) * \text{bsd} \tag{13}$$

where β is one of the other hyper-parameter ranges between 0 and 1.1 which sets the weight between the average of before value and the present value to compute the new weighted average. After calculating, exponentially weighted averages with updated parameters are followed as below and algorithm ends up at local optima with few iterations

$$\text{ws} = \text{ws}-\text{learning rate} * \text{Vwsd} \tag{14}$$

$$\text{bs} = \text{bs}-\text{learning rate} * \text{Vbsd} \tag{15}$$

These iterations finally minimize cost function, are a measure of how well the model is working and find the optimal parameters that yield the best model performance by attaining low latency introduced by the network.

2.2 Regularization to Achieve Rate of Processing Work

Regularization method solves the under-fitting problem when a model is small with high bias and in turn can achieve the rate of processing work. The sum of squared differences from the predicted value to the actual present value is referred to as loss function.

$$L(x, y) = \sum_{i=1}^{n} (y_i - f(x_i))^2 \tag{16}$$

$$f(x_i) = h_\theta x = \theta_0 + \theta_1 x_1 + \theta_2 x_2^2 \tag{17}$$

where here for a linear regression using two input variables, the loss function is specified and $i = 2$. Loss function computed with regularization term is defined as below.

$$L(x, y) = \sum_{i=1}^{n} (y_i - h_\theta(x_i))^2 \tag{18}$$

$$h_\theta x_i = \theta_0 + \theta_1 x_1 + \theta_2 x_2^2 \tag{19}$$

$$L(x, y) = \sum_{i=1}^{n} (y_i - h_\theta(x_i))^2 + \lambda \sum_{i=1}^{n} \theta_i^2 \tag{20}$$

where regularization term added is $\sum_{i=1}^{n} \theta_i^2$.

When λ is zero, then the loss function computed as

$$L(x, y) = \sum_{i=1}^{n} (y_i - h_\theta(x_i))^2 + 0 \text{ and } h_\theta x_i = \theta_0 \tag{21}$$

When λ is larger, then the weights are penalized and they become closer to zero. The under-fitting problem having a high bias can be resolved using with $L1$ regularization defined by the following equation:

$$L(x, y) = \sum_{i=1}^{n}(y_i - h_\theta(x_i))^2 + \lambda \sum_{i=1}^{n}|\theta_i| \tag{22}$$

3 Case Study of the Proposed Work

The proposed work discussed the performance optimization of real-time short video with image processing capabilities applied for the applications of image segmentation, and recognition is performed using deep learning technique and convolutional neural network. Frame interpolation using nonlinear method known as a bilinear interpolation is used to achieve the real-time frame rate by applying parametric ReLU for the further enhancement of frames. Further they can be estimated by the PSNR value to the original frames and the interpolated frames, calculated the errors between them and improved. Weight decay model hyper-parameter can be used for attaining the low latency introduced at network which is known to be transmission latency. The regularization technique is used to avoid under-fitting problem having high bias of the simple model when fewer parameters are to be processed and estimated using loss function further with optimizer error rates which are calculated. Overall, this model can be compressed and added on to the edge devices for the image processing of real-time video saving space and the time which is changeless state can optimize the performance of video processing capabilities.

References

1. Niklaus, S., Mai, L., Liu, F.: Video frame interpolation via adaptive convolution. In: Proceedings—30th IEEE Conference on Computer Vision and Pattern Recognition, CVPR 2017, pp. 2270–2279 (2017)
2. Parihar, A.S., Varshney, D., Pandya, K., Aggarwal, A.: A Comprehensive Survey on Video Frame Interpolation Techniques (2021). https://doi.org/10.1007/s00371-020-02016-y. https://doi.org/10.1007/s00371-020-02016-y
3. Ahn, H.E., Jeong, J., Kim, J.W., Kwon, S., Yoo, J.: A fast 4K video frame interpolation using a multi-scale optical flow reconstruction network. Symmetry **11**, 1–12 (2019). https://doi.org/10.3390/sym11101251
4. Niklaus, S., Mai, L., Liu, F.: Video Frame Interpolation via Adaptive Separable Convolution (2017)
5. Mathai, V., Baby, A., Sabu, A., Jose, J., Kuriakose, B.: Video Frame Interpolation using Deep Convolutional Neural Network. Springer International Publishing (2019). https://doi.org/10.1007/978-3-030-00665-5_82
6. PReLU activation: This Paper Introduced Both the PReLU. https://medium.com/@shoray.goel/prelu-activation-e294bb21fefa. Last accessed 2021/04/07
7. Jiang, H., Sun, D., Jampani, V., Yang, M.H., Learned-Miller, E., Kautz, J.: Super SloMo: high quality estimation of multiple intermediate frames for video interpolation. In: Proceedings of the IEEE Computer Society Conference on Computer Vision and Pattern Recognition, pp. 9000–9008 (2018). https://doi.org/10.1109/CVPR.2018.00938

8. Meyer, S., Djelouah, A., McWilliams, B., Sorkine-Hornung, A., Gross, M., Schroers, C.: PhaseNet for video frame interpolation. In: Proceedings of the IEEE Computer Society Conference on Computer Vision and Pattern Recognition, pp. 498–507 (2018). https://doi.org/10.1109/CVPR.2018.00059
9. Sanjar, K., Bekhzod, O., Kim, J., Kim, J., Paul, A., Kim, J.: Improved U-net: fully convolutional network model for skin-lesion segmentation. Appl. Sci. **10** (2020). https://doi.org/10.3390/app10103658
10. Nabi, J.: Hyper-Parameter Tuning Techniques in Deep Learning. https://towardsdatascience.com/hyper-parameter-tuning-techniques-in-deep-learning-4dad592c63c8. Last accessed 2021/04/05

Convolutional Neural Networks in Particle Classification

J. Tripathi and V. Bhatnagar

1 Introduction

HEP experiments aim for understanding the nature by studying the particle interactions recorded with particle detectors. An important part of understanding particle interactions involves correct categorization of particles produced in the detectors. Generally, the signature of particles left behind in the detector are first reconstructed into high-level objects such as showers, clusters, and tracks. The physics observables associated with these objects including length, energy, direction, and momentum are fed into machine learning (ML) algorithms such as boosted decision trees [1] or K-nearest neighbors [2] for correct identification of each particle. However, the features used to characterize the particles are known a priori. The goal of particle categorization in the HEP experiments is based on extracting typical features of a particle giving it a unique identification. This shares a great deal of similarity with the core problem of specific feature extraction in the field of computer vision. CNNs, being able to learn the features and patterns from images on their own, are well suited for the analyses of the images produced by the particle detectors. Over the last few years, CNNs have been quite successful in performing the key computer vision tasks such as face recognition, image recognition, and detection compared to the previous state-of-the-art machine learning methods.

2 Terminology

The following terms are used quite frequently in the field of machine learning in general and in this paper in particular.

J. Tripathi (✉) · V. Bhatnagar
Panjab University, Chandigarh, Panjab 160014, India
e-mail: tjyoti@pu.ac.in

V. Bhatnagar
e-mail: vipin@pu.ac.in

M. Dua et al. (eds.), *Proceedings of the International Conference on Paradigms of Communication, Computing and Data Sciences*, Algorithms for Intelligent Systems,
https://doi.org/10.1007/978-981-16-5747-4_65

(1) *Training and Testing*: Network should be trained on the training data and should be able to make good predictions for the testing dataset that it has not seen.
(2) *Hyperparameter*: refers to a variable that is set before initializing the training process, e.g., loss function, learning rate, batch size, etc.
(3) *Parameter*: refers to a variable that is learned during the process of training of a network, e.g., weights, biases, and kernel parameters.
(4) *Overfitting*: Model does not generalize well from observed to unobserved data, which is called overfitting. Because of existence of overfitting, the model performs perfectly on training set, while fitting poorly on the testing dataset [3].
(5) *Underfitting*: It happens when model is too simple and can neither model the training dataset nor generalize to new dataset.
(6) *Classification*: Assigning label to a given input. Classification is useful for making discrete predictions, e.g., mapping picture of an animal to its appropriate category.
(7) *Feature maps*: Set of features obtained by convolving trained kernels on each part of the image. Typically used in the field of image recognition.

3 Convolutional Neural Networks

CNNs are one of the well-known deep learning architectures widely used in pattern and image recognition, speech recognition, video analysis, and natural language processing problems. CNNs are inspired by the structure of the visual system and the visual cortex [4]. The *simple cells* in the visual cortex are sensitive to edge-like features within small regions of the retina whereas the *complex cells* are receptive to collections of simple cells and detect position-independent edge-like features [5]. CNNs mimic this using a layered structure for feature extraction from the input images. The first CNN was designed by Yan LeCun and his collaborators for document analysis in the year 1998 [6].

3.1 Architecture of a CNN

Artificial neurons form the basic building unit of CNNs and are inspired from the behavior of the human brain. A neuron receives inputs from external sources, where each input has an associated *weight* (assigned on the basis of its relative importance) which is summed with the *bias* of the neuron to produce output (shown in Fig. 1). The bias of a neuron is a measure of how easy it is to get the neuron activated. The weight and bias of a neuron get tuned during the training process of the network. A mathematical representation of a neuron is shown in Eq. (1) where x_j and w_j denote the set of input values and the associated weights, respectively, b is the bias of the neuron, and φ is the activation function. Activation functions are used in neural

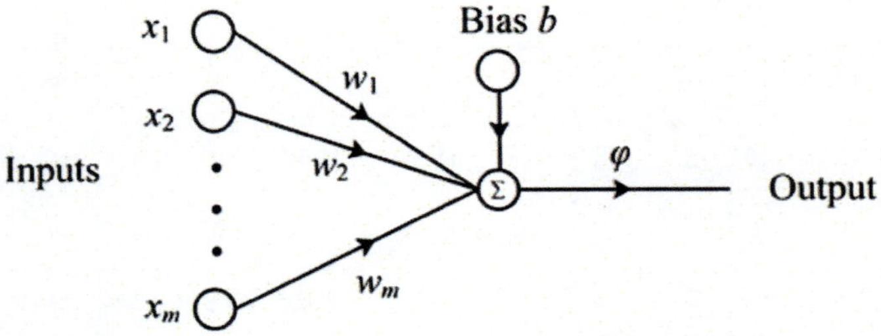

Fig. 1 Basic model of a neuron consisting of input signals (x), weights (w), bias (b), activation function (φ), and the output [7]

networks to compute the weighted sum of input and biases, of which is used to decide if a neuron can be fired or not.

$$y = \varphi\left(\sum_{j=1}^{m} w_j x_j + b\right) \tag{1}$$

A CNN typically comprises of three main types of layers, namely convolutional layer, pooling layer, and fully connected layer where each type of layer is made up of a set of neurons and plays a different role. The convolution and pooling layers perform feature extraction whereas a fully connected layer maps the extracted features into final output.

(i) *Convolution layers*: These layers mainly aim for feature extraction from the input. A convolution layer utilizes several convolution *kernels* which are used to compute different feature maps. A kernel refers to an mathematical operator with a grid of parameters applied to each part of the full image to extract different features. A kernel is an optimizable feature extractor making CNNs highly efficient for image processing. The initial convolution layers extract simple features like lines, edges, and corners whereas the higher level convolution layers are able to extract more abstract information from the images.
The output of the convolution layers goes through a linear or nonlinear trigger or activation function such as rectified linear units (ReLUs), hyperbolic tangent (tanh), and sigmoid (σ) functions to form the output feature maps. A ReLU implements the function $y = \max(x, 0)$ as shown in Fig. 2 and enables faster training of the network. An appropriate activation function must be chosen according to the task as it has a huge impact on the overall performance of the neural network.

(ii) *Pooling layers*: The convolution layers generate feature maps which are fed to the pooling layers to downsample their size. The number of input and output feature maps does not change. Pooling layers reduce the spatial dimensions (width $\times$ height) of the input feature maps reducing the memory requirement of the network [8]. The reduction in the size of the feature map leads to a loss

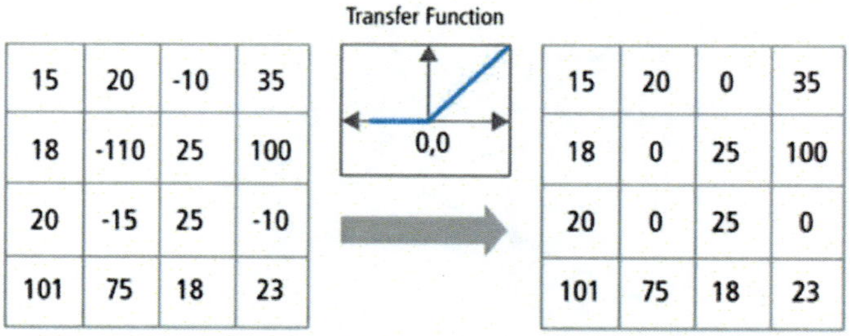

Fig. 2 ReLU activation function produces x for $x > 0$ and 0 as an output for $x < 0$ [11]

in the associated information, however such loss helps avoid overfitting of the network. The two most commonly used pooling operations are max pooling [9] and average pooling [10] as shown in Fig. 3.

(iii) *Fully connected layers*: These layers are often used as the final layers in the network where the high-level reasoning is performed. The neurons in the fully connected layers are connected to all the neurons in the previous layers, as implied by their name. This leads to the compilation of all the features extracted in previous layers for the calculation of the final output [11]. The extracted and downsampled feature maps from the convolution and pooling layers, respectively, are fed to the fully connected layers where they are mapped to the final outputs of the network such as probability scores to belong to each class in classification task [12].

Figure 4 shows a CNN architecture with the building blocks of the network including convolution, pooling, and fully connected layers for the task of computer vision. A typical CNN architecture having repeated stack of each type of layer is able to bring out the complicated information making it an useful particle classification technique in the field of HEP to sort the trajectories of the charged particles based on their type.

4 LArTPC Detector

Particle detectors in HEP are designed to identify the charged particle traversing through it after being produced in a collision or decay along with quantifying its kinematical properties. The charged particles ionize the material of the detecting medium while traversing the detector's volume, providing tiny signals that are recorded by the detector.

Investigating the details of each particle is a technical challenge as different particles exhibit different topology owing to the difference in their masses, charge, and the type of interactions they undergo [13].

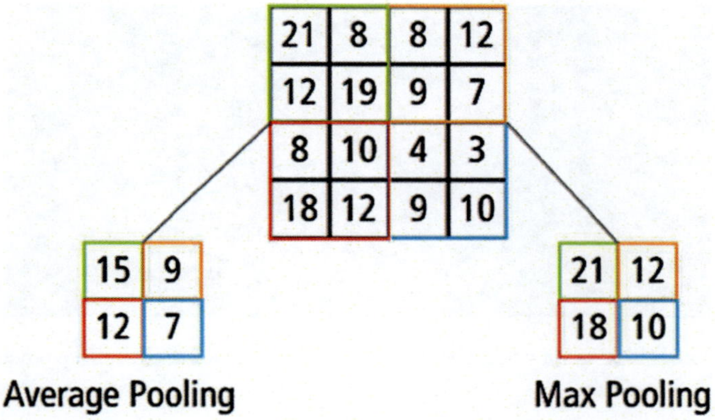

Fig. 3 Max pooling and average pooling operation on a feature map [11]

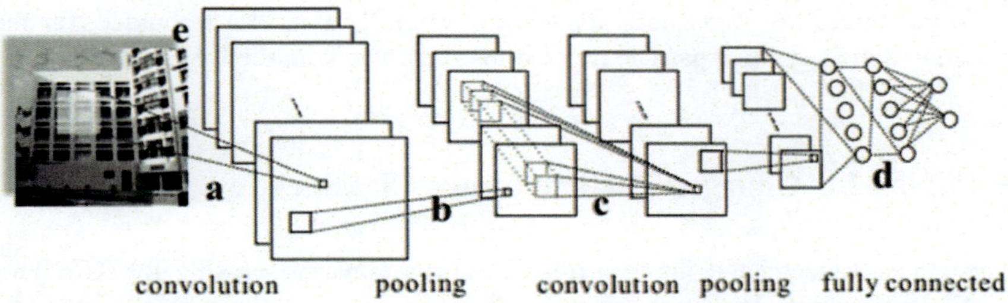

Fig. 4 CNN architecture with convolution, pooling, and fully connected layers for visual recognition [14]

(i) Charged hadrons such as pions and protons leave a track before depositing their energy in the detector.
(ii) The muons pass straight through the detector as they have low probability to interact and deposit energy mainly through ionization (Fig. 5).
(iii) The electrons and photons create electromagnetic showers with complex topological features with many branching trajectories.

This distinction between the energy deposition pattern helps the particle identification (PID) techniques to correctly identify each particle. LArTPC, a type of particle detector, is being increasingly used in the field of particle detection for its ability to provide efficient differentiation among the particles. It provides a great detail of the particle interactions producing high-resolution images of charged particles traveling in the detector. It consists of a huge volume of liquid argon bounded by a cathode and an anode plane. When a charged particle passes through the sensitive region of the detector, it produces ionization electrons and scintillation light along its trajectory [16]. The scintillation light is detected by an array of photomultiplier tubes (PMTs), providing a timing measurement whereas the ionization electrons drift

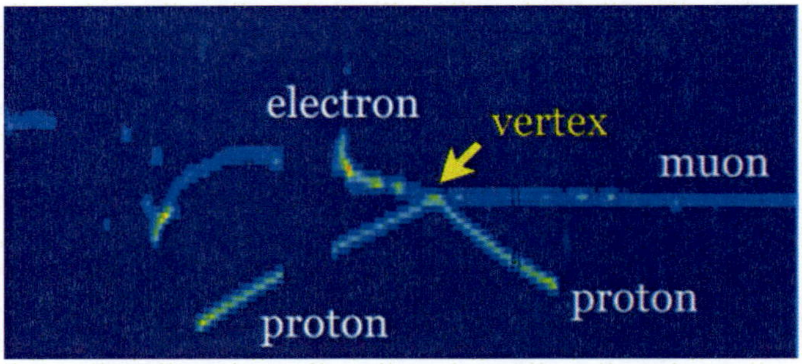

Fig. 5 An example image in which two protons, one electron, and one muon are produced [15]

toward the wire planes under an applied electric field, where their charge is collected and measured.

The next section studies the application of CNNs as a PID technique, for the classification of charged particle trajectories simulated with the LArTPC detector.

5 CNNs for Charged Particle Image Classification

The rapid advancement in the performance of the graphics processing unit (GPU) for parallel computing along with the availability of several high-quality labeled public datasets for training have paved the way to explore the potential of the CNNs.

The particle classification demonstrated in this study uses a publicly available dataset from deeplearnphysics.org. The contents of the dataset can be accessed using LARCV which is primarily written in C++ with an extensive Python API. LARCV is maintained by the Deep Learn Physics organization and software containers are distributed for algorithm development and data analysis use [15].

All files contain single-particle events which means each 2D image contain only one particle. There are five particle types including electron, gamma ray, muon, pion, and proton. An event fraction per particle type is equal for all samples. A particle's generation point is uniformly distributed within 5.12 m cubic volume and the generation direction is isotropic. A particle's momentum is uniformly distributed in the range specified per particle type:

(1) Electrons: 35.5–800 MeV/c
(2) Gamma rays: 35–800 MeV/c
(3) Muons: 90–800 MeV/c
(4) Pions: 105–800 MeV/c
(5) Protons: 275–800 MeV/c.

2.56 m cubic 3D volume is chosen to maximize the particle's trajectory within the volume and recorded in the file. The 2D images are created as 2D projections (xy, yz, zx)

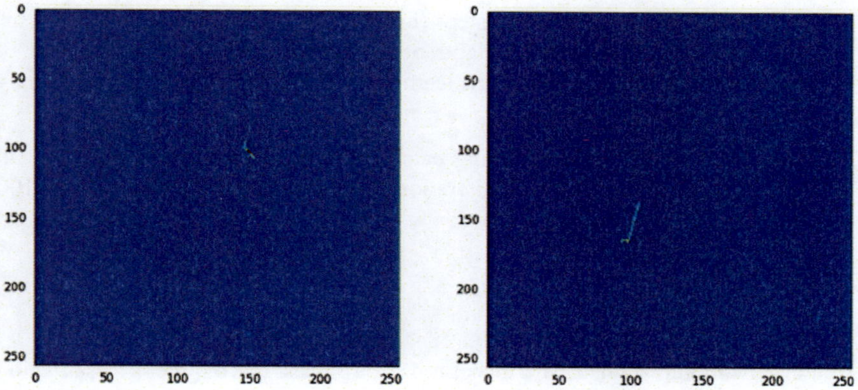

Fig. 6 Trajectory of a simulated proton recorded in the LArTPC detector's volume

of the 3D data. These are three channels of 2D images in each file. Events that contain any 2D projection image with less than 10 nonzero pixels are filtered out to remove empty and almost empty images from the set. Figure 6 shows the image of a proton traveling in the LArTPC detector.

5.1 Building and Training the Network

A sample of 10 layers deep CNN is constructed for the task of particle classification into 5 classes namely electron, gamma, muon, proton, and pion. It has 5×2 convolution layers with max pooling operation followed after every two layers with the last pooling layer using the average pooling operation and the ReLU function has been used as the activation function. The network is trained to find the value of the kernel parameters and the weights associated with each input to minimize the difference between the predicted and the actual output. The training of the network is achieved using TensorFlow framework which is an open-source platform for machine learning [17].

5.1.1 Configuration

The configuration variables are set before the training to enable the network divide the dataset into smaller sizes for the training and testing purpose.

(1) *Train batch size*: The entire dataset cannot be used for training at once so the data is sent in small collection of the labeled images called *batch* to the GPU.
(2) *Test batch size*: Small collection of images from the testing dataset. The performance of the network is monitored using the test batch.

(3) *Iterations*: The total number of steps (or batches) to train the network.
(4) *Weights Saving*: Saving the weights of the network after a certain number of training steps as the weights are learned by the network during the training process.

The network is trained with a training batch size and testing batch size of 50 and 100 images, respectively, with 5000 iterations where the weights are saved after 100 training steps.

5.1.2 Compilation

Neural networks are trained using an optimization process that requires a loss function to calculate the model error or the difference between the actual and the predicted output. These are added during the model's compilation step

(1) *Loss function*. A loss function predicts the *loss* which is the difference between the output predicted by the network and the actual output. The CNN is optimized by minimizing the loss function during its training by adjusting the parameters (weight and bias) of the network. Different types of loss functions are used depending on the problem in hand including mean-squared error, mean absolute error, and cross-entropy [18].
(2) *Optimization*. An optimizer updates the values of weights and biases to minimize the loss. Gradient-based algorithms such as Adagrad, RMSprop, and Adam are widely used for the optimization of the loss function. TensorFlow implements these algorithms with the help of subclasses of the *Optimizer* base class.
(3) *Learning rate*. Learning rate plays an important role in the training process. It is the step size considered for the loss function to reach a minimum during the training of the network.

In this study, the cross-entropy function is used as the objective to be minimized using the *RMSPropOptimizer* which is initialized with a learning rate of 0.0005 [19]. In the cross-entropy function, the loss increases as the predicted output value diverges from the actual input label. Since the network is getting trained to classify five types of particles, a separate loss is calculated for each label type and the result is summed up.

$$-\sum_{c=1}^{M} y_{oc} \log(p_{oc}) \tag{2}$$

where M is the number of classes (e, π, μ, etc.), log is the natural log, p is the predicted probability observation for o belonging to class c and y is the binary indicator (0 or 1) if class label c is the correct classification for observation o. The learnable parameters are updated according to the value of the calculated loss through the RMSPropOptimizer algorithm.

6 Results

The parameters including the kernel weights and biases of the network are optimized during the training process. The more the number of layers, the higher the response time needed to compute the response for a single image. Even addition of a single neuron in a layer leads to an increased amount of mathematical operations (e.g., matrix multiplication) at each iteration. Each particle's image in the training dataset is processed by the network, and the predicted output is compared with its actual label using the cross-entropy loss function.

Apart from providing the training sample, the same LArTPC simulation dataset gives us a testing sample as well to readily assess the performance of the trained network. The training and testing took about 24 h on Intel HD Graphics 620 with Intel Xeon E3-1200 v6/7th Gen Core Processor.

A common way to evaluate the performance of a network is to calculate *accuracy* and loss where accuracy is defined as fraction of instances when the method makes correct classification of the input data. The accuracy and loss as a function of the number of training iterations for the network implemented for particle classification with the LArTPC detector are shown in Fig. 7.

$$\text{Accuracy} = \frac{\text{Number of correct predictions}}{\text{Total number of predictions}} \tag{3}$$

This network reaches an accuracy of nearly 80% for the training data and 70% for the testing dataset after 5000 iterations. The loss curve reaches a minimum of 0.4 and 0.6 for the training and testing data, respectively, showing that the loss of the network falls with each iteration. The network learns topological features of the particles recorded in the detector from the training data. As a result, features associated to a fraction of testing data remain completely unseen to the network. Although the response of the CNN seems to understand the underlying physics processes involved in the signature left behind, it is quite complicated to bring out a clear distinction between each particle as there is significant overlap in their topological features. In the energy range of few hundred MeVs, muon and pion can look alike a lot in the detector which can confuse the network. Similarly, a neutral pion and an electron both create electromagnetic showers but there is difference in the depth of the shower created. The exploitation of main features of the particle track and shower development within any particular detecting medium in a given energy range can benchmark the processing of CNNs however that study is beyond the scope of this paper.

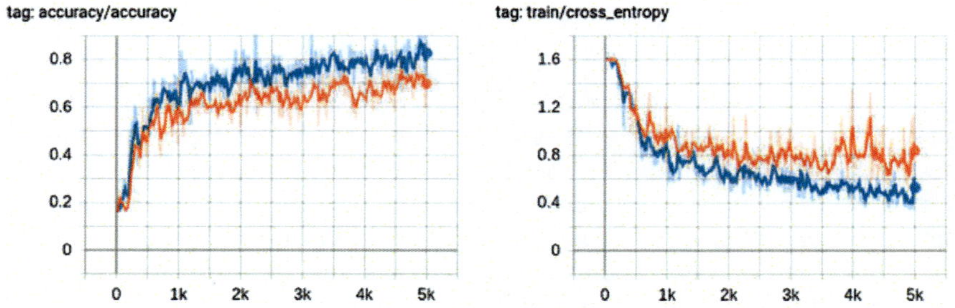

Fig. 7 Accuracy (left) and loss (right) curves for the training and testing samples. The orange line represents the metric measured on a test sample whereas the blue line represents the same for the training sample

7 Future Work

These studies demonstrate the potential of CNNs for the task of particle identification. After 5000 iterations, the accuracy of the network increases to 80% on the training dataset; however, it is comparatively lower on the testing dataset, nearly 70%, showing that the network has got slightly overfitted to the training data. The best way to avoid overfitting is using larger dataset for training, covering the entire range of inputs that the network might have to handle.

While the results show that CNNs can be used for image recognition of particles, there is lot more to explore. For making further improvements in the overall performance of the network, we plan to work on the following points:

- As already discussed in Sect. 5.1, loss is a measurement of error and the goal of network training is to minimize the loss by adjusting the parameters of the network. With the currently available dataset of fixed size, we plan on minimizing the loss with other optimizers like Adadelta, Adagrad, and SGD.
- Also, we need to carefully tune the structural parameters of the CNN network such as the number of convolution and pooling layers, number of maps per layer and the kernel sizes to see what best can be achieved from the CNN for the task of particle identification.

There are certainly other possible avenues to work on to improve the performance of the network. However, the proof-of-concept tests conducted in this study show that CNNs can deliver promising results and the given study can be used as a starting point for further exploration and development of completely new CNN network.

8 Conclusion

CNN is attracting huge amount of interest in the field of image recognition owing to its high-speed performance, easier and faster training, and fewer memory requirements. The CNN algorithms learn to make predictions based on the observations made. Being data-driven, the predictions are often much more accurate than those made by traditional methods of data analysis. CNNs are currently being used in several high-energy physics experiments including NOvA, MicroBOONE, and the upcoming DUNE experiment at Fermilab, USA for particle and event identification. The unique nature of the problems of HEP experiments has this capability of adding new knowledge to the field of neural networks and machine learning. The majority of industry applications work on real-world datasets where the labeled data of the same type is used for testing and training of the network. In contrast, HEP experiments usually train most of the network on simulated data that resembles the expected data. The detail to which these simulations are tunable is especially relevant to the study of machine learning algorithms. It provides the opportunity to study their behavior under controlled modifications in the training samples, which could greatly contribute to the challenge of explainability in and outside the field [20].

Acknowledgements The authors are thankful to the DeepLearnPhysics group as a whole for making the data available and contributing to the development of machine learning algorithms in the field of scientific research. The technique used is fully reproducible and the implementation is available on the Web site deeplearnphysics.org.

References

1. Altman, N.S.: An introduction to kernel and nearest-neighbor nonparametric regression. Am. Stat. **46**, 175–185 (1992). http://www.tandfonline.com/doi/pdf/10.1080/00031305.1992.10475879
2. Friedman, J.H.: Stochastic gradient boosting. Comput. Stat. Data Anal. **38**, 367–378 (2002)
3. https://elitedatascience.com/overfitting-in-machine-learning
4. Hubel, D., Wiesle, T.: Receptive fields and functional architecture of monkey striate cortex. J. Physiol. **195**, 215–243 (1968)
5. Aurisano, A., et al.: A convolutional neural network neutrino event classifier (2016). https://doi.org/10.1088/1748-0221/11/09/P09001
6. LeCun, Y., et al.: Gradient-based learning applied to document recognition. Proc. IEEE **86**(11), 2278–2324 (1998)
7. Shao, C.: A quantum model for multilayer perceptron (2018). https://arxiv.org/pdf/1808.10561.pdf
8. Voulodimos, A., et al.: Deep learning for computer vision: a brief review (2018). https://doi.org/10.1155/2018/7068349
9. Wu, H., Gu, X.: Max-pooling dropout for regularization of convolutional neural networks (2015). https://arxiv.org/pdf/1512.01400.pdf
10. Christlein, V. et al.: Deep generalized max pooling (2019). http://arxiv.org/org/pdf/1908.05040.pdfarxiv.org/pdf/1908.05040.pdf
11. Hijazi , S., et al.: Using convolutional neural networks for image recognition. https://ip.cadence.com/uploads/901/cnn_wp-pdf

12. Yamashita, R., et al.: Convolutional neural networks: an overview and application in radiology. https://link.springer.com/content/pdf/10.1007%2Fs13244-018-0639-9.pdf
13. Wingerter-Seez, I.: Particle physics instrumentation (2014). http://arxiv.org/abs/1804.11246arXiv:1804.11246
14. Ren, J.S.J., Xu, L.: On vectorization of deep convolutional neural networks for vision tasks (2015). https://arxiv.org/pdf/1501.07338.pdf
15. Adams, C.: et al.: PILArNet: public dataset for particle imaging liquid argon detectors in high energy physics (2020). https://arxiv.org/pdf/2006.01993.pdf
16. Adams, C., et al.: A deep neural network for pixel-level electromagnetic particle identification in the MicroBooNE liquid argon time projection chamber (2018). https://arxiv.org/pdf/1808.07269.pdf
17. A guide to tensorflow (part 1) (2018). https://codability.in/a-guide-to-tensorflow-part-1
18. ML Cheatsheet Documentation. https://readthedocs.org/projects/ml-cheatsheet/downloads/pdf/latest
19. Tieleman, T., Hinton, G.: Lecture 6.5-rmsprop: divide the gradient by a running average of its recent magnitude. In: COURSERA: Neural Networks for Machine Learning, vol. 4 (2012)
20. Psihas, F., et al.: A review on machine learning for neutrino experiments (2020). arXiv2008.01242.pdf

An Extensive Approach Towards Heart Stroke Prediction Using Machine Learning with Ensemble Classifier

Divya Paikaray and Ashok Kumar Mehta

1 Introduction

Heart stroke is one of the diseases which lead to death. It is better known as CVA (Cerebrovascular accident) in medical terms. An attempt has been made in this paper to analyze and predict heart stroke related disease using computer based analysis. The main objective of this paper is to employ machine learning algorithms for analysis of attributes at a given point of time in a patient. There are various machine learning algorithms that are going to be used in this paper. The main focus is on supervised machine learning algorithms, which will help in classification and predicting a model. The algorithm is used with the labeled dataset, which helps in classification and tries to find the accuracy of the model. The more accuracy leads to a better model. In an unsupervised approach, the dataset is unlabeled, in which the algorithm tries to extract the features and tries to draw the patterns on its own.

It has been observed that major causalities take place because of delayed treatment or diagnosis of stroke. The response time in case of a patient suffering from stroke is very less as it is mainly due to lack of blood flow to main body parts like the brain and heart. It is observed that timely prediction or diagnosis of strokes can avoid most severe cases. With the help of technology, mankind can be unfettered from the uncertainties of strokes and their suffering. Such an attempt has been made by using machine learning algorithms in this field.

In the past various studies have been done towards implementation of machine learning into medical treatment. Authors like T. Badriyah, S. Monisha, G. Çınarer, and many more have made an extensive study in this field which is discussed later in the paper.

D. Paikaray (✉) · A. K. Mehta
Department of Computer Applications, NIT Jamshedpur, Jamshedpur, India

A. K. Mehta
e-mail: akmehta.ca@nitjsr.ac.in

M. Dua et al. (eds.), *Proceedings of the International Conference on Paradigms of Communication, Computing and Data Sciences*, Algorithms for Intelligent Systems,
https://doi.org/10.1007/978-981-16-5747-4_66

The paper gives a brief insight into the previous work carried out in this field. Further, an attempt has been made to deduce an accurate methodology for prediction of heart stroke disease. Initially, data is collected consisting of various attributes which are needed for processing. The attributes of this disease that have been taken in this paper are gender, age, hypertension, heart disease, ever married, work type, residence type, average glucose level, BMI, smoking status. The various machine learning algorithms have been applied in this paper and the results are compared. An extensive study is carried out for determination of the result. The paper is concluded with a brief understanding of the importance of machine learning and various algorithms in the field of heart stroke analysis and its future prospects.

2 Background

Machine learning algorithms for stroke disease classification were proposed by Badriyah [1], in which eight algorithms were applied to the image dataset. The features were extracted, and algorithms were applied to it in which random forest algorithm gave the best accuracy compared to all algorithms.

Predictive analysis of student stress levels was proposed by Monisha [2], in which the Naïve Bayes technique is used to analyze the accurate facts and to organize the stress-causing factors based on probabilistic parameters and output are inferred by visualization tools.

Classification of Brain Tumors was proposed by Çınarer [3], which states that MRI is the popular technique that plays a vital role in the successful implementation of treatment with an accurate diagnosis. For this, various techniques and algorithms have been used for the classification of the brain tumor images in which SVM was found the best algorithm with 90% accuracy.

Prediction of Breast Cancer was proposed by Fatima et al. [4], in which the prediction of breast cancer has been analyzed by the comparative analysis of machine learning, deep learning, and data mining techniques. The main focus is to comparatively analyze different existing techniques in order to find out the most suitable method that will support the larger dataset with a better result in accuracy of prediction.

Using applications of Machine learning algorithms was proposed by Chang et al. [5] in which facial marks were traced by an ensemble algorithm of regression trees method (ERT). In this, the structural similarity between left and right faces was calculated. Histogram Oriented Gradients and support vector machine was adapted for face detection.

In this, Fang et al. [6] have emphasized ischemic strokes as one of the major mortality causing diseases. The feature was analyzed by the Shapiro–Wilk algorithm and Pearson correlation. The workflow method by the author includes primary feature selection followed by RFECV, which is used for features importance determination, and finally, the prediction and results are analyzed, which uses the random forest and other algorithms. In this, Monteiro [7] gives importance to ischemic stroke as it is not a risk-free treatment, and doctors only proceed with treatment when the potential benefits outweigh the perceived risk. This makes the prediction tool far more important for early diagnosis and prediction. The methodology is distributed into five main steps:

- Baseline
- Two hours after admission
- Twenty-four hours after admission
- Seven days after admission
- Discharge.

The results were compared in a tabular manner between random forests, Xg boost, Support vector machine, and so on. In this, Treib et al. [8] focuses on the patients who are having a good and healthy cardiovascular system but then also suffering from acute stroke. In this, they have measured the Cardiac Output and blood pressure of the patient with having no past history of heart disease.

Joo et al. [9] focus on the characteristic of machine learning and big data for predicting cardiovascular risk. Thus, it focuses on assessing the effectiveness of Machine learning algorithms in predicting the two years and ten years' risk of cardiovascular disease. The algorithm used is namely logistic regression, deep neural network, and random forests.

Guidi [10] has focused on the analysis of heart disease on the clinical decision support system. In this, the database is analyzed by comparing the performance of various algorithms such as neural networks, Support vector machine, and the system with fuzzy rules and random forest. Random forest performance was found to be the best in both the HF-Prediction and HF-Severity functions.

3 Proposed Framework

In this paper, the main focus is on predicting heart stroke, which is done on a dataset of few patients in which various machine learning algorithms are applied, including ensemble classifiers to test which algorithm gives the best accuracy and can be used for predicting stroke as fast as possible. The attributes play a very important role. Firstly, data preprocessing is done followed by the feature extraction after which different algorithms are applied to find the best predicted model for the particular dataset.

3.1 Collection of Data

In this, the data is collected of around ten thousand people of which around five thousand data are taken for the classification process. The dataset consists of attributes, namely age, gender, id, hypertension, heart disease, ever married, work type, residence type, and average glucose level, BMI, smoking status, which will predict the possibilities of stroke as positive or negative.

3.2 Data Preprocessing

Data Preprocessing is one of the important processes without which analysis will never generate results accurately. It includes data cleaning, data transformation, conversion of data, etc. Noise removal from data is essential as it gives accurate results. All the duplicate values are removed. The outliers are removed for better accuracy. For a better result, collecting relevant data from the specified raw dataset is very essential.

3.3 Feature Extraction

Feature Extraction is one of the important steps in which the important features are extracted according to which we can achieve the result with better accuracy and less error. Further, the contribution of the attribute are also determined in the same step.

3.4 Machine Learning Algorithms

Machine learning is the part of artificial intelligence that uses many advanced algorithms for better result and accuracy in any field. It works in data that should be preprocessed and if not then we need to preprocess the data so that it could work properly. Machine learning is of three types, i.e., supervised learning, unsupervised learning, and reinforcement learning. In this paper, the main focus is on supervised learning, which basically consists of a dependent variable that is predicted from the independent variable. Various algorithms are used for the classification that is:

- **Linear Discriminant analysis**—It can be defined as a method that is used in or applied in statistics that helps in finding the linear combination and can be used in linear classifiers.
- **Logistic regression**—It is another mathematical technique that is used in machine learning. In this, the probability of the target variable is predicted, which can be in the form of zero and one. It is another classification problem that is totally calculated using the concept of probability.
- **Naïve Baye's**—This Algorithm uses Baye's theorem for solving the classification problem. The naïve Bayes algorithm is a very effective classification algorithm as

it gives the result very fast. It is also calculated using the concept of probability which results in yes or no.

- **Support vector machine**—It is one of the most effective algorithms which can be used in classification and regression as well. This hyperplane plays an important role in classifying the data. It focuses on the more the margin and the more accurate the result will be. For enhancing the result, the hyperplane should be drawn in such a way that the margin will be more.

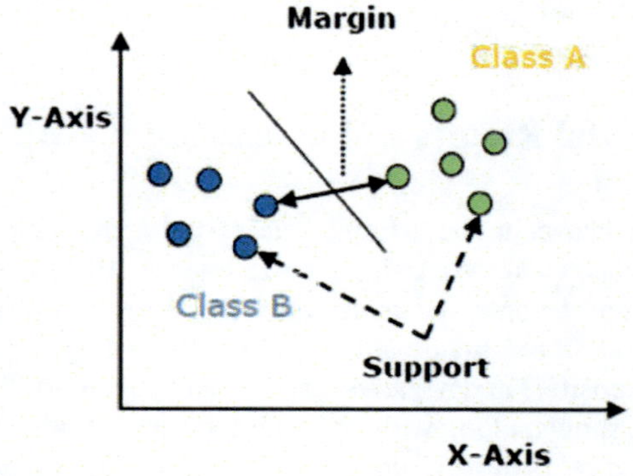

- **K-Nearest Neighbors classifier**—K-Nearest neighbors work on the nearest value, i.e., by finding the distance between the two by a very famous Euclidean distance formula. It classifies by using the nearest distance. This is one of the most used classifier algorithms. It is an algorithm that is a non-parametric classification method.

3.5 Ensemble Classifiers

An ensemble classifier does not depend on any single model to make the decision; instead, the method says to take a different number of models, and based on the model, the decision should be made so that the accuracy will be more. In this, different models are known as the base learner, and each base learner can use a different algorithm for them to maintain the diversity of their outputs.

- **Random Forest**—It is a kind of ensemble classifier that uses a decision tree in a randomized fashion. For this, we need to create a bootstrap dataset by sampling. Now using the bootstrap dataset, we create a decision tree in a random variable. Random forest is also used for ranking the importance of variables.
- **Bagging**—Bagging is also known as Bootstrap aggregation. In this, from the original dataset, we create a number of datasets randomly, selecting records from the original datasets. After this, each classifier is trained by the different bootstrap dataset, and all the classifiers are aggregated then one single classifier is developed with less error and more accuracy.

- **Ada Boost**—It is basically a sequential ensemble classifier. Ada boost is mainly used for boosting the performances of decision trees. In this, the weak learners are combined into a weighted sum that will represent the output of the boosted classifier.
- **Gradient Boosting Classifiers**—Gradient boosting classifier is one of the ensemble classifiers which combines together all the weak learner models to develop a good predictive model, which will result in good accuracy. The flowchart gives an idea of how the whole process takes place in developing a model with good accuracy.

4 Experimental Results and Discussion

In this paper, the dataset has been trained and tested with multiple algorithms. A total of 3065 data has been taken for training and a total of 2044 has been taken for testing purpose. The dataset contained many attributes shown in Fig. 1, which contains 41.4% female and 58.6% male patients, as shown in Fig. 2. In this, the stroke has been denoted as positive and negative, as shown in Fig. 3. In Fig. 4, the features that are extracted are shown, along with known important features. It is observed that the average glucose level is most important, followed by other features. Many machine learning algorithms have been applied. The graph in Fig. 5 shows the accuracy of all algorithms applied in the dataset. After applying the ensemble classifier, the random forest gives the highest accuracy among all, as shown in Fig. 6. The confusion matrix gives the precision value, sensitivity, and F1 score, as shown in Fig. 7. The table gives all the accuracy of different predictive models in Table 1.On the basis of this experiment, the logistic regression gives 95.04%, linear discriminant gives 93.47%, KNeighbhor gives 95%, Naive Bayes gives 94.6%, Support vector machine gives 95%, Ada Boost gives 94.74%, Bagging gives 94.91%, Gradient Boosting gives 94.82%, and random forest with the highest accuracy 95.10% (Fig. 8).

The confusion matrix for random forest describes the overall performance of the classification model on the set of test data and for which the true values are known. The precision value for this confusion matrix is 0.95, recall value is 1 and $f1$-score is 0.97 and the overall accuracy is 95%.

5 Conclusion

The paper focused on classifying the stroke dataset using various machine learning algorithms. Before classifying, the dataset has been preprocessed, cleaned, and the feature was extracted. In this paper, many algorithms were used, namely linear discriminant, logistic regression, and support vector machine, and so on. Based on this experiment, the random forest classifier gave the highest accuracy. In the future, the model can be utilized for comparison of accuracy by implementing a

Fig. 1 Flow diagram of training and testing model

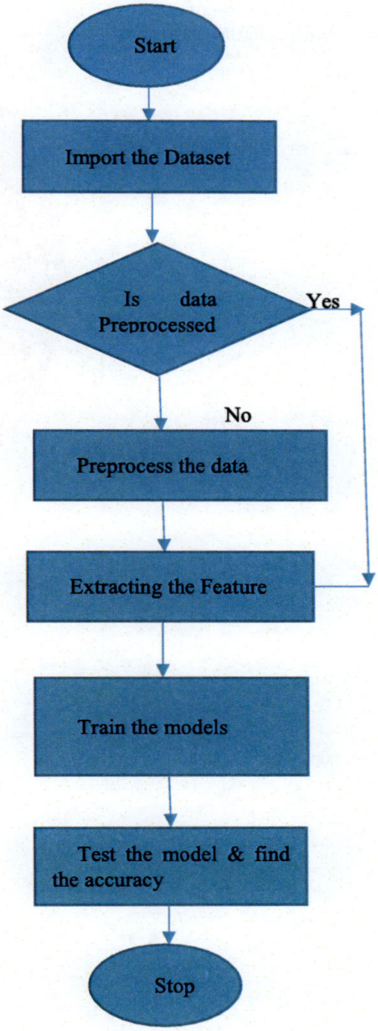

age	gender	id	hypertension	Heart_disease	Ever_married	Work_type	Residence_type	Avg_glucose_level	bmi	Smoking_status	stroke
75	0	9046	0	1	1	1	0	228.69	36.6	2	1
55	1	51676	0	0	1	0	1	202.21	25.8	0	1
.	.	.	.	.	.	.	.	.	.	.	.
.	.	.	.	.	.	.	.	.	.	.	.
.	.	.	.	.	.	.	.	.	.	.	.
95	1	44679	0	0	1	2	0	85.28	26.2	2	0

Fig. 2 Dataset used for the classification algorithm

Fig. 3 Population of males and females in the dataset

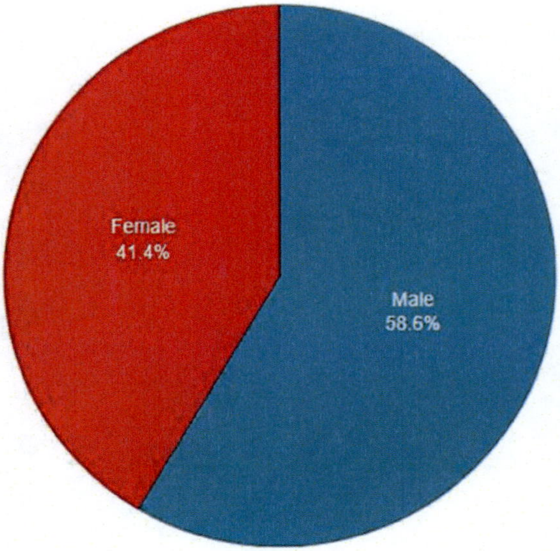

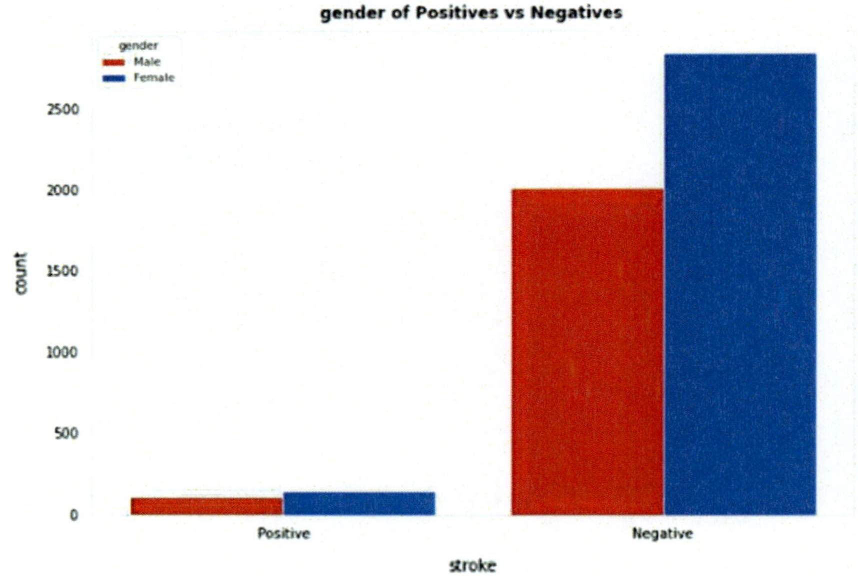

Fig. 4 Gender of positives versus negatives

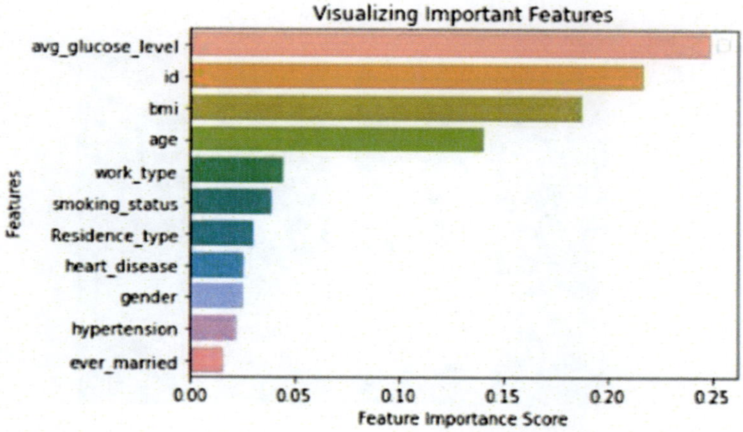

Fig. 5 Feature extraction

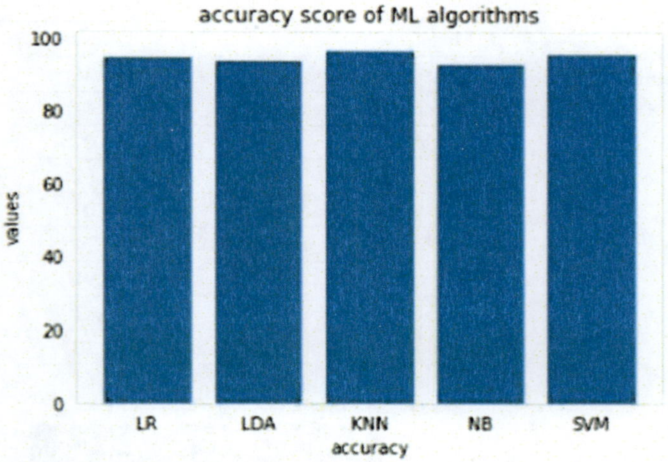

Fig. 6 Accuracy of ML algorithms

hybrid model and big data analytics. Further, an increase in accuracy will lead to an enhanced classification model with better appreciation and prediction of strokes.

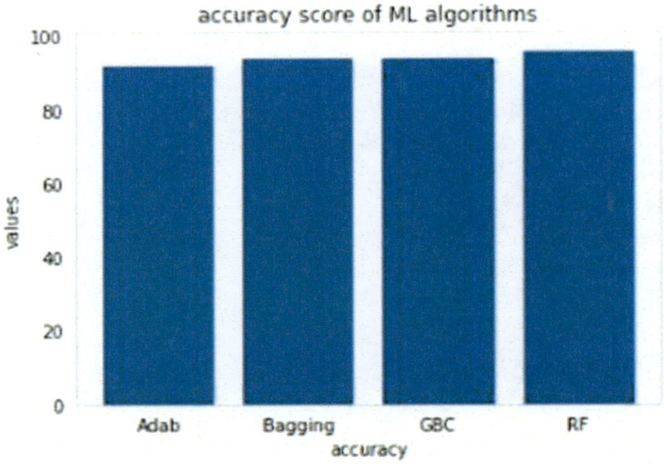

Fig. 7 Accuracy of ensemble classifiers algorithms

Table 1 Accuracy of different models using machine learning algorithms and ensemble classifier

Model	Accuracy (%)
Logistic regression	95.04
Linear discriminant analysis	93.47
Neighbor	95.0
Naïve Bayes	94.6
Support vector machine	95.0
Ada boost	94.74
Bagging	94.91
Gradient boosting	94.82
Random Forest	95.10

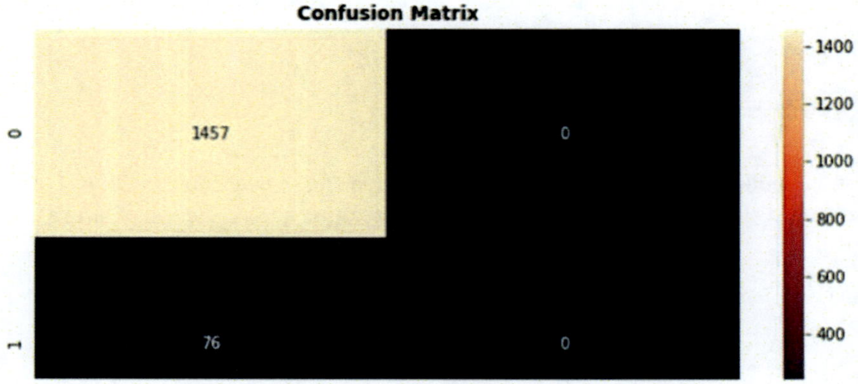

Fig. 8 Confusion matrix for random forest algorithm

References

1. Badriyah, T., Sakinah, N., Syarif, I., Syarif, D.R.: Machine learning algorithm for stroke disease classification. In: 2020 International Conference on Electrical, Communication, and Computer Engineering (ICECCE), Istanbul, Turkey, pp. 1–5 (2020). http://doi.org/10.1109/ICECCE49384.2020.9179307
2. Monisha, S., Meera, R., Vijay Swaminath, R., Arun Raj, L.: Predictive analysis of student stress level using Naïve Bayesian classification algorithm. In: 2020 International Conference on Computer Communication and Informatics (ICCCI), Coimbatore, India, pp. 1–7 (2020). http://doi.org/10.1109/ICCCI48352.2020.9104113
3. Çınarer, G., Emiroğlu, B.G.: Classification of brain tumors by machine learning algorithms. In: 2019 3rd International Symposium on Multidisciplinary Studies and Innovative Technologies (ISMS), Ankara, Turkey, pp. 1–4 (2019). http://doi.org/10.1109/ISMSIT.2019.8932878
4. Fatima, N., Liu, L., Hong, S., Ahmed, H.: Prediction of breast cancer, comparative review of machine learning techniques, and their analysis. IEEE Access **8**, 150360–150376 (2020). https://doi.org/10.1109/ACCESS.2020.3016715
5. Chang, C., Cheng, M., Ma, M.H.: Application of machine learning for facial stroke detection. In: 2018 IEEE 23rd International Conference on Digital Signal Processing (DSP), Shanghai, China, pp. 1–5 (2018). http://doi.org/10.1109/ICDSP.2018.8631568
6. Fang, G., Xu, P., Liu, W.: Automated ischemic stroke subtyping based on machine learning approach. IEEE Access **8**, 118426–118432 (2020). https://doi.org/10.1109/ACCESS.2020.3004977
7. Monteiro, M., et al.: Using machine learning to improve the prediction of functional outcome in ischemic stroke patients. IEEE/ACM Trans. Comput. Biol. Bioinf. **15**(6), 1953–1959 (2018). http://doi.org/10.1109/TCBB.2018.2811471
8. Treib, J., Haass, A., Krammer, I., et al.: Cardiac output in patients with acute stroke. J. Neurol. **243**, 575–578 (1996). https://doi.org/10.1007/BF00900944
9. Joo, G., Song, Y., Im, H., Park, J.: Clinical implication of machine learning in predicting the occurrence of cardiovascular disease using big data (nationwide cohort data in Korea). IEEE Access **8**, 157643–157653 (2020). https://doi.org/10.1109/ACCESS.2020.3015757
10. Guidi, G., Pettenati, M.C., Melillo, P., Iadanza, E.: A machine learning system to improve heart failure patient assistance. IEEE J. Biomed. Health Inform. **18**(6), 1750–1756 (2014). https://doi.org/10.1109/JBHI.2014.2337752

A Systematic Review of Stability Analysis for Memristor Neural Networks

M. S. Deepthi, H. R. Shashidhara, and R. Shruthi

1 Introduction

In addition to three passive circuit elements, Chua introduced memristor a new fourth circuit element in 1971 [1]. The HP labs in 2008 realized the practical model of memristor [1]. The two-terminal device memristor exhibit several characteristics similar to that of neuron present in the human brain [2]. It works same as synapse in the neural system and its conductance proportionally varies with respect to the amount of the current flowing through it over the period of time [3]. This has made the memristor an alternative to the resistors used in VLSI circuits and as led to development of the new neural network named memristive neural networks (MNNs) [1].

MNNs are promising candidate for the architecture in neuromorphic computing systems because of its high-density, nanoscale size, and non-volatility characteristics [4]. In MNNs, computation takes place in the domain of flux-charge, rather than in current-voltage field [5]. This gives potential to memristor to act as non-volatile memories and superior in characteristics compared to traditional neural networks (NNs) [5]. Many predominant outcomes of MNNs have been outlined in last three decades on nonlinear dynamic attributes such as associative memory, stability, dissipativity, attractivity, synchronization, and passivity [6]. MNN systems are widely used in various domains like pattern recognition, combinatorial optimization, and

M. S. Deepthi (✉) · H. R. Shashidhara · R. Shruthi
Department of Electronics and Communication Engineering, The National Institute of Engineering, Mysuru, Karnataka 570008, India
e-mail: deepthims@nie.ac.in

H. R. Shashidhara
e-mail: shashidharahr@nie.ac.in

R. Shruthi
e-mail: shruthir@nie.ac.in

M. Dua et al. (eds.), *Proceedings of the International Conference on Paradigms of Communication, Computing and Data Sciences*, Algorithms for Intelligent Systems,
https://doi.org/10.1007/978-981-16-5747-4_67

knowledge acquisition [3]. Such applications are dependent on the stability of the networks, hence studying and analysis of the stability of the MNNs is very important.

Stability of the MNNs has been extensively researched in past few years due to the wide spread of its applications [7] in several domains like signal processing, communication systems, optimization problems, pattern recognition, and other engineering areas [2]. In Wen et al. [8], the authors have investigated global exponential stability of MNNs with time-varying delays and general activation functions [8]. Recently fractional order MNNs have gained attention as they accurately describe the pinched hysteresis property of memristor than the integer order systems [6]. In Wu et al. [6] investigate the Mittag–Leffler global stabilization of fractional-order MNNs class using differential inclusions of fractional order and set-valued maps [6]. Complex-valued MNNs (CVMNNs) are also an emerging field of research where the inputs are complex-valued and the information is computed in complex domain. Analysis of CVMNNs is complicated and different compared to that of real-valued MNNs. There are several methods to ensure and analyze the stability of the CVMNNs. Some authors have discussed various stability analysis such as exponential stability, finite time stability, input-to-state stability, and global exponential stability [1, 7, 9–12]. Great number of stability analysis of variety of MNNs are available in previous literature but they all deal with the neural networks operating over infinite time interval. In Ali et al. [1] studied finite time analysis of complexed valued MNNs of fractional order characterized by propagation latencies. In [10], the finite time stability analysis of fractional order complex-valued MNNs has been investigated with both leakage and time-varying delays. Adequate condition for finite time stability was established using Gronwall-Bellman inequality, Holder inequality, and inequality scaling skills with fractional order $0 < \alpha < \frac{1}{2}$ and $\frac{1}{2} \leq \alpha \leq 1$. Two important ways in which finite time stability diverges from classical stability concept is, first it is concerned with system operating over fixed period of time. Secondly, finite time stability confines the system state variables over an assigned limited time interval [1, 7]. Reference [13] applied non-smooth analysis and control theory to derive sufficient criteria for the Lagrange stability of MNNs with discrete and distributed delays. Reference [14] proved robust stabilization of memristive Hopfield neural networks of fractional order with parameter disturbances, and under the frame work of Filippov sense, the uniqueness and existence of the equilibrium point are examined.

In the process of researching the stability of the neural networks, there is another important aspect that should not be ignored called time delay [15, 16]. Time delay is introduced in amplifiers because of limited switching speediness which is unavoidable and they cause negative effect on the systems causing oscillation along with instability. The investigation of the consequence of time delay on stability of the neural networks as gained wide attention in neural network field. For memristive complex-valued neural network with latency, adequate criteria for exponential, finite time, input-to-state, and global exponential stability dependent on the scale of latencies were recognized [7, 9, 10, 12]. Lately, the stability solutions for MNNs with delays, such as distributed delays, discrete delays, mixed time delays, proportional delays, probabilistic delays, and so on are obtained [1, 4, 13, 17].

Several investigations are made recently on the stability concerns of MNNs and are the active research area [7]. Hence, it is very necessary to have an exhaustive survey on the stability issues of MNNs. Even though currently there are some literature surveys on MNNs stability, there is a lack of comprehensive survey on issues related to stability of variety of MNNs with and without delays which stimulates us to present a detailed literature survey on this specific topic [15]. This paper concentrates on following MNNs: complex-valued MNNs, uncertain MNNs, fractional order MNNs, memristor BAM NNs, inertial MNNs, memristor recurrent neural networks, and discrete time stochastic MNNs.

This paper is assembled as follows. The background of MNNs is covered in Sect. 2. Section 3 provides a detailed overview of stability analysis and results of various MNNs. The summary and some future scopes are finally given in Sect. 4, which exhibits some favorable directions towards research in the stability analysis of MNNs.

2 Background of Memristive-Based Neural Networks

Artificial neural networks (ANN) are the new computational paradigm which imitates the function of the human brain. This property of the ANN as caused direct influence on the development in AI technology. The applications of the ANN include image processing and signal processing, computer vision, medicine, pattern recognition, financial systems, military systems, artificial intelligence, planning, control and search, human factors, power systems, and so on. In traditional ANN, the connection weights are realized by using resistors, where the resistor lack memory. Memristors are more suitable in implementing the synapse of the neural networks because of its characteristics like nanoscale size, non-volatility, passivity, and memory effect. Memristor shows hysteresis effect which incorporates non-volatile feature in it, making it well suited for neuro-morphic systems. In MNN, the resistors are replaced by new two-terminal device called memristor. MNNs open up the opportunity for parallelism and resist the complexities of implementing neural networks hardware circuitry.

MNNs without or with latency are mathematically modeled by differential or difference equations whose coefficients are state dependent [45]. Reference [18] proposes a preliminary MNN model with recurrent structure that approximately mimics the behavior of the human brain. Reference [1] models a fractional order complex-valued MNNs characterized by correlated delays using Caputo derivative and RL derivative case. Reference [19] models an uncertain MNNs with variable time delays and also considering the leakage term which has significant influence on the attributes of the system. The Hopfield neural network, a very popular neural network which is fitter than the fractional order neural network is mathematically modeled in [14] with the parameter disturbance. Some more research work establishing mathematical models of the MNNs are listed in the literature and most of them are relevant to the continuous domain. Most of the signals in the real world are digitized before and after the processing by the computer, hence discrete time MNNs have to be taken into account. The research on the discrete domain MNNs is still in infancy stage since it introduces lot of complexities in switching behavior of the system which is state dependent [17].

Up to now, there exist several works on the inspection of stability of MNNs using various mathematical techniques. The stability of the memristor recurrent neural networks (MRNNs) with remarkable capabilities like self-learning, fault tolerance, and nonlinear function approximation is analyzed using differential inclusions [20], Filippov solutions [3], impulsive delayed differential inequality and Lyapunov functions [21], and so on. The existing literature discusses the stability analysis of various other MNNs such as fractional order MNNs [1, 2, 6, 7, 10, 14, 16, 22–25], memristive Hopfield networks [14], memristive complex-valued neural networks [1, 7, 9, 10, 12], uncertain memristive neural networks [4, 19, 26, 27], inertial memristive neural networks [28], stochastic memristive neural networks. In precis, it is extremely important to study the stability analysis techniques of MNNs from both research, engineering and theoretical point of view.

3 Stability Analysis Techniques and Sufficient Conditions for Memristor-Based Neural Networks

Most of the research work with reference to MNNs is predominantly on the analysis of dynamic behaviors such as stability, synchronization, passivity, state estimation, and so on. There are several widely seen classes of MNNs such as MRNNs, CVMNNs, fractional order memristor neural networks (FOMNNs), memristor bidirectional associative MNNs, uncertain MNNs, inertial MNNs, discrete time stochastic MNNs to name a few, whose dynamic behaviors are analyzed and investigated.

Many paradigms have been proposed for the stability analysis of various MNNs such as linear matrix inequalities (LMI) method [8, 18, 25, 26], differential inclusions [6, 11, 20, 29, 46], set-valued maps [6, 24, 29, 30], Gronwall inequality [1, 7, 16, 24, 47], Holder in-equality [1, 10], non-smooth analysis [13, 30, 31, 50], Lyapunov function method [4, 8, 11, 12, 14, 16, 17, 19, 21, 22, 27–29, 32–34, 48, 49], Banach contraction principle [2], Brouwer's fixed-point theorem [9], quadratic convex combination method [35], free weighting matrix technique [8, 27], Halanay differential inequality [12, 28] to name but a few. In this section, we present lately advances in the stability analysis of both discrete and continuous time MNNs.

In [18], authors investigate the exponential stability and deduce adequate conditions for the same of MNNs. The following linear matrix inequality in Eq. (1) holds good if positive diagonal matrices P, C, and D exist for $\beta > 0$,

$$\begin{pmatrix} \Omega & \mathrm{PF} & \mathrm{PG} \\ \mathrm{FP} & -C & 0 \\ \mathrm{GP} & 0 & -D \end{pmatrix} < 0 \tag{1}$$

where $\Omega = 2\beta P - 2P - I_n + \tilde{A}C\tilde{A}^{\mathrm{T}} + \exp\{2\beta\tau\}\tilde{B}D\tilde{B}^{\mathrm{T}}$

In [20] an abridged scientific model of MRNN is projected as below in Eq. (2),

$$\begin{aligned}\dot{y}(t) &= P(y)\\ &= -Dy(t) + A(y)\hat{f}(y(t)) + B(y)\hat{g}(y(t-\tau(t))) + s \end{aligned} \tag{2}$$

It establishes global uniform asymptotic stability using differential inclusions since classical definitions of solutions cannot be applied for differential equations because of the discontinuous functions in the RHS of the memristor NN differential equation model. It has been proved that if $D-|A|_{\max}K-|B|_{\max}L$ is M-matrix, where $D, A, B, K,$ and L are matrices, then the MRNN is globally uniformly asymptotically stable.

The FOMNN [6] is investigated and is modeled using fractional differential equations. The problem of convergence and the base norms of dynamics of fractional order systems are better evaluated by fractional inequality of Caputo fractional derivative. The condition for global Mittag–Leffler stabilization of considered FOMNN is shown in Eq. (3), provided if constants $\gamma > 1$ and $\zeta > 0$ exists, where ζ is free weighing parameter,

$$\left(\sum_{i=1}^{n}|x_i(t)|^{\gamma}\right)^{\frac{1}{\gamma}} \le \left[\sum_{i=1}^{n}|x_i(0)|^{\gamma} E_{\alpha}(-\delta t^{\alpha})\right]^{\frac{1}{\gamma}}, \quad t \ge 0 \tag{3}$$

In [13], authors consider model of memristive neural networks which ratifies bounded and Lurie-type feedback functions. Multi-stability property of the system is explored considering MNNs satisfying Lagrange stability. Lagrange stability is established for both networks with bounded and Lurie-type feedback function by deriving different attractive sets using non-smooth analysis and control theory.

The uniform stability of complex and real-valued memristor neural networks was proposed in [2]. Since the solutions for differential equations with discontinuous right-hand side cannot be arrived using straight forward method, under the Filippov frame work differential inclusion theory is used to convert differential equations with discontinuous right-hand side into a differential inclusion. It is shown utilizing Banach contraction principle that if RVMNN and CVMNN satisfy certain Lipschitz and initial conditions then they are uniformly stable.

In [9], a model for memristor complex-valued neural network is defined and distinctiveness of equilibrium point is proved by showing that the activation functions of the network satisfy Lipschitz continuity condition in complex field. The MCVNN is declared to be global exponentially stable if $\Lambda - \Phi L_f - \Psi L_g$ is a M-matrix, where $\Lambda = \mathrm{diag}\{d_1, \ldots . d_n\}$, '$d_n$' indicates neuron self-inhibitions, $\Phi = (w_{pq})_{n\times n}$, $\Psi = (v)_{n\times n}$, $L_f = \mathrm{diag}\{l_1{}^f, \ldots . l_n{}^f\}$, and $L_g = \mathrm{diag}\{l_1{}^g, \ldots . l_n{}^g\}$.

The exponential stability of category of recurrent MNNs is studied in [3] and proved if the subjective solution $x(t)$ with $\varphi(s) \in C([t_0-\tau, t_0], \Re^n)$ satisfies Eq. (4)

$$|x_i(t)| \le \left(\sum_{j=1}^{n} \sup_{t_0-\tau \le s \le t_0} \left| \varphi(s)_j(s) \right| \right) \beta_i \exp\{-\alpha_i (t - t_0)\} \tag{4}$$

where β_i and α_i are positive constants.

In [4], authors have modeled an uncertain memristor recurrent neural network and the existence of exceptional equilibrium point is demonstrated using the theory of homeomorphism. The condition for global exponential stability of the uncertain MNNs by using functional Lyapunov method is given in Eq. (5),

$$\frac{\mathrm{d}}{\mathrm{d}t} V(t, y(t)) \le 0 \tag{5}$$

where $V(t, y(t))$ is Lyapunov function, $y = p - p^*$, 'p^*' represents the equilibrium point of the network.

The operation of the fractional order complex-valued MNNs limited to finite time interval with time delays is analyzed in [7]. The fractional order network is represented by the following differential equation model in Eq. (6),

$$\begin{aligned} D^{\alpha} z_p(t) = {} & -d_p z_p(t) + \sum_{q=1}^{m} a_{pq}\big(z_q(t)\big) f_q\big(z_q(t)\big) \\ & + \sum_{q=1}^{m} b_{pq}\big(z_q(t)\big) g_q\big(z_q(t - \tau(t))\big) + h_p \end{aligned} \tag{6}$$

where $t \ge 0$ 'm' indicates number of units in NN, $z(t)$ represents state variable which is complex-valued accomplice with pth neuron, $d_p > 0$ is a constant, h_p is exterior input vector, $f_q\big(z_q(t)\big)$, $g_q\big(z_q(t - \tau(t))\big)$ are nonlinear complex-valued activation functions and $b_{pq}\big(z_q(t)\big)$ are complex-valued connective memristor weights. Utilizing Gronwall inequality, Laplace transform and Mittag–Leffler, the finite time stability of the fractional order complex-valued MNN with latency when $1 < \alpha < 2$ is ensured with sufficient conditions.

In [10], the finite time stability of MNNs with fractional order and complex-valued variables, characterized with leakage and time-varying latencies are studied. The network is subjected to stability analysis for two values of α: $\frac{1}{2} \le \alpha \le 1$ and $0 < \alpha < \frac{1}{2}$ and sufficient conditions are obtained using Holder inequality and Gronwall inequality with respect to $(t_0, J, \delta, \varepsilon)$, where '$J$' denotes external input, 't_0' denotes initial time, 'δ' and 'ε' represent positive numbers.

A genre of fractional order MNNs is modeled in [16] and the existence of unique equilibrium point $x^* \in R^n$ is proved by contraction mapping principle such that $\varpi\big((d_i x_i)^*\big) = (d_i x_i)^*$, where $\varpi\big((d_i x_i)^*\big)$ is contraction mapping function, d_i is self-regulating parameter of neuron and x_i is equilibrium point. The condition for Mittag–Leffler global stability of the equilibrium point of the NN is given in Eq. (7),

$$\|x(t) - x^*\| \leq c\|x^0 - x^*\|E_q(-h(t - t_0)^q), \quad q > 0, t \geq t_0 \tag{7}$$

where x^* is the equilibrium point, 'c' and 'h' represent positive constants.

In [22], authors investigate the fractional order MNNs entailed with Caputo fractional derivative and establishes the occurrence and distinctiveness of the equilibrium point. The condition for Mittag Leffler global stability is given according to Eq. (8),

$$\|y(t) - y^*\| \leq M\|y(0) - y^*\|E_\alpha(-\lambda t^\alpha) \tag{8}$$

where y^* is the equilibrium point, 'M' is a nonsingular M-matrix, $E_\alpha(-\lambda t^\alpha)$ is a Mittag–Leffler function, α is fractional order of the NNs.

In [23], the distinctiveness, occurrence, and asymptotic global stability of equilibrium point of fractional order MNN are demonstrated under the condition that neuron activation functions f_i, g_i are enclosed and satisfy the Lipschitz condition with Lipschitz constant $F_i, G_i > 0$.

The exponential global stability of periodic solution for memristor BAM neural networks with latency and the occurrence of the equilibrium point is proved using methods like Yoshizawa-like theorem, functional differential inclusions theory, and inequality technique under the Filippov-framework. Also, the periodicity of the solution of the neural network with discontinuous right-hand sides was studied in [29].

In [35], authors develop new delay-dependent exponential stability criteria making use of generalized double-integral inequalities, which is a better method compared to existing ones. The disparities comprise free weighting matrices improving their flexible nature by reducing the conservatism of the stability criteria. This criterion is combined with quadratic convex combination technique to demonstrate the exponential stability of the MNNs.

The globally exponential stability of impulsive delayed recurrent MNNs using impulsive differential inequality and Lyapunov function were investigated in [21]. The adequate condition for the exponential global stability of equilibrium point u^* of the system is shown in Eq. (9),

$$\sum_{i=1}^{n}|v(t) - v^*| \leq \sup_{-\tau \leq s \leq 0}|\phi(s) - v^*|e^{-(\lambda-\delta)t} \tag{9}$$

where $v(t)$ is the state of the memristor unit, $\phi(s)$ denotes initial value of the system, u^* is the equilibrium point, λ is a constant, and δ is impulse jump operator.

In [32], authors explore the exponential stability of novel impulsive controlled memristor neural network model with variable latencies using Lyapunov–Razumikhin techniques and mathematics induction method and following stability criteria is derived and shown in Eq. (10),

$$\|x(t)\| \leq \beta\|\phi(t)\|_\tau e^{-\mu t}, \quad t \geq 0 \tag{10}$$

where $\beta = (\lambda_{\max}(P)/\lambda_{\min}(P))^{1/2}$ and $\mu = \lambda - \delta$, $\lambda_{\max}(P)$ and $\lambda_{min}(P)$ are maximum and minimum Eigen values of matrix P, $x(t)$ and $\phi(t)$ are the solution of the system and initial value of the memristor system, respectively.

[26] investigates the delay-dependent robust stability investigation of uncertain MNNs with delay state and norm-bounded uncertainty. Following are the conditions for robust stability based on LMIs given in Eq. (11),

$$\begin{aligned} &\Omega(P, R, \varepsilon, \overline{\tau}) < 0 \\ &I - \varepsilon_5 D^{\mathrm{T}} D > 0, I - \varepsilon_6 D^{\mathrm{T}} D > 0, I - \varepsilon_7 E_b^{\mathrm{T}} E_b > 0 \end{aligned} \tag{11}$$

where $\Omega(P, R, \varepsilon, \overline{\tau})$ is matrix function, P and R are matrices, ε is a positive scalar, $\overline{\tau}$ is constant time delay, D and E are real-valued matrices with relevant size.

A reaction diffusion uncertain memristor neural network with time variable latencies and leakage is defined in [27] and inspect asymptotic global stability by using Jensen integral inequality, Lyapunov stability theory, free weighting matrix, Schur complement Lemma approach together with LMI approach and claims that the outcome is less conservatism.

A novel model of multidirectional associative MNNs with time-varying delays was proposed in [33]. The conservativeness is reduced because of bounded time delays and it is not necessary for their derivatives to be differential. The equilibrium point of the MAMNN exists if activation functions $f_{ki}()$ and $g_{ki}()$ are bounded and continuous.

The equilibrium point of the system is globally exponentially stable if the following inequality in Eq. (12) is satisfied,

$$-\frac{1}{2D_{ki}} + \sum_{\substack{p=1 \\ p \neq k}}^{m} \sum_{j=1}^{n_p} \frac{1}{D_{pj}} \left[\left(A_{kipj}\sigma_{ki}\right)^2 + \left(B_{kipj}\rho_{ki}\right)^2 \right] < 0 \tag{12}$$

where

$$\begin{aligned} &D_{ki} = \min\left\{ \left|\underline{d}_{ki}\right|, \left|\overline{d}_{ki}\right| \right\}, A_{kipj} = \max\left\{ \left|\underline{a}_{kipj}\right|, \left|\overline{a}_{kipj}\right| \right\}, \\ &B_{pjki} = \max\left\{ \left|\underline{b}_{kipj}\right|, \left|\overline{b}_{kipj}\right| \right\}. \end{aligned}$$

In [24], authors prove the finite time stability of memristive-based fractional order neural networks of order $0 < \alpha < 1$, by using method of iteration, set-valued mapping, the theory of differential inclusion and contracting mapping principle. For fractional order memristive neural networks of order $1 < \alpha < 2$, finite time stability is proved based on the following inequality in Eq. (13),

$$\mathrm{e}^{-\omega t}(1+t)E_\alpha((\|A\|F + \mathrm{e}^{\omega\tau}\|B\|G)\Gamma(\alpha)t^\alpha) < \frac{\varepsilon}{\delta}, \quad t \in [0, T) \tag{13}$$

where $A = \left(A_{ij}\right)_{n\times n}$, $B = \left(B_{ij}\right)_{n\times n}$, $C = \text{diag}(c_1, \ldots, c_n)$, $F = \max_{1\le j\le n}\left\{F_j\right\}$, $G = \max_{1\le j\le n}\left\{G_j\right\}$, $\omega = \min_{1\le i\le n}\{c_i\}$, $E_\alpha()$ is Mittag–Leffler function, ε and δ are constants.

Paper [19] models an uncertain memristor-based NN with discrete delays as a switched neural network with undetermined parameters which are time-varying in nature. This paper establishes the LMIs-based stability criteria for system with both unstable and stable subsystems. Using mean dwell period and multiple Lyapunov-like function technique, the condition for USMNNs to be exponentially globally stable for switching signal $\sigma(t)$ with average dwell time satisfying $\alpha.T_a + \ln\beta$, where α and β are scalars and T_a is the average dwell time.

In [14], it investigates robust stability of fractional order memristor-based Hopfield neural networks with external interferences and also proves the occurrence and distinctiveness of the equilibrium point. Some conditions to achieve the robust stability based on fractional order Lyapunov direct method are presented. The equilibrium point x^* of the system is globally Mittag–Leffler stable and the FMHNNs with parameter disturbances are asymptotically globally stable under the following condition,

$$\left\|x(t) - x^*\right\| \le V\left(0, x(0) - x^*\right)E_\alpha\left(-\lambda t^\alpha\right) \tag{14}$$

where $x(t)$ is the solution of the system, x^* is the equilibrium point, $x(0)$ is the initial condition of the system, λ is positive constant, $\alpha \in (0, 1)$, $E_\alpha()$ is Mittag–Leffler function with one parameter.

[34] studies a group of NNs with memristors that are defined as dynamic MNNs. The significant feature of the DM-NNs is the processing takes place in charge-flux domain. The condition for global asymptotic stability of equilibrium point of the system is established dependent on the concept of Lyapunov diagonally stable (LDS) matrices. The equilibrium point is said to be asymptotically globally stable if it is locally stable in the sense of Lyapunov and is globally attractive for the solution of Eq. (15)

$$C\frac{\mathrm{d}\varphi(t)}{\mathrm{d}t} = -\hat{Q}(\varphi(t)) + G\varphi(t) + Q \quad \text{for } t \ge 0 \tag{15}$$

where the diagonal mapping

$$\hat{Q}(\varphi(t)) = \left(\hat{q}_1(\varphi_1(t))\right), \left(\hat{q}_2(\varphi_2(t))\right)\ldots\left(\hat{q}_n(\varphi_n(t))\right) : \mathbb{R}^n \to \mathbb{R}^n \tag{16}$$

and $Q \in \mathbb{R}^n$ is a constant vector, $\varphi(t)$ is vector of memristor fluxes.

In [25], a simplified mathematical model of memristor recurrent neural network in matrix form is proposed as below,

$$\begin{aligned} &D^\alpha x(t) = -Dx(t) + C(x(t))f(x(t)) + E(x(t))f(x(t - h(t))) + u(t), \quad t > 0 \\ &x(t) = \phi(t), \quad t \in [-h, 0] \end{aligned} \tag{17}$$

where $x(t)$ represents the system state vector, $\alpha \in (0, 1)$, 'D' stands for the ith neuron reset rate from a given potential toward the resting state. The function $C(x(t))$ and $E(x(t))$ represents the non-delayed and delayed memristive synaptic connection weights, respectively. $f(x(t))$ and $f(x(t-h(t))$ represent neuron activation function. Investigates the asymptotic stability of the system and derives the stability criteria with the help of FO Razumikhin theorem and linear matrix inequalities (LMI).

The input-to-state stability of memristor bidirectional associative memory neural networks with variable time delays was investigated in [30]. The condition for the input-to-state stability is derived based on non-smooth analysis and set-valued maps. The system is stable if $\|x(t; x_0, I(t))\| + \|y(t; x_0, J(t))\|$ is bounded, where $I(t)$ and $J(t)$ are bounded external inputs.

In [36], authors have developed a technique for stability analysis of memristor neural networks with time-varying latency based on segmentation of state space. The study shows that the exponential global stability of MNNs can be attained by reasonable selection of external inputs. The paper proves the global exponentially stability of distinctive equilibrium point of the memristive network located in the exponential attractor $\mathcal{D}$, where $\mathcal{D} = \prod_{i=1}^{n} \mathcal{D}_i = \mathcal{D}_1 \times \mathcal{D}_2 \times \cdots \times \mathcal{D}_n \subset \mathbb{R}^n$.

with $\mathcal{D}_i = \begin{cases} (1, +\infty), & i \in N_1 \\ [-1, 1], & i \in N_2 \\ (-\infty, -1), & i \in N_3 \end{cases}$ where N_1, N_2 and N_3 are index sets.

In [8], investigators have developed a delay-dependent criterion for exponential global stability of MNNs depended on free weighting matrix technique and Lyapunov Krasovskii functional scheme. The inequality to be satisfied for global exponential stability of memristor neural network is derived as follows,

$$-PQ_2^{-1}P \le -2P + Q_2 \tag{18}$$

where P and Q are matrices with appropriate dimensions.

The new inertial MNN model with variable latency and impulses is put forward in [28]. Sufficient conditions are derived on exponential global stability using extended Halanay differential inequality and Lyapunov functional technique.

[17] proposes a novel memristor neural network model, namely discrete time stochastic MNN where leakage and probabilistic delays are simultaneously considered using appropriate Lyapunov–Krasovskii function. The adequate condition for globally exponential stability in mean square is as below,

$$\mathbb{E}\left\{\|x(k)\|^2\right\} \le \alpha\beta^k \sup_{-\tau_K \le s \le 0} \mathbb{E}\left\{\|x(s)\|^2\right\} \tag{19}$$

holds for all $k \ge 0$, where $\tau_k = \max\{\tau_M, \ell\}$, $x(k)$ is neuron state vector, α and $\beta \in (0, 1)$ are scalars.

[31] investigates the input-to-output stability of memristive neural NNs since it plays an eminent part in revealing spike-timing-dependent-plasticity. The paper derives new criteria for input-to-output stability for the memristive neural network

using non-smooth analysis and control theory. The condition for the stability of the system is given as,

$$-1+\sum_{j=1}^{n}\left(\tilde{a}_{ij}+\tilde{b}_{ij}\right)l_i<0,\quad i=1,2\ldots,n \tag{20}$$

where $\tilde{a}_{ij}=\max\{|\hat{a}_{ij}|,|\check{a}_{ij}|\}$, $\tilde{b}_{ij}=\max\{|\hat{b}_{ij}|,|\check{b}_{ij}|\}$, $\hat{a}_{ij},\check{a}_{ij},\hat{b}_{ij}$ and $\check{b}_{ij}$are constants and l_i is Lipschitz constant.

[11] investigates the input to output stability of memristive complex-valued neural networks (MCVNN) class characterized by latencies.

The input-to-output stability condition for MCVNN utilizing set-valued maps, Lyapunov function method and differential inclusion theory is given by,

$$\|y(t;\,y_0,w(t))\|\le\beta(\|y_0\|_\infty,t)+\gamma\left(\|w\|_{\sup}\right),\quad t\ge 0 \tag{21}$$

where $y(t)$ is complex-valued state vector, z_0 is initial condition, $w(t)$ denotes external input vector.

[12] discusses and concentrates on the exponential global stability of stochastic memristor complex-valued neural networks (SMCVNN) with latency. The condition for the equilibrium point of the SMCVNN to be globally exponentially stable in the sense of mean square is every solution $z(t;\phi)$ should satisfy following condition

$$\mathbb{E}\left[|z(t;\phi)|^2\right]\le\gamma e^{-\beta t}\sup_{-\tau\le s\le 0}\mathbb{E}\left[|\phi(s)|^2\right] \tag{22}$$

where there should exist constants $\beta>0,\gamma>0$, $\mathbb{E}[.]$ represents correspondent expectation operator with respect to the given probability measure P.

[37] the author studies the associative memory based on MRNNs to solve the problem of storage capacity. Sufficient conditions for global stability and multi-stability of MRNN are derived using comparative principle and the existing stability criteria. If the coefficients of the MRNN satisfy that $C-|A|-|B|$ is a nonsingular M-matrix with $|A|=\left(|a_{ij}|\right)_{n\times n}$ and $|B|=\left(|b_{ij}|\right)_{n\times n}$ then the system is considered to attain global exponential stability. The parameters a_{ij} and b_{ij} depends on external inputs.

Paper [38] studies a new inertial memristor neural network model characterized by both proportional and distributed delays. An adequate and imperative criterion for exponential input-to-state stability are derived based on the concepts like inequality technique, differential inclusion, and Cauchy–Schwarz Inequality.

[39] investigates average square-based exponential stability for memristor stochastic neural networks characterized with leakage latency. Adequate and distinctive criteria are derived using appropriate matrix inequality technique, Lyapunov Krasovskii function, intermediate theorem of Schur complement and Itô's differential formula.

Table 1 Comparison of mathematical methods for stability analysis

Mathematical techniques for stability analysis	Characteristics	Typical references
Linear matrix inequalities [40]	Incorporating affine constraints, simple algebraic observation	[8, 18, 26, 25, 39]
Differential inclusion [41]	Mathematical tool for studying differential equations, large set of trajectories	[20, 29, 24, 38]
Lyapunov function	System operation over infinite interval of time, Concerned with specific bounds on the states	[1, 4, 16, 22, 29, 21, 27, 27–29, 11, 12]
Gronwall inequality [42]	Explicit bounds on solutions	[1, 7, 10, 16, 24]
Quadratic convex combination	Less conservative [43]	[35]
Free weighting matrix	More effective than basic inequality provides great freedom of the criteria [44]	[8, 27]
Non-smooth analysis [51]	Upper and lower limits, semicontinuity, differentiability	[13, 30, 31]

From the above conditions, we can observe that major stability analysis outcomes can be derived using linear matrix inequalities, set-valued maps, Lyapunov methods, and free weighting matrix techniques for continuous time MNNs. Among these literature reviews, there are mainly two research items in study on discrete MNNs [13, 17]. The differentiation of selected stability analysis methods is shown in Table 1.

4 Conclusions

In prior three decades, there is a fine tuning of MNNs stability with or without delays. Even so, there are still many problems to be solved related to dynamic behavior of MNNs. In this paper, the stability analysis techniques for several MNNs have been extensively reviewed including discrete time MNNs. Future scope concerned to the areas of MNNs networks based on the literature survey is as follows.

Analysis of synchronization in fixed period of fractional order MNNs with multiproportional latency, finite time analysis of MNNs employing pinning control strategy, analyzing the behavior of fractional-order MNNs considering the effect of external disturbances and latency, analyzing multiperiodicity and multi-stability of fractional order NNs stimulated by external inputs, designing fractional order MNNs-based associative memories with high capacity, analysis of synchronization and stability of stability of inertial MNNs under various schemes of control and so forth.

References

1. Ali, M.S., Narayanan, G., Orman, Z., Shekher, V., Arik, S.: Finite time stability analysis of fractional-order complex-valued memristive neural networks with proportional delays. Neural Process. Lett. **51**(1), 407–426 (2020)
2. Rakkiyappan, R., Velmurugan, G., Cao, J.: Stability analysis of memristor-based fractional-order neural networks with different memductance functions. Cogn. Neurodyn. **9**(2), 145–177 (2015)
3. Wen, S., Zeng, Z., Huang, T.: Exponential stability analysis of memristor-based recurrent neural networks with time-varying delays. Neurocomputing **97**, 233–240 (2012)
4. Wang, J., Liu, F., Qin, S.: Global exponential stability of uncertain memristor-based recurrent neural networks with mixed time delays. Int. J. Mach. Learn. Cybern. **10**(4), 743–755 (2019)
5. Di Marco, M., Forti, M., Pancioni, L.: Stability of memristor neural networks with delays operating in the flux-charge domain. J. Franklin Inst. **355**(12), 5135–5162 (2018)
6. Wu, A., Zeng, Z.: Global Mittag-Leffler stabilization of fractional-order memristive neural networks. IEEE Trans. Neural Netw. Learn. Syst. **28**(1), 206–217 (2015)
7. Rakkiyappan, R., Velmurugan, G., Cao, J.: Finite-time stability analysis of fractional-order complex-valued memristor-based neural networks with time delays. Nonlinear Dyn. **78**(4), 2823–2836 (2014)
8. Wen, S., Huang, T., Zeng, Z., Chen, Y., Li, P.: Circuit design and exponential stabilization of memristive neural networks. Neural Netw. **63**, 48–56 (2015)
9. Hou, P., Hu, J., Gao, J., Zhu, P.: Stability analysis for memristor-based complex-valued neural networks with time delays. Entropy **21**(2), 120 (2019)
10. Wang, L., Song, Q., Liu, Y., Zhao, Z., Alsaadi, F.E.: Finite-time stability analysis of fractional-order complex-valued memristor-based neural networks with both leakage and time-varying delays. Neurocomputing **245**, 86–101 (2017)
11. Liu, D., Zhu, S., Chang, W.: Input-to-state stability of memristor-based complex-valued neural networks with time delays. Neurocomputing **221**, 159–167 (2017)
12. Liu, D., Zhu, S., Chang, W.: Global exponential stability of stochastic memristor-based complex-valued neural networks with time delays. Nonlinear Dyn. **90**(2), 915–934 (2017)
13. Wu, A., Zeng, Z.: Lagrange stability of memristive neural networks with discrete and distributed delays. IEEE Trans. Neural Netw. Learn. Syst. **25**(4), 690–703 (2013)
14. Liu, S., Yu, Y., Zhang, S., Zhang, Y.: Robust stability of fractional-order memris-tor-based Hopfield neural networks with parameter disturbances. Phys. A **509**, 845–854 (2018)
15. Zhang, H., Wang, Z., Liu, D.: A comprehensive review of stability analysis of continuous-time recurrent neural networks. IEEE Trans. Neural Netw. Learn. Syst. **25**(7), 1229–1262 (2014)
16. Liu, W., Jiang, M., Yan, M.: Stability analysis of memristor-based time-delay fractional-order neural networks. Neurocomputing **323**, 117–127 (2019)
17. Liu, H., Wang, Z., Shen, B., Huang, T., Alsaadi, F.E.: Stability analysis for discrete-time stochastic memristive neural networks with both leakage and probabilistic delays. Neural Netw. **102**, 1–9 (2018)
18. Wu, A., Zeng, Z.: Exponential stabilization of memristive neural networks with time delays. IEEE Trans. Neural Netw. Learn. Syst. **23**(12), 1919–1929 (2012)

19. Li, X., She, K., Zhong, S., Shi, K., Kang, W., Cheng, J., Yu, Y.: Extended robust global exponential stability for uncertain switched memristor-based neural networks with time-varying delays. Appl. Math. Comput. **325**, 271–290 (2018)
20. Hu, J., Wang, J.: Global uniform asymptotic stability of memristor-based recurrent neural networks with time delays. In: The 2010 International Joint Conference on Neural Networks (IJCNN), pp. 1–8. IEEE, Barcelona, Spain (2010)
21. Wang, H., Duan, S., Li, C., Wang, L., Huang, T.: Exponential stability analysis of delayed memristor-based recurrent neural networks with impulse effects. Neural Comput. Appl. **28**(4), 669–678 (2017)
22. Chen, J., Zeng, Z., Jiang, P.: Global Mittag-Leffler stability and synchronization of memristor-based fractional-order neural networks. Neural Netw. **51**, 1–8 (2014)
23. Chen, L., Wu, R., Cao, J., Liu, J.B.: Stability and synchronization of memristor-based fractional-order delayed neural networks. Neural Netw. **71**, 37–44 (2015)
24. Chen, C., Zhu, S., Wei, Y.: Finite-time stability of delayed memristor-based fractional-order neural networks. IEEE Trans. Cybern. **50**(4), 1607–1616 (2018)
25. Chen, L., Huang, T., Machado, J.T., Lopes, A.M., Chai, Y., Wu, R.: Delay-dependent criterion for asymptotic stability of a class of fractional-order memristive neural networks with time-varying delays. Neural Netw. **118**, 289–299 (2019)
26. Wang, X., Li, C., Huang, T.: Delay-dependent robust stability and stabilization of uncertain memristive delay neural networks. Neurocomputing **140**, 155–161 (2014)
27. Li, R., Cao, J.: Stability analysis of reaction-diffusion uncertain memristive neural networks with time-varying delays and leakage term. Appl. Math. Comput. **278**, 54–69 (2016)
28. Zhang, W., Huang, T., He, X., Li, C.: Global exponential stability of inertial memristor-based neural networks with time-varying delays and impulses. Neural Netw. **95**, 102–109 (2017)
29. Li, H., Jiang, H., Hu, C.: Existence and global exponential stability of periodic solution of memristor-based BAM neural networks with time-varying delays. Neural Netw. **75**, 97–109 (2016)
30. Zhao, Y., Kurths, J., Duan, L.: Input-to-state stability analysis for memristive BAM neural networks with variable time delays. Phys. Lett. A **383**(11), 1143–1150 (2019)
31. Wu, A., Zeng, Z.: Input-to-state stability of memristive neural system with time delays. Circ. Syst. Signal Process. **33**(3), 681–698 (2014)
32. Duan, S., Wang, H., Wang, L., Huang, T., Li, C.: Impulsive effects and stability analysis on memristive neural networks with variable delays. IEEE Trans. Neural Netw. Learn. Syst. **28**(2), 476–481 (2016)
33. Wang, W., Yu, X., Luo, X., Li, L.: Stability analysis of memristive multidirectional associative memory neural networks and applications in information storage. Mod. Phys. Lett. B **32**(18), 1850207 (2018)
34. Di Marco, M., Forti, M., Pancioni, L.: New conditions for global asymptotic stability of memristor neural networks. IEEE Trans. Neural Netw. Learn. Syst. **29**(5), 1822–1834 (2017)
35. Wang, Z., Ding, S., Huang, Z., Zhang, H.: Exponential stability and stabilization of delayed memristive neural networks based on quadratic convex combination method. IEEE Trans. Neural Netw. Learn. Syst. **27**(11), 2337–2350 (2015)
36. Wu, A., Zeng, Z.: An improved criterion for stability and attractability of memristive neural networks with time-varying delays. Neurocomputing **145**, 316–323 (2014)
37. Bao, G., Chen, Y., Wen, S., Lai, Z.: Stability analysis for memristive recurrent neural network and its application to associative memory. J. Autom. **43**(12), 2244–2252 (2017)
38. Iswarya, M., Raja, R., Cao, J., Niezabitowski, M., Alzabut, J., Maharajan, C.: New results on exponential input-to-state stability analysis of memristor based complex-valued inertial neural networks with proportional and distributed delays. J. Math. Comput. Simul. (2021)
39. Wang, F., Chen, Y.: Mean square exponential stability for stochastic memristor-based neural networks with leakage delay. Chaos Solitons Fractals **146**, 110811 (2021).
40. Scherer, C., Weiland, S.: Linear matrix inequalities in control. Lecture Notes, vol. 3, no. 2. Dutch Institute for Systems and Control, Delft, The Netherlands (2000)

41. Aubin, J.P., & Cellina, A.: Differential Inclusions: Set-Valued Maps and Viability Theory, vol. 264. Springer Science & Business Media, Berlin (2012)
42. Ye, H., Gao, J., Ding, Y.: A generalized Gronwall inequality and its application to a fractional differential equation. J. Math. Anal. Appl. **328**(2), 1075–1081 (2007)
43. Zhang, H., Yang, F., Liu, X., Zhang, Q.: Stability analysis for neural networks with time-varying delay based on quadratic convex combination. IEEE Trans. Neural Netw. Learn. Syst. **24**(4), 513–521 (2013)
44. Zhang, C.K., He, Y., Jiang, L., Lin, W.J., Wu, M.: Delay-dependent stability analysis of neural networks with time-varying delay: a generalized free-weighting-matrix approach. Appl. Math. Comput. **294**, 102–120 (2017)
45. Liu, H., Ma, L., Wang, Z., Liu, Y., Alsaadi, F.E.: An overview of stability analysis and state estimation for memristive neural networks. Neurocomputing **391**, 1–12 (2020)
46. Rakkiyappan, R., Premalatha, S., Chandrasekar, A., Cao, J.: Stability and synchronization analysis of inertial memristive neural networks with time delays. Cogn. Neurodyn. **10**(5), 437–451 (2016)
47. Velmurugan, G., Rakkiyappan, R., Cao, J.: Finite-time synchronization of fractional-order memristor-based neural networks with time delays. Neural Netw. **73**, 36–46 (2016)
48. Zhang, G., Shen, Y., Yin, Q., Sun, J.: Global exponential periodicity and stability of a class of memristor-based recurrent neural networks with multiple delays. Inf. Sci. **232**, 386–396 (2013)
49. Li, R., Cao, J.: Finite-time stability analysis for Markovian jump memristive neural networks with partly unknown transition probabilities. IEEE Trans. Neural Netw. Learn. Syst. **28**(12), 2924–2935 (2016)
50. Wang, L., Zeng, Z., Zong, X., Ge, M.F.: Finite-time stabilization of memristor-based inertial neural networks with discontinuous activations and distributed delays. J. Franklin Inst. **356**(6), 3628–3643 (2019)
51. Ferrera, J.: An Introduction to Nonsmooth Analysis. Academic Press, Cambridge (2013)

Analysis of State-of-Art Attack Detection Methods Using Recurrent Neural Network

Priyanka Dixit and Sanjay Silakari

1 Introduction

Protecting computer networks from illegal access is one of the major challenges for the government and other commercial sectors. The threat minded actions generally targeting the vulnerable systems and create serious loss to the organizations, companies and other areas [1].

Cyber security is a collection of technology and wide grouping of distinct networks that provide protection, privacy and authentication against unauthorized access, fraud, data damages and other cyber crimes [1]. Cyber security is deals with broad range of problems such as attack detection, malware detection, security countermeasures, monitoring etc. Cyber attack detection is an important challenge of cyber security. Previously, researchers have used various techniques to design robust attack detection system but fail to achieve better performance and deals with real time systems. The attack detection system has broadly four important phases First, the type of attack (details of dataset used), secondly preprocessing of data it includes general steps like data cleaning, normalizing and third, feature selection or extraction which is most important phase, and finally classification in which classify normal or attack records.

Since the last few decades, several researchers have used all most many of machine learning approaches for the classification of cyber attacks without knowing their depth features. It seen that the conventional machine learning approaches are not found effective to solve the attack detection problem due to their complexity and limitations.

Now a day's machine learning techniques are in trend by simulating human brain architecture in the form of deep neural networks or deep learning. These deep networks are highly suitable to solve complex problems. Deep learning networks

P. Dixit (✉) · S. Silakari
Department of Computer Science and Engineering, University Institute of Technology, Rajiv Gandhi Proudyogiki Vishwavidyalaya , Bhopal, India

M. Dua et al. (eds.), *Proceedings of the International Conference on Paradigms of Communication, Computing and Data Sciences*, Algorithms for Intelligent Systems,
https://doi.org/10.1007/978-981-16-5747-4_68

are broadly classified in two categories are supervised and unsupervised. In supervised category the deep networks are convolution neural network (CNN), recurrent neural network (RNN), deep brief networks (DBN), dense neural network (DNN). In unsupervised are Generic neural network (GNN), Auto encoder and restricted Boltzmann machine (RBN). In this paper reviewing the application of RNN deep networks in attack detection systems.

The remaining paper is arranged as Sect. 2 Introduction of different parameters used for attack detection. Section 3 Focuses on the brief study of different datasets that are popularly preferred for cyber attack detection. Section 4 Literature review and comparative study among RNN networks. Section 5 Observations and future directions. Finally concluded.

1.1 Prominent Attacks of Cyber Security

Cyber security is deals with different types of attacks the major are denial of service attack (DoS), probe, malware, zero-day, phishing, user root, adversarial attacks, poisoning attack, evasive attack etc. In present scenario many of researchers have used deep learning methods for the detection of these attacks [2]. Here in this paper we focused on RNN network which one of the popular technique of deep learning, application on attack detection. The performance of the deep networks on the following types of network attacks such as Denial of service attack (Dos), Probe attack, User to root attack, Remote to local attack, Fuzzers, Backdoor, Exploit, Generic, Reconnaissance, Shellcode, Worms etc. are discussed.

1.2 Recurrent Neural Network

Recurrent Neural Network (RNNs) is one of deep network that easily deals with sequential data and used in both supervised and unsupervised learning. RNN are based on generative model architecture that also refers as graphical models because their visual representation is in the form of graph, where nodes are the random variables, edges are connection links among them. The generative models have unobservable hidden variables and for classification task the model goes for pre training stage. RNN consist memory cells that are able to store data of its previous input and network of feedback loop in which linking of multiple layer by layer [3, 4].

The RNN is work by well known basic two architectures namely Elman and Jordan RNNs. The Elman RNN architectural model has a simple feedback looping layer after layer. The Jordan RNN architectural model has a feedback looping of neurons within one layer to another layer. There is another feedback looping that connecting a neuron to itself.

Figures 1 and 2 are two base architectures of the RNN (Elman and Jordan) showing the context unit known to store information of the previous output of hidden layer.

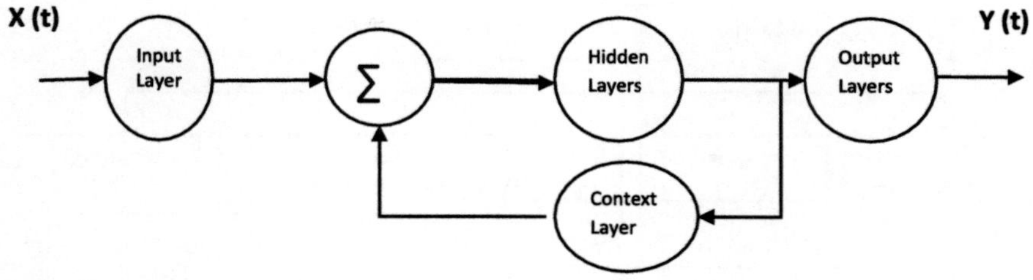

Fig. 1 Elman recurrent neural network

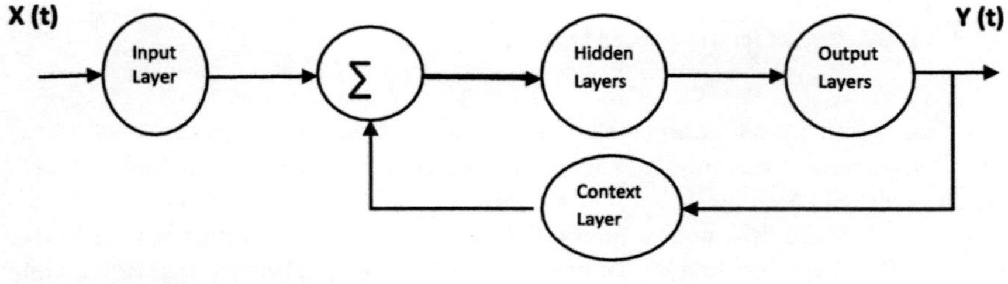

Fig. 2 Jordan recurrent neural network

RNN applications mainly in natural language processing, text recognition, speech recognition etc. due to modeling sequences by having cyclic connections. The general representation of RNN is shown below:

Here the notation for input vector sequence, the hidden vector sequence, and output vector sequence with X, H and Y, respectively.

$$X = (x_1, x_2, x_3, \ldots, x_n) \tag{1}$$

$$H = (h_1, h_2, h_3, \ldots, h_n) \tag{2}$$

$$Y = (y_1, y_2, y_3, \ldots, y_n) \tag{3}$$

In Eqs. 1, 2, and 3 X represents the input vector sequence, H represent hidden vector sequence and Y represent output vector sequence are calculated with $t = 1$ to T as follows:

$$h_t = \sigma(W_{xh}X_t + W_{hh}h_{t-1} + b_h) \tag{4}$$

$$y_t = W_{hy}h_t + b_y \tag{5}$$

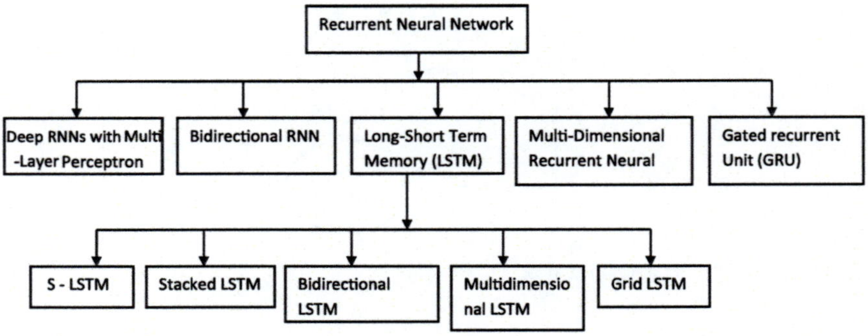

Fig. 3 Classification of recurrent neural network

In Eqs. 4 and 5 the function σ is represents activation function, x_t is input at time t, h_{t-1} is the state at time $t-1$, W and b is a weight and bias, Y_t is the output at each output layer [5].

In Fig. 3. Recurrent neural network is itself classify in various sub networks are Deep RNN with multilayer, Bidirectional RNN, Multi-Dimensional RNN, Gate Recurrent unit (GRU) and Long-short Term Memory (LSTM). LSTM itself sub classify as S-LSTM, Stacked LSTM, Bi-directional LSTM, Multi dimensional LSTM and Grid LSTM the detail description of above mentioned techniques are given [6–8].

The literature associated with different recurrent neural networks (RNN) architectures used for attack detection such as Generative Adversarial network based models (GANs), Long Short Term memory neural network (LSTM).

2 Performance Evaluation Parameters

In this section, the performance parameters for attack detection are discussed that shows the effectiveness of the model.

- True Positive (T_{p}) is equal to the correctly unaccepted, and can be denoted as the number of attack records that are determined as attack [9].
- False Positive (F_{p}) is the equal of incorrectly unaccepted, and can be denoted the number of normal records that are determined as attack [9].
- True Negative (T_{n}) is equal to the correctly accepted, and can be denoted as the number of normal records that are determined as normal [9].
- False Negative (F_{n}) is equal to the incorrectly accepted, and can be denoted as the number of attack records that are determined as normal [9].

1. **Accuracy (AC)**: the ratio of the total number of records classified correctly to the total the records shown in (6).

$$\mathrm{AC} = \frac{T_{\mathrm{p}} + T_{\mathrm{n}}}{T_{\mathrm{p}} + T_{\mathrm{n}} + F_{\mathrm{p}} + F_{\mathrm{n}}} \tag{6}$$

2. **True Positive Rate (TPR)**: also called as Detection Rate (DR), it is the ratio of total number of records identified correctly to the total number of attack records, as shown in (7).

$$\mathrm{TPR} = \frac{T_{\mathrm{p}}}{T_{\mathrm{p}} + F_{\mathrm{n}}} \tag{7}$$

3. **False Positive Rate (FPR)**: the ratio of percentage of the number of records rejected incorrectly is with the total number of normal records, as shown in (8).

$$\mathrm{FPR} = \frac{F_{\mathrm{p}}}{F_{\mathrm{p}} + T_{\mathrm{n}}} \tag{8}$$

3 Different Datasets of Cyber Attack

The following are the famous cyber security datasets used for the performance evaluation of attack detection system.

1. DARPA 98,99
   This dataset is generally consist network in and out packets logs records and is used in practice since 1998. Its training data have 7 weeks and the test data have 2 weeks of network attacks records in packet based format. It is not able to represent real time network traffic. It contains 41 features and the attacks are Dos, R2L, U2R, Probe attack [4, 10].
2. KDD Cup 1999
   This dataset is the base of DARPA 1998 assessment program and captures 7 weeks of network traffic records that contains approx 4,900,000 vectors. It consists of redundant and error records of data. This dataset has 41 features that are categories in following categories are basic features, traffic features, and content features. This dataset has following categories of attacks are Denial of service attack (DoS), Remote to Local attack (R2L), Probe attack and User to Root attack (U2R) [4].
3. NSL-KDD
   This dataset is the improved form of KDD Cup 99 and also consist 41 features that are categories in following categories are (1) basic features, (2) traffic features, and (3) content features. This dataset has following class of attacks are Denial of service attack (DoS), Remote to Local attack (R2L), Probe attack and User to Root attack (U2R) [4, 10].
4. ISCX2012
   This dataset is introduced in 2012 and captured 7 days of network traffic records. The network records such as infiltrating the inside network, hypertext transfer

protocol based DoS, DDoS and Brute Force Secure Shell hash. It consist two classes of profiles are first profiles focus on the attack scenario in an explicit way and second profiles summarize extracted behaviors of entities [4].

5. CIC DDoS 2019
   It contains the most up-to-date common Distributed Denial of Service attacks, which matches with true real-world data (PCAPs). This dataset has 85 features and the attacks are 12 Distributed Denial of Service attacks, MSSQL, SSDP while as UDP based attacks include CharGen, NTP and TFTP [4].
6. UNSW-NB15
   It is formed by using following tools are named as IXIA Perfect storm, Tcpdump, Argus, and Bro-ids tool. They are responsible for the following types of attacks such as Denial of service, Exploits, Generic, Reconnaissance, Shellcode, and Worms. It also contains approx two million and five lakhs and forty thousand and forty four vectors with forty nine features [4, 10].
7. CIC-IDS2017
   It contains data of a week from Mon July 03, 2017, to Fri, July 07, 2017. This dataset consist following attack categories are Brute Force secure shell hash, Denial of service, Heart bleed, Web based attack, infiltration, Botnet and Distributed denial of service attack, and Brute Force File transfer protocol. The CIC FlowMeter, which extract eightly features from the network traffic. Meanwhile this dataset mine behavior of twenty five users based on the protocols including File transfer protocol, Hyper text transfer protocol [4].
8. CSE-CICIDS-2018
   This dataset is introduced by Communications Security Establishment (CSE) and the Canadian Institute for Cyber security (CIC). It includes 7 different categories of attack such as Heart bleed, Brute-force, Denial of service attack, Distributed Denial of service attack, Web attacks, Botnet, and infiltration. It used to mine 80 features from the network traffic and created by using Network profiles in some standard way [4].

4 Related Work

Yin [9] defined recurrent neural network based intrusion detection model in both binary and multiclass classification applied on NSL KDD dataset and compared results with J48, naive Bayesian, and random forest, the performance parameters were accuracy rate and detection rate, false positive rate used. RNN performed outstanding in given parameters especially in multiclass classification.

Kim and Kim [11] presented a model, fusion of recurrent neural network with hessian-free optimization used for the intrusion detection. They used DARPA dataset KDDCUP99 to train and test the model. They select optimal features to train the model and analyzed results with different parameters. It is found that the results of this model is superior as compared with existing researches in prospective of performance. The model achieved 95.37% detection rate and 2.1% of false alarm rate.

Kim et al. [12] presented attack detection model based deep learning approach. They applied Long Short Term Memory (LSTM) architecture to a Recurrent Neural Network (RNN) and train the model using KDD Cup 1999 benchmark. The experimental results shown outstanding performance by achieved 98.88 detection rate, 10.04 FAR and 96.93 accuracy.

Krishnan and Raajan [13] presented attack detection model by using recurrent neural network shown result outstanding as compared with the conventional techniques such as decision tree, C4.5 and Conditional Random Field. The performance of the model is recorded as for Probe attack recall, precision is 97.8% and 88.34, Dos attack recall, precision is 97.05% and 99.9, U2R attack Recall, precision 62.7 and 56.12. R2L attack Recall, precision 28.81 and 94.1.

Fu et al. [5] introduced intelligent system based on hybrid of RNN-LSTM deep learning approach to detect network based attacks. The experimental results were evaluated on NSL KDD benchmark. The performance of the model found improved as compared with other methods such as GRNN, PNN, RBNN, KNN, SVM, and Bayesian. The result of the model is recorded as detection rate 98.85, FAR is 8.75 and Accuracy is 97.52.

Chen et al. [14] presented an effective attack detection model by using bidirectional long short term memory architecture. The experiment evaluations on UNSW-NB15 data set and performance is recorded as bi directional LSTM Precision is 0.93, Recall is 0.89 and $F1$ score is 0.91 with can compared with the performance of simple LSTM Precision is 0.85, Recall is 0.88 and $F1$ score is 0.86 The performance of Bidirectional LSTM is found improved.

In Table 1 presents a brief comparative study among the recurrent neural networks based on their performance for the detection of cyber attacks.

5 Observations and Future Directions

RNNs are able to effectively use text content or information between in and out sequences. The limitation of these networks that they can store only a limited range of information that creates diminishing influence of hidden layers. The Long short time memory neural network is able to improve long-term dependency and vanishing gradient problems. There are different LSTM variants used (Fig. 3) to enhance the classification performance.

In Table 1 it is analyzed that there are different RNN models were proposed in literature but due to the vanishing gradient effect of RNN model, LSTM, GRU which are RNNs variants shows better performance for cyber attack detection system.

Future scopes are RNN based deep networks can be applied to real time attack detection or some new attack benchmarks. We found in various research papers that RNNs based deep networks and its variants are able to performed better for attack detection system hence need to be more explored.

Table 1 Comparative study on the basis of performance among recurrent neural networks used for attack detection

Author and year	Technique	Performance	Datasets	Comparative techniques
Radford et al. [1]	RNN	**Accuracy** is 84%	ISCX IDS	–
Yin et al. [9]	RNN	**Accuracy**: Dos is 84.49% Probe is 83.40% R2L is 24.69% U2R is 11.50% **Detection rate**: Dos is 2.06% Probe is 2.16% R2L is 0.80% U2R is 0.70%	NSL KDD	–
Kim and Kim [11]	RNN	Detection rate is 95.37% FAR is 2.1%	KDDCUP99	–
Kim et al. [12]	RNN-LSTM	Detection rate is 98.88 FAR is 10.04 Accuracy is 96.93	KDDCUP 1999 and additional, original data	GRNN, PNN, RBNN, KNN, SVM, Bayesian
Staudemeyer [15]	RNN	Accuracy is 93.82% Cost is 22.13	KDD 1999	–
Krishnan and Raajan [13]	RNN	Probe attack Recall 97.8 Precision 88.34 DoS attack Recall 97.05 Precision 99.9 U2R attack Recall 62.7 Precision 56.12 R2L attack Recall 28.81 Precision 94.1	KDDCUP99	C4.5 (decision tree), conditional random field
Tchakoucht et al. [10]	RNN + ML-ESM	Detection rate is 97.90 FPR is 0.60	DARPA KDD'99,	RNN, ELM, SVM-PSO, association rule, ANN, CFA, RNN, ensemble learning
Tchakoucht et al. [10]	RNN + ML-ESM	Detection rate is 83 FPR is 3.30	NSLKDD	RNN, ELM, SVM-PSO, association rule, ANN, CFA, RNN, ensemble learning

(continued)

Table 1 (continued)

Author and year	Technique	Performance	Datasets	Comparative techniques
Tchakoucht et al. [10]	RNN + ML-ESM	Detection rate is 98 FPR is 5.10	UNSW NB 15	RNN, ELM, SVM-PSO, association rule, ANN, CFA, RNN, ensemble learning
Fu et al. [5]	RNN-LSTM	Detection rate 98.85 FAR is 8.75 Accuracy is 97.52	NSL KDD	GRNN, PNN, RBNN, KNN, SVM, BAYESIAN
Chen et al. [14]	Bi-LSTM	**Bi LSTM** Precision is 0.93 Recall is 0.89 $F1$ score is 0.91 **LSTM** Precision is 0.85 Recall is 0.88 $F1$ score is 0.86	UNSW-NB15	LSTM
Pranitha et al. [16]	GRU	Accuracy is 89.22%	NSL KDD	SVM, KNN, NB, RF, ANN, RNN, LSTM

6 Conclusion

This paper presents the literature review on recurrent neural network to design the architectures for attack detection systems. The application of Deep learning techniques in cyber security is very popular now days. Here discussing the latest literature articles of RNN and represent by comparative study about its contribution in detection of cyber attack. Also discussing the advanced datasets and different parameters to evaluate the performance of the attack detection model. It is found that the RNN variants networks are more effective for the attack detection system. It is also analyzed that due to the better performance of RNN networks can be explored more in cyber security domain.

References

1. Radford, B.J., Apolonio, L.M., Trias, A.J., Simpson, J.A.: Network traffic anomaly detection using recurrent neural networks. In: Proceedings of the 2017 National Symposium on Sensor Data and Fusion (2017)
2. Dixit, P., Silakari, S.: Deep learning algorithms for cybersecurity applications: a technological and status review. Comput. Sci. Rev. **39** (2021)
3. Hodo, E., Bellekens, X., Hamilton, A., Tachtatzis, C., Atkinson, R.: Shallow and Deep Networks Intrusion Detection System: A Taxonomy and Survey. Jan (2017). (Eprint) arxiv:1701.02145

4. Ferraga, M.A., Maglaras, L., Moschoyiannis, S., Janicke, H.: Deep learning for cyber security intrusion detection: approaches, datasets, and comparative study. J. Inf. Secur. Appl. **20** (2020)
5. Fu, Y., Lou, F., Meng, F., Tian, Z., Zhang, H., Jiang, F.: An intelligent network attack detection method based on RNN. In: 2018 IEEE Third International Conference on Data Science in Cyberspace
6. Salehinejad, H., Sankar, S., Barfett, J., Colak, E., Valaee, S.: Recent advances in recurrent neural networks (2018). arxiv:1801.01078
7. Kim, G., Yi, H., Lee, J., Paek, Y., Yoon, S.: LSTM-Based SystemCall Language Modeling and Robust Ensemble Method for Designing HostBased Intrusion Detection Systems (2016). arxiv:1611.01726
8. Yu, Y., Si, X., Hu, C., Zhang, J.: A review of recurrent neural networks: LSTM cells and network architectures. Neural Comput. **31**, 1235–1270 (2019)
9. Yin, C., Zhu, Y., Fei, J., He, X.: A Deep Learning Approach for Intrusion Detection Using Recurrent Neural Networks. IEEE (2017)
10. Tchakoucht, T.A., Ezziyyani, M.: Multilayered echo-state machine: a novel architecture for efficient intrusion detection. IEEE Access **6**, 72458–72468 (2018)
11. Kim, J., Kim, H.: Applying recurrent neural network to intrusion detection with Hessian free optimization. In: International Conference on Information Security Applications, Jeju Island, Korea, pp. 357–369 (2015)
12. Kim, J., Kim, J., Thu, H.L.T., Kim, H.: Long short term memory recurrent neural network classifier for intrusion detection. In: 2016 International Conference Platform Technology and Service (PlatCon), Jeju, Korea, 15–17 Feb 2016, pp. 1–5
13. Krishnan, R., Raajan, N.R.: An intellectual intrusion detection system model for attacks classification using RNN. Int. J. Pharm. Technol. **8**, 23157–23164 (2016)
14. Chen, W., Yang, S., Wang, X.A., Zhang, W., Zhang, J.: Network malicious behavior detection using bidirectional LSTM. In: CISIS (2018)
15. Staudemeyer, R.C.: Applying long short-term memory recurrent neural networks to intrusion detection. S. Afr. Comput. J. **56**, 136–154 (2015)
16. Pranitha, G., Kiran Mahesh Reddy, D., Deepika, B., Alekhya, G., Vennela, Ch.N.: Intrusion detection system using gated recurrent neural networks. Paideuma J. (2020)

Effects of Climate Change on Agriculture Productivity: An Exploratory Statistical Study with Small Data Set Neural Networks

Domenico Vito

1 Introduction

The Food and Agriculture Organization of the United Nations (FAO) calculated that 815 million people around the globe are daily affected by chronic hungry [1].

Almost of the 80% of the world low-income persons are considered to live in rural districts and rely mostly from agriculture, fisheries or forestry as main origin of income and food. On this scenario, the continuous rise of temperature due to climate change will threaten the progress towards eradicating hunger and ensuring the sustainability of our natural resource base to achieve the 2030 Agenda for Sustainable Development.

Agricultural adaptation has become crucial in hot spot regions of climate change, and agro-climatic model to estimate the crop productivity can help to address policies and intervention from national to local level. Samely, food security and nutrition are parts of the Sustainable Development Goals, particularly on SDG 2 and SDG 3, and they cannot be addressed without coping with climate change.

This work is proposing an exploratory study to estimate the impact of climate change on agricultural activities, particularly regarding the effects of temperature rise, comparing statistical and machine learning techniques. Such studies can support policies towards more sustainable development strategies.

D. Vito (✉)
Politecnico di Milano, Piazza Leonardo da Vinci 32, 20133 Milan, Italy
e-mail: domenico.vito@polimi.it

M. Dua et al. (eds.), *Proceedings of the International Conference on Paradigms of Communication, Computing and Data Sciences*, Algorithms for Intelligent Systems,
https://doi.org/10.1007/978-981-16-5747-4_69

2 Preliminary Concepts and Background

2.1 The Effects on Climate Change and Temperature on Agriculture

Climate change can have a harmful impact on food security at the global, regional and local level. It could affect food availability, reduce access to food and affect food quality. Standing to the special IPCC Special Report on the impacts of global warming of 1.5 °C [2], human activities are estimated to be the cause approximately 1.0 °C of global warming above pre-industrial levels (1990), with a rough range of 0.8–1.2 °C.

Global warming will go to reach 1.5 °C between 2030 and 2052 if it continues to increase at the current rate.

Increases in temperature and carbon dioxide can alter the normal production some crop yields in some places, furthermore it can contribute to spoilage and contamination [3–5].

Possible effects of climatic change on agriculture can be connected to the three main factors, namely

1. the increase of atmospheric CO_2 concentration,
2. the variations in temperature, precipitation and insolation and
3. the increase in sea levels with a significant reduction of the extension of agricultural areas and of the salinity of groundwater in coastal areas [6].

The impacts of climate change on agriculture affect not only the ecosystems but also the connected human activities: changes in the frequency and severity of droughts and floods hardly threaten farmers and ranchers regarding their food safety.

In general, climate change makes harder to grow crops and to cultivate livestocks ways and same places that were usual in the past.

2.2 Effect of Climate Change on Single Crops

For each particular crop, the increased temperature effects in a different way in relation to the crop's optimal temperature for growth and reproduction.

Warming may benefit in some zones, depending on the types of crops that are typically planted there, or allow farmers to shift to unconventional crops that are usually cultivated in warmer areas.

On the other hand, if a crop's optimum temperature is exceeded by the increased temperature, yields are going to decline [7]. Beside the temperature, also higher CO_2 levels can alter crop yields. Some laboratory experiments also suggested that plant growth levels can be increased by elevated CO_2 levels.

The effects of direct increases in atmospheric CO_2 would on the whole be positive, if they were not associated with the direct consequences of climatic variations.

In fact, a doubling of the CO_2 can determine an increase in the photosynthetic rate between 30 and 100%, depending on temperature levels and water availability [4, 6], but not for all species. Type C3 species (wheat, rice, soy, etc.) have a positive response to high concentrations of CO_2. Type C4 species (maize, sorghum, sugarcane, millet, etc.), although more photosynthetically efficient, are less responsive to the increase in CO_2 concentration. This difference in behaviour between C3 and C4 plants is due a different strategy in carbon fixation [8]. A further effect of the increase in CO_2 is on the efficiency of water use: in fact, an environment with a high concentration of CO_2 causes a decrease in the stomatal opening and consequently a reduction in transpiration per leaf area unit.

Other stressors and combination of them, such as changing temperatures, water and nutrient constraints and ozone can contrast such potential increases in yield.

For instance, if temperature exceeds a crop's optimal level in a condition of water scarcity, yield increases may be reduced or reversed. It has been demonstrated that elevated CO_2 is linked to reduced protein and nitrogen content in alfalfa and soybean plants [9], resulting in a loss of quality. Reduced grain and forage quality falls in to weaken the ability of pasture and rangeland to keep up grazing livestock.

Together being affected by climate crisis agriculture is, on the other hand, one of the sectors that can have an impact on GHG emission [10, 11].

Agriculture and forestry indeed could be both a source and a since of GHG emissions. Agriculture's greenhouse gas emissions are on continuous rising although not as fast as emissions from other human activities.

Globally, agriculture stands to face the triple challenge of increasing production in response to the growing food demand, adapting to climate crisis but in the same way having the potential in reducing agricultural greenhouse gas (GHG) emissions [11].

National agricultural policies have to incorporate these challenge as a mandate, having the tackle of climate change as cardinal point. On this objective, extensive studies on the macro-effect of climate change on high-scale productivity can help in better support decision-making processes.

3 Methods

The phrase "computational experiment" suddenly reminds to a comparison with classical "physical experiment": the computational experiment can be defined as simulation that are used as just extracted information such as data or models.

More precisely, a computational model is a model of the physical or rather complex system that first is expressed in a mathematical form to represent the main relationships among explanatory variables and then it is implemented in the form of a computer programme: it may be viewed as a function of inputs that when evaluated produced outputs [12]. The sake of this work has been to understand the impact of climate change on farming and cropping thought a computational experiment. The starting point has been the definition of the basic inputs which have relevant impacts on an output considered [12].

3.1 AquaCrop Model

AquaCrop is a water productivity model for the simulation of the aboveground biomass production in relationship to the water transpired exchanged by the crop [13].

The model gets local weather data [precipitation, minimum and maximum temperature, reference evapotranspiration (ETo)] and assesses the daily crop growth. The core equation that describes the of the AquaCrop growth model is

$$B_n = \mathrm{WP}^* \sum_{(i=1)}^{n} \frac{\mathrm{Tr}_i}{\mathrm{ET_o}i} \quad (1)$$

where B_n is the cumulative aboveground biomass production after n days; Tr_i is the daily crop transpiration (mm/day), $\mathrm{ET_o}i$ is the daily reference evapotranspiration, and i represents the sequential days of the period in which B_n is produced; WP* is the normalized crop water productivity (g m^{-2}).

Transpiration (T_r) is calculated as a function of the Canopy cover (CCi); for this reason, CC is a key variable for the yield estimation. AquaCrop model calculates CC as a function of four sets of main parameters and driving variables depending on weather data, crop type, soil and initial conditions of soil water content [14]. The main output is the crop yield at physiological maturity. This variable represents the culmination of all crop growth processes, in function of soil and climate characteristics. The result can be combined with the presence of different crops in the same cultivated area.

3.2 Data Sets

The input data for the computational models has been gathered mainly from FAOSTAT, The Food and Agriculture Organization Corporate Statistical Database (FAO), is a web-portal that collects and manages data from the Food and Agriculture Organization. FAOSTAT [15] allows free access to food and agriculture data for over 245 countries and territories. It collects all FAO regional groupings starting from year 1961 to the most recent year available. The repository contains country indicators, data on population, food production, investment, trade, agri-environmental indicators, forestry, emissions and so forth. All the data can be downloaded, and the functions for cross-crosscutting data analysis are integrated into the platform.

FAOSTAT database provided data related to productivity, productivity indexes and temperature rise that can be feasible to further analyses.

Production data on cereals crops pertains to the harvesting for dry grain only: hay, animal food and grazing components are excluded.

The FAO indices of agricultural production show the relative level of the aggregate volume of agricultural production for each year compared with the baseline period set to 2004–2006. These indexes are the sum of price-weighted quantities for each agricultural commodity that is produced after deductions of quantities used as seed and feed. These lasts are weighted in a similar manner. All the indices at the different governance level (country, regional and global) are calculated though the Laspeyres formula. Production quantities related to each commodity are weighted on the average international commodity prices for 2004–2006 and summed for each year. The temperature change data and agri-environmental indicators collect data on observed mean surface temperature changes by country, for the time lapse between 1961 and 2017 and are updated annually. The data relates on monthly, seasonal and annual mean temperature anomalies that are the temperature variations with respect to the baseline years that are 1951–1980. The used standard is GISTEMP, the Global Surface Temperature Change data of the National Aeronautics and Space Administration Goddard Institute for Space Studies (NASA-GISS). The temperature change domain indeed provides member countries with climate change relevant statistics in support of national and regional agri-environmental analysis. Data analysis has been conducted with the support of MATLAB and Orange software.

3.3 Agro-climatic Zones

The climatic and atmospheric conditions for the crop growth analysis have been evaluated considering the agro-climatic zones.

The agro-ecological zones (AEZ) methodology has been over the past 30 years by Food and Agriculture Organization of the United Nations (FAO) and the International Institute for Applied Systems Analysis (IIASA) for assessing agricultural land monitoring, potential and resources.

Global agro-ecological zones (GAEZ) are the methodology for assessing global land resources as well as a spatial database.

The spatial database contains 226,225 data layers in total at resolutions varying between 30 arc-seconds and 5 arc-minutes, several terabytes of data.

The access to data is granted through the GAEZ data portal, an interactive facility for free data access. The interface permits data visualization and provides various analysis outputs and download options to the user. The GAEZ portal also shows for the main crops the yield and production gaps, in terms of ratios and differences between actual yield and production and potentials reaching a number of 120 data layers as a whole. The platform also offers additional agro-climatic resource, crop suitability and potential yield themes with reference to the baseline period of 1961–1990 and individual historical years and future climate projections based on various GCMs and IPCC AR4 SRES emission scenarios.

3.4 Description of the Computational Experiment Algorithm

The main structure of the computational experiment has been organized taking inspiration from AquaCrop approach So taking into account the data of FAOSTAT, we considered the productivity of three important crops that characterize the diverse areas with diverse environment and on varying meteorological conditions.

The three elected crops are maize (*Zea mays* L.), winter wheat (*Triticum aestivum* L.) and rice (*Oryza sativa* L.). The productivity of these three crops has been analysed for three chosen countries corresponding to three main climatic areas and geographical zones, as described in Table 1.

A machine learning simulation on how the influence of climatic conditions could affect the productivity together with the temperature has been performed.

This simulation considered the problem of state productivity as a multivariate problem on which the input features are in the schema below (Fig. 1).

Table 1 Design of the analysis of temperature and productivity series for different climatic areas

Crop	Chosen county	Geographical zone	GAEZ
Winter wheat	Italy	Northwestern Europe	Subtropic, cool cold
Maize	Malawi	South East-Asia	Tropic, cold
Rice	Indonesia	Southern Africa	Tropic warm

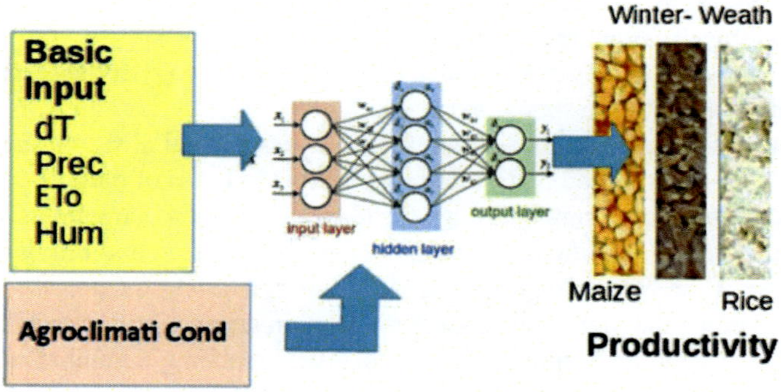

Fig. 1 Multivariate data flow for the analysis of crop productivity

3.5 NN Small Data Set Machine Learning Approach

The preliminary analysis was performed to consider a first orientation correlation.

Indeed it was more probable that the dependence among the input and the output features was nonlinear. To better describe the nonlinearities between the input features and the output features, an empirical data-driven machine learning approach based on neural networks was used.

More specifically, the model of "multilayer perceptron" with one hidden layer and one output layer proposed by Pasini et al. [16] has been implemented. The model is quite effective for studying cause–effect links in complex systems, with the aim of reconstructing the behaviour of a target variable by knowing the trends of other (causal) variables that for hypothesis are supposed to drive its evolution.

The basic elements of the network are the connections, each one having its associated weights (w_{jk} and W_{ij}), and the neurons of the hidden and the output layers, that are the single computational units.

In the single layer version, each neuron is connected to all others of the previous and the following layer, and there are not connections between neurons that belong to same layer. Mathematically, it can be considered as in (2).

$$O_i^{\mu} = f_i\left(\sum_i W_{ij} V_j^{\mu}\right) = f_i\left[\sum_j W_{ij} g_j\left(\sum_k w_{jk} I_k^{\mu}\right)\right] \quad (2)$$

4 Results

The small neural network prediction has been performed following the data flow in Fig. 3. Particularly, meteorological year mean temperature increase data, crop production and production indexes have been taken from FAOSTAT; indeed, reference evapotranspiration and precipitation data has been taken from global agro-ecological zones (GAEZ) (Fig. 2).

The NN has been implemented considering the following feature sets (Table 2).

The NN has been trained on the whole knowledge base of country data and tested on a subset of it with the small data set principles of Pasini et al. [16].

Indeed, the model was applied for the prediction of the national production data for each of the case studies. The software Orange@ [17] has been used to support the elaborations, in order to easily manage the elaborated data flow for the implementation. Table 3 resumes the scores for the model testing.

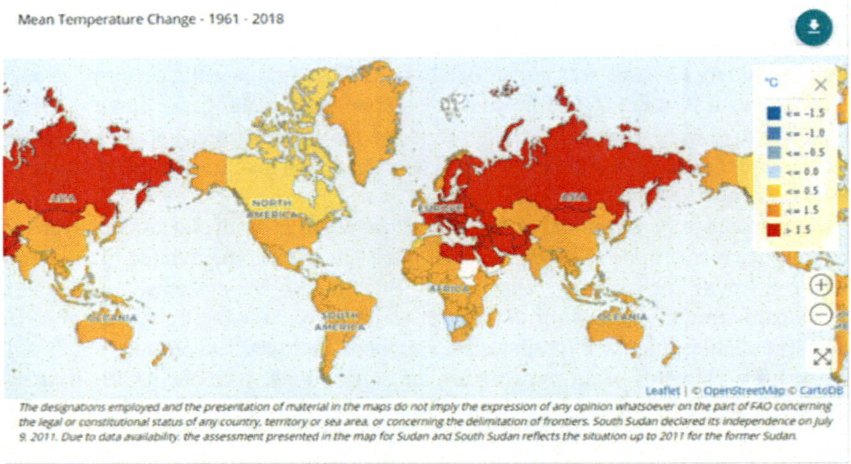

Fig. 2 Global mean temperature increase map

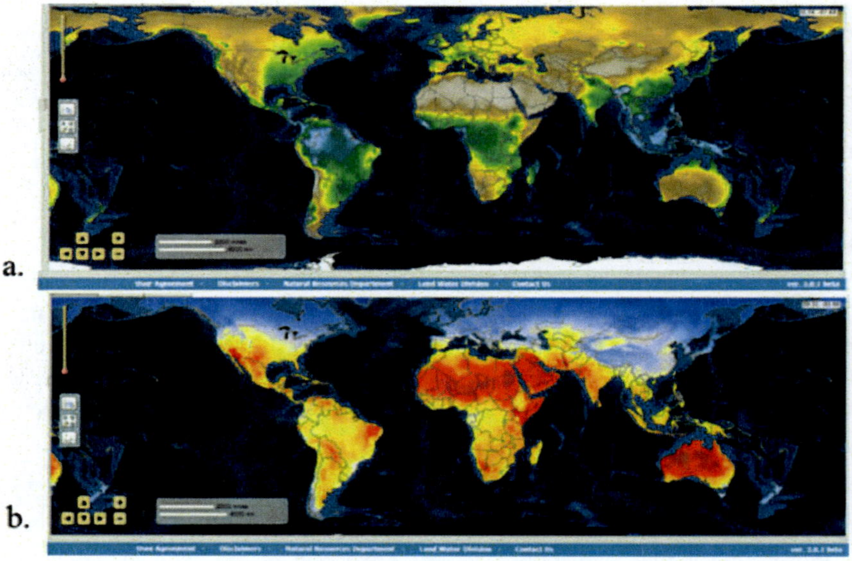

Fig. 3 Global precipitation (mm) (**a**) and ref. evapotranspiration (**b**) maps from GAEZ

Table 2 Feature set for the NN prediction

Input features	Output features
Meteorological year mean temperature	Crop production (maize, wheat or rice)
Precipitation weighted mean	Cereals indexes
Reference evotranspiration weighted mean	Crop indexes
Methodological year mean temperature	
Precipitation weighted mean	

5 Conclusions

Climate crisis is, at the same time, the burden and the challenge of our time. Agricultural activities are one of the most affected sectors by climate crisis.

As they are at the baseline of human societies, the primary effects on this sector will be reverberated to all other ones. Furthermore, on the whole chain of the agricultural production, farmers are one among the most vulnerable economically and socially, of this reverberation. To face the climate crisis indeed is essential to work on planning for the national to the local level with the best available science and technology. Today there a lot of knowledge that just working separately cannot effect policy decisions in a long-term impactful way. Data analysis and data-driven modelling can indeed support policy-makers to have a more reliable long-term vision which can be final results of their choices. Furthermore, the data-driven approach can help to integrate different dis-aggregate information into a more effective knowledge base, which offers inputs and outputs for the prediction of agricultural productivity at different scales. The presented study has offered an exploratory example on how to predict and analyse multivariate data sets in order to understand the effect of climate change phenomena on crop productivity. As exploratory study, it has a constrained field of applicability: several assumptions have been made like the invariance of crop growth parameters with time and condition as well as several simplifications in the machine learning model that is fitted to small data sets for the current application. The very scope indeed of this study has been to demonstrate and set up a "computational experiment" to better assess the effect of the rising temperatures on yield production and to assess a national level impact of the phenomena for the early stage of policy definition. As a first proof of concept, a great margin of improvement in the complexity of the set-up is open, even though reminding that data-driven models cannot stand alone, as they are just as a useful tool in support to cause–effect relationships. Policy implementation cannot prescind in any case from field studies and context evaluations.

Table 3 Validation of NN though accuracy (ACC) precision (PRC) RCL and AUC from the ROC curve

		A. Italy				B. Malawi				C. Indonesia			
	Production index	ACC%	PRC%	RCL%	AUC%	ACC%	PRC%	RCL%	AUC%	ACC%	PRC%	RCL%	AUC%
	Item	Wheat				Maize				Rice			
	Production	77.5	88.5	86	91	87.5	89.8	90	97	97.4	99.8	92	94
Cereals	Gross production index number	80.0	97	99	95	82.2	67	70	99	92.5	87	89	97
	Gross per capita production index number	90.0	90.7	80	20.2	93	94.8	85	90.5	95.0	96.8	95.8	97.5
	Net production index number	89.4	89.0	88.0	97	90.4	99.0	84.0	95	98.9	92.7	85.6	67
	Net per capita production index number	95.3	98.0	88.0	90.5	96.3	98.6	98.0	97.5	85.9	88.9	78.8	89.6
Crop	Gross production index number	98.3	89.2	89.3	98.2	78.3	89.7	97.3	99.2	99.3	97.7	94.7	93.3
	Gross per capita production index number	95.6	89.0	89.2	89.0	92.6	87.5	89.7	93.0	97.7	98.5	90.7	68.0
	Net production index number	67.9	88.5	89.7	60.4	97.9	90.5	91.7	95.4	98.8	98.0	89.9	95.6
	Net per capita production index number	90.9	99.9	90.9	89.3	96.9	98.9	97.5	90.8	89.9	68.9	77	89.0

References

1. FAO's work on climate change. United Nations Climate Change Conference 2017. http://www.fao.org/3/a-i8037e.pdf. Last accessed 02/03/2021
2. IPCC: Global warming of 1.5°C. An IPCC special report on the impacts of global warming of 1.5°C above pre-industrial levels and related global greenhouse gas emission pathways, in the context of strengthening the global response to the threat of climate change, sustainable development, and efforts to eradicate poverty (2018)
3. Wang, E., Martre, P., Zhao, Z., Ewert, F., Maiorano, A., Rötter, R.P., Asseng, S.: The uncertainty of crop yield projections is reduced by improved temperature response functions. Nat. Plants **3**(8), 1–13 (2017)
4. Liu, S., Mo, X., Lin, Z., Xu, Y., Ji, J., Wen, G., Richey, J.: Crop yield responses to climate change in the Huang-Huai-Hai Plain of China. Agric. Water Manag. **97**(8), 1195–1209 (2010)
5. Polley, H.W.: Implications of atmospheric and climatic change for crop yield and water use efficiency. Crop Sci. **42**(1), 131–140 (2002)
6. Pearch, R.W., Bjorkman, O.: Physiological effects. In: Lemon, E.R. (ed.) CO_2 and Plants, the Response of Plants to Rising Levels of Atmospheric CO_2. Westview Press, Boulder, Colorado (1983)
7. Climate impacts on agriculture and food supply. https://19january2017snapshot.epa.gov/climate-impacts/climate-impacts-agriculture-and-food-supply_.html. Last accessed 02/03/2021
8. Morrison, J.I.L.: Interactions between increasing CO2 concentration and temperature on plant growth. Plant Cell Environ. **22**(6), 659–682 (1999)
9. Alvar-Beltrán, J., Dao, A., Marta, A.D., Saturnin, C., Casini, P., Sanou, J., Orlandini, S.: Effect of drought, nitrogen fertilization, temperature, and photoperiodicity on quinoa plant growth and development in the Sahel. Agronomy **9**(10), 607 (2019)
10. Ziska, L., Crimmins, A., Auclair, A., DeGrasse, S., Garofalo, J.F., Khan, A.S., Loladze, I., Pérez de León, A.A., Showler, A., Thurston, J., Walls, I.: Ch. 7: food safety, nutrition, and distribution. In: The Impacts of Climate Change on Human Health in the United States: A Scientific Assessment. U.S. Global Change Research Program, Washington, DC (2016)
11. FAO (Food and Agriculture Organization): The water-energy-food nexus: a new approach in support of food security and sustainable agriculture. http://www.fao.org/policy-support/resources/resources-details/en/c/421718 (2014). Last accessed 02/03/2021
12. Morris, M.D.: Factorial sampling plans for preliminary computational experiments. Technometrics **33**(2), 161–174 (1991)
13. Silvestro, P.C., Casa, R., Pignatti, S., Castaldi, F., Yang, H., Yang, G.: Development of an assimilation scheme for the estimation of drought-induced yield losses based on multi-source remote sensing and the AquaCrop model. In: 2014 Dragon 3 Mid-Term Results International Symposium, 26–29 May, Chengdu, PR China (2014)
14. Steduto, P., Hsiao, T.C., Raes, D., Fereres, E.: AquaCrop—the FAO crop model to simulate yield response to water: I. Concepts and underlying principles. Agron. J. **101**(3), 426–437 (2009)
15. Olinelli, F., Gianinetto, M., Nana, E.: Potenziali impatti dei cambiamenti climatici a scala locale: il caso dell'Himalaya Nepalese (+ qualcosa sull'Italia) (2010)
16. Pasini, A.: Artificial neural networks for small dataset analysis. J. Thorac. Dis. **7**(5), 953–960 (2015)
17. Demšar, J., Zupan, B., Leban, G., Curk, T.: Orange: from experimental machine learning to interactive data mining. In: European Conference on Principles of Data Mining and Knowledge Discovery, pp. 537–539. Springer, Berlin (2004)

Prediction of Stock Value Using Recurrent Neural Network

Jayant Dhingra, Abhinav Sharma, and Rashmi Arora

1 Introduction

Timely prediction of stock prices can prove to be very profitable for individuals and organisations. Stock market is generally full of uncertainty and the stock prices tend to fall and rise all the time. There are a lot of variables in a price of a stock, due to which a lot of people fear before investing in the stock market. An individual or an organisation can gain a considerable amount of profits if they invest in the right stock at the right time.

Due to the complex nature of the stock market, machine learning can prove to be a very efficient way to find out which stocks can be profitable and which ones to avoid. It can price stock with an interesting efficacy. Due to this exceptional precision many researchers and stock market investors find machine learning techniques appealing when it comes to predicting the stock prices.

The dataset used in while predicting is an indispensable step because the accuracy of the stock price prediction mostly depends on the sturdiness of the dataset. A little alteration in the dataset can result in huge variations in the outcome of a ML model. In the proposed LSTM-RNN model, we used the dataset which was taken from a pharmaceutical company named Sun Pharmaceutical Industries Ltd. Some of the variables of this dataset include open value, close value, highest value, etc.

Regression and LSTM model were employed in calculating the stock price accurately. LSTM is one of the most successful RNN architectures and is capable of effectively neglecting memory that it considers to be not pertinent, making it very

J. Dhingra (✉) · A. Sharma
Department of Electronics and Communication, Guru Tegh Bahadur Institute of Technology, New Delhi, India

R. Arora
Department of Information and Technology, Guru Tegh Bahadur Institute of Technology, New Delhi, India

M. Dua et al. (eds.), *Proceedings of the International Conference on Paradigms of Communication, Computing and Data Sciences*, Algorithms for Intelligent Systems,
https://doi.org/10.1007/978-981-16-5747-4_70

effective. In the results, the graph of predicted values is plotted with the real values for comparison.

2 Related Works

Jayesh Potdar and Rejo Mathew put forward a review paper which compared various machine learning algorithms, some of them being support vector regression (SVR), improvised Levenberg Marquardt Algorithm (LMA) and others. Their analysis gave a precise price trend [1] which assisted in stock price prediction.

Sukhman Singh, Tarun Kumar Madan, Jitendra Kumar and Ashutosh Kumar Singh collectively presented distinct machine learning algorithms such as random forest, boosted decision tree and others along with a few hybrid methods to predict the stock prices for different stock exchanges [2] across the globe. Their work also covered various challenges that were encountered while building these models.

Ishita Parmar, Navanshu Agarwal and others presented a paper which focused on LSTM hinged machine learning and regression which helped in prediction of stock values quite accurately. Their approach considered a variety of stocks features such as open, close, high and volume. Their LSTM-based model resulted in a Train Score of 0.00106 MSE (0.03 RMSE) and a Test Score of 0.00875 MSE (0.09 RMSE) [3].

Meghan Misra, Ajay Prakash and Harkiran Kaur et al. have proposed methods which focus on categorising various methods which further help in predictive analysis in various domains [4]. The authors also suggested some enhancements and improvements that should be included to achieve overall better results and more robust models.

Wasiat Khan, Mustansar Ali Ghazanfar and others put forward a very invaluable work in which they compared various algorithms and applied them on social media and financial news data and discovered the impact of this data on stock market prediction accuracy for ten consecutive days. They compared a lot of classifiers to find the most efficient and robust classifier. They finally used deep learning classifiers which gave prediction accuracies of 80.53% and 75.16% are achieved using social media and financial news. Their work also suggested that the Random forest classifier is found to be consistent and the highest accuracy of 83.22% is achieved by its ensemble [5].

Naadun Sirimevan, I.G.U.H. Mamalgaha and others used an LSTM–RNN model to fill the information gap and predict an accurate future value. They experimented with Weighted Average and Differential Evolution techniques which gave extremely accurate predictions to one-day, seven-days, fifteen-days and 30 days for the future [6].

Ashish Sharma, Dinesh Bhuriya and Upendra Singh et al. have proposed a survey of a well-known efficient regression approach which successfully predicted the stock market price from the typical stock market data [7]. Their work can further be enhanced by using more variables.

Radu Iacomin et al. have proposed various studies with different ML algorithms such as ANN with different feature selections. Their result showed that support vector machines (SVM) with the help of feature selection principal component analysis (PCA) will have the success of making a profit [8].

Sumeet Sarode and others have put forward a unique approach in which they combined two distinct fields for analysis of the stock exchange. Their proposed system combines price prediction which is based on historical and real-time data. Lon Short-Term Memory is used for the prediction. Only the relevant news is collected and the filtered news is analysed to predict the sentiment around various companies. Their results provide responses which help in giving recommendations for future increases in profits [9].

Kunal Pahwa and Neha Agarwal et al. have proposed a model which uses pre-existing open source libraries and algorithms to help predict stock prices. Some of the techniques include Linear Regression and other supervised machine learning techniques [10].

Pawee Werawithayaset and Suratose Tritilanunt proposed an idea that reduced investment risks by predicting stock prices. Their experiment showed that Partial Least Square is the best algorithm of the three algorithms that were compared to predict the stock closing price [11].

2.1 Related Work Comparison and Conclusion

The purpose of this literature survey conclusion was to adumbrate the trend and advancement in the methods and accuracies which range from 68 to 83.22% and the method proposed in this paper attained accuracy of 92.6% (Table 1).

Table 1 Comparison of proposed method with literature

Reference number	Title	Method used	Inference
[3]	Stock market prediction using machine learning	Regression and LSTM based model	Test score of 0.00875 MSE (0.09 RMSE)
[5]	Stock market prediction using machine learning classifiers and social media, news	Random forest classifier and its ensemble	83.22% accuracy achieved
[8]	Stock market prediction	SVM with PCA	PCA with SVM gave 68% rate of recognition
Our model	Prediction of stock value using recurrent neural network	RNN and LSTM	Accuracy of 92.6% using tensorflow 2.0

Date	Open	High	Low	Close	Adj Close	Volume
2016-02-18	869.700012	876.500000	858.099976	870.450012	840.284668	2342525.0
2016-02-19	870.400024	870.400024	857.000000	864.200012	834.251343	1561649.0
2016-02-22	871.950012	885.799988	868.000000	882.450012	851.868835	2302969.0
2016-02-23	884.700012	898.450012	866.250000	871.099976	840.912170	2470030.0
2016-02-24	866.349976	870.000000	849.599976	854.549988	824.935608	1716644.0

Fig. 1 Dataset table of sun pharmaceutical industries

3 Dataset

The dataset used in the model is of a pharmaceutical company named Sun Pharmaceutical Industries Ltd. which is an Indian multinational company and is registered in both National Stock Exchange (NSE) and Bombay Stock Exchange (BSE). The following dataset is fetched from the National Stock Exchange (NSE) website [12], which contains 7 columns and 1145 rows. Figure 1 shows some of the rows and columns of the data used in this model.

4 Proposed System

The proposed stock market prediction system was designed with the help of Long Short-Term Memory (LSTM) network. According to the designed model the dataset was first cleaned and pre-processed and visualisation was performed for the better understanding of the data. The data was divided into train and test set accordingly. After that five layers of LSTM were formed, and training data was compiled to train the machine and finally the results were noted and a comparison graph was formed is shown in Fig. 2.

5 Implementation

The implementation of this model is pretty straightforward as the dataset was first fetched from NSE website of Sun Pharmaceutical which is an Indian multinational company registered in NSE. The historical data of the past 5 years was fetched and analysed and visualised using various python libraries and it is cleaned afterwards. The dataset was further divided into test and train sets and it is scaled further using MinMax scaler. After that LSTM layers were formed, and the training data was

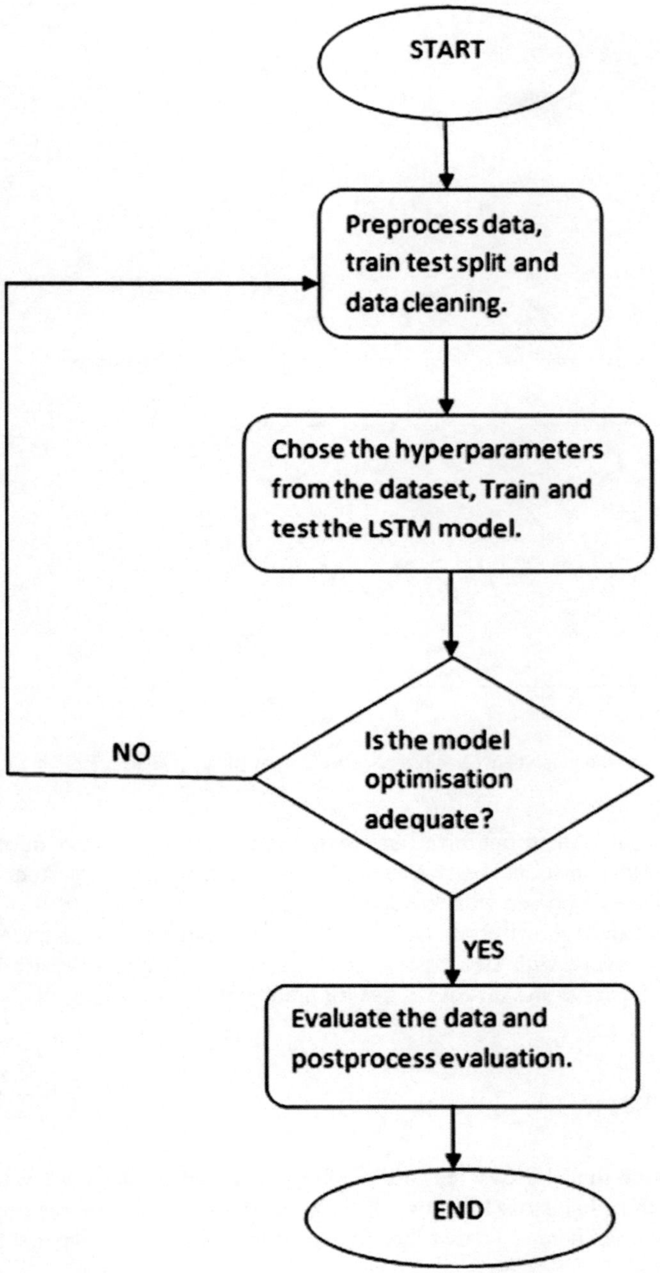

Fig. 2 Block diagram of LSTM model

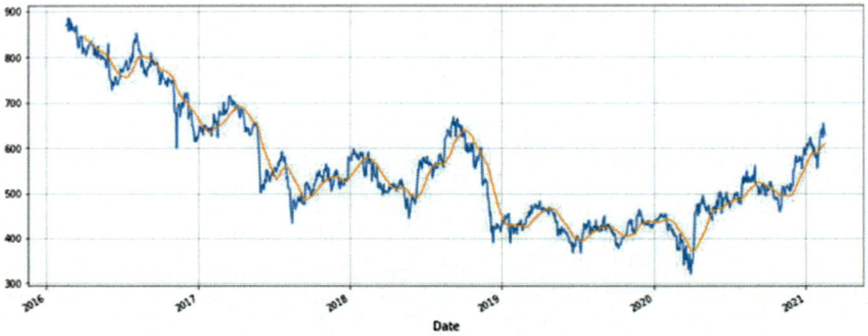

Fig. 3 Comparison graph between open price and rolling mean of close price

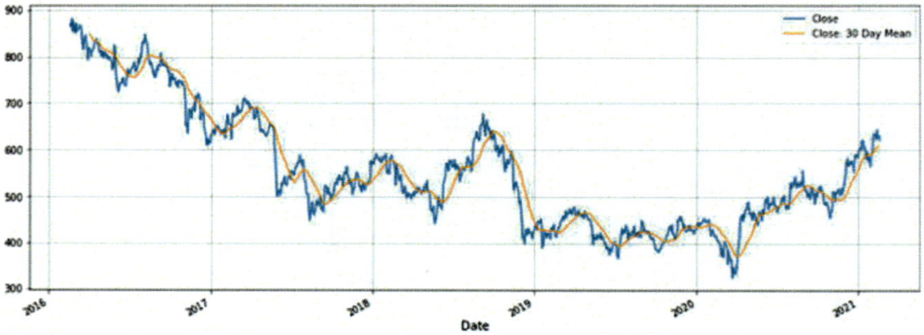

Fig. 4 Comparison graph between close price and rolling mean of close price

compiled using Adam optimizer and losses were calculated using mean squared error with 100 epochs and batch size of 32. In the end, the predicted stock price was calculated and compared with real stock price.

Figures 3 and 4 shown above are basically the comparison of Open price and close price, respectively, with their past 30 days rolling mean which is basically used to smoothen the curve and for ease in use for prediction.

6 Results

The proposed model uses a tensorflow library and predicts the stock value of Sun Pharmaceutical Industries Ltd. Long Short-Term Memory, a type of Recurrent Neural Network is used here to predict the value of stock price. The comparison graph is shown in Fig. 5.

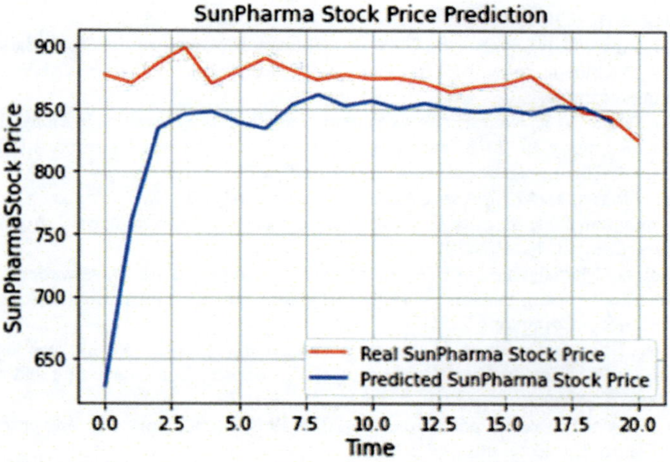

Fig. 5 Comparison curve between real stock price and predicted stock price

It is quite evident by Fig. 5 that the model shows minimal signs of overfitting and it is pretty usable for evaluation, and moreover, the model gives out the accuracy of 92.6% by taking 100 epochs and using tensorflow == 2.0.

7 Conclusion

The complex nature of the stock market makes it look petrifying for the investors. Its volatility makes people really apprehensive before investing in the stock market and choosing their stocks. Advancements in the field of machine learning can be used to solve this problem for all the people looking to invest in the stock market. Various machine learning algorithms can be used to predict the stock market prices quite accurately. The technique used in this paper is the robust and accurate LSTM-RNN model which was applied on the dataset which was taken from a pharmaceutical company named Sun Pharmaceutical Industries Limited which gave an impressive accuracy of 92.6% with minimal signs of overfitting and underfitting.

References

1. Potdar, J., Mathew, R.: Machine learning algorithms in stock market prediction. In: Proceeding of the International Conference on Computer Networks, Big Data and IoT ICCBI—2018, pp.192–197
2. Singh, S., Madan, T.K., Kumar, J., Singh, A.K., Singh, S., Madan, T.K., Kumar, J., Singh, A.K.: 2019 2nd International Conference on Intelligent Computing, Instrumentation and Control

Technologies (ICICICT)
3. Parmar, I., Agarwal, N., Saxena, S., Gupta, S., Dhiman, H., Chouhan, L.: Stock market prediction using machine learning. In: First International Conference on Secure Cyber Computing and Communications
4. Misra, M., Yadav, A.P., Kaur, H.: Stock market prediction using machine learning algorithms: a classification study. In: 2018 International Conference on Recent Innovations in Electrical, Electronics & Communication Engineering (ICRIEECE)
5. Khan, W., Ghazanfar, M.A., Azam, M.A., Karami, A., Alyoubi, K.H., Alfakeeh, A.S.: Stock market prediction using machine learning classifiers and social media, news. J. Ambient Intell. Humanized Comput. 1–24 (2020)
6. Sirimevan, N., Mamalgaha, I.G.U.H., Jayasekara, C., Mayuran, Y.S., Jayawardena, C.: Stock market prediction using machine learning techniques. In: 2019 International Conference on Advancements in Computing (ICAC)
7. Sharma, A., Bhuriya, D., Singh, U.: Survey of stock market prediction using machine learning approach. In: 2017 International Conference of Electronics, Communication and Aerospace Technology (ICECA)
8. Iacomin, R.: Stock market prediction. In: 2015 19th International Conference on System Theory, Control and Computing (ICSTCC)
9. Sarode, S., Tolani, H.G., Kak, P., Lifna, C.S.: Stock price prediction using machine learning techniques. In: 2019 International Conference on Intelligent Sustainable Systems (ICISS)
10. Pahwa, K.: Stock market analysis using supervised machine learning. In: 2019 International Conference on Machine Learning, Big Data, Cloud and Parallel Computing (COMITCon)
11. Werawithayaset, P., Tritilanunt, S.: Stock closing price prediction using machine learning. In: 2019 17th International Conference on ICT and Knowledge Engineering (ICT&KE)
12. National Stock Exchange India: https://www.nseindia.com/get-quotes/equity?symbol=SUNPHARMA. Accessed on 20/02/2021

Credit Card Fraud Detection: An Exploration of Different Sampling Methods to Solve the Class Imbalance Problem

Mythili Krishnan and Madhan Kumar Srinivasan

1 Introduction

Credit card fraud as the term denotes can refer to both credit card and debit card fraud, i.e., all payment cards. Any unauthorized transaction made through a credit/debit card without the knowledge of the cardholder is termed as fraud. Recent data from FTC has shown that credit card fraud has increased by 105% from Q1 2019 to Q1 2020 [1] and around $232 million was returned to customers as compensation for fraud in the USA. In India, a news report stated that India lost 128 crores in Fraud between Oct 2019 and Dec 2019 [2]. At an overall level, global payment losses have tripled to about $32.39 Billion from 2011 to 2020 and is projected to increase by 25% by 2027 [3]. Fraud through credit card can be conducted in many ways with the presence of the physical card (lost/stolen card or counterfeit cards) and card not being present, i.e., through Internet or virtual methods when the card might not be present with the fraudster physically. Fraudsters also took advantage of the COVID-19 situation and with an increase in Internet transactions there was also an increase in the ransomware attacks by 25% from Q4 2019 to Q4 2020 [3]. This type of fraud is mainly transactional as opposed to an identity take-over fraud where the fraudster impersonates the card holder and commits fraud. Hence, we will analyze transactional data mainly along with some demographic information for the detection methods.

The impact of fraud and its loss to business and customers is manifold. While customers can lose money, companies can incur financial loss, suffer decrease in customer satisfaction and loss of reputation. Lack of standard metrics, imbalanced

M. Krishnan (✉) · M. K. Srinivasan
Accenture, Bangalore, India
e-mail: mythili.krishnan@accenture.com

M. K. Srinivasan
e-mail: madhan.srinivasan@accenture.com

M. Dua et al. (eds.), *Proceedings of the International Conference on Paradigms of Communication, Computing and Data Sciences*, Algorithms for Intelligent Systems,
https://doi.org/10.1007/978-981-16-5747-4_71

data and cost of fraud detection outweighing the actual transaction amount and inefficiencies in dealing with new fraudulent patterns makes fraud detection a challenging problem. In this paper, we will illustrate different algorithms and compare their efficacy for credit card fraud detection, specifically we will look at the different sampling methods that can be used for fraud detection and discuss their merits and demerits.

2 Literature Review

In traditional or canonical machine learning algorithms, the inherent assumption is that the number of objects or data points will be similar in all classes. One possible way to mitigate this in machine learning is through backward propagation in Neural Networks. Bartosz Krawczyk [4] has illustrated three ways of tackling this—data level methods, i.e., modify the sample, algorithm level methods, i.e., modify the existing algorithms or hybrid method—a combination of the former two. In his paper, Bartosz also talks about the data imbalance problem and its evolution with the advent of big data and extension of machine learning. Fraud detection is an extreme imbalanced problem where the imbalance can range from 1:1000 to 1:5000 where minority class is poorly represented because frauds are too few in numbers compared to the genuine transactions.

The data-based methods and model/algorithm-based methods are also cited by Abdi et al. [5] where they have discussed a new model-based method centering on the concept of Mahalanobis distance. By this method, synthetic samples are generated, while keeping the covariance structure of the minority class intact. Another important aspect in fraud detection is the misclassification rate and a successful/good model will have lower false positives and false negatives. Abdi et al. [5] also touch upon this and illustrate how the risk of overlapping is reduced with their method of using Mahalnobis distance. They also elucidate the various methodologies that can be deployed to deal with the class imbalance problem using concepts like active learning and also discuss the other related problems of small sample and class overlap which is especially a deterrent in classification problems like ours. The problem can be looked at as a supervised as well as an anomaly detection problem. Yu and Wang [6] talk about outlier detection as a method for fraud detection, and they have used distance sum algorithms for this. The distance is calculated between observed value and pre-determined values for attributes that capture a customer's behavior.

Fraud detection methods using Neural Networks is also a popular method as shown by Ghosh and Reilly [7] where they have used the ANN framework to create a fraud detection framework for a banking dataset. But Neural Networks are computationally more expensive, and resource consumption is much more. Kazemi and Zarrabi [8] and Dhankhad et al. [9] have illustrated that classical models can perform as well as Neural Networks in this scenario. Also, as illustrated by Varmedja et al. [10] that by using SMOTE as the sampling technique other than Neural Network, Logistic Regression and Random Forest algorithms are equally efficient in detecting fraud with high accuracy. Jemima Jebaseeli et al. [11] have also illustrated the supremacy of Random Forest algorithms, which due to high-weightage attained greater accuracy than other methods because of randomization by bootstrapping the data. Also, Rai and Dwivedi [12] found that Random Forest performed better than the two competing algorithms Logistic Regression and Naïve Bayes in their experiment of credit card fraud detection.

Pumsirirat [13] illustrated that Neural Network models are optimal in some scenarios with large datasets. The tipping point finally lies at how much data is at hand, and the use case one is considering because Neural Networks can adapt well in different domains.

Other than the usual under-sampling, over-sampling methods, some research also has been done to enhance these methods further. Park and Park [14] explored a combination of over-sampling and under-sampling method based on slow-start algorithm and their method performed better than SMOTE. Vaishnavi et al. [15] talk about the concept of drift problems where imbalance in the data can arise when a variable change over time in unforeseen ways. They have used clustering method to separate out high, medium and low transactions and have tried different methods like SMOTE and Mathew Coefficient Classifier for dealing with class imbalance. Yeşilkanat et al. [16] introduced a sliding window approach for generating training data that has been referred to by Vaishnavi et al. [15] as well to adapt to changes in fraudulent behavior and capture the unknown patterns. They have also included a character level word embedding model on merchant names and automated generation of transaction data using Gradient Boosting. Somasundaram and Reddy [17] also talk about the concept drift problem, they proposed a Transaction Window Bagging model, and their model have exhibited higher performance on Brazilian Bank's dataset. Zhou et al. [18] have compared different algorithms and have chosen XGBOOST as the preferred algorithm for credit card fraud detection, and their focus is on mobile payment fraud detection.

In this paper, we will discuss how by using a combination of ensemble algorithms, and sampling methods it is possible to capture more fraudulent transactions. We will illustrate how ensemble techniques can be used for fraud detection. Ensemble methods work well because they combine the results of different modeling algorithms and can produce more accurate and robust results.

3 Business Problem

To illustrate the business problem, let us first introduce the concept of Type I and Type II errors. Table 1 illustrates the misclassification metrics with TP (true positive)/TN (true negative) being the cases of correct predictions, i.e., predicting fraud cases as fraud (TP) and predicting non-fraud cases as non-fraud (TN). While FP (false positive) refers to those cases that are actually non-fraud, but the model predicted as Fraud and FN (false negative) are those cases that are actually fraud but the model predicted as non-fraud. The modeling performance metrics like accuracy, F-score

Table 1 Type I and Type II errors

Predicted transactions	True/actual transactions	
	Fraud	Non-fraud
Fraud	TP	FP (Type I error)
Non-fraud	FN (Type II error)	TN

can be derived using these concepts as illustrated by Branco et al. [19]. Refer to Table 1 for the definitions.

Type I error is defined as the error of predicting the non-frauds as fraud, i.e., FP or false positives and Type II refers to the error in predicting the frauds as non-frauds, i.e., FN or false negatives. The objective of our fraud detection problem is to minimize the errors be it Type I or Type II, but it is also imperative in any machine learning problem, both Type I and Type II error cannot be minimized at the same time. So, any bank or financial organization would have to take a hit in one of the aspects—so we have to either minimize Type I error keeping Type II error at a certain level or minimize Type II error keeping Type I error at a certain level. In our case of credit card fraud detection, let us discuss in detail which error should take priority. Before that let us define some important metrics using the Type I and Type II errors, i.e., the concepts of false positives and false negatives that we will consider later in the paper (Table 2).

Table 2 Model metrics definition–precision, recall, accuracy, *F*-score

Model metrics	Definition
Precision	TP/(TP + FP)
Recall	TP/(TP + FN)
Accuracy	(TP + TN)/(TP + FP + FN + TN)
F-score	2 * [(Precision * Recall)/(Precision + Recall)]

Precision measures how accurately the model can capture fraud, i.e., out of the total predicted fraud cases, how many turned out to be fraud in actual. Recall measures out of all the actual fraud cases that has occurred, how many could the model predict as fraud. Accuracy is the measure of the overall model adequacy which is defined as how many of the majority and minority classes could be predicted correctly. *F*-score is a balance between precision, and recall because precision and recall are inversely related.

The business problem in case of a bank or financial organization is to detect as many fraudulent transactions as possible because fraud is expensive and can result in not only financial loss for the bank but also for the customer along with customer dissatisfaction and loss of reputation. Hence, the metric that the banks should be more concerned with is recall that maximizes the correct prediction of fraud and in turn if the false positives are a bit high it will mean blocking a few genuine customers the impact of which is not that high. However, a balance should be maintained because blocking too many genuine transactions can also lead to customer dissatisfaction and penalizing the genuine cases.

4 Approach

4.1 Dataset

We have used the Kaggle credit card data that contains credit card transactions on which Principal Component Analysis (PCA) has already been performed to maintain the confidentiality of the dataset. The dataset consists of 31 variables named V1 to V28, amount, time and class. Class is the variable which takes the value 0 for genuine transactions and 1 for fraudulent transactions. This will be the target variable or the dependent variable in our model. There are 284,807 observations in the data. This data will be used for conducting the analysis further.

As per Table 3, the dataset is highly imbalanced with the number of frauds being

Table 3 Percentage of fraud and non-fraud transactions in the dataset

Non-fraud	Fraud
0	1
99.83%	0.17%

only 0.17% of the data.

We can visualize the class imbalance with the help of a scatter plot as given in Fig. 1.

Class imbalance is a deterrent because most of the algorithms in machine learning focus on learning from the occurrences/observation points that occur frequently, i.e., the majority class. This is called the frequency bias. So, in cases of imbalanced

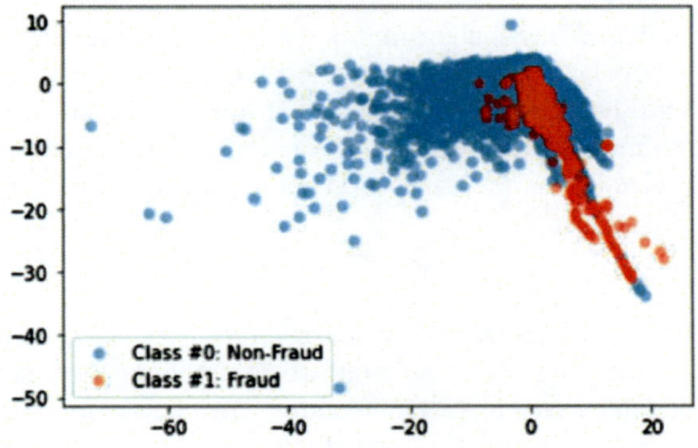

Fig. 1 Class imbalance using scatter plot

dataset, these algorithms might not work well. It has been observed that few techniques that will work well are tree-based algorithms or anomaly detection algorithms. Traditionally, in fraud detection problems, business rule-based methods are often used. Tree-based methods also work well because a tree creates rule-based hierarchy that can separate both the classes. Decision Trees tend to over-fit the data, and to eliminate this possibility, we will consider Random Forest and an ensemble method.

4.2 Methodology

Decision Tree models have typically performed well on fraud detection datasets because of their high flexibility, easier to understand and easy implementation criteria. Branco et al. [19] have compared different fraud detection techniques both supervised and unsupervised and have proved that Decision Tree techniques can work as well and sometimes better in an imbalanced dataset. As seen in the paper by Niveditha et al. [20], they could predict fraud using Random Forest model with 98.6% accuracy. In this paper, we will be testing the Random Forest model and compare it with an ensemble model technique. As we will see, a simple Random Forest model with tuned parameters can also work as well as an ensemble model technique.

A combination of under-sampling the majority class and over-sampling the minority class which can achieve better results has been discussed in Chawla et al. [21]. We will also conduct an experiment and test the ensemble and Random Forest models by applying different sampling techniques and then compare the results. The ensemble algorithm has been constructed with Logistic Regression, Random Forest and Naïve Bayes as the chosen algorithms.

Random Forest works by building multiple Decision Tree predictors and the final selected class or output is the mode of the classes of the individual Decision Trees. This method works like voting where voting is done based on the most popular class. For example, if rule 1 is dominated by 2 trees each of which predicts a particular transaction as fraud versus rule 2 where only 1 tree predicts the same transaction as non-fraudulent, then rule 1 will take precedence here and final prediction will be fraud.

Naïve Bayes algorithm is a probabilistic classifier based on Bayes' theorem. It is popular because of its highly scalable feature, easy to interpret, and also it is not sensitive to irrelevant features. It is Naïve because it assumes that the features in the model is independent of each other. Given the feature set, it will compute the probability of the transaction being fraud or non-fraud.

$$P(y/X) = [P(X/y)P(y)]/P(X) \tag{1}$$

where y is the class variable fraud/non-fraud and X is the feature set $(x_1, x_2, \ldots, x_n)$.

Logistic Regression is one of the simplest supervised algorithm to predict a target variable based on certain features. The dependent variable as is in our case is binary in nature, i.e., fraud or non-fraud. It works well on large datasets.

Let us now consider the different sampling methods that we will use. The sampling method that will be considered in this paper are:

1. Random under-sampling: For random under-sampling, random samples are drawn from the majority sample; here, this means the non-fraud/genuine transactions such that the sample size is equivalent to the fraudulent observations. Yen and Lee [22] have illustrated how cluster based under-sampling on back-propagation of neural network can reduce the impact of imbalanced class distribution and increase accuracy of the model. One drawback of under-sampling as the name suggests is—as we are discarding some observations there might be some loss of information. Figure 2 provides a pictorial depiction of this method.
2. Random over-sampling: This methodology as the name suggests is opposite of under-sampling, and here, the minority class, i.e., the fraudulent observations is duplicated and the number of observations in the minority class is increased to get a balanced dataset. One limitation of this method is creation of duplicates due to which there can be potential over-fitting at times. Figure 3 will provide a pictorial depiction of this method.
3. SMOTE: Synthetic minority over-sampling technique or SMOTE is a sampling method that uses synthetic data with KNN. Gulowaty and Ksieniewicz [23] have illustrated the usage of SMOTE as an over-sampling algorithm for imbalanced data streams. SMOTE does not use duplicate data. Each minority class sample along with their k-nearest neighbors is considered. Then, synthetic examples are created along the line segments that join any/all the minority class examples and their k-nearest neighbors. This is illustrated in Fig. 4.

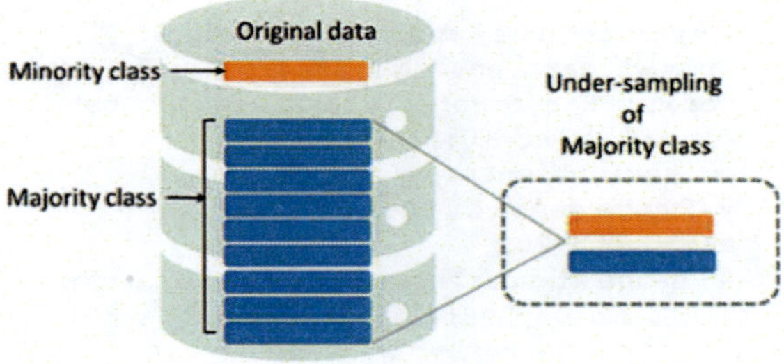

Fig. 2 Pictorial depiction of random under-sampling

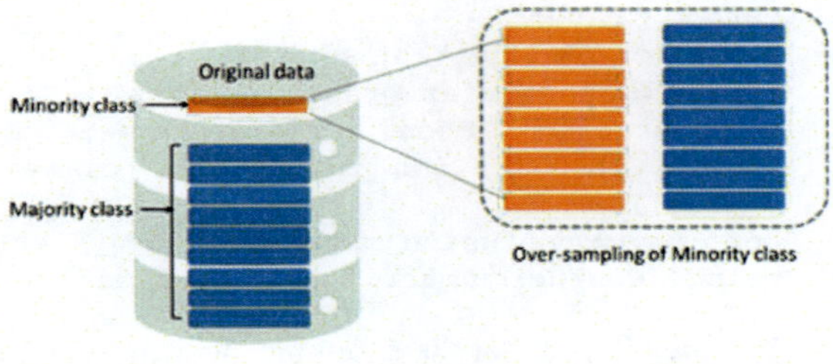

Fig. 3 Pictorial depiction of random over-sampling

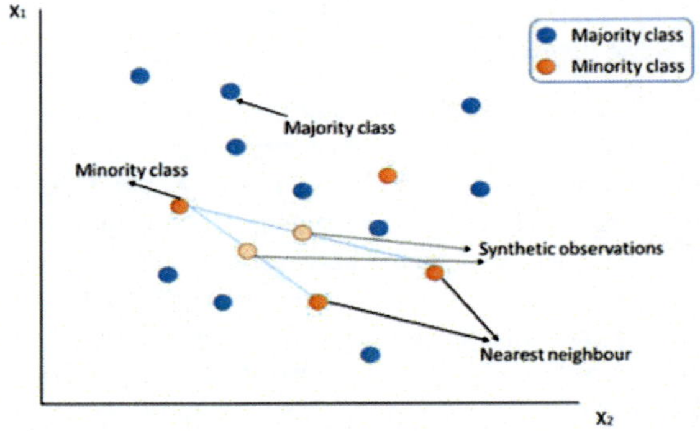

Fig. 4 SMOTE sampling methodology

4.3 Results

The Random Forest algorithm and an ensemble algorithm constructed with Logistic Regression, Random Forest and Naïve Bayes were chosen as the algorithms to be tested in the experiment, and the results were compared. The dataset was split into 80:20 training and test samples with this pseudo code:

```
from sklearn.model_selection import train_test_split
x_train,x_test,y_train,y_test = train_test_split (x,y,
test_size=0.2, random_state=0)
```

The Random Forest algorithm along with the tuned parameters can be implemented with this pseudo code:

```
# Training the Random forest model
from sklearn.ensemble import RandomForestClassifier
model = RandomForestClassifier(bootstrap=True,
class_weight={0:1, 1:12},
criterion='entropy',
# Change depth of model
max_depth=10,
# Change the number of samples in leaf nodes
min_samples_leaf=10,
# Change the number of trees to use
n_estimators=20,
n_jobs=-1,
random_state=5)
# Predict Y on the test set
y_pred = classifier.predict(x_test)
# Computation of model performance metrics
print('Classifcation report:\n', classification_report(y_test,
y_pred))
conf_mat = confusion_matrix(y_true=y_test, y_pred=y_pred)
print('Confusion matrix:\n', conf_mat)
```

The pseudo code for the ensemble algorithm can be written as:

```
# Import the libraries
from sklearn.ensemble import RandomForestClassifier, VotingClas-
sifier
from sklearn.linear_model import LogisticRegression
from sklearn.naive_bayes import GaussianNB
# Define Models
clf1 = LogisticRegression(random_state=1)
clf2 = RandomForestClassifier(random_state=1)
clf3 = GaussianNB()
# Combine the different algorithms into an ensemble model
ensemble_model = VotingClassifier(estimators=[('lr', clf1),
('rf', clf2), ('gnb', clf3)], voting='hard')
# Fit and predict the ensemble model with the test data
ensemble_model.fit(x_train, y_train)
ensemble_model.predict(x_test)
```

The models can be implemented after applying the sampling techniques. For this, the pipeline module was used. The pseudo code of one of the sampling methods, i.e., SMOTE with the ensemble model is provided for reference:

```
# Import the pipeline module from imblearn
from imblearn.over_sampling import SMOTE
# Define which resampling method and which ML model to use in the
pipeline. In our case this is the ensemble model
resampling = SMOTE(sampling_strategy='auto',random_state=0)
model = VotingClassifier(estimators=[('lr', clf1), ('rf', clf2),
('gnb', clf3)], voting='hard')
# Define the pipeline and combine SMOTE with the ensemble model
pipeline = Pipeline([('SMOTE', resampling), ('EM', model)])
# Fit the model
pipeline.fit(x_train, y_train)
predicted = pipeline.predict(x_test)
# Obtain the results from the classification report and confusion
matrix
print('Classifcation report:\n', classification_report(y_test,
predicted))
conf_mat = confusion_matrix(y_true=y_test, y_pred=predicted)
print('Confusion matrix:\n', conf_mat)
```

Table 4 illustrates the comparison between all the different model combinations:

As we can observe from Table 4, if we consider all the parameters and performance metrics then we should choose the Random Forest model with the tuned parameters because the accuracy is almost 100% with the precision, recall and *F*-score all being in the range of >80%. Table 4 illustrates the number of frauds that the model is able to capture along with the false positives and false negatives. But if we look at the ensemble model with SMOTE then 5 more frauds are getting captured, but the false positives, i.e., genuine transactions sacrificed is a bit high. While we are considering the model performance metrics let us also consider the actual number of frauds, false positives and false negatives. This will provide a more holistic view of the different algorithms/models. Table 5 will be used to illustrate this.

In this case, we need to look at the value of the transactions captured and compare them with the value of the false positive transactions. If the value of the extra 5

Table 4 Comparison of different model results

Model performance metrics	Precision	Recall	*F*-score	Accuracy (%)
Random Forest with no tuning	0.94	0.75	0.84	99.95
Random Forest with tuned parameters	0.82	0.83	0.82	99.94
Random Forest with under-sampling	0.01	0.95	0.02	86
Random Forest with over-sampling	0.18	0.86	0.3	99
Random Forest with SMOTE	0.07	0.9	0.13	98
Ensemble model with no sampling	0.89	0.77	0.83	99.94
Ensemble model with under-sampling	0.06	0.92	0.12	98
Ensemble model with over-sampling	0.23	0.88	0.37	99
Ensemble model with SMOTE	0.28	0.88	0.43	99.58

Table 5 Number of frauds, false positives (Type I Error) and false negatives (Type II Error)

	True positives	False positives	False negatives
Random Forest with no tuning	76	5	25
Random Forest with tuned parameters	84	11	17
Random Forest with under-sampling	96	7838	5
Random Forest with over-sampling	87	384	14
Random Forest with SMOTE	91	1254	10
Ensemble model with no sampling	78	10	23
Ensemble model with under-sampling	93	1405	8
Ensemble model with over-sampling	89	295	12
Ensemble model with SMOTE	89	227	12

fraudulent transactions outweighs, the genuine transactions by a large extent and the business is ready to take the hit, and then it is wise to go ahead with this model. Though Random Forest with SMOTE and ensemble model with under-sampling has high recall (this is the metric of importance in our use case for fraud prediction), but the false positives is too high hence these models were not selected. Table 6 indicates our 2 selected models.

We can further test the robustness and stability of these models by performing tenfold cross validation. Table 6 provides the cross validation results of our selected models. Tenfold cross validation on the ensemble model with SMOTE yields 99.57%

Table 6 Selected models

Model performance metrics	Precision	Recall	*F*-score	Accuracy (%)
Random Forest with tuned parameters	0.82	0.83	0.82	99.94
Ensemble model with SMOTE	0.28	0.88	0.43	99.57

accuracy with 0.09% variation which means the model is extremely stable and consistent. The Random Forest model with the tuned parameters has an accuracy of 99.94% and standard deviation of 0.01%, and hence, this model is also very stable and consistent.

5 Deployment

Here, we will briefly discuss the deployment option of the fraud detection algorithms. As discussed earlier, it is not possible to minimize both the Type I and Type II errors, and hence, the banks will try to minimize any one of the two, while keeping the other at a certain threshold. In this credit card fraud detection case, banks will try to minimize the Type II error so that recall is high by keeping the Type I error at a certain tolerable limit. It is important to maintain a tolerable limit of Type I error, or else a lot of genuine transactions might be sacrificed in the effort of catching fraudulent transactions.

For implementation, a lot of banks or financial organizations also deploy the threshold approach. This threshold approach is created in conjunction with the machine learning algorithms along with certain business rules. The algorithm alone is not the decision maker, rather the model is augmented with certain business judgments in the form of rules. The final decision of tagging the transaction as fraud or genuine can be over-written at times by the business rules. This is illustrated in Fig. 5.

The business rules can help in determining the appropriate action that needs to be taken on a real time basis with minimum human intervention.

6 Conclusion

Credit card fraud detection continues to allude all researchers by being a classic data imbalance problem. The far-reaching benefits of accurately predicting fraud both in business related aspects and customer related aspects cannot be undermined. In this paper, we looked at the problem of fraud detection for credit cards using decision tree-based algorithms like Random Forest which has proved to be equally successful along with an ensemble technique using a combination of algorithms. We have established the results using pseudo codes and model performance metrics like accuracy, precision and recall with recall being the metric of choice in this particular use case. We have also discussed in details how different sampling techniques can alter/enhance the results and presented a comparison of the different sampling techniques and algorithms. This work hasn't considered variable selection. But in future, extensive research can be conducted to focus on how to alter the variables to achieve better results with the similar algorithms.

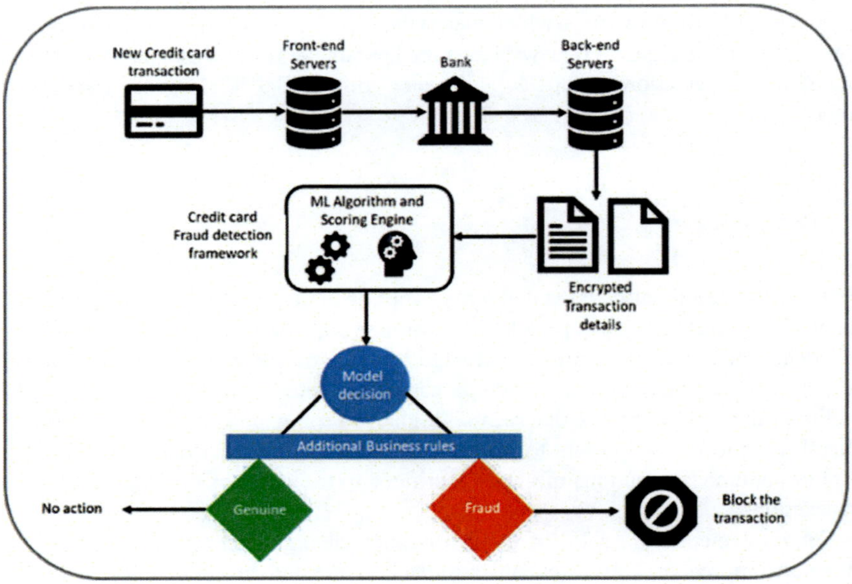

Fig. 5 Deployment of the machine learning algorithm with business rules

References

1. Federal Trade Commission (FTC): Compare identity theft report types. http://tableaupublic@ftc.gov. Last updated 2020/10/17
2. Reserve Bank of India Report (RBI): https://timesofindia.indiatimes.com/business/india-business/in-92-days-india-lost-rs-128-crore-in-card-online-fraud/articleshow/74571025.cms. Last updated 2020/03/11
3. Merchantsavvy Homepage: http://www.merchantsavvy.co.uk. Last updated 2020/10
4. Krawczyk, B.: Learning from imbalanced data: open challenges and future directions. In: Progress in Artificial Intelligence, vol. 5. Springer, Berlin, pp 221–232 (2016)
5. Abdi, L., Hashemi, S.: To combat multi-class imbalanced problems by means of over-sampling techniques. Engineering **28**(1) (2016). http://doi.org/10.1109/TKDE.2015.2458858
6. Yu, W.-F., Wang, N.: Research on credit card fraud detection model based on distance sum. In: 2009 International Joint Conference on Artificial Intelligence. IEEE, China (2009). http://doi.org/10.1109/JCAI.2009.146
7. Ghosh, S., Reilly, D.L.: Credit card fraud detection with a neural network. In: Proceedings of the Twenty-Seventh Hawaii International Conference on System Sciences, Wailea, HI, USA, pp. 621–630 (1994). http://doi.org/10.1109/HICSS.1994.323314
8. Kazemi, Z., Zarrabi, H.: Using deep networks for fraud detection in the credit card transactions. In: 2017 IEEE 4th International Conference on Knowledge-Based Engineering and Innovation (KBEI), pp. 630–633. IEEE (2017)
9. Dhankhad, S., Mohammed, E., Far, B.: Supervised machine learning algorithms for credit card fraudulent transaction detection. In: A Comparative Study, 2018 IEEE International Conference on Information Reuse and Integration (IRI), pp. 122–125. IEEE (2018)

10. Varmedja, D., Karanovic, M., Sladojevic, S., Arsenovic, M., Anderla, A.: Credit card fraud detection—machine learning methods. In: 18th International Symposium INFOTEH-JAHORINA (INFOTEH), pp. 1–5 (2019). http://doi.org/10.1109/INFOTEH.2019.8717766 (East Sarajevo, Bosnia and Herzegovina 2019)
11. Jemima Jebaseeli, T., Venkatesan, R., Ramalakshmi, K.: Fraud detection for credit card transactions using random forest algorithm. In: Peter, J., Fernandes, S., Alavi, A. (eds.) Intelligence in Big Data Technologies—Beyond the Hype. Advances in Intelligent Systems and Computing, vol. 1167. Springer, Singapore (2021). http://doi.org/10.1007/978-981-15-5285-4_18
12. Rai, A.K., Dwivedi, R.K.: Fraud detection in credit card data using machine learning techniques. In: Bhattacharjee, A., Borgohain, S., Soni, B., Verma, G., Gao, X.Z. (eds.) Machine Learning, Image Processing, Network Security and Data Sciences. MIND 2020. Communications in Computer and Information Science, vol. 1241. Springer, Singapore (2020). http://doi.org/10.1007/978-981-15-6318-8_31
13. Pumsirirat, A., Yan, L.: Credit card fraud detection using deep learning based on auto-encoder and restricted Boltzmann machine. Int. J. Adv. Comput. Sci. Appl. **9**(1), 18–25 (2018)
14. Park, S., Park, H.: Combined oversampling and undersampling method based on slow-start algorithm for imbalanced network traffic. Computing (2020). https://doi.org/10.1007/s00607-020-00854-1(2020)
15. Dornadula, V.N., Geetha, S.: Credit card fraud detection using machine learning algorithms. Procedia Comput. Sci. **165**, 631–641 (2019). ISSN 1877-0509. http://doi.org/10.1016/j.procs.2020.01.057
16. Yeşilkanat, A., Bayram, B., Köroğlu, B., Arslan, S.: An adaptive approach on credit card fraud detection using transaction aggregation and word embeddings. In: Maglogiannis, I., Iliadis, L., Pimenidis, E. (eds.) Artificial Intelligence Applications and Innovations. AIAI 2020. IFIP Advances in Information and Communication Technology, vol. 583. Springer, Cham (2020). http://doi.org/10.1007/978-3-030-49161-1_1
17. Somasundaram, A., Reddy, S.: Parallel and incremental credit card fraud detection model to handle concept drift and data imbalance. Neural Comput. Appl. **31**, 3–14 (2019). https://doi.org/10.1007/s00521-018-3633-8
18. Zhou, H., Chai, H., Qiu, M.: Fraud detection within bankcard enrollment on mobile device based payment using machine learning. Front. Inf. Technol. Electronic. Eng. **19**, 1537–1545 (2018). http://doi.org/10.1631/FITEE.1800580
19. Branco, P., Torgo, L., Ribeiro, R.P.: A survey of predictive modelling under imbalanced distributions. In: ACM Computing Surveys, Article no 31 (2016). http://doi.org/10.1145/2907070
20. Niveditha, G., Abarna, K., Akshaya, G.: Credit card fraud detection using random forest algorithm. Int. J. Sci. Res. Comput. Sci. Eng. Inf. Technol. (2019). http://doi.org/10.32628/CSEIT195261
21. Chawla, N.V., Bowyer, K.W., Hall, L.O., Kegelmeyer, W.P.: SMOTE: synthetic minority over-sampling technique. J. Artif. Intell. Res. **16**(1) (2002)
22. Yen, S.J., Lee, Y.S.: Under-sampling approaches for improving prediction of the minority class in an imbalanced dataset. In: Huang, D.S., Li, K., Irwin, G.W. (eds.) Intelligent Control and Automation. Lecture Notes in Control and Information Sciences, vol. 344. Springer, Berlin (2006). http://doi.org/10.1007/978-3-540-37256-1_89
23. Gulowaty, B., Ksieniewicz, P.: SMOTE algorithm variations in balancing data streams. In: Yin, H., Camacho, D., Tino, P., Tallón-Ballesteros, A., Menezes, R., Allmendinger, R. (eds.) Intelligent Data Engineering and Automated Learning—IDEAL 2019. IDEAL 2019. Lecture Notes in Computer Science, vol. 11872. Springer, Cham (2019). http://doi.org/10.1007/978-3-030-33617-2_31

An Overview of Pulmonary Tuberculosis Detection and Classification Using Machine Learning and Deep Learning Algorithms

Priyanka Saha and Sarmistha Neogy

1 Introduction

Pulmonary Tuberculosis (TB) is a potential noxious disease that is spread by a bacterium and mainly affects the lungs. It is considered a worldwide leading respiratory disease. Pulmonary Tuberculosis can be life threatening if not detected at right time followed by proper treatment. According to worldwide survey, China, India, and Indonesia have the highest number of cases every year [1]. In India, around 2.69 million cases are reported in 2018 and around 2.4 million cases are registered in 2019 [2]. Hence, early detection and prevention are the only keys to help in decreasing the number of cases. In the modern era of Artificial Intelligence, Computer Aided Diagnosis System (CAD) is rapidly revolving around the pattern of disease diagnosis and treatment. With the help of Learning algorithms, strong modeling of Radiographic Images is helpful for analysis of the data thus helping in proper classification and prediction.

A bacterium named Mayobacterium Tuberculosis which spreads through air is the main reason for Pulmonary Tuberculosis. Detection of it is done by examining microscopic images of sputum for acid-fast bacilli, i.e., observing pus samples or sputum for confirming the presence of mycobacterium TB. A designated highly equipped Laboratory is required to examine the presence of bacteria in this process. It is a time-consuming system and thus requires more resource and expertise to analyze. But most of the countries lack the infrastructure to do these acid-fast bacilli tests. Sputum Smear Microscopy is another feasible solution for pulmonary tuberculosis detection [3]. It is fast and a straight-forward approach and popular technique. As the number of Tb cases increasing, Chest X-Ray is an effective, hassle free, and affordable solution these days. Other methods that are used for Tuberculosis detection are Tuberculin Skin test, culture test, Interferon Gamma Release Assay (IGRA),

P. Saha (✉) · S. Neogy
Department of Computer Science and Engineering, Jadavpur University, Kolkata 700032, India

M. Dua et al. (eds.), *Proceedings of the International Conference on Paradigms of Communication, Computing and Data Sciences*, Algorithms for Intelligent Systems,
https://doi.org/10.1007/978-981-16-5747-4_72

GeneXpert, etc. [4]. Generally, a sputum smear image dataset contains three types of images with single bacilli, images with touching or overlapping bacilli, and debris. A touching bacilli image consists of several single bacilli objects that are overlapped with each other. To detect the severity of TB, detection of all the bacilli is important. But most of the research works are carried out single bacilli object [5, 6].

So, an automated system is required to analyze faster, accurate decision making and controlling the number of cases. As manual detection is a tedious job, different Intelligent Diagnosis Models are proposed for Pulmonary Tuberculosis Detection. Models based on Learning Algorithms have shown proper accuracies and many researchers have proposed models for automatic Pulmonary Tuberculosis detection using chest X-Ray (CXR) images. Here in this paper, we will discuss the methods that have been proposed and their efficiency and prediction probabilities for detecting TB. It gives an overview of the research works done for implementing an automated model for TB data analysis. Section 2 of this paper focuses on the process of data collection and preparation of the data for learning algorithms model implementation. Section 3 discusses how different learning algorithms are implemented to derive an automated model for TB detection. Section 4 compares the performances of the models and the accuracy and error rate of each model.

2 Data Collection and Preprocessing

The first and foremost requirement for any data analysis model is to collect adequate amount of data. Data collection is the main bottleneck for any Data analysis project. The steps involved in making an automated TB detection and decision-making system involve image collection and preprocessing, image segmentation, feature extraction, and finally classification. The majority amount of time for analyzing a machine learning or deep learning model is spent on data preprocessing, i.e., data collection, data resizing, data cleaning, data analysis, data visualization, and feature engineering. In 1998, Veropoulos et al. [7] first proposed an image processing technique to detect TB bacilli from microscopic sputum smear images. An adaptive threshold-based image segmentation technique was proposed for Costa et al. [8] for ZN-stained sputum smear images. A threshold-based segmentation considering Cr, a plane of YCbCr, and lab color space was used for classification of bacilli from sputum smear images [9]. Zhei et al. used a two-stage segmentation method for image segmentation for better classification accuracy [10]. Surgirtha and Murugesan [10] represented a color segmentation method for sputum smear image classification. As deep learning algorithms are becoming more popular, adequate amount of data is required for training a model. Data collection can be done in three important steps—Data Acquisition, Data Labeling, and Improving Existing data. The below figure shows a high label research landscape for data collection (Fig. 1).

After data collection is complete, the next important step is to preprocess the data. A deep learning models' accuracy and prediction power depend on the quality of data. The idea of data processing is to aggregate more information from the data

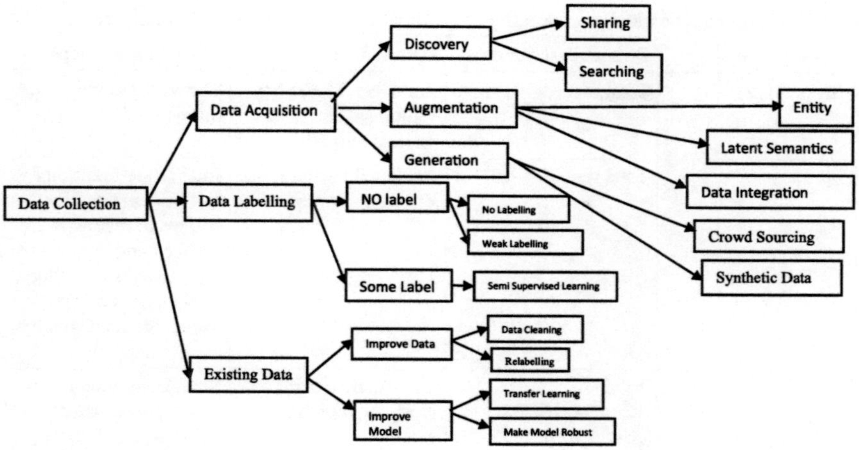

Fig. 1 Data collection framework [11]

and discards any unimportant data that may affect the analysis result. As Pulmonary tuberculosis data is dependent on chest X-ray images and smear sputum images, preprocessing these images is the first and foremost step for implementation. As images are collected from different sources like smartphones, digital microscope, and cameras, adjusting the contrast of the images are of major concern. Image preprocessing steps can be classified as Image Denoising, Image segmentation, and Image Augmentation. Image Denoising can be done with an Image Encoder, Using Principal Component Analysis (PCA), Discrete Wavelet Transform (DWT), and Discrete Fourier Transform (DFT), with Anisotropic Diffusion, etc. Many researchers have implemented preprocessing steps based on their dataset structure. Table 1 describes the data collection and data preparation steps involved in different models.

3 Machine Learning, Deep Learning, and Ensemble Based Learning Methods for Pulmonary Tuberculosis Detection

After applying various techniques for preprocessing images and extracting relevant features like texture, edge, shape, and obtaining different views of the images in horizontal, vertical, rotated format, next step is to apply a classifier to detect the occurrence of pulmonary tuberculosis. Data preprocessing is done in earlier step so that classifier can predict accurately. Hence proper dataset must be obtained for a powerful machine learning, deep learning classifier application. Classifiers that have been applied by the researchers in their corresponding work are discussed here.

Table 1 Data collection and preprocessing steps involved in TB dataset in different studies

Author's name	Dataset description	Data source	Preprocessing steps
Muyama [12]	Total records—148	Ziehl–Neelsen sputumsmear microscopic images	Converted to jpeg and normalized
Li [13]	Total records—501	Affiliated Hospital of Zhejiang University	Images are resampled to $1 \times 1 \times 1$ mm^3. Irrelevant regions were reduced and, 5 types of PTB images—Military, Infiltrative, Caseous, Tuberculoma, Cavitary are obtained
Chang [14]	Total records—16,503 images form 1727 samples	Tao-Yuan General Hospital, Taiwan	SMOTE is adopted for training data balancing. For image of minority class, images are synthesized by *k* times where *k* defines a ratio of majority class to minority class in training
Oloko-Oba [15]	Total records—662	Shenzhen tuberculosis dataset [16]	Dataset augmented to size 10,000 with zoom random of 0.5, flip top bottom of 0.5, flip left right with a probability of 0.5, with an area of 0.8, rotate left right of 5
Ayaz [17]	Not mentioned	Montgomery and Shenzhen Dataset available at U.S. National Library of Medicine (NLM), National Institutes of Health (NIH)	Images were resized to 300×300 pixels for Gabor filter-based feature extraction
Yugaswara et al. [18]	Total records—81	Public Health Center, Jakarta	Not discussed
Ali [19]	Not available	Gulab Davi Hospital	Mean or mode for replacing missing qualities and elimination of presence of having missing qualities

(continued)

Table 1 (continued)

Author's name	Dataset description	Data source	Preprocessing steps
Raju [20]	Dataset 1—138 images Dataset 2—662 images Dataset 3—113 images	Montgomery County, Maryland, Shenzhen Hospital in Shenzhen, A medical college in India	Intensified edges of foreground portion of cropped images are resized to 128 × 128, 256 × 256, 512 × 512, 1024 × 1024 pixels with pixel normalization
El-Melegy [21]	Total records—500	ZNSM-iDB Dataset [22]	Data Augmentation was used for dataset size increase
Hernández [23]	Total records—800	Montgomery County's Tuberculosis screening program	Image duplication, CLAHE application, resized to 224 × 224 pixels. Gaussian noise added and image normalized by 1/255 factor and then cropped to 99 pixels
Kant [24]	Total records—202 from first 4 set of datasets 3	ZiehlNeelsen Sputum smear Microscopy image DataBase (ZNSM-iDB)	Imageset with 20 × 20-pixel microscopic field view patch was used for classification
Abiyev [25]	Total records—112,120	National Institutes of Health—Clinical Center [26]	Not discussed
Karnkawinpong [27]	Total records—3310	National Library of Medicine (NLM) and the Ministry of Public Health, Thailand	−10-to-10-degree random rotation for training set, 10 to 10, second test set rotated from −30 to 30
Sathitratanacheewin [28]	Dataset 1—662 Dataset 2—112,120	Shenzhen No. 3 Hospital in Shenzhen, Guangdong providence, China, NIH Clinical Center, Bethesda, Maryland, USA	Not discussed
López [29]	Total records—492 9770 patches are extracted from focus smear microscopic images	[30]	Resized to 40 × 40 pixels Data Augmentation technique rotation in two different direction (90 and 180°) of all 9770 patches results in 29,310 total patches

(continued)

Table 1 (continued)

Author's name	Dataset description	Data source	Preprocessing steps
Santosh [31]	Dataset 1—138 Dataset 2—682 Dataset 3—156	Montgomery County Hospital collection; Shenzhen Hospital, China collection; and Indian (IN) collection, New Delhi	Graph Cut Algorithm is used for ROI selection. Lung region symmetry analyzed using multi scale shape features and edge plus texture feature
Lakhani [32]	Total records—1007	Montgomery County, Shenzhen, China Belarus TB Public Health Program, Thomas Jefferson Hospital	Image augmentation, cropping to 227 × 227, pixels, mean subtraction, and mirror image
Jaeger [33]	Dataset 1—138 Dataset 2—615 Dataset 3—247	Department of Health and Human Services of Montgomery County (MC), Maryland Shenzhen Hospital, China Japanese Society of Radiological Technology (JSRT) Dataset	Graph Cut Segmentation Method is used for Lung region
Priya [34]	Total records—100	Not discussed	Images were segmented using active contour segmentation method Hu's and Zernike moments are used for feature extraction PCA and KPCA are used to identify most significant features
van Ginneken [35]	Dataset 1—616 Dataset 2—200	Political asylum in The Netherlands University of Chicago Hospitals Maryland, USA	Active Shape Model Segmentation helps to find required region within lungs

3.1 Types of Machine Learning Algorithms

Machine learning models can be divided into two categories such as Interactive Model and Automated Model [36].

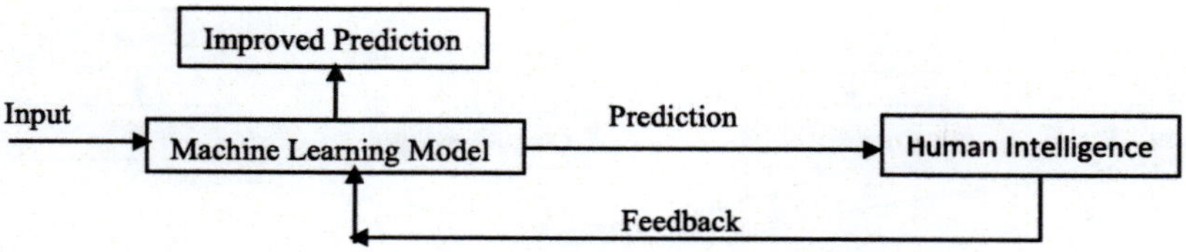

Fig. 2 Flowchart of interactive machine learning

3.1.1 Interactive Models of Machine Learning

These models involve humans in building machine learning models. By incorporating human feedback during the learning process, a model can be trained faster and more efficiently to produce accurate results. Among the dominating machine learning approaches such as supervised learning and unsupervised learning, human feedback is not involved in the training process. As healthcare data is vast and in different formats, it will be more effective to involve human understanding while training. Researchers proposed Transparent Boosting Tree Algorithm which combines machine learning expertise and human intelligence to improve the learning process. Below flowchart shows how human interaction can be involved [37] (Fig. 2).

3.1.2 Automated Models of Machine Learning

On contrast to interactive machine learning model, Automated machine learning models do not correlate with humans as they are completely self-automated. As the number of pulmonary tuberculosis is higher in every country, the amount of data involving in detection of tuberculosis is also high. So, automation is desirable as dealing with huge datasets while involving human interaction is not possible. Parameters of these models are pre-assigned by the developers to enhance the efficiency of the model. Fine tuning of these parameters is also available for efficiency enhancement.

3.2 Ensemble Machine Learning Algorithm

Ensemble Machine Learning algorithm is an approach to combine different machine learning algorithms for better prediction power than a single classifier's power. Ensemble learning aims to solve the problem of bias and variance. By combining many models, we can reduce the ensemble error, while retaining individual model's complexities. Ensemble methods can be classified as NON-generative methods and Generative Methods.

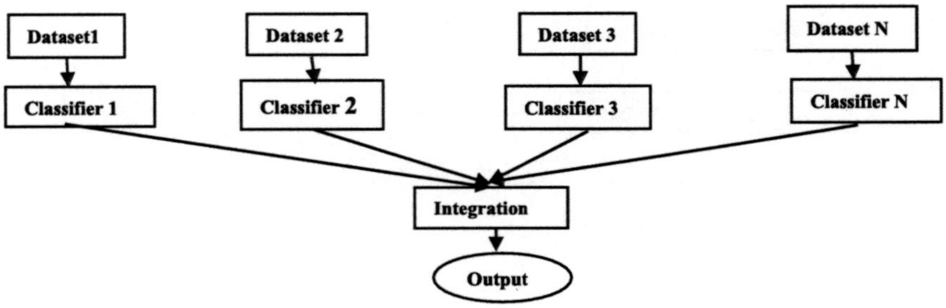

Fig. 3 Bagging ensemble framework

3.2.1 Non-generative Methods

Voting

The simplest form of ensemble is majority voting that allows the combination of multiple base learner's prediction. The Base Learners are considered as voters and class is considered a contender. The algorithm takes vote to a contender as a winner.

Stacking

Stacking is a form of meta-learning. The base learners are used to generate metadata for the problem's dataset and then utilize another learner called a meta-learner to process metadata. Base learners are considered as Label 0 and meta learners are considered as Label 1.

3.2.2 Generative Methods

Bagging

It is the most widely used ensemble technique and can be considered as a non-hybrid classifier. Bagging or bootstrap aggregation is a generative ensemble learning that reduces the variance problem. It creates several base learners with same classifier by sub sampling the original training set (Fig. 3).

Boosting

It is a variation of the bagging technique where converts a weak learner into a string learner sequentially and trying to correct make its predecessor prediction. While bagging is a parallel model where each model is trained independently, in boosting current model use the architecture of its previous classifier. Boosting is able to reduce bias and variance also. Weights are assigned to the dataset at the beginning and are updated after each training step which helps subsequent learners to focus more on wrongly classified data that are now assigned with higher weights.

3.3 Deep Learning Algorithms for Pulmonary Tuberculosis Detection

Deep Learning is a subset of Machine learning Algorithms that consists of statistical analysis algorithms that train data repeatedly to make predictions. It utilizes different layers of non-linear processing units for feature extraction. Output of previous layer acts as an input layer to the present layer. It can learn in many ways supervised, unsupervised and semi-supervised. Different architectures are present like Deep Neural Network, Recurrent Neural Network, Convolutional Neural Network, Deep Belief Network, etc. Deep learning is basically an advancement of Artificial Neural Network which consists of multiple hidden layers with great labels of abstraction. The popularity of deep learning in healthcare sector is driven large dataset availability, advancement of GPU for better performance with graphics and videos.

A variant of deep learning model is Convolutional Neural network and one of the widely used deep learning algorithms for computer vision tasks. It consists of three important layers—convolutional layer, pooling layer, and a fully connected layer. Convolution a pooling layer are used for feature learning process and fully connected layer is used for classification purposes. It has emerged as a huge success when it comes to medical image analysis, image segmentation, image annotation, etc. In the field of health image analysis, there are some famous CNN architectures available to increase the prediction capacity. Some of the following architectures are:

3.3.1 VGG Net

VGG Net is a straight-forward CNN architecture [38]. All filters are a size off 3×3 and default image input size is 224×224. VGG architecture consists of a stack of convolutional layer s with increasing filter size. VGG16 and VGG19 are two popular VGG architectures.

3.3.2 ResNet (Residual Network)

ResNet is a deeper network than VGGNet with total 152 layers [39]. In this network, every layer has a residual connection with the previous network. Here current layer is not only connected to the previous layer but also connected layers behind the previous layer. Training of this network is done by using batch normalization layers

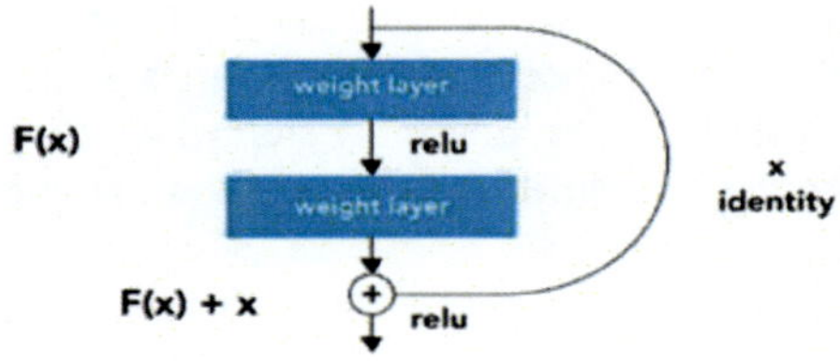

Fig. 4 ResNet building block [39]

after every convolutional layer. Batch normalization will give a boost to the weights and thus higher learning rate can be achieved while training and will reduce vanishing gradient problem. Many variants of ResNets are available by adding some Conv2D layers (Fig. 4).

3.3.3 Dense Net

In contrast to ResNet, where current layer has a residual connection with the layer before the previous layer, all layers preceding to current layer are concatenated and fed as input to the current layer. Vanishing gradient problem can be minimized more here as all layers are connected to the output layer. It has more layer connections than ResNet [40].

3.3.4 Inception Net

Inception Net is based on the idea of making a wider network. A wider network can be obtained by parallel connection of multiple layers with different filters and lastly concatenating all those parallel paths to pass to next layer.

Earlier different automatic TB detection approaches have been proposed to identify TB bacteria from microscopic sputum smear images. The basic steps involved in these automated systems are autofocusing of microscope, image acquisition and preprocessing, segmentation of TB bacilli, bacilli classification, and finally predicting a TB bacillus. Different auto-focusing methods have been experimented by researchers in different studies such as Gradient Based Operator like Gaussian Derivative [41, 42], Tenengrad (TEN) [41], Absolute Tenengrad (ATEN) [41, 43, 44], Total Variation [41], Laplacian based operator such as Image Laplacian (LAP) [41], Sum modified Laplacian [43], Wavelet Based Operator like Ratio of the sum of wavelet coefficient (WAVR) [41], DCT based operator like Discrete Cosine Transform (DCT) [41], Mid-Frequency DCT [41], statistics Based Operator like Standard Deviation [45], Histogram Entropy [45, 46], Variance of Log Histogram (VLH) [42]. In [4], TB detection methods from sputum smear microscopic images with image processing techniques are discussed. Accuracy of these automated systems for TB detection has increased over the years.

With the advancement of AI, researchers have opted for various machine learning classifiers such as Support Vector Machine (SVM), Neural Network (NN), Random Forest (RF), Bayesian Network to classify the image dataset as normal or abnormal [18, 19, 35]. Since the popularity of deep learning algorithms increases and results of deep learning applications have promising results, application of deep learning in tuberculosis detection has made prominent progress. Among all the deep learning algorithms, Deep Convolution Neural Network (DCNN) is a popular supervised model for TB detection and classification [13, 25, 27, 28]. Different algorithms of DCNN have performed well on TB datasets and accuracy obtained nearly 90%. Table 2 lists the machine learning and deep learning architectures used by researchers in their models for pulmonary Tuberculosis detection.

4 Performance Evaluation

After successful application of data collection, data preprocessing, feature engineering, feature extraction, machine learning or deep learning model implementation, parameter tuning of these models for better prediction, and finally followed by some prediction based on some class or labels or probability, the most important concern of any research work is how much accurate and effective the model is based on some model's parameter evaluation. Different performance metrics are there to evaluate the machine learning or deep learning models. Accuracy of any model is measured by calculating the number of false predictions made on the test dataset. Evaluation makes it easier to generalize a model and helps in enhancing the predictive power of any model by parameter tuning. There are different performance evaluation metrics available based on statistical analysis. They are Confusion Matrix, Accuracy, Sensitivity, specificity, precision, recall, $F1$ Score, Receiver Operation Characteristics (ROC), and Area Under Curve (AUC). While detecting pulmonary tuberculosis, researchers have performed different accuracy measures to discuss their models' performance and effectiveness.

4.1 Confusion Matrix

A confusion matrix is a table that often uses as a measure to project the performance of a model on a test set. It is the easiest and most intuitive metric to determine the performance of binary as well as a multi class classification problem. The table contains four different combinations of predicted and actual values. The matrix combinations are True Positive, False Positive, True Negative, and False Negative. The matrix can be defined as follows (Fig. 5):

Table 2 Machine learning classifiers and deep learning architectures in different studies

Author's name	Model description
Muyama [12]	Two pretrained CNNs'—VGGNet and GoogLeNet Inception v3 were used considering two scenarios—fast feature extraction without data augmentation, fast feature extraction using data augmentation, and fine tuning
Li [13]	(i) DenseVoxNet, 3D U-Net, and V-Net networks are used for image segmentation (ii) Region Proposal Network is used for detecting Region of interest (ROI) with 3D boundary boxes (iii) Four 3D CNN models with different feature extraction methods followed by same RPN output layer were used for evaluation
Chang [14]	An Encoder-Classifier model approach has been adopted where an input image x is encoded as a latent vector *v*. A fully connected. network (FCN) with its parameter set θf *is* used as a classifier to classify *v* into one of the *n* classes
Oloko-Oba [15]	A ConvNet consisting 6 convolutional. layers, batch normalization, and dense layer. ReLU activation function is used except in 2nd dense layer where softmax activation function is used
Ayaz [17]	An ensemble method that combines hand-crafted features with deep features (convolutional neural network-based). Hand-crafted features were extracted by Gabor Feature and deep features were extracted via pretrained deep learning models
Yugaswara [18]	Logistic Regression (LR), K-Nearest Neighbor (KNN), Naive Bayes (NB), Support Vector Machine (SVM), Random Forest (RF), Neural Network (NN), and Linear Discriminant Analysis (LDA) were implemented on the preprocessed dataset with parameter fine tuning
Raju [20]	Model 1 implements a Deep Residual Network with ReLU and Batch Norm as pre-activation, weights are initialized with orthogonal weight initialization OxfordNet is implemented in Model 2 with orthogonal weight initialization with learning rate 0.003
Ali [19]	Using WEKA toolkit, SVM, Logistic Regression, Naïve Bayes, and C5.0 were implemented
El-Melegy [21]	An automatic method based on faster R-CNN was proposed which combines a CNN, a Region Proposal Network, a region of Interest pooling Layer, and a classifier. VGG16 was used for CNN part
Hernández [23]	Pretrained models ResNet50, InceptionV3, VGG16, VGG19 were used as classifiers in the ensemble method. Voting by a majority (VM) and Sum of Probabilities was used for the ensemble method
Kant [24]	A five-layer full convoluted Neural Network (CNN) was used without any fully connected units. Cascading of Network is used to rectify the errors of initial network and to differentiate false positive and true positive cases
Abiyev [25]	BPNN, CpNN, CNN was used as a classifier

(continued)

Table 2 (continued)

Author's name	Model description
Karnkawinpong [27]	AlexNet, VGG-16, CapsNet were implemented with parameter tuning to classify tuberculosis lesion from CXR image
Sathitratanacheewin [28]	InceptionV3 and pretrained DCNN are used to classify TB CXR images
López [29]	Three different CNN architectures were implemented: Model 1—one convolution layer Model 2—2 convolution layers Model 3—3 convolution layers Models were trained in 3 patch versions—RGB, R-G, greyscale
Santosh [31]	Ensemble Voting based classifier with three base classifiers such as Bayesian Network, MLP, and RF were used
Lakhani [32]	AlexNet and GoogleNet were used for classification. Ensembles were performed on best performing algorithm
Jaeger [33]	For the segmented lung field, texture and shape features are computed and fed as input to a binary classifier. Using a decision rule and thresholds, classification is made for TB positive and Tb negative
Priya [34]	Tuberculosis (TB) digital image classification has been done using active contour method and Differential Evolution based Extreme Learning Machines (DE-ELM). Further, significant features are derived using PCA and KPCA (Kernel PCA)
van Ginneken [35]	Texture features are obtained from the overlapped segmented regions. KNN is applied for each training set of each region. A weighted multiplier is used to combine the classification results of the region

		Actual Value	
		TB Detected	TB not detected.
Predicted Value	TB detected.	**TP**	**FP**
	not detected	**FN**	**TN**

Fig. 5 Confusion matrix: rows indicate actual value or known value; column indicates predicted value by the model

4.2 Accuracy

Accuracy is the most common and initial metric used as an evaluation metric of a classifier. It is a measure of calculating all the correctly identified cases, i.e., all correctly identified positive and all correctly identified negative cases. It basically considers all the classes with equal importance.

$$\text{Accuracy} = \frac{(\text{TP} + \text{TN})}{(\text{TP} + \text{FP} + \text{TN} + \text{FN})} \tag{1}$$

4.3 Precision

Precision Metric quantifies the no of correct predictions made. It is calculated by a ratio of total correctly predicted positive values by total number of positive values predicted by the model.

$$\text{Precision} = \frac{\text{TP}}{\text{TP} + \text{FP}} \tag{2}$$

4.4 Sensitivity/Recall

Recall or Sensitivity defines the number of correct positive predictions made from all positive predictions that could have been made.

$$\text{Recall} = \frac{\text{TP}}{\text{TP} + \text{FN}} \tag{3}$$

4.5 Specificity

Specificity indicates the ability of a classifier to identify negative result correctly.

$$\text{Specificity} = \frac{\text{TN}}{\text{TN} + \text{FP}} \tag{4}$$

4.6 F1 Score

$F1$ score embodies a harmonic mean of Precision and Recall. It provides a better overview of the incorrectly classified data points than the accuracy metric.

$$F1\,\text{Score} = 2 * \frac{(\text{Precision} * \text{Recall})}{(\text{Precision} + \text{Recall})} \tag{5}$$

Fig. 6 AUC area

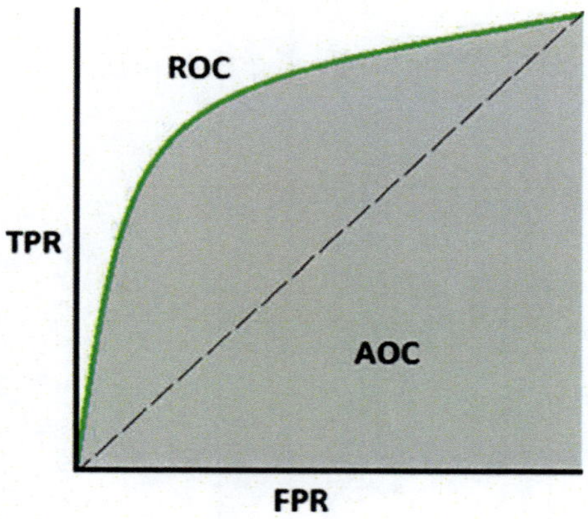

4.7 ROC Curve

Receiver Operating Characteristic Curve graph represents the overall performance of a classification model considering all classification thresholds. In the graph, the x-axis represents false positive rate and y-axis represents true positive rate. ROC graph presents a clear summary view of all the information. The following insights can be gain from ROC curve:

1. Point at (0, 0), indicates zero true positive and zero false positives.
2. A diagonal line represents true positive rate and false positive rate is equal. A classifier has predicted correctly classified TB samples and incorrectly classified TB data points with the same proportion.
3. A point at (1, 1) signifies a model's successful prediction of all correct TB datapoints but non-TB datapoints are misinterpreted.

4.8 AUC (Area Under Curve)

In AUC, the two-dimensional area under the ROC curve starting from (0, 0) to (1, 1) is considered. AUC projects the performance across all possible classification thresholds. AUC helps in analyzing the probability that a random TB negative data point ranks lower than a random TB positive data point (Fig. 6; Table 3).

5 Conclusion

Though Medical Datasets are complicated, and all aspects of examinations are important for proper diagnosis, but often medical datasets are limited as all phenomena are not included which is the main bottleneck in developing any automated TB

Table 3 Different performance metrics' analysis and comparison in different studies

Author's name	Classifier used	Accuracy	Precision	Recall	Sensitivity	Specificity	$F1$ score	ROC
Muyama [12]	VGGNet, InceptionV3	79.6%, 76.8%	79.5%, 74.8%	78.9%, 80.8%	–	–	–	–
Li [13]	Four 3D CNN classifiers		93.7%	98.7%				
Chang [14]	An Encoder-Classifier model		99%	99%				
Oloko-Oba [15]	ConvNet with 6 conv layer	87.8%	–	–	–	–	–	–
Ayaz [17]	CNN-Ensemble	93.47%, 90.6% in LR, 97.59% in CNN as level 0 classifier	–	–	–	–	–	99%, 97%
Yugaswara [18]	LDA, LR, KNN, SVM, RF, NN, NB	Lies in 97– 99%	–	–	Lies in 91–98%	96–100%	95–99%	–
Raju [20]	Deep Residual Network, Oxford Net		–	–	82.08%, 84.91%	93.80%, 93.02%	–	–
Ali [19]	SVM, NB, C5.0, Logistic Regression	81–84%	–	–	–	–	–	–
El-Melegy [21]	Faster R-CNN	–	82.6%	98.3%	–	–	89.7%	–
Hernández [23]	ResNet50, InceptionV3 VGG16, VGG19 with voting ensemble	79–86%	–	–	–	–	–	–
Kant [24]	Cascaded CNN	–	67.55%	83.78%	–	–	–	–

(continued)

Table 3 (continued)

Author's name	Classifier used	Accuracy	Precision	Recall	Sensitivity	Specificity	$F1$ score	ROC
Abiyev [25]	CNN, BPNN, CpNN	80.04–92.4%	–	–	–	–	–	–
Karnkawinpong [27]	AlexNet, VGG16, CapsNet	86.86–90.63%	–	–	85.43–89.07%	88.06–91.94%	–	–
Sathitratanacheewin [28]	Inception V3, DCNN	98.45%, 85.02%	–	–	72%	82%	–	–
López [29]	Three CNN architecture	97%	–	–	–	–	–	–
Santosh [31]	Voting Ensemble with Bayes N/W, MLP, RF	91%	–	–	–	–	–	96%
Jaeger [33]	A binary classifier with texture and shape features	First set—78.3%, Second set—90%	–	–	–	–	–	First set—87%, 2nd set—90%
Lakhani [32]	Ensemble of ALexNet, GoogleNet	–	–	–	–	–	–	99%
Priya [34]	DE-ELM Classifier, KPCA derived Hu's and Zernike moments	97.5%, 95%	–	–	–	–	–	–
van Ginneken [35]	KNN on all segmented region	–	–	–	86%, 97%	50%, 90%	–	82%, 98.6%

detection system. Pulmonary tuberculosis from radiography images has significant importance. As the number of cases increases each year with a steady rate, detection rate is required to increase to improve reporting efficiency and maintain a steady, smooth workflow to mitigate any delay concerning TB detection and any missing diagnosis point. Using an automated system will help them to detect any anomaly easily and will reduce doctor's fatigue from extreme pressure of examining this large amount of data. Various pulmonary tuberculosis detection and classification have been proposed by different researchers which starts from collecting data, data preprocessing, feature engineering, classifier implementation and finally measuring accuracy of their proposed work. The most common problem faced by researchers was unavailability of labeled data. Labeling of collected data requires enough time and expert supervision. So, this is a major concern while building a model for pulmonary tuberculosis detection. In this study, different ideas and perspectives of researchers are discussed that can help in building a model for pulmonary tuberculosis detection in the future.

References

1. GBD Tuberculosis Collaborators: The global burden of tuberculosis: results from the global burden of disease study 2015. Lancet Infect. Dis. **18**(3), 261–284 (2018)
2. https://tbfacts.org/tb-statistics-india/
3. Chang, J., Arbeláez, P., Switz, N., Reber, C., Tapley, A., Davis, J.L., et al.: Automated tuberculosis diagnosis using fluorescence images from a mobile microscope. In: Proceedings of Medical Image Computing and Computer-Assisted Intervention—MICCAI, vol. 7512, pp. 345–352 (2012)
4. Panicker, R.O., Soman, B., Saini, G., Rajan, J.: A review of automatic methods based on image processing techniques for tuberculosis detection from microscopic sputum smear images. J. Med. Syst. **40**(1), 1–13 (2016)
5. Khutlang, R., Krishnan, S., Dendere, R., Whitelaw, A., Veropoulos, K., Learmonth, G., et al.: Classification of mycobacterium tuberculosis in images of ZN stained sputum smears. IEEE Trans. Inf. Technol. Biomed. **14**(4), 949–957 (2010)
6. Khutlang, R., Krishnan, S., Whitelaw, A., Douglas, T.S.: Automated detection of tuberculosis in Ziehl-Neelsen stained sputum smears using two one-class classifiers. J. Microsc. **237**(1), 96–102 (2010)
7. Veropoulos, K., Campbell, C., Learmonth, G.: Image processing and neural computing used in the diagnosis of tuberculosis. In: Proceedings of IEE Colloquium on Intelligent Methods in Healthcare and Medical Applications (Digest No. 1998/514), pp. 8/1–8/4 (1998)
8. Costa, M.G., Costa Filho, C.F.F., Sena, J.F., Salem, J., Lima, M.O.: Automatic identification of Mycobacterium tuberculosis with conventional light microscopy. In: Proceedings of 30th Annual International IEEE Engineering in Medicine and Biology Society, pp. 382–385 (2008)
9. Sotaquirá, M., Rueda, L., Narvaez, R.: Detection and quantification of bacilli and clusters present in sputum smear samples: a novel algorithm for pulmonary tuberculosis diagnosis. In: Proceedings of International Conference on Digital Image Processing, pp. 117–121 (2009)
10. Surgitha, G.E., Murugesan, G.: Detection of tuberculosis bacilli from microscopic sputum smear images. In: ICBSII. IEEE Press, Chennai, India (2017). http://doi.org/10.1109/ICBSII.2017.8082271
11. Roh, Y., Heo, G., Whang, S.E.: A survey on data collection for machine learning: a big data—AI integration perspective. IEEE Trans. Knowl. Data Eng. http://doi.org/10.1109/TKDE.2019.2946162

12. Muyama, L., Nakatumba-Nabende, J., Mudali, D.: Automated detection of tuberculosis from sputum smear microscopic images using transfer learning techniques. In: Abraham, A., Siarry, P., Ma, K., Kaklauskas, A. (eds.) Intelligent Systems Design and Applications. ISDA 2019. Advances in Intelligent Systems and Computing, vol. 1181. Springer, Cham (2021). http://doi.org/10.1007/978-3-030-49342-4_6
13. Li, X., Zhou, Y., Du, P., et al.: A deep learning system that generates quantitative CT reports for diagnosing pulmonary tuberculosis. Appl. Intell. (2020). https://doi.org/10.1007/s10489-020-02051-1
14. Chang, R.I., Chiu, Y.H., Lin, J.W.: Two-stage classification of tuberculosis culture diagnosis using convolutional neural network with transfer learning. J. Supercomput. **76**, 8641–8656 (2020). http://doi.org/10.1007/s11227-020-03152-x
15. Oloko-Oba, M., Viriri, S.: Tuberculosis abnormality detection in chest X-rays: a deep learning approach. In: Chmielewski, L.J., Kozera, R., Orłowski, A. (eds.) Computer Vision and Graphics. ICCVG 2020. Lecture Notes in Computer Science, vol. 12334. Springer, Cham (2020). http://doi.org/10.1007/978-3-030-59006-2_11
16. https://lhncbc.nlm.nih.gov/publication/pub9931
17. Ayaz, M., Shaukat, F., Raja, G.: Ensemble learning based automatic detection of tuberculosis in chest X-ray images using hybrid feature descriptors. Phys. Eng. Sci. Med. (2021). https://doi.org/10.1007/s13246-020-00966-0
18. Yugaswara, H., Fathurahman, M., Suhaeri: Experimental analysis of tuberculosis classification based on clinical data using machine learning techniques. In: Ghazali, R., Nawi, N., Deris, M., Abawajy, J. (eds.) Recent Advances on Soft Computing and Data Mining. SCDM 2020. Advances in Intelligent Systems and Computing, vol. 978. Springer, Cham (2020). http://doi.org/10.1007/978-3-030-36056-6_15
19. Ali, M., Arshad, W.: Prediction of tuberculosis using supervised learning techniques under Pakistani patients. In: Khanna, A., Gupta, D., Bhattacharyya, S., Snasel, V., Platos, J., Hassanien, A. (eds.) International Conference on Innovative Computing and Communications. Advances in Intelligent Systems and Computing, vol. 1087. Springer, Singapore (2020). http://doi.org/10.1007/978-981-15-1286-5_4
20. Raju, M., Aswath, A., Kadam, A., Pagidimarri, V.: Automatic detection of tuberculosis using deep learning methods. In: Laha, A. (ed.) Advances in Analytics and Applications. Springer Proceedings in Business and Economics. Springer, Singapore (2019). http://doi.org/10.1007/978-981-13-1208-3_11
21. El-Melegy, M., Mohamed, D., ElMelegy, T.: Automatic detection of tuberculosis bacilli from microscopic sputum smear images using faster R-CNN, transfer learning and augmentation. In: Morales, A., Fierrez, J., Sánchez, J., Ribeiro, B. (eds.) Pattern Recognition and Image Analysis. IbPRIA 2019. Lecture Notes in Computer Science, vol. 11867. Springer, Cham (2019). http://doi.org/10.1007/978-3-030-31332-6_24
22. Shah, M.I., et al.: Ziehl-Neelsen sputum smear microscopy image database: a resource to facilitate automated bacilli detection for tuberculosis diagnosis. J. Med. Imaging **4**(2) (2017)
23. Hernández, A., Panizo, Á., Camacho, D.: An ensemble algorithm based on deep learning for tuberculosis classification. In: Yin, H., Camacho, D., Tino, P., Tallón-Ballesteros, A., Menezes, R., Allmendinger, R. (eds.) Intelligent Data Engineering and Automated Learning—IDEAL 2019. IDEAL 2019. Lecture Notes in Computer Science, vol. 11871. Springer, Cham (2019). http://doi.org/10.1007/978-3-030-33607-3_17
24. Kant, S., Srivastava, M.M.: Towards automated tuberculosis detection using deep learning. In: 2018 IEEE Symposium Series on Computational Intelligence (SSCI), Bangalore, India, pp. 1250–1253 (2018). http://doi.org/10.1109/SSCI.2018.8628800
25. http://doi.org/10.1155/2018/4168538
26. Wang, X., Peng, Y., Lu, L., Lu, Z., Bagheri, M., Summers, R.M.: Chest X-ray 8: hospital-scale chest x-ray database and benchmarks on weakly-supervised classification and localization of common thorax diseases. In: Proceedings of IEEE CVPR 2017, Honolulu, HI, USA (2017)
27. Karnkawinpong, T., Limpiyakorn, Y.: Chest X-ray analysis of tuberculosis by convolutional neural networks with affine transforms. In: ACM International Conference Proceedings Series, pp. 90–93 (2018). https://doi.org/10.1145/3297156.3297251

28. Sathitratanacheewin, S., Pongpirul, K.: Deep learning for automated classification of tuberculosis-related chest x-ray: dataset specificity limits diagnostic performance generalizability. arXiv preprint arXiv: 1811.07985 (2018)
29. López, Y.P., Costa Filho, C.F.F., Aguilera, L.M.R., Costa, M.G.F.: Automatic classification of light field smear microscopy patches using convolutional neural networks for identifying mycobacterium tuberculosis. In: 2017 CHILEAN Conference on Electrical, Electronics Engineering, Information and Communication Technologies (CHILECON), Pucon, pp. 1–5 (2017). http://doi.org/10.1109/CHILECON.2017.8229512
30. Costa, M.G.F., et al.: A sputum smear microscopy image dataset for automatic bacilli detection in conventional microscopy. Presented at the 36th Annual International Conference of the IEEE Engineering in Medicine and Biology Society, Chicago (2015)
31. Santosh, K.C., Antani, S.: Automated chest X-ray screening: can lung region symmetry help detect pulmonary abnormalities? IEEE Trans. Med. Imaging **37**(5), 1168–1177 (2018). https://doi.org/10.1109/TMI.2017.2775636
32. Lakhani, P., Sundaram, B.: Deep learning at chest radiography: automated classification of pulmonary tuberculosis by using convolutional neural networks. Radiology **284**, 574–582 (2017). https://doi.org/10.1148/radiol.2017162326
33. Jaeger, S., Karargyris, A., Candemir, S., Folio, L., Siegelman, J., Callaghan, F., Xue, Z., Palaniappan, K., Singh, R.K., Antani, S., Thoma, G., Wang, Y.-X., Lu, P.-X., McDonald, C.J.: Automatic tuberculosis screening using chest radiographs. IEEE Trans. Med. Imaging **33**(2), 233–245 (2014). http://doi.org/10.1109/TMI.2013.2284099. Epub 2013 Oct 1. PMID: 24108713
34. Priya, E., Srinivasan, S., Ramakrishnan, S.: Classification of tuberculosis digital images using hybrid evolutionary extreme learning machines. In: Nguyen, N.T., Hoang, K., Jędrzejowicz, P. (eds.) Computational Collective Intelligence. Technologies and Applications. ICCCI 2012. Lecture Notes in Computer Science, vol. 7653. Springer, Berlin (2012). http://doi.org/10.1007/978-3-642-34630-9_28
35. van Ginneken, B., Katsuragawa, S., ter Haar Romeny, B.M., Doi K., Viergever, M.A.: Automatic detection of abnormalities in chest radiographs using local texture analysis. IEEE Trans. Med. Imaging **21**(2), 139–149 (2002). http://doi.org/10.1109/42.993132
36. Karmani, P., Chandio, A.A., Korejo, I.A., Chandio, M.S.: A review of machine learning for healthcare informatics specifically tuberculosis disease diagnostics. In: Bajwa, I., Kamareddine, F., Costa, A. (eds.) Intelligent Technologies and Applications. INTAP 2018. Communications in Computer and Information Science, vol. 932. Springer, Singapore (2019). http://doi.org/10.1007/978-981-13-6052-7_5
37. https://arxiv.org/abs/1610.05463
38. https://arxiv.org/abs/1409.1556
39. He, K., Zhang, X., Ren, S., Sun, J.: Deep residual learning for image recognition. In: 2016 IEEE Conference on Computer Vision and Pattern Recognition (CVPR), Las Vegas, NV, pp. 770–778 (2016). http://doi.org/10.1109/CVPR.2016.90
40. Huang, G., Liu, Z., Van Der Maaten, L., Weinberger, K.Q.: Densely connected convolutional networks. In: 2017 IEEE Conference on Computer Vision and Pattern Recognition (CVPR), Honolulu, HI, pp. 2261–2269 (2017). http://doi.org/10.1109/CVPR.2017.243
41. Mateos-Pérez, J.M., Redondo, R., Nava, R., Valdiviezo, J.C., Cristóbal, G., Escalante-Ramérez, B., Ruiz-Serrano, M.J., Pascau, J., Desco, M.: Comparative evaluation of autofocus algorithms for a real-time system for automatic detection of Mycobacterium tuberculosis. Cytometry A **81**(3), 213–221 (2012). http://doi.org/10.1002/cyto.a.22020
42. Russell, M.J., Douglas, T.S.: Evaluation of autofocus algorithms for tuberculosis microscopy. In: Proceedings of 29th Annual International Conference of the IEEE Engineering in Medicine and Biology Society (EMBS), pp. 3489–3492 (2007). http://doi.org/10.1109/IEMBS.2007.4353082
43. Osibote, O.A., Dendere, R., Krishnan, S., Douglas, T.S.: Automated focusing in bright-field microscopy for tuberculosis detection. J. Microsc. **240**(2), 155–163 (2010). https://doi.org/10.1111/j.1365-2818.2010.03389.x

44. Zhai, Y., Liu, Y., Zhou, D., Liu, S.: Automatic identification of Mycobacterium tuberculosis from ZN-stained sputum smear: algorithm and system design. In: Proceedings of IEEE International Conference on Robotics and Biomimetics (ROBIO), Tianjin, pp. 41–46 (2010). http://doi.org/10.1109/ROBIO.2010.5723300
45. Costa Filho, C.F.F., Costa, M.G.F., Júnior, A.K.: Autofocus functions for tuberculosis diagnosis with conventional sputumsmear microscopy. In: Méndez-Vilas, A. (ed.) Proceedings of Current Microscopy Contributions to Advances in Science and Technology. Formatex Research Center, pp. 13–20 (2012)
46. Junior, A.K., Costa, M.G., Costa Filho, C.F.F., Fujimoto, L.B.M., Salem, J.: Evaluation of autofocus functions of conventional sputum smear microscopy for tuberculosis. In: Proceedings of Annual International Conference of the IEEE Engineering in Medicine and Biology Society, Argentina, pp. 3041–3044 (2010). http://doi.org/10.1109/IEMBS.2010.5626143
47. Panicker, R.O., Kalmady, K.S., Rajan, J., Sabu, M.K.: Automatic detection of tuberculosis bacilli from microscopic sputum images using deep learning methods. Biocybern. Biomed. Eng. **38**, 691–699 (2018). https://doi.org/10.1016/j.bbe.2018

Explorative Study of Explainable Artificial Intelligence Techniques for Sentiment Analysis Applied for English Language

Rohan Kumar Rathore and Anton Kolonin

1 Introduction

It is evident that artificial intelligence applications resulting from data-driven machine learning models can automate numerous decision-making tasks. Particularly, artificial neural network (ANN)-based models are giving breakthrough solution performances in various industries [1]. One of the widespread use cases is the sentiment analysis of text for the English language. A sentiment analysis model could be a standalone application or it could be a component of a more complex application. Sentiment analysis is a machine learning problem of classification type where text needs to be classified on a set of sentiment polarities like positive sentiment or negative sentiment. The solution of the sentiment analysis problem involves the following broad steps—text pre-processing, featurization, classification algorithm and evaluation method. Due to the maturity of this use case, different algorithms have been optimized to achieve breakthrough performances. Artificial neural network-based models are known to produce superior results over classical models like the support vector machine (SVM) model for sentiment analysis task [2]. In spite of the success of artificial neural network-based models, they lack transparency in their decision-making process. This leads to categorizing them as black-box models, resulting in not being deployed in real-world scenarios.

Several techniques have been introduced to address the problem of extracting the decision-making process of artificial neural networks. All techniques can be broadly classified as either pedagogical or decompositional. Pedagogical techniques treat models as black-box and estimate their decision-making process by studying the

R. K. Rathore (✉) · A. Kolonin
Novosibirsk State University, Novosibirsk 630090, Russia
e-mail: r.rathore@g.nsu.ru

A. Kolonin
e-mail: akolonin@aigents.com

M. Dua et al. (eds.), *Proceedings of the International Conference on Paradigms of Communication, Computing and Data Sciences*, Algorithms for Intelligent Systems,
https://doi.org/10.1007/978-981-16-5747-4_73

relation between model inputs and their corresponding outputs. Decompositional techniques focus on extracting the rules of the models at the level of individual units [3]. In the case of artificial neural networks, the information stored in weights and biases of individual nodes of layers is utilized in estimating the decision-making process of the model. Apart from these two categories, there have been numerous attempts to build novel explainable models or modified black-box models to make them explainable. Han Liu et al. built an explainable sentiment analysis model by fuzzy information granulation approach [4].

The next major concern is gauging the effectiveness of explainable artificial intelligence techniques which is a field of active research. One identified method is the conducting of simulatability tests in experimental settings. A model is said to be simulatable when a person can predict its behaviour on new inputs. These tests are human subject tests measuring the relative performance of human judgment of predicting the model's behaviour with or without seeing the explainable artificial intelligence technique's explanations. It consists of two tests—Forward simulation test where given an input and an explanation, users must predict what the model would output for the given input. The second is Counterfactual simulation test where given an input, the model's output for that input, and an explanation of that output, and then the user must predict what the model will output when given a perturbation of the original input. The explanation itself is algorithmically generated by the XAI technique for explaining the model [5].

In this work, we explore one pedagogical XAI technique (LIME: Local Interpretable Model-Agnostic Explanations) and one decompositional XAI technique (LRP: Layer-wise Relevance Propagation) on artificial neural network (black-box)-based sentiment analysis model and compare their performance quantitatively based on the simulatability test. The dataset in use is the publicly available IMDB movie reviews dataset. The result analysis highlights that these techniques are capable of explaining the model's decision-making process, and these human subject tests are capable of determining the effectiveness of XAI techniques. This provides a comprehensive understanding of our sentiment analysis AI application. This approach can be extended to other use cases and other XAI techniques.

2 Exploration of Techniques

In this section, we review LIME and LRP techniques and explore these on artificial neural network-based sentiment analysis models.

2.1 Local Interpretable Model-Agnostic Explanation (LIME)

LIME is an algorithm that explains the predictions of classification or regression models by approximating them locally with an interpretable model. In classification tasks, it identifies an interpretable model over the interpretable representation that is locally faithful to the classifier. Interpretable representation is a binary vector indicating the presence or absence of a word. The original representation of the text instance that is being explained is denoted by $x \in R^d$, and the binary vector for interpretable representation is denoted by $x' \in \{0, 1\}^{d'}$ [6].

The explanation of a model is defined as an interpretable model $g \in G$ where G is a family of interpretable models. $\Omega(g)$ measures the complexity of the explanation $g \in G$. Assuming $f: R^d \rightarrow R$ as a binary classifier, $f(x)$ is the probability that x belongs to a certain class. $\pi_x(z)$ measures the proximity between an instance z to x. Overall, L is a measure of the unfaithfulness of g in approximating classifier f. The explanation produced by LIME is obtained as given in Eq. (1):

$$\xi(x) = \arg\min L(f, g, \pi_x) + \Omega(g); \quad g \in G \tag{1}$$

To learn the local behaviour of f, $L(f, g, \pi_x)$ is approximated by drawing samples that are weighted by π_x. Instances are sampled around x' by drawing nonzero elements of x' uniformly at random. For a given perturbed sample $z' \in \{0, 1\}^{d'}$, the sample is recovered in the original representation $z \in R^d$ and subsequently labelled for the explanation model as $f(z)$. Given this dataset Z of perturbed samples with the associated labels, Eq. (1) is optimized to get an explanation $\xi(x)$ [6].

If G belongs to the class of linear models, such that $g(z') = w_g.z'$, the locally weighted square loss is used as L, as defined in Eq. (2), where $\pi_x(z) = \exp(-D(x, z)^2/\sigma^2)$ is an exponential kernel defined on some distance function D (cosine distance for text) with width σ [6].

$$L(f, g, \pi_x) = \sum \pi_x(z)(f(z) - g(z'))^2; \quad z, z' \in Z \tag{2}$$

For text classification, to ensure that the explanation is interpretable, interpretable representations are chosen as bag of words, and by setting a limit K on the number of words, i.e. $\Omega(g) = \infty 1[\|w_g\|_0 > K]$. K can be adapted to be as big as the user can handle, or could have different values of K for different instances [6].

In our work, the explanation from LIME technique is shown (Fig. 1) for the following input text below. We modified the default colour scheme of the technique output to make it better align with our simulatability tests control settings.

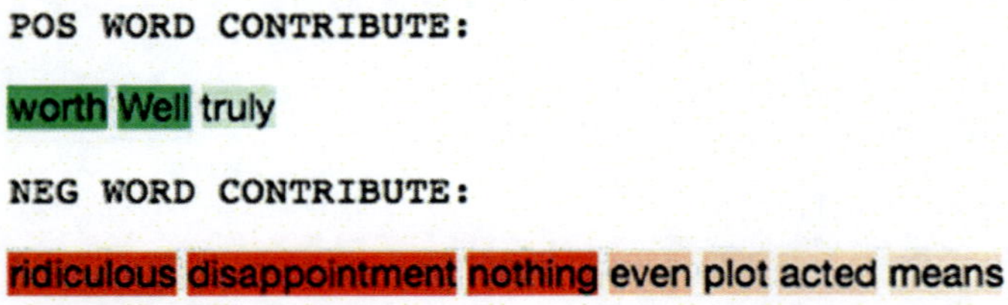

Fig. 1 LIME explanation of a movie review

Input text: "*This movie was beyond disappointment. Well acted story that means nothing. The plot is ridiculous and even what story there is goes absolutely nowhere. It truly isn't worth a nickel, buffalo or otherwise..pun intended!.*"

2.2 Layer-Wise Relevance Propagation (LRP)

LRP proposes relevance scores to neurons of artificial neural networks and the propagation procedure implemented by LRP is subject to the conservation property. Let j and k be neurons at two consecutive layers of the artificial neural network. The quantity z_{jk} models the extent to which neuron j has contributed to making neuron k relevant where the denominator is responsible for conservation property. Propagating relevance scores $(R_k)_k$ at a given layer onto neurons of the lower layer is achieved by applying the rule given in Eq. (3). The propagation procedure terminates once the input features are reached. This rule can be interpreted within the Deep Taylor Decomposition (DTD) framework where each step of propagation procedure can be modelled as an own Taylor decomposition performed over local quantities in the network graph. Therefore, LRP is viewed as a succession of Taylor expansions performed locally at each neuron of artificial neural network [7].

$$R_j = \sum_k \frac{zjk}{\sum_j zjk} R_k \tag{3}$$

Deep neural networks with rectifier (ReLU) nonlinearities are composed of neurons of the type Eq. (4):

$$a_k = \max\left(0, \sum_{0,j} aj.zjk\right) \tag{4}$$

The application of LRP to deep rectifier networks can be implemented based on the following generic rule Eq. (5):

$$R_j = \sum_k \frac{aj.\rho(wjk)}{\varepsilon + \sum_{0,k} aj.\rho(wjk)} R_k \tag{5}$$

The computation of this propagation rule can be decomposed in four steps:
(forward pass) $\forall_k : z_k = \varepsilon + \sum_{0,j} aj.\rho(wjk).sk.$
(element-wise division) $\forall_k$: $s_k = R_k/z_k$.
(backward pass) $\forall_j : c_j = \sum_k \rho(wjk).sk.$
(element-wise product) $\forall_j$: $R_j = a_j.c_j$.

WORD CONTRIBUTE:

ridicul disappoint noth absolut worth well act even plot mean

Fig. 2 LRP explanation of a movie review

In LRP, a_0 is set as 1 and w_{0k} is defined as the neuron bias. The first step is a forward pass on a copy of the layer where the weights and biases are applied to the map $\theta \rightarrow \rho(\theta)$ and further added by the small increment. The second and fourth steps are simple element-wise operations. In the third step, c_j can be expressed as the gradient computation Eq. (6):

$$c_j = \left[\nabla \left(\sum_k zk(a).sk \right) \right]_j \tag{6}$$

where $a = (a_j)_j$ is the vector of lower-layer activations, where z_k is a function of it and s_k is treated as constant [7].

In our work, explanation from LRP technique is shown (Fig. 2) on the same input text above. Here also, we modified the default colour scheme of the technique output to make it better align with our simulatability tests control settings.

3 Experimental Design

3.1 Dataset

We used IMDB movies reviews dataset. It has 50,000 movie reviews having equal distribution of positive and negative sentiments.

3.2 Model

We trained our artificial neural network on preprocessed and Tf-Idf vectorized movie reviews text, split into training and testing dataset as 80% and 20%, respectively. We obtained accuracy of 88.0 and 84.7% on training and testing dataset, respectively.

3.3 Simulatability Tests

A model is simulatable when a person can predict its behaviour on new inputs [5]. In this experiment, there are two tests—Forward simulation test and Counterfactual simulation test. First, we ask human subjects to predict model behaviour on test cases without seeing the technique's explanation, and later, we ask the same while seeing the technique's explanation. All the other control settings for the experiment were kept the same. Therefore, any change in the accuracy of the human prediction (with respect to model prediction) is due to the effect of seeing the explanation. This shows the effectiveness of the XAI technique.

We followed the procedure given by Hase et al. [5] in a simplified control setting. There are two tests in the procedure—Forward simulation test and Counterfactual simulation test.

Forward simulation test: This test is conducted in two phases—prediction phase pre and prediction phase post. In the prediction phase pre, human subjects see the test cases and are asked to tag the labels for each. Here the human subjects attempt to simulate the model behaviour without seeing the explanations from the XAI techniques. Since no other information is given, these are human subjects' best estimates to the test cases. Accuracy of human subject prediction with respect to model prediction is computed (pre simulation accuracy). In the prediction phase post, human subjects again see the same test cases along with seeing the explanations from the XAI techniques and are asked to tag the labels for each. Accuracy of human subject prediction with respect to model prediction is computed (post simulation accuracy).

Counterfactual simulation test: This test is conducted in the similar two phases, prediction phase pre, and prediction phase post. In the prediction phase pre, human subjects see the test cases, the model predicted labels, the perturbed instances of the test cases and are asked to tag the labels for each perturbed instance. Here the human subjects attempt to simulate the model behaviour on the perturbed instances without seeing the explanations from the XAI techniques. Accuracy of human subject prediction with respect to model prediction is computed (pre simulation accuracy). In the prediction phase post, human subjects again see the same test cases, the model predicted labels, the perturbed instances of the test cases along with seeing the explanations from the XAI techniques and are asked to tag the labels for each perturbed instance. Accuracy of human subject prediction with respect to model prediction is computed (post simulation accuracy).

Each technique was passed through both these tests. In our experiment, one human subject was involved in the experiment. A total of 120 data points were collected. We kept the test cases for pre- and post-human subject tests the same. Our perturbation method is the random switching of words from the training dataset vocabulary at 10% random places. We changed the colour scheme of the techniques output to make the human decision-making efforts for simulating the model behaviour unbiased across different test cases. Figures 1 and 2 show the explanation by LIME and LRP technique, respectively, for a case highlighting the importance of words with the density of the colour. In LIME, two lists of bag of words are generated for each label; while in LRP, only one list of bag of words is generated. The code for experimental design is publicly available in GitHub—https://github.com/rohancode/simulatibility_test.

Table 1 Change in human subject accuracy after being given explanations of model behaviour

XAI phase	Forward simulation (%)	Counterfactual simulation (%)	Total (%)
LIME—Pre	90.0	65.0	77.5
LIME—Post	90.0	90.0	90.0
LIME—Change	**0.0**	**25.0**	**12.5.0**
LRP—Pre	90.0	65.0	77.5
LRP—Post	95.0	85.0	90.0
LRP—Change	**5.0**	**20.0**	**12.5.0**

4 Results

The results of the experiment are summarized in Table 1. We observe that both LIME and LRP techniques improve the accuracy of model prediction capability of the human subject by 12.5%. The major changes were observed in the Counterfactual Tests for both techniques where for LIME, the accuracy improved from 65 to 90% and for LRP, the accuracy improved from 65 to 85%.

The results highlight that XAI techniques increase the human understanding of our black-box sentiment analysis model. It should be noted that these results are specific to our experimental control settings.

5 Conclusion

This study provides an overview of LIME and LRP explainable artificial intelligence techniques applied on artificial neural network-based sentiment analysis models trained on IMDB movie reviews dataset. The effectiveness of these techniques is measured quantitatively by simulatability tests. The analysis of the results highlights that both techniques increase the human understanding of the model.

References

1. Dargan, S., Kumar, M., Ayyagari, M.R., et al.: A survey of deep learning and its applications: a new paradigm to machine learning. Arch. Comput. Methods Eng. **27**, 1071–1092 (2014)
2. Medhat, W., Hassan, A., Korashy, H.: Sentiment analysis algorithms and applications: a survey. Ain Shams Eng. J. **5**(4), 1093–1113 (2014)
3. Andrews, R., Diederich, J., Tickle, A.B.: Survey and critique of techniques for extracting rules from trained artificial neural networks. Knowl.-Based Syst. **8**(6), 373–389 (1995)
4. Liu, H., Cocea, M.: Fuzzy information granulation towards interpretable sentiment analysis. Granular Comput. **2**, 289–302 (2017)
5. Hase, P., Bansal, M.: Evaluating explainable AI: which algorithmic explanations help users predict model behavior? arXiv:2005.01831 (2020)

6. Ribeiro, M.T., Singh, S., Guestrin, C.: "Why should i trust you?": explaining the predictions of any classifier. In: Proceedings of the 22nd ACM SIGKDD International Conference on Knowledge Discovery and Data Mining (KDD '16), pp. 1135–1144. Association for Computing Machinery, New York (2014)
7. Montavon, G., Binder, A., Lapuschkin, S., Samek, W., Müller, K.R.: Layer-wise relevance propagation: an overview. In: Samek, W., Montavon, G., Vedaldi, A., Hansen, L., Müller, K.R. (eds.) Explainable AI: Interpreting, Explaining and Visualizing Deep Learning, LNCS, vol. 11700, pp. 193–209. Springer, Cham (2019)

Author Index

M. Dua et al. (eds.), *Proceedings of the International Conference on Paradigms of Communication, Computing and Data Sciences*, Algorithms for Intelligent Systems,
https://doi.org/10.1007/978-981-16-5747-4